CRITICAL CARE NEUROLOGY
PART II

HANDBOOK OF CLINICAL NEUROLOGY

Series Editors

MICHAEL J. AMINOFF, FRANÇOIS BOLLER, AND DICK F. SWAAB

VOLUME 141

CRITICAL CARE NEUROLOGY PART II

Series Editors

MICHAEL J. AMINOFF, FRANÇOIS BOLLER, AND DICK F. SWAAB

Volume Editors

EELCO F.M. WIJDICKS AND ANDREAS H. KRAMER

VOLUME 141

3rd Series

ELSEVIER

Radarweg 29, PO Box 211, 1000 AE Amsterdam, Netherlands
The Boulevard, Langford Lane, Kidlington, Oxford OX5 1GB, United Kingdom
50 Hampshire Street, 5th Floor, Cambridge, MA 02139, United States

© 2017 Elsevier B.V. All rights reserved.

No part of this publication may be reproduced or transmitted in any form or by any means, electronic or mechanical, including photocopying, recording, or any information storage and retrieval system, without permission in writing from the publisher. Details on how to seek permission, further information about the Publisher's permissions policies and our arrangements with organizations such as the Copyright Clearance Center and the Copyright Licensing Agency, can be found at our website: www.elsevier.com/permissions.

This book and the individual contributions contained in it are protected under copyright by the Publisher (other than as may be noted herein).

Notices
Knowledge and best practice in this field are constantly changing. As new research and experience broaden our understanding, changes in research methods, professional practices, or medical treatment may become necessary.

Practitioners and researchers must always rely on their own experience and knowledge in evaluating and using any information, methods, compounds, or experiments described herein. In using such information or methods they should be mindful of their own safety and the safety of others, including parties for whom they have a professional responsibility.

With respect to any drug or pharmaceutical products identified, readers are advised to check the most current information provided (i) on procedures featured or (ii) by the manufacturer of each product to be administered, to verify the recommended dose or formula, the method and duration of administration, and contraindications. It is the responsibility of practitioners, relying on their own experience and knowledge of their patients, to make diagnoses, to determine dosages and the best treatment for each individual patient, and to take all appropriate safety precautions.

To the fullest extent of the law, neither the Publisher nor the authors, contributors, or editors, assume any liability for any injury and/or damage to persons or property as a matter of products liability, negligence or otherwise, or from any use or operation of any methods, products, instructions, or ideas contained in the material herein.

British Library Cataloguing-in-Publication Data
A catalogue record for this book is available from the British Library

Library of Congress Cataloging-in-Publication Data
A catalog record for this book is available from the Library of Congress

ISBN: 978-0-44-463599-0

For information on all Elsevier publications
visit our website at https://www.elsevier.com/

Publisher: Mara Conner
Editorial Project Manager: Kristi Anderson
Production Project Manager: Sujatha Thirugnana Sambandam
Cover Designer: Alan Studholme

Typeset by SPi Global, India

Handbook of Clinical Neurology 3rd Series

Available titles

Vol. 79, The human hypothalamus: basic and clinical aspects, Part I, D.F. Swaab, ed. ISBN 9780444513571
Vol. 80, The human hypothalamus: basic and clinical aspects, Part II, D.F. Swaab, ed. ISBN 9780444514905
Vol. 81, Pain, F. Cervero and T.S. Jensen, eds. ISBN 9780444519016
Vol. 82, Motor neurone disorders and related diseases, A.A. Eisen and P.J. Shaw, eds. ISBN 9780444518941
Vol. 83, Parkinson's disease and related disorders, Part I, W.C. Koller and E. Melamed, eds. ISBN 9780444519009
Vol. 84, Parkinson's disease and related disorders, Part II, W.C. Koller and E. Melamed, eds. ISBN 9780444528933
Vol. 85, HIV/AIDS and the nervous system, P. Portegies and J. Berger, eds. ISBN 9780444520104
Vol. 86, Myopathies, F.L. Mastaglia and D. Hilton Jones, eds. ISBN 9780444518996
Vol. 87, Malformations of the nervous system, H.B. Sarnat and P. Curatolo, eds. ISBN 9780444518965
Vol. 88, Neuropsychology and behavioural neurology, G. Goldenberg and B.C. Miller, eds. ISBN 9780444518972
Vol. 89, Dementias, C. Duyckaerts and I. Litvan, eds. ISBN 9780444518989
Vol. 90, Disorders of consciousness, G.B. Young and E.F.M. Wijdicks, eds. ISBN 9780444518958
Vol. 91, Neuromuscular junction disorders, A.G. Engel, ed. ISBN 9780444520081
Vol. 92, Stroke – Part I: Basic and epidemiological aspects, M. Fisher, ed. ISBN 9780444520036
Vol. 93, Stroke – Part II: Clinical manifestations and pathogenesis, M. Fisher, ed. ISBN 9780444520043
Vol. 94, Stroke – Part III: Investigations and management, M. Fisher, ed. ISBN 9780444520050
Vol. 95, History of neurology, S. Finger, F. Boller and K.L. Tyler, eds. ISBN 9780444520081
Vol. 96, Bacterial infections of the central nervous system, K.L. Roos and A.R. Tunkel, eds. ISBN 9780444520159
Vol. 97, Headache, G. Nappi and M.A. Moskowitz, eds. ISBN 9780444521392
Vol. 98, Sleep disorders Part I, P. Montagna and S. Chokroverty, eds. ISBN 9780444520067
Vol. 99, Sleep disorders Part II, P. Montagna and S. Chokroverty, eds. ISBN 9780444520074
Vol. 100, Hyperkinetic movement disorders, W.J. Weiner and E. Tolosa, eds. ISBN 9780444520142
Vol. 101, Muscular dystrophies, A. Amato and R.C. Griggs, eds. ISBN 9780080450315
Vol. 102, Neuro-ophthalmology, C. Kennard and R.J. Leigh, eds. ISBN 9780444529039
Vol. 103, Ataxic disorders, S.H. Subramony and A. Durr, eds. ISBN 9780444518927
Vol. 104, Neuro-oncology Part I, W. Grisold and R. Sofietti, eds. ISBN 9780444521385
Vol. 105, Neuro-oncology Part II, W. Grisold and R. Sofietti, eds. ISBN 9780444535023
Vol. 106, Neurobiology of psychiatric disorders, T. Schlaepfer and C.B. Nemeroff, eds. ISBN 9780444520029
Vol. 107, Epilepsy Part I, H. Stefan and W.H. Theodore, eds. ISBN 9780444528988
Vol. 108, Epilepsy Part II, H. Stefan and W.H. Theodore, eds. ISBN 9780444528995
Vol. 109, Spinal cord injury, J. Verhaagen and J.W. McDonald III, eds. ISBN 9780444521378
Vol. 110, Neurological rehabilitation, M. Barnes and D.C. Good, eds. ISBN 9780444529015
Vol. 111, Pediatric neurology Part I, O. Dulac, M. Lassonde and H.B. Sarnat, eds. ISBN 9780444528919
Vol. 112, Pediatric neurology Part II, O. Dulac, M. Lassonde and H.B. Sarnat, eds. ISBN 9780444529107
Vol. 113, Pediatric neurology Part III, O. Dulac, M. Lassonde and H.B. Sarnat, eds. ISBN 9780444595652
Vol. 114, Neuroparasitology and tropical neurology, H.H. Garcia, H.B. Tanowitz and O.H. Del Brutto, eds. ISBN 9780444534903
Vol. 115, Peripheral nerve disorders, G. Said and C. Krarup, eds. ISBN 9780444529022
Vol. 116, Brain stimulation, A.M. Lozano and M. Hallett, eds. ISBN 9780444534972
Vol. 117, Autonomic nervous system, R.M. Buijs and D.F. Swaab, eds. ISBN 9780444534910
Vol. 118, Ethical and legal issues in neurology, J.L. Bernat and H.R. Beresford, eds. ISBN 9780444535016
Vol. 119, Neurologic aspects of systemic disease Part I, J. Biller and J.M. Ferro, eds. ISBN 9780702040863
Vol. 120, Neurologic aspects of systemic disease Part II, J. Biller and J.M. Ferro, eds. ISBN 9780702040870
Vol. 121, Neurologic aspects of systemic disease Part III, J. Biller and J.M. Ferro, eds. ISBN 9780702040887
Vol. 122, Multiple sclerosis and related disorders, D.S. Goodin, ed. ISBN 9780444520012
Vol. 123, Neurovirology, A.C. Tselis and J. Booss, eds. ISBN 9780444534880

AVAILABLE TITLES (Continued)

Vol. 124, Clinical neuroendocrinology, E. Fliers, M. Korbonits and J.A. Romijn, eds. ISBN 9780444596024
Vol. 125, Alcohol and the nervous system, E.V. Sullivan and A. Pfefferbaum, eds. ISBN 9780444626196
Vol. 126, Diabetes and the nervous system, D.W. Zochodne and R.A. Malik, eds. ISBN 9780444534804
Vol. 127, Traumatic brain injury Part I, J.H. Grafman and A.M. Salazar, eds. ISBN 9780444528926
Vol. 128, Traumatic brain injury Part II, J.H. Grafman and A.M. Salazar, eds. ISBN 9780444635211
Vol. 129, The human auditory system: Fundamental organization and clinical disorders, G.G. Celesia and G. Hickok, eds. ISBN 9780444626301
Vol. 130, Neurology of sexual and bladder disorders, D.B. Vodušek and F. Boller, eds. ISBN 9780444632470
Vol. 131, Occupational neurology, M. Lotti and M.L. Bleecker, eds. ISBN 9780444626271
Vol. 132, Neurocutaneous syndromes, M.P. Islam and E.S. Roach, eds. ISBN 9780444627025
Vol. 133, Autoimmune neurology, S.J. Pittock and A. Vincent, eds. ISBN 9780444634320
Vol. 134, Gliomas, M.S. Berger and M. Weller, eds. ISBN 9780128029978
Vol. 135, Neuroimaging Part I, J.C. Masdeu and R.G. González, eds. ISBN 9780444534859
Vol. 136, Neuroimaging Part II, J.C. Masdeu and R.G. González, eds. ISBN 9780444534866
Vol. 137, Neuro-otology, J.M. Furman and T. Lempert, eds. ISBN 9780444634375
Vol. 138, Neuroepidemiology, C. Rosano, M.A. Ikram and M. Ganguli, eds. ISBN 9780128029732
Vol. 139, Functional neurologic disorders, M. Hallett, J. Stone and A. Carson, eds. ISBN 9780128017722
Vol. 140, Critical care neurology Part I, E.F.M. Wijdicks and A.H. Kramer, eds. ISBN 9780444636003

Foreword

Modern hospitals in the developed countries have changed remarkably in character over the last quarter-century, no longer serving as a hospice for the chronically sick. Instead, their focus is now primarily on surgical patients requiring perioperative care, patients requiring a procedural intervention, and patients with critical illnesses requiring care in the intensive care unit because of the complexity of their disorders. In the same manner as many other medical disciplines, neurology has become for the most part an outpatient specialty. Patients requiring surgery or with complex neurologic disorders necessitating a multidisciplinary approach and constant monitoring now make up a large component of the patients admitted to hospital and seen by neurologists. It was with this in mind that we felt the need to include critical care neurology within the embrace of the *Handbook of Clinical Neurology* series. To this end, we approached two leaders in the field to develop the subject, and are delighted that they agreed to do so and with what they have achieved.

Eelco Wijdicks is professor of neurology and chair of the division of critical care neurology at the Mayo Clinic College of Medicine, Rochester, Minnesota, and is a well-known author and the founding editor of the journal *Neurocritical Care*. Andreas H. Kramer is a clinical associate professor in the departments of critical care medicine and clinical neurosciences at the Hotchkiss Brain Institute of the University of Calgary, in Alberta, Canada. Both are leaders in the field of neurointensive care, with wide experience in patient management and an international record in developing evidence-based guidelines for optimizing patient care. Together they have developed two volumes of the *Handbook* to cover the pathophysiology and treatment of patients with acute neurologic or neurosurgical disorders requiring care in the intensive care unit (Volume 140), or with neurologic complications that have arisen in the setting of a medical or surgical critical illness (Volume 141).

Forty-one chapters deal with all aspects of these disorders, including ethical and prognostic considerations. Many of the management issues that are discussed in these pages are among the most difficult ones faced by contemporary clinicians, and the availability of these authoritative reviews – buttressed by the latest advances in medical science – will increase physician confidence by providing the most up-to-date guidelines for improving patient care. We are grateful to Professors Wijdicks and Kramer, and to the various contributors whom they enlisted as coauthors, for crafting two such comprehensive volumes that will be of major utility both as reference works for all practitioners and as practical guides for those in the front line.

As series editors, we reviewed all of the chapters in these volumes, making suggestions for improvement as needed. We believe that all who are involved in the care of critically ill patients in the hospital setting will find them a valuable resource. The availability of the volume electronically on Elsevier's Science Direct site should increase their accessibility and facilitate searches for specific information.

As always, we extend our appreciation to Elsevier, our publishers, for their continued support of the *Handbook* series, and warmly acknowledge our personal indebtedness to Michael Parkinson in Scotland and to Mara Conner and Kristi Anderson in California for their assistance in seeing these volumes to fruition.

Michael J. Aminoff
François Boller
Dick F. Swaab

Preface

New subspecialties in neurology continue to germinate, and critical care neurology (also known as neurocritical care) is one of the more recent ones. The field has matured significantly over the last two decades, and a neurointensivist is a recognizable and legitimate specialist. The field involves primarily the care of patients with an acute neurologic or neurosurgical disorder. These disorders are life-threatening because the main injury may damage critical structures and often affects respiration and even the circulation. A neurologic complication may also appear *de novo* in the setting of a medical or surgical critical illness. These two clinical situations form the pillars of this field and therefore justify two separate volumes. In these two books we include traditional sections focused on epidemiology and pathophysiology, but others are more tailored towards management of the patient, sections we think are informative to the general neurologist. Therapeutic interventions and acute decisions are part of a daily commitment of a neurointensivist. We assumed that a focus on management (and less on diagnostics) will be most useful for the reader of this handbook series. The immediacy of management focuses on prevention of further intracranial complications (brain edema and brain tissue shift, increased intracranial pressure, and seizures) and systemic (cardiopulmonary) insults.

We have written extensively on many of these topics but in these two volumes we let other practitioners write about their practice, experiences, and research. They have all made a name for themselves and we are pleased they were able to contribute to this work. Although the major topics are reviewed, we realize some may have been truncated or not covered because we tried to avoid a substantial overlap with other volumes in the series.

This is a contributed book with all its inherent quirks, stylistic mismatches, and inconsistencies, but we hope we have edited a text that is more than the sum of its parts. We appreciate the fact that the series editors of the *Handbook of Clinical Neurology* recognized this field of neurology. Herein, we are making the argument that delivery of care by a neurointensivist is an absolute requirement and its value for the patient is undisputed. Still, the best way to achieve this is through integrated care, and neurointensivists can only function in a multidisciplinary cooperative practice. The new slate of neurointensivists in the USA can be certified in neurology, neurosurgery, internal medicine, anesthesiology, or other critical care specialties and time will tell if this all-inclusiveness will dilute or strengthen the specialty. One fact is clear: our backgrounds are different and this significantly helped in shaping this volume.

We thank the editors of the series – Michael Aminoff, Francois Boller, and Dick Swaab – for inviting us three years ago to prepare these volumes. We must particularly thank Michael Parkinson and Sujatha Thirugnana Sambandam, who steered the books to fruition.

I—Eelco Wijdicks—know the series very well and when I did my neurology residency in Holland in the early 1980s it was known as "Vinken and Bruyn," and residents and staff would always look there first to find a solution for a difficult patient, to read up on an usual disorder or to understand a mechanism. I admired the beautiful covers and authoritative reviews and I remember it had a special place in our library. I was thrilled to see the complete series in the Mayo Neurology library when I arrived in the USA.

We are both honored to have contributed to this renowned series of clinical neurology books.

Eelco F.M. Wijdicks
Andreas H. Kramer

Contributors

N. Badjatia
Department of Neurology, University of Maryland School of Medicine, Baltimore, MD, USA

J. Ch'ang
Neurological Institute, Columbia University, New York, NY, USA

J. Claassen
Neurological Institute, Columbia University, New York, NY, USA

R. Dhar
Division of Neurocritical Care, Department of Neurology, Washington University, St. Louis, MO, USA

M. Diringer
Department of Neurology, Washington University, St. Louis, MO, USA

I.R.F. da Silva
Neurocritical Care Unit, Americas Medical City, Rio de Janeiro, Brazil

J.A. Frontera
Neurological Institute, Cleveland Clinic, Cleveland, OH, USA

J.E. Fugate
Department of Neurology, Mayo Clinic, Rochester, MN, USA

R.G. Geocadin
Neurosciences Critical Care Division, Department of Anesthesiology and Critical Care Medicine and Departments of Neurology and Neurosurgery, Johns Hopkins University School of Medicine, Baltimore, MD, USA

G. Hermans
Department of General Internal Medicine, UZ Leuven, Leuven, Belgium

J. Horn
Department of Intensive Care, Academic Medical Center, Amsterdam, The Netherlands

R.M. Jha
Department of Critical Care Medicine, University of Pittsburgh, Pittsburgh, PA, USA

J.T. Jo
Neuro-Oncology Center, University of Virginia, Charlottesville, VA, USA

E.J.O. Kompanje
Department of Intensive Care, Erasmus MC University Medical Center, Rotterdam, The Netherlands

A.H. Kramer
Departments of Critical Care Medicine and Clinical Neurosciences, Hotchkiss Brain Institute, University of Calgary and Southern Alberta Organ and Tissue Donation Program, Calgary, AB, Canada

M.A. Kumar
Departments of Neurology, Neurosurgery, Anesthesiology and Critical Care, Perelman School of Medicine, Hospital of the University of Pennsylvania, Philadelphia, PA, USA

M.D. Levine
Department of Emergency Medicine, Mayo Clinic, Phoenix, AZ, USA

M. Mulder
Department of Critical Care and the John Nasseff Neuroscience Institute, Abbott Northwestern Hospital, Allina Health, Minneapolis, MN, USA

E. Nourollahzadeh
Division of Neurocritical Care and Emergency Neurology, Department of Neurology, Yale New Haven Hospital, New Haven, CT, USA

B. Pfausler
Neurocritical Care Unit, Department of Neurology, Medical University Innsbruck, Innsbruck, Austria

D. Schiff
Neuro-Oncology Center, University of Virginia, Charlottesville, VA, USA

E. Schmutzhard
Neurocritical Care Unit, Department of Neurology, Medical University Innsbruck, Innsbruck, Austria

K.N. Sheth
Division of Neurocritical Care and Emergency Neurology, Department of Neurology, Yale New Haven Hospital, New Haven, CT, USA

L. Shutter
Department of Critical Care Medicine, University of Pittsburgh, Pittsburgh, PA, USA

A.J.C. Slooter
Department of Intensive Care Medicine, University Medical Center Utrecht, Utrecht, The Netherlands

M. Toledano
Department of Neurology, Mayo Clinic, Rochester, MN, USA

S.J. Traub
Department of Emergency Medicine, Mayo Clinic, Phoenix, AZ, USA

M. van der Jagt
Department of Intensive Care, Erasmus MC University Medical Center, Rotterdam, The Netherlands

R.R. van de Leur
Department of Intensive Care Medicine, University Medical Center Utrecht, Utrecht, The Netherlands

J.D. VanDerWerf
Department of Neurology, Perelman School of Medicine, Hospital of the University of Pennsylvania, Philadelphia, PA, USA

E.F.M. Wijdicks
Division of Critical Care Neurology, Mayo Clinic and Neurosciences Intensive Care Unit, Mayo Clinic Campus, Saint Marys Hospital, Rochester, MN, USA

W.L. Wright
Neuroscience Intensive Care Unit, Emory University Hospital Midtown, Atlanta, GA, USA

I.J. Zaal
Department of Intensive Care Medicine, University Medical Center Utrecht, Utrecht, The Netherlands

Contents of Part II

Foreword vii
Preface ix
Contributors xi

SECTION 2 Neurologic complications of critical illness

24. **The scope of neurology of critical illness**
    E.F.M. Wijdicks (Rochester, USA)   443

25. **Delirium in critically ill patients**
    A.J.C. Slooter, R.R. van de Leur, and I.J. Zaal (Utrecht, The Netherlands)   449

26. **Posterior reversible encephalopathy in the intensive care unit**
    M. Toledano and J.E. Fugate (Rochester, USA)   467

27. **Acute neurotoxicology of drugs of abuse**
    S.J. Traub and M.D. Levine (Phoenix, USA)   485

28. **Seizures in the critically ill**
    J. Ch'ang and J. Claassen (New York, USA)   507

29. **Intensive care unit-acquired weakness**
    J. Horn and G. Hermans (Amsterdam, The Netherlands and Leuven, Belgium)   531

30. **Neurologic complications of transplantation**
    R. Dhar (St. Louis, USA)   545

31. **Neurologic complications of cardiac and vascular surgery**
    K.N. Sheth and E. Nourollahzadeh (New Haven, USA)   573

32. **Neurology of cardiopulmonary resuscitation**
    M. Mulder and R.G. Geocadin (Minneapolis and Baltimore, USA)   593

33. **Therapeutic hypothermia protocols**
    N. Badjatia (Baltimore, USA)   619

34. **Neurologic complications of polytrauma**
    R.M. Jha and L. Shutter (Pittsburgh, USA)   633

35. **Neurologic complications in critically ill pregnant patients**
    W.L. Wright (Atlanta, USA)   657

36. **Neurologic complications of sepsis**
    E. Schmutzhard and B. Pfausler (Innsbruck, Austria)   675

37. **Neurologic complications of acute environmental injuries**
    I.R.F. da Silva and J.A. Frontera (Rio de Janeiro, Brazil and Cleveland, USA) 685

38. **Neurologic manifestations of major electrolyte abnormalities**
    M. Diringer (St. Louis, USA) 705

39. **Management of neuro-oncologic emergencies**
    J.T. Jo and D. Schiff (Charlottesville, USA) 715

40. **Management of neurologic complications of coagulopathies**
    J.D. VanDerWerf and M.A. Kumar (Philadelphia, USA) 743

41. **Prognosis of neurologic complications in critical illness**
    M. van der Jagt and E.J.O. Kompanje (Rotterdam, The Netherlands) 765

Index I-1

Contents of Part I

Foreword vii
Preface ix
Contributors xi

SECTION 1 Care in the neurosciences intensive care unit

1. **The history of neurocritical care**
   E.F.M. Wijdicks (Rochester, USA) — 3

2. **Airway management and mechanical ventilation in acute brain injury**
   D.B. Seder and J. Bösel (Portland and Boston, USA and Heidelberg, Germany) — 15

3. **Neuropulmonology**
   A. Balofsky, J. George, and P. Papadakos (Rochester, USA) — 33

4. **Neurocardiology**
   N.D. Osteraas and V.H. Lee (Chicago, USA) — 49

5. **Principles of intracranial pressure monitoring and treatment**
   M. Czosnyka, J.D. Pickard, and L.A. Steiner (Cambridge, UK and Basel, Switzerland) — 67

6. **Multimodal neurologic monitoring**
   G. Korbakis and P.M. Vespa (Los Angeles, USA) — 91

7. **Continuous EEG monitoring in the intensive care unit**
   G.B. Young and J. Mantia (London and Ontario, Canada) — 107

8. **Management of the comatose patient**
   E.F.M. Wijdicks (Rochester, USA) — 117

9. **Management of status epilepticus**
   M. Pichler and S. Hocker (Rochester, USA) — 131

10. **Critical care in acute ischemic stroke**
    M. McDermott, T. Jacobs, and L. Morgenstern (Ann Arbor, USA) — 153

11. **Management of intracerebral hemorrhage**
    A.M. Thabet, M. Kottapally, and J.C. Hemphill III (San Francisco and Miami, USA) — 177

12. **Management of aneurysmal subarachnoid hemorrhage**
    N. Etminan and R.L. Macdonald (Mannheim, Germany and Toronto, Canada) — 195

13. **Management of acute neuromuscular disorders**
    E.F.M. Wijdicks (Rochester, USA) — 229

14. **Critical care management of traumatic brain injury** 239
    D.K. Menon and A. Ercole (Cambridge, UK)

15. **Management of acute traumatic spinal cord injuries** 275
    C.D. Shank, B.C. Walters, and M.N. Hadley (Birmingham, USA)

16. **Decompressive craniectomy in acute brain injury** 299
    D.A. Brown and E.F.M. Wijdicks (Rochester, USA)

17. **Diagnosis and management of spinal cord emergencies** 319
    E.P. Flanagan and S.J. Pittock (Rochester, USA)

18. **Diagnosis and management of acute encephalitis** 337
    J.J. Halperin (Philadelphia, USA)

19. **Management of bacterial central nervous system infections** 349
    M.C. Brouwer and D. van de Beek (Amsterdam, The Netherlands)

20. **Management of infections associated with neurocritical care** 365
    L. Rivera-Lara, W. Ziai, and P. Nyquist (Baltimore, USA)

21. **Determinants of prognosis in neurocatastrophes** 379
    K. Sharma and R.D. Stevens (Baltimore, USA)

22. **Family discussions on life-sustaining interventions in neurocritical care** 397
    M.M. Adil and D. Larriviere (New Orleans, USA)

23. **Organ donation protocols** 409
    C.B. Maciel, D.Y. Hwang, and D.M. Greer (New Haven, USA)

Index I-1

Section 2

Neurologic complications of critical illness

Chapter 24

The scope of neurology of critical illness

E.F.M. WIJDICKS*
Division of Critical Care Neurology, Mayo Clinic and Neurosciences Intensive Care Unit, Mayo Clinic Campus, Saint Marys Hospital, Rochester, MN, USA

Abstract

Critical illness increases the probability of a neurologic complication. There are many reasons to consult a neurologist in a critically ill patient and most often it is altered alertness with no intuitive plausible explanation. Other common clinical neurologic problems facing the intensive care specialist and consulting neurologist in everyday decisions are coma following prolonged cardiovascular surgery, newly perceived motor asymmetry, seizures or other abnormal movements, and generalized muscle weakness. Assessment of long-term neurologic prognosis is another frequent reason for consultation and often to seek additional information about the patient's critical condition by the attending intensivist. Generally speaking, consultations in medical or surgical ICU's may have a varying catalog of complexity and may involve close management of major acute brain injury.

This chapter introduces the main principles and scope of this field. Being able to do these consults effectively–often urgent and at any hour of the day–requires a good knowledge of general intensive care and surgical procedures. An argument can be made to involve neurointensivists or neurohospitalists in these complicated consults.

The subspecialty of critical care neurology has two major pillars – the care of patients with critical neurologic illness and the care of critically ill patients who during the most treacherous phase of their clinical course, or soon thereafter, develop a neurologic complication. This second part of the volume *Critical Care Neurology* is about the second category of patients, namely those admitted to medical and surgical intensive care units (ICUs) presenting with a *de novo* neurologic problem. These patients are seen in consultation by neurologists for diagnosis and management – often expediently – but remain under the care of intensivists and surgeons. The complications observed may be quite specific and neurologists immediately appreciate that a neurologic complication in a critically ill patient often occurs in a complex, rapidly changing clinical situation. Moreover, most intensivists feel uncomfortable in handling this new neurologic condition themselves, and request not only assistance with identification of the neurologic disorder, but also in management. Many general neurologists who work on the hospital consultation service – often not more than a few weeks a year – feel uncomfortable seeing these medically unstable patients, who every so often cannot even leave the ICU for neuroimaging. It is therefore common that neurologists or neurohospitalists request formal assistance by a neurointensivist, if these individuals are among the staff. Furthermore, once the patient is assessed, continuous attention may be needed, which may involve prolonged bedside care and later calls at night by nursing staff or attending intensivists to help direct management.

More than in any place in the hospital, consulting in the ICU involves questions about de-escalating care. The attending team and family may consider withdrawing life-sustaining interventions, or at least a do-not-resuscitate status, and therefore need a neurologist's input. Such involvement is partly a reflection of the high prevalence of neurologic catastrophes in patients with a critical illness. Frequently the clinical situation is clear, such as, for example, in persistently comatose survivors

---

*Correspondence to: Eelco F.M. Wijdicks, Department of Neurology, Mayo Clinic, 200 First Street SW, Rochester MN 55905, USA. E-mail: wijde@mayo.edu

following prolonged cardiopulmonary arrest and in patients with polytrauma and severe traumatic brain injury–in other situations the degree of brain injury may be far more difficult to ascertain.

Neurologists are asked to participate in family conferences and they can be helpful in clarifying the bigger picture. Sometimes the neurologic complication is a defining moment and little can be done to help the patient. In such situations, neurologists could be conclusive if there is an undeniable poor outcome and thereby keep the managing team from treating a patient in a futile situation. However, although it is important to decisively prognosticate when certain, another fundamental rule of ICU consultation is to hold back when information is incomplete or the clinical situation is not fully understood. Some neurologic manifestations (e.g., coma in progressive multiorgan failure) are a direct manifestation of a relentless downward spiral. At the other end of this spectrum of severity are transient manifestations (e.g., briefly altered consciousness or twitching). Many of these passing manifestation have no impact on outcome and may remain unexplained or attributed to a probable drug effect.

Critical illness increases the probability of a neurologic complication, and current best estimates are that approximately 10% of patients will develop some sort of neurologic manifestation (Bleck et al., 1993; Howard, 2007; Wijdicks, 2012). Neurology of critical illness is an important field, which requires renewed attention and research. The rationale for this expertise is summarized in Table 24.1 and shows common clinical neurologic problems facing the intensive care specialist and consulting neurologist in everyday decisions.

Most ICU consults are relatively urgent or emergent consults. The urgency is often determined by the inability to understand the full clinical picture and when the situation is particularly concerning at face value. Examples are consults in the ICU for acutely impaired consciousness that require a rapid but comprehensive assessment of the cause of coma and whether it can be immediately reversed (Wijdicks, 2016b). In this ICU practice we can expect three clinical scenarios: acute loss of consciousness, failure of patients to fully awaken after recuperation from a major surgical procedure, and, occasionally, coma in a developing but as yet undiagnosed critical illness. These latter situations are the most challenging both for the neurologists coming to the bedside and the intensivist trying to grasp the situation.

Another common issue is the patient with "altered mental status." This category of neurologic deficits – patients who are agitated and less responsive – may in comparison appear less concerning. Patients are confused and may not respond quickly, rarely fixate on objects, and cannot follow simple commands. Some are able to speak, others are unable to respond. We assume that, in most situations, patients will have acute brain dysfunction from sepsis-associated encephalopathy, the effects of medications, or from new-onset acute renal or liver failure, or both.

Unfortunately, for many years, neurologists had the tendency to call any patient with an encephalopathy "multifactorial metabolic encephalopathy," followed by listing the abnormalities that make up the patient's critical illness. None of this would advance understanding of these complicated patients. More experience in examining and following such patients has resulted in a better effort to try to understand the true nature of acute brain dysfunction. One principle is to set apart the major driver of neurologic manifestations, but equally common now is to consider other possible explanations, such as structural injury (Lockwood, 1987; Iacobone et al., 2009; Hughes et al., 2012). We now know that many patients termed "encephalopathic" really had a structural brain injury, including those with electroencephalogram (EEG) patterns (e.g., diffuse slowing or triphasic waves) traditionally associated with "metabolic dysfunction." Posterior reversible encephalopathy syndrome is so prevalent that it is often placed high in the differential diagnosis, and if the circumstances are right, should be investigated with magnetic resonance imaging.

Acute confusional state or delirium may trigger a consult, but many intensivists do recognize this entity and treat it appropriately (Brown, 2014). The most difficult situation is to assess a patient with decreased or increased arousal, abnormal perception, abnormal attention, and incoherent language. Within this category are patients with apraxia and aphasia. Abulia is suggestive of a frontal syndrome, but may be misinterpreted as so-called "hypoactive delirium," although most patients with delirium have no new structural central nervous

Table 24.1

Why neurology consults in the ICU matter?

Neurologic consultation in the ICU requires a broad base of medical knowledge
Neurologic consultation provides diagnostic, therapeutic, and prognostic advice
Neurologic consultation may detect an unsuspected neurologic disorder
Neurologic consultation in the ICU may change approach to the patient
Neurologic consultation involves end-of-life decisions for some patients

ICU, intensive care unit.
From Wijdicks (2016a) Solving Critical Consults. In: Core Principles of Acute Neurology Series. Used with permission from Mayo Foundation for Medical Education and Research.

system lesion (Devinsky and D'Esposito, 2004). A large proportion of ICU patients with "sundowning" or agitated delirium have pre-existing cognitive decline or prior undiagnosed advanced dementia. An unexplained observation is that delirium is associated with prolonged ICU stay and increased mortality (Ely et al., 2004; Pandharipande et al., 2013), yet none of the clinical trials that have aggressively treated patients with delirium have shown an improved mortality rate (Zaal and Slooter, 2012; Flannery and Flynn 2013). Most physicians feel that treatment should be swift with potent sedative drugs because there are few other options to calm the patient and ensure safety (Makii et al., 2010).

Consults for new-onset seizures or new movement abnormalities are also comparatively frequent. A new focal finding (i.e., hemiparesis or marked asymmetry) is less commonly, but consults are often for newly perceived asymmetries. A major challenge is to recognize an acute stroke during a critical illness or after a major vascular procedure. Patients may have a delayed presentation or recognition of neurologic deficits, particularly when anesthetic drugs have been and are still being metabolized in the postoperative phase of surgery. The challenge here is early recognition to allow an endovascular intervention because intravenous thrombolysis is usually contraindicated. Acute ischemic stroke may warrant endovascular treatment if the situation allows, although the computed tomography scan may already show an established infarct. In unclear situations we often perform a CT angiogram with or without CT perfusion. This gives a good sense of injury and what tissue is at risk.

Consults in surgical and trauma ICUs are often related to diagnostic evaluation of new spinal cord and traumatic brain injury. In most instances, other specialties have already been involved (i.e., neurosurgery). A special category is consultation in the transplant recipient, which may have already started before transplantation (e.g., fulminant hepatic failure) The neurologist's presence is often appreciated by the attending intensivist or surgeon if care involves management of increased intracranial pressure (brain edema in fulminant hepatic failure or traumatic diffuse axonal injury),recognition of neurotoxicity and co-management of opportunistic CNS infections. Another special category is the patient admitted with a left ventricular assist device and new neurologic symptoms. Decisions on discontinuation or modification of anticoagulation often involve a neurologist.

In the surgical ICU, consults may involve sudden appearance of paraplegia after awakening from anesthesia. Acute infarction of the spinal cord could allow immediate placement of a lumbar drain to reduce cerebrospinal fluid spinal pressure and possibly blood pressure augmentation to improve residual spinal blood flow. In each of these scenarios, prompt decisions are warranted that could lead to improved outcome if appropriate measures are taken. Urgent consultation for a possible complication of carotid artery surgery involves assessment for ischemic stroke or management of blood pressure and heart rate instability (the latter is mostly managed by a neurointensivist, but a general neurologist should be aware of this major complication involving damage to the baroreceptors).

Generalized weakness in the ICU is very common and nearly always prompts a neurologic consult (Maramattom and Wijdicks, 2006). Most neurologists will expect (and diagnose) critical illness polyneuropathy, critical illness myopathy, or both. These are the most common causes of weakness in the ICU setting. The prevalence of ICU-acquired weakness is high in survivors of critical illness and will likely increase further as more patients survive sepsis, multiorgan failure, and other fulminant conditions.

Failure to wean off the ventilator (or unexplained reintubation) is another trigger for a comprehensive neurologic assessment and a neurologic disorder other than critical illness polyneuropathy may be found.

Finally, consults may involve explanation of neuroimaging findings or interpretation of an abnormal EEG in a patient with an undefined repetitive movement (Firosh Khan et al., 2005; Oddo et al., 2009; Young, 2009; Claassen et al., 2013).

Any consult in a critically ill neurologic patient must proceed with the steps outlined in Table 24.2. A neurologic consult in a critically ill patient may lead to a diagnosis not initially considered by the managing team and is frequent in our experience (Mittal et al., 2015). These recognized neurologic disorders may all have major consequences – diagnostic, prognostic, and therapeutic. Consultations may have a varying catalog of complexity and may involve management of major acute neurologic injury (Fig. 24.1). Consults in ICUs are complex by

Table 24.2

Essentials of a neurology consult in the intensive care unit

Assess details on severity of critical illness
Assess blood pressure and extent of blood pressure support
Assess drug administration over 5–7 days
Verify onset of symptoms with nursing staff
Assess major confounders (therapeutic hypothermia, extracorporeal membrane oxygenation, acute metabolic derangements, and acid–base abnormalities)
Inquire about possible movements, or twitching
Assess for drugs strongly related to delirium, movement disorders

From Wijdicks (2016a) Solving Critical Consults. In: Core Principles of Acute Neurology Series. Used with permission from Mayo Foundation for Medical Education and Research.

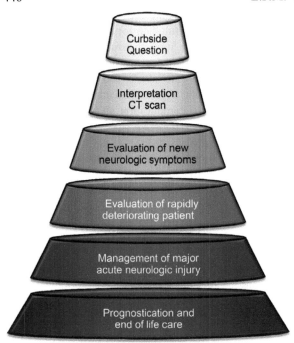

Fig. 24.1. Complexity of neurologic consults in the intensive care unit. CT, computed tomography. From Wijdicks (2016a) Solving Critical Consults. In: Core Principles of Acute Neurology Series. Used with permission from Mayo Foundation for Medical Education and Research.

nature, and neurologists should expect not only to solve a diagnostic problem, but to be actively involved in ongoing management of the neurologic considerations. Situations in which this may occur include acute ischemic stroke after cardiovascular surgery, recurrent seizures in acute hyponatremia, posterior reversible encephalopathy syndrome, and immunosuppression neurotoxicity.

CONCLUSION

ICU consults are the most challenging in the hospital because: (1) decisions may have to be made in an evolving situation; (2) the primary diagnosis may be unclear; (3) neurologic examination can be compromised when patients are markedly edematous, jaundiced, immobile, bruised, and have major operative wounds or an open abdomen or chest; and (4) neuroimaging and electrophysiology may not be particularly helpful. One could argue for a separate hospital service staffed by experienced neurohospitalists or neurointensivists. Still many of us are caught unaware by a variety of presentations, and as long as experience is gained, it is ideally gained by a specialized group that will be able to apply this knowledge to future patients. We have seen a number of conditions emerge more clearly as a result of us taking all ICU consults in both Mayo-affiliated hospital (Table 24.3). We suspect that telemedicine, would be ideal for such consults in the future (Wilcox and Adhikari, 2012; Lilly et al., 2014a, b).

Table 24.3

Some initially unrecognized neurologic conditions in the intensive care unit

Posterior reversible encephalopathy syndrome
Serotonin syndrome
Cefepime neurotoxicity
Nonconvulsive status epilepticus
Amyotrophic lateral sclerosis
Hyperammonemic stupor
Central pontine myelinolysis

REFERENCES

Bleck TP, Smith MC, Pierre-Louis SJ et al. (1993). Neurologic complications of critical medical illnesses. Crit Care Med 21: 98–103.

Brown CH (2014). Delirium in the cardiac surgical ICU. Curr Opin Anaesthesiol 27: 117–122.

Claassen J, Taccone FS, Horn P et al. (2013). Recommendations on the use of EEG monitoring in critically ill patients: consensus statement from the neurointensive care section of the ESICM. Intensive Care Med 39: 1337–1351.

Devinsky O, D'Esposito M (2004). Neurology of Cognitive and Behavioral Disorders. Oxford University Press, Oxford.

Ely EW, Shintani A, Truman B et al. (2004). Delirium as a predictor of mortality in mechanically ventilated patients in the intensive care unit. JAMA 291: 1753–1762.

Firosh Khan S, Ashalatha R, Thomas SV et al. (2005). Emergent EEG is helpful in neurology critical care practice. Clin Neurophysiol 116: 2454–2459.

Flannery AH, Flynn JD (2013). More questions than answers in ICU delirium: pressing issues for future research. Ann Pharmacother 47: 1558–1561.

Howard RS (2007). Neurological problems on the ICU. Clin Med 7: 148–153.

Hughes CG, Patel MB, Pandharipande PP (2012). Pathophysiology of acute brain dysfunction: what's the cause of all this confusion? Curr Opin Crit Care 18: 518–526.

Iacobone E, Bailly-Salin J, Polito A et al. (2009). Sepsis-associated encephalopathy and its differential diagnosis. Crit Care Med 37: S331–S336.

Lilly CM, McLaughlin JM, Zhao H et al. (2014a). A multicenter study of ICU telemedicine reengineering of adult critical care. Chest 145: 500–507a.

Lilly CM, Zubrow MT, Kempner KM et al. (2014b). Critical Care Telemedicine: Evolution and State of the Art. Crit Care Med 42: 2429–2436.

Lockwood AH (1987). Metabolic encephalopathies: opportunities and challenges. J Cereb Blood Flow Metab 7: 523–526.

Makii JM, Mirski MA, Lewin 3rd JJ (2010). Sedation and analgesia in critically ill neurologic patients. J Pharm Pract 23: 455–469.

Maramattom BV, Wijdicks EF (2006). Acute neuromuscular weakness in the intensive care unit. Crit Care Med 34: 2835–2841.

Mittal MK, Kashyap R, Herasevich V et al. (2015). Do patients in a medical or surgical ICU benefit from a neurologic consultation? Int J Neurosci 125: 512–520.

Oddo M, Carrera E, Claassen J et al. (2009). Continuous electroencephalography in the medical intensive care unit. Crit Care Med 37: 2051–2056.

Pandharipande PP, Girard TD, Jackson JC et al. (2013). Long-term cognitive impairment after critical illness. N Engl J Med 369: 1306–1316.

Wijdicks EFM (2012). Neurologic Complications of Critical Illness. 3rd edn. Oxford, New York.

Wijdicks EFM (2016a). Solving critical consults. In: Core Principles of Acute Neurology Series, Oxford University Press, Oxford.

Wijdicks EFM (2016b). Why you may need a neurologist to see a comatose patient in the ICU. Crit Care 20: 193.

Wilcox ME, Adhikari NK (2012). The effect of telemedicine in critically ill patients: systematic review and meta-analysis. Crit Care 16: R127.

Young GB (2009). Continuous EEG, monitoring in the ICU: challenges and opportunities. Can J Neurol Sci 36 (Suppl 2): S89–S91.

Zaal IJ, Slooter AJ (2012). Delirium in critically ill patients: epidemiology, pathophysiology, diagnosis and management. Drugs 72: 1457–1471.

Chapter 25

Delirium in critically ill patients

A.J.C. SLOOTER*, R.R. VAN DE LEUR, AND I.J. ZAAL
Department of Intensive Care Medicine, University Medical Center Utrecht, Utrecht, The Netherlands

Abstract

Delirium is common in critically ill patients and associated with increased length of stay in the intensive care unit (ICU) and long-term cognitive impairment. The pathophysiology of delirium has been explained by neuroinflammation, an aberrant stress response, neurotransmitter imbalances, and neuronal network alterations. Delirium develops mostly in vulnerable patients (e.g., elderly and cognitively impaired) in the throes of a critical illness. Delirium is by definition due to an underlying condition and can be identified at ICU admission using prediction models. Treatment of delirium can be improved with frequent monitoring, as early detection and subsequent treatment of the underlying condition can improve outcome. Cautious use or avoidance of benzodiazepines may reduce the likelihood of developing delirium. Nonpharmacologic strategies with early mobilization, reducing causes for sleep deprivation, and reorientation measures may be effective in the prevention of delirium. Antipsychotics are effective in treating hallucinations and agitation, but do not reduce the duration of delirium. Combined pain, agitation, and delirium protocols seem to improve the outcome of critically ill patients and may reduce delirium incidence.

INTRODUCTION

The term delirium is derived from the Latin word *delirare*, meaning "to go out of the furrow while ploughing a field" which implies to deviate from the straight track (Adamis et al., 2007). Despite its long history, delirium in the intensive care unit (ICU) has not received much attention. In 1980, the third edition of the *Diagnostic and Statistical Manual of Mental Disorders* (DSM-III) brought cohesion in the various terms under the umbrella of delirium (American Psychiatric Association, 1980). Delirium has recently been redefined in the DSM-5 as an acute disturbance in attention and awareness, with additional disturbances in cognition, not explained by a pre-existing neurocognitive disorder, and caused by another medical condition (American Psychiatric Association, 2013; Zaal and Slooter, 2014). Delirium has a heterogeneous etiology, but homogeneous presentation. In ICU patients, delirium usually occurs in the setting of multiorgan failure and in the sickest patients.

A related condition to delirium is encephalopathy, for which there is no uniformly accepted definition and which may just be another nondistinct term. Acute or subacute encephalopathy has been defined as the manifestations of widespread failure of cerebral metabolism due to metabolic, toxic, or infectious disease. These manifestations typically involve a disorder of consciousness or awareness, such as delirium, stupor, or coma, as well as a variety of focal neurologic signs and sometimes seizures (Lipowski, 1990). Most neurologists use the term encephalopathy to describe diffuse brain pathology that manifests as delirium or in more severe cases as coma, with occasionally focal neurologic signs.

Delirium together with more severe disorders of consciousness has been named by some as "acute brain failure" (Morandi et al., 2008), analogous to other heterogeneous conditions in ICU patients, such as acute renal failure or acute heart failure. The term acute brain failure further expresses a continuum of pathology, whereas the term delirium may falsely suggest a dichotomy. Finally,

---

*Correspondence to: Arjen J.C. Slooter, Department of Intensive Care Medicine, University Medical Center Utrecht, Utrecht, The Netherlands. E-mail: a.slooter-3@umcutrecht.nl

the term subsyndromal delirium has been used to describe patients with some delirium features, who do not fulfill diagnostic criteria for delirium (Ouimet et al., 2007b).

DSM-5 criteria for delirium are broad and include numerous other acute neurologic conditions. The typical critically ill patient with delirium has been exposed to a variety of both predisposing and precipitating risk factors at the same time, and it is usually impossible to assign one specific cause for delirium (Ely et al., 2001a). The term delirium in the context of acute structural brain injury is usually used to describe secondary deterioration after the acute event due to systemic disorders, such as adverse effects of drugs, metabolic alterations, or infectious diseases. The terminology in delirious patients remains ambiguous, reflecting the patient's symptomatology.

EPIDEMIOLOGY

It has been recently appreciated that delirium is a common phenomenon in the ICU. A meta-analysis of 42 studies and 16 595 critically ill patients showed an occurrence rate of delirium of 31.8% (Salluh et al., 2015). The rates varied between 9.2% in nonventilated surgical intensive care patients (Serafim et al., 2012) and 91% in a group of mechanically ventilated cancer patients (Almeida et al., 2014). This variation is presumably caused by a different case mix, as the proportion of elective surgical patients (versus acutely admitted patients) and risk factors for delirium is not equally distributed between study populations. Furthermore, the use of different screening instruments and the subsequent different interpretation of subsyndromal delirium could have caused variability (Ouimet et al., 2007b) and is all indicative of a definitional problem.

The median duration of ICU delirium was found to be 3 days (interquartile range 2–7 days) and the median time to onset 2 days (interquartile range 1–4 days), although this can vary strongly between patients (Pisani et al., 2009b; Zaal et al., 2015c). In some cases, delirium can be present for weeks.

Over 100 different risk factors have been described for delirium. In ICU patients, an average of 11 of these factors has been reported to be present at the same time (Ely et al., 2001a) and may imply an interaction between predisposing, precipitating, and triggering factors. In practice, this means that a young and healthy person will only develop delirium when seriously critically ill, while an old and demented person may already become delirious from fever and a urinary tract infection (Fig. 25.1). A semiquantitative best-evidence approach may identify risk factors for ICU delirium based on the number, quality, and consistency of previous studies (Zaal et al., 2015b).

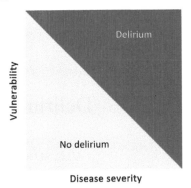

Fig. 25.1. The interaction between disease severity and vulnerability in the pathogenesis of delirium.

The risk of delirium increases with 2% with every additional year once a patient reached the age of 65 (Pandharipande et al., 2006). Dementia doubles the risk to develop delirium in the ICU (McNicoll et al., 2003; Van Rompaey et al., 2009). It has been suggested that there could be a genetic predisposition to delirium (Ely et al., 2007; Stoicea et al., 2014).

Precipitating factors in the ICU can be classified into three domains: acute illness, exposure to medications, and environmental effects. First, acute illness as reflected in the Acute Physiological and Chronic Health Evaluation (APACHE) II score, emergency surgery, mechanical ventilation, metabolic acidosis, (poly)trauma, and acute systemic inflammation were precipitating factors. Each additional point on the APACHE II score on admission increased the risk of delirium by 5–6%. A score of 18 and higher did not increase the risk any further (Pandharipande et al., 2006; Ouimet et al., 2007a; Van Rompaey et al., 2009; Wolters et al., 2015b; Zaal et al., 2015b). Second, benzodiazepine use, especially with continuous infusion, causes a dose-dependent increase in the risk of delirium (Pandharipande et al., 2006, 2008; Zaal et al., 2015a). Associations have also been found between use of opiates or corticosteroids and development of ICU delirium, but these findings have been inconsistent (Schreiber et al., 2014; Kamdar et al., 2015; Wolters et al., 2015a; Zaal et al., 2015b). Environmental factors such as lack of daylight, ICU sound level, and interruptions increased the risk of delirium (Van Rompaey et al., 2009; Caruso et al., 2014). A summary of predisposing and precipitating risk factors for delirium in ICU patients is shown in Figures 25.2 and 25.3.

NEUROPATHOLOGY

The pathophysiology is multifactorial and thus remains poorly understood. A variety of hypotheses on the pathophysiology of delirium have been described, which may be complementary, rather than competing (Maldonado, 2013). The main hypotheses focus on neuroinflammation,

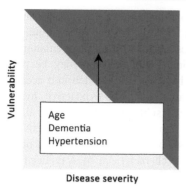

Fig. 25.2. Predisposing factors in the pathogenesis of intensive care unit delirium.

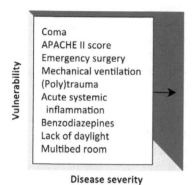

Fig. 25.3. Precipitating factors in the pathogenesis of intensive care unit delirium. APACHE II, Acute Physiological and Chronic Health Evaluation II.

an aberrant stress response, neurotransmitter imbalances, and neuronal network alterations. In addition, oxidative stress and a disturbance of circadian integrity may play a role in the pathogenesis of delirium (Maldonado, 2013).

Critical illness is often associated with an inflammatory response, particularly in trauma or sepsis. A peripheral proinflammatory cytokine signal can be transmitted to the brain through afferents of the vagus nerve, transport across the blood–brain barrier (BBB), or entry at the circumventricular region, where the BBB is nonexistent or discontinuous. During sepsis increased levels of interleukin (IL)-1, IL-6, and tumor necrosis factor (TNF)-α have been found in the cerebrospinal fluid (Waage et al., 1989). This could cause activation of microglia, leading to further cytokine release and neuronal dysfunction (Van Gool et al., 2010). Interestingly, increased brain TNF-α levels persist in rodents for months after administration of lipopolysaccharide, used to simulate sepsis (Qin et al., 2007). Microglia seem to be inhibited by acetylcholine, and in elderly patients with a neurodegenerative disorder, the inhibitory function of acetylcholine may be impaired. This could thus explain that age and cognitive impairment are risk factors for delirium and that delirium causes further long-term cognitive impairment (Ebersoldt et al., 2007; Van Gool et al., 2010; Field et al., 2012; Pandharipande et al., 2013). Levels of IL-6, IL-8, procalcitonin, and C-reactive protein are elevated during delirium (De Rooij et al., 2007; McGrane et al., 2011). Postmortem studies of brains of delirious patients showed increased activity of microglia, astrocytes, and IL-6 and a high frequency of ischemic lesions in different parts of the brain, especially in the hippocampus (Janz et al., 2010; Munster et al., 2011). Other possible effects of sepsis in the pathogenesis of delirium are impaired perfusion pressure and ischemia due to systemic hypotension and hypoxemia, as well as microcirculatory alterations, including endothelial dysfunction (Ebersoldt et al., 2007; Pfister et al., 2008).

The aberrant stress response hypothesis points out the possible adverse effects of acute stress. Glucocorticoids, the central hormones in the human stress response, are regulated through the limbic–hypothalamic–pituitary–adrenal axis (LHPA axis). Several stressors, such as surgery, systemic inflammation, and pain, cause the brain to activate the LHPA axis and thereby increase the levels of cortisol. In healthy individuals, this response is adaptive and has adequate feedback regulation. Cognitive decline and aging are associated with impaired feedback regulation of the stress response pathway and these patients may develop sustained high cortisol levels. Glucocorticoids, in turn, can pass the BBB, and sustained high cortisol levels are known to cause impairment in cognition and attention (Maclullich et al., 2008). High cortisol levels were indeed associated with delirium (Mu et al., 2010; Plaschke et al., 2010; Van den Boogaard et al., 2011). One study showed that systemic corticosteroid administration was related to the development of delirium (Schreiber et al., 2014), but another study could not confirm this (Wolters et al., 2015a).

It is further presumed that delirium is associated with reduced cholinergic activity. The acetylcholine neurotransmitter system plays a central role in attention and consciousness, which are particularly affected in delirium (Hshieh et al., 2008). Secondly, the use of anticholinergic drugs is associated with an increased severity of the symptoms of delirium in non-ICU patients (Han et al., 2001). Exposure to anticholinergic drugs in the ICU, however, was not associated with a greater risk of the development of delirium (Wolters et al., 2015b). Furthermore, cholinesterase inhibitors did not decrease the risk for delirium (Van Eijk et al., 2010). The cholinergic system is thought to be balanced with dopamine and serotonin and an excess of these two neurotransmitters may be associated with delirium. Dopaminergic drugs, like levodopa, may induce delirium and dopamine antagonists, such as haloperidol, may alleviate certain delirium symptoms. Dopamine and serotonin influence arousal, motor function, and the sleep–wake cycle (Hshieh et al., 2008). Moreover, dopamine has a regulatory influence on the cholinergic system (Trzepacz, 1999). Lastly, release of gamma-aminobutyric acid (GABA) is presumed to be involved in the development of delirium, as benzodiazepines, that increase GABA activity, increase the risk of delirium (Pandharipande et al., 2006; Zaal et al., 2015a).

Recently, numerous studies have been performed on neuronal networks in various neurologic and psychiatric diseases. A functional connectivity study using electroencephalography (EEG) and a global approach for network analysis, found delirium to be associated with a more random functional network and loss of functional connectivity (Van Dellen et al., 2014). Delirium could therefore be considered as a disconnection syndrome. Using a seed-based approach and functional magnetic resonance imaging, reduced autocorrelation was found between the executive network and the default-mode network that is involved in attention, as well as disruptions in interregional connectivity in subcortical regions that are involved in arousal (Choi et al., 2012). Considering the aforementioned multifactorial etiology of delirium, it is assumed that the interaction of different pathways leads to disruption of large-scale neuronal networks, and that this disruption causes attention deficits, a reduced level of consciousness, and other features of delirium.

In addition, ischemia seems to play a role in the pathogenesis of delirium. Studies with preoperative MRI of the brain found white-matter hyperintensities, carotid stenosis, and new cortical ischemic lesions, acquired during surgery, to be risk factors for postoperative delirium (Shioiri et al., 2010; Root et al., 2013; Omiya et al., 2015). During a delirium episode reduced cerebral blood flow was found in mainly the parietal and frontal regions, with recovery of blood flow after symptom resolution (Yokota et al., 2003; Alsop et al., 2006; Fong et al., 2006). In an investigation where ICU patients received an MRI scan at hospital discharge and 3 months later, the duration of delirium was found to be associated with white-matter disruption and smaller brain volumes (Gunther et al., 2012; Morandi et al., 2012; Jackson et al., 2015). It remains unclear whether these were consequences of delirium, or predisposing factors.

CLINICAL PRESENTATION

A key characteristic of delirium is a decreased level of consciousness and disturbance of attention. Attention disturbances usually manifest with difficulty sustaining attention, but distraction by irrelevant stimuli also often occurs. A disturbed sleep–wake rhythm is almost always present, as well as disorientation in time and place. Patients with delirium may be restless or not and some have categorized delirium according to psychomotor alterations into hypoactive, hyperactive, and mixed motor subtypes (Meagher et al., 2000). The hypoactive subtype is the most frequent type of delirium but is nondistinct and may be underdiagnosed because it is characterized by a reduced alertness and a reduced amount of motor activity (i.e., hypokinesia) and speech. Hypoactive delirium and metabolic encephalopathy (already an umbrella term for many metabolic disturbances) cannot be clinically distinguished. The hyperactive subtype, and for most neurologists and psychiatrists considered the prototype, is characterized by an increased and inappropriate quantity of motor activity (i.e., hyperkinesia, albeit usually with bradykinesia), with restlessness and sometimes agitation. This form of delirium is more often associated with alcohol withdrawal (Zaal et al., 2015c). The remaining patients have the mixed motor subtype, and alternate between the hyper- and hypoactive type (Peterson et al., 2006; Pandharipande et al., 2007a; Meagher, 2009). It should be noted, however, that classifications of motor subtypes are usually based on brief observations in time, and that there is no uniform classification of motor subtypes.

Other features of delirium are disorganized thinking and memory deficiency but emotional manifestations such as anxiety, sadness, and irritation are seen. Perceptual disturbances (i.e., hallucinations and delusions) have traditionally been associated with a delirium, but occur in less than 40% of cases (Marquis et al., 2007). Hallucinations are usually complex visual (landscapes, animals), and acoustic hallucinations (sounds and voices) are rare. Autonomic features, such as tachycardia and hyperventilation, are common and may cause additional distress to the patient (i.e., increased myocardial demand). Delirium severity often fluctuates during the course of the day, with classically restlessness at night ("sundowning"), and sleepiness during the day.

NEURODIAGNOSTICS AND IMAGING

The diagnosis of delirium is not difficult in predisposed patients with an acute illness who develop a slightly decreased level of consciousness with disturbed attention in the course of hours to a few days. Patients with a more severe decreased level of consciousness (i.e., no response to voice) are considered unarousable or comatose. There are numerous bedside tests to assess attention in nonmechanically ventilated patients. Of these, repeating the months of the year backward appeared to be most sensitive to detect delirium in non-ICU patients (Adamis et al., 2015). In intubated or tracheostomized patients, the Attention Screening Examination can be used, which is part of the Confusion Assessment Method for the ICU (CAM-ICU) (Ely et al., 2001b). The examiner reads a series of 10 letters and the patient is instructed to squeeze the hand of the examiner when he or she hears the letter A. Disturbed attention with this test is defined as two or more errors. Patients with a paresis, for example, due to ICU-acquired weakness, can close their eyes instead of squeezing the examiner's hand. Other important features of delirium, such as disorientation, emotional problems, and hallucinations, can be assessed with close observation in patients who cannot communicate verbally. Assessment of other core features of delirium such as a disturbed sleep–wake rhythm and psychomotor disturbances (bradykinesia, hypo- or hyperkinesia) can also only be based on observations, although it should be noted that estimations of sleep based on inspection consistently overestimate sleep time (Aurell and Elmqvist, 1985).

The diagnostic evaluation of patients with possible delirium should include tests to detect "red flags": symptoms and signs that suggest a specific neurologic syndrome and that are not compatible with the most common type of ICU delirium, which is due to multiple interacting predisposing and precipitating factors, as described above. Important "red flags" are focal neurologic signs, an atypical presentation (e.g., predominance of hallucinations with preserved attention and consciousness), an atypical onset (either very acute or chronic), and a paucity of delirium risk factors. Agitation with dilated pupils, clonus, hyperthermia, and hyperactive bowel sounds may indicate a serotonin syndrome (Boyer and Shannon, 2005). Hyperthermia is also a prominent feature of neuroleptic malignant syndrome, with extreme rigidity and an elevated serum creatine kinase level. Wernicke encephalopathy should be considered in malnourished patients with ophthalmoplegia, nystagmus, pupillary abnormalities, or limb ataxia (Sechi and Serra, 2007).

EEG changes, including diffuse slowing of background activity, periodic discharges such as triphasic waves, and polymorphic delta activity, can support the diagnosis of delirium (Kaplan and Rossetti, 2011). In addition, EEG may show electrographic seizures and periodic epileptiform discharges in some patients with delirium, particularly in the context of sepsis (Oddo et al., 2009). EEG is a sensitive, but less specific, technique to detect delirium. Neuroimaging has been used in several studies to improve understanding of the pathophysiology of delirium (Soiza et al., 2008), but the diagnostic yield in typical cases of delirium is low, and neuroimaging is not recommended in typical cases.

Screening

Delirium is often underdiagnosed in the ICU and up to 70% of delirium cases are missed by ICU physicians (Van Eijk et al., 2009). A delay of more than 24 hours in the treatment of ICU delirium was found to be associated with impaired outcome (Heymann et al., 2010). To improve recognition of delirium in the ICU, various screening instruments have been developed. Of these, the two most commonly used and investigated instruments are the CAM-ICU and the Intensive Care Delirium Screening Checklist (ICDSC) (Bergeron et al., 2001; Ely et al., 2001b; Patel et al., 2009).

The CAM-ICU is an assessment at one moment in time with a binary outcome (delirium or no delirium). It is an adaption of the CAM for patients who cannot communicate verbally and can be completed in approximately 2 minutes. It consists of four features: (1) acute onset of mental status changes, or a fluctuating course; (2) inattention; (3) disorganized thinking; and (4) an altered level of consciousness (Table 25.1). The ICDSC is an eight-item screening tool based on observations during a nursing shift, with an outcome that can range from 0 to 8 points. Hence, a patient can be scored as no delirium (0 points), subsyndromal delirium (1–3 points), or delirium (>4 points) (Ouimet et al., 2007b). The eight features are an altered level of consciousness, inattention, disorientation, hallucinations or delusions, psychomotor agitation and retardation, inappropriate speech or mood, sleep–wake cycle disturbances, and symptom fluctuations (Table 25.2).

In two meta-analyses, the CAM-ICU and the ICDSC had a similar pooled sensitivity (75.5–80.0% vs. 74–80.1%), but the CAM-ICU had a higher specificity (95.9% vs. 74.6–81.9%) (Gusmao-Flores et al., 2012; Neto et al., 2012). It should be noted that almost all of these studies were performed in a research setting. In routine daily practice, the sensitivity of the CAM-ICU was 47% overall, and 31% for the hypoactive subtype (Van Eijk et al., 2011).

Various other approaches are currently explored to develop more sensitive delirium monitoring with objective tools. These include objective assessments of attention deficits (Tieges et al., 2015), which require active collaboration of the patient. Alternatively, monitoring could be based on physiologic alterations, such as temperature variability, eye movements, and blinks (Van der Kooi et al., 2013, 2014). A promising approach seems to be a brief EEG recording with a limited number of electrodes and automated processing (Van der Kooi et al., 2012).

Table 25.1

The Confusion Assessment Method for the Intensive Care Unit (CAM-ICU)*

CAM-ICU

Criterium

1. Acute onset <u>and/or</u> fluctuating course
2. Inattention
   Read the following series of 10 letters and let patient squeeze on the letter 'A'
   S A V E A H A A R T[†] *Positive if more than two errors.*
3. Altered level of consciousness
   Positive if RASS score other than zero.
4. Disorganized thinking
   Positive if more than 1 mistake on the following questions or commands.

   Ask the following questions:[‡]
   - Will a stone float on water?
   - Are there fish in the sea?
   - Does one pound weigh more than two pounds?
   - Can you use a hammer to pound a nail?

   Command
   Say to patient: "Hold op this many fingers" (hold two fingers in front of the patient), "Now do the same with the other hand" (without repeating the number of fingers)

Positive CAM-ICU: Criterium 1 <u>plus</u> 2 <u>and</u> either criterium 3 <u>or</u> 4

Adapted from Ely et al. (2001b).
*The CAM-ICU can only be administered when the RASS score > −3.
[†]If the patient has a neuromuscular disease and squeezing is impossible, eye blinks can be used.
[‡]An alternative set of questions are available.

Table 25.2

The Intensive Care Delirium Screening Checklist (ICDSC)*

ICDSC

Criterium

1. Altered level of consciousness
   Score if RASS score other than zero (without recent sedative use)
2. Inattention
   Score if inadequate response to the following test:
   Read the following series of 10 letters and let patient squeeze on the letter 'A'
   S A V E A H A A R T[†]
3. Disorientation
   Score if disoriented in name, place and/or date
4. Hallucination, delustion or psychosis
   Score if present
5. Psychomotor agitation or retardation
   Either hyperactive, requiring sedatives or restraints, or hypoactive
6. Inappropriate speech or mood
   Score if present
7. Sleep-wake cycle disturbance
   Either frequent awakening/ <4 hours of sleep or sleeping during most of the day
8. Symptom fluctuation
   Fluctuation of any of the above symptoms over a 24 hour period

Score 1 point for every item present:
0 points: no delirium
1–3 points: subsyndromal delirium
4–8 points: delirium

Adapted from Bergeron et al. (2001) and Ouimet et al. (2007b).
*It is only possible to assess the ICDSC if the RASS score > −3. The first four criteria are based on a bedside assessment, the other four on observations throughout the entire shift.
[†]If the patient has a neuromuscular disease and squeezing is impossible, eye blinks can be used.

HOSPITAL COURSE AND MANAGEMENT

Treatment of delirium is first management of the underlying condition. A delay in treatment of ICU delirium was found to be associated with impaired outcome (Heymann et al., 2010). Secondly, symptomatic treatment aims to reduce symptoms such as agitation and hallucinations) (Table 25.3).

Prediction

Based on 10 characteristics that are known within the first 24 hours of ICU admission, it can be predicted which patients will develop delirium during their stay in the ICU (Van den Boogaard et al., 2012a). The pooled area under the receiver operating characteristics curve (AUROC) of this PRE-DELIRIC model was 0.85, which was significantly higher than the AUROC for nurses' and physicians' predictions (0.59) (Van den Boogaard et al., 2012a). The same authors developed another model, based on characteristics at ICU admission, called E-PRE-DELIRIC, which consists of the following nine factors: age, history of cognitive impairment, history of alcohol abuse, admission category (surgery, medical, trauma, neurology/neurosurgery), urgent admission, mean arterial pressure, use of corticosteroids, respiratory failure, and blood urea nitrogen. This model can be used at ICU admission to calculate the risk for delirium during the complete ICU stay and has an AUROC of 0.75 (Van den Boogaard et al., 2014; Wassenaar et al., 2015).

Table 25.3

Hospital course and management of delirium in the intensive care unit (ICU)

Prediction	
E-PRE-DELIRIC model	Wassenaar et al. (2015)
Non-pharmacological prevention and treatment	
Evaluation of precipitating risk factors	
Multicomponent prevention protocols	Inouye et al. (1999); Hshieh et al. (2015)
Early physical and occupational therapy	Schweickert et al. (2009)
Sleep promotion program:	
– Earplugs during the night	Van Rompaey et al. (2012)
– Reduced exposure to light and noise during the night	Zaal et al. (2013); Kamdar et al. (2013)
– Improved exposure to daylight	Zaal et al. (2013); Kamdar et al. (2013)
– Single-room ICU	Zaal et al. (2013)
Reorientation program	Colombo et al. (2012)
Pharmacological prevention	
Low-dose haloperidol (in high risk patients)	Van den Boogaard et al. (2013a)
Risperidone	Prakanrattana and Prapaitrakool (2007); Hakim et al. (2012)
Avoid continuous infusion of benzodiazepines	Pandharipande et al. (2006); Zaal et al. (2015b)
Dexmedetomidine for sedation	Maldonado et al. (2009); Riker et al. (2009)
Pharmacological treatment	
Haloperidol (for acute agitated patients)	Jacobi et al. (2002); Page et al. (2013)
Olanzapine (if haloperidol is contraindicated)	Skrobik et al. (2004)
Quetiapine (in addition to haloperidol)	Devlin et al. (2010)
Dexmedetomidine	Riker et al. (2009); Reade et al. (2009); Barr et al. (2013)
Combined pain, agitation and delirium protocols	Dale et al. (2014); Balas et al. (2014)

Nonpharmacologic prevention and treatment

The focus in the prevention of ICU delirium should be on minimizing modifiable precipitating risk factors and treatment of underlying conditions (Hsieh et al., 2013).

Multicomponent nonpharmacologic prevention approaches in non-ICU patients resulted in a significant decrease in the odds for delirium of about 50%, as shown in a meta-analysis of 11 studies (Hshieh et al., 2015). Several practical interventions were used, such as early mobilization, enhancing sleep through a warm drink and relaxing music at bedtime, quieter hallways and fewer sleep interruptions for medication and procedures, visual and hearing aids, and reorientation with a card of care team members and a day schedule. The prevention program was especially effective in the primary prevention, as it had no effect on the risk of recurrence (Inouye et al., 1999). Other effective interventions included a proactive daily geriatrician consultation and the assistance of family members, who placed familiar objects in the room and helped the patient with reorientation (Marcantonio et al., 2001; Martinez et al., 2012).

There is little evidence on the effects of nonpharmacologic prevention. In a randomized controlled trial (RCT) on early physical and occupational therapy compared to usual care, early mobilization was found to be safe and associated with a decrease of the median delirium duration from 4 days to 2 days (Schweickert et al., 2009).

Another RCT, comparing 69 ICU patients with earplugs during the night with 67 patients without earplugs, showed that earplugs decreased the risk for delirium and mild confusion by more than 50%. This improvement is entirely due to a decline in mild confusion on the Neelon and Champagne Confusion Scale (NEECHAM), as the frequency of moderate to severe confusion did not differ between the two groups (Van Rompaey et al., 2008, 2012).

In a before-and-after study on several sleep quality improvement interventions (prevention of daytime sleeping with more daylight and more daytime activities and reduction of nighttime noise and light), a lower incidence of delirium and coma was found. Sleep ratings, however, did not improve after the intervention and it is therefore not possible to attribute the lower delirium incidence specifically to sleep (Kamdar et al., 2013).

Another before-and-after study showed that moving from a conventional ICU with wards to a single-room ICU with a view, designed to reduce noise and improve daylight exposure, led to a shorter duration of delirium (mean 0.4 days). The incidence of delirium was however similar in both groups (Zaal et al., 2013).

An ICU reorientation strategy halved the occurrence rate of delirium in a third before-and-after study. Nurses stimulated the memory of the patients, provided them with information about their location and illness progression, and asked if patients wanted to read or listen to music. The rooms were equipped with a clock in front of every bed and nighttime noises were reduced to a minimum (Colombo et al., 2012).

In summary, multicomponent nonpharmacologic strategies have been shown to be effective in non-ICU patients. Considering the fact that ICU patients are exposed to numerous risk factors for delirium, a multifactorial approach is preferable. The few studies in ICU patients suggest that implementing an early mobilization program, the use of earplugs, reductions of nighttime light and noise, with increased exposure to daylight and a reorientation program, can reduce the frequency of delirium substantially.

Pharmacologic prevention

Based on the presumed pathogenesis of delirium, several groups of medications could be avoided in the prevention and treatment of delirium: antipsychotics, alpha-2-receptor agonists, statins, corticosteroids, and cholinesterase inhibitors. The number of trials on pharmacologic prevention of delirium in the ICU is limited and most of these trials concern elective surgery patients (Serafim et al., 2015).

Only two randomized placebo-controlled trials have been conducted on antipsychotics as prophylaxis and early treatment for delirium in ICU patients. In the Hope-ICU trial, treatment with haloperidol, a first-generation antipsychotic, 2.5 mg intravenously every 8 hours or placebo was initiated within 72 hours after ICU admission in 141 patients. Patients in the haloperidol group spent about the same number of days without delirium and without coma, as did patients in the placebo group (median 5 days vs. 6 days (Page et al., 2013). The study has been criticized for not discriminating between delirious and nondelirious patients at baseline and thereby failing to make a distinction between prophylaxis and treatment.

In a recent multicenter placebo-controlled RCT, low-dose haloperidol (1 mg every 6 hours) was studied for the prevention of delirium in 68 ICU patients with subsyndromal delirium (ICDSC 1–3). A similar number of patients given haloperidol (35%) and placebo (23%) developed delirium ($p = 0.29$). Haloperidol did however reduce agitation (Al-Qadheeb et al., 2016).

A recent observational study suggested even a detrimental effect of haloperidol on the occurrence of delirium. The cumulative dose of haloperidol and next-day diagnosis of delirium was studied in 93 older medical ICU patients. Each additional cumulative milligram of haloperidol was associated with 5% higher odds of next-day delirium in nonintubated cases, although no association was found in intubated patients (Pisani et al., 2015).

By contrast, a beneficial effect of haloperidol has been observed in a placebo-controlled RCT on 457 noncardiac surgery ICU patients. Prophylactic low-dose haloperidol (0.5 mg bolus followed by 0.1 mg/hour) during the first 12 hours resulted in a decreased incidence of delirium of 15%, compared to 23% in the placebo group ($p = 0.03$). With a low mean APACHE II score and median length of stay in the ICU shorter than 24 hours, this study population can however not be considered as critically ill (Wang et al., 2012).

In addition, a strong positive effect of haloperidol prophylaxis was found in a before-and-after study. Patients with a predicted risk of more than 50% or a history of dementia or alcohol abuse received a prophylactic haloperidol dose of 1 mg every 8 hours intravenously during the intervention period. Delirium incidence decreased from 75% to 65% and mortality from 12% to 7.3% (Van den Boogaard et al., 2013a). Based on this study, haloperidol prophylaxis is currently studied in a placebo-controlled RCT in 2145 ICU patients (Van den Boogaard et al., 2013b).

Another strategy for the prevention of delirium is to minimize the effects of pharmacologic risk factors, which may induce or prolong ICU delirium. Especially benzodiazepines seem to be associated with delirium in the ICU, but there is also evidence that opioids, anticholinergic drugs, and corticosteroids are risk factors (Han et al., 2001; Schreiber et al., 2014; Zaal et al., 2015b; Wolters et al., 2015a, b).

Benzodiazepines appear to increase the risk for transitioning into delirium in a dose-dependent way. In one study, the risk of delirium was raised from 60% without lorazepam to a plateau of 100% for doses larger than 20 mg of lorazepam (Pandharipande et al., 2006). Another study confirmed these results with other more common benzodiazepines in the ICU, such as midazolam and oxazepam, and showed that for every 5 mg midazolam equivalent, the risk increased 4% with a plateau for doses of 150 mg and more. The association was only found in awake patients who received benzodiazepines by continuous intravenous infusion and the increased risk continued to exist for 2 days after the exposure. Intermittent administration of benzodiazepines was not associated with an increased risk (Zaal et al., 2015a).

As mentioned earlier, it is unclear if opioids are a risk factor for delirium, including positive and negative associations (Zaal et al., 2015b). Since undertreated severe pain is a presumed risk factor for delirium, adequate pain management is important (Morrison et al., 2003). Concerning drugs with anticholinergic side-effects and corticosteroids, little and conflicting evidence is available and it is therefore not possible to make recommendations (Schreiber et al., 2014; Wolters et al., 2015a, b). Anticholinergic drugs possibly increase the symptom severity of delirium, but not the frequency (Han et al., 2001). Currently, over 600 drugs are presumed to have anticholinergic effects, and additive effects of these drugs seem to result in side-effects (Tune, 2001). It should however be noted that there are various scales available to assess anticholinergic effects, and that there is no consensus which scale represents anticholinergic load best (Carnahan et al., 2006; Rudolph, 2008; Durán et al., 2013).

Dexmedetomidine, a highly selective α_2-receptor agonist with sedative and analgesic effects, is an acceptable alternative for both benzodiazepines and opiates. One double-blind multicenter RCT with 375 medical and surgical ICU patients compared continuous infusion of dexmedetomidine with midazolam. The drugs were equally efficient in maintaining a Richmond Agitation and Sedation Scale (RASS) between –2 and +1, but the dexmedetomidine group had significantly lower delirium prevalence (54% vs. 76.6%) and more delirium-free days (mean of 2.5 vs. 1.7 days) than the midazolam group (Riker et al., 2009). Another RCT in the same population investigated lorazepam instead of midazolam and showed a similar delirium prevalence and delirium-free days (Pandharipande et al., 2007b). The higher costs of dexmedetomidine in comparison to propofol and midazolam might be compensated by reductions in ICU costs, due to shorter duration of mechanical ventilation and ICU stay (Turunen et al., 2015). The application of clonidine, another α_2-receptor agonist, has been suggested to reduce costs as well (Gagnon et al., 2015). It is unclear whether dexmedetomidine is superior to clonidine.

Statins may have, besides inhibiting cholesterol synthesis, an anti-inflammatory and neuroprotective effect and could therefore be beneficial in delirium. A prospective cohort study compared patients who used statins at home and continued using them in the ICU with patients who did not. Statin use was associated with more delirium-free days and a lower C-reactive protein (Morandi et al., 2014). Based on this study, and on a possible beneficial effect in cardiac surgery patients (Katznelson et al., 2009), a placebo-controlled RCT has started on prophylaxis and early treatment with simvastatin (Casarin et al., 2015).

Based on the hypothesis that reduced cholinergic activity plays a central role in the pathogenesis of delirium, prophylactic use of rivastigmine, a cholinesterase inhibitor, has been investigated in one RCT in cardiac surgery patients. Rivastigmine or a placebo was administered from 1 day preoperatively to 6 days after the procedure, and no difference was found in the incidence or duration of delirium between the groups (Gamberini et al., 2009).

Sleep–wake cycle disruption is often observed in delirium and some studies found lower melatonin levels in delirious patients (Mo et al., 2015). As melatonin plays an important role in the sleep–wake cycle, exogenous melatonin administration is suggested for the prevention of delirium. The only RCT on this subject compared ramelteon, a melatonin receptor agonist, at night with a placebo in 67 nonintubated ICU and acute ward patients. Delirium incidence was significantly lower in the ramelteon group (Hatta et al., 2014). This trial was however small ($n=67$) and the benefit was exceptionally large and questionable (risk of delirium 3% vs. 32%).

In summary, the evidence on the prophylactic use of antipsychotics is contradictory. Findings in postoperative patients cannot always be extrapolated to all critically ill patients. Avoiding delirogenic drugs can be an effective strategy in the prevention of delirium. Infusion of benzodiazepines has the strongest evidence as being a risk factor for delirium, whereas alpha-2 agonists seem to be a good alternative.

Pharmacologic treatment

The pharmacologic treatment of delirium in the ICU is still mostly based on clinical experience, as there is little high-quality evidence for its efficacy (Serafim et al., 2015).

The rationale for treatment of delirium with antipsychotics is that delirium may be associated with increased dopaminergic activity and decreased cholinergic activity. However, evidence is limited (Meagher et al., 2013). Antipsychotics can be divided into two classes, typical (or first-generation) versus atypical (or second-generation) antipsychotics, which differ in receptor binding and side-effects. The typical antipsychotics, like haloperidol, were developed in the 1950s and are high-affinity dopamine receptor antagonists. This may lead to acute extrapyramidal side-effects, such as parkinsonism, akathisia, and dystonia. Clozapine, quetiapine, and risperidone are examples of atypical antipsychotics and have been shown to cause fewer extrapyramidal symptoms than haloperidol (Leucht et al., 2009). This is believed to be the result of broader receptor binding, including serotonergic, adrenergic, and acetylcholinergic receptors, dopaminergic inactivation, and a faster dissociation from dopamine receptors (Gross and Geyer, 2012). A rare, but serious, adverse effect of some antipsychotics (e.g., haloperidol, quetiapine, risperidone, and ziprasidone) is a dose-dependent blockade of cardiac potassium channels, resulting in electrocardiogram

QT-interval prolongation, which subsequently can cause *torsades de pointes* and ventricular arrhythmia (Gross and Geyer, 2012). Guidelines therefore advise not to use antipsychotics in patients with existing QT-interval prolongation or who are using other QT-interval-prolonging drugs (Barr et al., 2013).

Haloperidol is commonly used for the treatment of ICU delirium (Devlin et al., 2011), although the current guidelines conclude that no evidence is available that haloperidol reduces the duration of delirium and that haloperidol use should be restricted to ICU delirium with agitated or psychotic symptoms (Martin et al., 2010; Young et al., 2010; Barr et al., 2013). Intravenous administration is favorable in critically ill patients, as they often have changed enteral absorption. To treat acutely agitated delirious patients, the following dosage scheme has been described: a loading dose of 2 mg intravenously, which can be repeated every 20 minutes if agitation persists, with a lower dose in elderly patients. When delirium is controlled, a maintenance dose can be given regularly every 4–6 hours for a few days, titrated according to symptoms. The therapy should then be tapered for several days (Jacobi et al., 2002; Young et al., 2010). In the above-mentioned Hope-ICU trial, the group receiving a dose of 2.5 mg haloperidol intravenously every 8 hours, supplemented with doses of 2.5–5 mg if agitation developed, had a lower proportion of agitated patients (RASS score of +2 and more) (Page et al., 2013).

One double-blind multicenter RCT compared haloperidol, ziprasidone, and placebo in 103 critically ill patients. There was no difference in delirium duration, coma/delirium-free days and extrapyramidal symptoms between the groups, suggesting that antipsychotics have little effect on the course of delirium. The small sample size and lack of standardized sedation protocols could have hindered the detection of significant differences. As the trial did not differentiate between positive symptoms, including agitation and hallucinations, and negative symptoms, including inattention, it is possible that antipsychotics treated some of the positive symptoms (Girard et al., 2010b).

Atypical antipsychotics have been compared to haloperidol in two other RCTs in mixed ICU patients. The first trial randomized 73 patients with delirium to receive either haloperidol (2.5–5 mg every 8 hours orally) or olanzapine (5 mg daily orally). The drugs were equally effective in treating delirium symptoms, but fewer extrapyramidal side-effects were reported in the olanzapine group. This suggests that olanzapine can be useful for patients in whom haloperidol is contraindicated (Skrobik et al., 2004). In the second trial, an RCT with 36 patients, quetiapine or placebo was administered with as-needed haloperidol to delirious patients. Patients receiving quetiapine had a faster resolution of delirium and less agitation, whereas the incidence of adverse effects was similar between groups (Devlin et al., 2010). Unfortunately, olanzapine and quetiapine are only available in tablet form.

In view of the acetylcholine deficiency hypothesis and previous studies in non-ICU patients, cholinesterase inhibitors have been studied as treatment of delirium (Fischer, 2001; Hshieh et al., 2008; Overshott et al., 2010). In a multicenter, double-blind, randomized placebo-controlled RCT, the cholinesterase inhibitor rivastigmine was added to usual care with haloperidol. The study was halted after 140 of the planned 440 patients were included, because of increased mortality (12 vs. 4 patients) and delirium duration (5 vs. 3 days) in the rivastigmine group (Van Eijk et al., 2010). In a small pilot RCT in postoperative hip surgery patients another cholinesterase inhibitor, donepezil, was tested. Donepezil did not reduce the presence or severity of delirium, but did result in more side-effects (Marcantonio et al., 2011).

Several other drugs have been tried. A small open-label randomized pilot trial in 20 patients investigated the use of dexmedetomidine instead of haloperidol for patients who could not be extubated exclusively due to hyperactive delirium. Dexmedetomidine shortened the time to extubation and ICU discharge (Reade et al., 2009). These results were confirmed in a subsequent randomized controlled trial comparing dexmedetomidine (starting at 0.5 μg/kg/hr, maximum dose 1.5 μg/kg/hr) with placebo among 74 patients in whom agitated delirium precluded safe extubation. Ventilator-free days were increased with use of dexmedetomidine and there was more rapid resolution of delirium (Reade et al., 2016). Other small studies and case reports pointed at the use of methylphenidate for hypoactive delirium and exogenous melotonin in ICU delirium, but further research is needed on these drugs (Gagnon et al., 2005; Mo et al., 2015).

In summary, haloperidol is commonly prescribed as treatment of ICU delirium. While there is no evidence that it decreases the duration of delirium, it has been found to reduce agitation. There is limited evidence that atypical antipsychotics have fewer side-effects than haloperidol, but atypical antipsychotics cannot be administered intravenously. It is not recommended to use cholinesterase inhibitors to treat delirium.

Combined pain, agitation, and delirium protocols

Several of the measures above described have been implemented in combined pain, agitation, and delirium (PAD) protocols, as there are many interconnections between the management of these conditions. Most of these protocols include measures such as validated monitoring and screening tools for PAD, daily spontaneous awakening and breathing trials, adequate nonpharmacologic and pharmacologic management of delirium and pain, and early physical therapy (Trogrlić et al., 2015). One before-and-after study with 296 patients showed that, after the implementation of the Awakening and Breathing Coordination, Delirium monitoring/management, and Early exercise/mobility (ABCDE) protocol, the duration of mechanical ventilation decreased by 3 days and delirium incidence from 62% to 49% (Balas et al., 2014). Another before-and-after study with a similar PAD protocol in surgical ICU patients also resulted in reduced delirium, mechanical ventilation, and ICU stay duration (Dale et al., 2014). Although evidence is not based on an RCT, implementation of the PAD protocols seem thus to be important to improve outcome in ICU patients.

CLINICAL TRIALS AND GUIDELINES

Over the years, several guidelines for the management of delirium have been published. The summary of the guideline of the Society of Critical Care Medicine, is shown in Table 25.4 (Barr et al., 2013). Other influential

Table 25.4

"Delirium" in the "Statements and recommendations" section of the "Clinical practice guidelines for the management of pain, agitation, and delirium in adult patients in the intensive care unit" (Barr et al., 2013)

Outcomes associated with delirium
Delirium is associated with increased mortality in adult ICU patients (A)
Delirium is associated with prolonged ICU and hospital LOS in adult ICU patients (A)
Delirium is associated with the development of post-ICU cognitive impairment in adult ICU patients (B)

Detecting and monitoring delirium
We recommend routine monitoring of delirium in adult ICU patients (+1B)
The Confusion Assessment Method for the ICU (CAM-ICU) and the Intensive Care Delirium Screening Checklist (ICDSC) are the most valid and reliable delirium-monitoring tools in adult ICU patients (A)
Routine monitoring of delirium in adult ICU patients is feasible in clinical practice (B)

Delirium risk factors
Four baseline risk factors are positively and significantly associated with the development of delirium in the ICU: pre-existing dementia, history of hypertension and/or alcoholism, and a high severity of illness at admission (B)
Coma is an independent risk factor for the development of delirium in ICU patients (B)
Conflicting data surround the relationship between opioid use and the development of delirium in adult ICU patients (B)
Benzodiazepine use may be a risk factor for the development of delirium in adult ICU patients (B)
There are insufficient data to determine the relationship between propofol use and the development of delirium in adult ICU patients (C)
In mechanically ventilated adult ICU patients at risk of developing delirium, dexmedetomidine infusions administered for sedation may be associated with a lower prevalence of delirium compared to benzodiazepine infusions (B)

Delirium prevention
We recommend performing early mobilization of adult ICU patients whenever feasible to reduce the incidence and duration of delirium (+1B)
We provide no recommendation for using a pharmacologic delirium prevention protocol in adult ICU patients, as no compelling data demonstrate that this reduces the incidence or duration of delirium in these patients (0,C)
We provide no recommendation for using a combined nonpharmacologic and pharmacologic delirium prevention protocol in adult ICU patients, as this has not been shown to reduce the incidence of delirium in these patients (0,C)
We do not suggest that either haloperidol or atypical antipsychotics be administered to prevent delirium in adult ICU patients (−2C)
We provide no recommendation for the use of dexmedetomidine to prevent delirium in adult ICU patients, as there is no compelling evidence regarding its effectiveness in these patients (0,C)

Delirium treatment
There is no published evidence that treatment with haloperidol reduces the duration of delirium in adult ICU patients (no evidence)
Atypical antipsychotics may reduce the duration of delirium in adult ICU patients (C)
We do not recommend administering rivastigmine to reduce the duration of delirium in ICU patients (−1B)
We do not suggest using antipsychotics in patients at significant risk for *torsades de pointes* (i.e., patients with baseline prolongation of QTc interval, patients receiving concomitant medications known to prolong the QTc interval, or patients with a history of this arrhythmia) (−2C)
We suggest that in adult ICU patients with delirium unrelated to alcohol or benzodiazepine withdrawal, continuous intravenous infusions of dexmedetomidine rather than benzodiazepine infusions be administered for sedation to reduce the duration of delirium in these patients (+2B)

ICU, intensive care unit; LOS, length of stay.

guidelines include the British National Institute for Health and Clinical Excellence (NICE) guideline "Diagnosis, prevention, and management of delirium" (Young et al., 2010) and the guidelines of the American Geriatrics Society on postoperative delirium (American Geriatrics Society Expert Panel on Postoperative Delirium in Older Adults, 2014).

After the MENDS trial had showed that dexmedetomidine sedation resulted in more delirium-free and coma-free days than lorazepam (Pandharipande et al., 2007b), the MENDS II trial will compare dexmedetomidine with propofol in a double-blind RCT. The trial's study population will consist of 530 ventilated septic patients and it has delirium/coma-free days as primary endpoint (Vanderbilt University, 2014a).

As a meta-analysis of small trials in postoperative patients (Fok et al., 2015) and a before-and-after study in ICU patients (Van den Boogaard et al., 2013a) suggested a beneficial effect of antipsychotics, a large ($n=2145$) multicenter double-blind, placebo-controlled RCT is currently performed to study whether haloperidol can prevent mortality (primary endpoint) and delirium (Radboud University, 2013).

Based on a lower risk of ICU delirium in statin users in observational studies (Morandi et al., 2014; Page et al., 2014), a single-center placebo-controlled RCT has started to investigate the use of simvastatin as prophylactic drug for delirium in the ICU (Casarin et al., 2015). Another large-scale placebo-controlled RCT investigates treatment with haloperidol versus ziprasidone (Vanderbilt University, 2014b).

COMPLEX CLINICAL DECISIONS

Delirium is often associated with substantial anxiety that does not respond sufficiently to nonpharmacologic measures such as reassurance. In those cases where anxiety is extreme, it is unclear whether a trial should be started with a benzodiazepine, which has anxiolytic properties but may prolong delirium as well (Pandharipande et al., 2006; Zaal et al., 2015a). It should be noted that anxiety in delirium may be caused by hallucinations and delusions, but these are difficult to assess in patients who cannot verbally communicate, for example because of mechanical ventilation, and who do not respond to closed questions. It is important to realize that psychosis is not limited to hyperactive delirious patients, and a trial with an antipsychotic may be warranted.

Further, it is controversial whether disturbed sleep is an early sign of delirium, a risk factor for delirium, or both. Benzodiazepines are often prescribed to promote sleep. However, a benzodiazepine may induce delirium in these cases, and although benzodiazepines may promote light sleep (nonrapid-eye-movement (REM) sleep stage 1), it suppresses deep sleep (non-REM stage 3) and REM sleep, which appears to be crucial for recovery. Theoretically, alpha-2 agonists, such as dexmedetomidine and clonidine, may promote sleep (Brown et al., 2010), but this has not yet been studied extensively.

Finally, benzodiazepines are often prescribed to prevent or treat symptoms of alcohol withdrawal. This practice is supported by a meta-analysis of RCTs that showed that benzodiazepines were superior to placebo in the treatment of alcohol withdrawal (Holbrook et al., 1999). Patients with isolated alcohol withdrawal syndrome are rarely admitted to the ICU, but alcohol withdrawal often contributes to delirium in ICU patients who are exposed to a variety of other delirium risk factors. As benzodiazepines also increase the risk of ICU delirium, it is currently unclear whether benzodiazepines in alcoholic ICU patients are superior to other approaches, such as alpha-2 agonists and anticonvulsant drugs.

OUTCOME PREDICTION

Delirium has, for a long time, been considered as an "inconvenient" presentation of critical illness and considered completely reversible after treatment of the underlying disorder (Van Eijk and Slooter, 2010). Over the last two decades, numerous studies have consistently shown that delirium in ICU patients is associated with a worse prognosis (Salluh et al., 2015).

The experience of delirium is very distressing for the patient, but also for family members and caregivers (Breitbart et al., 2002). Yet, many patients have no recollection of an episode of delirium (Breitbart et al., 2002). The rate of self-extubation and removal of catheters is obviously increased in patients with delirium (Dubois et al., 2001; Van den Boogaard et al., 2012c). Delirium is incontrovertibly associated with an increased ICU length of stay and duration of hospital admission. This is also reflected in the duration of mechanical ventilation and also significantly longer in patients with delirium (Zhang et al., 2013; Salluh et al., 2015). Primarily due to increased duration of admission, delirium results in higher ICU and hospital expenses (Milbrandt et al., 2004). Early delirium research showed a two to three times increased risk of dying during ICU admission for patients who developed delirium (Salluh et al., 2015), whereas the underlying pathophysiologic mechanism remained unclear. In more recent research, with adjustments for disease severity until delirium onset, delirium was no longer associated with increased mortality in the ICU (Klein Klouwenberg et al., 2014).

With regard to long-term mortality, every additional day of delirium was found to increase mortality by 10% at 6 months and 1 year after discharge (Ely et al.,

2004; Pisani et al., 2009a). However, when follow-up was restricted to the period after ICU discharge and more extensive adjustments were made, no increased 1-year mortality was observed (Wolters et al., 2014).

Cognitive impairment

Delirium in the ICU was consistently found to be associated with impaired cognitive function at 3, 12, and 18 months after discharge, also when patients with pre-existing cognitive impairment were excluded and adjustments were made for confounders such as severity of illness (Girard et al., 2010a; Van den Boogaard et al., 2012b; Pandharipande et al., 2013, Wolters et al., 2014). In addition to cognitive problems, delirium was found to be associated with more problems in activities of daily living and worse scores on sensorimotor function tests at follow-up (Brummel et al., 2014). Anxiety, depression, and posttraumatic stress disorder are common in survivors of critical illness (Parker et al., 2015), and related to delirium in non-ICU patients (Davydow, 2009). However, only very few relatively small studies have investigated the association between ICU delirium and subsequent psychiatric morbidity in ICU survivors, and most of these could not establish a relationship. One larger study found an association between prolonged ICU delirium and depression, but not PTSD, 1 year after ICU discharge (Jackson et al., 2014).

Future studies will focus on cognitive impairment as this appears to be independently and consistently associated with delirium. Alternative explanations for the association of delirium and long-term cognitive impairment are that delirium is a marker of vulnerability to cognitive impairment, and that delirium is an intermediate factor in the development of cognitive impairment (Fong et al., 2015).

Arguments that delirium may cause cognitive impairment using the Bradford Hill criteria for causality are shown in Table 25.5 (Hill, 1965; Slooter, 2013). The relationship between delirium and long-term cognitive impairment is relatively strong (odds ratio 2–3: Wolters et al., 2014). It is consistent, as delirium increases the risk of cognitive impairment in all previous studies in both ICU and non-ICU patients, and delirium is a risk factor for acceleration of existing dementia (Fong et al., 2015). The relationship shows a biologic gradient, as more delirium days are associated with more cognitive impairment (Pandharipande et al., 2013). "Plausibility and coherence" are supported by observations that neuroinflammation may persist and increase also in cases of transient systemic inflammation (Van Gool et al., 2010). Patient-related factors that link delirium directly to impaired outcome are lethargy (with aspiration pneumonia, pressure ulcers, thrombosis) and agitation (resulting in falls, and complications of antipsychotics, sedatives, and physical restraints).

By contrast, two criteria for causality seem not to be fulfilled. First, temporality, as delirium does not always precede cognitive impairment, and pre-existing cognitive impairment is also a risk factor for delirium. Second, the criterion of experiment is not fulfilled as there is no RCT that shows that treatment of delirium improves long-term cognitive dysfunction. It should, however, be noted that multicomponent geriatric interventions resulted in less cognitive impairment in non-ICU patients (Pitkälä et al., 2006). We cannot definitively conclude that delirium may have a causal role in the development of long-term cognitive impairment after critical illness.

REFERENCES

Adamis D, Treloar A, Martin FC et al. (2007). A brief review of the history of delirium as a mental disorder. Hist Psychiatry 18: 459–469.

Adamis D, Meagher D, Murray O et al. (2015). Evaluating attention in delirium: a comparison of bedside tests of attention. Geriatr Gerontol Int. http://dx.doi.org/10.1111/ggi.12592.

Almeida ICT, Soares M, Bozza FA et al. (2014). The impact of acute brain dysfunction in the outcomes of mechanically ventilated cancer patients. PLoS One 9: e85332.

Al-Qadheeb NS, Skrobik Y, Schumaker G et al. (2016). Preventing ICU subsyndromal delirium conversion to delirium with low-dose IV haloperidol: a double-blind, placebo-controlled pilot study. Crit Care Med 44: 583–591.

Alsop DC, Fearing MA, Johnson K et al. (2006). The role of neuroimaging in elucidating delirium pathophysiology. J Gerontol A Biol Sci Med Sci 61: 1287–1293.

American Geriatrics Society Expert Panel on Postoperative Delirium in Older Adults (2014). Postoperative delirium in older adults: best practice statement from the American Geriatrics Society. J Am Coll Surg 220: 136–148.e1.

American Psychiatric Association (1980). Diagnostic and Statistical Manual of Mental Disorders, 3rd edn. American Psychiatric Publishing, Washington, DC.

American Psychiatric Association (2013). Diagnostic and Statistical Manual of Mental Disorders, 5th edn. American Psychiatric Publishing, Arlington, VA.

Table 25.5

Bradford Hill criteria for causality

Strength of association	✓
Consistency	✓
Temporality	✗
Biological gradient	✓
Plausibility and coherence	✓
Experiment	?

Adapted from Hill (1965).

Aurell J, Elmqvist D (1985). Sleep in the surgical intensive care unit: continuous polygraphic recording of sleep in nine patients receiving postoperative care. BMJ 290: 1029–1032.

Balas MC, Vasilevskis EE, Olsen KM et al. (2014). Effectiveness and safety of the awakening and breathing coordination, delirium monitoring/management, and early exercise/mobility bundle. Crit Care Med 42: 1024–1036.

Barr J, Fraser GL, Puntillo K et al. (2013). Clinical practice guidelines for the management of pain, agitation, and delirium in adult patients in the intensive care unit. Crit Care Med 41: 263–306.

Bergeron N, Dubois M-J, Dumont M et al. (2001). Intensive care delirium screening checklist: evaluation of a new screening tool. Intensive Care Med 27: 859–864.

Boyer EW, Shannon M (2005). The serotonin syndrome. N Engl J Med 352: 1112–1120.

Breitbart W, Gibson C, Tremblay A (2002). The delirium experience: delirium recall and delirium-related distress in hospitalized patients with cancer, their spouses/caregivers, and their nurses. Psychosomatics 43: 183–194.

Brown EN, Lydic R, Schiff ND (2010). General anesthesia, sleep, and coma. N Engl J Med 363: 2638–2650.

Brummel NE, Jackson JC, Pandharipande PP et al. (2014). Delirium in the ICU and subsequent long-term disability among survivors of mechanical ventilation. Crit Care Med 42: 369–377.

Carnahan RM, Lund BC, Perry PJ et al. (2006). The Anticholinergic Drug Scale as a measure of drug-related anticholinergic burden: associations with serum anticholinergic activity. J Clin Pharmacol 46: 1481–1486.

Caruso P, Guardian L, Tiengo T et al. (2014). ICU architectural design affects the delirium prevalence. Crit Care Med 42: 2204–2210.

Casarin A, McAuley DF, Alce TM et al. (2015). Evaluating early administration of the hydroxymethylglutaryl-CoA reductase inhibitor simvastatin in the prevention and treatment of delirium in critically ill ventilated patients (MoDUS trial): study protocol for a randomized controlled trial. Trials 16: 218.

Choi S-HH, Lee H, Chung T-SS et al. (2012). Neural network functional connectivity during and after an episode of delirium. Am J Psychiatry 169: 498–507.

Colombo R, Corona A, Praga F et al. (2012). A reorientation strategy for reducing delirium in the critically ill. Results of an interventional study. Minerva Anestesiol 78: 1026–1033.

Dale CR, Kannas DA, Fan VS et al. (2014). Improved analgesia, sedation, and delirium protocol associated with decreased duration of delirium and mechanical ventilation. Ann Am Thorac Soc 11: 367–374.

Davydow DS (2009). Symptoms of depression and anxiety after delirium. Psychosomatics 50: 309–316.

De Rooij SE, van Munster BC, Korevaar JC et al. (2007). Cytokines and acute phase response in delirium. J Psychosom Res 62: 521–525.

Devlin JW, Roberts RJ, Fong JJ et al. (2010). Efficacy and safety of quetiapine in critically ill patients with delirium: a prospective, multicenter, randomized, double-blind, placebo-controlled pilot study. Crit Care Med 38: 419–427.

Devlin JW, Bhat S, Roberts RJ et al. (2011). Current perceptions and practices surrounding the recognition and treatment of delirium in the intensive care unit: a survey of 250 critical care pharmacists from eight states. Ann Pharmacother 45: 1217–1229.

Dubois MJ, Bergeron N, Dumont M et al. (2001). Delirium in an intensive care unit: a study of risk factors. Intensive Care Med 27: 1297–1304.

Durán CE, Azermai M, Vander Stichele RH (2013). Systematic review of anticholinergic risk scales in older adults. Eur J Clin Pharmacol 69: 1485–1496.

Ebersoldt M, Sharshar T, Annane D (2007). Sepsis-associated delirium. Intensive Care Med 33: 941–950.

Ely EW, Gautam S, Margolin R et al. (2001a). The impact of delirium in the intensive care unit on hospital length of stay. Intensive Care Med 27: 1892–1900.

Ely EW, Margolin R, Francis J et al. (2001b). Evaluation of delirium in critically ill patients: validation of the Confusion Assessment Method for the Intensive Care Unit (CAM-ICU). Crit Care Med 29: 1370–1379.

Ely EW, Shintani A, Truman B et al. (2004). Delirium as a predictor of mortality in mechanically ventilated patients in the intensive care unit. JAMA 291: 1753–1762.

Ely EW, Girard TD, Shintani AK et al. (2007). Apolipoprotein E4 polymorphism as a genetic predisposition to delirium in critically ill patients. Crit Care Med 35: 112–117.

Field RH, Gossen A, Cunningham C (2012). Prior pathology in the basal forebrain cholinergic system predisposes to inflammation-induced working memory deficits: reconciling inflammatory and cholinergic hypotheses of delirium. J Neurosci 32: 6288–6294.

Fischer P (2001). Successful treatment of nonanticholinergic delirium with a cholinesterase inhibitor. J Clin Psychopharmacol 21: 118.

Fok MC, Sepehry AA, Frisch L et al. (2015). Do antipsychotics prevent postoperative delirium? A systematic review and meta-analysis. Int J Geriatr Psychiatry 30: 333–344.

Fong TG, Bogardus ST, Daftary A et al. (2006). Cerebral perfusion changes in older delirious patients using 99mTc HMPAO SPECT. J Gerontol A Biol Sci Med Sci 61: 1294–1299.

Fong TG, Davis D, Growdon ME et al. (2015). The interface between delirium and dementia in elderly adults. Lancet Neurol 14: 823–832.

Gagnon B, Low G, Schreier G (2005). Methylphenidate hydrochloride improves cognitive function in patients with advanced cancer and hypoactive delirium: a prospective clinical study. J Psychiatry Neurosci 30: 100–107.

Gagnon DJ, Riker RR, Glisic EK et al. (2015). Transition from dexmedetomidine to enteral clonidine for ICU sedation: an observational pilot study. Pharmacother J Hum Pharmacol Drug Ther 35: 251–259.

Gamberini M, Bolliger D, Lurati Buse GA et al. (2009). Rivastigmine for the prevention of postoperative delirium in elderly patients undergoing elective cardiac surgery – a randomized controlled trial. Crit Care Med 37: 1762–1768.

Girard TD, Jackson JC, Pandharipande PP et al. (2010a). Delirium as a predictor of long-term cognitive impairment in survivors of critical illness. Crit Care Med 38: 1513–1520.

Girard TD, Pandharipande PP, Carson SS et al. (2010b). Feasibility, efficacy, and safety of antipsychotics for intensive care unit delirium: the MIND randomized, placebo-controlled trial. Crit Care Med 38: 428–437.

Gross G, Geyer MA (2012). Current antipsychotics, Springer, Berlin.

Gunther ML, Morandi A, Krauskopf E et al. (2012). The association between brain volumes, delirium duration, and cognitive outcomes in intensive care unit survivors: the VISIONS cohort magnetic resonance imaging study. Crit Care Med 40: 2022–2032.

Gusmao-Flores D, Salluh JIF, Chalhub RÁ et al. (2012). The confusion assessment method for the intensive care unit (CAM-ICU) and intensive care delirium screening checklist (ICDSC) for the diagnosis of delirium: a systematic review and meta-analysis of clinical studies. Crit Care Med 16: R115.

Hakim SM, Othman AI, Naoum DO (2012). Early treatment with risperidone for subsyndromal delirium after on-pump cardiac surgery in the elderly. Anesthesiology 116: 987–997.

Han L, McCusker J, Cole M et al. (2001). Use of medications with anticholinergic effect predicts clinical severity of delirium symptoms in older medical inpatients. Arch Intern Med 161: 1099–1105.

Hatta K, Kishi Y, Wada K et al. (2014). Preventive effects of ramelteon on delirium. JAMA Psychiat 71: 397.

Heymann A, Radtke F, Schiemann A et al. (2010). Delayed treatment of delirium increases mortality rate in intensive care unit patients. J Int Med Res 38: 1584–1595.

Hill AB (1965). The environment and disease: association or causation? Proc R Soc Med 58: 295–300.

Holbrook AM, Crowther R, Lotter A et al. (1999). Meta-analysis of benzodiazepine use in the treatment of acute alcohol withdrawal. CMAJ 160: 649–655.

Hshieh TT, Fong TG, Marcantonio ER et al. (2008). Cholinergic deficiency hypothesis in delirium: a synthesis of current evidence. J Gerontol A Biol Sci Med Sci 63: 764–772.

Hshieh TT, Yue J, Oh E et al. (2015). Effectiveness of multicomponent nonpharmacological delirium interventions. JAMA Intern Med 175: 512.

Hsieh SJ, Ely EW, Gong MN (2013). Can intensive care unit delirium be prevented and reduced? Lessons learned and future directions. Ann Am Thorac Soc 10: 648–656.

Inouye SK, Bogardus ST, Charpentier PA et al. (1999). A multicomponent intervention to prevent delirium in hospitalized older patients. N Engl J Med 340: 669–676.

Jackson JC, Pandharipande PP, Girard TD et al. (2014). Depression, post-traumatic stress disorder, and functional disability in survivors of critical illness in the BRAIN-ICU study: a longitudinal cohort study. Lancet Respir Med 2: 369–379.

Jackson JC, Morandi A, Girard TD et al. (2015). Functional brain imaging in survivors of critical illness: a prospective feasibility study and exploration of the association between delirium and brain activation patterns. J Crit Care 30: 653. e1–653.e7.

Jacobi J, Fraser GL, Coursin DB et al. (2002). Clinical practice guidelines for the sustained use of sedatives and analgesics in the critically ill adult. Crit Care Med 30: 119–141.

Janz DR, Abel TW, Jackson JC et al. (2010). Brain autopsy findings in intensive care unit patients previously suffering from delirium: a pilot study. J Crit Care 25: 538.e7–538. e12.

Kamdar BB, King LM, Collop NA et al. (2013). The effect of a quality improvement intervention on perceived sleep quality and cognition in a medical ICU. Crit Care Med 41: 800–809.

Kamdar BB, Niessen T, Colantuoni E et al. (2015). Delirium transitions in the medical ICU: exploring the role of sleep quality and other factors. Crit Care Med 43: 135–141.

Kaplan PW, Rossetti AO (2011). EEG patterns and imaging correlations in encephalopathy: encephalopathy part II. J Clin Neurophysiol 28: 233–251.

Katznelson R, Djaiani GN, Borger MA et al. (2009). Preoperative use of statins is associated with reduced early delirium rates after cardiac surgery. Anesthesiology 110: 67–73.

Klein Klouwenberg PMC, Zaal IJ, Spitoni C et al. (2014). The attributable mortality of delirium in critically ill patients: prospective cohort study. BMJ 349: g6652.

Leucht S, Corves C, Arbter D et al. (2009). Second-generation versus first-generation antipsychotic drugs for schizophrenia: a meta-analysis. Lancet 373: 31–41.

Lipowski Z (1990). Delirium: acute confusional states, Oxford University Press, New York.

Maclullich AMJ, Ferguson KJ, Miller T et al. (2008). Unravelling the pathophysiology of delirium: a focus on the role of aberrant stress responses. J Psychosom Res 65: 229–238.

Maldonado JR (2013). Neuropathogenesis of delirium: review of current etiologic theories and common pathways. Am J Geriatr Psychiatry 21: 1190–1222.

Maldonado JR, Wysong A, van der Starre PJA et al. (2009). Dexmedetomidine and the reduction of postoperative delirium after cardiac surgery. Psychosomatics 50: 206–217.

Marcantonio ER, Flacker JM, Wright RJ et al. (2001). Reducing delirium after hip fracture: a randomized trial. J Am Geriatr Soc 49: 516–522.

Marcantonio ER, Palihnich K, Appleton P et al. (2011). Pilot randomized trial of donepezil hydrochloride for delirium after hip fracture. J Am Geriatr Soc 59: 282–288.

Marquis F, Ouimet S, Riker R et al. (2007). Individual delirium symptoms: do they matter? Crit Care Med 35: 2533–2537.

Martin J, Heymann A, Bäsell K et al. (2010). Evidence and consensus-based German guidelines for the management of analgesia, sedation and delirium in intensive care – short version. Ger Med Sci 8. Doc02.

Martinez FT, Tobar C, Beddings CI et al. (2012). Preventing delirium in an acute hospital using a non-pharmacological intervention. Age Ageing 41: 629–634.

McGrane S, Girard TD, Thompson JL et al. (2011). Procalcitonin and C-reactive protein levels at admission as predictors of duration of acute brain dysfunction in critically ill patients. Crit Care Med 15: R78.

McNicoll L, Pisani MA, Zhang Y et al. (2003). Delirium in the intensive care unit: occurrence and clinical course in older patients. J Am Geriatr Soc 51: 591–598.

Meagher DJ (2009). Motor subtypes of delirium: Past, present and future. Int Rev Psychiatry 21: 59–73.

Meagher DJ, O'Hanlon D, O'Mahony E et al. (2000). Relationship between symptoms and motoric subtype of delirium. J Neuropsychiatry Clin Neurosci 12: 51–56.

Meagher DJ, McLoughlin L, Leonard M et al. (2013). What do we really know about the treatment of delirium with antipsychotics? Ten key issues for delirium pharmacotherapy. Am J Geriatr Psychiatry 21: 1223–1238.

Milbrandt EB, Deppen S, Harrison PL et al. (2004). Costs associated with delirium in mechanically ventilated patients. Crit Care Med 32: 955–962.

Mo Y, Scheer CE, Abdallah GT (2015). Emerging role of melatonin and melatonin receptor agonists in sleep and delirium in intensive care unit patients. J Intensive Care Med: 1–5.

Morandi A, Pandharipande P, Trabucchi M et al. (2008). Understanding international differences in terminology for delirium and other types of acute brain dysfunction in critically ill patients. Intensive Care Med 34: 1907–1915.

Morandi A, Rogers BP, Gunther ML et al. (2012). The relationship between delirium duration, white matter integrity, and cognitive impairment in intensive care unit survivors as determined by diffusion tensor imaging. Crit Care Med 40: 2182–2189.

Morandi A, Hughes CG, Thompson JL et al. (2014). Statins and delirium during critical illness: a multicenter, prospective cohort study. Crit Care Med 42: 1899–1909.

Morrison RS, Magaziner J, Gilbert M et al. (2003). Relationship between pain and opioid analgesics on the development of delirium following hip fracture. J Gerontol A Biol Sci Med Sci 58: 76–81.

Mu D-L, Wang D-X, Li L-H et al. (2010). High serum cortisol level is associated with increased risk of delirium after coronary artery bypass graft surgery: a prospective cohort study. Crit Care Med 14: R238.

Neto AS, Nassar AP, Cardoso SO et al. (2012). Delirium screening in critically ill patients: a systematic review and meta-analysis. Crit Care Med 40: 1946–1951.

Oddo M, Carrera E, Claassen J et al. (2009). Continuous electroencephalography in the medical intensive care unit. Crit Care Med 37: 2051–2056.

Omiya H, Yoshitani K, Yamada N et al. (2015). Preoperative brain magnetic resonance imaging and postoperative delirium after off-pump coronary artery bypass grafting: a prospective cohort study. Can J Anesth 62: 595–602.

Ouimet S, Kavanagh BP, Gottfried SB et al. (2007a). Incidence, risk factors and consequences of ICU delirium. Intensive Care Med 33: 66–73.

Ouimet S, Riker R, Bergeon N et al. (2007b). Subsyndromal delirium in the ICU: evidence for a disease spectrum. Intensive Care Med 33: 1007–1013.

Overshott R, Vernon M, Morris J et al. (2010). Rivastigmine in the treatment of delirium in older people: a pilot study. Int Psychogeriatr 22: 812–818.

Page VJ, Ely EW, Gates S et al. (2013). Effect of intravenous haloperidol on the duration of delirium and coma in critically ill patients (Hope-ICU): a randomised, double-blind, placebo-controlled trial. Lancet Respir Med 1: 515–523.

Page VJ, Davis D, Zhao XB et al. (2014). Statin use and risk of delirium in the critically ill. Am J Respir Crit Care Med 189: 666–673.

Pandharipande PP, Shintani A, Peterson J et al. (2006). Lorazepam is an independent risk factor for transitioning to delirium in intensive care unit patients. Anesthesiology 104: 21–26.

Pandharipande PP, Cotton BA, Shintani A et al. (2007a). Motoric subtypes of delirium in mechanically ventilated surgical and trauma intensive care unit patients. Intensive Care Med 33: 1726–1731.

Pandharipande PP, Pun BT, Herr DL et al. (2007b). Effect of sedation with dexmedetomidine vs lorazepam on acute brain dysfunction in mechanically ventilated patients: the MENDS randomized controlled trial. JAMA 298: 2644–2653.

Pandharipande PP, Cotton BA, Shintani A et al. (2008). Prevalence and risk factors for development of delirium in surgical and trauma intensive care unit patients. J Trauma 65: 34–41.

Pandharipande PP, Girard TD, Jackson JC et al. (2013). Long-term cognitive impairment after critical illness. N Engl J Med 369: 1306–1316.

Parker AM, Sricharoenchai T, Raparla S et al. (2015). Posttraumatic stress disorder in critical illness survivors: a metaanalysis. Crit Care Med 43: 1121–1129.

Patel RP, Gambrell M, Speroff T et al. (2009). Delirium and sedation in the intensive care unit: survey of behaviors and attitudes of 1384 healthcare professionals. Crit Care Med 37: 825–832.

Peterson JF, Pun BT, Dittus RS et al. (2006). Delirium and its motoric subtypes: a study of 614 critically ill patients. J Am Geriatr Soc 54: 479–484.

Pfister D, Siegemund M, Dell-Kuster S et al. (2008). Cerebral perfusion in sepsis-associated delirium. Crit Care Med 12: R63.

Pisani MA, Kong SYJ, Kasl SV et al. (2009a). Days of delirium are associated with 1-year mortality in an older intensive care unit population. Am J Respir Crit Care Med 180: 1092–1097.

Pisani MA, Murphey TE, Araujo KLB et al. (2009b). Benzodiazepine and opioid use and the duration of ICU delirium in an older population. Crit Care Med 37: 177–183.

Pisani MA, Araujo KLB, Murphy TE (2015). Association of cumulative dose of haloperidol with next-day delirium in older medical ICU patients. Crit Care Med 43: 996–1002.

Pitkälä KH, Laurila JV, Strandberg TE et al. (2006). Multicomponent geriatric intervention for elderly inpatients with delirium: a randomized, controlled trial. J Gerontol A Biol Sci Med Sci 61: 176–181.

Plaschke K, Fichtenkamm P, Schramm C et al. (2010). Early postoperative delirium after open-heart cardiac surgery is associated with decreased bispectral EEG and increased cortisol and interleukin-6. Intensive Care Med 36: 2081–2089.

Prakanrattana U, Prapaitrakool S (2007). Efficacy of risperidone for prevention of postoperative delirium in cardiac surgery. Anaesth Intensive Care 35: 714–719.

Qin L, Wu X, Block ML et al. (2007). Systemic LPS causes chronic neuroinflammation and progressive neurodegeneration. Glia 55: 453–462.

Radboud University (2013). Effects of prophylactic use of haloperidol in critically ill patients with a high risk for delirium. In: ClinicalTrials.gov. https://clinicaltrials.gov/show/NCT01785290. Accessed 4 Oct 2015.

Reade MC, Eastwood GM, Bellomo R et al. (2016). Effect of dexmedetomidine added to standard care on ventilator-free time in patients with agitated delirium: a randomized clinical trial. JAMA 315: 1460–1480.

Reade MC, O'Sullivan K, Bates S et al. (2009). Dexmedetomidine vs. haloperidol in delirious, agitated, intubated patients: a randomised open-label trial. Crit Care Med 13: R75.

Riker RR, Shehabi Y, Bokesch PM et al. (2009). Dexmedetomidine vs midazolam for sedation of critically ill patients: a randomized trial. JAMA 301: 489–499.

Root JC, Pryor KO, Downey R et al. (2013). Association of pre-operative brain pathology with post-operative delirium in a cohort of non-small cell lung cancer patients undergoing surgical resection. Psychooncology 22: 2087–2094.

Rudolph JL (2008). The anticholinergic risk scale and anticholinergic adverse effects in older persons. Arch Intern Med 168: 508.

Salluh JIF, Wang H, Schneider EB et al. (2015). Outcome of delirium in critically ill patients: systematic review and meta-analysis. BMJ 350: h2538.

Schreiber MP, Colantuoni E, Bienvenu OJ et al. (2014). Corticosteroids and transition to delirium in patients with acute lung injury. Crit Care Med 42: 1480–1486.

Schweickert WD, Pohlman MC, Pohlman AS et al. (2009). Early physical and occupational therapy in mechanically ventilated, critically ill patients: a randomised controlled trial. Lancet 373: 1874–1882.

Sechi G, Serra A (2007). Wernicke's encephalopathy: new clinical settings and recent advances in diagnosis and management. Lancet Neurol 6: 442–455.

Serafim RB, Dutra MF, Saddy F et al. (2012). Delirium in postoperative nonventilated intensive care patients: risk factors and outcomes. Ann Intensive Care 2: 51.

Serafim RB, Bozza FA, Soares M et al. (2015). Pharmacologic prevention and treatment of delirium in intensive care patients: A systematic review. J Crit Care 30: 799–807.

Shioiri A, Kurumaji A, Takeuchi T et al. (2010). White matter abnormalities as a risk factor for postoperative delirium revealed by diffusion tensor imaging. Am J Geriatr Psychiatry 18: 743–753.

Skrobik YK, Bergeron N, Dumont M et al. (2004). Olanzapine vs haloperidol: treating delirium in a critical care setting. Intensive Care Med 30: 444–449.

Slooter AJC (2013). Neurocritical care: critical illness, delirium and cognitive impairment. Nat Rev Neurol 9: 666–667.

Soiza RL, Sharma V, Ferguson K et al. (2008). Neuroimaging studies of delirium: a systematic review. J Psychosom Res 65: 239–248.

Stoicea N, McVicker S, Quinones A et al. (2014). Delirium-biomarkers and genetic variance. Front Pharmacol 5: 75.

Tieges Z, Stíobhairt A, Scott K et al. (2015). Development of a smartphone application for the objective detection of attentional deficits in delirium. Int Psychogeriatr 27: 1251–1262.

Trogrlić Z, van der Jagt M, Bakker J et al. (2015). A systematic review of implementation strategies for assessment, prevention, and management of ICU delirium and their effect on clinical outcomes. Crit Care Med 19: 157.

Trzepacz PT (1999). Update on the neuropathogenesis of delirium. Dement Geriatr Cogn Disord 10: 330–334.

Tune LE (2001). Anticholinergic effects of medication in elderly patients. J Clin Psychiatry 62 (Suppl 2): 11–14.

Turunen H, Jakob SM, Ruokonen E et al. (2015). Dexmedetomidine versus standard care sedation with propofol or midazolam in intensive care: an economic evaluation. Crit Care Med 19: 67.

Van Dellen E, van der Kooi AW, Numan T et al. (2014). Decreased functional connectivity and disturbed electroencephalography of intensive care unit patients with delirium after cardiac surgery. Anesthesiology 121: 328–335.

Van den Boogaard M, Kox M, Quinn KL et al. (2011). Biomarkers associated with delirium in critically ill patients and their relation with long-term subjective cognitive dysfunction; indications for different pathways governing delirium in inflamed and noninflamed patients. Crit Care Med 15: R297.

Van den Boogaard M, Pickkers P, Slooter AJC et al. (2012a). Development and validation of PRE-DELIRIC (PREdiction of DELIRium in ICu patients) delirium prediction model for intensive care patients: observational multicentre study. BMJ 344: e420.

Van den Boogaard M, Schoonhoven L, Evers AWM et al. (2012b). Delirium in critically ill patients: impact on long-term health-related quality of life and cognitive functioning. Crit Care Med 40: 112–118.

Van den Boogaard M, Schoonhoven L, van der Hoeven JG et al. (2012c). Incidence and short-term consequences of delirium in critically ill patients: a prospective observational cohort study. Int J Nurs Stud 49: 775–783.

Van den Boogaard M, Schoonhoven L, van Achterberg T et al. (2013a). Haloperidol prophylaxis in critically ill patients with a high risk for delirium. Crit Care Med 17: R9.

Van den Boogaard M, Slooter AJ, Brüggemann RJ et al. (2013b). Prevention of ICU delirium and delirium-related outcome with haloperidol: a study protocol for a multicenter randomized controlled trial. Trials 14: 400.

Van den Boogaard M, Schoonhoven L, Maseda E et al. (2014). Recalibration of the delirium prediction model for ICU patients (PRE-DELIRIC): a multinational observational study. Intensive Care Med 40: 361–369.

Van der Kooi AW, Leijten FS, Van der Wekken RJ et al. (2012). Electroencephalography-based monitoring of delirium in the ICU: what are the opportunities? Crit Care Med 16: P338.

Van der Kooi AW, Kappen TH, Raijmakers RJ et al. (2013). Temperature variability during delirium in ICU patients: an observational study. PLoS One 8: e78923.

Van der Kooi AW, Rots ML, Huiskamp G et al. (2014). Delirium detection based on monitoring of blinks and eye movements. Am J Geriatr Psychiatry 22: 1575–1582.

Van Eijk MM, Slooter AJC (2010). Delirium in intensive care unit patients. Semin Cardiothorac Vasc Anesth 14: 141–147.

Van Eijk MMJ, van Marum RJ, Klijn IAM et al. (2009). Comparison of delirium assessment tools in a mixed intensive care unit. Crit Care Med 37: 1881–1885.

Van Eijk MM, Roes KC, Honing ML et al. (2010). Effect of rivastigmine as an adjunct to usual care with haloperidol on duration of delirium and mortality in critically ill patients: a multicentre, double-blind, placebo-controlled randomised trial. Lancet 376: 1829–1837.

Van Eijk MM, van den Boogaard M, van Marum RJ et al. (2011). Routine use of the confusion assessment method for the intensive care unit: a multicenter study. Am J Respir Crit Care Med 184: 340–344.

Van Gool WA, van de Beek D, Eikelenboom P (2010). Systemic infection and delirium: when cytokines and acetylcholine collide. Lancet 375: 773–775.

van Munster BC, Aronica E, Zwinderman AH et al. (2011). Neuroinflammation in delirium: a postmortem case-control study. Rejuvenation Res 14: 615–622.

Van Rompaey B, Schuurmans MJ, Shortridge-Baggett LM et al. (2008). A comparison of the CAM-ICU and the NEECHAM Confusion Scale in intensive care delirium assessment: an observational study in non-intubated patients. Crit Care Med 12: R16.

Van Rompaey B, Elseviers MM, Schuurmans MJ et al. (2009). Risk factors for delirium in intensive care patients: a prospective cohort study. Crit Care Med 13: R77.

Van Rompaey B, Elseviers MM, Van Drom W et al. (2012). The effect of earplugs during the night on the onset of delirium and sleep perception: a randomized controlled trial in intensive care patients. Crit Care Med 16: R73.

Vanderbilt University (2014a). The MENDSII study, maximizing the efficacy of sedation and reducing neurological dysfunction and mortality in septic patients with acute respiratory failure. In: ClinicalTrials.gov. https://clinicaltrials.gov/show/NCT01739933. Accessed 4 Oct 2015.

Vanderbilt University (2014b). The Modifying the Impact of ICU-Associated Neurological Dysfunction-USA (MIND-USA) study. In: ClinicalTrials.gov.

Waage A, Halstensen A, Shalaby R et al. (1989). Local production of tumor necrosis factor alpha, interleukin 1, and interleukin 6 in meningococcal meningitis. Relation to the inflammatory response. J Exp Med 170: 1859–1867.

Wang W, Li H-L, Wang D-X et al. (2012). Haloperidol prophylaxis decreases delirium incidence in elderly patients after noncardiac surgery: a randomized controlled trial. Crit Care Med 40: 731–739.

Wassenaar A, van den Boogaard M, van Achterberg T et al. (2015). Multinational development and validation of an early prediction model for delirium in ICU patients. Intensive Care Med: 1048–1056.

Wolters AE, van Dijk D, Pasma W et al. (2014). Long-term outcome of delirium during intensive care unit stay in survivors of critical illness: a prospective cohort study. Crit Care Med 18: R125.

Wolters AE, Veldhuijzen DS, Zaal IJ et al. (2015a). Systemic corticosteroids and transition to delirium in critically ill patients. Crit Care Med 1.

Wolters AE, Zaal IJ, Veldhuijzen DS et al. (2015b). Anticholinergic medication use and transition to delirium in critically ill patients: a prospective cohort study. Crit Care Med 43: 1846–1852.

Yokota H, Ogawa S, Kurokawa A et al. (2003). Regional cerebral blood flow in delirium patients. Psychiatry Clin Neurosci 57: 337–339.

Young J, Murthy L, Westby M et al. (2010). Diagnosis, prevention, and management of delirium: summary of NICE guidance. BMJ 341: c3704.

Zaal IJ, Spruyt CF, Peelen LM et al. (2013). Intensive care unit environment may affect the course of delirium. Intensive Care Med 39: 481–488.

Zaal IJ, Slooter AJ (2014). Light levels of sedation and DSM-5 criteria for delirium. Intensive Care Med 40: 300–301.

Zaal IJ, Devlin JW, Hazelbag M et al. (2015a). Benzodiazepine-associated delirium in critically ill adults. Intensive Care Med 41: 2130–2137.

Zaal IJ, Devlin JW, Peelen LM et al. (2015b). A systematic review of risk factors for delirium in the ICU. Crit Care Med 43: 40–47.

Zaal IJ, Tekatli H, van der Kooi AW et al. (2015c). Classification of daily mental status in critically ill patients for research purposes. J Crit Care 30: 375–380.

Zhang Z, Pan L, Ni H (2013). Impact of delirium on clinical outcome in critically ill patients: a meta-analysis. Gen Hosp Psychiatry 35: 105–111.

Chapter 26

Posterior reversible encephalopathy in the intensive care unit

M. TOLEDANO AND J.E. FUGATE*
Department of Neurology, Mayo Clinic, Rochester, MN, USA

Abstract

Posterior reversible encephalopathy syndrome (PRES) is increasingly diagnosed in the emergency department, and medical and surgical intensive care units. PRES is characterized by acute onset of neurologic symptoms in the setting of blood pressure fluctuations, eclampsia, autoimmune disease, transplantation, renal failure, or exposure to immunosuppressive or cytotoxic drugs, triggers known to admit patients to the intensive care unit (ICU). Although the exact pathophysiology remains unknown, there is growing consensus that PRES results from endothelial dysfunction. Because of the heterogeneous nature of the disorder, it is probable that different mechanisms of endothelial injury are etiologically important in different clinical situations. The presence of bilateral vasogenic edema on brain imaging, particularly in parieto-occipital regions, is of great diagnostic utility but PRES remains a clinical diagnosis. Although largely reversible, PRES can result in irreversible neurologic injury and even death. The range of clinical and radiographic manifestations of the syndrome is probably broader than previously thought, and it is imperative that clinicians become familiar with the full spectrum of the disorder, as prompt recognition and elimination of an inciting factor improve outcome. PRES may be the most frequent toxic-metabolic encephalopathy seen in the ICU.

INTRODUCTION

A reversible syndrome associated with posterior subcortical white-matter changes and characterized by alterations in consciousness, headaches, seizures, and visual disturbances was first described by Hinchey and colleagues in 1996. Initially termed reversible posterior leukoencephalopathy syndrome, a new name, posterior reversible encephalopathy syndrome (PRES), was proposed, as it became clear that cortical areas were also commonly affected (Casey et al., 2000). Initially described in patients with hypertensive encephalopathy, eclampsia, chemotherapy exposure, or following allogeneic bone marrow and solid-organ transplantation (SOT) (Hinchey et al., 1996), PRES has been linked with a growing list of medical conditions and medications (Table 26.1). Associations with autoimmune diseases, metabolic disorders, and sepsis/shock have also been described, and immunosuppressive or cytotoxic drugs are frequently implicated.

Brain imaging usually reveals bilateral vasogenic edema predominantly affecting the parieto-occipital regions. Controversy remains regarding the pathophysiology of PRES, but there is growing consensus that the syndrome is triggered by endothelial injury leading to blood–brain barrier disruption and vasogenic edema. The initial insult to the endothelium can have multiple causes but is likely secondary to abrupt blood pressure changes and/or direct effects from circulating inflammatory cytokines or cytotoxic drugs. Prognosis is favorable but failure to recognize the syndrome and initiate treatment in a timely manner can result in irreversible cytotoxic edema and neurologic injury.

In this review, we describe the epidemiology, pathophysiology, clinical and radiographic features, approach to diagnosis and treatment, as well as the prognosis of PRES. We emphasize the importance of recognizing the entire range of radiographic and clinical presentations, which is likely broader than previously considered,

---

*Correspondence to: Jennifer E. Fugate, D.O., Mayo Clinic, 200 First St., Rochester MN 55905, USA. Tel: +1-507-284-4741, E-mail: Fugate.Jennifer@mayo.edu

Table 26.1

Conditions and medications associated with posterior reversible encephalopathy syndrome

Pre-eclampsia/eclampsia
Marked elevation or fluctuations in blood pressure
Hypertensive encephalopathy
Dysautonomia (e.g., Guillain–Barré syndrome, spinal cord injury)
Induced hypertension (e.g., treatment for vasospasm in aneurysmal subarachnoid hemorrhage)
Autoimmune disease
Connective tissue inflammatory disease
Scleroderma
Sjögren's disease
Systemic lupus erythematosus
Vasculitis
Cryoglobulinemia
Granulomatosis with polyangiitis
Polyarteritis nodosa
Inflammatory bowel disease
Crohn's disease
Ulcerative colitis
Other
Hashimoto thyroiditis
Neuromyelitis optica
Primary sclerosing cholangitis
Thrombocytopenia purpura/hemolytic uremic syndrome
Infection/sepsis/shock
Immunosuppressive, immunomodulatory, and chemotherapeutic drugs
Combination chemotherapy
Angiogenesis inhibitors
Bevacizumab
Tyrosine kinase inhibitors: pazopanib, sorafenib, sunitinib
Calcineurin inhibitors
Cyclosporine A
Tacrolimus (rarely sirolimus)
Bortezomib
Cisplatin and other platinum-based agents
Cytarabine/gemcitabine
Interferon-alpha
Methothrexate
Rituximab
Vincristine
Disorders of metabolism
Porphyria (acute intermittent porphyria)
Ornithine transcarbamylase deficiency
Miscellaneous
Acute and chronic renal failure
Amphetamine/cocaine use
Blood transfusion
Hypercalcemia, hypomagnesemia
Intravenous immunoglobulin

and which is of relevance both to patient outcome and in future study design. The more severe presentations of PRES that may require care in an intensive care unit (ICU) are described. We also highlight some areas of ongoing controversy as well as possible directions for future research.

EPIDEMIOLOGY

Although studies of PRES have increased exponentially over the last decade, these consist almost exclusively of isolated case reports or case series. These reports have been instrumental in enhancing our understanding of the clinical and radiographic manifestations of the disease, as well as its associated conditions and risk factors. Nonetheless, epidemiologic data remain limited. The syndrome appears to affect patients in all age groups, including infancy (Kummer et al., 2010). There also appears to be a female predominance, even when patients with eclampsia are excluded (Bartynski and Boardman, 2007; Fugate et al., 2010; Liman et al., 2012). However, the incidence of PRES in the general population is not known.

Studies have attempted to estimate the incidence in select populations known to be at risk. A recent retrospective review of 2588 admissions to a pediatric ICU identified 10 patients with PRES, resulting in an estimated incidence of 0.4% in this patient population (Raj et al., 2013). The incidence of PRES in adult ICU patients, however, is not known. The reported incidence of PRES following allogeneic bone marrow transplant (BMT) in adults varies between 2.7 and 25%, depending on the underlying hematologic malignancy and on the myeloablative preconditioning regimen used (Reece et al., 1991; Bartynski et al., 2004, 2005). The incidence following SOT is lower and varies between 0.4 and 0.6% (Bartynski et al., 2008; Wu et al., 2010). Interestingly, PRES appears to occur earlier following liver transplantation (often in the setting of severe infection) compared with kidney transplantation (Singh et al., 2000). In a retrospective observational study of incident end-stage kidney disease patients in southwest Ireland over a 10-year period, 5 out of 592 (0.84%) patients developed PRES (Canney et al., 2015). A single-center retrospective study of 3746 patients with systemic lupus erythematous (SLE) estimated a prevalence of 0.69% in this patient population (Lai et al., 2013).

PATHOPHYSIOLOGY AND NEUROPATHOLOGY

The pathogenesis of PRES remains incompletely understood, but it appears to be related to a breakdown of normal cerebral autoregulation and endothelial cell function. This breakdown can be precipitated by several mechanisms, with hypertension being the most commonly described. Because of the heterogeneous nature of the disorder, it is probable that different mechanisms are etiologically important in different clinical situations.

Autoregulatory failure and hypertension

Rapidly developing hypertension with failed autoregulation causing hyperperfusion and associated breakdown of the blood–brain barrier remain the leading theory of the pathophysiologic changes underlying PRES. Autoregulation is an intrinsic function of the cerebral vasculature, designed to maintain a stable blood flow despite fluctuations in cerebral perfusion pressure (calculated as mean arterial blood pressure minus intracranial pressure) (Guyton, 2006; Budohoski et al., 2013). This process is largely driven by changes in arteriolar wall diameter with vasodilation occurring when blood pressure drops and vasoconstriction occurring when pressure increases (Budohoski et al., 2013; Kowianski et al., 2013). Chemical signals from neurons, endothelial cells, and astrocytes all contribute to the control of cerebral autoregulation (Kowianski et al., 2013), but vascular tone is primarily modulated by the endothelium via release of vasodilators (nitric oxide, prostacyclin, hydrogen sulfide, and endothelium-derived hyperpolarizing factor), and vasoconstrictors (thromboxane A_2, endothelin 1, and angiotensin II).

In humans, the lower and upper limits of autoregulation are approximately 50–150 mmHg, respectively (Budohoski et al., 2013). Reduced blood pressure below the lower limit of autoregulation can result in hypoperfusion and potential ischemia (Kontos et al., 1978). Conversely, abrupt and severe increases in blood pressure above the upper limit of autoregulation result in breakthrough with passive arteriolar dilatation, hyperperfusion, and injury to the capillary bed allowing the interstitial extravasation of plasma and macromolecules (vasogenic edema) (MacKenzie et al., 1976; Auer, 1978; Leopold, 2013). Sympathetic stimulation, which can be heightened during episodes of acute hemodynamic stress, increases the upper limit of autoregulation in animal studies (Waldemar et al., 1989; Guyton, 2006). The upper limit is also known to be increased in the setting of chronic hypertension (Strandgaard and Paulson, 1984).

The hypertension/hyperperfusion theory is supported by studies showing that acute hypertension commonly accompanies PRES and that prompt treatment of hypertension is associated with both clinical and radiologic improvement (Fugate et al., 2010; Ni et al., 2011; Cruz et al., 2012). The preferential involvement of posterior brain regions in PRES may also be explained by this theory in that the relative lack of sympathetic innervation in the posterior fossa likely renders the area more susceptible to hyperperfusion injury.

An earlier theory had postulated that PRES results from autoregulatory vasoconstriction in response to hypertension, leading to hypoperfusion, ischemia, and subsequent edema. According to proponents of this theory, vasogenic edema may result from cerebral hypoxia and associated production of hypoxemia-inducible factor-1α, which in turn upregulates vascular endothelial growth factor (VEGF), leading to increased vascular permeability (Bartynski, 2008b). Blood vessel irregularities consistent with vasoconstriction have occasionally been demonstrated on both noninvasive (computed tomography (CT) or magnetic resonance angiography (MRA)), as well as catheter cerebral angiography (CA) (Will et al., 1987; Trommer et al., 1988; Singhal, 2004; Bartynski and Boardman, 2008). Additionally, perfusion studies in patients with PRES have mostly demonstrated reduced perfusion. Of a combined 78 patients reported in the literature with perfusion imaging, 75 (96%) showed radiographic evidence of hypoperfusion (Schwartz et al., 1992; Naidu et al., 1997; Engelter et al., 1999; Apollon et al., 2000; Casey et al., 2004; Brubaker et al., 2005; Bartynski and Boardman, 2008). However, most patients with PRES do not have demonstrable vascular narrowing on imaging, and ischemic infarction is only rarely demonstrated on imaging. Also, the timing of perfusion studies varied substantially amongst individual patients in the aforementioned studies and blood pressure values at the time of imaging were not recorded. Importantly, perfusion imaging was undertaken after treatment of hypertension, sometimes many days after symptom onset. As such, they are not necessarily representative of the actual state of cerebral perfusion at the time of presentation.

Some authors have questioned the etiologic role of hypertension in PRES by noting that 15–20% of patients are normotensive or even hypotensive at presentation (Rabinstein et al., 2012), and that even when hypertension is present, less than 50% have a documented mean arterial pressure exceeding the upper limit of cerebral blood flow autoregulation (Mueller-Mang et al., 2009; Li et al., 2012). Although an important observation, these findings should be interpreted with caution, as most of the data comes from retrospective studies in which the most extreme readings might not have been measured or reported. Moreover, the upper limit of autoregulation varies among individuals, making it difficult to ascertain the etiologic significance of single recordings (van Beek et al., 2008). It is possible that, in certain patients, acute hypertension could lead to endothelial dysfunction and breakdown of the blood–brain barrier through different mechanisms, even when the extent of hypertension does not exceed the upper limit of autoregulation. Certainly, factors other than hypertension can contribute to endothelial dysfunction and may render patients more susceptible to changes in blood pressure. The alternative hypothesis, that hypertension is itself a reaction to insufficient brain perfusion caused by endothelial dysfunction from systemic toxic effects, fails to explain the finding that hypertension typically precedes the development of PRES (Rabinstein et al., 2012). Wide fluctuations in blood pressure, rather than absolute rise in blood pressure, may be a precipitant of the syndrome

(Liman et al., 2012), and even patients with sepsis and hypotension can develop PRES (Bartynski et al., 2006). Ultimately, the patient's mean baseline blood pressure, the proportional rise and rapidity with which changes take place, as well as the presence of pronounced fluctuations in blood pressure are all important factors in disrupting the blood–brain barrier, resulting in vasogenic edema.

Other causes of endothelial dysfunction

Although PRES can occur with uncontrolled hypertension, it commonly arises in the setting of a significant systemic/metabolic process or exposure to certain immunosuppressive/cytotoxic drugs (Table 26.1). In addition to injury caused by hypertension, endothelial dysfunction may result from direct effects of excessive inflammatory cytokines and/or toxicity from certain drugs (Marra et al., 2014). During an inflammatory response, activation of lymphocytes and monocytes results in cytokine release (e.g., interleukin (IL)-1, IL-6, tumor necrosis factor (TNF)-α, and interferon (IFN)-γ). Subsequent endothelial cell activation results in increased production of adhesion molecules (e.g., vascular cell adhesion molecule 1 (VCAM-1), intracellular adhesion molecule 1 (ICAM-1), E-selectin) that facilitate interaction with circulating leukocytes. Additionally, TNF-α induces the expression of VEGF, which induces endothelial cell swelling and increases vascular permeability, leading to brain edema (Leopold, 2013).

Pre-eclampsia and eclampsia are associated with a strong inflammatory cytokine response. Subsequent endothelial cell activation induces expression of adhesion molecules followed by cell swelling and increased permeability in the setting of high levels of VEGF-A (Postma et al., 2014).

The higher incidence of PRES following BMT compared to SOT may be due in part to the preconditioning regimens used in the former (marrow ablative chemotherapy, total body irradiation) (Lamy et al., 2014). And indeed, a greater frequency of PRES is reported in patients receiving higher-dose myoablative regimens compared to nonmyoablative ones (Bartynski et al., 2004, 2005). Myoablative regimens likely lead to endothelial and tissue injury and induce production of inflammatory cytokines. Graft-versus-host disease, a complication of BMT, is probably a more severe immunoreactive process than solid-organ rejection, also partially accounting for the discrepancy in incidence between the two (Lamy et al., 2014). Similarly, PRES appears to occur earlier following liver transplantation compared to renal patients possibly secondary to the poor health of patients requiring emergent liver transplantation (Singh et al., 2000). Nonetheless, transplant rejection and infection are present in most SOT patients who develop PRES (Bartynski et al., 2008).

PRES has also been associated with infection and sepsis. Although the nature of this association remains incompletely understood, it is likely mediated by cytokine production. Microbes cause a host immune response leading to increased release of TNF-α and IL-1, and upregulation of VCAM-1 and ICAM-1 (Pruitt et al., 1995). During an infection, polymorphonuclear cells become activated and marginated and eventually adhere to the vascular endothelium, resulting in the release of mediators and increased permeability (Parent and Eichacker, 1999). This can lead to interstitial edema, which could manifest as PRES in the brain.

Nearly half of patients with PRES have pre-existing autoimmune disease (Bartynski, 2008a; Fugate et al., 2010; Lamy et al., 2014). Specific disorders include, but are not limited to, scleroderma, rheumatoid arthritis, SLE, Sjögren's syndrome, polyarteritis nodosa, granulomatosis with polyangiitis, neuromyelitis optica, Crohn's disease, ulcerative colitis, primary sclerosis cholangitis, thrombotic thrombocytopenic purpura, and hypothyroidism (Burrus et al., 2009; Fugate et al., 2010; Lai et al., 2013; Shaharir et al., 2013). Many of these patients are on immunosuppressant drugs, which could exert direct toxic effects, but endothelial activation and cytokine production in the setting of chronic or acute inflammation are likely important drivers behind the association with PRES.

Immunosuppressive or cytotoxic drugs used after BMT or SOT, or to treat malignancies, are strongly associated with PRES. The effect appears to be time- and dose-independent; patients can develop PRES months after initiation and despite normal serum drug concentrations (Schwartz et al., 1995). Drugs often implicated in PRES include calcineurin inhibitors such as cyclosporine and tacrolimus (but only rarely sirolimus) (Schwartz et al., 1995; Wong et al., 2003; Bodkin and Eidelman, 2007). Cyclosporine neurotoxic effects might be facilitated by the endogenous vasoactive substance endothelin, hypertension, and hypomagnesemia (Schwartz et al., 1995). Chemotherapeutic agents include vincristine, cisplatin, gemcitabine, bortezomib, and cytarabine (Hurwitz et al., 1988; Ito et al., 1997; Rajasekhar and George, 2007; Kelly et al., 2008; Vaughn et al., 2008), as well as antiangiogenic drugs that antagonize the action of VEGF, such as sunitinib, bevacizumab, and sorafenib (Tlemsani et al., 2011; Seet and Rabinstein, 2012).

Although renal failure is classically associated with PRES (present in up to 55% cases) (Mueller-Mang et al., 2009; Fugate et al., 2010; Liman et al., 2012), it remains unclear whether renal dysfunction is an independent risk factor or a marker of comorbid hypertension, autoimmune disease, or another systemic condition. Similarly, other metabolic derangements such as hyponatremia, hypomagnesemia, hypercalcemia, and hepatic dysfunction may be independent risk factors or markers of systemic disease.

Histopathology

Biopsy samples of PRES patients show reactive endothelial changes, increased VEGF expression, as well as intravascular and perivascular T lymphocytes without evidence of inflammation, ischemia, or neuronal damage (Horbinski et al., 2009). These findings support the hypothesis that activation and dysfunction of the endothelium underlie PRES.

CLINICAL PRESENTATION

The classic neurologic manifestations of PRES include encephalopathy, seizures, visual disturbances, headache, and focal neurologic deficits (Burnett et al., 2010; Fugate et al., 2010; Cordelli et al., 2012; Li et al., 2012; Liman et al., 2012). Because most of the literature on the syndrome is comprised of retrospective observational studies, the frequency of symptoms varies depending on the study population or sample size. The frequency of the main presenting symptoms is shown in Table 26.2.

The clinical presentation of PRES can be acute or subacute with symptoms developing over 24–48 hours. Continued progression of symptoms for many weeks is very uncommon. Encephalopathy ranges in severity from mild confusion, stupor, or even coma in rare cases (Keswani and Wityk, 2002). Seizures occur in about 60–75% of cases and can be generalized tonic-clonic or focal (Burnett et al., 2010; Fugate et al., 2010; Li et al., 2012; Liman et al., 2012). Although uncommon, status epilepticus can occur, and clinicians should have a low threshold to obtain an electroencephalogram (EEG) in patients with PRES with persistent and unexplained altered consciousness (Kozak et al., 2007; Burnett et al., 2010; Fugate et al., 2010). Visual disturbances such as decreased visual acuity, field deficits, cortical blindness, and hallucinations are common (Tallaksen et al., 1998; Kahana et al., 2005). Papilledema with flame-shaped retinal hemorrhages and exudates can be present in the setting of hypertension, but fundoscopic examination is often unremarkable (Dinsdale, 1982). Headache, when present, is usually dull, diffuse, and gradual in onset. A thunderclap-onset headache in the context of PRES should raise suspicion for associated reversible cerebral vasoconstriction syndrome (RCVS), and prompt cerebrovascular imaging (Benziada-Boudour et al., 2009). Focal findings, such as aphasia or hemiparesis, are present in 5–15% of patients (McKinney et al., 2007; Burnett et al., 2010; Fugate et al., 2010; Li et al., 2012). Radiographic and clinical evidence of myelopathy has been rarely documented (de Havenon et al., 2014).

NEURODIAGNOSTICS AND IMAGING

The clinical presentation of PRES is often nonspecific and the symptoms and signs of PRES, such as encephalopathy, seizures, visual disturbances, and headache, can be seen in many other disorders. Transient elevations in blood pressure are difficult to interpret in the acute setting, as pronounced hypertension can be a physiologic response to cerebral ischemia or an underlying systemic process such as infection. The differential diagnosis is wide and includes various neurologic conditions, including other vasculopathies, infectious/inflammatory encephalitides, malignancy, and toxic/metabolic encephalopathies (Fig. 26.1).

EEG can help confirm the presence of ictal or epileptogenic activity and often demonstrates focal or multifocal sharp waves (Kozak et al., 2007; Lee et al., 2008b). Focal or diffuse slowing is also commonly demonstrated. Although EEG findings are nonspecific, the presence of bilateral occipital sharp waves in patients with status epilepticus should raise suspicion for PRES (Kozak et al., 2007). Cerebrospinal fluid (CSF) analysis is likewise nonspecific but commonly shows an elevated protein (rarely > 100 mg/dL) (Lee et al., 2008b; Datar et al., 2015b). Although mild CSF pleocytosis may occur, its presence should prompt consideration of a different diagnosis (Datar et al., 2015b).

Brain imaging is very useful both to rule out other conditions and to confirm the diagnosis. However, even when brain magnetic resonance imaging (MRI) shows

Table 26.2

Prevalence of clinical symptoms and signs in patients with posterior reversible encephalopathy syndrome

Encephalopathy (50–80%) (Li et al., 2012; Liman et al., 2012; Brewer et al., 2013)
Seizure (60–75%) (Burnett et al., 2010; Fugate et al., 2010; Liman et al., 2012; Brewer et al., 2013)
Headache (50%) (Tlemsani et al., 2011; Legriel et al., 2012; Li et al., 2012)
Visual disturbances (33%) (Burnett et al., 2010; Legriel et al., 2012; Liman et al., 2012; Brewer et al., 2013)
Focal neurological deficit (10–15%) (Burnett et al., 2010; Cruz et al., 2012; Liman et al., 2012)
Status epilepticus (5–15%) (Burnett et al., 2010; Fugate et al., 2010; Li et al., 2012)

Modified from (Fugate and Rabinstein, 2015).

Clinical mimic

Ischemic/hemorrhagic stroke
- Hyperacute in onset
- Findings on MRI can be unilateral

Osmotic demyelinating syndrome
- Does not preferentially affect the parieto-occipital lobes
- History of rapid normalization of sodium or glucose concentration (can not always be confirmed)
- Characteristic central pontine signal abnormality in bat-wing shape

Toxic leukoencephalopathy
- History of substance use and/or positive drug screen test
- Presence of characteristic toxidrome
- MRI abnormalities tend to be symmetric

CNS vasculitis
- Subacute clinical presentation
- CSF pleocytosis
- Cytotoxic edema not in topographic distribution typical of PRES

Autoimmune or paraneoplastic encephalitis
- History of malignancy or tumor
- Presence of neural-specific antibody in serum or CSF
- CSF pleocytosis
- Findings on MRI can be unilateral. Limbic structures are commonly involved

Infectious encephalitis
- Fever
- CSF pleocytosis
- Positive CSF Gram stain or culture/positive microbial serology or PCR
- Imaging findings can be unilateral

Acute demyelinating encephalomyelitis
- Mostly a disease of children
- Preceded by vaccination or viral/bacterial illness

Mitochondrial encephalomyositis
- Family history
- Presence of hearing loss, opthalmoplegia, short stature, myopathy
- Magnetic spectroscopy can show abnormally raised lactate and decreased N-acetyl-asparate concentrations

Progressive multifocal leukoencephalopathy
- Subacute-to-chronic clinical presentation

Malignancy or tumor (lymphoma, gliomatosis cerebri, metastatic disease)
- Subacute-to-chronic clinical presentation
- History of unintentional weight loss
- Abnormal CSF cytology
- MRI findings on MRI can be unilateral and fail to resolve

Leukoariosis
- No acute clinical presentation
- Periventricular, mostly confluent signal abnormality on MRI

Radiographic mimic

Fig. 26.1. Differential diagnosis in posterior reversible encephalopathy syndrome (PRES). MRI, magnetic resonance imaging; CNS, central nervous system; CSF, cerebrospinal fluid; PCR, polymerase chain reaction. (Modified from Fugate and Rabinstein, 2015.)

regions of T2 signal abnormality suggestive of PRES, the differential remains wide (Fig. 26.1). Overreliance on imaging can lead to misdiagnosis and both the clinical context and the judgment of the clinician are crucial in establishing the correct diagnosis.

Imaging

Although neuroradiographic abnormalities of PRES may be apparent on noncontrast CT, they are better demonstrated on brain MRI. T2-weighted sequences such as fluid-attenuated inversion recovery (FLAIR) are the most sensitive (Bartynski and Boardman, 2007). Specificity is not known as there is no accepted diagnostic gold standard in PRES.

Classically, brain imaging demonstrates relatively symmetric vasogenic edema involving the parieto-occipital white matter of both cerebral hemispheres (Fig. 26.2) but other brain regions are commonly involved. The frontal and temporal lobes, for example,

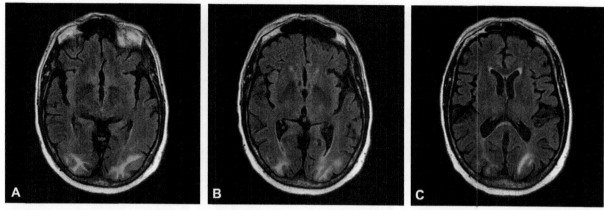

Fig. 26.2. (A–C) Classic radiographic presentation of posterior reversible encephalopathy syndrome. Axial T2 fluid-attenuated inversion recovery sequences show T2 signal abnormality that is mostly subcortical and involves the bilateral occipital lobes predominantly.

can be affected in up to 75% of cases (Fugate et al., 2010; Cruz et al., 2012; Legriel et al., 2012). Brainstem and basal ganglia involvement can be seen in up to a third of cases (Mueller-Mang et al., 2009; Fugate et al., 2010; Liman et al., 2012), and cerebellar involvement in up to half (Fugate et al., 2010). Rare cases demonstrating signal change in the spinal cord have also been described (de Havenon et al., 2014). However, edema in these regions rarely occurs in isolation and almost always coincides with parieto-occipital involvement. Cortical lesions are not uncommon and, although vasogenic edema is almost always bilateral, it can be asymmetric. Strictly unilateral lesions or isolated brainstem, cerebellar, or spinal cord involvement should prompt consideration of a different diagnosis.

Three dominant MRI patterns have been described in the radiology literature and are present in up to 70% of PRES patients (Bartynski and Boardman, 2007): a parieto-occipital pattern, holohemispheric watershed pattern, and superior frontal sulcus pattern. These patterns are supportive of the diagnosis, although not pathognomonic. Interestingly, neither the pattern nor the extent of brain edema is strongly correlated with the severity of clinical presentation (Mueller-Mang et al., 2009; Fugate et al., 2010). T2 signal changes affecting regions outside these dominant patterns are not uncommon.

Diffusion-weighted imaging demonstrates restricted diffusion in 15–30% of cases (Covarrubias et al., 2002; Burnett et al., 2010; Fugate et al., 2010; Cruz et al., 2012; Li et al., 2012, 2013; Liman et al., 2012). Usually small areas of restricted diffusion are embedded within larger regions of vasogenic edema, although rarely, more extensive confluent regions of restricted diffusion can occur (Fig. 26.3). Such lesions can be hard to distinguish from those caused by ischemic infarction, although in PRES the distribution of abnormalities is usually not confined to a single vascular territory (Lamy et al., 2004). Although lesions demonstrating restricted diffusion can rarely be reversible, their presence is usually associated with irreversible structural damage and residual neurologic deficit (Moon et al., 2013).

Gyriform signal enhancement – likely reflecting disruption of the blood–brain barrier – can be seen in up to 20% of cases following administration of gadolinium (Fugate et al., 2010; Kastrup et al., 2015) (Fig. 26.4), but no study has systematically assessed the effect of timing on the presence or absence of contrast enhancement.

Between 20 and 25% of cases of PRES are complicated by intracranial hemorrhage (Bartynski and Boardman, 2007; Hefzy et al., 2009; Mueller-Mang et al., 2009; Burnett et al., 2010; Fugate et al., 2010; Sharma et al., 2010). Intraparenchymal hemorrhage (IPH) is most common, followed by sulcal subarachnoid hemorrhage (SAH), although both types can co-occur in 18–30% of patients (Hefzy et al., 2009; Sharma et al., 2010) (Fig. 26.5). In one retrospective study of 151 patients with PRES, a single hemorrhage type was found in 16 out of 23 (70%) patients who bled, compared to 30% who had multiple types (Hefzy et al., 2009). Isolated IPH was found in 13 out of the 16 patients with a single hemorrhage type (81%), compared to 3 (19%) who had isolated SAH (Hefzy et al., 2009). In another retrospective study of 263 patients, sulcal SAH was found in 14 out of 60 (23%) who bled, compared to 46 (77%) who had IPH (Sharma et al., 2010). Amongst those patients who bleed, the most common risk factors are intrinsic coagulopathy and ongoing therapeutic anticoagulation (Hefzy et al., 2009). The risk of intracranial bleed may be greatest following allogeneic BMT (Hefzy et al., 2009). One study reported a high rate of microhemorrhages (58%) on susceptibility-weighted imaging (McKinney et al., 2007), but the clinical significance of this finding remains unclear.

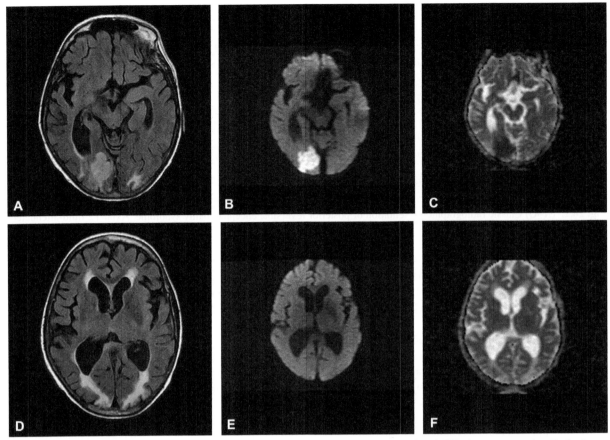

Fig. 26.3. Restricted diffusion in posterior reversible encephalopathy syndrome. Restricted diffusion can occasionally be large and homogeneous, and difficult to distinguish from ischemic infarction (A–C), despite being surrounded by more typical areas of vasogenic edema (D–F).

Diffuse or focal vasoconstriction, vasodilatation, and string-on-a bead appearance, consistent with vasospasm, have been documented in patients with PRES on either noninvasive (CTA or MRA), or on CA. In a series of 47 patients with PRES who underwent catheter CA or MRA, reversible vasculopathy was observed in 40 (85%) (Bartynski and Boardman, 2008). Although this may suggest that vasospasm is common in PRES, the finding should be interpreted with caution, as more than half of the 116 patients with PRES evaluated at the treatment center during the duration of the study did not undergo angiography and it is likely that selection bias resulted in overestimation (Bartynski and Boardman, 2008). Other studies report lower rates of vasoconstriction (15–30%) in patients with PRES who undergo angiography (although still less than half of patients undergo vessel imaging in these studies) (Burnett et al., 2010; Fugate et al., 2010; Li et al., 2012).

The imaging features seen in angiography in patients with PRES are similar to those observed in patients with RCSV, a condition of reversible vasospasm that can be seen postpartum or after exposure to certain vasoactive substances (Ducros et al., 2007; Singhal et al., 2011; Fugate et al., 2012). The topographic distribution of ischemic infarcts seen in RCVS is similar to that seen in PRES (Singhal et al., 2011), and PRES has been reported in 17–38% of patients with RCVS (Ducros et al., 2010; Singhal et al., 2011; Fugate et al., 2012). This overlap suggests that a clinical and pathophysiologic continuum may exist between these two entities but more studies are needed to elucidate the nature of their relationship.

HOSPITAL COURSE AND MANAGEMENT

Prompt recognition of PRES and elimination or treatment of a precipitating cause usually result in a favorable outcome. However, no randomized control trials have been conducted assessing individual therapeutic interventions and management is largely guided by expert consensus.

Seizures and status epilepticus are common indications for PRES patients to be admitted to an ICU. In one multicenter study of patients admitted to an ICU for "severe"

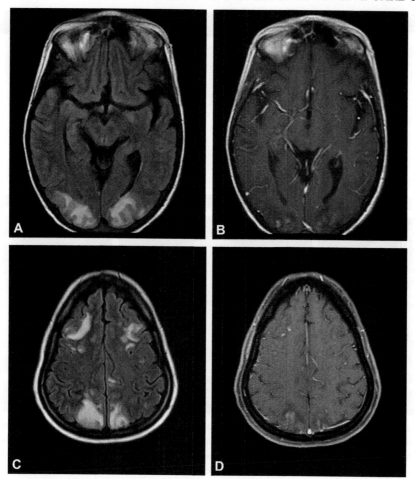

Fig. 26.4. Gadolinium enhancement in posterior reversible encephalopathy syndrome. Axial T2 fluid-attenuated inversion recovery sequences (**A**, **C**) show T2 signal change in frontal and parieto-occipital regions with associated contrast enhancement in axial T1 gadolinium sequences (**B**, **D**).

PRES, 31 of 70 (44%) had status epilepticus (Legriel et al., 2012). Seizures are treated with antiepileptic drugs (AEDs) as they would be in any acute neurologic condition, and status epilepticus may require anesthetic agents and prolonged continuous EEG monitoring. However, no prospective randomized studies are available to guide choice of drug or duration of treatment. Periodic sharp waves in the bioccipital head regions may suggest PRES and may not necessarily need to be treated with AEDs if there are no clinical correlates. AEDs can usually be discontinued several weeks after the initial presentation and, in the absence of residual brain lesions; long-term treatment with AEDs is rarely necessary (Lee et al., 2008b).

Magnesium sulfate is indicated to prevent further seizures in women with eclampsia, which can be associated with PRES. Magnesium has cerebral vasodilatory effects, and reduces blood vessel permeability. The role of magnesium supplementation in treating or preventing seizures in PRES outside the setting of eclampsia is not known.

Hypertensive encephalopathy is treated with immediate hypertension reduction, and the choice of a specific antihypertensive drug is left to the discretion of the physician. Suitable agents include nicardipine, labetalol, hydralazine, and sodium nitroprusside (Mancia et al., 2014). Acute, severe hypertension is one of the most common reasons for admission of a PRES patient to the ICU (Legriel et al., 2012). Most experts recommend that mean arterial blood pressure should not be lowered by more than 25% within the first few hours so as to avoid the theoretic risk of cerebral, renal, or coronary ischemia (Mancia et al., 2014). Pronounced fluctuations in blood pressure should also be avoided and continuous infusion of intravenous drugs may be needed to achieve this goal (Mancia et al., 2014).

There are a few unique presentations of PRES due to blood pressure fluctuation that may be seen in the neuroscience ICU. Neurointensivists should be aware that patients with aneurysmal SAH who become markedly hypertensive (either spontaneously or by induction with

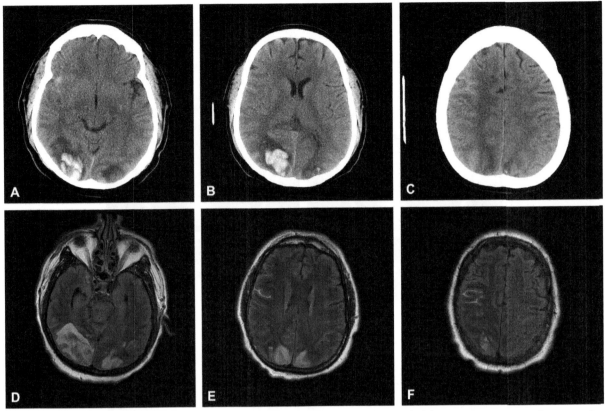

Fig. 26.5. Intracranial hemorrhage in posterior reversible encephalopathy syndrome. Axial noncontrast computed tomography showing acute intraparenchymal hemorrhage surrounded by vasogenic edema in the bilateral posterior regions, associated with sulcal subarachnoid hemorrhage shown as hyperdensity over the right frontal lobe (A–C). Axial T2 fluid-attenuated inversion recovery sequences in the same patient showing associated vasogenic edema in the posterior regions and right frontal subarachnoid blood seen as hyperintensity (D–F).

vasopressors) can develop PRES (Giraldo et al., 2011). Thus PRES can be a late – albeit rare – cause of secondary neurologic deterioration in aneurysmal SAH patients and should be considered after more common causes (e.g., delayed cerebral ischemia, infarction, hyponatremia, hydrocephalus) are excluded. Additionally, patients with dysautonomia may be at risk for PRES. Patients in the ICU with Guillain–Barré syndrome often have dysautonomia and new-onset encephalopathy or seizures in a patient with Guillain–Barré syndrome should prompt consideration of PRES.

If PRES is caused by a specific medication, the offending agent should be discontinued in the acute setting if possible (Tlemsani et al., 2011). Failure to discontinue can impede recovery (Junna and Rabinstein, 2007), although cases are reported where symptoms improve without discontinuation of the offending agent (Schwartz et al., 1995). When a substitute agent is initiated, patients should be followed closely and monitored for recurrence of symptoms. It is not advisable to rechallenge the patient with the same agent as recurrence has been documented in this setting (Covarrubias et al., 2002; Serkova et al., 2004). Other systemic conditions such as sepsis or flareups of autoimmune disorders should be treated according to existing guidelines.

Rarely, patients may develop cerebellar or brainstem edema severe enough to cause obstructive hydrocephalus by compression of the cerebral aqueduct or the fourth ventricle (Fig. 26.6). If the hydrocephalus becomes symptomatic, it may need to be treated with an external ventricular drain. Because the edema in PRES is reversible, long-term placement of a ventriculoperitoneal shunt is almost never necessary. There are some cases of severe PRES reported in which the edema was so extensive and rapidly progressive that it caused significant mass effect and elevated intracranial pressure (Facchini et al., 2013). Refractory intracranial hypertension can occur in such cases, requiring ICP monitoring and barbiturate infusions, but aggressive treatment is fully justified as some patients may have near-complete or complete recoveries (Facchini et al., 2013).

OUTCOME PREDICTION

Most of the literature suggests that PRES is a benign disorder. In the majority of cases, clinical improvement can be observed within a period of days to weeks, after

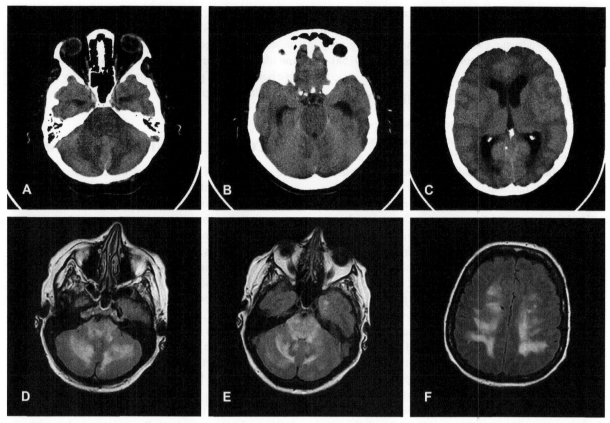

Fig. 26.6. Severe brainstem and cerebellar edema in posterior reversible encephalopathy syndrome. Axial noncontrast computed tomography showing effacement of the fourth ventricle (**A**), hydrocephalus causing temporal tip dilation (**B**), and rounding of the frontal horns of lateral ventricles (**C**). Axial T2 fluid-attenuated inversion recovery sequences better demonstrate extensive cerebellar (**D**), brainstem (**E**), and subcortical edema (**F**).

removal of the inciting cause (Roth and Ferbert, 2010; Brewer et al., 2013), although occasionally patients can take a few weeks to recover (Tlemsani et al., 2011). Radiologic improvement lags behind clinical recovery by a few weeks. However, despite its name, PRES is not always reversible. The extent of reversibility varies in the scientific literature, in part depending on how PRES is defined. Some investigators have required clinical and radiographic reversibility as part of their inclusion criteria (Hefzy et al., 2009; Mueller-Mang et al., 2009; Fugate et al., 2010; Cruz et al., 2012; Gao et al., 2012), whereas others have not (Burnett et al., 2010; Liman et al., 2012, 2014). Not surprisingly, studies that use reversibility to define PRES report almost universal recovery. On the other hand, studies that do not exclude patients with clinical and radiographic sequelae report persistent imaging abnormalities or neurologic deficits in up to 10–25% of cases (Burnett et al., 2010; Moon et al., 2013; Shaharir et al., 2013; Liman et al., 2014). Although not well characterized, epilepsy, persistent hemiparesis, decreased visual acuity, and dizziness have been reported. Intracranial hemorrhage and presence of restricted diffusion on MRI have been associated with incomplete recovery (Covarrubias et al., 2002; Moon et al., 2013; Shaharir et al., 2013). Time from onset to control of the inciting factor and hyperglycemia were reported to be independently associated with poor outcomes in one study of 70 patients with severe PRES requiring admission to an ICU (Legriel et al., 2012).

In more severe cases death can occur and mortality rates of up to 3–6% have been reported in some case series (Mueller-Mang et al., 2009; Legriel et al., 2012; Moon et al., 2013). Severe brain injury and mortality, when they do occur, can be attributed to intracranial hemorrhage, posterior fossa edema with brainstem compression and herniation or hydrocephalus, as well as diffuse brain edema and increased intracranial pressure (Lee et al., 2008a). This subset of patients with significant morbidity and mortality highlights the importance of prompt recognition and treatment of the causative factor of PRES.

Recurrent PRES has been reported in about 5–10% of cases and appears to be more common in patients with uncontrolled hypertension as the precipitating cause of their syndrome (Li et al., 2013). Recurrent seizures occur in 10–15% of patients during the first few years following PRES (Datar et al., 2015a). However, in the vast majority of cases these can be attributed to provoking factors such as recurrent PRES rather than the development of epilepsy. In the absence of recurrent PRES, recurrent seizures are extremely rare after resolution of the syndrome, even in cases that had originally presented with status epilepticus (Datar et al., 2015a).

CLINICAL TRIALS AND GUIDELINES

Although studies of PRES have increased exponentially since its original description more than two decades ago, these consist almost exclusively of cases series and case reports. Management is guided by expert consensus, as no prospective randomized control interventional studies have ever been conducted. No animal model exists and pathophysiologic studies are scarce. Consequently our understanding of the mechanisms leading to endothelial dysfunction and increased blood–brain barrier permeability remains limited. Treatment consists of correction or elimination of an inciting factor but no interventions targeting proximal causes that could possibly prevent PRES in patients known to be at risk have been proposed. Large case series have enhanced our understanding of the clinical and radiologic spectrum of the disorder, its risk factors, and prognosis. However, it remains unclear why specific individuals within populations at risk develop PRES while others do not. Clustering of risk factors (e.g., a BMT patient on tacrolimus with renal failure) likely accounts for some, but not all, of this difference. Genetic susceptibility is a possibility, and one that could potentially open the way for targeted prevention strategies in the future, but at present no epidemiologic data or genetic studies exist to support this hypothesis. Based on the co-occurrence of PRES and neuromyelitis optica (Magana et al., 2009), we did undertake genotypic analysis in patients with PRES but did not find associations with aquaporin-4 pleomorphisms (unpublished work).

As previously discussed, many studies have utilized reversibility of clinical and radiographic manifestation of PRES as part of their inclusion criteria, thus cementing the conception of PRES as a strictly reversible syndrome and possibly excluding cases with less favorable outcomes. Similarly, most studies have mandated the presence of vasogenic edema on brain imaging as part of their diagnostic criteria. Consequently, a diagnosis of PRES relies heavily on radiologic studies in clinical practice. We have seen patients with hypertensive encephalopathy presenting with clinical characteristics of PRES but no evidence of vasogenic edema on MRI. It is highly likely that these patients have brain injury that cannot be visualized utilizing current MRI sequences but whose symptoms are caused by the same underlying pathophysiologic mechanism as PRES. Conversely, we have seen patients with large confluent regions of restricted diffusion but only scant vasogenic edema occurring in a topographic pattern highly suggestive of PRES. Elsewhere (Fugate and Rabinstein, 2015), we have proposed an algorithm for the diagnosis of PRES that does not require reversibility or the presence of vasogenic edema (Fig. 26.7). This algorithm can identify patients falling outside the artificially restrictive boundaries of previous definitions and may be used in future study designs.

COMPLEX CLINICAL DECISIONS

The diagnosis of PRES relies on clinical and radiographic findings of limited specificity and the differential is wide. Reliance on extensive ancillary testing to rule out competing diagnoses is rarely necessary. Most patients with PRES present with one or more characteristic neurologic symptoms in the setting of at least one known risk factor. Brain imaging showing vasogenic edema in a topographic distribution typical of PRES helps confirm the diagnosis. Nonetheless, the diagnosis can be challenging at times. Clinicians need to be familiar with atypical presentations and remain vigilant for "red flags" in order to minimize unnecessary testing or misdiagnosis.

Fever in the setting of encephalopathy should prompt consideration of a lumbar puncture, which becomes mandatory if seizures or focal deficits are also present (in which case, brain imaging should be obtained before performing a lumbar puncture). Persistent encephalopathy in spite of correction of the presumed inciting cause should raise suspicion for nonconvulsive seizure activity, and an EEG should be performed. Similarly, progressive neurologic deficits days to weeks after treatment should prompt further imaging. Although focal neurologic signs can occur with PRES, acute onset of a focal deficit should raise suspicion for ischemic infarction or intracerebral hemorrhage. Marked hypertension is also a feature of acute stroke where it occurs as a physiologic response and is necessary to maintain perfusion. Most of the time this scenario presents no difficulty but patients presenting with isolated aphasia can be challenging. Superimposed encephalopathy, if present, is suggestive of PRES, but this may not always be apparent. MRI may be helpful but vascular and perfusion imaging may be necessary if doubts remain. Similarly, patients with PRES can present with acute brainstem dysfunction and stupor mimicking "top of the basilar" syndrome. Brainstem edema can be present in cases of PRES but further imaging may be necessary to rule out acute ischemia.

Absence of a known precipitating factor should prompt more careful screening. PRES is usually a manifestation of underlying systemic disease and has been described as the initial presentation of an autoimmune disease such as SLE (Kur and Esdaile, 2006) or a metabolic disorder such as porphyria (Utz et al., 2001). Screening should be guided by history, as well as

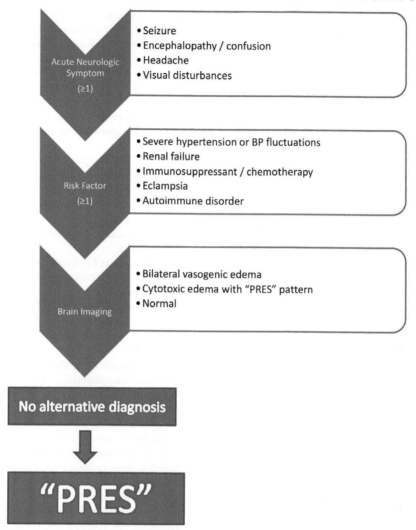

Fig. 26.7. Proposed algorithm for the diagnosis of posterior reversible encephalopathy syndrome (PRES). BP, blood pressure. (Reproduced with permission from Fugate and Rabinstein, 2015.)

additional clinical and paraclinical data, but a wider search may be warranted in certain cases.

NEUROREHABILITATION

Although reversible in most cases, a small but important subset of patients with PRES can be left with permanent neurologic sequelae. Patients whose course is complicated by hemorrhage or ischemic infarction are at higher risk. Residual focal neurologic symptoms such as hemiparesis and permanent visual disturbances have been reported, but more global dysfunction such as cognitive difficulties and dizziness can also occur. Neurorehabilitation should be targeted to the patient's specific deficit and referral to a physical medicine and rehabilitation specialist or, if needed, a speech therapist or neuropsychologist should be considered.

CONCLUSIONS

PRES is a heterogeneous clinical syndrome characterized by acute neurologic symptoms arising in the setting of severe abrupt hypertension or blood pressure fluctuations, autoimmune disorders, disorders of metabolism, eclampsia, transplantation, or exposure to immunosuppressant drugs. Although some questions remain regarding the details of the underlying pathophysiology, PRES is caused by endothelial cell dysfunction. This can occur secondary to changes in blood pressure or as a direct effect of circulating inflammatory cytokines or cytotoxic drugs. Brain imaging showing vasogenic edema in one of the characteristic patterns (parieto-occipital, watershed zones, or frontal sulcus pattern) can be very useful, but presence of edema on imaging is not necessary to make the diagnosis. Prognosis is favorable, but neurologic sequelae and even death can occur, particularly when cerebral ischemia or hemorrhage occurs. Prompt initiation of treatment may improve outcome and clinicians should be familiar with both typical and atypical presentations of the disorder.

Although endothelial dysfunction underlies the pathophysiology of this disorder, further studies are needed to unravel the exact mechanisms of injury. A reliable animal model would be very helpful in this regard and may open the way for pathophysiologically guided therapies. Large, multicenter, prospective studies will be needed to further define populations at risk, as well as delineate the clinical and radiographic boundaries of the disorder.

REFERENCES

Apollon KM, Robinson JN, Schwartz RB et al. (2000). Cortical blindness in severe preeclampsia: computed tomography, magnetic resonance imaging, and single-photon-emission computed tomography findings. Obstet Gynecol 95: 1017–1019.

Auer LM (1978). The pathogenesis of hypertensive encephalopathy. Experimental data and their clinical relevance with special reference to neurosurgical patients. Acta Neurochir Suppl (Wien) 27: 1–111.

Bartynski WS (2008a). Posterior reversible encephalopathy syndrome, part 1: fundamental imaging and clinical features. AJNR Am J Neuroradiol 29: 1036–1042.

Bartynski WS (2008b). Posterior reversible encephalopathy syndrome, part 2: controversies surrounding pathophysiology of vasogenic edema. AJNR Am J Neuroradiol 29: 1043–1049.

Bartynski WS, Boardman JF (2007). Distinct imaging patterns and lesion distribution in posterior reversible encephalopathy syndrome. AJNR Am J Neuroradiol 28: 1320–1327.

Bartynski WS, Boardman JF (2008). Catheter angiography, MR angiography, and MR perfusion in posterior reversible encephalopathy syndrome. AJNR Am J Neuroradiol 29: 447–455.

Bartynski WS, Zeigler ZR, Shadduck RK et al. (2004). Pretransplantation conditioning influence on the occurrence of cyclosporine or FK-506 neurotoxicity in allogeneic bone marrow transplantation. AJNR Am J Neuroradiol 25: 261–269.

Bartynski WS, Zeigler ZR, Shadduck RK et al. (2005). Variable incidence of cyclosporine and FK-506 neurotoxicity in hematopoeitic malignancies and marrow conditions after allogeneic bone marrow transplantation. Neurocrit Care 3: 33–45.

Bartynski WS, Boardman JF, Zeigler ZR et al. (2006). Posterior reversible encephalopathy syndrome in infection, sepsis, and shock. AJNR Am J Neuroradiol 27: 2179–2190.

Bartynski WS, Tan HP, Boardman JF et al. (2008). Posterior reversible encephalopathy syndrome after solid organ transplantation. AJNR Am J Neuroradiol 29: 924–930.

Benziada-Boudour A, Schmitt E, Kremer S et al. (2009). Posterior reversible encephalopathy syndrome: a case of unusual diffusion-weighted MR images. J Neuroradiol 36: 102–105.

Bodkin CL, Eidelman BH (2007). Sirolimus-induced posterior reversible encephalopathy. Neurology 68: 2039–2040.

Brewer J, Owens MY, Wallace K et al. (2013). Posterior reversible encephalopathy syndrome in 46 of 47 patients with eclampsia. Am J Obstet Gynecol 208: 468.e461–468.e466.

Brubaker LM, Smith JK, Lee YZ et al. (2005). Hemodynamic and permeability changes in posterior reversible encephalopathy syndrome measured by dynamic susceptibility perfusion-weighted MR imaging. AJNR Am J Neuroradiol 26: 825–830.

Budohoski KP, Czosnyka M, Kirkpatrick PJ et al. (2013). Clinical relevance of cerebral autoregulation following subarachnoid haemorrhage. Nat Rev Neurol 9: 152–163.

Burnett MM, Hess CP, Roberts JP et al. (2010). Presentation of reversible posterior leukoencephalopathy syndrome in patients on calcineurin inhibitors. Clin Neurol Neurosurg 112: 886–891.

Burrus TM, Wijdicks EF, Rabinstein AA (2009). Brain lesions are most often reversible in acute thrombotic thrombocytopenic purpura. Neurology 73: 66–70.

Canney M, Kelly D, Clarkson M (2015). Posterior reversible encephalopathy syndrome in end-stage kidney disease: not strictly posterior or reversible. Am J Nephrol 41: 177–182.

Casey SO, Sampaio RC, Michel E et al. (2000). Posterior reversible encephalopathy syndrome: utility of fluid-attenuated inversion recovery MR imaging in the detection of cortical and subcortical lesions. AJNR Am J Neuroradiol 21: 1199–1206.

Casey SO, McKinney A, Teksam M et al. (2004). CT perfusion imaging in the management of posterior reversible encephalopathy. Neuroradiology 46: 272–276.

Cordelli DM, Masetti R, Bernardi B et al. (2012). Status epilepticus as a main manifestation of posterior reversible encephalopathy syndrome after pediatric hematopoietic stem cell transplantation. Pediatr Blood Cancer 58: 785–790.

Covarrubias DJ, Luetmer PH, Campeau NG (2002). Posterior reversible encephalopathy syndrome: prognostic utility of quantitative diffusion-weighted MR images. AJNR Am J Neuroradiol 23: 1038–1048.

Cruz Jr RJ, DiMartini A, Akhavanheidari M et al. (2012). Posterior reversible encephalopathy syndrome in liver transplant patients: clinical presentation, risk factors and initial management. Am J Transplant 12: 2228–2236.

Datar S, Singh T, Rabinstein AA et al. (2015a). Long-term risk of seizures and epilepsy in patients with posterior reversible encephalopathy syndrome. Epilepsia 56: 564–568.

Datar S, Singh TD, Fugate JE et al. (2015b). Albuminocytologic dissociation in posterior reversible encephalopathy syndrome. Mayo Clin Proc 90: 1366–1371.

de Havenon A, Joos Z, Longenecker L et al. (2014). Posterior reversible encephalopathy syndrome with spinal cord involvement. Neurology 83: 2002–2006.

Dinsdale HB (1982). Hypertensive encephalopathy. Stroke 13: 717–719.

Ducros A, Boukobza M, Porcher R et al. (2007). The clinical and radiological spectrum of reversible cerebral vasoconstriction syndrome. A prospective series of 67 patients. Brain 130: 3091–3101.

Ducros A, Fiedler U, Porcher R et al. (2010). Hemorrhagic manifestations of reversible cerebral vasoconstriction syndrome: frequency, features, and risk factors. Stroke 41: 2505–2511.

Engelter ST, Petrella JR, Alberts MJ et al. (1999). Assessment of cerebral microcirculation in a patient with hypertensive encephalopathy using MR perfusion imaging. AJR Am J Roentgenol 173: 1491–1493.

Facchini A, Magnoni S, Civelli V et al. (2013). Refractory intracranial hypertension in posterior reversible encephalopathy syndrome. Neurocrit Care 19: 376–380.

Fugate JE, Rabinstein AA (2015). Posterior reversible encephalopathy syndrome: clinical and radiological manifestations, pathophysiology, and outstanding questions. Lancet Neurol 14: 914–925.

Fugate JE, Claassen DO, Cloft HJ et al. (2010). Posterior reversible encephalopathy syndrome: associated clinical and radiologic findings. Mayo Clin Proc 85: 427–432.

Fugate JE, Ameriso SF, Ortiz G et al. (2012). Variable presentations of postpartum angiopathy. Stroke 43: 670–676.

Gao B, Liu FL, Zhao B (2012). Association of degree and type of edema in posterior reversible encephalopathy syndrome with serum lactate dehydrogenase level: initial experience. Eur J Radiol 81: 2844–2847.

Giraldo EA, Fugate JE, Rabinstein AA et al. (2011). Posterior reversible encephalopathy syndrome associated with hemodynamic augmentation in aneurysmal subarachnoid hemorrhage. Neurocrit Care 14: 427–432.

Guyton AC (2006). Cerebral blood flow, cerebrspinal fluid and brain metabolism. In: AC Guyton (Ed.), Textbook of medical physiology, 11th edn. Elsevier Saunders, Philadelphia.

Hefzy HM, Bartynski WS, Boardman JF et al. (2009). Hemorrhage in posterior reversible encephalopathy syndrome: imaging and clinical features. AJNR Am J Neuroradiol 30: 1371–1379.

Hinchey J, Chaves C, Appignani B et al. (1996). A reversible posterior leukoencephalopathy syndrome. N Engl J Med 334: 494–500.

Horbinski C, Bartynski WS, Carson-Walter E et al. (2009). Reversible encephalopathy after cardiac transplantation: histologic evidence of endothelial activation, T-cell specific trafficking, and vascular endothelial growth factor expression. AJNR Am J Neuroradiol 30: 588–590.

Hurwitz RL, Mahoney Jr DH, Armstrong DL et al. (1988). Reversible encephalopathy and seizures as a result of conventional vincristine administration. Med Pediatr Oncol 16: 216–219.

Ito Y, Niwa H, Iida T et al. (1997). Post-transfusion reversible posterior leukoencephalopathy syndrome with cerebral vasoconstriction. Neurology 49: 1174–1175.

Junna MR, Rabinstein AA (2007). Tacrolimus induced leukoencephalopathy presenting with status epilepticus and prolonged coma. J Neurol Neurosurg Psychiatry 78: 1410–1411.

Kahana A, Rowley HA, Weinstein JM (2005). Cortical blindness: clinical and radiologic findings in reversible posterior leukoencephalopathy syndrome: case report and review of the literature. Ophthalmology 112: e7–e11.

Kastrup O, Schlamann M, Moenninghoff C et al. (2015). Posterior reversible encephalopathy syndrome: the spectrum of MR imaging patterns. Clin Neuroradiol 25: 161–171.

Kelly K, Kalachand R, Murphy P (2008). Bortezomib-induced reversible posterior leucoencephalopathy syndrome. Br J Haematol 141: 566.

Keswani SC, Wityk R (2002). Don't throw in the towel! A case of reversible coma. J Neurol Neurosurg Psychiatry 73: 83–84.

Kontos HA, Wei EP, Navari RM et al. (1978). Responses of cerebral arteries and arterioles to acute hypotension and hypertension. Am J Physiol 234: H371–H383.

Kowianski P, Lietzau G, Steliga A et al. (2013). The astrocytic contribution to neurovascular coupling – still more questions than answers? Neurosci Res 75: 171–183.

Kozak OS, Wijdicks EF, Manno EM et al. (2007). Status epilepticus as initial manifestation of posterior reversible encephalopathy syndrome. Neurology 69: 894–897.

Kummer S, Schaper J, Mayatepek E et al. (2010). Posterior reversible encephalopathy syndrome in early infancy. Klin Padiatr 222: 269–270.

Kur JK, Esdaile JM (2006). Posterior reversible encephalopathy syndrome – an underrecognized manifestation of systemic lupus erythematosus. J Rheumatol 33: 2178–2183.

Lai CC, Chen WS, Chang YS et al. (2013). Clinical features and outcomes of posterior reversible encephalopathy syndrome in patients with systemic lupus erythematosus. Arthritis Care Res (Hoboken) 65: 1766–1774.

Lamy C, Oppenheim C, Meder JF et al. (2004). Neuroimaging in posterior reversible encephalopathy syndrome. J Neuroimaging 14: 89–96.

Lamy C, Oppenheim C, Mas JL (2014). Posterior reversible encephalopathy syndrome. Handb Clin Neurol 121: 1687–1701.

Lee SY, Dinesh SK, Thomas J (2008a). Hypertension-induced reversible posterior leukoencephalopathy syndrome causing obstructive hydrocephalus. J Clin Neurosci 15: 457–459.

Lee VH, Wijdicks EF, Manno EM et al. (2008b). Clinical spectrum of reversible posterior leukoencephalopathy syndrome. Arch Neurol 65: 205–210.

Legriel S, Schraub O, Azoulay E et al. (2012). Determinants of recovery from severe posterior reversible encephalopathy syndrome. PLoS One 7: e44534.

Leopold J (2013). The endothelium. In: JB Creager, J Loscalzo (Eds.), Vascular medicine: a companion to Braunwald's heart disease, 2nd edn. Elsevier Saunders, Philadelphia.

Li Y, Gor D, Walicki D et al. (2012). Spectrum and potential pathogenesis of reversible posterior leukoencephalopathy syndrome. J Stroke Cerebrovasc Dis 21: 873–882.

Li R, Mitchell P, Dowling R et al. (2013). Is hypertension predictive of clinical recurrence in posterior reversible encephalopathy syndrome? J Clin Neurosci 20: 248–252.

Liman TG, Bohner G, Heuschmann PU et al. (2012). The clinical and radiological spectrum of posterior reversible encephalopathy syndrome: the retrospective Berlin PRES study. J Neurol 259: 155–164.

Liman TG, Bohner G, Endres M et al. (2014). Discharge status and in-hospital mortality in posterior reversible encephalopathy syndrome. Acta Neurol Scand 130: 34–39.

MacKenzie ET, Strandgaard S, Graham DI et al. (1976). Effects of acutely induced hypertension in cats on pial arteriolar caliber, local cerebral blood flow, and the blood-brain barrier. Circ Res 39: 33–41.

Magana SM, Matiello M, Pittock SJ et al. (2009). Posterior reversible encephalopathy syndrome in neuromyelitis optica spectrum disorders. Neurology 72: 712–717.

Mancia G, Fagard R, Narkiewicz K et al. (2014). 2013 ESH/ESC practice guidelines for the management of arterial hypertension. Blood Press 23: 3–16.

Marra A, Vargas M, Striano P et al. (2014). Posterior reversible encephalopathy syndrome: the endothelial hypotheses. Med Hypotheses 82: 619–622.

McKinney AM, Short J, Truwit CL et al. (2007). Posterior reversible encephalopathy syndrome: incidence of atypical regions of involvement and imaging findings. AJR Am J Roentgenol 189: 904–912.

Moon SN, Jeon SJ, Choi SS et al. (2013). Can clinical and MRI findings predict the prognosis of variant and classical type of posterior reversible encephalopathy syndrome (PRES)? Acta Radiol 54: 1182–1190.

Mueller-Mang C, Mang T, Pirker A et al. (2009). Posterior reversible encephalopathy syndrome: do predisposing risk factors make a difference in MRI appearance? Neuroradiology 51: 373–383.

Naidu K, Moodley J, Corr P et al. (1997). Single photon emission and cerebral computerised tomographic scan and transcranial Doppler sonographic findings in eclampsia. Br J Obstet Gynaecol 104: 1165–1172.

Ni J, Zhou LX, Hao HL et al. (2011). The clinical and radiological spectrum of posterior reversible encephalopathy syndrome: a retrospective series of 24 patients. J Neuroimaging 21: 219–224.

Parent C, Eichacker PQ (1999). Neutrophil and endothelial cell interactions in sepsis. The role of adhesion molecules. Infect Dis Clin North Am 13: 427–447.x.

Postma IR, Slager S, Kremer HP et al. (2014). Long-term consequences of the posterior reversible encephalopathy syndrome in eclampsia and preeclampsia: a review of the obstetric and nonobstetric literature. Obstet Gynecol Surv 69: 287–300.

Pruitt JH, Copeland 3rd EM, Moldawer LL (1995). Interleukin-1 and interleukin-1 antagonism in sepsis, systemic inflammatory response syndrome, and septic shock. Shock 3: 235–251.

Rabinstein AA, Mandrekar J, Merrell R et al. (2012). Blood pressure fluctuations in posterior reversible encephalopathy syndrome. J Stroke Cerebrovasc Dis 21: 254–258.

Raj S, Overby P, Erdfarb A et al. (2013). Posterior reversible encephalopathy syndrome: incidence and associated factors in a pediatric critical care population. Pediatr Neurol 49: 335–339.

Rajasekhar A, George Jr TJ (2007). Gemcitabine-induced reversible posterior leukoencephalopathy syndrome: a case report and review of the literature. Oncologist 12: 1332–1335.

Reece DE, Frei-Lahr DA, Shepherd JD et al. (1991). Neurologic complications in allogeneic bone marrow transplant patients receiving cyclosporin. Bone Marrow Transplant 8: 393–401.

Roth C, Ferbert A (2010). Posterior reversible encephalopathy syndrome: long-term follow-up. J Neurol Neurosurg Psychiatry 81: 773–777.

Schwartz RB, Jones KM, Kalina P et al. (1992). Hypertensive encephalopathy: findings on CT, MR imaging, and SPECT imaging in 14 cases. AJR Am J Roentgenol 159: 379–383.

Schwartz RB, Bravo SM, Klufas RA et al. (1995). Cyclosporine neurotoxicity and its relationship to hypertensive encephalopathy: CT and MR findings in 16 cases. AJR Am J Roentgenol 165: 627–631.

Seet RC, Rabinstein AA (2012). Clinical features and outcomes of posterior reversible encephalopathy syndrome following bevacizumab treatment. QJM 105: 69–75.

Serkova NJ, Christians U, Benet LZ (2004). Biochemical mechanisms of cyclosporine neurotoxicity. Mol Interv 4: 97–107.

Shaharir SS, Remli R, Marwan AA et al. (2013). Posterior reversible encephalopathy syndrome in systemic lupus erythematosus: pooled analysis of the literature reviews and report of six new cases. Lupus 22: 492–496.

Sharma A, Whitesell RT, Moran KJ (2010). Imaging pattern of intracranial hemorrhage in the setting of posterior reversible encephalopathy syndrome. Neuroradiology 52: 855–863.

Singh N, Bonham A, Fukui M (2000). Immunosuppressive-associated leukoencephalopathy in organ transplant recipients. Transplantation 69: 467–472.

Singhal AB (2004). Postpartum angiopathy with reversible posterior leukoencephalopathy. Arch Neurol 61: 411–416.

Singhal AB, Hajj-Ali RA, Topcuoglu MA et al. (2011). Reversible cerebral vasoconstriction syndromes: analysis of 139 cases. Arch Neurol 68: 1005–1012.

Strandgaard S, Paulson OB (1984). Cerebral autoregulation. Stroke 15: 413–416.

Tallaksen CM, Kerty E, Bakke S (1998). Visual hallucinations in a case of reversible hypertension-induced brain oedema. Eur J Neurol 5: 615–618.

Tlemsani C, Mir O, Boudou-Rouquette P et al. (2011). Posterior reversible encephalopathy syndrome induced by anti-VEGF agents. Target Oncol 6: 253–258.

Trommer BL, Homer D, Mikhael MA (1988). Cerebral vasospasm and eclampsia. Stroke 19: 326–329.

Utz N, Kinkel B, Hedde JP et al. (2001). MR imaging of acute intermittent porphyria mimicking reversible posterior leukoencephalopathy syndrome. Neuroradiology 43: 1059–1062.

van Beek AH, Claassen JA, Rikkert MG et al. (2008). Cerebral autoregulation: an overview of current concepts and methodology with special focus on the elderly. J Cereb Blood Flow Metab 28: 1071–1085.

Vaughn C, Zhang L, Schiff D (2008). Reversible posterior leukoencephalopathy syndrome in cancer. Curr Oncol Rep 10: 86–91.

Waldemar G, Paulson OB, Barry DI et al. (1989). Angiotensin converting enzyme inhibition and the upper limit of cerebral blood flow autoregulation: effect of sympathetic stimulation. Circ Res 64: 1197–1204.

Will AD, Lewis KL, Hinshaw Jr DB et al. (1987). Cerebral vasoconstriction in toxemia. Neurology 37: 1555–1557.

Wong R, Beguelin GZ, de Lima M et al. (2003). Tacrolimus-associated posterior reversible encephalopathy syndrome after allogeneic haematopoietic stem cell transplantation. Br J Haematol 122: 128–134.

Wu Q, Marescaux C, Wolff V et al. (2010). Tacrolimus-associated posterior reversible encephalopathy syndrome after solid organ transplantation. Eur Neurol 64: 169–177.

Chapter 27

Acute neurotoxicology of drugs of abuse

S.J. TRAUB* AND M.D. LEVINE
Department of Emergency Medicine, Mayo Clinic, Phoenix, AZ, USA

Abstract

Many substances can affect the central nervous system, and may cause patients to become critically ill. Acute central neurotoxicologic syndromes associated with drugs of abuse are usually caused by an overdose of sedative-hypnotic agents (including alcohol) or opioids, withdrawal from sedative-hypnotic agents, or an overdose of anticholinergic or sympathomimetic agents. Clinical findings are often syndromic, making physical examination the most important diagnostic tool in the approach to the patient with an unknown ingestion. Treatment focusses on supportive care as the most important intervention for all such patients, augmented by antidotal therapy when appropriate.

INTRODUCTION

Toxicology is the study of adverse interactions between xenobiotics and living organisms. Many toxins directly affect the central nervous system (CNS), and many patients poisoned by such toxins become acutely and critically ill.

In this chapter, we focus on acute central neurotoxicologic syndromes of drugs of abuse. When considering drug-induced alterations in mental status in critically ill patients, the decision as to which neurotoxicologic syndromes to include and which to exclude is challenging. We have chosen to exclude subacute and chronic conditions (such as the personality changes that may accompany mercury poisoning), conditions that manifest predominantly or exclusively with peripheral findings (such as botulism, tetanus, or strychnine poisoning), conditions whose presentations are often dominated by nonneurologic derangements (such as aspirin or tricyclic antidepressant poisoning), and conditions with CNS findings that rarely, if ever, produce critical illness (such as toxicity from marijuana or hallucinogens). This allows us to focus on six major syndromes that are of particular interest to those who care for critically ill neurologic patients: sedative-hypnotic toxicity, opioid toxicity, sympathomimetic toxicity, anticholinergic toxicity, dissociative agent toxicity, and sedative-hypnotic withdrawal.

In the first section, we discuss each of the syndromes, with particular attention to pathophysiology, presentation, and treatment. We depart from the structure of other chapters in this series in that we omit neurorehabilitation, as complete neurologic recovery from these syndromes is the rule unless a secondary complication – such as intracranial hemorrhage after cocaine use or anoxic brain damage after opioid use – has occurred. Of note, when discussing laboratory testing and imaging in the first section, we address only laboratory testing that is important when the diagnosis of agent- or class-specific poisoning is established.

In the second section, we present an approach to the undifferentiated patient with a suspected acute central neurotoxicologic syndrome related to one of the aforementioned drugs of abuse. We focus primarily on the importance of physical examination to facilitate diagnosis, mention additional crucial diagnostic interventions to perform, and discuss pitfalls.

MAJOR NEUROTOXICOLOGIC SYNDROMES

Sedative-hypnotic toxicity

EPIDEMIOLOGY

Sedative-hypnotic agents refer to a class that includes ethanol, pharmaceutical agents, and many drugs of abuse

---

*Correspondence to: Stephen Traub, MD, Department of Emergency Medicine, Mayo Clinic in Arizona, 6335 E. Monterra Way, Scottsdale AZ 85266, USA. Tel: +1-480-766-9663, E-mail: traub.stephen@mayo.edu

Table 27.1

Commonly abused sedative-hypnotic agents

Ethanol
Benzodiazepines
Barbiturates
Gamma-hydroxybutyrate
Chloral hydrate
Meprobamate
Carisoprodol
Zolpidem
Zaleplon
Zopiclone
Eszopiclone

(Table 27.1). Sedative-hypnotic abuse and dependence occur worldwide, and are widespread. In the US, the lifetime rate of alcohol abuse is almost 20%, and the lifetime rate of dependence is 10–15% (Hasin et al., 2007). In France, one study found that approximately half of all patients using prescription sedative hypnotics were physically dependent on them (Guerlais et al., 2015). Sedative-hypnotic toxicity is a problem that is encountered by virtually every clinician who works with the critically ill.

Pharmacology

Most sedative-hypnotic agents act directly or indirectly at the gamma-aminobutyric acid (GABA) receptor to hyperpolarize neurons. Two distinct subtypes of GABA receptors play roles in sedative-hypnotic toxicity: $GABA_A$ receptors (the site of action of benzodiazepines and barbiturates) modulate neuronal activity via chloride conductance (Olsen et al., 1986), whereas $GABA_B$ receptors (the site of action of gamma-hydroxybutyrate and baclofen) exert their effects via modulation of potassium and calcium conductance (Emson, 2007; Marshall, 2008). The exact mechanism of action of ethanol is not entirely known; specific ethanol receptors (if they exist) have eluded detection and characterization. It is possible that ethanol does not act via direct receptors, but rather by inducing conformational changes in proteins via substitution at key water-binding sites (Harris et al., 2008). Regardless of the exact mechanism, ethanol likely exerts much of its effect at the level of $GABA_A$ receptors (Valenzuela and Jotty, 2015) and through interference with stimulatory N-methyl-D-aspartate (NMDA) neurotransmission (He et al., 2013).

Clinical presentation

The dominant finding in sedative-hypnotic toxicity is CNS depression, which exists on a continuum that is generally dose-dependent. Early and mild toxicity manifests with some degree of slurred speech, ataxia, and delayed response time to stimuli. As toxicity progresses, patients often become lethargic; in severe toxicity, lethargy may progress to coma.

It is worth noting, however, that sedative-hypnotic agents are not monolithic in their effects. There are important clinical differences between drugs that work predominantly at the $GABA_A$ receptor, and general class differences between those that work at the $GABA_A$ versus the $GABA_B$ receptor.

Although both barbiturate and benzodiazepine poisoning may present with profoundly depressed mental status, barbiturate poisoning may also manifest with apnea or circulatory collapse (Hadden et al., 1969; Spear and Protass, 1973) whereas such presentations are rare with benzodiazepine poisoning (Gaudreault et al., 1991). It is for this reason that benzodiazepines have essentially replaced barbiturates as the sedative-hypnotic agent of choice in medical practice. Newer sedative-hypnotic agents (such as zolpidem), often used as sleep aids, also demonstrate excellent safety profiles (Garnier et al., 1994).

Similar to the $GABA_A$ agents, $GABA_B$ drugs profoundly decrease mental status in overdose. In contradistinction to $GABA_A$ agents, however, $GABA_B$ agents may also produce increased muscular activity or seizures (Leung et al., 2006; Schep et al., 2012).

Peripheral physical examination findings are generally unremarkable in most cases of sedative-hypnotic toxicity. Benzodiazepines rarely produce abnormalities of peripheral physiology, and the term "coma with normal vital signs" is frequently used in reference to the clinical presentation of benzodiazepine overdose. Barbiturates may be associated with bradycardia, hypothermia, and skin blistering, while gamma-hydroxybutyrate may be associated with bradycardia as well. Traumatic injuries should prompt clinicians to exclude alternative etiologies of altered mental status, such as subdural hemorrhage.

Laboratory testing

Patients with known uncomplicated sedative-hypnotic intoxication require minimal, if any, laboratory testing other than rapid serum glucose testing to exclude hypoglycemia.

The use of blood or breath ethanol levels to assess the degree of alcohol intoxication, while appealing, is unreliable. Although many legal jurisdictions throughout the world define blood ethanol levels (such as 80 mg/dL) at which certain activities (such as driving) are prohibited, it is not the case that all individuals exhibit the same degree of intoxication at a given blood ethanol level. Significant intoxication may occur at levels below legal limits in alcohol-naive individuals, and chronic ethanol users with markedly elevated blood levels may appear to be minimally affected (Sullivan et al., 1987). For these reasons, ethanol intoxication is a clinical, and not a numeric, diagnosis.

CLINICAL COURSE AND MANAGEMENT

The treatment of sedative-hypnotic poisoning is almost entirely supportive. We recommend endotracheal intubation for airway protection or apnea, supplemental oxygen as needed to maintain normal oxygen saturation, and intravenous fluids followed by vasopressor agents to treat hypotension, noting that these interventions are rarely needed when the sedative-hypnotic agent involved is a benzodiazepine or a so-called "z drug" – zolpidem, zopiclone (or its S-enantiomer, eszoplocine), or zaleplon.

Thiamine deficiency is prevalent not only among chronic alcoholics but among patients who present to the emergency department with ethanol intoxication (Li et al., 2008). Clinicians must maintain a high index of suspicion for Wernicke encephalopathy or Korsakoff psychosis (together, Wernicke–Korsakoff syndrome), as the rate of initial misdiagnosis of these conditions is high (Harper et al., 1986). Thiamine (100 mg intravenously (IV)) is appropriate for any patient who presents with an ethanol-related disorder in whom any suspicion for Wernicke–Korsakoff syndrome exists.

The use of analeptic (CNS arousal) agents in the treatment of sedative-hypnotic toxicity should be mentioned only to be condemned. In an era when barbiturate intoxication was common and the use of analeptics (such as amphetamines) prevalent, a landmark study showed that the use of supportive care alone produced superior clinical results (Clemmesen and Nilsson, 1961). This "Scandanavian method" has become the cornerstone of treatment of most poisoned patients, but particularly those with sedative-hypnotic toxicity.

There is no antidotal therapy for most sedative-hypnotic agents; benzodiazepines are an exception. However, the use of specific reversal agents in the treatment of acute benzodiazepine toxicity is controversial (see section on complex clinical decisions, below). Patients with presumed sedative-hypnotic toxicity who fail to improve over time should undergo a more rigorous workup for altered mental status

COMPLEX CLINICAL DECISIONS

While patients with sedative-hypnotic toxicity may be critically ill, most treatment decisions are relatively straightforward, and outlined above. Exceptions to this rule are the decision to intubate patients with severe gamma-hydroxybutyrate ingestion and the decision to administer flumazenil (a specific benzodiazepine reversal agent) to patients with benzodiazepine toxicity.

Gamma-hydroxybutyrate toxicity is often associated with a waxing and waning mental status. Patients who appear comatose may become alert with stimulation, only to lapse back to an apparently comatose state when the stimulation is removed. In such patients, the decision to proceed with advanced airway management is complicated. Clinicians with significant experience in airway management as well as the management of gamma-hydroxybutyrate ingestion may feel comfortable observing such patients without intubation; in one series, 87% of gamma-hydroxybutyrate-poisoned patients with a low Glasgow Coma Scale score (3–8) were managed conservatively, and none had adverse outcomes (Munir et al., 2008). We suggest, however, that clinicians with less experience should have a lower threshold to intubate such patients.

Flumazenil is a specific antagonist of benzodiazepines at the benzodiazepine receptor. Although its use in benzodiazepine poisoning is intellectually attractive, there are significant potential pitfalls. In a patient who is habituated to benzodiazepines, even at low doses, administration of flumazenil may precipitate benzodiazepine withdrawal (Mintzer et al., 1999). While flumazenil use at low doses may be safe, particularly in those in whom habituation may not have had time to occur (Moore et al., 2014), we recommend against its routine use as acute benzodiazepine toxicity is rarely if ever life-threatening, and can almost always be managed safely and effectively with supportive care alone.

Opioids

EPIDEMIOLOGY

Opioids are naturally occurring or synthetic agents that are widely used for their ability to relieve pain and abused for their euphoria-inducing effects. Most opioids are structural analogs of morphine, a naturally occurring substance from the opium poppy, *Papaver somniferum*.

Opioid abuse in the US is a significant and growing problem. Exposure to opioids often occurs at a young age, with one in four adolescents reporting some type of use (Boyd et al., 2014). Prescription opioid diversion and abuse in the US have increased dramatically (Dart et al., 2015). In recent years, opioid overdose deaths from prescription medications have grown faster than those from heroin, although heroin deaths have increased as well (Rudd et al., 2014). The current US opioid epidemic has led to novel interventions in an attempt to stem it, including the use of controlled-substance monitoring databases to help prescribers identify opioid abusers (McAllister et al., 2015) and the distribution of naloxone, a specific narcotic antagonist, to laypersons (Wheeler et al., 2015).

PHARMACOLOGY

Opioids are exogenous xenobiotics that mimic the actions of endorphins (so named because the are endogenous

Table 27.2

Traditional opioid receptor names and functions

Receptor name	Functions
Mu	Analgesia
	Sedation
	Euphoria
	Respiratory depression
	Decreased gastrointestinal motility
Kappa	Analgesia (spinal and suprasinal)
	Miosis
	Dysphoria
Delta	Analgesia (spinal and supraspinal)
Nociceptin/orphanin	Analgesia
	Anxiolysis

morphine-like substances) – metenkephalin, leuenkephalin, β-endorphin and dynorphin. There are several different classification schemasta for opioid receptors: the classic receptor-naming scheme, with the most significant clinical findings related to each receptor, is shown in Table 27.2. Opioid receptors are members of the G superfamily of proteins (Waldhoer et al., 2004), with cellular mechanisms of action that may involve adenylate cyclase, calcium channels, or potassium channels.

The rewarding properties of opioids are a function of mu-receptor agonism (Lutz and Kieffer, 2013). The cellular neuropathology of the most lethal aspect of opioid toxicity, respiratory depression, is also a result of mu-receptor agonism in the brainstem (Boom et al., 2012). Respiratory effects are caused predominantly by a depression in both the hypercarbic and hypoxic drive to breathe (Weil et al., 1975), although the former plays a larger role.

CLINICAL PRESENTATION

The most prominent finding in opioid toxic patients is an alteration in mental status, and patients with opioid toxicity present in a stereotypical fashion. Patients may be euphoric, but are often subdued; with significant toxicity, they generally present tired-appearing or lethargic. The term narcotic itself derives from a Greek term meaning "to make stiff or numb." Patients may lose their train of thought or speech, or even "nod off" during the medical interview.

Several peripheral physical examination findings support the diagnosis of opioid toxicity. Respiratory rate and depth are slowed, often to dangerous levels. Hypercapnia may contribute to the depressed level of consciousness. Pupils are constricted (Knaggs et al., 2004), often to a degree described as "pinpoint," and bowel sounds are markedly decreased or absent.

The acute respiratory distress syndrome (ARDS) may occur in opioid-poisoned patients. It often comes to clinical attention immediately after the patient's ventilatory status has improved, frequently after administration of the narcotic antagonist naloxone (Narcan) (Schwartz and Koenigsberg, 1987).

Individual opioids may, as a result of their unique structure, also cause cardiac or neurologic effects via nonopioid pathways. Propoxyphene and meperidine may both cause neuromuscular hyperactivity, including seizures (Hagmeyer et al., 1993; Basu et al., 2009). Propoxyphene may cause QRS prolongation, particularly in overdose (Afshari et al., 2005), while methadone is associated with QTc prolongation, even in patients receiving doses in the lower level of therapeutic (Roy et al., 2012).

LABORATORY TESTING

When opioid toxicity is known, minimal testing is appropriate. We recommend a fingerstick glucose in any patient with an alteration in mental status, and electrocardiogram (ECG) to exclude any toxin-induced conduction abnormality (particularly as some opioids have this effect), a pregnancy test in women of childbearing age, and testing for salicylates and acetaminophen in any patient with known or suspected suicidal ideation.

If patients have been immobile for an extended period, a serum creatine kinase and creatinine are indicated to assess for rhabdomyolysis and pigment-induced nephropathy, respectively.

CLINICAL COURSE AND MANAGEMENT

Opioid toxicity often presents with respiratory failure, presenting the treating physician with an immediate question regarding airway management and correction of hypoventilation. In most patients who are unresponsive and apneic, endotracheal intubation is the appropriate intervention. In opioid-poisoned patients, however, administration of a specific narcotic antagonist such as naloxone may restore consciousness, airway-protective mechanisms, and ventilation.

In all patients with opioid-induced respiratory depression, bag-valve mask ventilation should be started and continued until naloxone has restored appropriate spontaneous ventilation. Recommendations regarding the initial dosing of naloxone vary. We recommend starting with 0.04 mg IV, a dose that is lower than often advocated, but less likely to precipitate opiate withdrawal. Although opiate withdrawal *per se* is not life-threatening, opioid withdrawal may unmask severe pain syndromes, particularly in patients with malignancies (Boland et al., 2013). If there is no response to the initial dose in 2–3 minutes, we recommend an escalating dosing strategy of 0.4 mg/2 mg/4 mg/10 mg/15 mg, similar to that which has been advocated by others (Boyer, 2012). If a patient has no response to the cumulative dose from this protocol, then uncomplicated opioid toxicity is almost assuredly not the cause of the patient's mental status changes and respiratory depression.

Because of the relatively brief duration of action of naloxone, some patients (particularly those who have ingested long-acting opioid agonists, such as sustained-release oxycodone or methadone) will require additional naloxone dosing. When an additional bolus dose is required due to a recurrence of respiratory depression, we recommend that the patient be placed on a continuous naloxone infusion. An appropriate initial dose for continuous infusion is approximately two-thirds of the cumulative dose to which the patient responded, infused per hour (Goldfrank et al., 1986). For example, if the patient responded to a cumulative dose of 0.44 mg, an appropriate starting dose for the naloxone continuous infusion would be 0.3 mg/hour.

Acute complications from opioid use include hypoxic coma (if the patient has had a prolonged period of opioid-induced hypoventilation prior to restoration of appropriate ventilation), ARDS, and rhabdomyolysis or compartment syndrome (if the patient has had a prolonged period of immobility).

COMPLEX CLINICAL DECISIONS

Clinicians who treat patients with opioid toxicity may be faced with complex decisions. These include the decision to use naloxone in patients with mild alterations in mental status and/or respiratory insufficiency, and the disposition decision for patients who have a clear sensorium after the administration of naloxone and who wish to leave against medical advice.

The use of naloxone in patients with relatively mild depression in mental status or respiratory effort deserves particular discussion. It must be considered in the context of the appropriate goal of naloxone administration, which is to restore appropriate spontaneous ventilation. In patients with normal oxygenation (without supplemental oxygen) and mild decreases in mentation or respiratory rate, naloxone administration may be unnecessary. In patients who have mild iatrogenic reduction in ventilation after administration of known quantities of opioids for severe pain syndromes, small amounts of supplemental oxygen may be appropriate, provided that the patient is closely monitored. However, in patients who are hypoxemic in the setting of self-induced opioid toxicity (often with agents or doses that are unknown), naloxone is the most appropriate intervention. While it is tempting to simply apply supplemental oxygen to such patients rather than risk turning a cooperative patient into a combative one, patients treated in this fashion are at risk for developing significant hypercapnia. Of note, this strategy applies to patients without confounding conditions – such as pneumonia or opioid-induced ARDS – which require the administration of supplemental oxygen.

The patient whose sensorium clears after the administration of naloxone who wishes to leave the hospital represents a particular challenge. Such patients often have a normal mental status, but are at risk for recurrent toxicity, particularly if they have ingested a long-acting opioid whose duration of action exceeds that of naloxone. Although several studies have failed to find evidence of significant rates of death after opioid-toxic patients leave against medical advice after naloxone administration (Vilke et al., 1999, 2003; Wampler et al., 2011), most of the patients in these studies used relatively short-acting opioids (such as heroin), and the data cannot be immediately extrapolated to longer-acting agents. Whenever possible, such patients should be strongly encouraged to remain for a period of observation for several hours.

Dissociative agents

EPIDEMIOLOGY

Dissociative agents have been used medicinally for decades, with phencyclidine (PCP) first having been used as an anesthetic in the 1950s. Early reports noted a profound state of anesthesia, in which major surgeries were performed without loss of respiratory drive; however, severe emergence reactions (including psychosis and hallucinations) limited PCP's use (Greifenstein et al., 1958). More recently, ketamine has been introduced for sedation and general anesthesia, and as an adjunct to treat refractory unipolar major depressive disorder, posttraumatic stress disorder, and acute suicidal ideation (Green and Krauss, 2004; Green and Sherwin, 2005; Potter and Choudhury, 2014; Schwartz et al., 2016). Dextromethorphan is a widely available over-the-counter cough suppressant that has dissociative properties at higher doses.

Dissociative agents are abused recreationally. PCP ("angel dust") abuse reached epidemic proportions in the 1970s (Stillman and Petersen, 1979), but its use today is sporadic. Ketamine, often referred to as "special K," "vitamin K," or "super K," has become popular at rave parties and night clubs (Jansen, 1993). Dextromethorphan abuse, which became popular in the early 2000s, is often referred to as "robo tripping" or "skittling." The former term refers to the well-known dextromethorphan-containing cough medication Robitussin, whereas the latter refers to the resemblance of many dextromethorphan-containing cough and cold medications to the candy Skittles.

Dissociative agents may be ingested via many routes. PCP is often consumed orally, as thermal degradation results in inactivation of nearly 50% of the drug when smoked (National Highway Traffic Safety Administration, n.d.). Ketamine can be consumed either orally or parenterally, although poor oral bioavailability makes insufflation the preferred method. Although dextromethorphan is typically consumed orally, inhalational use has been reported (Hendrickson and Cloutier, 2007).

Pharmacology

Phencyclidine, ketamine, and dextromethorphan all share common mechanisms of action. These agents bind to the calcium channel on the NMDA receptor (Siu and Drachtman, 2007; Lodge and Mercier, 2015), resulting in NMDA antagonism. These agents also inhibit both peripheral and central catecholamine uptake (Akunne et al., 1991; Steinmiller et al., 2003; Boyer, 2004) and have effects on opioid sigma receptors (Wolfe and De Souza, 1993). These cellular effects mediate the clinical effects of pain control (Hillhouse and Negus, 2016) and psychosis (Jodo, 2013), and cause the dissociative effects observed in high doses.

Clinical presentation

All dissociative agents may produce a dose-dependent psychosis and impairment of sensory input, whereby perception and motor activity are separated from each other. Patients may appear calm, intoxicated, or agitated. Hallucinations may occur, and when present, are frequently auditory. Horizontal, vertical, or rotatory nystagmus may be observed (McCarron et al., 1981a, b).

Catecholamine reuptake inhibition may result in sympathomimetic effects, including hypertension and tachycardia. However, unlike other sympathomimetic agents, hyperthermia is relatively uncommon (McCarron et al., 1981b: 1000 cases; Ng et al., 2010).

Because of its effects on serotonin, dextromethorphan abuse may result in serotonin syndrome, either as an isolated ingestion (Ganetsky et al., 2007) or when taken in the setting of chronic serotonin selective reuptake inhibitor use (Schwartz et al., 2008).

Respiratory depression may occur with dextromethorphan, particularly in children (Shaul et al., 1977; Henretig, 1994).

The clinical presentation of dissociative agent toxicity may vary, due in large part to differences in dosing from patient to patient. Ketamine, for example, produces little more than analgesia at low doses, sensory distortions at moderate doses, and complete dissociation and anesthesia at higher doses (Peltoniemi et al., 2016). The route of administration may also alter the observed clinical effects; for a given dose of ketamine, insufflation will cause more symptoms than oral ingestion. If coingestants are present (as is frequently the case with PCP; Dominici et al., 2015), a mixed clinical picture may result.

Laboratory testing

When the diagnosis of dissociative agent toxicity is known, minimal testing is appropriate. Depending on the scenario, it may also be appropriate to assess serum glucose, obtain an ECG, perform a pregnancy test, ensure there has been no dangerous coingestion, and rule out rhabdomyolysis.

Management

The first step in the management of dissociative agent toxicity is to ensure that the patient's airway is patent, and that breathing and circulation adequate. Patients who are agitated should, if possible, be placed in a quiet, secluded room to reduce sensory stimuli.

Benzodiazepines are appropriate for agitation, and for the treatment of dissociative-induced pyschosis. If such symptoms prove refractory to benzodiazepines, butyrophenones are a reasonable alternative (Giannini et al., 1984).

When present, rhabdomyolysis should be treated with intravenous fluids per usual protocols.

Patients may develop emergence reactions as their acute toxicity resolves. Such reactions, which are characterized most frequently by hallucinations and confusion, are best treated with benzodiazepines.

Complex clinical decisions

The decision to use naloxone to reverse the decrease in mental status and respiratory depression associated with dextromethorphan may be difficult. In patients who are habituated to narcotics, the use of naloxone will precipitate opioid withdrawal, further complicating assessment and management. In patients who are opioid-naive, however, there is little down side. Of note, higher than usual doses of naloxone (on the order of 10 mg IV) may be required (Wolfe and Caravati, 1995; Chyka et al., 2007).

Urinary acidification has been advocated in the past as a means of enhancing PCP elimination through ion trapping. However, the risks of systemic acidemia are far greater than the benefits of slightly increased elimination, and such an approach is no longer recommended.

Sedative-hypnotic withdrawal

Epidemiology

Sedative-hypnotic agents (the class that includes ethanol) are the most commonly abused drugs in the world. In many patients, chronic and habitual use leads to a state of physiologic dependence; in such cases, abrupt cessation of drug use produces significant derangements in homeostasis, referred to as withdrawal.

In societies in which ethanol use is common, lifetime ethanol dependence amongst drinkers and former drinkers is likely on the order of 10% (Grant and Dawson, 1997); in the US in 2014, over 1.5 million people with alcohol use disorder were treated in specialized facilities (Substance Abuse and Mental Health Services Administration, 2014).

Pharmacology

When ethanol is ingested chronically, complex changes (including alterations in gene expression and receptor morphology) lead to a reduction in GABA activity and an increase in glutamic NMDA activity (Littleton, 1998). In such patients who continue to ingest ethanol, the ongoing CNS-depressant effects of ethanol counteract these changes, and the result is a relative homeostasis.

If such patients abruptly cease ingesting ethanol, however, this CNS-depressant input is lost, and patients' endogenous state – decreased baseline GABA activity and increased NMDA activity – produces the signs and symptoms of ethanol withdrawal.

Clinical presentation

Ethanol withdrawal may begin as early as hours or as late as days after cessation of ethanol ingestion. Conceptually, ethanol withdrawal may be divided into four syndromes: tremulousness, hallucinosis, withdrawal seizures, and delirium tremens. It is important to note that there is no "standard" progression of ethanol withdrawal, and it is the rule (rather than the exception) that patients will not pass through all four syndromes. In clinical practice, nursing protocols have been considered useful and in the US CIWA protocol is instituted in patients with a known or suspicious history of alcohol dependence (Table 27.3).

Alcoholic tremulousness

Early ethanol withdrawal is characterized by sympathetic hyperactivity; anxiety is the most prominent symptom, whereas hypertension, tachycardia, and diaphoresis are prominent signs (Blondell, 2005). In patients with alcoholic tremulousness, the sensorium remains clear and the patient often craves ethanol; self-administration of ethanol usually abolishes the condition.

Alcohol withdrawal seizures

Alcohol withdrawal seizures may occur after cessation of ethanol consumption, and a classic study clearly implicates the causal link between ethanol cessation and seizures in humans (Isbell et al., 1955). Alcohol withdrawal seizures may occur at any time after the cessation of ethanol intake, but the occurrence peaks in the 12–24-hour range (Victor and Brausch, 1967). Alcohol withdrawal seizures are usually characterized by one or two generalized tonic-clonic events, with a clear sensorium after a brief postictal state, although status epilepticus is reported (Alldredge and Lowenstein, 1993).

Alcoholic hallucinosis

Alcoholic hallucinosis is characterized by a perception disorder that may be auditory, visual, or tactile in nature, and the hallucinations are similar to those experienced by schizophrenics (Soyka, 1990). In one study, the prevalence of alcoholic hallucinosis in patients being treated for ethanol withdrawal was approximately 10% (Tsuang et al., 1994). Patients with alcoholic hallucinosis have a clear sensorium and stable vital signs, which distinguishes them from patients with delirium tremens (see below).

Delirium tremens

Delirium tremens begins several days after the cessation of ethanol, and is a life-threatening condition characterized by profound autonomic instability (severe tachycardia, hypertension, and diaphoresis) in conjunction with a marked alteration in consciousness (such as marked confusion and agitation) (Isbell et al., 1955). Patients with concurrent medical illness, a history of delirium tremens, and a longer period of abstinence at presentation are at increased risk of developing delirium tremens (Ferguson et al., 1996; Lee et al., 2005).

Laboratory testing

In the patient with a clinical diagnosis of ethanol withdrawal, there are no laboratory tests to either confirm the disease state or grade its severity. Any patient with ethanol withdrawal and a clouded sensorium should undergo rapid glucose testing to exclude hypoglycemia.

A high index of suspicion must exist for concurrent medical conditions that may have precipitated abstinence. We recommend urinalysis and chest X-ray to exclude obvious infectious causes, a complete blood count and serum electrolytes; and testing of liver function and serum lipase.

Clinical course and management

In patients with ethanol withdrawal, the initial goal of treatment is to approximate physiologic homeostasis via agents that have pharmacologic cross-reactivity with ethanol. Benzodiazepines have an excellent safety profile, even when used in high doses, and for this reason are superior to barbiturates. Benzodiazepines are also clearly superior to other agents, such as phenothiazines, that have been used historically (Kaim et al., 1969).

Table 27.3

Clinical institute withdrawal of alcohol scale, revised (CIWA - Ar)

Nausea and vomiting
Ask: "Do you feel sick to your stomach? Have you vomited?"
0   no nausea and no vomiting
1   mild nausea with no vomiting
2
3
4   intermittent nausea with dry heaves
5
6
7   constant nausea, frequent dry heaves and vomiting

Tremor
Arms extended and fingers spread apart
0   no tremor
1   not visible, but can be felt fingertip to fingertip
2
3
4   moderate, with patient's arms extended
5
6
7   severe, even with arms not extended

Paroxysmal sweats
Degree of sweating
0   no sweat visible
1   barely perceptible sweating, palms moist
2
3
4   beads of sweat obvious on forehead
5
6
7   drenching sweats

Anxiety
Ask: "Do you feel nervous?"
0   No anxiety, at ease
1   Mildly anxious
2
3
4   Moderately anxious, or guarded, so anxiety is inferred
5
6
7   Equivalent to acute panic states as seen in severe delirium or acute schizophrenic reactions

Agitation
0   Normal activity
1   Somewhat more activity than normal activity
2
3
4   Moderately fidgety and restless
5
6
7   Paces back and forth during most of the interview, or constantly thrashes about

Tactile disturbances
Ask: "Have you any itching, pins and needles sensations, any burning, any numbness, or do you feel bugs crawling on or under your skin?"
0   None
1   Very mild itching, pins and needles, burning, or numbness
2   Mild itching, pins and needles, burning or numbness
3   Moderate itching, pins and needles, or numbness
4   Moderately severe hallucinations
5   Severe hallucinations
6   Extremely severe hallucinations
7   Continuous hallucinations

Auditory disturbances
Ask: "Are you more aware of sounds around you? Are they harsh? Do they frighten you? Are you hearing anything that is not disturbing to you? Are you hearing things you know are not there?"
0   Very mild harshness or ability to frighten
1   Mild harshness or ability to frighten
2   Moderate harshness or ability to frighten
3   Moderate itching, pins and needles, or numbness
4   Moderately severe hallucinations
5   Severe hallucinations
6   Extremely severe hallucinations
7   Continuous hallucinations

Visual disturbances
Ask: "Does the light appear to be too bright? Is its color different? Does it hour your eyes? Are you seeing anything that is disturbing to you? Are you seeing things you know are not there?"
0   Not present
1   Very mild sensitivity
2   Mild sensitivity
3   Moderate sensitivity
4   Moderately severe hallucinations
5   Severe hallucinations
6   Extremely severe hallucinations
7   Continuous hallucinations

Headache/fullness in head
Ask: "Does your head feel different? Does it feel like there is a band around your head?" Do not rate for dizziness, or lightheadedness. Otherwise, reate severity
0   Not present
1   Very mild
2   Mild
3   Moderate
4   Moderately severe
5   Severe
6   Very severe
7   Extremely severe

Orientation/clouding of sensorium
Ask: "What day is this? Where are you? Who am I?"
0   Oriented, can do serial additions
1   Can't do serial additions or is uncertain about date
2   Disoriented for date by no more than 2 calendar days
3   Disoriented for date by more than 2 calendar days
4   Disoriented to place or person

In mild ethanol withdrawal, as is seen with alcoholic tremulousness, small doses of oral benzodiazepines will often suffice. Patients who require treatment in a critical care setting, however, usually manifest moderate to severe symptoms of ethanol withdrawal; such patients are commonly treated with intravenous benzodiazepines.

While all benzodiazepines possess the requisite pharmacodynamic profile to achieve sedation, there are important differences in pharmacokinetics. The ideal agent would be an intravenous medication with a rapid onset and a relatively long duration of action, as moderate to severe ethanol withdrawal usually resolves over the course of several days. For this indication, we believe that diazepam is superior to lorazepam: the onset of action is faster (Greenblatt et al., 1989), and diazepam has active metabolites that prolong its duration of action. In patients with liver failure, however, lorazepam may be a better choice as the use of diazepam may result in prolonged sedation due to reduced clearance.

Regardless of the agent chosen, the initial clinical goal is rapid sedation. Dosing is empiric, and tailored to patient response; while diazepam 10 mg IV is a reasonable initial choice, rapid escalation is often appropriate (particularly in patients with a history of significant ethanol consumption) and hundreds of milligrams may be necessary. Once initial sedation has been achieved, however, further dosing should be based on symptoms, not a fixed timetable. Symptom-triggered therapy results in a shorter duration of therapy with less benzodiazepine use than a fixed-dosing approach, and has been validated in both the emergency department and inpatient settings (Daeppen et al., 2002; Cassidy et al., 2012; Sachdeva et al., 2014). We recommend the use of the Clinical Institute for the Withdrawal of Alcohol scale (Table 27.3) to guide therapy.

We recommend the administration of thiamine (100 mg) for all patients with delirium tremens and for any other patient in whom mental status changes raise the concern of Wernicke–Korsakoff syndrome.

Complex clinical decisions

Two complex clinical decisions may confront the intensivist treating ethanol withdrawal: how to treat ethanol withdrawal that is refractory to benzodiazepines, and whether to use dexmetetomidine as adjunctive treatment.

Some patients with ethanol withdrawal have persistent symptoms despite large doses of appropriate benzodiazepines. The reasons why the majority of patients respond well to benzodiazepine therapy but others do not are unclear, although ethanol induces changes in GABA receptors and GABA transmission (Liang et al., 2006; Follesa et al., 2015) and patient-to-patient variation in these changes may play a role. One small study suggests that patients requiring more than 50 mg of intravenous diazepam in the first hour of therapy are more likely to develop ethanol withdrawal that is refractory to benzodiazepines (Hack et al., 2006).

Patients with benzodiazepine-refractory ethanol withdrawal require additional treatment with agents that are cross-tolerant with ethanol. The addition of phenobarbital may provide adequate sedation (Gold et al., 2007), although the concurrent use of benzodiazepines and phenobarbital increases the risk of severe respiratory depression requiring mechanical ventilation. Propofol is another option in such patients (Lorentzen et al., 2014), and also requires mechanical ventilation.

Dexmedetomidine is a centrally acting alpha-adrenergic agonist that is used to treat agitation in critically ill patients; it has also been used in the treatment of ethanol withdrawal (Muzyk et al., 2013). In one small study, however, the addition of dexmedetomidine did not improve outcomes in ethanol withdrawal, and was associated with more adverse drug reactions (Mueller et al., 2014). At this time, we continue to advocate for the use of medications that are cross-tolerant with ethanol, and do not advocate for the routine use of dexmedetomidine, even as adjunctive treatment.

Anticholinergic toxicity

Epidemiology

Anticholinergic xenobiotics are abundant in modern medicine. Some are used specifically for their anticholinergic side-effects (such as tricyclic antidepressants to treat nocturnal enuresis), whereas others possess anticholinergic properties as undesirable side-effects (as is the case with antiparkinsonian and antipsychotic drugs). Anticholinergic medications and plants are also abused for their ability to induce delirium, as a "legal high" (Thomas et al., 2009; Sutter et al., 2014). Anticholinergic xenobiotics may be encountered in the form of dermal preparations (scopolamine), inhalational agents (ipratropium), or oral medications (diphenhydramine).

The use of medications with anticholinergic properties is common, with some series involving elderly patients estimating a prevalence of nearly 10% (Machado-Alba et al., 2016). Overdoses on medications with anticholinergic effects are responsible for 15–20% of all poisoning admissions and up to 40% of those to an intensive care unit (Dawson and Buckley, 2015). In 2014, more than 100 000 ingestions involving anticholinergic agents were reported to US poison control centers (Mowry et al., 2015), numbers that likely underestimate true prevalence due to underrecognition and lack of mandatory reporting.

Pharmacology

Anticholinergic agents block the effects of acetylcholine through competitive inhibition at the binding site.

Table 27.4

Common xenobiotics with anticholinergic properties

Belladonna alkaloids	Antihistamines	Antipsychotics	Antispasmodics	Miscellaneous	Plants and mushrooms
Atropine	Brompheniramine	Chlorpromazine	Dicyclomine	Amantadine	*Datura stramomium*
Belladonna	Dimenhydrinate	Clozapine	Oxybutynin	Benztropine	*Mandrigora officinarum*
Glycopyrolate	Diphenhydramine	Loxapine		Carbamazepine	*Hyoscyamine niger*
Homatropine	Doxylamine	Mesoridazine		Cyproheptadine	*Amanita muscaria*
Hyosciamine	Hydroxyzine	Olanzapine		Cyclobenzaprine	*Amanita pantherina*
Ipatropium	Meclizine	Quetiapine		Tricyclic antidepressants	
Scopolamine		Thioridazine			

Generally speaking, "anticholinergic toxicity" refers to inhibition of muscarinic acetylcholine receptors. There are five unique muscarinic receptors (M1–M5) throughout both the peripheral nervous system and CNS, all of which are G-protein-coupled receptors. Activation of M2 and M4 receptors inhibits adenylate cyclase activity and activates potassium channels, resulting in hyperpolarization (Eglen, 2006; Ishii and Kurachi, 2006); activation of M1, M3, and M5 receptors results in activation of phospholipase C, subsequently generating 1,4,5-triphophate and 1,2-diacylglycerol, ultimately increasing intracellular calcium and activation of protein kinase C (Eglen, 2006).

Most antimuscarinic agents contain either a tertiary or quaternary amine. Quaternary amines (e.g., glycopyrrolate) are charged, limiting gastrointestinal absorption and penetration of the blood–brain barrier. Uncharged tertiary amines (e.g., atropine), in contrast, are well absorbed and have significant CNS penetration.

A list of commonly encountered xenobiotics with anticholinergic properties is presented in Table 27.4.

Clinical presentation

Anticholinergic toxicity classically produces both central and peripheral manifestations. However, this syndrome is heterogeneous, and patients may present with a spectrum of clinical manifestations. Perhaps owing to the variability of presentation, anticholinergic toxicity is underrecognized (Corallo et al., 2009; Levine, 2016).

Central manifestations of anticholinergic toxicity include encephalopathy, agitation, mutism, and coma. Seizures may also occur. Carphologia, a striking condition in which patients pick at imaginary objects, is often present (Levine et al., 2011). A quiet, mumbling speech,

Table 27.5

Grading system for anticholinergic-induced central nervous system effects (Burns et al., 2000)

Grade	Clinical findings
0	Relaxed and cooperative
1	Anxious, irritable
2	Intermittently/mildly disoriented, confused, hallucinations, moderate agitation
3	Incomprehensible speech, marked agitation
4	Seizure, coma

often characterized as a "mouthful of marbles" and nearly or frankly incomprehensible, is common with severe toxicity (Levine et al., 2011; Dawson and Buckley, 2015). A grading system has been proposed to assess the severity of CNS effects (Table 27.5).

Peripheral manifestations include mydriasis, tachycardia, urinary retention, reduced gastrointestinal motility (resulting in hypoactive bowel sounds, as well as delayed drug absorption and toxicity: Dawson and Buckley, 2015), hyperthermia, and dry, flushed skin. A palpable bladder may be appreciated on abdominal examination. Peripheral effects may be dose-related, with lower doses producing drying of the mucosal membranes and higher doses producing anhydrosis, tachycardia, and mydriasis (Longo, 1966; Gowdy, 1972). Mydriasis may not be present in cases of mixed ingestion or with ingestions of drugs which antagonize multiple receptors. Absence of bowel sounds is often a late finding. Fever may occur due to neuromuscular excitation, agitation, and impaired heat dissipation (lack of appropriate sweating) (Dawson and Buckley, 2015).

Medications with anticholinergic properties may produce findings in addition to those in the anticholinergic toxidrome. ECG changes may include QRS widening (from the sodium channel blockade seen with diphenhydramine) and QT prolongation (from the blockade of potassium efflux seen with quetiapine). Rhabdomyolysis may occur as a consequence of agitation, or possibly due to direct myotoxicity (Levine, 2016). Acute angle closure glaucoma (Lai and Gangwani, 2012) and urinary retention (particularly in patients with a predisposing condition, such as benign prostatic hyperplasia) (Vande Griend and Linnebur, 2012) are reported. The use of anticholinergic agents in elderly patients has been associated with an increased risk of falls (Richardson et al., 2015; Chatterjee et al., 2016; Marcum et al., 2016), which may be associated with significant morbidity. Elderly patients and those with underlying organic brain disease may be at increased risk of developing mild central anticholinergic toxicity, particularly confusion, following therapeutic drug administration.

Laboratory testing

When anticholinergic toxicity is known, minimal further testing is required. A rapid bedside glucose should be performed to eliminate hypoglycemia as an etiology of any alteration in mental status, and an ECG should be obtained to assess for conduction system abnormalities. Additional testing that should be considered is as described for other poisonings.

Clinical course and management

The initial management for patients with anticholinergic toxicity should focus on airway, breathing, and circulation. Endotracheal intubation should be performed in patients who are unable to protect their airway and in those with rapidly declining mental status, especially if there is concurrent concern for possible aspiration.

Seizures are best treated with benzodiazepines (e.g., 5–10 mg of intravenous diazepam, or 1–2 mg of intravenous lorazepam); rectal diazepam is an alternative. Intramuscular absorption of benzodiazepines is erratic, but this route is also acceptable if intravenous access cannot be established. Patients who are not voiding spontaneously, especially those with a palpable, distended bladder, should have a urinary catheter placed. Rhabdomyolysis should be treated as per usual practice.

Complex clinical decisions

Mechanistically, the ideal treatment for anticholinergic-induced agitation or delirium would be a selective agonist of the M1 receptor, but such an antidote does not currently exist (Dawson and Buckley, 2015). The two primary pharmacologic options for treatment of anticholinergic-induced delirium are the administration of GABA agonists, such as benzodiazepines, or the administration of a cholinesterase inhibitor, such as physostigmine.

Benzodiazepines are nonspecific CNS depressants, which can be used to treat agitation from many etiologies, including anticholinergic-induced delirium. As an agonist of the $GABA_A$ receptor, benzodiazepines result in increased release of inhibitory neurotransmitters.

Cholinesterase inhibitors such as physostigmine temporarily prevent the breakdown of acetylcholine in the synaptic cleft, thereby countering the effects of anticholinergic agents (Shannon, 1998). Physostigmine (from an extract of the Calabar bean) was first used in 1863 to reverse atropine-induced mydriasis, and the following year to treat atropine-induced delirium (Nickalls and Nickalls, 1988). There is substantial variability in the use of physostigmine to treat anticholinergic toxicity, even among medical toxicologists (Dawson and Buckley, 2015).

A typical starting dose of physostigmine is 1–2 mg for adults, and 0.02 mg/kg in pediatric patients. The drug should be administered over at least 5 minutes. It is contraindicated in patients with evidence of abnormal cardiac conduction, including intraventricular conduction delay or evidence of atrioventricular block.

Safe use of physostigmine relies on careful patient selection. Adverse effects include "overcorrection" of the anticholinergic state, with development of cholinergic symptoms (salivation, lacrimation, urination, diarrhea, vomiting, bradycardia, pulmonary secretions, bronchospasm). Seizures are reported but rare (Watkins et al., 2015), and their occurrence is usually directly related to a rapid rate of administration (Shannon, 1998). Some toxicologists administer benzodiazepines immediately prior to, or concurrently with, physostigmine in order to raise the seizure threshold.

There are no prospective studies directly comparing physostigmine with benzodiazepines. One retrospective registry of anticholinergic toxicity treated by medical toxicologists suggested that the intubation rate in patients receiving benzodiazepines was lower than in those receiving physostigmine. In another study, physostigmine was significantly more effective than benzodiazepines in treating agitation and reversing delirium, and physostigmine was associated with both a reduced incidence of complications and a shorter time to recovery (Burns et al., 2000). Regardless of whether a benzodiazepine or physostigmine is used, it is imperative that patients have frequent reassessment, with additional medications titrated based on the presence of significant symptoms. Because of its short half-life, physostigmine may require redosing if anticholinergic symptoms return after the first dose.

Sympathomimetic toxicity

EPIDEMIOLOGY

Sympathomimetic agents – those whose actions tend to augment or exaggerate the effects of the sympathetic nervous system – have a long history in modern medicine. In the 1920s, ephedrine was extracted from *Ephedra* species, and amphetamine became the first drug marketed specifically to treat depression (Rasmussen, 2006). Since that time, amphetamine and its derivatives have seen many uses in medicine, including treatment of narcolepsy and postencephalitic parkinsonism, and have been used to promote weight loss (Rasmussen, 2008). Military pilots have used amphetamines for decades to combat fatigue during long missions (Emonson and Vanderbeek, 1995; Rasmussen, 2008).

Currently, prescription amphetamines and amphetamine derivatives are widely used for the treatment of attention deficit hyperactivity disorder (ADHD). The abuse of stimulants is currently a significant public health concern: diversion of prescription stimulants used to treat ADHD has been reported in up to 10% of high school students, and up to 35% of college students are prescribed such medicines (Clemow, 2015).

While diversion of legitimate prescriptions is common (Chen et al., 2016) and remains a public health concern, most recreational abuse of stimulants involves illicit drugs, including cocaine, amphetamines, and synthetic cathinones ("bath salts"). In recent years, the abuse of phenythylamines and synthetic cathinones has increased significantly (Baumann et al., 2014).

Cathinones (Table 27.6) are currently gaining in popularity. This class was originally derived from the plant *Catha edulis*, common in the Middle East and parts of Africa. Cathinone is found only in fresh leaves, within a few days of harvest. Synthetic cathinones, which are phenylalkylamines or beta-keto amphetamine derivatives, are designed to produce similar effects, but with a longer "shelf-life." These drugs are often referred to as "plant food" or "bath salts" in an attempt to circumvent US drug laws and allow for legal sale. Illicit synthetic use was not widely reported until 2007 (Prosser and Nelson, 2012).

PHARMACOLOGY

Generally speaking, stimulants such as cocaine, amphetamines, and cathinones exert their effects by increasing the synaptic concentrations of excitatory neurotransmitters.

Cocaine

The principal effects of cocaine are to inhibit the reuptake of several biologic amines, including norepinephrine, epinephrine, and dopamine (Tella et al., 1992; Hall et al., 2004), and to increase the levels of the excitatory amino acids aspartate and glutamate (Smith et al., 1995). Cocaine also blocks sodium channels, which accounts for its use as a topical anesthetic in minor surgical procedures of the head and neck; at high systemic levels, however, this may lead to effects on cardiac conduction, such as QRS widening (Wood et al., 2009).

Amphetamines and phenylethylamines

Most amphetamines exert their actions via release of norepinephrine and dopamine from presynaptic nerve terminals. The precise mechanism of dopamine release is dose-dependent: at low doses, it is released by exchange diffusion via the dopamine uptake transporter; at high doses, amphetamines alkalinize the presynaptic dopamine-containing vesicles, resulting in release of dopamine from the vesicles and into the synapse via reverse transport (Suzler et al., 1995). The majority of the recognized peripheral effects of amphetamines are the result of stimulation of alpha- and beta-adrenergic receptors.

Most designer amphetamines have affinity for the 5HT2 serotonin receptors (Nelson et al., 2014). Their activity at the serotonin receptor varies, with some acting as agonists and others acting as antagonists (Dean et al., 2013). Amphetamines can also block the reuptake of norepinephrine, dopamine, and serotonin. Some amphetamines increase the release of monoamines, such as serotonin and norepinephrine (Nagai et al., 2007), while others (such as 2C-B) may also function as direct alpha-adrenergic agonists.

Table 27.6

Common synthetic cathinones

3-Fluoromethcathinone	Ethcathinone
4-Fluoromethathinone (flephedrone)	Ethylone
Alpha-pyrrolidinovalerophenone	Mephedrone
Butylone	Methedrone
Brephedrone	Methcathinone
Butanamine (BDB)	Methylenedioxypyrovalerone
Dimethylecathinone	Naphyrone

Synthetic cathinones

Synthetic cathinones are heterogeneous with respect to mechanism of action. Butylone, ethylone, mephedrone, methylone, and naphyrone function in a manner similar to cocaine, serving as nonselective inhibitors of monoamine reuptake. Butylone, ethylone, mephedrone, and methylone cause serotonin release. Cathinone, methcathinone, and flephedrone act in a manner similar to amphetamines, preferentially causing the release of dopamine and inhibiting dopamine and norepinephrine reuptake (Simmler et al., 2013). Pyrovalerone and 3,4-methylenedioxypyrovalerone (MDPV) inhibit dopamine, serotonin, and norepinephrine transport (Schifano et al., 2016), but – unlike amphetamines – do not cause the release of monoamines (Simmler et al., 2013).

CLINICAL PRESENTATION

Virtually all stimulants produce hyperadrenergic effects. Central neurologic manifestations include euphoria, elevated alertness, and often a feeling of grandiosity or manic behavior. Psychomotor agitation is common. Hypertension, tachycardia, mydriasis, and diaphoresis are typical, owing to a combination of alpha- and beta-adrenergic-mediated effects.

Although various sympathomimetic agents can produce a common toxicity pattern, there are some differences in the toxicity profile for individual classes of sympathomimetic agents.

Cocaine abuse can result in numerous cardiovascular events. Chronic use of cocaine is associated with accelerated formation of atherosclerosis and left ventricular hypertrophy. Acutely, cocaine toxicity may manifest with myocardial ischemia or infarction (Mittleman et al., 1999; Pozner et al., 2005), the risk of which is greatest in the first hour following use (Mittleman et al., 1999). Additional cardiovascular events following cocaine consumption include aortic dissection and ventricular arrhythmias, the latter owing to the drug's sodium channel-blocking properties (Pozner et al., 2005).

Cocaine abuse may also produce prominent cerebrovascular effects, including vascular narrowing associated with vasoconstriction or vasculitis; increased platelet aggregation (resulting in thrombosis); and complications attributable to severe hypertension. Consequently, current or previous cocaine abuse may be a factor contributing to both ischemic and hemorrhagic stroke in young individuals (Treadwell and Robinson, 2007; Cheng, 2016). Cocaine has also rarely been associated with the development of toxic leukoencephalopathy; it is possible, however, that this finding is related to adulterants present in cocaine samples, such as levamisole (Buchanan et al., 2011; Blanc et al., 2012).

There are several pulmonary manifestations of cocaine, including noncardiogenic pulmonary edema, pulmonary hypertension, bronchiolitis obliterans with organizing pneumonia, alveolar hemorrhage, eosinophilic lung disease, and aspiration pneumonia (De Almeida et al., 2014). In addition, the use of cocaine has been associated with severe asthma exacerbations (Levine et al., 2005).

Seizures, choreoathetoid movements, abruptio placentae, and perforated gastrointestinal ulcers have all also been associated with cocaine use.

Compared with cocaine, amphetamines cause more neuropsychiatric manifestations and less catastrophic cardiovascular effects, although hypertension and tachycardia are still common. The higher rate of acute psychosis may be related to the increased dopaminergic effects of amphetamines compared to cocaine (Gold et al., 1989). However, both ischemic and hemorrhagic stroke may also occur with amphetamines (Ho et al., 2009), and there is also a possible association with cerebral vasculitis (Margolis and Newton, 1971).

Many amphetamines, including methylenedioxymethamphetamine (MDMA, or "ecstasy"), are serotonergic. Hyponatremia is associated with serotonergic amphetamines, and may be due to a combination of fluid loss (via sweating) with free-water replacement and the syndrome of inappropriate antidiuretic hormone (SIADH) (Traub et al., 2002). Recurrent seizures and serotonin syndrome have both been associated with the use of select amphetamines (Bosak et al., 2013).

Cathinones

While all cathinones are associated with increased motor activity, some (MDPV and mephedrone) may produce more motor activity than others (methedrone). In contrast to the euphoria associated with cocaine and amphetamines, the altered sensorium of cathinone use is more frequently negative, ranging from anxiety and restlessness to overt agitation and psychosis (Schifano et al., 2016).

Rhabdomyolysis is relatively common (O'Connor et al., 2015) and compartment syndrome may occur in unique areas (e.g., paraspinal musculature) (Levine et al., 2013). Hyponatremia, although relatively uncommon, may be due to SIADH (Wood et al., 2010).

LABORATORY TESTING

When sympathomimetic toxicity is known, minimal further testing is needed, apart from the general workup described above for other poisonings (such as rapid glucose testing and an ECG).

Rhabdomyolysis is relatively common with stimulant toxicity. In one study, rhabdomyolysis occurred most often with the synthetic cathinones (63%), and was somewhat less common with cocaine (33%) (O'Connor et al., 2015).

Patients who present with sympathomimetic toxicity but with lethargy as opposed to CNS stimulation should be strongly considered for CNS imaging. Neuroimaging may demonstrate ischemic or hemorrhagic stroke or vascular abnormalities in such cases. Characteristic magnetic resonance imaging (MRI) findings with cocaine-induced leukoencephalopathy may include increased signal intensity in the globus pallidus, splenium, and other white-matter regions, in some cases with restricted diffusion (De Roock et al., 2007). Such patients should also undergo an assessment of serum sodium, particularly when a serotonergic amphetamine (such as MDMA or a 2C amphetamine) is implicated, to rule out hyponatremia as an etiology for their decreased mental status.

Patients with chest discomfort should, in addition to electrocardiographic testing, have a chest radiograph performed and be considered for serial cardiac biomarker (i.e., troponin) testing.

Management

The first step in the management of stimulant toxicity is to ensure adequate airway, breathing, and circulation. Agitation should be treated aggressively, as mortality from sympathomimetic toxicity may be due to excited delirium and hyperthermia. We recommend benzodiazepines (e.g., 1–2 mg IV lorazepam for adults; 0.1 mg/kg for pediatric patients); multiple repeated doses, often higher than usually used, may be required. If hyperthermia persists despite repeated doses of benzodiazepines and active external cooling, we recommend neuromuscular relaxation with a nondepolarizing agent.

While agitation should be treated with benzodiazepines, acute psychosis may show some improvement with the use of a low-dose D_2 receptor agonist, such as haloperidol. Such use may be especially relevant in pediatric patients with apparent hallucinations following amphetamine exposure (Ruha and Yarema, 2006).

Seizures should be initially treated with benzodiazepines. Refractory seizures may require barbiturates (e.g., 15–20 mg/kg of phenobarbital) or propofol. In general, patients who require phenobarbital or propofol to control their seizures should be intubated. Rhabdomyolysis should be treated with intravenous fluids, as per normal practice guidelines.

Cocaine can cause sodium channel blockade, which may manifest as intraventricular conduction delay and widening of the QRS complex on the ECG. This is best treated with sodium bicarbonate (150 mEq IV bolus for adults; 2 mEq/kg IV bolus for pediatric patients).

Patients are commonly hypertensive and tachycardic. Benzodiazepines are the first choice for control of hypertension, but if additional blood pressure control is needed, we recommend nitrates or calcium channel blockers. Short-acting, titratable medications are preferred.

Beta-blockade should not be used in the setting of cocaine toxicity. Human cardiac catheterization studies (Lange et al., 1990) and case reports (Sand et al., 1991; Fareed et al., 2007) suggest that mixed β_1/β_2-adrenergic antagonists, as well as β_1-selective antagonists, are either not effective or potentially harmful. Labetalol, a mixed α/β-adrenergic antagonist, did not reverse vasospasm in a cardiac catheterization study, but did lower blood pressure (Boehrer et al., 1993). However, because intravenous labetalol is predominantly a β-adrenergic antagonist (with a ratio of $\beta:\alpha$ effect of 7:1) (Richards et al., 1977), we believe that the proscription against β-adrenergic antagonists in the setting of cocaine should apply to labetalol as well.

Complex clinical decisions

Although the decision to not give β-adrenergic antagonists in sympathomimetic toxicity is logical, and there is significant supporting evidence for not using it in the setting of cocaine toxicity, there is less data with respect to the use of β-adrenergic antagonists in amphetamine toxicity. Cocaine and amphetamines are different classes of molecules, and although both produce a hyperadrenergic state, it does not necessarily follow that the use of β-adrenergic antagonists will result in the same outcomes in both conditions. Human data regarding the use of β-blockers in the setting of amphetamine use are scant, and limited to case reports and volunteer studies using small doses of drug. Some advocate for the use of β-adrenergic antagonists in the setting of amphetamine use (Richards et al., 2015); however, we believe that the limited available data are insufficient to conclude that they are safe in this setting, and continue to advocate for alternative therapies (such as phentolamine, vasodilators, and calcium channel blockers) to manage hemodynamic abnormalities.

APPROACH TO THE PATIENT WITH AN UNKNOWN INGESTION

The patient in whom a toxic ingestion is strongly suspected, but the offending agent is unknown, may challenge the diagnostic abilities of the most senior physicians. In such cases, meticulous attention to history (when available) and physical examination will often provide the most important clues to facilitate diagnosis and treatment; ancillary testing, while important, often plays only a supporting role.

Patients suffering from a toxic ingestion are often confused and unable to provide a history. Even in such patients, however, environmental clues (often available from family members or first responders) may provide useful information that identifies an individual toxin. Examples include patients with syringes or other drug paraphernalia near them, and patients found comatose with an empty bottle next to them when the prescription was filled just days ago. Even when such relatively obvious clues are absent, however, ancillary information (such as pill bottles found in the patient's medicine cabinet or a medication list found in the patient's wallet) may provide valuable information.

Most patients with acute central neurotoxicologic syndromes present with alterations in mental status. The initial comprehensive physical examination should be performed while considering other etiologies of alterations in mental status, particularly traumatic. Any concern about trauma should raise the suspicion of traumatic brain injury. Certain findings, such as Battle's sign (ecchymosis posterior to the pinna over the mastoid), raccoon's eyes (bilateral orbital ecchymoses) or hemotympanum are concerning for a basilar skull fracture. Skin findings such as "track marks" (the presence of scarring or small linearly arranged eschars over a vein) strongly suggest the use of intravenous drugs.

The toxidrome-oriented physical examination

A syndrome is a constellation of symptoms and signs that is characteristic of a disease process; a toxidrome is a syndrome caused by a given toxin or class of toxins. As part of the physical examination, the clinician should specifically look for findings that, when part of a larger pattern, provide strong evidence that a patient has been poisoned by a certain substance or class. The crucial data to consider in such patients are vital signs, as well as findings related to mental status, the eyes (pupillary findings as well as nystagmus), mucous membranes, skin, and abdomen (bowel sounds).

The central findings (affect) of patients with acute central neurotoxicologic syndromes caused by drugs of abuse can be quickly triaged into depressed or unresponsive vs. elevated or delirious. Although some patients with ethanol toxicity may initially present as boisterous or stimulated, the majority of patients with significant ethanol toxicity are depressed or unresponsive, as are patients with opioid and dissociative-agent poisoning. In contrast, patients with sympathomimetic or anticholinergic toxicity and ethanol withdrawal all typically present with a heightened state of awareness or delirium.

Most findings related to vital signs, pupil diameter, mucous membranes, skin, and bowel sounds can be understood through the effect of the drug on the sympathetic and parasympathetic nervous system. Other key physiologic points to consider include the effect of opioids on respiratory drive (at the level of the brainstem), with resultant bradypnea; the ability of stimulants and ethanol withdrawal to generate heat; the ability of anticholinergic medications to interfere with heat dissipation, resulting in hyperpyrexia; and the ability of dissociative agents to produce nystagmus.

The physical examination findings for the toxidromes of each of the central neurotoxicologic syndromes caused by drugs of abuse discussed in this chapter are listed in Table 27.7. Not all findings need to be present

Table 27.7

Characteristics of toxidromes

Finding	Sedative-Hypnotics	Opioids	Sympatho-Mimetics	Anti-Cholinergic	Dissociative Agents	Sedative Withdrawal
Pulse	Normal or Slightly Decreased	Normal or Slightly Decreased	Increased	Increased	Increased	Increased
Blood Pressure	Normal or Slightly Decreased	Normal or Slightly Decreased	Increased	Normal or Slightly Increased	Increased	Increased
Respirations	Normal or Slightly Decreased	Markedly Decreased	Normal	Normal	Normal	Normal
CNS	Sedated	Sedated	Agitated	Agitated Delirium	Confusion to Unresponsiveness	Agitated, Tremulous
Mucous Membranes	Normal	Normal	Normal	Dry	Salivating (with Ketamine)	Normal
Eyes	Variable	Constricted Pupils	Dilated pupils	Dilated Pupils	Nystagmus (any direction)	Dilated Pupils
Skin	Normal	Normal	Sweating	Dry	Normal	Sweating
Lungs	Normal	Normal	Normal	Normal	Normal	Normal
Bowel Sounds	Normal	Decreased	Normal	Decreased	Normal	Normal

to make a diagnosis; bowel sounds may be present in patients with anticholinergic toxicity, and patients with sympathomimetic toxicity may have a relatively normal blood pressure.

When faced with a patient with a suspected toxic ingestion and classic toxidrome findings, the patient is much less of a diagnostic dilemma. One study found that the accuracy of the toxidrome-oriented physical examination was on the order of 80–90% in assessing poisoned patients (Nice et al., 1988).

Laboratory testing and imaging

Most patients with a suspected but unknown toxic ingestion present with alterations in mental status as their chief complaint. Central nervous system imaging is generally not needed in uncomplicated patients. We advocate for judicious use of CNS imaging, particularly when the diagnosis is unclear or the presentation raises concerns of a central nervous system complication (such as ischemic or hemorrhagic stroke in the setting of cocaine use).

In all patients with alteration of mental status, we recommend obtaining a rapid serum glucose level immediately. When an acute central neurotoxicologic syndrome caused by a drug of abuse is suspected, we also recommend obtaining an ECG to assess for toxin-induced abnormalities such as QRS or QTc prolongation. When suicidal intent cannot be excluded, we recommend serum testing for salicylates and acetaminophen. Simple blood work may identify an anion gap or osmolar gap, which may be a clue regarding the possibility of other ingestions.

Additional diagnostic testing is then determined by the certainty of the clinical diagnosis. Limited or no testing is necessary when clinical circumstances strongly suggest a certain toxin or class of toxins, even if the patient himself or herself does not readily admit to the ingestion. For example, a patient who presents with agitated delirium, tachycardia, dilated pupils, dry mucous membranes, dry skin, and absent bowel sounds, whose clinical picture improves dramatically after the administration of physostigmine, likely requires no further diagnostic workup to determine the etiology of the alteration in mental status.

When a toxic ingestion is one of several possibilities in the differential diagnosis, however, a broader approach is necessary. We recommend a workup that includes computed tomography (CT) of the head, a complete blood count, and tests of renal and hepatic function. In patients in whom there is reason to suspect hyperammonemia from a medical disorder (such as liver failure) or an acquired condition (such as valproic acid therapy), a serum ammonia level is appropriate. The decision to perform a lumbar puncture is based on clinical presentation, although clinicians should be aware that classic history and physical examination findings alone are insufficiently accurate to reliably exclude CNS infection (Brouwer et al., 2012).

The reflexive use of urine drug-of-abuse testing (colloquially, "tox screens") deserves specific discussion, and three points in particular bear consideration. First, these tests are not measures of intoxication, but rather an indicator of whether a patient was exposed to a certain substance. While it is true that patients intoxicated with cocaine will almost always have a urine drug screen that is positive for cocaine, the converse – that patients with a urine drug screen that is positive for cocaine will almost always be intoxicated with cocaine – is false. The clinical effects of cocaine usually last on the order of minutes to hours (depending on route of ingestion), whereas cocaine metabolites may persist in the urine for several days. Second, urine drug-of-abuse tests suffer from both false-negative (Bertol et al., 2013) and false-positive (Brahm et al., 2010; Rengarajan and Mullins, 2013) results, potentially steering clinicians away from correct diagnoses or towards erroneous ones. Third, there is little evidence that urine drug-of-abuse screening tests alter patient management (Tenenbein, 2009). For all of these reasons, we recommend against the routine use of urine drug-of-abuse tests.

CONCLUSION

Patients who ingest drugs of abuse may present with varied acute central neurotoxicologic findings. In all patients, initial care requires attention to the airway, breathing, and circulation. Empiric testing should usually include a fingerstick blood glucose level, an ECG, a pregnancy test in women of childbearing age, as well as testing for salicylates and acetaminophen in any patient with known or suspected suicidal ideation. Additional testing is performed based on the patient's symptoms and the certainty (or lack thereof) of the toxin-related diagnosis. Management is generally supportive, although antidotal therapy may be appropriate in some cases. When the toxin is not known, historic clues may be of some help, but the physical examination – and not indiscriminate testing – will provide the most useful clinical information.

REFERENCES

Afshari R, Maxwell S, Dawson A et al. (2005). ECG abnormalities in co-proxamol (paracetamol/dextropropoxyphene) poisoning. Clin Toxicol (Phila) 43 (4): 255–259.

Akunne HC, Reid AA, Thurkuf A et al. (1991). [^{3}H]1-[2-(2-thienyl) cyclohexyl]-piperidine labeled two high-affinity binding sites associated with the biogenic amine reuptake complex. Synapse 8: 289–300.

Alldredge BK, Lowenstein DH (1993). Status epilepticus related to alcohol abuse. Epilepsia 34 (6): 1033–1037.

Basu D, Banerjee A, Harish T et al. (2009). Disproportionately high rate of epileptic seizure in patients abusing dextropropoxyphene. Am J Addict 18 (5): 417–421.

Baumann MH, Solis Jr E, Watterson LR et al. (2014). Bath salts, spice, and related designer drugs: the science behind the headlines. J Neurosci 34: 15150–15158.

Bertol E, Vaiano F, Borsotti M et al. (2013). Comparison of immunoassay screening tests and LC-MS-MS for urine detection of benzodiazepines and their metabolites: results of a national proficiency test. J Anal Toxicol 37 (9): 659–664.

Blanc PD, Chin C, Lynch KL (2012). Multifocal inflammatory leukoencephalopathy associated with cocaine abuse: is levamisole responsible? Clin Toxicol (Phila) 50 (6): 534–535.

Blondell RD (2005). Ambulatory detoxification of patients with alcohol dependence. Am Fam Physician 71 (3): 495–502.

Boehrer JD, Moliterno DJ, Willard JE et al. (1993). Influence of labetalol on cocaine-induced coronary vasoconstriction in humans. Am J Med 94: 608.

Boland J, Boland E, Brooks D (2013). Importance of the correct diagnosis of opioid-induced respiratory depression in adult cancer patients and titration of naloxone. Clin Med (Lond) 13 (2): 149–151.

Boom M, Niesters M, Sarton E et al. (2012). Non-analgesic effects of opioids: opioid-induced respiratory depression. Curr Pharm Des 18 (37): 5994–6004.

Bosak A, LoVecchio F, Levine M (2013). Recurrent seizures and serotonin syndrome following "2C-I" ingestion. J Med Toxicol 9: 196–198.

Boyd CJ, Young A, McCabe SE (2014). Psychological and drug abuse symptoms associated with nonmedical use of opioid analgesics among adolescents. Subst Abus 35 (3): 284–289.

Boyer EW (2004). Dextromethorphan abuse. Pediatr Emerg Care 20: 858–863.

Boyer EW (2012). Management of opioid analgesic overdose. N Engl J Med 367 (2): 146–155.

Brahm NC, Yeager LL, Fox MD et al. (2010). Commonly prescribed medications and potential false-positive urine drug screens. Am J Health Syst Pharm 67: 1344–1350.

Brouwer MC, Thwaites GE, Tunkel AR et al. (2012). Dilemmas in the diagnosis of acute community-acquired bacterial meningitis. Lancet 380 (9854): 1684–1692.

Buchanan JA, Heard K, Burbach C et al. (2011). Prevalence of levamisole in urine toxicology screens positive for cocaine in an inner-city hospital. JAMA 305 (16): 1657–1658.

Burns MJ, Linden CH, Graudins A et al. (2000). A comparison of physostigmine and benzodiazpeines for the treatment of anticholinergic poisoning. Ann Emerg Med 35: 374–381.

Cassidy EM, O'Sullivan I, Bradshaw P et al. (2012). Symptom-triggered benzodiazepine therapy for alcohol withdrawal syndrome in the emergency department: a comparison with the standard fixed dose benzodiazepine regimen. Emerg Med J 29 (10): 802–804.

Chatterjee S, Bali V, Carnahan RM et al. (2016). Anticholinergic medication use and risk of dementia among elderly nursing home residents with depression. Am J Geriatr Psychiatry. pii:S1064-S17481.

Chen LY, Crum RM, Strain EC et al. (2016). Prescriptions, nonmedicinal use, and emergency department visitsinvolving prescription stimulants. J Clin Psychiatry 77: e297–e304.

Cheng YC, Ryan KA, Qadwai SA et al. (2016). Cocaine Use and Risk of Ischemic Stroke in Young Adults. Stroke 47 (4): 918–922.

Chyka PA, Erdman AR, Manoguerra AS et al. (2007). Dextromethorphan poisoning: an evidence-based consensus guideline for out-of-hospital management. Clin Toxicol (Phila) 45 (6): 662–677.

Clemmesen C, Nilsson E (1961 Mar-Apr). Therapeutic trends in the treatment of barbiturate poisoning. The Scandinavian method. Clin Pharmacol Ther 2: 220–229.

Clemow DB (2015). Misuse of methylphenidate. Curr Top Behav Neurosci. e-pub ahead of print.

Corallo CE, Whitfield A, Wu A (2009). Anticholinergic syndrome following an unintentional overdose of scopolamine. Ther Clin Risk Manag 5: 719–723.

Daeppen JB, Gache P, Landry U et al. (2002). Symptom-triggered vs fixed-schedule doses of benzodiazepine for alcohol withdrawal: a randomized treatment trial. Arch Intern Med 162 (10): 1117–1121.

Dart RC, Severtson SG, Bucher-Bartelson B (2015). Trends in opioid analgesic abuse and mortality in the United States. N Engl J Med 372 (16): 1573–1574.

Dawson AH, Buckley NA (2015). Pharmacological management of anticholinergic delirium-theory, evidence, and practice. Br J Clin Pharmacol 81: 516–524.

De Almeida RR, de Souza LS, Mancano AD et al. (2014). High-resolution computed tomographic findings of cocaine-induced pulmonary disease: a state of the art review. Lung 192: 225–233.

Dean BV, Stellpflug SJ, Burnett AM et al. (2013). 2C or not 2C: phenethylamine designer drug review. J Med Toxicol 9: 172–178.

De Roock S, Hantson P, Laterre PF et al. (2007). Extensive pallidal and white matter injury following cocaine overdose. Intensive Care Med 33 (11): 2030–2031.

Dominici P, Kopec K, Manur R et al. (2015). Phencyclidine intoxication case series study. J Med Toxicol 11: 321–325.

Eglen RW (2006). Muscarinic receptor subtypes in neuronal and non-neuronal cholinergic function. Auton Autacoid Pharmacol 26: 219–233.

Emonson DL, Vanderbeek RD (1995). The use of amphetamines in U.S. Air Force tactical operations during Desert Shield and Storm. Aviat Space Environ Med 66: 260–263.

Emson PC (2007). GABA(B) receptors: structure and function. Prog Brain Res 160: 43–57.

Fareed FN, Chan G, Hoffman RS (2007). Death temporally related to the use of a beta adrenergic receptor antagonist in cocaine associated myocardial infarction. J Med Toxicol 3: 169.

Ferguson JA, Suelzer CJ, Eckert GJ et al. (1996). Risk factors for delirium tremens development. J Gen Intern Med 11 (7): 410–414.

Follesa P, Floris G, Asuni GP et al. (2015 Nov 9). Chronic intermittent ethanol regulates hippocampal GABA(A) receptor delta subunit gene expression. Front Cell Neurosci 9: 445.

Ganetsky M, Babu KM, Boyer EW (2007). Serotonin syndrome in dextromethorphan ingestion responsive to propofol therapy. Pediatr Emerg Care 23 (11): 829–831.

Garnier R, Guerault E, Muzard D et al. (1994). Acute zolpidem poisoning – analysis of 344 cases. J Toxicol Clin Toxicol 32 (4): 391–404.

Gaudreault P, Guay J, Thivierge RL et al. (1991). Benzodiazepine poisoning. Clinical and pharmacological considerations and treatment. Drug Saf 6 (4): 247–265.

Giannini AJ, Eighan MS, Loiselle RH et al. (1984). Comparison of haloperidol and chlorpromazine in the treatment of phencyclidine psychosis. J Clin Pharmacol 24 (4): 202–204.

Gold LH, Geyer MA, Koob GF (1989). Neurochemical mechanisms involved in behavioral effects of amphetamines and related designer drug. NIDA Res MOnogr 94: 101–126.

Gold JA, Rimal B, Nolan A et al. (2007). A strategy of escalating doses of benzodiazepines and phenobarbital administration reduces the need for mechanical ventilation in delirium tremens. Crit Care Med 35 (3): 724–730.

Goldfrank L, Weisman RS, Errick JK et al. (1986). A dosing nomogram for continuous infusion intravenous naloxone. Ann Emerg Med 15 (5): 566–570.

Gowdy JM (1972). Stramonium intoxication: review of the symptomatology in 212 cases. JAMA 221: 585–587.

Grant BF, Dawson DA (1997). Age at onset of alcohol use and its association with DSM-IV alcohol abuse and dependence: results from the National Longitudinal Alcohol Epidemiologic Survey. J Subst Abuse 9: 103–110.

Green SM, Krauss B (2004). Ketamine is a safe, effective, and appropriate technique for emergency department paediatric procedural sedation. Emerg Med J 21: 271–272.

Green SM, Sherwin TS (2005). Incidence and severity of recovery agitation after ketamine sedation in young adults. Am J Emerg Med 23: 142–144.

Greenblatt DJ, Ehrenberg BL, Gunderman J et al. (1989). Kinetic and dynamic study of intravenous lorazepam: comparison with intravenous diazepam. J Pharmacol Exp Ther 250 (1): 134–140.

Greifenstein FE, Devault M, Yoshitake J et al. (1958). A study of 1-aryl cyclo hexyl amine for anesthesia. Anesth Analg 37: 283–294.

Guerlais M, Grall-Bronnec M, Feuillet F et al. (2015). Dependence on prescription benzodiazepines and Z-drugs among young to middle-aged patients in France. Subst Use Misuse 50 (3): 320–327.

Hack JB, Hoffmann RS, Nelson LS (2006). Resistant alcohol withdrawal: does an unexpectedly large sedative requirement identify these patients early? J Med Toxicol 2 (2): 55–60.

Hadden J, Johnson K, Smith S et al. (1969). Acute barbiturate intoxication. Concepts of management. JAMA 209 (6): 893–900.

Hagmeyer KO, Mauro LS, Mauro VF (1993). Meperidine-related seizures associated with patient-controlled analgesia pumps. Ann Pharmacother 27 (1): 29–32.

Hall FS, Sora I, Drgonova J et al. (2004). Molecular mechanisms underlying the rewarding effects of cocaine. Ann N Y Acad Sci 1025: 47–56.

Harper CG, Giles M, Finlay-Jones R (1986). Clinical signs in the Wernicke–Korsakoff complex: a retrospective analysis of 131 cases diagnosed at necropsy. J Neurol Neurosurg Psychiatry 49 (4): 341–345.

Harris RA, Trudell JR, Mihic SJ (2008). Ethanol's molecular targets. Sci Signal 1 (28): re7.

Hasin DS, Stinson FS, Ogburn E et al. (2007). Prevalence, correlates, disability, and comorbidity of DSM-IV alcohol abuse and dependence in the United States: results from the National Epidemiologic Survey on Alcohol and Related Conditions. Arch Gen Psychiatry 64 (7): 830–842.

He Q, Titley H, Grasselli G et al. (2013). Ethanol affects NMDA receptor signaling at climbing fiber-Purkinje cell synapses in mice and impairs cerebellar LTD. J Neurophysiol 109 (5): 1333–1342.

Hendrickson RG, Cloutier RL (2007). "Crystal dex:" free-base dextromethorphan. J Emerg Med 32: 393–396.

Henretig FM (1994). Special considerations in the poisoned pediatric patient. Emerg Med Clin North Am 12: 549.

Hillhouse TM, Negus SS (2016 Feb 23). Effects of the non-competitive N-methyl-D-aspartate receptor antagonists ketamine and MK-801 on pain-stimulated and pain-depressed behaviour in rats. Eur J Pain. http://dx.doi.org/10.1002/ejp.847.

Ho EL, Josephson SA, Lee HS et al. (2009). Cerebrovascular complications of methamphetamine abuse. Neurocrit Care 10: 295–305.

Isbell H, Fraser HF, Wikler A et al. (1955). An experimental study of the etiology of rum fits and delirium tremens. Q J Stud Alcohol 16 (1): 1–33.

Ishii M, Kurachi Y (2006). Muscarinic acetylcholine receptors. Curr Pharm Des 12: 3573–3581.

Jansen KL (1993). Non-medical use of ketamine. BMJ 306: 601–602.

Jodo E (2013). The role of the hippocampo-prefrontal cortex system in phencyclidine-induced psychosis: a model for schizophrenia. J Physiol Paris 107 (6): 434–440.

Kaim SC, Klett CJ, Rothfeld B (1969). Treatment of the acute alcohol withdrawal state: a comparison of four drugs. Am J Psychiatry 125 (12): 1640–1646.

Knaggs RD, Crighton IM, Cobby TF et al. (2004). The pupillary effects of intravenous morphine, codeine, and tramadol in volunteers. Anesth Analg 99 (1): 108–112.

Lai JS, Gangwani RA (2012). Medication-induced acute angle closure attack. Hong Kong Med J 18: 139–145.

Lange RA, Cigarroa RG, Flores ED et al. (1990). Potentiation of cocaine-induced coronary vasoconstriction by beta-adrenergic blockade. Ann Intern Med 112: 897–903.

Lee JH, Jang MK, Lee JY et al. (2005). Clinical predictors for delirium tremens in alcohol dependence. J Gastroenterol Hepatol 20 (12): 1833–1837.

Leung NY, Whyte IM, Isbister GK (2006). Baclofen overdose: defining the spectrum of toxicity. Emerg Med Australas 18 (1): 77–82.

Levine M (2016). Pediatric toxicological emergencies in the ICU. In: DA Turner, JS Kilinger (Eds.), Current Concepts in Pediatric Critical Care, Society of Critical Care Medicine, Mt. Prospect, IL, pp. 73–80.

Levine M, Iliescu ME, Margellos-Anast H et al. (2005). The effects of cocaine and heroin use on intubation rates and hospital utilization in patients with acute asthma exacerbations. Chest 128: 1951–1957.

Levine M, Brooks DE, Truitt CA et al. (2011). Toxicology in the ICU, part 1: general overview and approach to treatment. Chest 140: 795–806.

Levine M, Levitan R, Skolnik A (2013). Compartment syndrome after "bath salts" use: a case series. Ann Emerg Med 61: 480–483.

Li SF, Jacob J, Feng J et al. (2008). Vitamin deficiencies in acutely intoxicated patients in the ED. Am J Emerg Med 26 (7): 792–795.

Liang J, Zhang N, Cagetti E et al. (2006). Chronic intermittent ethanol-induced switch of ethanol actions from extrasynaptic to synaptic hippocampal GABAA receptors. J Neurosci 26 (6): 1749–1758.

Littleton J (1998). Neurochemical mechanisms underlying alcohol withdrawal. Alcohol Health Res World 22 (1): 13–24.

Lodge D, Mercier MS (2015). Ketamine and phencyclidine: the good, the bad and the unexpected. Br J Pharmacol 172 (17): 4254–4276.

Longo VG (1966). Behavioral and electroencephalographic effects of atropine and related compounds. Pharmacol Rev 18: 965–996.

Lorentzen K, Lauritsen AØ, Bendtsen AO (2014). Use of propofol infusion in alcohol withdrawal-induced refractory delirium tremens. Dan Med J 61 (5): A4807.

Lutz PE, Kieffer BL (2013). The multiple facets of opioid receptor function: implications for addiction. Curr Opin Neurobiol 23 (4): 473–479.

Machado-Alba JE, Castro-Rodriguez A, Alzate-Piedrahita JA et al. (2016). Anticholinergic risk and frequency of anticholinergic drug prescriptions in a population over 65. J Am Med Dire Assoc 17: 275. E1–275. E4.

Marcum ZA, Wirtz HS, Pettinger M et al. (2016). Anticholinergic medication use and falls in postmenopausal women: findings from the women's health initiative cohort study. BMC Geriatr 16: 76.

Margolis MT, Newton TH (1971). Methamphetamine ("speed") arteritis. Neuroradiology 2: 179–182.

Marshall FH (2008). The role of GABA(B) receptors in the regulation of excitatory neurotransmission. Results Probl Cell Differ 44: 87–98.

McAllister MW, Aaronson P, Spillane J et al. (2015). Impact of prescription drug-monitoring program on controlled substance prescribing in the ED. Am J Emerg Med 33 (6): 781–785.

McCarron MM, Schulze BW, Thompson GA et al. (1981a). Acute phencyclidine intoxication: incidence of clinical findings in 1,000 cases. Ann Emerg Med 10: 237–242.

McCarron MM, Schulze BW, Thompson GA et al. (1981b). Acute phencyclidine intoxication: clinical patterns, complications, and treatment. Ann Emerg Med 10: 290–297.

Mintzer MZ, Stoller KB, Griffiths RR (1999). A controlled study of flumazenil-precipitated withdrawal in chronic low-dose benzodiazepine users. Psychopharmacology (Berl) 147 (2): 200–209.

Mittleman MA, Mintzer D, Maclure M et al. (1999). Triggering of myocardial infarction by cocaine. Circulation 99: 2737–2741.

Moore PW, Donovan JW, Burkhart KK et al. (2014). Safety and efficacy of flumazenil for reversal of iatrogenic benzodiazepine-associated delirium toxicity during treatment of alcohol withdrawal, a retrospective review at one center. J Med Toxicol 10 (2): 126–132.

Mowry JB, Spyker DA, Brooks DE et al. (2015). 2015 Annual report of the American Association of Poison Control Centers' National Poison Data System (NPDS): 32nd annual report. Clin Toxicol 53: 962–1146.

Mueller SW, Preslaski CR, Kiser TH et al. (2014). A randomized, double-blind, placebo-controlled dose range study of dexmedetomidine as adjunctive therapy for alcohol withdrawal. Crit Care Med 42 (5): 1131–1139.

Munir VL, Hutton JE, Harney JP et al. (2008). Gamma-hydroxybutyrate: a 30 month emergency department review. Emerg Med Australas 20 (6): 521–530.

Muzyk AJ, Kerns S, Brudney S et al. (2013). Dexmedetomidine for the treatment of alcohol withdrawal syndrome: rationale and current status of research. CNS Drugs 27 (11): 913–920.

Nagai F, Nonaka R, Satoh K (2007). The effects of non-medically used psychoactive drugs on monoamine neurotransmission in rat brain. Eur J Pharmacol 559: 132–137.

National Highway Traffic Safety Administration (n.d.). Drugs and human performance fact sheets: Phencyclidine (PCP). http://www.nhtsa.gov/people/injury/research/job185drugs/phencyclidine.htm. Accessed 20 May, 2016.

Nelson ME, Bryant SM, Aks SE (2014). Emerging drugs of abuse. Emerg Med Clin N Am 32: 1–28.

Ng SH, Tse ML, Ng HW et al. (2010). Emergency department presentation of ketamine abusers in Hong Kong: a review of 233 cases. Hong Kong Med J 16: 6–11.

Nice A, Leikin JB, Maturen A et al. (1988). Toxidrome recognition to improve efficiency of emergency urine drug screens. Ann Emerg Med 17 (7): 676–680.

Nickalls RW, Nickalls EA (1988). The first use of physostigmine in the treatment of atropine poisoning. A translation of Kleinwachter's paper entitled "Observations on the effect of Calabar bean extract as an antidote to atropine poisoning". Anaesthesia 43: 776–779.

O'Connor AD, Padilla-Jones A, Gerkin RD et al. (2015). Prevalence of rhabdomyolysis in sympathomimetic toxicity: a comparison of stimulants. J Med Toxicol 11: 195–200.

Olsen RW, Yang J, King RG et al. (1986). Barbiturate and benzodiazepine modulation of GABA receptor binding and function. Life Sci 39 (21): 1969–1976.

Peltoniemi MA, Hagelberg NM, Olkkola KT et al. (2016). Ketamine: a review of clinical pharmacokinetics and pharmacodynamics in anesthesia and pain therapy. Clin Pharmacokinetc. E pub ahead of print. PMID: 27028535.

Potter DE, Choudhury M (2014). Ketamine: repurposing and redefining a multifaceted drug. Drug Discov Today 19: 1848–1854.

Pozner CN, Levine M, Zane R (2005). The cardiovascular effects of cocaine. J Emerg Med 29: 173–178.

Prosser JM, Nelson LS (2012). The toxicology of bath salts: a review of synthetic cathinones. J Med Toxicol 8: 33–42.

Rasmussen N (2006). Making the first anti-depressant: amphetamine in American medicine, 1929–1950. J Hist Med Allied Sci 61: 288–323.

Rasmussen N (2008). America's first amphetamine epidemic 1929-1971. Am J Public Health 98: 974–985.

Rengarajan A, Mullins ME (2013). How often do false-positive phencyclidine urine screens occur with use of common medications? Clin Toxicol 51: 493–496.

Richards DA, Prichard BN, Boakes AJ et al. (1977). Pharmacological basis for antihypertensive effects of intravenous labetalol. Br Heart J 39 (1): 99–106.

Richards JR, Albertson TE, Derlet RW et al. (2015). Treatment of toxicity from amphetamines, related derivatives, and analogues: a systematic clinical review. Drug Alcohol Depend 150: 1–13.

Richardson K, Bennett K, Maidment ID et al. (2015). Use of medications with anticholinergic activity and self-reported injurious falls in older community-dwelling adults. J Am Geriatr Soc 63: 1561–1569.

Roy AK, McCarthy C, Kiernan G et al. (2012). Increased incidence of QT interval prolongation in a population receiving lower doses of methadone maintenance therapy. Addiction 107 (6): 1132–1139.

Rudd RA, Paulozzi LJ, Bauer MJ et al. (2014). Increases in heroin overdose deaths – 28 states, 2010 to 2012. MMWR Morb Mortal Wkly Rep 63 (39): 849–854.

Ruha AM, Yarema MC (2006). Pharmacologic treatment of acute pediatric methamphetamine toxicity. Pediatr Emerg Care 22: 782–785.

Sachdeva A, Chandra M, Deshpande SN (2014). A comparative study of fixed tapering dose regimen versus symptom-triggered regimen of lorazepam for alcohol detoxification. Alcohol Alcohol 49 (3): 287–291.

Sand IC, Brody SL, Wrenn KD et al. (1991). Experience with esmolol for the treatment of cocaine-associated cardiovascular complications. Am J Emerg Med 9: 161.

Schep LJ, Knudsen K, Slaughter RJ et al. (2012). The clinical toxicology of γ-hydroxybutyrate, γ-butyrolactone and 1,4-butanediol. Clin Toxicol (Phila) 50 (6): 458–470.

Schifano F, Papanti GD, Orsolini L et al. (2016). Novel psychoactive substances: the pharmacology of stimulants and hallucinogens. Expert Rev Clin Pharmacol 4: 1–12.

Schwartz JA, Koenigsberg MD (1987). Naloxone-induced pulmonary edema. Ann Emerg Med 16 (11): 1294–1296.

Schwartz AR, Pizon AF, Brooks DE (2008). Dextromethorphan-induced serotonin syndrome. Clin Toxicol 46: 771–773.

Schwartz J, Murrough JW, Iosifescu DV (2016). Ketamine for treatment-resistant depression: recent developments and clinical applications. Evid Based Ment Health. http://dx.doi.org/10.1136/eb-2016-102355 (epub ahead of print).

Shannon M (1998). Toxicology reviews: Physostigmine. Ped Emerg Care 14: 224–226.

Shaul WL, Wandell M, Robertson WO (1977). Dextromethorphan toxicity: reversal by naloxone. Pediatrics 59: 117.

Simmler LD, Buser TA, Donzelli M et al. (2013). Pharmacological characterization of designer cathinones in vitro. Br J Pharmacol 168: 458–470.

Siu A, Drachtman R (2007 Spring). Dextromethorphan: a review of N-methyl-D-aspartate receptor antagonist in the management of pain. CNS Drug Rev 13 (1): 96–106.

Smith JA, Mo Q, Guo H et al. (1995). Cocaine increases extraneuronal levels of aspartate and glutamate in the nucleus accumbens. Brain Res 683 (2): 264–269.

Soyka M (1990). Psychopathological characteristics in alcohol hallucinosis and paranoid schizophrenia. Acta Psychiatr Scand 81 (3): 255–259.

Spear PW, Protass LM (1973). Barbiturate poisoning – an endemic disease. Five years' experience in a municipal hospital. Med Clin North Am 57 (6): 1471–1479.

Steinmiller CL, Maisonneuve IM, Glick SD (2003). Effects of dextromethorphan on dopamine release in the nucleus accumbens: interactions with morphine. Pharmacol Biochem Behav 74: 803–810.

Stillman R, Petersen RC (1979). The paradox of phencyclidine (PCP) abuse. Ann Intern Med 90: 428–429.

Substance Abuse and Mental Health Services Administration (SAMHSA) (2014). National Survey on Drug Use and Health (NSDUH). Accessed January 15, 2016 at www.SAMHSA.gov.

Sullivan Jr JB, Hauptman M, Bronstein AC (1987). Lack of observable intoxication in humans with high plasma alcohol concentrations. J Forensic Sci 32 (6): 1660–1665.

Sutter ME, Chenoweth J, Albertson TE (2014). Alternative drugs of abuse. Clin Rev Allergy Immunol 46: 3–18.

Suzler D, Chen TK, Lau YY et al. (1995). Amphetamine redistributes dopamine from synaptic vesicles to the cytosol and promotes reverse transport. J Neurosci 15 (5 Pt 2): 4102–4108.

Tella SR, Schindler CW, Goldberg SR (1992). Cardiovascular effects of cocaine in conscious rats: relative significance of central sympathetic stimulation and peripheral neuronal monoamine uptake and release mechanisms. J Pharmacol Exp Ther 262 (2): 602–610.

Tenenbein M (2009). Do you really need that emergency drug screen? Clin Toxicol (Phila) 47 (4): 286–291.

Thomas A, Nallur DG, Jones N et al. (2009). Diphenhydramine abuse and detoxification: a brief review and case report. J Psychopharmacol 23: 101–105.

Traub SJ, Hoffman RS, Nelson LS (2002). The "ecstasy" hangover: hyponatremia due to 3,4-methylenedioxymethamphetamine. J Urban Health 79 (4): 549–555.

Treadwell SD, Robinson TG (2007). Cocaine use and stroke. Postgrad Med J 83 (980): 389–394.

Tsuang JW, Irwin MR, Smith TL et al. (1994). Characteristics of men with alcoholic hallucinosis. Addiction 89 (1): 73–78.

Valenzuela CF, Jotty K (2015). Mini-review: effects of ethanol on GABAA receptor-mediated neurotransmission in the cerebellar cortex – recent advances. Cerebellum 14 (4): 438–446.

Vande Griend JP, Linnebur SA (2012). Inhaled anticholinergic agnets and acute urinary retention in men with lower urinary tract symptoms or benign prostatic hyperplasia. Ann Pharmacother 46: 1245–1249.

Victor M, Brausch C (1967). The role of abstinence in the genesis of alcoholic epilepsy. Epilepsia 8 (1): 1–20.

Vilke GM, Buchanan J, Dunford JV et al. (1999). Are heroin overdose deaths related to patient release after prehospital treatment with naloxone? Prehosp Emerg Care 3 (3): 183–186.

Vilke GM, Sloane C, Smith AM et al. (2003). Assessment for deaths in out-of-hospital heroin overdose patients treated with naloxone who refuse transport. Acad Emerg Med 10 (8): 893–896.

Waldhoer M, Bartlett SE, Whistler JL (2004). Opioid receptors. Annu Rev Biochem 73: 953–990.

Wampler DA, Molina DK, McManus J et al. (2011). No deaths associated with patient refusal of transport after naloxone-reversed opioid overdose. Prehosp Emerg Care 15 (3): 320–324.

Watkins JW, Schwarz ES, Arroyo-Plasencia AM (2015). The use of physostigmine by toxicologists in anticholinergic toxicity. J Med Toxicol 11: 179–184.

Weil JV, McCullough RE, Kline JS et al. (1975). Diminished ventilatory response to hypoxia and hypercapnia after morphine in normal man. N Engl J Med 292 (21): 1103–1106.

Wheeler E, Jones TS, Gilbert MK et al. (2015). Opioid overdose prevention programs providing naloxone to laypersons – United States, 2014. MMWR Morb Mortal Wkly Rep 64 (23): 631–635.

Wolfe TR, Caravati EM (1995). Massive dextromethorphan ingestion and abuse. Am J Emerg Med 13 (2): 174–176.

Wolfe Jr SA, De Souza EB (1993). Sigma and phencyclidine receptors in the brain-endocrine-immune axis. NIDA Res Monogr 133: 95–123.

Wood DM, Dargan PI, Hoffman RS (2009). Management of cocaine-induced cardiac arrhythmias due to cardiac ion channel dysfunction. Clin Toxicol (Phila) 47 (1): 14–23.

Wood DM, Davies S, Greene SL et al. (2010). Case series of individuals with analytically confirmed acute mephedrone toxicity. Clin Toxicol 48: 924–927.

Chapter 28

Seizures in the critically ill

J. CH'ANG AND J. CLAASSEN*
Neurological Institute, Columbia University, New York, NY, USA

Abstract

Critically ill patients with seizures are either admitted to the intensive care unit because of uncontrolled seizures requiring aggressive treatment or are admitted for other reasons and develop seizures secondarily. These patients may have multiorgan failure and severe metabolic and electrolyte disarrangements, and may require complex medication regimens and interventions. Seizures can be seen as a result of an acute systemic illness, a primary neurologic pathology, or a medication side-effect and can present in a wide array of symptoms from convulsive activity, subtle twitching, to lethargy. In this population, untreated isolated seizures can quickly escalate to generalized convulsive status epilepticus or, more frequently, nonconvulsive status epileptics, which is associated with a high morbidity and mortality. Status epilepticus (SE) arises from a failure of inhibitory mechanisms and an enhancement of excitatory pathways causing permanent neuronal injury and other systemic sequelae. Carrying a high 30-day mortality rate, SE can be very difficult to treat in this complex setting, and a portion of these patients will become refractory, requiring narcotics and anesthetic medications. The most significant factor in successfully treating status epilepticus is initiating antiepileptic drugs as soon as possible, thus attentiveness and recognition of this disease are critical.

INTRODUCTION

Critically ill patients with seizures are either admitted to the intensive care unit (ICU) because of uncontrolled seizures requiring aggressive treatment or are admitted for other reasons and develop seizures secondarily. Seizures may be seen in the context of an acute systemic illness, a primary neurologic pathology, or as a medication side-effect. In this population, untreated isolated seizures can quickly escalate to generalized convulsive status epilepticus (GCSE) or, more frequently, nonconvulsive status epileptics (NCSE) which is associated with a high morbidity and mortality (Mirski and Varelas, 2008). Carrying a high 30-day mortality rate, status epilepticus (SE) can be very difficult to treat in a complex ICU setting, and a portion of these patients will become refractory, requiring narcotics and anesthetic medications (Chen and Wasterlain, 2006; Knake et al., 2009; Claassen et al., 2012).

SE is defined as clinical and/or electrographic seizure activity (seen on electroencephalogram (EEG)) lasting greater than 5 minutes or recurrent seizures in a 5-minute interval without return to neurologic baseline (Lowenstein and Alldredge, 1998; Lowenstein et al., 1999; Meldrum, 1999; Alldredge et al., 2001; Chen and Wasterlain, 2006; Knake et al., 2009). SE is classified as generalized convulsive (overt jerking movements) or nonconvulsive (seizure activity seen only on EEG without overt jerking movements). It can also be classified as partial status, which is manifested as focal motor convulsions, focal sensory symptoms, or focal impairments in function, such as aphasia without alternations of consciousness. The definition of SE was partly based on research showing that repetitive seizures in animals became self-sustaining and pharmacoresistant within 15–30 minutes (Vicedomini and Nadler, 1987; Mazarati et al., 1998a). In adults, generalized convulsive seizures were found to last approximately 60 seconds for both behavioral and EEG changes, with none of these seizures lasting more than 2 minutes (Theodore et al., 1994). A retrospective study from the Richmond group found that 60% of seizures lasting 10–30 minutes required antiepileptic drug (AED)

---

*Correspondence to: Jan Claassen, MD, PhD, Neurological Institute, Columbia University, 177 Fort Washington Avenue, MHB 8 Center, Room 300, New York NY 10032, USA. Tel: +1-212-305-7236, Fax: +1-212-305-2792, E-mail: jc1439@columbia.edu

treatment and, in that time, irreversible neuronal damage would ensue (Bleck, 1991; DeLorenzo et al., 1999). Thus, if a seizure lasted 5 minutes or more, it was most likely not going to abort spontaneously (Theodore et al., 1994; Shinnar et al., 2001; Jenssen et al., 2006).

Patients with NCSE do not have overt signs of convulsive activity but electrographic seizures are recorded on EEG. Two distinct phenotypes of NCSE should be differentiated: those patients with seizures in the acute brain injury setting, often encountered in the ICU, which is typically associated with coma or severely impaired consciousness, and the "wandering confused" patient that can been see in the elderly. These two groups of patients do not share the same underlying etiology or prognosis but both may have SE without accompanying convulsions. The ICU patients with acute brain injury and NCSE are often comatose, requiring intubation and continuous drips of intravenous (IV) antiepileptic medications. NCSE may present as the "wandering confused" patient in the emergency department with inattention and behavioral changes, who is found to have ongoing seizure activity without associated convulsions (Shorvon, 2007). These patients have a relatively good prognosis and may have a chronic epileptic syndrome (Brophy et al., 2012). The management of this form of NCSE is less aggressive than those with acute brain injury and not the focus of the ensuing discussion.

Focal motor status (Fig. 28.1), also termed epilepsia partialis continua, involves repetitive movements or neurologic symptoms confined to restricted body regions (i.e., left-thumb twitching). These seizures are frequently caused by a focal structural brain lesion or nonketotic hyperglycemia and are very resistant to treatment, often requiring neurosurgical intervention if due to a structural lesion (Singh and Strobos, 1980). Detailed discussions on subtypes have been published (Singh and Strobos, 1980; Shorvon, 2007). This chapter will focus on the diagnosis and management of different types of SE in patients presenting in the emergency department or ICU. This chapter complements Chapter 7 and Chapter 9 in part one of these volumes.

EPIDEMIOLOGY

SE is a major neurologic emergency, with an overall incidence of 41 per 100 000 individuals per year. Age of onset shows a bimodal distribution, with the highest incidences during the first year of life and a second peak amongst those over 60 years of age, with a combined incidence of 86 per 100 000 individuals per year (DeLorenzo et al., 1996; Chin et al., 2006). SE accounts for 20% of all neurologic problems seen in emergency departments, 1% of all emergency department visits, and presents as the patient's first seizure in 50% of cases (Epilepsy Foundation of America's Working Group on Status Epilepticus, 1993; Pallin et al., 2008; Farhidvash et al., 2009). Almost half (48%) of patients with convulsive SE will have ongoing electrographic seizure activity and 14% are in NCSE (DeLorenzo et al., 1998). In the absence of a large epidemiologic study including EEG monitoring, accurate incidence and prevalence numbers are not available for NCSE or nonconvulsive seizures (NCSz). In the general population, the incidence is 1.5 per 100 000 per year and, of patients presenting in SE, approximately 5% of these will be in NCSE (Tomson et al., 1992; Towne et al., 2000). However, in the ICU setting, seizures are nonconvulsive in 90% of patients with seizures (Claassen et al., 2004). NCSz and NCSE are more common in the critically ill and are associated with sepsis, cardiac or respiratory arrest, coma, history of epilepsy, convulsive seizures, traumatic brain injury, intracerebral hemorrhages (ICH), and certain EEG patterns like periodic lateralized epileptiform discharges, generalized periodic epileptiform discharges, and burst suppression (DeLorenzo et al., 1996; Towne et al., 2000; Varelas et al., 2003; Vespa et al., 2003; Claassen et al., 2004; Jirsch and Hirsch, 2007; Oddo et al., 2009; Trinka and Leitinger, 2015) (Fig. 28.2). Intensivists should have a high index of suspicion for seizures, as 8% of patients in medical, cardiac, and surgical ICUs and 61% of patients in neurologic ICUs will have seizures if they undergo continuous electroencephalograph (cEEG) monitoring (Claassen et al., 2004; Oddo et al., 2009; Kurtz et al., 2014). In the medical ICU, especially those with sepsis, 10% of patients (without brain injury) have NCSz and NCSE, and up to 34% in the neurologic ICU (Oddo et al., 2009; Kurtz et al., 2014).

SE that does not respond to two or more AEDs is labeled as refractory status epilepticus (RSE), regardless of the time elapsed (Brophy et al., 2012). Although there has never been a study to determine the incidence of RSE, it is estimated to occur in approximately 2000–6000 cases per year; however, this estimate is likely an underestimation as it was made prior to the wide use of cEEGs (Jagoda and Riggio, 1993). Approximately 30% of patients with SE in the ICU setting will have RSE and NCSE has a higher likelihood of becoming refractory (Treiman et al., 1998; Mayer et al., 2002). NCSE or focal motor seizures present at onset were independent risk factors for the development of RSE (Mayer et al., 2002). RSE is further more likely to occur in patients with severe brain injuries, including trauma, infection, and stroke (Hocker et al., 2014). Interestingly, cause of seizures, duration of seizures before treatment, and APACHE II scores (Acute Physiology and Chronic Health Evaluation II scores) were

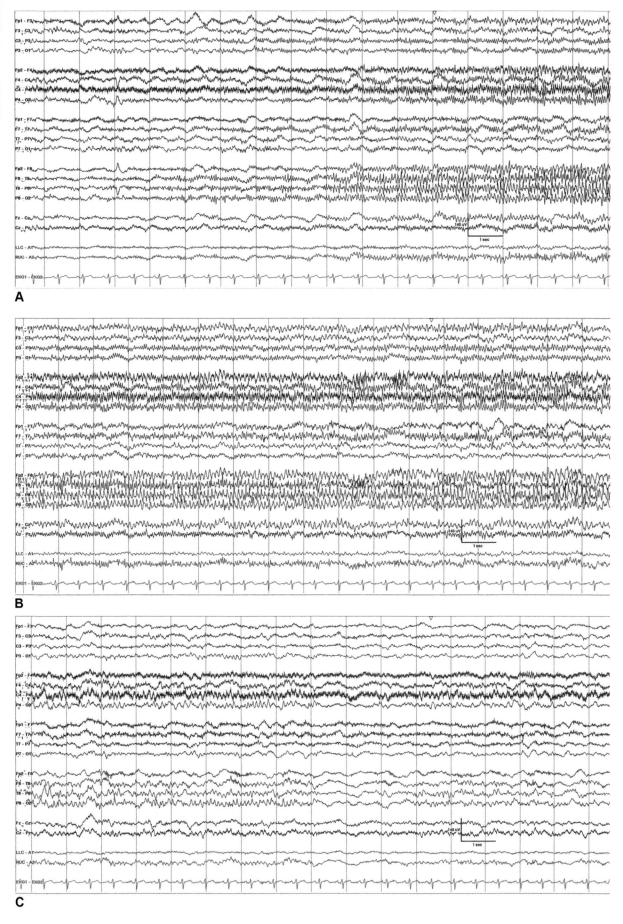

Fig. 28.1—Cont'd

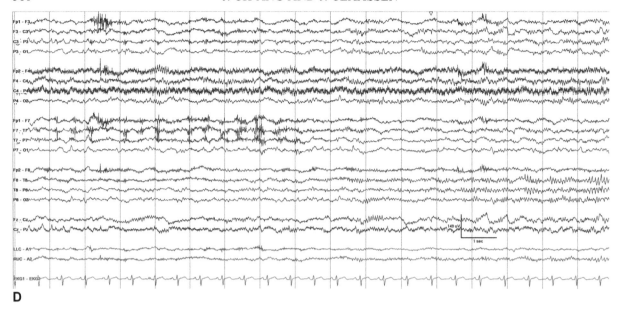

Fig. 28.1—Cont'd (panels **A–D** consecutive pages of 15 seconds of electroencephalogram). A 26-year-old woman with N-methyl-D-aspartate encephalitis was admitted for seizures and orofacial movements. She was found to be in partial status epilepticus arising from the right temporoparietal region with spread to the left hemisphere. Clinical symptoms included behavioral arrest and left ocular deviation, facial tonic-clonic contraction, followed by a brief period of confusion. She was treated with rituximab and seizures were controlled with levetiracetam and phenytoin. On discharge follow-up at 4 months, the patient's seizures were controlled and she was finishing the college semester.

not associated (Mayer et al., 2002). Generally, patients in RSE are comatose or have subtle signs of seizure similar to other patients in NCSE (Husain et al., 2003). Among patients with RSE, 10–15% will fail to respond to third-line therapy and are considered to have superrefractory SE (SRSE) (Novy et al., 2010). In a small Chinese study, approximately 68% of SRSE were seen in patients with encephalitis other reports of SRSE were in patients with autoimmune or paraneoplastic encephalitis (Hocker et al., 2014; Tian et al., 2015).

NEUROPATHOLOGY

SE occurs when patients' intrinsic mechanisms that help inhibit or terminate seizures are overwhelmed and cause the enhancement of excitatory pathways (Mazarati et al., 1998a). Isolated seizures are terminated through inhibitory feedback, resulting in a postictal state during which the brain is more refractory to develop seizure activity (Mazarati et al., 1998a). In the first milliseconds to seconds, neurotransmitters and modulators are released and $GABA_A$ receptors become desensitized (Naylor et al., 2005). In the following minutes to hours, there are maladaptive changes with increased expression of proconvulsive neuropeptides such as substance P (which also increases glutamate release), the internalization of $GABA_A$ receptors, and the depletion of inhibitory neuropeptides (Sperk et al., 1986; Mazarati et al., 1998b; Liu et al., 1999; Naylor et al., 2005). Also seen is the activation of N-methyl-D-aspartate (NMDA) receptors (hence the effectiveness of ketamine), an increase in α-amino-3-hydroxy-5-methyl-4-isoxazolepropionic acid (AMPA) and NMDA receptors, and increased presynaptic modulators of glutamate release, all of which suggest the potentiation of excitatory pathways (Mazarati and Wasterlain, 1997, 1999; Naylor et al., 2005; Chen and Wasterlain, 2006). This excessive neuronal firing, through the activation of both intracellular proteases and nitric oxide and the generation of free radicals, causes necrosis and mitochondrial dysfunction, leading to neuronal death (Fujikawa, 1996; Cock et al., 2002). In the next hours to weeks, seizure-provoked neuronal death results in long-term changes in gene expression and neuronal reorganization (Chen and Wasterlain, 2006).

The mechanisms underlying NCSE remain poorly understood; however, electrographic seizures can damage the brain in the absence of clinical convulsions. In animal models and at autopsy of patients without pre-existing epilepsy in NCSE, necrosis in the cortex and hippocampus has been seen (Wasterlain et al., 1993). Neuronal damage in NCSE is further suggested by elevations of neuron-specific enolase following SE, a critical enzyme for energy metabolism (Rabinowicz et al., 1995).

Pharmacoresistance progressively develops during SE from these dynamic changes in neurotransmitters,

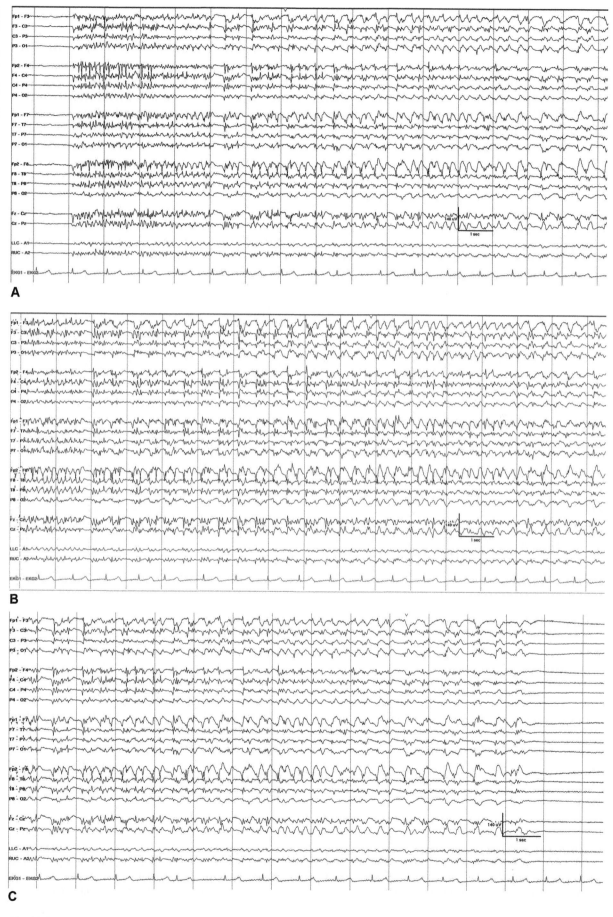

Fig. 28.2—Cont'd

neuropeptides, and receptors mentioned above. The endocytosis of $GABA_A$ receptors may partially explain the loss of benzodiazepine responsiveness (Naylor et al., 2005). The time course parallels the loss of benzodiazepine responsiveness, which falls between 10 and 45 minutes (Kapur and Macdonald, 1997). Benzodiazepine efficacy was found to decrease 20-fold within 30 minutes of SE onset (Kapur and Macdonald, 1997). Phenytoin was found to fail later than diazepam and because of its actions on voltage-gated Na channels, its failure is independent of the GABA receptor changes (Mazarati et al., 1998a). In a SE animal model, when phenytoin was given 40 minutes after 30 minutes of SE or 10 minutes after 60 minutes of SE, phenytoin's efficacy was greatly decreased (Mazarati et al., 1998a).

CLINICAL PRESENTATION

The initial clinical presentation of patients with seizures depends on the neuroanatomy which is seizing, the underlying cause seizure, and the presence of abovementioned convulsions (GCSE vs. NCSE). Clinical findings in patients with ongoing seizure activity (especially RSE and SRSE) will further depend on the effects of medications and complications. There are positive clinical signs, including twitching, automatisms, and rhythmic jerking, and negative signs, including confusion, aphasia, and staring (Table 28.1). As discussed above, seizures associated with overt rhythmic jerking for greater than 5 minutes or more than two convulsions in a 5-minute interval without return to baseline are called GCSE. After a seizure has ended, patients may have transient postictal motor weakness, known as Todd's paralysis, postictal confusion, or lethargy. If the level of consciousness does not improve by 20 minutes after the movements have stopped or the patient continues to have persistent alteration in neurologic function from baseline 30–60 minutes later, NCSz or NCSE should be considered. In the critically ill, it can be especially difficult to distinguish whether or not someone is having a seizure. Often these patients have altered consciousness secondary to systemic disease or medication effect and do not have seizures. Also in the ICU population, it is difficult to recognize seizures in patients receiving paralytics, and autonomic signs seen in seizures (tachycardia, hypoxia, pupillary dilation, blood pressure fluctuation) could also be accounted for by their systemic illness, pain, or inadequate sedation.

NEURODIAGNOSTICS AND IMAGING

SE is a neurologic emergency, particularly if generalized convulsive SE, treatment is started prior to an extensive diagnostic workup. When the clinical diagnosis of GCSE is made or there is concern for NCSE, patients should be assessed first for their airway, breathing, and circulation (ABC). Treatment as well as basic diagnostic studies and stabilization of vital sign parameters should be initiated in parallel to terminate clinical and electrographic seizures immediately, which may include treatment of the underlying cause (i.e., SE secondary to hypoglycemia). Generally, the first AED should be given simultaneously. When there is rapid control of SE and the patient is hemodynamically stable, the diagnostic work up should be guided by the clinical picture (Table 28.2) with consideration of a wide range of underlying causes (Table 28.3). In the ICU, the etiology of SE can be different from those who are admitted from the community. In a general population-based study, low levels of AEDs (34%), prior remote neurologic insults (24%), and strokes (22%) account for most cases of SE, with other important etiologies being hypoxia (13%), metabolic (15%), and alcohol withdrawal (13%) (DeLorenzo et al., 1996). In the ICU,

Table 28.1

Semiologic spectrum of nonconvulsive seizures and nonconvulsive status epilepticus

Negative symptoms	Positive symptoms	Other
Anorexia	Agitation/aggression	Laughter
Aphasia/mutism	Automatism	Nausea/vomiting
Amnesia	Blinking	Nystagmus/eye deviation
Catatonia	Crying	Perseveration
Coma	Delirium	Psychosis
Confusion	Delusions	Tremulousness
Lethargy	Echolalia	
Staring	Facial twitching	

Reproduced from Jirsch and Hirsch (2007).

Fig. 28.2—Cont'd (panels **A–C** consecutive pages of 15 seconds of electroencephalogram (EEG)). A 82-year-old man admitted for ventricular fibrillation arrest (return to spontaneous circulation 6 minutes), who during rewarming developed very frequent cyclic seizures (each lasting 10–15 seconds, occurring every 1–5 minutes) qualifying for nonconvulsive status epilepticus. The EEG was characterized by generalized spike and polyspike discharges. There was no clinical correlate. Treatment included levetiracetam and valproic acid. Burst suppression was achieved with midazolam and propofol infusions. The patient ultimately lost brainstem reflexes and the family opted to withdraw care.

Table 28.2

Diagnostic tests and imaging in status epilepticus

The diagnostic workup should be completed simultaneously with treatment.

All patients
Fingerstick glucose
Monitor vital signs
Head computed tomography scan (appropriate for most cases)
Laboratory tests: blood glucose, complete blood count, basic metabolic panel, calcium (total and ionized), magnesium, antiepileptic drug levels (phenytoin, valproate, carbamazepine), coagulation studies, pregnancy test (if female), blood culture
Continuous electroencephalograph monitoring

Tailored to clinical presentation
Brain magnetic resonance imaging
Lumbar puncture
Comprehensive toxicology panel, including toxins that frequently cause seizures (i.e., isoniazid, tricyclic antidepressants, theophylline, cocaine, sympathomimetics, alcohol, organophosphates, and cyclosporine)
Other laboratory tests: urine ethyl glucuronide, inborn errors of metabolism, heavy metals

Additional studies
Bacterial and fungal cultures, AFB smears and cultures. Coxsackie complement fixation
Encephalitis panel (PCR for HSV, VZV, CMV, EBV, enterovirus, SLE, EEE, CA encephalitis, Powassan and WNV), ELISA for WNV, Lyme titer and Western blot, cryptococcal antigen, cytology, flow cytometry, paraneoplastic panel, anti-NMDA receptor antibodies, anti-VGKC complex antibodies
Serum and fecal test

Autoimmune
Anti-RBC antibodies, Rh factor, ANA, ANCA, anti-ENA, anti-DNA, ACE

Viral
Dengue immunoglobulin, hepatitis A, B, C panel, NY state encephalitis panel (PCR for HSV, VZV, CMV, EBV, enterovirus, SLE, EEE, CA encephalitis, Powassan and WNV)

Bacterial
Anaerobes and AFB smear and cultures from brain biopsy, VDRL, Lyme titers, *Legionella*, *Haemophilus* GPB, *Streptococcus pneumoniae*, streptococcus group B, meningo A, Y, B/E, C, W135, *Bartonella* titers

Parasitic and fungal
Stool for ova, parasites, protozoa, *Cyclospora*, *Cryptosporidium*, *Isospira*. Blood for *Echinococcus*, *Histoplasma*, *Blastomyces*, *Aspergillus*

Adapted from Brophy et al. (2012).
AFB, acid-fast bacilli; PCR, polymerase chain reaction; HSV, herpes simplex virus; VZV, varicella-zoster virus; CMV, cytomegalovirus; EBV, Epstein–Barr virus; SLE, systemic lupus erythematosus; EEE, Eastern equine encephalitis; CA encephalitis, California encephalitis; WNV, West Nile virus; ELISA, enzyme-linked immunosorbent assay; RBC, red blood cell; Rh, rhesus; ANCA, antineutrophil cytoplasmic antibody; anti-ENA, antibodies to extractable nuclear antigens; ACE, angiotensin-converting enzyme; GPB, Gram-positive bacteria; VDRL, Venereal Disearch Research Laboratory.

the majority of seizures are nonconvulsive (90%) and of those found to be in NCSE, hypoxia/anoxia was the most common etiology (42%), followed by stroke (22%), infection (5%), head trauma (5%), metabolic disorders (5%), and alcohol withdrawal or low AEDs (5%) (Towne et al., 2000; Claassen et al., 2004). In the neurologic ICU, the highest rate of NCsz or NCSE was found in patients with encephalitis and brain hemorrhages (Hauser, 1990). In other critically ill subpopulations, 29% of patients with intracerebral hemorrhage had NCSz, 26% with central nervous system (CNS) infection, 23% with brain tumor, 22% with severe head trauma, 18% with subarachnoid hemorrhage (SAH), and 31% with a prior history of epilepsy (Vespa et al., 1999, 2003; Dennis et al., 2002; Claassen et al., 2004).

All patients presenting to the emergency room (ER) with SE should receive glucose fingerstick, basic laboratory measures (including complete blood count, basic metabolic panel, ionized and total calcium, magnesium), pertinent AED levels, and computed tomography (CT) scan of the head (appropriate in the majority of cases) (Brophy et al., 2012; Claassen et al., 2015). Additional diagnostic tests may be required based on the clinical scenario. Those with possible exposure or ingestion as the potential cause for SE should have a comprehensive toxicology panel to include substances that frequently cause seizures, such as isoniazid, tricyclic antidepressants, theophylline, cocaine, sympathomimetics, alcohol, organophosphates, and cyclosporine (Brophy et al., 2012). If the patient's presentation is concerning for a CNS infection

Table 28.3

Underlying etiologies of status epilepticus

The diagnostic workup should be completed simultaneously with treatment.

Acute processes

Metabolic disturbances: electrolyte abnormalities – hyponatremia, hypocalcemia, hypomagnesemia, hypophosphatemia, hypoglycemia, hyperglycemia with hyperosmolar state, renal failure

Sepsis

Central nervous system infection: meningitis, encephalitis, abscess

Stroke
  Ischemic stroke, intracerebral hemorrhage, subarachnoid hemorrhage, cerebral sinus thrombosis

Head trauma with or without epidural or subdural hematoma

Drug issues

Drug toxicity
  Withdrawal from opioid, benzodiazepine, barbiturate, or alcohol
  Noncompliance with AEDs

Hypoxia, cardiac arrest

Hypertensive encephalopathy, posterior reversible encephalopathy syndrome

Autoimmune encephalitis (i.e., anti-NMDA receptor antibodies, anti-VGKC complex antibodies), paraneoplastic syndromes

Chronic processes

Pre-existing epilepsy: breakthrough seizures or discontinuation of AEDs

Chronic ethanol abuse in setting of ethanol intoxication or withdrawal

CNS tumors

Remote CNS pathology (e.g., stroke, abscess, TBI, cortical dysplasia)

Special considerations in children

Prolonged febrile seizures are the most frequent cause of SE in children

CNS infections, especially bacterial meningitis, inborn errors of metabolism, ingestion of substances

Adapted from Brophy et al. (2012).
AEDs, antiepileptic drugs; NMDA, *N*-methyl-D-aspartate; VGKC, voltage-gated potassium channel; CNS, central nervous system; TBI, traumatic brain injury; SE, status epilepticus.

(fever, leukocytosis, nuchal rigidity), then a lumbar puncture should be completed (preferably after a CT scan of the head) and empiric antibiotics and antivirals should be started immediately. Empiric treatment should be continued until the cerebrospinal fluid profile and other supportive tests come back negative.

For those patients already in the ICU, SE may have a multifactorial cause from neurologic complications (i.e., brain hemorrhages), multiorgan failure (i.e., development of sepsis or renal failure), metabolic disarray (i.e., hepatic failure), and medications (i.e., administration of seizure-lowering medications such as certain antibiotics or antipsychotics). Magnetic resonance imaging (MRI) can be used on a case-by-case basis to identify stroke, posterior reversible encephalopathy syndrome, or abscesses. Ongoing seizures may cause restricted diffusion (seen on diffusion-weighted imaging on apparent diffusion coefficient) in the hippocampus, thalamus, especially the pulvinar, and the cerebral cortex (Kim et al., 2001; Farina et al., 2004; Szabo et al., 2005).

cEEG monitoring is required to detect NCSz/NCSE and to direct treatment in SE, particularly if refractory to initial interventions. Ideally, cEEG should be started within 1 hour of SE onset (Brophy et al., 2012). Delay to diagnosis and seizure duration are associated with increased mortality (Young et al., 1996). Specific indications for cEEG include patients who: (1) do not return to preconvulsive neurologic baseline; (2) are suspected to have NCSz (coma, altered mental status); (3) have epileptiform activity or periodic discharges on initial 30 minutes of EEG; or (4) have intracranial hemorrhage (Brophy et al., 2012). Since almost 50% of patients continue to have electrographic seizures after control of GCSE, those who do not have improvement in their mental status by 20 minutes should be evaluated with cEEG. The diagnosis of NCSE and detection of RSE require EEG.

Criteria for NCSE are seen in Table 28.4. It must be noted that if patients do not have these criteria fulfilled, it does not mean NCSE can be ruled out. Further monitoring and a benzodiazepine trial should be done for patients with neurologic impairment and rhythmic or periodic focal or generalized epileptiform discharges on EEG (Table 28.5) (Jirsch and Hirsch, 2007; Claassen, 2009). Comatose patients should undergo a

Table 28.4

Criteria for nonconvulsive seizures on electroencephalogram (EEG)

Any pattern lasting at least 10 seconds satisfying any one of the three primary criteria:
1. Repetitive generalized or focal spikes, sharp waves, spike-and-wave complexes at ≥3 seconds
2. Repetitive generalized or focal spikes, sharp waves, spike-and-wave or sharp-and-slow wave complexes at <3 seconds and the secondary criterion
3. Sequential rhythmic, periodic, or quasiperiodic waves at ≥1 second and unequivocal evolution in frequency (gradually increasing or decreasing by at least 1 second), morphology, or location (gradual spread into or out of a region involving at least two electrodes). Evolution in amplitude alone is not sufficient. Change in sharpness without other change in morphology is not enough to satisfy evolution in morphology

Secondary criterion

Significant improvement in clinical state or appearance of previously absent normal EEG patterns (such as posterior dominant alpha rhythm) temporally coupled to acute administration of a rapidly acting antiepileptic drug. Resolution of the epileptiform discharges leaving diffuse slowing without clinical improvement and without appearance of previously absent normal EEG patterns would not satisfy the secondary criterion

Adapted from Jirsch and Hirsch (2007).

Table 28.5

Benzodiazepine trial for the diagnosis of nonconvulsive status epilepticus

If patients have rhythmic or periodic focal or generalized epileptiform discharges on EEG with neurologic impairment, then:
Monitoring
EEG, pulse oximetry, blood pressure, ECG, respiratory rate
Antiepileptic drug trial
Sequential small doses of rapidly acting short-duration benzodiazepine such as midazolam at 1 mg/dose
Between doses, perform clinical and EEG assessment
Trial is stopped after any of the following:
  Persistent resolution of EEG pattern (and exam repeated)
  Definitive clinical improvement
  Respiratory depression, hypotension, or other adverse effects.
  A maximum dose is reached (such as 0.2 mg/kg midazolam)
Test is considered positive if there is resolution of the potential ictal EEG pattern and an improvement in either the clinical state or the appearance of previously absent normal EEG patterns. If EEG improves, but patient does not, the result is equivocal.

Reproduced from Jirsch and Hirsch (2007).
EEG, electroencephalogram; ECG, electrocardiogram.

minimum of 48 hours' cEEG and those with epileptiform discharges should also have prolonged monitoring, since these discharges have the potential to develop into seizures (Claassen et al., 2004). Those who are not comatose may only need 24 hours evaluating a sleep-and-wake cycle. If the patient is in RSE, more diagnostic studies should be ordered for further investigation, prior studies may need to be repeated, and the differential diagnosis should be broadened (Table 28.2).

HOSPITAL COURSE AND MANAGEMENT

Seizure duration and time to first treatment are major determinants of morbidity and mortality in patients with SE (Towne et al., 1994; Young et al., 1996; DeLorenzo et al., 1998). Thus, the most critical step in stopping clinical and electrographic seizure activity involves prompt administration of AEDs and in adequate amounts. Based on randomized controlled trials reflected in the Neurocritical Care Society management guidelines for SE, first a benzodiazepine is given as emergent rescue therapy and then, unless the precipitating cause of seizures is corrected and the patient has stopped seizing, a second AED, called the control therapy, is required (Brophy et al., 2012). As outlined above in the neurodiagnostics and imaging section, conceptually the initial approach involves out-of-hospital and in-hospital stabilization of vital signs (ABCs of life support), identifying the underlying cause of SE, and detecting NCSz/SE, while following a standardized protocol of AED administration (outlined below).

Emergent control therapy

Lorazepam emerged as the most efficacious initial agent, with a 52% success rate of stopping seizure activity within 20 minutes in a landmark randomized controlled trial by Treiman and colleagues (Treiman et al., 1998; Alldredge et al., 2001). Compared to other benzodiazepines, lorazepam has a longer duration of antiseizure effect (12–24 hours compared to 15–30 minutes, for example, for diazepam) (Epilepsy Foundation of America's Working Group on Status Epilepticus, 1993). The reason is that diazepam is even more lipophilic than lorazepam and is quickly redistributed to other fatty tissues, causing both brain and serum concentrations to decrease quickly, allowing SE to recur

Table 28.6

Antiepileptic drug dosing and pharmacokinetics

Lorazepam
Loading dose: 0.1 mg/kg IV up to 4 mg per dose, may repeat once in 5 minutes
Onset of action: 3–10 minutes
Duration of effect: 12–24 hours
Adverse effects: respiratory depression, hypotension, sedation
Considerations: dilute 1:1 with saline. Use midazolam 10 mg IM if no IV is present

Phenytoin
Loading dose: 20 mg/kg IV, maximal infusion rate of 50 mg/min, 25 mg/min in elderly, patients with pre-existing cardiac disease. May give an additional 5–10 mg/kg 10 minutes after loading infusion if still seizing
Target serum level: total 15–25 µg/mL, free level 2–3 µg/mL. Total level adjustment for albumin (alb): total level/(alb $\times$ 0.1)+0.1. Measure serum phenytoin levels 2 hours after IV or 4 hours after IM. Monitor the free level when on valproic acid, benzodiazepines, or other medications that are highly protein-bound, and in low-albumin states
Maintenance dosing: 5–7 mg/kg/day in 2–3 divided doses
Onset of action: 10–25 minutes
Duration of effect: 24 hours
Adverse effects: arrhythmias, hypotension, purple-glove syndrome, Stevens–Johnson syndrome, pancytopenia, hepatotoxicity
Considerations: induces hepatic metabolism of other medications and may displace other protein-bound drugs, thus increasing levels. Metabolized by cytochrome p450 enzyme. Precipitation if given with potassium, insulin, heparin, norepinephrine, cephalosporin, dobutamine

Fosphenytoin
Loading dose: 20 mg/kg IV, maximal infusion rate of 150 mg/min, may give an additional 5–10 mg/kg dose 10 minutes after loading infusion if still seizing
Target serum level: see phenytoin
Maintenance dosing: phenytoin 5–7 mg/kg/day in 2–3 divided doses
Onset of action: 10–25 minutes
Adverse effects: see phenytoin. Less infusion-site reactions such as phlebitis, soft-tissue damage
Considerations: measure serum phenytoin levels 2 hours after IV or 4 hours after IM

Valproic acid
Loading dose: 20–40 mg/kg IV at rate of 3–6 mg/kg/min; may give an additional 20 mg/kg IV dose 10 minutes after loading infusion if still seizing
Target serum levels: total: 80–140 µg/mL, free: 4–11 µg/mL (only consider if toxicity is suspected). Serum concentration levels may be obtained immediately following loading-dose infusion
Maintenance dosing: 1 gram IV q6 hours
Adverse effects: hyperammonemic encephalopathy (consider L-carnitine 33 mg/kg q 8 hours), pancreatitis, rare liver failure, thrombocytopenia and qualitative platelet defect, hypofibrinogenemia, tremor
Considerations: interacts with phenytoin, must follow the free phenytoin level. Hepatic enzyme inhibitor. Meropenem and amikacin decrease valproate serum concentrations (accelerated renal excretion). No sedation and rare hypotension make it a drug of choice in patients with a do-not-intubate status.

Levetiracetam
Loading dose: 20 mg/kg IV or 1–3 gram IV over 15 minutes
Target serum levels: 12–46 µg/mL. Levels may not be readily available by the lab
Maintenance dosing: 1 gram IV q12 hours
Adverse effects: psychosis, behavioral agitation
Considerations: not hepatically metabolized. Dose adjustments needed in poor renal function, dialysis, and continuous renal replacement therapy. Minimal drug interactions and side-effects

Lacosamide
Loading dose: 200–400 mg IV over 15 minutes
Maintenance dosing: 100–200 mg q12 hours
Adverse effects: PR prolongation, hypotension, nausea
Considerations: minimal drug interactions and side-effects. Limited studies in using it in treatment of status epilepticus

Phenobarbital
Loading dose: 20 mg/kg IV, infusion rate of 50–100 mg/min, may give an additional 5–10 mg/kg 10 minutes after loading infusion if still seizing
Target trough serum level: 30–50 µg/mL
Maintenance dosing: 1–3 mg/kg/day in 2–3 divided doses
Onset of action: 20–30 minutes
Duration of effect: >48 hours
Adverse effects: hypotension, respiratory depression, sedation
Considerations: prolonged half-life in adults, ranging from 50 to 150 hours. Powerful sedative effect may contribute to coma

IV, intravenous; IM, intramuscular.

unless another agent is given (Epilepsy Foundation of America's Working Group on Status Epilepticus, 1993). Animal studies have shown that the longer the duration until treatment is initiated, the more likely it is that pharmacoresistance will develop (Kapur and Macdonald, 1997). A crucial case series by Lowenstein and Alldredge (1993) found that, when an AED was initiated within 30 minutes of seizure onset, 80% of patients responded, and only 40% responded when a medication was given more than 2 hours after seizure onset.

In the first major randomized controlled trial for SE, the Veterans Affairs Cooperative Study by Treiman and colleagues (1998) further supported the notion that time to treatment may be at least as important as finding the best first-line drug. Patients who were in the earliest stage of overt convulsions in SE had a 75% chance of successful termination with the initial AED, then, when in a waxing-and-waning pattern, success with the first agent fell to 30% (Treiman et al., 1998). When the patient had progressed to the stage of continuous ictal activity, 25% responded to the initial AED and finally, when there was a suppression burst or brief suppression pattern, 7–8% respectively responded (Treiman et al., 1998).

These observations led to several trials exploring the out-of-hospital initiation of SE therapy, which determined that lorazepam, when given by emergency medical services (EMS) prior to reaching the hospital, established 59% seizure control on arrival to the ER, compared to the 43% treated with diazepam and 21% given placebo (Alldredge et al., 2001). Prior to this trial concerns were raised that giving benzodiazepines in an out-of-hospital setting might be associated with respiratory complications, but Alldredge and colleagues (2001) found that ongoing seizure activity caused more respiratory complications than benzodiazepines. In fact, 23% of patients in the placebo arm required bag-masked ventilation and intubation, compared to 10–11% in the lorazepam and diazepam arms respectively (Alldredge et al., 2001). Although there always is a risk of respiratory depression and hypotension from rapid administration of benzodiazepines, this helped dispel the notion that administering benzodiazepines to a seizing patient would contribute to respiratory compromise.

Administration of IV lorazepam may be challenging, particularly in the EMS setting, as establishing IV access can be difficult in a seizing patient and IV lorazepam has a short shelf-life when not refrigerated (Gottwald et al., 1999). Intramuscular (IM) midazolam is being increasingly used after one study showed that IM midazolam was at least as effective and safe as IV lorazepam in the prehospital setting (Silbergleit et al., 2012). Other forms of benzodiazepines exist when IV lorazepam is not possible, such as nasal, buccal midazolam and rectal diazepam (Brophy et al., 2012). Most recently, a study from Europe demonstrated that add-on therapy of levetiracetam (LEV) to clonazepam did not increase the seizure control rate in convulsive SE but that high doses of clonazepam in the control arm were observed to have a control rate of more than 80% (Navarro et al., 2016). This study raises the question of possibly exploring higher initial doses of benzodiazepines in addition to cutting down on the time to treatment initiation (Claassen, 2016). If not given prior to ER arrival, benzodiazepines should be administered emergently upon arrival to the ER.

A patient's airway and ventilation should be monitored and, if the airway is not secure from seizures, a postictal state, or escalation of seizure treatment to IV AEDs, then intubation should be considered early. Strict vitals monitoring (blood pressure, body temperature, telemetry, oxygenation) and IV access should be continued throughout AED administration as both AEDs and ongoing seizure activity may be associated with life-threatening hypotension and arrhythmias. Cerebral autoregulation is severely impaired and cerebrovascular resistance falls in SE as well as in many acute brain injuries; therefore, the cerebral perfusion pressure is directly dependent on the systemic blood pressure and even relative hypotension may require the use of vasopressors (Epilepsy Foundation of America's Working Group on Status Epilepticus, 1993). While assessing the patient for ABCs, a fingerstick should be done (if not done by EMS) and if low <60 mg/dL, administer D50W 50 mL IV and thiamine 100 mg IV (Epilepsy Foundation of America's Working Group on Status Epilepticus, 1993; Brophy et al., 2012). Hypoglycemic seizures will only terminate with glucose administration. Throughout all of this, as mentioned in the previous section, diagnostic studies should be done to identify any etiologies that must be urgently attended to or can be easily reversed. Obtaining a CT scan of the head is appropriate in most cases and, when concerned for a CNS infection, lumbar puncture should be done and antibiotics should be started immediately.

Urgent control therapy

After a benzodiazepine has been given, patients should receive a second AED unless the cause of seizures is definitively corrected, such as hypoglycemia (Brophy et al., 2012). Benzodiazepines are fast acting, but are not a good maintenance therapy to prevent the recurrence of seizures or SE. For those patients whose seizures responded to a benzodiazepine, a loading dose of a second AED causes a quick rise to therapeutic blood levels and then this AED is continued for maintenance dosing. If the patient, however, did not respond to a benzodiazepine, then the goal of the urgent control AED is to stop SE. The agent of choice for urgent control therapy is based mostly on observational data, small trials, and *post hoc* analyses of clinical trials (Epilepsy Foundation of America's Working Group on Status Epilepticus, 1993; Lowenstein and Alldredge, 1998). Guidelines recommend IV administration of fosphenytoin/phenytoin or valproate sodium (Brophy et al., 2012). Alternative AEDs which have been reported include: phenobarbital, LEV, or a continuous-infusion antiepileptic such as midazolam. See Table 28.6 for medication details. There are not enough data to support a clear second agent; however, many expert neurologists still prefer fosphenytoin or phenytoin (Claassen et al., 2003). A recent meta-analysis found that valproic acid, LEV, and phenobarbital were more effective than phenytoin; however, these conclusions were based on data using varying definitions of SE, several possible confounding variables, and a very heterogeneous patient population, making the data difficult to interpret (Yasiry and Shorvon, 2014). When choosing the second AED, each clinical scenario should be assessed, including comorbidities, but the most important goal is to administer an AED promptly to attain therapeutic serum levels. In patients with known epilepsy who had been taking an AED, it is reasonable to give an IV bolus of that AED before moving on to a new agent (Brophy et al., 2012).

Refractory status epilepticus

Patients who continue to seize clinically or electrographically after both the initial and second agent have been given are considered to be in RSE regardless of elapsed time. Additional diagnostic investigations or a repetition of tests previously obtained may be required if the underlying cause of seizures has not yet been identified (see additional studies in Table 28.2). After two standard AEDs have been given, only 2% and 5% of GCSE and NCSE respectively will respond to a third standard agent (Treiman et al., 1990). Because of this poor response, most experts recommend rapid treatment escalation to continuous-infusion AEDs (cIV AEDs), most commonly midazolam, propofol, and in the past also pentobarbital (Table 28.7). Respiratory compromise and hypotension should be expected when using anesthetic drips. To use these drips, patients should generally be intubated and, commonly, these patients require vasopressors and central-line placement. IV valproic acid or IV phenobarbital may be alternatives for patients with a "do not intubate" status.

There is a paucity of evidence regarding which cIV AED is preferred. Pentobarbital had been the traditional agent for the last 50 years but has fallen out of favor and is now primarily used for SRSE (see below). A systematic review found pentobarbital to be more effective than midazolam and propofol at controlling seizures but concluded that mortality was not different when comparing pentobarbital to midazolam or propofol infusions (Claassen et al., 2002a). This review was based on combining many case reports with many inherent limitations, an overall small sample size, inconsistent use of cEEG, particularly for patients treated with pentobarbital, and a heterogeneous patient population with different AED combinations preceding anesthetic drip administration. However, the low doses of midazolam and propofol used at that time are most likely the main reason for the differences in efficacy.

Pentobarbital has an extremely long half-life and is associated with a number of systemic complications, including respiratory depression, myocardial depression, sedation, hypotension, thrombocytopenia, refractory acidosis, ileus, decreased clearance of bronchial secretions (leading to mucus plugging and pneumonia), and immunosuppression. These side-effects make midazolam and propofol more favorable choices for the treatment of RSE (Jagoda and Riggio, 1993; Yaffe and Lowenstein, 1993; Devlin et al., 1994; Abou Khaled and Hirsch, 2006; Rossetti et al., 2011; Hocker et al., 2014). The only randomized controlled trial for RSE, which compared propofol to barbiturate infusions, was terminated early to poor recruitment but, in its limited data, expectedly longer duration of mechanical ventilation was seen in patients treated with barbiturates (Rossetti et al., 2011).

Midazolam and propofol drips are now more commonly used because they are shorter acting and cause fewer hemodynamic disturbances (Table 28.7). Even though a small study showed that propofol and midazolam did not differ in their seizure control, propofol was associated with a higher mortality (although it was not statistically significant: 57% propofol and 17% midazolam) (Prasad et al., 2001). Given other mixed data regarding propofol's safety profile, including propofol infusion syndrome, midazolam has become the more popular choice (Brophy et al., 2012; Riviello et al., 2013). Midazolam also acts on the $GABA_A$ receptor with a rapid onset of action. Side-effects include hypotension, respiratory depression, and tachyphylaxis (Abou Khaled and Hirsch, 2006). In a single-center comparison of two protocols of low- (0.2 mg/kg/h) versus high- (0.4 mg/kg/h) dose cIV midazolam, high-dose cIV midazolam was safely tolerated and had a lower seizure withdrawal rate and lower mortality at discharge (Fernandez et al., 2014). Expectedly, hypotension was more frequent in the high-dose group but, when given in a controlled setting such as the ICU, it was not associated with worse outcome. In patients with refractory NCSE being treated with cIV midazolam, 18% had acute treatment failure and 56% had breakthrough seizures (Claassen et al., 2001).

Propofol, a $GABA_A$ receptor agonist, has a rapid onset of action (<3 minutes) and easy reversibility

Table 28.7

Antiepileptic drug dosing and pharmacokinetics for refractory status epilepticus and superrefractory status epilepticus

Midazolam
Loading dose: 0.2 mg/kg with 0.2–0.4 mg/kg boluses repeated every 5 minutes until seizures stop, up to maximum 2 mg/kg
Maintenance dosing: initial rate 0.05 mg/kg/h. If recurrent seizures, give 0.1–0.2 mg/kg bolus, then increase maintenance rate by 0.05–0.1 mg/kg/h every 3–4 hours or by approximately 20%
Dose range: 0.05–2.9 mg/kg/h titrated to burst suppression or seizure control
Onset of action: minutes, less than 1 hour to stop status epilepticus
Half-life: 1.5–3.5 hours; with prolonged use there may be tachyphylaxis and prolongation of half-life up to days
Duration of effect: minutes to hours
Adverse effects: respiratory depression, hypotension, sedation
Considerations: renally eliminated. Rapid redistribution. Does not contain propylene glycol

Propofol
Loading dose: 1–2 mg/kg with 1–2 mg/kg boluses repeated every 3–5 minutes until seizures stop, up to maximum of 10 mg/kg
Maintenance dosing: initial rate 20 μg/kg/min. If recurrent seizures, increase maintenance rate 5–10 μg/kg/min every 5 minutes or 1 mg/kg bolus plus increasing maintenance rate
Dose range: 20–200 μg/kg/min. Do not exceed a dosage of 5 mg/kg/h for >48 h (increased risk of propofol infusion syndrome), titrated to burst suppression or seizure control
Onset of action: less than 3 minutes
Duration of effect: 5–10 minutes after discontinuation
Adverse effects: respiratory depression, pancreatitis, hypertriglyceridemia, hypotension, bradycardia, and propofol infusion syndrome: metabolic acidosis, cardiac failure, rhabdomyolysis, hypotension, and death
Considerations: risk factors for propofol infusion syndrome: infusion >48 hours at high doses (>5 mg/kg/h), severe head injury, lean mass, and concurrent use of catecholamines or steroids. Follows creatine phosphokinase, triglycerides, acid–base status. Must adjust daily caloric intake (1.1 kcal/min)

Pentobarbital
Loading dose: 5 mg/kg, infusion rate 25–50 mg/min with repeated 5 mg/kg boluses until seizures stop
Target trough serum level: 30–45 μg/mL
Maintenance dosing: initial rate 1 mg/kg/h, titrate to EEG suppression burst pattern. If recurrent seizures, 5 mg/kg bolus followed by increase in the maintenance rate by 0.5–1 mg/kg/h q 12 hours
Dose range: 0.5–10 mg/kg/h
Taper: after 24 hours of EEG control of seizure activity at a rate of 0.5–1 mg/kg/h q 4–6 hours
Onset of action: 15–20 minutes
Half-life: 15–60 hours
Adverse effects: hypotension, thrombocytopenia, potential immunosuppression, Stevens–Johnson syndrome, myocardial depression, metabolic acidosis (diluted in propylene glycol), ileus
Considerations: while weaning off continuous IV pentobarbital, consider adding phenobarbital as a maintenance agent (serum levels > 100 μg/mL may be needed) to prevent recurrent seizure activity. At high doses, there is complete loss of neurologic function

Ketamine
Loading dose: 1–2 mg/kg IV over 1 minute with 1.5 mg/kg bolus every 3–5 minutes until seizures stop, up to a maximum of 4.5 mg/kg
Maintenance dosing: initial infusion rate is 20 μg/kg/min. If recurrent seizures, a bolus should be followed by increase of maintenance rate by 10–20 μg/kg/min until seizure control
Dose range: 5–125 μg/kg/min
Adverse effects: elevated blood pressure
Considerations: caution in patients with elevated intracranial pressure, traumatic brain injury, ocular injuries, hypertension, chronic congestive heart failure, myocardial infarction, tachyarrhythmias, and in patients with a history of alcohol abuse. Consider using with benzodiazepines for possible synergism.

EEG, electroencephalogram.

(Abou Khaled and Hirsch, 2006). Propofol infusion syndrome, a dreaded complication, may be encountered but is rare. This syndrome consists of metabolic acidosis, cardiac failure, rhabdomyolysis, hypotension, and possibly death. Risk factors include prolonged infusion (>48 hours), high doses (>5 mg/kg/h), severe head injury, lean mass, and concurrent use of steroids or catecholamines (Kumar et al., 2005). The mechanism of injury is thought to be disruption of mitochondrial function and fatty-acid oxidation, leading to lactic acidosis and organ dysfunction (Kumar et al., 2005). In fatal cases, it is recommended that autopsy should include electron microscopy of cardiac and skeletal muscle for mitochondrial dysfunction (Kumar et al., 2005). A retrospective single-center study looked at 27 cases and cIV infusions of propofol were safe at mean rates of 4.8 mg/kg/h for 3 days with 67% permanent seizure control and no major adverse effects (Rossetti et al., 2004). Still, caution must be exercised when using propofol at high doses for prolonged periods of time. Acid–base status must also be monitored if used with other carbonic anhydrase inhibitors, topiramate or zonisamide (Abou Khaled and Hirsch, 2006).

No consensus has been reached regarding the intensity and duration of cIV AED treatments for RSE. In a retrospective analysis of 63 patients who were treated with cIV AEDs, these authors found that patients had more infections and increased relative risk of death, raising awareness of the risks of using cIV AEDs and also questioning to what extent SE should be treated with these cIV agents (Sutter et al., 2014). However, this observation of a small cohort of patients should be taken with caution given that it was based on a retrospective case series where there is an inability to adequately control for bias of why certain patients were selected for cIV agents and there remains a concern about generalizability (Sutter et al., 2014).

Given the little data surrounding electrographic targets and their duration, expert opinion prefers seizure suppression as a goal, but others prefer EEG burst suppression or complete background suppression (Brophy et al., 2012). A retrospective review suggested that mortality and return to functional baseline were independent of which cIV AED was used and the extent of electrographic burst suppression (Rossetti et al., 2005). The European Federation of the Neurological Societies of 2010 recommends a goal of burst suppression on EEG if propofol or barbiturates are used and seizure suppression if midazolam is used (Meierkord et al., 2010). Once the electrographic goal has been reached, cIV AED therapy is continued for 24–48 hours and then a very gradual weaning of the medication should be done while conventional maintenance AEDs are optimized. Therapeutic concentrations may exceed published target concentrations but dosing should be individualized to seizure control with monitoring for toxicity. The withdrawal of these infusions should be done while under cEEG to monitor for recurrent electrographic seizures. There are no data to dictate how to best wean cIV AEDs. If patients have withdrawal seizures, they should be restarted on the cIV AED at the rate prior to initiating the weaning (which presumably had controlled the seizures) and maintained for at least 24 hours. Prior to tapering the anesthetic drip again, at least one of the nonanesthetic AEDs that the patient is already receiving should be optimized (achieve therapeutic levels or increase the dose) or a new nonanesthetic AED should be started (Brophy et al., 2012).

Superrefractory status epilepticus

Seizures that are refractory to third-line agents are rare and termed SRSE. Formally, it is defined as SE that continues for 24 hours or more after the initiation of a cIV AED, including those patients who had attained seizure control with cIV AED but recurred during weaning (Hocker et al., 2014). These patients may have severe acute brain injury but, in a significant portion, no overt cause can be identified (Cuero and Varelas, 2015). These patients are not well studied and, with no randomized clinical trials, the treatment may vary from center to center. Generally, adding an additional cIV agent while optimizing other AEDs or starting alternative treatments is practiced. A recent study reported seizure termination in 90% of SRSE patients with pentobarbital infusions (Pugin et al., 2014). These investigators also reported the success of using phenobarbital while weaning pentobarbital to prevent seizure relapses (Pugin et al., 2014).

SPECIFIC ANTIEPILEPTIC MEDICATIONS

Each of the commonly used drugs is discussed here in more detail. Drug dosing and pharmacokinetics are summarized in Table 28.6.

Phenytoin and fosphenytoin

Phenytoin has been used for half a century and is still extremely popular for the management of SE. It blocks the membrane channels that sodium moves through during depolarization, suppressing repetitive firing. While lorazepam was most efficacious in controlling SE as a first-line agent, phenytoin emerged as an alternative choice, and it is one of two recommended second-line AEDs (Treiman et al., 1998; Brophy et al., 2012). The recommended loading dose is 20 mg/kg IV (or IM) at a maximal rate of 50 mg/min. In some patients, as much as 30 mg/kg may be needed to stop SE (Osorio and Reed, 1989). Phenytoin is approximately 90% bound to plasma proteins, primarily albumin, which makes the unbound 10% pharmacologically active. Thus, low-albumin states such as renal failure or medications that bind to albumin and displace phenytoin from its binding site can cause elevations in phenytoin levels. Since the cytochrome P450 enzymes metabolize phenytoin, medications which alter the function of these enzymes can cause toxicity or inadequate levels. Loading doses should not be adjusted for renal or hepatic insufficiency. In hepatic and renal disease, phenytoin levels should be monitored and dosing should be adjusted for albumin level or unbound phenytoin levels ("free levels") should be obtained.

Valproic acid, for example, displaces phenytoin from plasma protein-binding sites and enhances the systemic clearance of the total drug. It also inhibits the cytochrome P450 enzymes, thereby inhibiting phenytoin metabolism, thus increasing the total concentration of free drug in the serum (Perucca et al., 1980). Because of these conflicting mechanisms, unbound phenytoin levels should be strictly monitored and may be difficult to interpret when valproic acid is given (Lai and Huang, 1993). The maximal rate is limited to 50 mg/min because phenytoin administered at high rates may cause cardiovascular complications such as arrhythmias and hypotension. Patients who are no longer in SE may be loaded with phenytoin at a slower rate. It takes 20–25 minutes for phenytoin to reach its maximal effect after a loading dose (Wilder, 1983). Phenytoin at its maximal rate of 50 mg/kg causes hypotension in 28–50% of patients and bradycardia and ectopic beats in 2% (Cranford et al., 1978; Wilder, 1983). Phenytoin increases the conduction in myocardial junctions and shortens the duration of action potentials in cardiac tissue (York and Coleridge, 1988). It causes hypotension through peripheral vasodilatation and a negative ionotropic effect (Conn et al., 1967; Cranford et al., 1978). Cardiac complications are more common in patients over 50 years old and with pre-existing heart disease and are due to the phenytoin itself and the diluent, propylene glycol (Cranford et al., 1978). The parenteral preparation of phenytoin is not water-soluble and thus must be dissolved in a solvent solution of propylene glycol, which has a variety of adverse effects, including hypotension and cardiac arrhythmias, such as bradycardia and asystole (Louis et al., 1967). Propylene glycol is suspected to enhance vagal activity and depress myocardial activity (Louis et al., 1967). Fosphenytoin was then created because it was more water-soluble and could be administered at a more rapid rate. Importantly, these side-effects of phenytoin such as QT prolongation can be mitigated by slowing or stopping the infusion (Cranford et al., 1978). Additionally, phenytoin is highly caustic to veins and any extravasation may lead to subsequent tissue necrosis (also known as purple-hand syndrome).

Fosphenytoin is a prodrug of phenytoin and dosing is expressed in phenytoin equivalents (amount of phenytoin released from the prodrug). It is approved for administration at a maximal rate of 150 mg/min. Hypotension and arrhythmia may still occur; therefore, it is safer to start the infusion at a slower rate and then increase the rate as tolerated. Due to the difference in solvents, fosphenytoin has less infusion-site reactions such as phlebitis and soft-tissue damage, even when extravasation occurs. Usually transient pruritus occurs.

Valproic acid

Valproic acid is a potent first-line and urgent control agent, aborting 66% and 79% of seizures after drug infusion respectively (Misra et al., 2006; Brophy et al., 2012). Its efficacy has also been attributed to synergistic effects with previously given antiepileptic medications (Misra et al., 2006). Valproic acid is loaded at 20–40 mg/kg IV and given over 10 minutes. An additional 20 mg/kg IV over 5 minutes can be given if the patient is still seizing. Target levels are 80–140 µg/mL and levels may be obtained following the loading dose. Valproate has been associated with liver dysfunction, encephalopathy, rarely hypotension and if avoidable should not be given to any patient with active bleeding because it can cause qualitative and quantitative platelet defects (Misra et al., 2006). Respiratory suppression is less frequent than with anesthetic infusions but may be seen.

Levetiracetam

LEV is often used off-label in SE as an urgent control agent. LEV has a distinct mechanism of action. In the animal model, it was found to inhibit neuronal hypersynchronization in the hippocampus and the development of electric kindling in the amygdala (Loscher et al., 1998; Niespodziany et al., 2003). The main target has been identified to be the SV2A receptor, a membrane protein on all synaptic vesicles which modulates vesicle fusion (Lynch et al., 2004). It is thought that LEV binding to this receptor enhances its function to inhibit abnormal epileptic bursts (Lynch et al., 2004). In 18 patients with SE refractory to benzodiazepines, LEV was able to control SE in 16 subjects (Knake et al., 2008). The loading dose of LEV is 20 mg/kg IV (up to 3 grams) over 15 minutes. In a small retrospective chart review, LEV terminated SE in 69% of patients after they had failed at least one AED (Moddel et al., 2009). No cardiac side-effects were seen; only nausea and vomiting during loading (Moddel et al., 2009). Renal function needs to be considered when choosing maintenance dosing as LEV is primarily cleared through the kidneys. Supplemental doses need to be given after dialysis. LEV is not metabolized by the liver, making it a good agent for patients with hepatic failure. LEV can be used in the treatment of SRSE in acute intermittent porphyria (Cuero and Varelas, 2015). LEV did not succeed as add-on therapy to benzodiazepines for out-of-hospital treatment of GCSE (Navarro et al., 2016).

Lacosamide

Lacosamide is becoming increasingly used for patients with SE and loaded at 200–400 mg over 3–5 minutes (Albers et al., 2011). Lacosamide selectively enhances slow inactivation of voltage-gated sodium channels without affecting fast inactivation and it also binds to a protein, collapsin-response mediator protein 2, involved in the modulation of NMDA receptor subunits (Errington et al., 2006; Beyreuther et al., 2007). Being one of the more recently introduced AEDs, it appears to have a low side-effect profile and low potential for pharmacokinetic drug–drug interactions. The most common side-effects are dizziness, headache, diplopia, nausea, vertigo, vomiting, and abnormal coordination (Halasz et al., 2009). Notably, it can produce a dose-related increase in the PR interval (Halasz et al., 2009). Lacosamide is not well studied as a treatment for SE, with most evidence coming from case reports and small studies.

Phenobarbital

Phenobarbital is approved by the US Food and Drug Administration for the treatment of SE, but has fallen out of favor as an early treatment agent for adults with SE given the many alternative agents available. Of note, in the largest randomized controlled trial comparing different agents to treat SE, phenobarbital had the second highest success rate for seizure control at 58%, following lorazepam at 65% (Treiman et al., 1998). At very high doses respiratory depression, loss of consciousness, and hypotension need to be considered. Phenobarbital is loaded 20 mg/kg IV at a rate of 50–100 mg/min. An additional load of 5–10 mg/kg can be given if the patient is still seizing. It can also be used as a maintenance agent while weaning off cIV pentobarbital to prevent recurrent seizure activity (see pentobarbital) (Pugin et al., 2014).

Ketamine

Upregulation of NMDA receptors occurs in ongoing SE, making ketamine, an NMDA receptor antagonist, appear to be a logical agent for RSE and SRSE. In a multicenter case series of 60 episodes of RSE and in a systematic literature review of 110 adult patients, ketamine achieved permanent control in approximately 57% of RSE (Gaspard et al., 2013; Zeiler et al., 2014). When treating SRSE, ketamine achieved 60% of SE control and overall dosing was 0.9–10 mg/kg/h (Gaspard et al., 2013). Mortality was found to be lower in the subjects who responded to ketamine, but the investigators suggested that this may have reflected a lower-severity SE (Gaspard et al., 2013). Benzodiazepines should be given concomitantly with ketamine infusions as it seems ketamine has a synergistic effect (Martin and Kapur, 2008; Hsieh et al., 2010). Ketamine has not been associated with cardiac depression; however, it does cause elevation in blood pressure. It should be used cautiously in patients with elevated intracranial pressure, traumatic brain injury, ocular injuries, hypertension, chronic congestive heart failure, myocardial infarction, tachyarrhythmias, and in patients with a history of alcohol abuse (Mewasingh et al., 2003).

Alternative treatments for SRSE

Topiramate (300–1600 mg/day per nasogastric tube) has multiple mechanisms of action affecting multiple receptors and ion channels (Cuero and Varelas, 2015). Similar to phenytoin, it blocks sodium channels and, when used in combination with other AEDs, it is synergistic at sodium channel blockade, GABA potentiation, calcium channel inhibition, and AMPA/kainite receptor inhibition. This medication is administered via nasogastric tube and has been found to be effective in aborting RSE while tapering cIV AEDs at doses of 300–1600 mg/day (Towne et al., 2003).

Isoflurane, an inhalational anesthetic agent, was shown to be efficacious in stopping RSE when titrated to burst suppression patterns (Kofke et al., 1989; Mirsattari et al., 2004). Unfortunately, upon discontinuation, seizures frequently recurred. Side-effects included hypotension, atelectasis, and ileus. Desflurane is also used.

Immunomodulators, steroids (methylprednisolone 1 g/day IV for 5 days, followed by prednisone 1 mg/kg/day for 1 week), IV immunoglobulins (0.4 g/kg/1 day IV for 5 days), plasmapheresis (five sessions), and adrenocorticotropic hormone are recommended in SRSE on a case-by-case basis (Cuero and Varelas, 2015). These interventions have been reported in select cases to help control seizures in syndromes with underlying immune mechanisms, such as Rasmussen's encephalitis, limbic encephalitis, acute disseminated encephalomyelitis, and paraneoplastic disorders. These alternative agents have not been adequately studied (Prasad et al., 1996).

Lidocaine, when used as 1.5–2 mg/kg bolus over 2 minutes and maintenance dose 3–4 mg/kg/h, suppressed seizures in 75% of RSE with the first bolus (Pascual et al., 1992). It has a very short half-life and can be used safely in patients when respiratory depression or worsening mental status is undesirable (Pascual et al., 1992). It can be neurotoxic and epileptogenic at high doses; thus it has a narrow pharmacologic range. Very few data are available to support this medication in the treatment of SE and it is generally not used.

Hypothermia has been shown to suppress electrographic seizures experimentally, in several case reports and in a small case series (Corry et al., 2008). It is recommended to cool to 33–35°C for 24–48 hours and rewarm by 0.1–0.2°C/h, but the data are sparse (Cuero and Varelas, 2015). Hypothermia may emerge as a promising adjunct to anesthetic agents due to its neuroprotective effects, but side-effects include shivering, coagulopathy, venous thromboembolism, immunosuppression, and electrolyte abnormalities. These may exacerbate complications seen with anesthetic agents.

Other medications used include gabapentin (in acute intermittent porphyria), magnesium 4 grams bolus IV and 2–6 g/h infusion (likely only helpful in eclampsia; keep serum levels <6 mEq/L), pyridoxine 100–600 mg/day IV or via nasogastric tube (Cuero and Varelas, 2015). Other nonpharmacologic therapies include the ketogenic diet, neurosurgical resection of an epileptogenic focus, if any, electroconvulsive therapy, vagal nerve stimulation, deep-brain stimulation, and transcranial magnetic stimulation (Brophy et al., 2012; Cuero and Varelas, 2015).

SE causes multiple systemic physiologic changes and requires ICU-level care. Systemic complications can be from ongoing seizure activity or can result from the AEDs themselves. During the first 30 minutes, there is a sympathetic overdrive, with increased catecholamines altering homeostatic mechanisms (Simon, 1985). Ten minutes after the offset of a single generalized seizure, norepinephrine concentrations were elevated to 12 times normal and epinephrine concentrations to 40 times normal (Simon, 1985). Life-threatening arrhythmias and stress cardiomyopathy are likely secondary to this sympathetic surge. On cardiac pathology, contraction band necrosis is seen; this may be the pathophysiologic correlate in SE-related death (Manno et al., 2005). When assessing electrocardiograms, 58.3% of patients had abnormalities, the most frequent being ischemic changes (Boggs et al., 1993). Patients also have an early increase in systemic blood pressure, with later normalization. The pulmonary vascular bed reacts differently during SE. Pulmonary arterial pressures continue to increase with every seizure, resulting in pulmonary edema (Simon, 1985). Patients with SE typically are not able to protect their airway and may develop hypoxia and aspiration. These patients are at risk of infection and may have infection as an underlying cause. During all of these systemic changes, fever can be from sustained muscle activity; leukocytosis can be from demargination; metabolic acidosis (secondary to excess anaerobic activity) and an elevated lactate may also be seen, but infection should always be ruled out first. SE can also result in renal failure from increased creatine kinase levels causing rhabdomyolysis. Lastly, electrolyte abnormalities such as hyperkalemia from muscle necrosis and metabolic acidosis and hyperglycemia from sympathetic overdrive can be seen. These complications can aggravate the underlying etiology, may cause further seizures, and complicate treatment.

Myoclonic status epilepticus

Myoclonic status epilepticus (MSE) is defined as a prolonged period of myoclonic jerks that are continuous or clustered, usually greater than 30 minutes, correlated with generalized spike wave or polyspike and wave on cEEG (Fig. 28.3). This is most frequently seen after

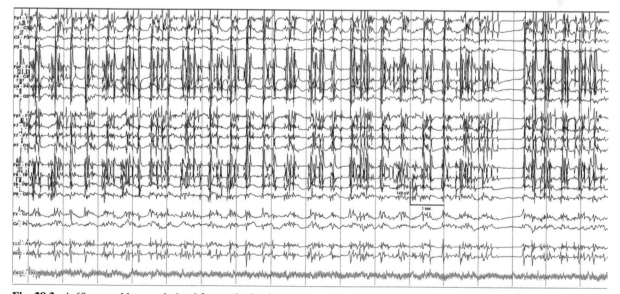

Fig. 28.3. A 69-year-old man admitted for septic shock complicated by hypoxic-ischemic injury and right thalamic stroke who exhibited multiple myoclonic seizures lasting 20 minutes or more. At times these seizures originated in the right hemisphere with secondary spread and were clinically associated with left facial jerking and chewing movements. Seizures were controlled with levetiracetam and a midazolam drip, resulting in suppression burst and later complete background suppression of the electroencephalogram. The patient ultimately died of medical complications.

cardiac arrest, associated with hypoxia, and associated with poor outcome if seen early after cardiac arrest (Wijdicks et al., 2006). The prognostic significance of MSE alone after hypothermia for cardiac arrest is less well established as recoveries have been reported but post-hypoxic SE still remains a poor predictor of outcomes in this population. (Bernard et al., 2002; Rossetti et al., 2010).

Seizure prophylaxis in the critically ill

After a patient has had a single seizure in the ICU, treatment to prevent recurrent seizures is needed as these patients are at increased risk of refractory seizures and SE. If a metabolic or physiologic abnormality can be resolved definitively, prophylactic treatment may not always be necessary. However, if a seizure is secondary to a structural lesion in the CNS, seizure recurrence is higher and prophylaxis is commonly given. Hemorrhagic stroke is associated with a twofold increase in the risk for seizure compared to ischemic infarction (Arboix et al., 1997). Currently, ICH guidelines do not recommend AED prophylaxis for all patients (Hemphill et al., 2015). For those ICH patients who have a depressed mental status out of proportion to the degree of brain injury, cEEG monitoring is indicated to rule out NCSE (Hemphill et al., 2015). In patients with aneurysmal SAH, the current guidelines from the American Stroke Association state that the use of prophylactic anticonvulsants may be considered in the immediate posthemorrhagic period because seizures early on could cause more damage or rebleeding from an unsecured aneurysm (Connolly et al., 2012). In patients with moderate to severe closed-head injuries, many clinicians will give prophylaxis, especially for those who suffer substantial cortical injury. In a double-blind study, phenytoin was found to be beneficial in reducing seizures during the first week after severe head injury, but continuation of phenytoin beyond 1 week did not benefit patients (Temkin et al., 1990). Thus, the American Academy of Neurology recommends giving prophylactic phenytoin during the first 7 days after traumatic head injury (Chang et al., 2003; Brain Trauma Foundation et al., 2007). Many institutions have now switched to LEV.

CLINICAL TRIALS AND GUIDELINES

Several landmark trials have brought about important changes in the treatment of SE, including Leppik et al. (1983), Treiman et al. (1998), Alldredge et al. (2001), Silbergleit et al. (2012), and Navarro et al.(2016). However, overall there is a paucity of data to base many of the management decisions on when treating SE. There are no data to recommend choices for second-line treatment, how to best treat RSE or SRSE, and no or few data for most of the newer AEDs. There are currently several trials examining new agents and comparing conventional agents. A neurosteroid, allopregnanolone, has been proposed as a new treatment for SE. In animal models, it was found to terminate seizure activity after benzodiazepines had failed (Rogawski et al., 2013). It acts as a positive allosteric modulator, potentiating synaptic $GABA_A$ receptors and enhancing extrasynaptic $GABA_A$ receptors which are believed to mediate tonic inhibition (Rogawski et al., 2013). In 2 pediatric patients, allopregnanolone allowed the weaning of cIV AEDs with resolution of SE and there continue to be clinical trials studying this new potential treatment (Broomall et al., 2014). Another neurosteroid, SGE-102, is being studied for the treatment of RSE while being administered with midazolam (Reddy et al., 2013). Finally, as other AEDs emerge as possible urgent control therapies after benzodiazepines have failed, there is going to be a randomized controlled trial over 50 centers internationally to study the efficacy between phenytoin, LEV, and valproic acid (Cock and ESETT Group, 2011). The Neurocritical Care Society published guidelines in 2012 on how to treat SE and discussed the level of evidence behind each treatment (Brophy et al., 2012). Similarly, the European Federation of Neurological Societies published guidelines for the management of SE (Meierkord et al., 2010). Recommendations on EEG monitoring for patients with or without primary brain injury in the ICU have been proposed by the European Society of Intensive Care Medicine (Claassen et al., 2013b).

COMPLEX CLINICAL DECISIONS

Management of patients with SE is highly complex and challenging, especially if there is no response to initial treatments. One must consider the underlying etiology, the sequelae of ongoing seizure activity, and the side-effects of interventions in order to adequately control seizures. The large number of possible interventions and the paucity of hard evidence to base choices on further complicate the management of the patient with SE. Even with guidelines, there are many decisions that must be made on a case-by-case basis, with the ultimate goal of terminating seizures as quickly as possible. One of the biggest challenges is that seizures are difficult to diagnose as most do not have overt clinical signs and EEG findings may be ambiguous. Recorded EEG patterns may not be clearly seizures but concurrently are not normal and highly epileptiform. These are termed the ictal–interictal continuum, for which treatment is even more controversial than for definitive NCSz or NCSE.

It is also difficult to know how aggressive to be with treatment regarding what electrographic target to treat to, seizure suppression versus burst suppression or complete background suppression, and when to decide if further medication is futile. The Neurocritical Care Society guidelines state it is appropriate to continue prolonged therapy for young patients with little comorbidity, self-limited disease processes, and the absence of poor prognostic intracranial lesions (Brophy et al., 2012).

OUTCOME PREDICTION

SE is a disease that carries a high morbidity and mortality rate. The neurotoxicity of prolonged or repetitive seizures will cause neuronal death and is associated with a higher mortality and worse clinical outcomes (Claassen et al., 2001, 2012). For SE, at hospital discharge, mortality was 9–21% (Alldredge et al., 2001; Claassen et al., 2002b; Novy et al., 2010). The 30-day mortality ranged between 19 and 27% (Logroscino et al., 1997; Treiman et al., 1998). Intermittent SE had a lower mortality rate when compared to continuous SE (20 vs. 31%) (Waterhouse et al., 1999). Etiology was found to be the most significant predictor of mortality; others included older age, medical comorbidities, high initial APACHE score, and acute symptomatic seizures (Prasad et al., 2001; Claassen et al., 2002b; Koubeissi and Alshekhlee, 2007). Prolonged hospitalization and acute symptomatic seizures were predictors of functional disability (Claassen et al., 2002b). In patients with stroke, aneurysmal SAH, CNS infections, traumatic brain injury, and intracranial hemorrhage, NCSE was independently associated with poor outcome (Geocadin et al., 2002; Claassen et al., 2007). A standardized score to comprehensively assess factors found to be predictive of poor outcome in patients with SE has been proposed and is called the Status Epilepticus Severity Score (Rossetti et al., 2008). Factors included in this score are age, history of seizures, seizure type, and the extent of consciousness impairment. Importantly, for adults with SE following SAH, every hour of seizure on cEEG was associated with worse functional and cognitive outcome at 3 months (De Marchis et al., 2016).

Patients who had NCSz were found to have long-term hippocampal atrophy on MRI (Vespa et al., 2010). After acute brain injury such as SAH, increased metabolism and blood flow have been shown with the onset of NCSz. Interestingly, brain oxygen may drop, intracranial pressure may rise rapidly, but regional cerebral blood flow may increase only minutes and not seconds after the onset of the seizure (Ko et al., 2011; Claassen et al., 2013a). One possible explanation for these observations may be that seizures cause more damage in acutely brain-injured patients than in those with epilepsy, damaging intrinsic defense mechanisms such as vasoreactivity.

RSE has a mortality rate between 23 and 61% (Young et al., 1980; Rashkin et al., 1987; Stecker et al., 1998; Krishnamurthy and Drislane, 1999; Claassen et al., 2001, 2002a; Rossetti et al., 2005; Novy et al., 2010) and this poor outcome is independent of what treatments are chosen (Rossetti et al., 2005). RSE was associated with more medical complications, longer neurologic ICU and hospital stays, and increased disability at discharge (Mayer et al., 2002). SRSE also has been seen to have a very high mortality rate of near 50% (Tian et al., 2015). A population-based study, which followed patients for 10 years, found that in approximately one-third of patients SE would recur (Hesdorffer et al., 2007).

NEUROREHABILITATION

For patients who survive SE, depending on their etiology and other predictors of poor cognitive and functional outcome, neurorehabilitation has a prominent role in restoring patients' functional outcome, quality of life, and cognitive function. Particularly patients with RSE or SRSE have prolonged mechanical ventilation and prolonged ICU stays which are associated with critical illness myopathy, critical illness polyneuropathy, and ICU-acquired weakness, all of which require intensive neurorehabilitative efforts (Dangayach et al., 2016).

REFERENCES

Abou Khaled KJ, Hirsch LJ (2006). Advances in the management of seizures and status epilepticus in critically ill patients. Crit Care Clin 22: 637–659. abstract viii.

Albers JM, Moddel G, Dittrich R et al. (2011). Intravenous lacosamide–an effective add-on treatment of refractory status epilepticus. Seizure 20: 428–430.

Alldredge BK, Gelb AM, Isaacs SM et al. (2001). A comparison of lorazepam, diazepam, and placebo for the treatment of out-of-hospital status epilepticus. N Engl J Med 345: 631–637.

Arboix A, Garcia-Eroles L, Massons JB et al. (1997). Predictive factors of early seizures after acute cerebrovascular disease. Stroke 28: 1590–1594.

Bernard SA, Gray TW, Buist MD et al. (2002). Treatment of comatose survivors of out-of-hospital cardiac arrest with induced hypothermia. N Engl J Med 346: 557–563.

Beyreuther BK, Freitag J, Heers C et al. (2007). Lacosamide: a review of preclinical properties. CNS Drug Rev 13: 21–42.

Bleck TP (1991). Convulsive disorders: status epilepticus. Clin Neuropharmacol 14: 191–198.

Boggs JG, Painter JA, Delorenzo RJ (1993). Analysis of electrocardiographic changes in status epilepticus. Epilepsy Res 14: 87–94.

Brain Trauma Foundation, American Association of Neurological Surgeons, Congress of Neurological Surgeons, Joint Section on Neurotrauma and Critical

Care Bratton SL, Chestnut RM, Ghajar J, et al. (2007). Guidelines for the management of severe traumatic brain injury. XIII. Antiseizure prophylaxis. J Neurotrauma 24 (Suppl 1): S83–S86.

Broomall E, Natale JE, Grimason M et al. (2014). Pediatric super-refractory status epilepticus treated with allopregnanolone. Ann Neurol 76: 911–915.

Brophy GM, Bell R, Claassen J et al. (2012). Guidelines for the evaluation and management of status epilepticus. Neurocrit Care 17: 3–23.

Chang BS, Lowenstein DH, Quality standards subcommittee of the american academy of Neurology (2003). Practice parameter: antiepileptic drug prophylaxis in severe traumatic brain injury: report of the Quality Standards Subcommittee of the American Academy of Neurology. Neurology 60: 10–16.

Chen JW, Wasterlain CG (2006). Status epilepticus: pathophysiology and management in adults. Lancet Neurol 5: 246–256.

Chin RF, Neville BG, Peckham C et al. (2006). Incidence, cause, and short-term outcome of convulsive status epilepticus in childhood: prospective population-based study. Lancet 368: 222–229.

Claassen J (2009). How I treat patients with EEG patterns on the ictal–interictal continuum in the neuro ICU. Neurocrit Care 11: 437–444.

Claassen J (2016). Dr No: double drug fails to eliminate status epilepticus. Lancet Neurol 15: 23–24.

Claassen J, Hirsch LJ, Emerson RG et al. (2001). Continuous EEG monitoring and midazolam infusion for refractory nonconvulsive status epilepticus. Neurology 57: 1036–1042.

Claassen J, Hirsch LJ, Emerson RG et al. (2002a). Treatment of refractory status epilepticus with pentobarbital, propofol, or midazolam: a systematic review. Epilepsia 43: 146–153.

Claassen J, Lokin JK, Fitzsimmons BF et al. (2002b). Predictors of functional disability and mortality after status epilepticus. Neurology 58: 139–142.

Claassen J, Hirsch LJ, Mayer SA (2003). Treatment of status epilepticus: a survey of neurologists. J Neurol Sci 211: 37–41.

Claassen J, Mayer SA, Kowalski RG et al. (2004). Detection of electrographic seizures with continuous EEG monitoring in critically ill patients. Neurology 62: 1743–1748.

Claassen J, Jette N, Chum F et al. (2007). Electrographic seizures and periodic discharges after intracerebral hemorrhage. Neurology 69: 1356–1365.

Claassen J, Silbergleit R, Weingart SD et al. (2012). Emergency neurological life support: status epilepticus. Neurocrit Care 17 (Suppl 1): S73–S78.

Claassen J, Perotte A, Albers D et al. (2013a). Nonconvulsive seizures after subarachnoid hemorrhage: Multimodal detection and outcomes. Ann Neurol 74: 53–64.

Claassen J, Taccone FS, Horn P et al. (2013b). Recommendations on the use of EEG monitoring in critically ill patients: consensus statement from the neurointensive care section of the ESICM. Intensive Care Med 39: 1337–1351.

Claassen J, Riviello Jr JJ, Silbergleit R (2015). Emergency neurological life support: status epilepticus. Neurocrit Care 23 (Suppl 2): 136–142.

Cock HR, ESETT Group (2011). Established status epilepticus treatment trial (ESETT). Epilepsia 52 (Suppl 8): 50–52.

Cock HR, Tong X, Hargreaves IP et al. (2002). Mitochondrial dysfunction associated with neuronal death following status epilepticus in rat. Epilepsy Res 48: 157–168.

Conn RD, Kennedy JW, Blackmon JR (1967). The hemodynamic effects of diphenylhydantoin. Am Heart J 73: 500–505.

Connolly Jr ES, Rabinstein AA, Carhuapoma JR et al. (2012). Guidelines for the management of aneurysmal subarachnoid hemorrhage: a guideline for healthcare professionals from the American Heart Association/American Stroke Association. Stroke 43: 1711–1737.

Corry JJ, Dhar R, Murphy T et al. (2008). Hypothermia for refractory status epilepticus. Neurocrit Care 9: 189–197.

Cranford RE, Leppik IE, Patrick B et al. (1978). Intravenous phenytoin: clinical and pharmacokinetic aspects. Neurology 28: 874–880.

Cuero MR, Varelas PN (2015). Super-refractory status epilepticus. Curr Neurol Neurosci Rep 15: 74.

Dangayach NS, Smith M, Claassen J (2016). Electromyography and nerve conduction studies in critical care: step by step in the right direction. Intensive Care Med 42: 1168–1171.

De Marchis GM, Pugin D, Meyers E et al. (2016). Seizure burden in subarachnoid hemorrhage associated with functional and cognitive outcome. Neurology 86: 253–260.

DeLorenzo RJ, Hauser WA, Towne AR et al. (1996). A prospective, population-based epidemiologic study of status epilepticus in Richmond, Virginia. Neurology 46: 1029–1035.

DeLorenzo RJ, Waterhouse EJ, Towne AR et al. (1998). Persistent nonconvulsive status epilepticus after the control of convulsive status epilepticus. Epilepsia 39: 833–840.

DeLorenzo RJ, Garnett LK, Towne AR et al. (1999). Comparison of status epilepticus with prolonged seizure episodes lasting from 10 to 29 minutes. Epilepsia 40: 164–169.

Dennis LJ, Claassen J, Hirsch LJ et al. (2002). Nonconvulsive status epilepticus after subarachnoid hemorrhage. Neurosurgery 51: 1136–1143. discussion 1144.

Devlin EG, Clarke RS, Mirakhur RK et al. (1994). Effect of four i.v. induction agents on T-lymphocyte proliferations to PHA in vitro. Br J Anaesth 73: 315–317.

Epilepsy Foundation of America's Working Group on Status Epilepticus (1993). Treatment of convulsive status epilepticus. Recommendations of the Epilepsy Foundation of America's Working Group on Status Epilepticus. JAMA 270: 854–859.

Errington AC, Coyne L, Stohr T et al. (2006). Seeking a mechanism of action for the novel anticonvulsant lacosamide. Neuropharmacology 50: 1016–1029.

Farhidvash F, Singh P, Abou-Khalil B et al. (2009). Patients visiting the emergency room for seizures: insurance status and clinic follow-up. Seizure 18: 644–647.

Farina L, Bergqvist C, Zimmerman RA et al. (2004). Acute diffusion abnormalities in the hippocampus of children with new-onset seizures: the development of mesial temporal sclerosis. Neuroradiology 46: 251–257.

Fernandez A, Lantigua H, Lesch C et al. (2014). High-dose midazolam infusion for refractory status epilepticus. Neurology 82: 359–365.

Fujikawa DG (1996). The temporal evolution of neuronal damage from pilocarpine-induced status epilepticus. Brain Res 725: 11–22.

Gaspard N, Foreman B, Judd LM et al. (2013). Intravenous ketamine for the treatment of refractory status epilepticus: a retrospective multicenter study. Epilepsia 54: 1498–1503.

Geocadin RG, Sherman DL, Christian Hansen H et al. (2002). Neurological recovery by EEG bursting after resuscitation from cardiac arrest in rats. Resuscitation 55: 193–200.

Gottwald MD, Akers LC, Liu PK et al. (1999). Prehospital stability of diazepam and lorazepam. Am J Emerg Med 17: 333–337.

Halasz P, Kalviainen R, Mazurkiewicz-Beldzinska M et al. (2009). Adjunctive lacosamide for partial-onset seizures: efficacy and safety results from a randomized controlled trial. Epilepsia 50: 443–453.

Hauser WA (1990). Status epilepticus: epidemiologic considerations. Neurology 40: 9–13.

Hemphill 3rd JC, Greenberg SM, Anderson CS et al. (2015). Guidelines for the management of spontaneous intracerebral hemorrhage: a guideline for healthcare professionals from the American Heart Association/American Stroke Association. Stroke 46: 2032–2060.

Hesdorffer DC, Logroscino G, Cascino GD et al. (2007). Recurrence of afebrile status epilepticus in a population-based study in Rochester, Minnesota. Neurology 69: 73–78.

Hocker S, Tatum WO, Laroche S et al. (2014). Refractory and super-refractory status epilepticus – an update. Curr Neurol Neurosci Rep 14: 452.

Hsieh CY, Sung PS, Tsai JJ et al. (2010). Terminating prolonged refractory status epilepticus using ketamine. Clin Neuropharmacol 33: 165–167.

Husain AM, Horn GJ, Jacobson MP (2003). Non-convulsive status epilepticus: usefulness of clinical features in selecting patients for urgent EEG. J Neurol Neurosurg Psychiatry 74: 189–191.

Jagoda A, Riggio S (1993). Refractory status epilepticus in adults. Ann Emerg Med 22: 1337–1348.

Jenssen S, Gracely EJ, Sperling MR (2006). How long do most seizures last? A systematic comparison of seizures recorded in the epilepsy monitoring unit. Epilepsia 47: 1499–1503.

Jirsch J, Hirsch LJ (2007). Nonconvulsive seizures: developing a rational approach to the diagnosis and management in the critically ill population. Clin Neurophysiol 118: 1660–1670.

Kapur J, Macdonald RL (1997). Rapid seizure-induced reduction of benzodiazepine and Zn^{2+} sensitivity of hippocampal dentate granule cell GABAA receptors. J Neurosci 17: 7532–7540.

Kim JA, Chung JI, Yoon PH et al. (2001). Transient MR signal changes in patients with generalized tonicoclonic seizure or status epilepticus: periictal diffusion-weighted imaging. AJNR Am J Neuroradiol 22: 1149–1160.

Knake S, Gruener J, Hattemer K et al. (2008). Intravenous levetiracetam in the treatment of benzodiazepine refractory status epilepticus. J Neurol Neurosurg Psychiatry 79: 588–589.

Knake S, Hamer HM, Rosenow F (2009). Status epilepticus: a critical review. Epilepsy Behav 15: 10–14.

Ko SB, Ortega-Gutierrez S, Choi HA et al. (2011). Status epilepticus-induced hyperemia and brain tissue hypoxia after cardiac arrest. Arch Neurol 68: 1323–1326.

Kofke WA, Young RS, Davis P et al. (1989). Isoflurane for refractory status epilepticus: a clinical series. Anesthesiology 71: 653–659.

Koubeissi M, Alshekhlee A (2007). In-hospital mortality of generalized convulsive status epilepticus: a large US sample. Neurology 69: 886–893.

Krishnamurthy KB, Drislane FW (1999). Depth of EEG suppression and outcome in barbiturate anesthetic treatment for refractory status epilepticus. Epilepsia 40: 759–762.

Kumar MA, Urrutia VC, Thomas CE et al. (2005). The syndrome of irreversible acidosis after prolonged propofol infusion. Neurocrit Care 3: 257–259.

Kurtz P, Gaspard N, Wahl AS et al. (2014). Continuous electroencephalography in a surgical intensive care unit. Intensive Care Med 40: 228–234.

Lai ML, Huang JD (1993). Dual effect of valproic acid on the pharmacokinetics of phenytoin. Biopharm Drug Dispos 14: 365–370.

Leppik IE, Derivan AT, Homan RW et al. (1983). Double-blind study of lorazepam and diazepam in status epilepticus. JAMA 249: 1452–1454.

Liu H, Mazarati AM, Katsumori H et al. (1999). Substance P is expressed in hippocampal principal neurons during status epilepticus and plays a critical role in the maintenance of status epilepticus. Proc Natl Acad Sci U S A 96: 5286–5291.

Logroscino G, Hesdorffer DC, Cascino G et al. (1997). Short-term mortality after a first episode of status epilepticus. Epilepsia 38: 1344–1349.

Loscher W, Honack D, Rundfeldt C (1998). Antiepileptogenic effects of the novel anticonvulsant levetiracetam (ucb L059) in the kindling model of temporal lobe epilepsy. J Pharmacol Exp Ther 284: 474–479.

Louis S, Kutt H, McDowell F (1967). The cardiocirculatory changes caused by intravenous Dilantin and its solvent. Am Heart J 74: 523–529.

Lowenstein DH, Alldredge BK (1993). Status epilepticus at an urban public hospital in the 1980s. Neurology 43: 483–488.

Lowenstein DH, Alldredge BK (1998). Status epilepticus. N Engl J Med 338: 970–976.

Lowenstein DH, Bleck T, Macdonald RL (1999). It's time to revise the definition of status epilepticus. Epilepsia 40: 120–122.

Lynch BA, Lambeng N, Nocka K et al. (2004). The synaptic vesicle protein SV2A is the binding site for the antiepileptic drug levetiracetam. Proc Natl Acad Sci U S A 101: 9861–9866.

Manno EM, Pfeifer EA, Cascino GD et al. (2005). Cardiac pathology in status epilepticus. Ann Neurol 58: 954–957.

Martin BS, Kapur J (2008). A combination of ketamine and diazepam synergistically controls refractory status epilepticus induced by cholinergic stimulation. Epilepsia 49: 248–255.

Mayer SA, Claassen J, Lokin J et al. (2002). Refractory status epilepticus: frequency, risk factors, and impact on outcome. Arch Neurol 59: 205–210.

Mazarati AM, Wasterlain CG (1997). Loss of hippocampal inhibition and enhanced LTP after self-sustaining status epilepticus. Epilepsia 38: 178.

Mazarati AM, Wasterlain CG (1999). N-methyl-D-aspartate receptor antagonists abolish the maintenance phase of self-sustaining status epilepticus in rat. J Neurosci 256: 187–190.

Mazarati AM, Baldwin RA, Sankar R et al. (1998a). Time-dependent decrease in the effectiveness of antiepileptic drugs during the course of self-sustaining status epilepticus. Brain Res 814: 179–185.

Mazarati AM, Liu H, Soomets U et al. (1998b). Galanin modulation of seizures and seizure modulation of hippocampal galanin in animal models of status epilepticus. J Neurosci 18: 10070–10077.

Meierkord H, Boon P, Engelsen B et al. (2010). EFNS guideline on the management of status epilepticus in adults. Eur J Neurol 17: 348–355.

Meldrum BS (1999). The revised operational definition of generalised tonic-clonic (TC) status epilepticus in adults. Epilepsia 40: 123–124.

Mewasingh LD, Sekhara T, Aeby A et al. (2003). Oral ketamine in paediatric non-convulsive status epilepticus. Seizure 12: 483–489.

Mirsattari SM, Sharpe MD, Young GB (2004). Treatment of refractory status epilepticus with inhalational anesthetic agents isoflurane and desflurane. Arch Neurol 61: 1254–1259.

Mirski MA, Varelas PN (2008). Seizures and status epilepticus in the critically ill. Crit Care Clin 24: 115–147. ix.

Misra UK, Kalita J, Patel R (2006). Sodium valproate vs phenytoin in status epilepticus: a pilot study. Neurology 67: 340–342.

Moddel G, Bunten S, Dobis C et al. (2009). Intravenous levetiracetam: a new treatment alternative for refractory status epilepticus. J Neurol Neurosurg Psychiatry 80: 689–692.

Navarro V, Dagron C, Elie C et al. (2016). Levetiracetam and clonazepam in status epilepticus: a prehospital double-blind randomised trial. Lancet Neurol 15: 47–55.

Naylor DE, Liu H, Wasterlain CG (2005). Trafficking of GABA(A) receptors, loss of inhibition, and a mechanism for pharmacoresistance in status epilepticus. J Neurosci 25: 7724–7733.

Niespodziany I, Klitgaard H, Margineanu DG (2003). Desynchronizing effect of levetiracetam on epileptiform responses in rat hippocampal slices. Neuroreport 14: 1273–1276.

Novy J, Logroscino G, Rossetti AO (2010). Refractory status epilepticus: a prospective observational study. Epilepsia 51: 251–256.

Oddo M, Carrera E, Claassen J et al. (2009). Continuous electroencephalography in the medical intensive care unit. Crit Care Med 37: 2051–2056.

Osorio I, Reed RC (1989). Treatment of refractory generalized tonic-clonic status epilepticus with pentobarbital anesthesia after high-dose phenytoin. Epilepsia 30: 464–471.

Pallin DJ, Goldstein JN, Moussally JS et al. (2008). Seizure visits in US emergency departments: epidemiology and potential disparities in care. Int J Emerg Med 1: 97–105.

Pascual J, Ciudad J, Berciano J (1992). Role of lidocaine (lignocaine) in managing status epilepticus. J Neurol Neurosurg Psychiatry 55: 49–51.

Perucca E, Hebdige S, Frigo GM et al. (1980). Interaction between phenytoin and valproic acid: plasma protein binding and metabolic effects. Clin Pharmacol Ther 28: 779–789.

Prasad AN, Stafstrom CF, Holmes GL (1996). Alternative epilepsy therapies: the ketogenic diet, immunoglobulins, and steroids. Epilepsia 37 (Suppl 1): S81–S95.

Prasad A, Worrall BB, Bertram EH et al. (2001). Propofol and midazolam in the treatment of refractory status epilepticus. Epilepsia 42: 380–386.

Pugin D, Foreman B, De Marchis GM et al. (2014). Is pentobarbital safe and efficacious in the treatment of super-refractory status epilepticus: a cohort study. Crit Care 18: R103.

Rabinowicz AL, Correale JD, Bracht KA et al. (1995). Neuron-specific enolase is increased after nonconvulsive status epilepticus. Epilepsia 36: 475–479.

Rashkin MC, Youngs C, Penovich P (1987). Pentobarbital treatment of refractory status epilepticus. Neurology 37: 500–503.

Reddy K, Reife R, Cole AJ (2013). SGE-102: a novel therapy for refractory status epilepticus. Epilepsia 54 (Suppl 6): 81–83.

Riviello Jr JJ, Claassen J, Laroche SM et al. (2013). Treatment of status epilepticus: an international survey of experts. Neurocrit Care 18: 193–200.

Rogawski MA, Loya CM, Reddy K et al. (2013). Neuroactive steroids for the treatment of status epilepticus. Epilepsia 54 (Suppl 6): 93–98.

Rossetti AO, Reichhart MD, Schaller MD et al. (2004). Propofol treatment of refractory status epilepticus: a study of 31 episodes. Epilepsia 45: 757–763.

Rossetti AO, Logroscino G, Bromfield EB (2005). Refractory status epilepticus: effect of treatment aggressiveness on prognosis. Arch Neurol 62: 1698–1702.

Rossetti AO, Logroscino G, Milligan TA et al. (2008). Status Epilepticus Severity Score (STESS): a tool to orient early treatment strategy. J Neurol 255: 1561–1566.

Rossetti AO, Oddo M, Logroscino G et al. (2010). Prognostication after cardiac arrest and hypothermia: a prospective study. Ann Neurol 67: 301–307.

Rossetti AO, Milligan TA, Vulliemoz S et al. (2011). A randomized trial for the treatment of refractory status epilepticus. Neurocrit Care 14: 4–10.

Shinnar S, Berg AT, Moshe SL et al. (2001). How long do new-onset seizures in children last? Ann Neurol 49: 659–664.

Shorvon S (2007). What is nonconvulsive status epilepticus, and what are its subtypes? Epilepsia 48 (Suppl 8): 35–38.

Silbergleit R, Durkalski V, Lowenstein D et al. (2012). Intramuscular versus intravenous therapy for prehospital status epilepticus. N Engl J Med 366: 591–600.

Simon RP (1985). Physiologic consequences of status epilepticus. Epilepsia 26 (Suppl 1): S58–S66.

Singh BM, Strobos RJ (1980). Epilepsia partialis continua associated with nonketotic hyperglycemia: clinical and biochemical profile of 21 patients. Ann Neurol 8: 155–160.

Sperk G, Wieser R, Widmann R et al. (1986). Kainic acid induced seizures: changes in somatostatin, substance P and neurotensin. Neuroscience 17: 1117–1126.

Stecker MM, Kramer TH, Raps EC et al. (1998). Treatment of refractory status epilepticus with propofol: clinical and pharmacokinetic findings. Epilepsia 39: 18–26.

Sutter R, Marsch S, Fuhr P et al. (2014). Anesthetic drugs in status epilepticus: risk or rescue? A 6-year cohort study. Neurology 82: 656–664.

Szabo K, Poepel A, Pohlmann-Eden B et al. (2005). Diffusion-weighted and perfusion MRI demonstrates parenchymal changes in complex partial status epilepticus. Brain 128: 1369–1376.

Temkin NR, Dikmen SS, Wilensky AJ et al. (1990). A randomized, double-blind study of phenytoin for the prevention of post-traumatic seizures. N Engl J Med 323: 497–502.

Theodore WH, Porter RJ, Albert P et al. (1994). The secondarily generalized tonic-clonic seizure: a videotape analysis. Neurology 44: 1403–1407.

Tian L, Li Y, Xue X et al. (2015). Super-refractory status epilepticus in West China. Acta Neurol Scand 132: 1–6.

Tomson T, Lindbom U, Nilsson BY (1992). Nonconvulsive status epilepticus in adults: thirty-two consecutive patients from a general hospital population. Epilepsia 33: 829–835.

Towne AR, Pellock JM, Ko D et al. (1994). Determinants of mortality in status epilepticus. Epilepsia 35: 27–34.

Towne AR, Waterhouse EJ, Boggs JG et al. (2000). Prevalence of nonconvulsive status epilepticus in comatose patients. Neurology 54: 340–345.

Towne AR, Garnett LK, Waterhouse EJ et al. (2003). The use of topiramate in refractory status epilepticus. Neurology 60: 332–334.

Treiman DM, Walton NY, Collins JF et al. (1990). Treatment of status epilepticus if first drug fails [abstract]. Epilepsia 40: 243.

Treiman DM, Meyers PD, Walton NY et al. (1998). A comparison of four treatments for generalized convulsive status epilepticus. Veterans Affairs Status Epilepticus Cooperative Study Group. N Engl J Med 339: 792–798.

Trinka E, Leitinger M (2015). Which EEG patterns in coma are nonconvulsive status epilepticus? Epilepsy Behav 49: 203–222.

Varelas PN, Spanaki MV, Hacein-Bey L et al. (2003). Emergent EEG: indications and diagnostic yield. Neurology 61: 702–704.

Vespa PM, Nuwer MR, Nenov V et al. (1999). Increased incidence and impact of nonconvulsive and convulsive seizures after traumatic brain injury as detected by continuous electroencephalographic monitoring. J Neurosurg 91: 750–760.

Vespa PM, O'Phelan K, Shah M et al. (2003). Acute seizures after intracerebral hemorrhage: a factor in progressive midline shift and outcome. Neurology 60: 1441–1446.

Vespa PM, Mcarthur DL, Xu Y et al. (2010). Nonconvulsive seizures after traumatic brain injury are associated with hippocampal atrophy. Neurology 75: 792–798.

Vicedomini JP, Nadler JV (1987). A model of status epilepticus based on electrical stimulation of hippocampal afferent pathways. Exp Neurol 96: 681–691.

Wasterlain CG, Fujikawa DG, Penix L et al. (1993). Pathophysiological mechanisms of brain damage from status epilepticus. Epilepsia 34 (Suppl 1): S37–S53.

Waterhouse EJ, Garnett LK, Towne AR et al. (1999). Prospective population-based study of intermittent and continuous convulsive status epilepticus in Richmond, Virginia. Epilepsia 40: 752–758.

Wijdicks EF, Hijdra A, Young GB et al. (2006). Practice parameter: prediction of outcome in comatose survivors after cardiopulmonary resuscitation (an evidence-based review): report of the Quality Standards Subcommittee of the American Academy of Neurology. Neurology 67: 203–210.

Wilder BJ (1983). Efficacy of phenytoin in treatment of status epilepticus. Adv Neurol 34: 441–446.

Yaffe K, Lowenstein DH (1993). Prognostic factors of pentobarbital therapy for refractory generalized status epilepticus. Neurology 43: 895–900.

Yasiry Z, Shorvon SD (2014). The relative effectiveness of five antiepileptic drugs in treatment of benzodiazepine-resistant convulsive status epilepticus: a meta-analysis of published studies. Seizure 23: 167–174.

York RC, Coleridge ST (1988). Cardiopulmonary arrest following intravenous phenytoin loading. Am J Emerg Med 6: 255–259.

Young GB, Blume WT, BOLTON CF et al. (1980). Anesthetic barbiturates in refractory status epilepticus. Can J Neurol Sci 7: 291–292.

Young GB, Jordan KG, Doig GS (1996). An assessment of nonconvulsive seizures in the intensive care unit using continuous EEG monitoring: an investigation of variables associated with mortality. Neurology 47: 83–89.

Zeiler FA, Teitelbaum J, Gillman LM et al. (2014). NMDA antagonists for refractory seizures. Neurocrit Care 20: 502–513.

Chapter 29

Intensive care unit-acquired weakness

J. HORN[1]* AND G. HERMANS[2]

[1]*Department of Intensive Care, Academic Medical Center, Amsterdam, The Netherlands*

[2]*Department of General Internal Medicine, UZ Leuven, Leuven, Belgium*

Abstract

When critically ill, a severe weakness of the limbs and respiratory muscles often develops with a prolonged stay in the intensive care unit (ICU), a condition vaguely termed intensive care unit-acquired weakness (ICUAW). Many of these patients have serious nerve and muscle injury. This syndrome is most often seen in surviving critically ill patients with sepsis or extensive inflammatory response which results in increased duration of mechanical ventilation and hospital length of stay. Patients with ICUAW often do not fully recover and the disability will seriously impact on their quality of life. In this chapter we discuss the current knowledge on the pathophysiology and risk factors of ICUAW. Tools to diagnose ICUAW, how to separate ICUAW from other disorders, and which possible treatment strategies can be employed are also described. ICUAW is finally receiving the attention it deserves and the expectation is that it can be better understood and prevented.

With an increasing survival rate in patients admitted to the intensive care unit (ICU), the long-term consequences of surviving critical illness have become more apparent. Surviving critical illness is often with injury to multiple organ systems, including in some the central and peripheral nervous system. The combination of physical and psychologic sequela seen in patients with a prolonged ICU admission has been called the post intensive care syndrome (PICS) (Needham et al., 2012). Generalized muscle weakness, caused by a combination of muscle and nerve injury, is one of the main issues in PICS. In this chapter we discuss the current knowledge on the pathophysiology and risk factors of ICU-acquired weakness (ICUAW). Tools to diagnose ICUAW, how to separate ICUAW from other disorders, and which possible treatment strategies can be employed are also described. Efforts to minimize this complication may impact on quality of life in ICU survivors. Intensivists and neurologists are often uncertain about what to do when these unfortunate patients are seen and effective neurorehabilitation needs to be developed.

EPIDEMIOLOGY

One of the first description of patients developing weakness during ICU admission was by Bolton and colleagues over 30 years ago (Bolton et al., 1984). Five relatively young patients admitted to the ICU for different reasons developed profound limb weakness and decreased tendon reflexes on physical examination. In 4 patients electrophysiologic studies showed an axonal polyneuropathy and myopathic changes with needle electromyography. Strength improved gradually in these patients (Bolton et al., 1984).

Over the years the syndrome of weakness in the ICU, then called critical illness polyneuropathy (CNP), was recognized and further details were given by many groups in many countries. The incidence reportedly ranged from 50 to 100% of ICU patients, indicating that it is a frequently occurring problem (De Jonghe et al., 2002; Bercker et al., 2005; Stevens et al., 2007; Ali et al., 2008; Sharshar et al., 2009; Mirzakhani et al., 2013). The wide range in incidence rates depends on the group of ICU

---

*Correspondence to: Janneke Horn, MD, Department of Intensive Care, Academic Medical Center, Meibergdreef 9, 1105 AZ Amsterdam, The Netherlands. Tel: +31-205662509, E-mail: j.horn@amc.uva.nl

patients studied and the criteria used to diagnose axonal polyneuropathy. For example, in critically ill patients with sepsis the incidence of generalized weakness is much higher than in patients with other illnesses (Bolton, 2005). Moreover, it became apparent that electrophysiologic abnormalities are more frequent than clinically detectable weakness.

Many different terms have been used in the last few decades to describe this neurologic syndrome or entity in the sickest ICU patients (Stevens et al., 2009). Terminology could be only descriptive and whether the cause of the weakness was primarily an injury to the nerve, muscle, or a combination. Terms used in the past were: CIP, acute quadriplegic myopathy, acute necrotizing myopathy of intensive care, and critical illness neuromyopathy. These differences in terminology seriously hindered studies on the topic. In 2009 a consensus meeting was held which introduced a more general term but also criteria for ICUAW (Stevens et al., 2009). ICUAW is most often seen in severely ill patients with multiorgan failure and ICUAW has been considered to be part of the multiorgan failure syndrome, involving nerves and muscles (Witt et al., 1991; De Letter et al., 2001; Garnacho-Montero et al., 2001; Bednařík et al., 2005; Nanas et al., 2008; Hermans et al., 2013). Nevertheless, severe weakness is also found sometimes in patients who have no other signs of multiorgan failure (Campellone et al., 1998). ICUAW is a clinical assessment, mostly based on strength examination in ICU patients, but more detailed neurologic examination (reflex pattern, tone, muscle bulk, and sensation) and electrophysiologic assessment are required to diagnose a peripheral nervous system or primary muscle disorder such as CIP or critical illness myopathy (CIM). Each of these disorders has definitional characteristics.

NEUROPATHOLOGY

Muscle weakness results from a complex interplay of factors that culminates in the loss of muscle mass and contractility. A summary of factors is shown in Figure 29.1. Muscle weakness may be brought about by structural damage to the nerves or muscles, but structural damage may not be easily demonstrated in all cases. Investigation of the pathophysiologic mechanisms leading to ICUAW in patients is hampered by several factors. Ideally, nerve and muscle tissue should be investigated at different time points of disease progression.

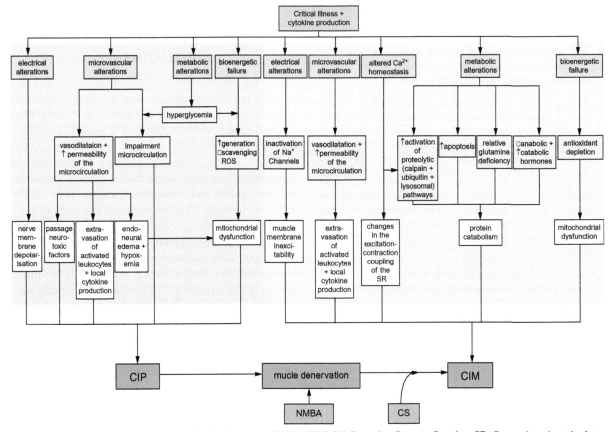

Fig. 29.1. Summary of factors involved in development of ICUAW. ROS, Reactive Oxygen Species; SR, Sarcoplasmic reticulum; CIP, Critical Illness Polyneuropathy; NMBA, NeuroMuscular Blocking Agent; CS, Corticosteroids; CIM, Critical Illness Myopathy. (Reproduced from Hermans et al., 2014a, with permission from John Wiley.)

Moreover, extensive biopsies may be problematic in patients with a coagulation disorder which is often present in critically ill patients and this procedure may be ill advised if it does not result in change in therapy (Latronico et al., 1996). When nerve biopsies are performed, they usually comprise pure sensory nerves such as the sural nerve or superficial peroneal nerve. This is suboptimal, as ICUAW is primarily a syndrome with problems in strength. Motor nerves have been rarely systematically investigated and often postmortem, thus skewing results to the more severe cases (Zochodne et al., 1987).

Some insights are derived from pathophysiologic mechanisms in ICUAW in animal experiments. Several sepsis models have been used to study ICUAW, but many of these models do not resemble the clinical situation (Witteveen et al., 2014). For example, the period of sepsis induced in these animals is usually short, animals are young, and strength assessments have not been systematically performed. Despite these limitations several factors have been identified that play a role in the development of CIP and CIM.

In CIP, axonal damage is found in the peripheral as well as in the phrenic nerves (Latronico et al., 1996). Sepsis induces microvascular changes in the endoneurium, as evidenced by enhanced expression of E-selectin on the vascular endothelium (Fenzi et al., 2003). This is considered to induce increased vascular permeability, allowing toxic factors to penetrate the nerve ends (Bolton, 2005). Additionally, increased permeability may induce endoneural edema, compromising energy delivery to the axon, and ultimately resulting in axonal injury. Mitochondrial dysfunction due to stress-induced hyperglycemia may further aggravate this process (Van den Berghe, 2004; Vanhorebeek et al., 2005; Hermans et al., 2007). Animal experiments also documented rapidly reversible nerve dysfunction (without actual structural damage) caused by Na channelopathy (Novak et al., 2009). So far, this mechanism has not been studied in critically ill patients and its clinical relevance remains to be seen (Batt et al., 2013b).

CIM has been investigated more extensively. Loss of muscle mass occurs early and, as expected, is more pronounced in multiple-organ failure as compared to single-organ failure (Puthucheary et al., 2013). Atrophy results from increased protein degradation not compensated by protein synthesis (Batt et al., 2013b). Protein wasting predominantly involves myosin, causing thick filament myopathy. A detailed overview of pathways governing catabolic signaling is outside the scope of this chapter but has been published (Friedrich et al., 2015). The main proteolytic system involved is the ubiquitin proteasome system. Key players initiating the catabolic process include (Derde et al., 2012;

Puthucheary, 2013; Wollersheim et al., 2014) inflammation, immobilization, the endocrine stress responses, the rapidly developing nutritional deficit, impaired microcirculation and denervation (Bloch et al., 2012; Batt et al., 2013b; Weber-Carstens et al., 2013). Possibly also deficient autophagy, a catabolic process involved in clearing large protein aggregates and cellular debris, contributes to poor muscle cell quality control in ICUAW (Hermans et al., 2013). In addition to structural changes muscles can be dysfunctional due to several other factors. These include muscle membrane inexcitability due to an acquired sodium channelopathy (Ackermann et al., 2014), mitochondrial dysfunction, and bioenergetic failure (Brealey et al., 2002), but also impaired excitation–contraction coupling (Rossignol et al., 2008; Zink et al., 2008).

RISK FACTORS FOR ICU WEAKNESS

Risk factors for neuromuscular complications in critically ill patients have been reported in several studies on ICU populations (Hermans and Van den Berghe, 2015). Bedrest in itself is known to have deleterious effect on musculoskeletal system and muscle mass disappears quickly (Convertino et al., 1997; Parry and Puthucheary, 2015). Muscle immobilization rapidly induces loss of muscle mass and strength. Though immobility itself is insufficient to explain the pronounced weakness in critically ill patients, it is a contributing factor and early mobilization can effectively reduce ICUAW (Patel and Pohlman, 2014). This is also in line with the observation that several surrogate markers of immobility, such as duration of ICU stay (Witt et al., 1991; Van den Berghe et al., 2005), mechanical ventilation prior to awakening (De Jonghe et al., 2002), time until awakening, and clinical evaluation of muscle strength (Hermans et al., 2013), were identified as predictors of ICUAW (Fan et al., 2014b).

Furthermore, sepsis, the systemic inflammatory response syndrome (SIRS), and multiple-organ failure are considered to be central players. This is based on the high incidence of sepsis in the earliest reports of patients with CIP (Bolton et al., 1984; Zochodne et al., 1987), as well as the high prevalence of neuromuscular complications in specific populations of patients with sepsis and organ failure (Witt et al., 1991; Garnacho-Montero et al., 2001; Bednarík et al., 2005). This observation is consistent with the identification of sepsis (Hermans et al., 2014a), bacteremia (Van den Berghe et al., 2001; Nanas et al., 2008), SIRS (De Letter et al., 2001; Bednarík et al., 2005), multiple-organ failure (De Jonghe et al., 2002), use of vasopressors (Van den Berghe et al., 2001, 2005) or aminoglycosides (Nanas et al., 2008), certain mediators of inflammation (Weber-Carstens et al., 2010), and the presence of septic encephalopathy (Garnacho-Montero et al., 2001) as risk factors for neuromuscular complications in the ICU. Both severity and duration of SIRS and organ failure were identified as risk factors (Campellone et al., 1998; De Letter et al., 2001; Garnacho-Montero et al., 2001; Van den Berghe et al., 2001, 2005; Nanas et al., 2008; Weber-Carstens et al., 2010; Hermans et al., 2013). Hyperglycemia, frequently present in critically ill patients, is also an independent risk factor for both electrophysiologic (Witt et al., 1991) and clinical (Nanas et al., 2008) manifestations of ICUAW. In fact, it seems that possibly controlling hyperglycemia to a normal range using insulin reduced the incidence of ICUAW (Van den Berghe et al., 2005; Hermans et al., 2007; Patel and Pohlman, 2014).

Several case reports have suggested a link between weakness and corticosteroids. Detrimental effects of corticosteroids (Campellone et al., 1998; De Jonghe et al., 2002; Hermans et al., 2013) and prolonged administration of neuromuscular blocking agents (Garnacho-Montero et al., 2001; Hermans et al., 2007, 2013) on the neuromuscular system of critically ill patients were described and were further supported by animal experimental data (Rouleau et al., 1987; Massa et al., 1992; Rich et al., 1998). Several prospective studies, including data from randomized controlled trials (RCTs), could not confirm these findings (De Letter et al., 2001; Garnacho-Montero et al., 2001; Bednarík et al., 2005; Bercker et al., 2005; Van den Berghe et al., 2005; Steinberg et al., 2006; Hermans et al., 2007; Nanas et al., 2008; Papazian et al., 2010; Weber-Carstens et al., 2010; Fan et al., 2014b; Patel and Pohlman, 2014). These conflicting data may suggest there may be more complex interplay of factors and that the neuromuscular effects of corticosteroids may vary according to dose, timing, and glycemic control (Hermans et al., 2007). In addition, though corticosteroids could have negative effects directly on the muscle, corticosteroids may reduce duration of shock and mechanical ventilation, thereby reducing one of the most important risk factors (Annane et al., 2002; Hough et al., 2009). Age is also independently related to ICUAW (Hermans et al., 2013; Patel and Pohlman, 2014), as well as parenteral nutrition (Garnacho-Montero et al., 2001), female sex (De Jonghe et al., 2002), hyperosmolality (Garnacho-Montero et al., 2001), hypoalbuminemia (Witt et al., 1991), and renal replacement therapy (Van den Berghe et al., 2001).

Respiratory weakness, weakness of the diaphragm, has been poorly studied. This muscle is difficult to assess clinically, electrophysiologically, and histologically. Available literature indicates that respiratory muscle weakness is associated with infection or sepsis (De Jonghe et al., 2007; Demoule et al., 2013; Supinski and Callahan, 2013), disease severity (Demoule et al., 2013), and peripheral weakness (De Jonghe et al., 2007). Indeed, 80% of patients with ICUAW demonstrated diaphragm weakness when sophisticated nonvolitional methods were used (Jung et al., 2016). Also, in patients with CIP, electrophysiologic data from the phrenic nerve and diaphragm show similar abnormalities as those documented in the peripheral nerves and muscles (Zifko et al., 1998). However, mechanical ventilation itself, through immobilization of the diaphragm, may contribute to atrophy of the diaphragm and diaphragmatic muscle weakness in critically ill patients; this is known as ventilator-induced diaphragmatic dysfunction and a major confounder in assessment for neuromuscular respiratory failure (De Jonghe et al., 2007; Levine et al., 2008; Hermans et al., 2010; Jaber et al., 2011a, b; Demoule et al., 2013).

CLINICAL PRESENTATION

The patient with a typical presentation of ICUAW is a critically ill patient who recovers (and may wake up from sedation) and is found to have a severe weakness of all extremities and may even move nothing at all. Weakness affects all limb musculature but generally is more pronounced proximally (De Jonghe et al., 2002; Hermans et al., 2012). The facial muscles are usually spared (Bolton et al., 1986). In most patients the tendon reflexes are reduced, but they can also be normal. The respiratory muscles and more specifically the diaphragm seem to be extremely vulnerable in critically ill patients (De Jonghe et al., 2007; Supinski and Callahan, 2013). This leads to a prolonged need for mechanical ventilation and often a cautious, step-by-step weaning from respiratory support. A comprehensive neurologic examination should further include inspection for fasciculations or myokymias, assessment of pattern of weakness and findings on examination that could suggest an underlying neurologic disorder. Tone, muscle strength, and sensory examination are all important to examine although, in most patients, flaccid, immobile extremities are found with normal cranial nerve examination. It is important to assess whether the patient is truly "neurologically weak" or deconditioned. Furthermore, is the weakness of peripheral or central origin, and thus could spinal cord damage be implicated? Equally important is to question whether a more chronic systemic illness might have predisposed the patient to sepsis and may be the cause of the neuropathy. Severe weakness in some patients – particularly if atrophy is substantial, tongue fasciculations are seen, and reflex pattern is high – may indicate amyotrophic lateral sclerosis.

It should be pointed out that neurologic examination of these patients is difficult when level of consciousness is diminished or altered by sedation (Bercker et al., 2005; Hough et al., 2011). As assessment of strength and sensation is often unreliable except when limbs are immobile, flaccid, and do not move with nailbed compression.

ICUAW is a syndrome of generalized limb weakness that develops while the patient is critically ill and for which there is no alternative explanation other than the critical illness itself (Stevens et al., 2009; Fan et al., 2014a). There is still no universally accepted "gold standard" for diagnosing ICUAW. The recent American Thoracic Society clinical practice guidelines however emphasize the importance of clinical examination using the Medical Research Council (MRC) scale as the current reference standard.

Weakness is graded with the MRC scale, a subjective 5-point rating scale that scores for maximal force produced by voluntary muscle contraction. The current consensus is that for a diagnosis of ICUAW MRC scores for six different bilaterally tested muscle groups should be summed (the so-called MRC sum score or MRC-SS) (Stevens et al., 2009; Fan et al., 2014a; Sharshar et al., 2014). An MRC-SS <48 was defined as the cutoff for diagnosing ICUAW. Both proximal and distal muscle groups should be tested whenever possible, including deltoids, biceps, wrist extensors, iliopsoas, quadriceps, and ankle dorsiflexor muscles. In patients in whom it is impossible to test all these six muscle groups, an average MRC score <4 per muscle group can be used.

For optimal physical examination using the MRC scale an awake and attentive patient is needed. This can be a limitation in the ICU, as many ICU patients are sedated for shorter or longer periods. Delirium may also jeopardize the reliability of the MRC scale testing (Hough et al., 2011). Despite these limitations MRC testing in critically ill patients seems feasible and reliable (Ali et al., 2008; Hermans et al., 2012; Vanpee et al., 2014). A complete examination of all six bilateral muscle groups may be very demanding for recovering ICU patients. In those patients, a simple surrogate marker for overall strength could be obtained by assessment of handgrip strength (Ali et al., 2008; Hermans et al., 2012). Whether this test is reliable is still being investigated and cutoff values need to be validated (Parry et al., 2015a).

Whether the autonomic dysfunction, which is often seen in critically ill patients, is caused by a neuropathy of the autonomic nervous system remains to be established. Axonal degeneration of the sympathetic chain and vagal nerve has been noted in selected patients with CIP and suggests the autonomic system might be involved in the process (Zochodne et al., 1987). Others found abnormal heart rate variability in all ICU patients studied but no correlation with ICUAW (Wieske et al., 2013a). In a small pilot study investigating other aspects of the autonomic nervous system, abnormal heart rate variability, cold face testing, and skin wrinkle testing were found (Wieske et al., 2013b).

NEURODIAGNOSTICS

Several modalities are used in the diagnostic process of ICUAW. Physical examination, electromyography, and nerve conduction studies (NCS) have been used most often (Fan et al., 2014a). The diagnosis of ICUAW can be made with physical examination alone but electrophysiologic studies such as needle electromyography and NCS may be additionally useful in uncooperative patients (Lacomis, 2013).

To differentiate between CIP and CIM electrophysiologic testing and biopsies are potentially useful (Stevens et al., 2009). NCS can be used for investigation of the nerves and differentiate between axonal and demyelinating neuropathies. Furthermore, NCS have the advantage that they can be used in patients who cannot be scored with the MRC scale and thus are used to diagnose ICUAW early (Woittiez et al., 2001; Weber-Carstens et al., 2009). Due to interference of electric devices in the ICU, NCS can be technologically demanding in the ICU setting. Therefore, the use of a simple, one-nerve examination has been evaluated and seems to be a promising tool (Latronico et al., 2014; Moss et al., 2014). One study group examined whether the results of a single-peroneal-nerve NCS corresponded to the results of a complete NCS combined with needle electromyography and found a good correlation between tests. Others investigated single NCS in a small study in patients with and without ICUAW (Wieske et al., 2015b). In both groups abnormal NCS results were found, suggesting better definitional values are needed.

Generally, a reduction in the amplitude of the compound muscle action potential (CMAP) is used as a marker of nerve or muscle dysfunction. A normal CMAP on screening electrophysiologic testing, 8 days after ICU admission, excludes a neurologic cause for ICUAW with a high predictive value (Hermans et al., 2015). This study also showed that a reduced CMAP in the first week of ICU admission was an independent predictor for 1-year mortality.

Direct muscle stimulation has been used to further differentiate between problems in the nerve or the muscle (Rich et al., 1997; Weber-Carstens et al., 2009). A first CMAP is obtained by conventional stimulation of the nerve. A second CMAP is obtained by direct stimulation of the muscle by inserting a needle into it. By comparing both CMAPs the problem can be localized in the nerve or the muscle. In a neuropathy, decreased CMAPs are found with nerve stimulation and normal CMAPs with muscle stimulation. Decreased CMAPs with both types of stimulation are found in myopathy. The nerve-to-muscle CMAP ratio of <0.5 is indicative of an isolated neuropathy; a ratio >0.5 is indicative of either a myopathy or a combination of neuropathy and myopathy

(Lefaucheur et al., 2006). Repetitive stimulation can be used to examine the neuromuscular junction. In ICUAW the neuromuscular junction is not affected and test results should be normal. Abnormalities in the neuromuscular junction are indicative of ongoing effects of neuromuscular blocking agents or previously undiagnosed myasthenia gravis (Lacomis, 2013).

Assessment of the morphology and recruitment pattern of motor unit action potentials (MUAPs) can be helpful to diagnose myopathy (Leijten et al., 1996; Zifko et al., 1998; Tennilä et al., 2000). Characteristically, with myopathy, MUAPs have a short duration and low amplitude. For optimal electromyography an awake and attentive patient is needed to activate the muscle and to reliably evaluate MUAPs. This of course limits its applicability in sedated, or otherwise unconscious or delirious critically ill patients.

Muscle tissue can be obtained via open biopsy or needle biopsy techniques. It carries the risk of bleeding or wound infection. In cases of unclear clinical or electrophysiologic findings or when a strong suspicion of a myopathy other than CIM exists, muscle biopsy may be helpful. In ICUAW total muscle mass is reduced by atrophy or necrosis of muscle fibers. Furthermore, loss of myosin, the so-called thick filaments, is a histologic hallmark of CIM (Faragher et al., 1996; Lacomis et al., 1996; Larsson et al., 2000; Sander et al., 2002). Histologic changes in muscles can often be found early after the onset of critical illness (Ahlbeck et al., 2009).

Other new techniques may be helpful but experience is limited. Muscle ultrasound is a promising diagnostic tool for ICUAW (Connolly et al., 2014; Bunnell et al., 2015). It can provide information on muscle mass and structure in a simple and noninvasive way (Tillquist et al., 2014). Cross-sectional muscle area decreases rapidly with critical illness, but there is paucity of data linking this to development of ICUAW, although small studies have been reported (Puthucheary et al., 2010; Parry et al., 2015b). Further research is needed to validate reproducibility, reliability for clinical, electrophysiologic, and pathologic changes as well as to determine association of ultrasound findings with important patient-centered outcomes. Ultrasound may also be used to visualize peripheral nerves, but systematic studies of patients with ICUAW are not available (Beekman and Visser, 2004).

Biomarkers may also be of help in diagnosing muscle or nerve injury in critically ill patients. The ideal biomarker should be easy to assess and discriminate between ICUAW and other causes of weakness and help to differentiate between CIP and CIM. Although creatinine kinase levels in blood are increased in patients with ICUAW, it is not a good biomarker (De Jonghe et al., 2002). Plasma neurofilament levels were found to be increased in patients with ICUAW as compared to patients without ICUAW but neurofilaments are nonspecific markers of axonal injury and levels can be increased in many other illnesses (Wieske et al., 2014).

HOSPITAL COURSE AND MANAGEMENT

ICUAW is associated with a prolonged need for mechanical ventilation and increased length of stay on the ICU and in the hospital (De Jonghe et al., 2002, 2004, 2007; Ali et al., 2008; Sharshar et al., 2009). Furthermore, mortality is increased in these patients during and after the ICU admission (Wieske et al., 2015a). As the exact cause of the weakness is unknown, treatment remains primarily aimed at the primary critical illness, such as sepsis or trauma. As soon as the patient stabilizes, further worsening of the weakness might be prevented by mobilization and muscle exercise through physiotherapy (Burtin et al., 2009; Calvo-Ayala et al., 2013).

Early mobilization is started as soon as possible, potentially within 24–48 hours of ICU admission. It is safe and feasible in critically ill patients while they remain on the mechanical ventilator (Morris et al., 2008). Early physical and occupational therapy combined with minimizing sedation shortened duration of mechanical ventilation and increased the rate of return to an independent functional status at hospital discharge, in addition to other favorable outcomes (Schweickert et al., 2009). Moreover functional status, quality of life, as well as quadriceps strength improved with bed cycling in addition to regular physiotherapy (Burtin et al., 2009). Despite beneficial effects of early rehabilitation, practices vary and may not be optimal (Barber et al., 2015; Dafoe et al., 2015).

Next to mobilization by physiotherapists, electric stimulation of muscles in ICU patients has been investigated (Gerovasili et al., 2009; Kho et al., 2015). Studies have been small and results were conflicting and rarely used in ICUs, if ever (Parry et al., 2013).

Critical illness represents a hypercatabolic state. Additionally, malnutrition occurs rapidly and frequently in critically ill patients because a dysfunctional gastrointestinal tract often hampers adequate enteral feeding. Moreover, early supplementation of a caloric deficit during the first week in ICU did not prevent muscle atrophy, suggesting that such supplementation cannot avert the early catabolic state (Hermans et al., 2013). Furthermore, patients with no supplementation developed less muscle weakness and recovered more rapidly from weakness. These findings are in line with the observation that increased protein delivery during the first week was associated with more pronounced muscle atrophy, although the mechanism is not fully understood (Puthucheary et al., 2013). Full versus trophic enteral feeding in the first 6 days of ICU did not result in any functional difference when assessed up to 12 months after ICU admission (Needham et al., 2013). Optimal time and dosing of nutrients however remain a matter of debate and require further study (Casaer, 2015; Casaer and Ziegler, 2015).

Correcting hyperglycemia, identified as a risk factor for neuromuscular complications in the ICU, resulted in decreased incidence of electrophysiologic abnormalities and need for prolonged mechanical ventilation in medical and surgical patients in a single center. If confirmed, this may be the only available therapeutic intervention (Hermans et al., 2007).

ICUAW is clearly associated with worse outcome and it independently predicts prolonged weaning, ICU and hospital stay, as well as ICU and hospital mortality (De Jonghe et al., 2004; Ali et al., 2008; Sharshar et al., 2009). This is not surprising given the observation that sepsis and multiple-organ failure are central risk factors for ICUAW. ICUAW might contribute to morbidity (Hermans et al., 2014b). Several mechanisms can be put forward, including associated respiratory muscle weakness (Jung et al., 2016), possibly contributing to recurrent respiratory failure and increased mortality (Demoule et al., 2013; Supinski and Callahan, 2013). Swallowing dysfunction is often present in patients with ICUAW and may result in pneumonia, acute respiratory distress syndrome (ARDS), respiratory failure, and other complications that contribute to poor outcome (Mirzakhani et al., 2013; Ponfick et al., 2015). Functional status at hospital discharge – as measured with the 6-minute walk distance – is significantly decreased for weak patients as compared to matched not weak patients (Hermans et al., 2014b). Similarly, in a matched set of ICUAW and non-ICUAW patients, hospitalization costs are higher and disposition is different (fewer patients discharged to home and more often to rehabilitation units). Weakness gradually recovers in weeks to months. Clinical recovery occurs in the majority of patients and was present in 86–96% of ARDS patients at 1-year and 91% at 2-year follow-up, but electrophysiologic abnormalities may remain for a long time in a substantial amount of patients (Fletcher et al., 2003; Fan et al., 2014b; Needham et al., 2014). Herridge and Tansey (2011) found that limited physical function and quality of life in relatively young and previously active ARDS survivors were mainly determined by ICUAW. Several studies have investigated the late effects of ICUAW. ICUAW independently contributes to 6-month and 1-year mortality and a decrease in physical functioning (Hermans et al., 2014b; Wieske et al., 2015a). Furthermore, ICUAW at any time during a 2-year follow-up was found to be associated with reduced functional status and quality of life (Fan et al., 2014b).

ICUAW is therefore recognized as a crucial component in PICS.

CLINICAL TRIALS AND GUIDELINES

Besides the clinical trials on early physical therapy and early mobilization as described above, several studies have been performed aiming at optimal treatment of the underlying disease. As the specific cause of ICUAW is currently unclear, optimal treatment of the underlying disease is thought to possibly limit the development of ICUAW. ICU treatment strategies aiming at a prevention of ICUAW development were recently summarized in a Cochrane review by Hermans et al. (2014a). They included all RCTs investigating the effect of an intervention on the occurrence of ICUAW or CIP/CIM. The number of studies available was limited to five (since this study, two new studies have been identified) (Van den Berghe et al., 2005; Steinberg et al., 2006; Hermans et al., 2007, 2013; Schweickert et al., 2009; Papazian et al., 2010; Yosef-Brauner et al., 2015). Table 29.1 summarizes these studies.

In December 2014 a guideline of the American Thoracic Society on diagnosis of ICUAW in adults was published (Fan et al., 2014a). Thirty-one studies were identified, of which 28 were prospective cohort studies and only three were RCTs. The median number of patients included in the studies was 43 (interquartile range 25–72), showing that most studies were relatively small.

Based on the available literature the authors of this guideline proposed three recommendations:

1. We recommend well-designed, adequately powered and executed randomized controlled trials comparing physical rehabilitation or other alternative treatments with usual care in patients with ICUAW that measure and report patient-important outcomes. (strong recommendation, very low-quality evidence)
2. We recommend clinical research to determine the role of prior patient disability in the development of and recovery from ICUAW. (strong recommendation, very low-quality evidence)
3. We recommend clinical research that determines whether or not patients would want to know if they have ICUAW even though no specific therapy currently exists and how patient preferences influence medical decision-making or the perception of prognosis. (strong recommendation, very low-quality evidence)

The paucity of clinical material, standardization of outcomes, and neurologic expertise is obvious and reflected in these guidelines. Other guidelines on ICUAW from the European and American societies for critical care

Table 29.1

Randomized trials to prevent or treat intensive care unit-acquired weakness (ICUAW)

Study	Population	Intervention	Primary endpoint/relevant endpoint	ICUAW/CIP/CIM diagnosis	Effect intervention
Van den Berghe et al. (2005)	Mechanically ventilated adults: $n=405$ (of 1548 in main study)	Strict glucose control (80–110 mg/dL) versus conventional care (180–210 mg/dL)	Death from any cause during ICU stay/development of CIP	Electromyography on day 7 at the ICU, repeated weakly. Definition CIP: abundant spontaneous activity such as positive sharp waves and fibrillation potentials	Reduction of development of CIP, 49% of population in conventional treatment group versus 25% in strict glucose control group: $p<0.0001$
Steinberg et al. (2006)	ARDS patients on mechanical ventilation for >7 days: $n=180$	Methylprednisolone versus placebo	Mortality after 60 days/occurrence of neuromyopathy	Presence of the terms "myopathy," "myositis," "neuropathy," "paralysis," or "unexplained weakness" in the medical record	30% of patients in methylprednisolone group with neuromyopathy versus 22% in placebo group (not significant)
Hermans et al. (2007)	Adult patients admitted to a medical ICU assumed to be admitted for 3 days: $n=420$ (of 1200 in main study)	Strict glucose control (80–110 mg/dL) versus conventional care (180–210 mg/dL)	Death from any cause in the hospital/development of CIP/CIM	Electromyography on day 7 at the ICU, repeated weakly. Definition CIP: abundant spontaneous activity such as positive sharp waves and fibrillation potentials	Reduction of development of CIP/CIM: 50.5% of population in conventional treatment group versus 38.9% in strict glucose control group, $p=0.02$
Schweickert et al. (2009)	Adults ICU patients on mechanical ventilation <72 hours: $n=104$	Exercise and mobilization versus standard care	Return to independent functional status/ICU-acquired paresis. Both at hospital discharge	MRC assessment in three muscle groups in upper and lower limbs. ICUAW when total score was <48	ICU-acquired paresis in 31% of patients in intervention group versus 49% in control group (not significant). Median MRC score not different between groups
Papazian et al. (2010)	Adult ARDS patients on mechanical ventilation: $n=340$	Neuromuscular blocker (cisatracurium) versus placebo	Mortality before hospital discharge and within 90 days/ICU-acquired paresis and MRC scores at day 28 and hospital discharge	ICU-acquired paresis defined as overall MRC score <48	No difference in development of ICU-acquired paresis or MRC scores at day 28 or hospital discharge
Hermans et al. (2013)	Medical and surgical ICU patients, NRS >3 $n=600$ (of 4640 in main study)	Late parenteral nutrition (not in the first week), versus early parenteral nutrition (within 48 hours) to complement insufficient enteral feeding	Duration of dependency on intensive care/ICUAW and recovery hereof	MRC sum score <48 at first measurement (from D8 screened three times weekly for awakening)	Reduced incidence of weakness with late parenteral nutrition (105/305 or 34%) compared with early parenteral nutrition (127/295 or 43%), $p=0.030$. Also enhanced recovery of weakness in the late parenteral nutrition group $p=0.021$
Yosef-Brauner et al. (2015)	Mechanical ventilation for 48 hours expected to need another 48 hours of mechanical ventilation and ICUAW: $n=18$	Physical therapy twice a day, conventional physical therapy once a day	Muscle strength indices	MRC sum score <48	Improvement in strength examination with MRC and dynamometry

CIP, critical illness polyneuropathy; CIM, critical illness myopathy; ICU, intensive care unit; ARDS, adult respiratory distress syndrome; MRC, Medical Research Council; NRS, Nutritional Risk Screening.

COMPLEX CLINICAL DECISIONS

Despite the fact that ICUAW is considered the most prevalent cause for weakness in ICU patients, other causes have to be considered (Maramattom and Wijdicks, 2006b). The differential diagnosis is broad and exclusion of conditions like Guillain–Barré syndrome or myasthenia gravis is necessary as this can have therapeutic consequences or prognostic implications. A careful medical history (when did weakness develop?) and physical examination (pattern of weakness, focal neurologic deficits, and brainstem functions) should eliminate most differential diagnoses (Maramattom and Wijdicks, 2006a). If necessary additional diagnostic tests can be performed to exclude other causes. ICUAW can only be diagnosed if weakness develops after the onset of critical illness. The mnemonic MUSCLES can be used to evaluate other causes for weakness in the ICU (Table 29.2) (Maramattom and Wijdicks, 2006a).

OUTCOME PREDICTION

Different trajectories of recovery from weakness and functional implications are increasingly being recognized. This heterogeneity is suggested to represent various clinical phenotypes and results from interaction of ICUAW with other factors such as age, comorbidities, pre-existing neuromuscular problems, and ICU length of stay (Batt et al., 2013a). Also cognitive dysfunction and patient's as well as caregiver's mental state may affect outcomes. One study of acute lung injury survivors found that the number of days of bedrest was the only independent predictor of weakness at any time during a 2-year follow-up period (Fan et al., 2014b). Additionally, increasing evidence points to possible differences in recovery of patients with neuropathy versus myopathy. Case series indicate that recovery from CIM may be faster, more complete, and less likely to yield severe persisting disability, and coexistent CIP in patients with CIM may hamper recovery (Guarneri et al., 2008; Intiso et al., 2011; Koch et al., 2011, 2014). Finally, the likelihood of mortality during the first year following ICU admission in patients with ICUAW at awakening is affected by whether or not patients recovered from ICUAW by the end of ICU discharge as well as by the severity of persisting weakness at that time (Hermans et al., 2014b).

NEUROREHABILITATION

Does rehabilitation after ICU and hospital discharge improve the functional recovery and outcome in patients with ICUAW? In contrast to the many papers that have been published on the effectiveness of early physical therapy and mobilization in ICU patients, carefully conducted studies on the effects of physical rehabilitation after hospital discharge are not available. A Cochrane review on this topic could not identify one properly conducted RCT, quasiRCT, or controlled cross-over trial that included the population of interest (Mehrholz et al., 2015).

A Cochrane review specifically on exercise rehabilitation programs initiated after ICU discharge was reported in 2015 (Connolly et al., 2015). Six trials, five RCTs and one controlled trial, were included in the review (Jones et al., 2003; Porta et al., 2005; Salisbury et al., 2010; Elliott et al., 2011; Batterham et al., 2014). The total patient population consisted of 483 adult survivors of critical illness who had received mechanical ventilation for at least 24 hours. The interventions investigated in the included studies ranged from a 6-week rehabilitation manual without any supervisory input (Jones et al., 2003) to an 8-week physiotherapist-supervised cycle ergometer exercise program (Batterham et al., 2014) to a 12-week rehabilitation program consisting of cognitive, physical, and functional components, supervised by an exercise physiologist, with support from a trained social worker once at home (Jackson et al., 2012). The outcome measurements used were very diverse. The overall quality of the evidence was rated as very low.

The authors of the review concluded that they were unable to determine an overall effect of an exercise-based intervention after ICU discharge on functional exercise capacity or health-related quality of life. The effectiveness of exercise-based rehabilitation after ICU discharge remains to be studied.

Table 29.2

Differential diagnosis of weakness on the intensive care unit

Category		Example
M	Medications	Steroids, neuromuscular blockers
U	Undiagnosed neuromuscular disorder	Myasthenia, mitochondrial myopathy
S	Spinal cord disease	Trauma, ischemia
C	Critical illness	Critical illness polyneuropathy, critical illness myopathy, critical illness neuromyopathy
L	Loss of muscle mass	Cachectic myopathy, rhabdomyolysis
E	Electrolyte disorders	Hypokalemia, hypophosphatemia
S	Systemic illness	Vasculitis, paraneoplastic

Reproduced from Maramattom and Wijdicks (2006a).

REFERENCES

Ackermann KA, Bostock H, Brander L et al. (2014). Early changes of muscle membrane properties in porcine faecal peritonitis. Crit Care 18: 484.

Ahlbeck K, Fredriksson K, Rooyackers O et al. (2009). Signs of critical illness polyneuropathy and myopathy can be seen early in the ICU course. Acta Anaesthesiol Scand 53: 717–723.

Ali NA, O'Brien JM, Hoffmann SP et al. (2008). Acquired weakness, handgrip strength, and mortality in critically ill patients. Am J Resp Crit Care Med 178: 261–268.

Annane D, Sebille V, Charpentier C et al. (2002). Effect of treatment with low doses of hydrocortisone and fludrocortisone on mortality in patients with septic shock. JAMA 288: 862–871.

Barber EA, Everard T, Holland AE et al. (2015). Barriers and facilitators to early mobilisation in intensive care: a qualitative study. Aust Crit Care 28: 177–182.

Batt J, Dos Santos CC, Cameron JI et al. (2013a). Intensive-care unit acquired weakness (ICUAW): clinical phenotypes and molecular mechanisms. Am J Resp Crit Care Med 187: 238–246.

Batt J, Santos CCD, Herridge MS (2013b). Muscle injury during critical illness. JAMA 310: 1569–1570.

Batterham AM, Bonner S, Wright J et al. (2014). Effect of supervised aerobic exercise rehabilitation on physical fitness and quality-of-life in survivors of critical illness: an exploratory minimized controlled trial (PIX study). Br J Anaesth 113: 130–137.

Bednarík J, Vondracek P, Dusek L et al. (2005). Risk factors for critical illness polyneuromyopathy. J Neurol 252: 343–351.

Beekman R, Visser LH (2004). High-resolution sonography of the peripheral nervous system – a review of the literature. Eur J Neurol 11: 305–314.

Bercker S, Weber-Carstens S, Deja M et al. (2005). Critical illness polyneuropathy and myopathy in patients with acute respiratory distress syndrome. Crit Care Med 33: 711–715.

Bloch S, Polkey MI, Griffiths M et al. (2012). Molecular mechanisms of intensive care unit-acquired weakness. Eur Resp J 39: 1000–1011.

Bolton CF (2005). Neuromuscular manifestations of critical illness. Muscle Nerve 32: 140–163.

Bolton CF, Gilbert JJ, Hahn AF et al. (1984). Polyneuropathy in critically ill patients. J Neurol Neurosurg Psychiatry 47: 1223–1231.

Bolton CF, Laverty DA, Brown JD et al. (1986). Critically ill polyneuropathy: electrophysiological studies and differentiation from Guillain–Barré syndrome. J Neurol Neurosurg Psychiatry 49: 563–573.

Brealey D, Brand M, Hargreaves I et al. (2002). Association between mitochondrial dysfunction and severity and outcome of septic shock. Lancet 360: 219–223.

Bunnell A, Ney J, Gellhorn A et al. (2015). Quantitative neuromuscular ultrasound in intensive care unit-acquired weakness: a systematic review. Muscle Nerve 52: 701–708.

Burtin C, Clerckx B, Robbeets C et al. (2009). Early exercise in critically ill patients enhances short-term functional recovery. Crit Care Med 37: 2499–2505.

Calvo-Ayala E, Khan BA, Farber MO et al. (2013). Interventions to improve the physical function of ICU survivors: a systematic review. Chest 144: 1469–1480.

Campellone JV, Lacomis D, Kramer DJ et al. (1998). Acute myopathy after liver transplantation. Neurology 50: 46–53.

Casaer MP (2015). Muscle weakness and nutrition therapy in ICU. Curr Opin Clin Nutr Metab Care 18: 162–168.

Casaer MP, Ziegler TR (2015). Nutritional support in critical illness and recovery. Lancet Diabetes Endocrinol 3: 734–745.

Connolly B, MacBean V, Crowley C et al. (2014). Ultrasound for the assessment of peripheral skeletal muscle architecture in critical illness: a systematic review. Crit Care Med: 1–10.

Connolly B, Salisbury L, O'Neill B et al. (2015). Exercise rehabilitation following intensive care unit discharge for recovery from critical illness. Cochrane Database Syst Rev 6. CD008632.

Convertino VA, Bloomfield SA, Greenleaf JE (1997). An overview of the issues: physiological effects of bed rest and restricted physical activity. Med Sci Sports Exerc 29: 187–190.

Dafoe S, Chapman MJ, Edwards S et al. (2015). Overcoming barriers to the mobilisation of patients in an intensive care unit. Anaesth Intensive Care 43: 719–727.

De Jonghe B, Sharshar T, Lefaucheur J-P et al. (2002). Paresis acquired in the intensive care unit: a prospective multicenter study. JAMA 288: 2859–2867.

De Jonghe B, Bastuji-Garin S, Sharshar T et al. (2004). Does ICU-acquired paresis lengthen weaning from mechanical ventilation? Int Care Med 30: 1117–1121.

De Jonghe B, Bastuji-Garin S, Durand MC et al. (2007). Respiratory weakness is associated with limb weakness and delayed weaning in critical illness. Crit Care Med 35: 2007–2015.

De Letter MACJ, Schmitz PIM, Visser LH et al. (2001). Risk factors for the development of polyneuropathy and myopathy in critically ill patients. Crit Care Med 29: 2281–2286.

Demoule A, Jung B, Prodanovic H et al. (2013). Diaphragm dysfunction on admission to ICU: prevalence, risk factors and prognostic impact – a prospective study. Am J Resp Crit Care Med 188: 213–219.

Derde S, Hermans G, Derese I et al. (2012). Muscle atrophy and preferential loss of myosin in prolonged critically ill patients. Crit Care Med 40: 79–89.

Elliott D, McKinley S, Alison J et al. (2011). Health-related quality of life and physical recovery after a critical illness: a multi-centre randomised controlled trial of a home-based physical rehabilitation program. Crit Care 15: R142.

Fan E, Cheek F, Chlan L et al. (2014a). An official American Thoracic Society clinical practice guideline: the diagnosis of intensive care unit-acquired weakness in adults. Am J Resp Crit Care Med 190: 1437–1446.

Fan E, Dowdy DW, Colantuoni E et al. (2014b). Physical complications in acute lung injury survivors: a two-year longitudinal prospective study. Crit Care Med 42: 849–859.

Faragher MW, Sc BMED, Day BJ (1996). Critical care myopathy: an electrophysiological and histological study. Muscle Nerve 19: 516–518.

Fenzi F, Latronico N, Refatti N et al. (2003). Enhanced expression of E-selectin on the vascular endothelium of peripheral nerve in critically ill patients with neuromuscular disorders. Acta Neuropathol 106: 75–82.

Fletcher SN, Kennedy DD, Ghosh IR et al. (2003). Persistent neuromuscular and neurophysiologic abnormalities in long-term survivors of prolonged critical illness. Crit Care Med 31: 1012–1016.

Friedrich O, Reid MB, Van den Berghe G et al. (2015). The sick and the weak: neuropathies/myopathies in the critically ill. Physiol Rev 95: 1025–1109.

Garnacho-Montero J, Madrazo-Osuna J, García-Garmendia JL et al. (2001). Critical illness polyneuropathy: risk factors and clinical consequences. A cohort study in septic patients. Int Care Med 27: 1288–1296.

Gerovasili V, Stefanidis K, Vitzilaios K et al. (2009). Electrical muscle stimulation preserves the muscle mass of critically ill patients: a randomized study. Crit Care 13: R161.

Guarneri B, Bertolini G, Latronico N (2008). Long-term outcome in patients with critical illness myopathy or neuropathy: the Italian multicentre CRIMYNE study. J Neurol Neurosurg Psychiatry 79: 838–841.

Hermans G, Van den Berghe G (2015). Clinical review: intensive care unit acquired weakness. Crit Care 19: 274.

Hermans G, Wilmer A, Meersseman W et al. (2007). Impact of intensive insulin therapy on neuromuscular complications and ventilator dependency in the medical intensive care unit. Am J Resp Crit Care Med 175: 480–489.

Hermans G, Agten A, Testelmans D et al. (2010). Increased duration of mechanical ventilation is associated with decreased diaphragmatic force: a prospective observational study. Crit Care 14: R127.

Hermans G, Clerckx B, Vanhullebusch T et al. (2012). Interobserver agreement of Medical Research Council sum-score and handgrip strength in the intensive care unit. Muscle Nerve 45: 18–25.

Hermans G, Casaer MP, Clerckx B et al. (2013). Effect of tolerating macronutrient deficit on the development of intensive-care unit acquired weakness: a subanalysis of the EPaNIC trial. Lancet Respir Med 1: 621–629.

Hermans G, De Jonghe B, Bruyninckx F et al. (2014a). Interventions for preventing critical illness polyneuropathy and critical illness myopathy. Cochrane Database Syst Rev 1. CD006832.

Hermans G, Van Mechelen H, Clerckx B et al. (2014b). Acute outcomes and 1-year mortality of intensive care unit-acquired weakness. A cohort study and propensity-matched analysis. Am J Resp Crit Care Med 190: 410–420.

Hermans G, Van Mechelen H, Bruyninckx F et al. (2015). Predictive value for weakness and 1-year mortality of screening electrophysiology tests in the ICU. Intensive Care Med 41: 2138–2148.

Herridge MS, Tansey CM (2011). Functional disability 5 years after acute respiratory distress syndrome. N Engl J Med 364: 1293–1304.

Hough CL, Steinberg KP, Taylor Thompson B et al. (2009). Intensive care unit-acquired neuromyopathy and corticosteroids in survivors of persistent ARDS. Int Care Med 35: 63–68.

Hough CL, Lieu BK, Caldwell ES (2011). Manual muscle strength testing of critically ill patients: feasibility and interobserver agreement. Crit Care 15: R43.

Intiso D, Amoruso L, Zarrelli M et al. (2011). Long-term functional outcome and health status of patients with critical illness polyneuromyopathy. Acta Neurol Scand 123: 211–219.

Jaber S, Jung B, Matecki S et al. (2011a). Clinical review: ventilator-induced diaphragmatic dysfunction – human studies confirm animal model findings! Crit Care 15: 206.

Jaber S, Petrof BJ, Jung B et al. (2011b). Rapidly progressive diaphragmatic weakness and injury during mechanical ventilation in humans. Am J Resp Crit Care Med 183: 364–371.

Jackson JC, Ely EW, Morey MC et al. (2012). Cognitive and physical rehabilitation of intensive care unit survivors: results of the RETURN randomized controlled pilot investigation. Crit Care Med 40: 1088–1097.

Jones C, Skirrow P, Griffiths RD et al. (2003). Rehabilitation after critical illness: a randomized, controlled trial. Crit Care Med 31: 2456–2461.

Jung B, Moury PH, Mahul M et al. (2016). Diaphragmatic dysfunction in patients with ICU-acquired weakness and its impact on extubation failure. Intensive Care Med 42: 853–861.

Kho ME, Truong AD, Zanni JM et al. (2015). Neuromuscular electrical stimulation in mechanically ventilated patients: a randomized, sham-controlled pilot trial with blinded outcome assessment. J Crit Care 30: 32–39.

Koch S, Spuler S, Deja M et al. (2011). Critical illness myopathy is frequent: accompanying neuropathy protracts ICU discharge. J Neurol Neurosurg Psychiatry 82: 287–293.

Koch S, Wollersheim T, Bierbrauer J (2014). Long term recovery in critical illness myopathy is complete, contrary to polyneuropathy. Muscle Nerve 50: 431–436.

Lacomis D (2013). Electrophysiology of neuromuscular disorders in critical illness. Muscle Nerve 47: 452–463.

Lacomis D, Giuliani MJ, Cott AV et al. (1996). Acute myopathy of intensive care: clinical, electromyographic, and pathological aspects. Ann Neurol 40: 645–654.

Larsson L, Li X, Eriksson LI (2000). Acute quadriplegia and loss of muscle myosin in patients treated with nondepolarizing neuromuscular blocking agents and corticosteroids: mechanisms at the cellular and molecular levels. Crit Care 28: 34–45.

Latronico N, Fenzi F, Recupero D et al. (1996). Critical illness myopathy and neuropathy. Lancet 347: 1579–1582.

Latronico N, Nattino G, Guarneri B et al. (2014). Validation of the peroneal nerve test to diagnose critical illness polyneuropathy and myopathy in the intensive care unit: the multicentre Italian CRIMYNE-2 diagnostic accuracy study. F1000Res 3: 127.

Lefaucheur JP, Nordine T, Rodriguez P et al. (2006). Origin of ICU acquired paresis determined by direct muscle stimulation. J Neurol Neurosurg Psychiatry 77: 500–506.

Leijten FS, De Weerd AW, Poortvliet DC et al. (1996). Critical illness polyneuropathy in multiple organ dysfunction syndrome and weaning from the ventilator. Int Care Med 22: 856–861.

Levine S, Nguyen T, Taylor N et al. (2008). Rapid disuse atrophy of diaphragm fibers in mechanically ventilated humans. N Engl J Med 358: 1327–1335.

Maramattom BV, Wijdicks EFM (2006a). Acute neuromuscular weakness in the intensive care unit. Crit Care Med 34: 2835–2841.

Maramattom BV, Wijdicks EFM (2006b). Neuromuscular disorders in medical and surgical ICUs: case studies in critical care neurology. Neurol Clin 24: 371–383.

Massa R, Carpenter S, Holland P et al. (1992). Loss and renewal of thick myofilaments in glucocorticoid-treated rat soleus after denervation and reinnervation. Muscle Nerve 15: 1290–1298.

Mehrholz J, Pohl M, Kugler J et al. (2015). Physical rehabilitation for critical illness myopathy and neuropathy: an abridged version of Cochrane Systematic Review. Eur J Phys Rehabil Med 51: 655–661.

Mirzakhani H, Williams JN, Mello J et al. (2013). Muscle weakness predicts pharyngeal dysfunction and symptomatic aspiration in long-term ventilated patients. Anesthesiology 119: 389–397.

Morris PE, Goad A, Thompson C et al. (2008). Early intensive care unit mobility therapy in the treatment of acute respiratory failure. Crit Care Med 36: 2238–2243.

Moss M, Yang M, Macht M et al. (2014). Screening for critical illness polyneuromyopathy with single nerve conduction studies. Int Care Med 40: 683–690.

Nanas S, Kritikos K, Angelopoulos E et al. (2008). Predisposing factors for critical illness polyneuromyopathy in a multidisciplinary intensive care unit. Acta Neurol Scand 118: 175–181.

Needham DM, Davidson J, Cohen H et al. (2012). Improving long-term outcomes after discharge from intensive care unit: report from a stakeholders' conference. Crit Care Med 40: 502–509.

Needham DM, Dinglas VD, Bienvenu OJ et al. (2013). One year outcomes in patients with acute lung injury randomised to initial trophic or full enteral feeding: prospective follow-up of EDEN randomised trial. BMJ 346: f1532.

Needham DM, Wozniak AW, Hough CL et al. (2014). Risk factors for physical impairment after acute lung injury in a national, multicenter study. Am J Respir Crit Care Med 189: 1214–1224.

Novak KR, Nardelli P, Cope TC et al. (2009). Inactivation of sodium channels underlies reversible neuropathy during critical illness in rats. J Clin Invest 119: 1150–1158.

Papazian L, Forel JM, Gacouin A et al. (2010). Neuromuscular blockers in early acute respiratory distress syndrome. N Engl J Med 363: 1107–1116.

Parry SM, Puthucheary ZA (2015). The impact of extended bed rest on the musculoskeletal system in the critical care environment. Extrem Physiol Med 4: 16.

Parry SM, Berney S, Granger CL et al. (2013). Electrical muscle stimulation in the intensive care setting: a systematic review. Crit Care Med 41: 2406–2418.

Parry SM, Berney S, Granger CL et al. (2015a). A new two-tier strength assessment approach to the diagnosis of weakness in intensive care: an observational study. Crit Care 19: 52.

Parry SM, El-Ansary D, Cartwright MS et al. (2015b). Ultrasonography in the intensive care setting can be used to detect changes in the quality and quantity of muscle and is related to muscle strength and function. J Crit Care 30: 1151.

Patel BK, Pohlman AS (2014). Impact of early mobilization on glycemic control and intensive care unit-acquired weakness in mechanically ventilated critically ill patients. CHEST J: 1–17.

Ponfick M, Linden R, Nowak DA (2015). Dysphagia – a common, transient symptom in critical illness polyneuropathy: a fiberoptic endoscopic evaluation of swallowing study. Crit Care Med 43: 365–372.

Porta R, Vitacca M, Gile LS et al. (2005). Supported arm training in patients recently weaned from mechanical ventilation. Chest 128: 2511–2520.

Puthucheary ZA (2013). Acute skeletal muscle wasting in critical illness. JAMA: 1–10.

Puthucheary Z, Montgomery H, Moxham J et al. (2010). Structure to function: muscle failure in critically ill patients. J Physiol 588: 4641–4648.

Puthucheary ZA, Rawal J, McPhail M et al. (2013). Acute skeletal muscle wasting in critical illness. JAMA 310: 1591–1600.

Rich MM, Bird SJ, Raps EC et al. (1997). Direct muscle stimulation in acute quadriplegic myopathy. Muscle Nerve 20: 665–673.

Rich MM, Teener JW, Raps EC et al. (1998). Muscle inexcitability in patients with reversible paralysis following steroids and neuromuscular blockade. Muscle Nerve 21: 1231–1232.

Rossignol B, Gueret G, Pennec J-P et al. (2008). Effects of chronic sepsis on contractile properties of fast twitch muscle in an experimental model of critical illness neuromyopathy in the rat. Crit Care Med 36: 1855–1863.

Rouleau G, Karpati G, Carpenter S et al. (1987). Glucocorticoid excess induces preferential depletion of myosin in denervated skeletal muscle fibers. Muscle Nerve 10: 428–438.

Salisbury LG, Merriweather JL, Walsh TS (2010). The development and feasibility of a ward-based physiotherapy and nutritional rehabilitation package for people experiencing critical illness. Clin Rehabil 24: 489–500.

Sander HW, Golden M, Danon MJ (2002). Quadriplegic areflexic ICU illness: selective thick filament loss and normal nerve histology. Muscle Nerve 26: 499–505.

Schweickert WD, Pohlman MC, Pohlman AS et al. (2009). Early physical and occupational therapy in mechanically ventilated, critically ill patients: a randomised controlled trial. Lancet 373: 1874–1882.

Sharshar T, Bastuji-Garin S, Stevens RD et al. (2009). Presence and severity of intensive care unit-acquired paresis at time of awakening are associated with increased intensive care unit and hospital mortality. Crit Care Med 37: 3047–3053.

Sharshar T, Citerio G, Andrews PJD et al. (2014). Neurological examination of critically ill patients: a pragmatic approach. Report of an ESICM expert panel. Int Care Med 40: 495.

Steinberg KP, Hudson LD, Goodman RB et al. (2006). Efficacy and safety of corticosteroids for persistent acute respiratory distress syndrome. N Engl J Med 354: 1671–1684.

Stevens RD, Dowdy DW, Michaels RK et al. (2007). Neuromuscular dysfunction acquired in critical illness: a systematic review. Int Care Med 33: 1876–1891.

Stevens RD, Marshall SA, Cornblath DR et al. (2009). a framework for diagnosing and classifying intensive care unit-acquired weakness. Crit Care Med 37: S299–S308.

Supinski GS, Callahan LA (2013). Diaphragm weakness in mechanically ventilated critically ill patients. Crit Care 17: R120.

Tennilä A, Salmi T, Pettilä V et al. (2000). Early signs of critical illness polyneuropathy in ICU patients with systemic inflammatory response syndrome or sepsis. Int Care Med 26: 1360–1363.

Tillquist M, Kutsogiannis DJ, Wischmeyer PE et al. (2014). Bedside ultrasound is a practical and reliable measurement tool for assessing quadriceps muscle layer thickness. JPEN 38: 886–890.

Van den Berghe G (2004). How does blood glucose control with insulin save lives in intensive care? J Clin Invest 114: 1187–1195.

Van den Berghe G, Wouters P, Weekers F et al. (2001). Intensive insulin therapy in critically ill patients. N Engl J Med 345: 1359–1367.

Van den Berghe G, Schoonheydt K, Becx P et al. (2005). Insulin therapy protects the central and peripheral nervous system of intensive care patients. Neurology 64: 1348–1353.

Vanhorebeek I, Vos RD, Mesotten D et al. (2005). Protection of hepatocyte mitochondrial ultrastructure and function by strict blood glucose control with insulin in. Lancet 53–59.

Vanpee G, Hermans G, Segers J et al. (2014). Assessment of limb muscle strength in critically ill patients: a systematic review. Crit Care Med 42: 701–711.

Weber-Carstens S, Koch S, Spuler S et al. (2009). Nonexcitable muscle membrane predicts intensive care unit-acquired paresis in mechanically ventilated, sedated patients. Crit Care Med 37: 2632–2637.

Weber-Carstens S, Deja M, Koch S et al. (2010). Risk factors in critical illness myopathy during the early course of critical illness: a prospective observational study. Crit Care 14: R119.

Weber-Carstens S, Schneider J, Wollersheim T et al. (2013). Critical illness myopathy and GLUT4: significance of insulin and muscle contraction. Am J Respir Crit Care Med 187: 387–396.

Wieske L, Chan Pin Yin DR, Verhamme C et al. (2013a). Autonomic dysfunction in ICU-acquired weakness: a prospective observational pilot study. Int Care Med 39: 1610–1617.

Wieske L, Kiszer ER, Schultz MJ et al. (2013b). Examination of cardiovascular and peripheral autonomic function in the ICU: a pilot study. J Neurol 260: 1511–1517.

Wieske L, Witteveen E, Petzold A et al. (2014). Neurofilaments as a plasma biomarker for ICU-acquired weakness: an observational pilot study. Crit Care 18: R18.

Wieske L, Dettling-Ihnenfeldt DS, Verhamme C et al. (2015a). Impact of ICU-acquired weakness on post-ICU physical functioning: a follow-up study. Crit Care 19: 196.

Wieske L, Verhamme C, Witteveen E et al. (2015b). Feasibility and diagnostic accuracy of early electrophysiological recordings for ICU-acquired weakness: an observational cohort study. Neurocrit Care 22: 385–394.

Witt N, Zochodne D, Bolton CF et al. (1991). Peripheral nerve function in sepsis and multiple organ failure. Chest 99: 176–184.

Witteveen E, Wieske L, Verhamme C et al. (2014). Muscle and nerve inflammation in intensive care unit-acquired weakness: a systematic translational review. J Neurol Sci 345: 15–25.

Woittiez AJJ, Veneman TF, Rakic S (2001). Critical illness polyneuropathy in patients with systemic inflammatory response syndrome or septic shock [1]. Int Care Med 27: 613.

Wollersheim T, Woehlecke J, Krebs M et al. (2014). Dynamics of myosin degradation in intensive care unit-acquired weakness during severe critical illness. Int Care Med 40: 528–538.

Yosef-Brauner O, Adi N, Ben Shahar T et al. (2015). Effect of physical therapy on muscle strength, respiratory muscles and functional parameters in patients with intensive care unit-acquired weakness. Clin Respir J 9: 1–6.

Zifko UA, Zipko HT, Bolton CF (1998). Clinical and electrophysiological findings in critical illness polyneuropathy. J Neurol Sci 159: 186–193.

Zink W, Kaess M, Hofer S et al. (2008). Alterations in intracellular Ca 2-homeostasis of skeletal muscle fibers during sepsis. Crit Care Med 36: 5–9.

Zochodne DW, Bolton CF, Wells GA et al. (1987). Critical illness polyneuropathy: a complication of sepsis and multiple organ failure. Brain 110: 819–841.

Chapter 30

Neurologic complications of transplantation

R. DHAR*
Division of Neurocritical Care, Department of Neurology, Washington University, St. Louis, MO, USA

Abstract

Major neurologic morbidity, such as seizures and encephalopathy, complicates 20–30% of organ and stem cell transplantation procedures. The majority of these disorders occur in the early posttransplant period, but recipients remain at risk for opportunistic infections and other nervous system disorders for many years. These long-term risks may be increasing as acute survival increases, and a greater number of "sicker" patients are exposed to long-term immunosuppression. Drug neurotoxicity accounts for a significant proportion of complications, with posterior reversible leukoencephalopathy syndrome, primarily associated with calcineurin inhibitors (i.e., cyclosporine and tacrolimus), being prominent as a cause of seizures and neurologic deficits. A thorough evaluation of any patient who develops neurologic symptoms after transplantation is mandatory, since reversible and treatable conditions could be found, and important prognostic information can be obtained.

INTRODUCTION

Organ and hematopoietic stem cell transplantation (HSCT) have revolutionized the management of patients with end-stage organ failure, hematologic malignancies, and other life-threatening diseases. The last few decades have witnessed major breakthroughs in transplant techniques and immunosuppression regimens that have greatly improved graft and patient survival. These interventions are often life-saving and improve quality of life for transplant recipients (Tome et al., 2008). However, they have resulted in an increasing number of unstable patients with organ dysfunction and multiple comorbidities undergoing complex transplant procedures, followed by a prolonged period of immunosuppression. In fact, over 25 000 transplants occurred in the USA in 2014, with an estimated 110 000 worldwide (White et al., 2014; Israni et al., 2015). Transplants, broken down by organ, with common indications and respective 1- and 5-year survival rates are shown in Table 30.1.

Transplant procedures expose sick recipients to hemodynamic and physiologic perturbations, as well as multiple potentially toxic medications and infectious agents. This complex milieu places these patients at high risk for a variety of neurologic complications, some occurring primarily in the immediate posttransplant period, and others developing in a delayed fashion. Acute morbidity often overlaps with systemic and graft-related derangements in those with a complicated postoperative course. Much of this morbidity is shared between various transplant types, while some are more common with, or specific to, certain transplant situations (e.g., embolic cerebrovascular events after heart transplant or central pontine myelinolysis (CPM) after liver transplantation). Table 30.1 also summarizes some of these organ-specific considerations.

HSCT involves transfer of stem cells from either one person to another (i.e., allogeneic) or, after procurement and storage, back to the donor (i.e., autologous) (Copelan, 2006). This allows reconstitution of marrow and immune function after ablation of the native bone marrow (and hopefully the underlying malignant/disease process). These toxic myeloablative conditioning regimens vary across centers, consisting of high-dose chemotherapy, sometimes in combination with total-body irradiation (TBI). Patients lack immune or hematopoietic function, and are highly susceptible to infections and

---

*Correspondence to: Rajat Dhar, MD, FRCP(C), Department of Neurology (Division of Neurocritical Care), Campus Box 8111, 660 S Euclid Avenue, Saint Louis MO 63110, USA. Tel: +1-314-362-2999, E-mail: dharr@neuro.wustl.edu

Table 30.1

Details of solid-organ and hematopoietic transplant procedures

Organ	Kidney	Liver	Lungs	Heart	Intestinal	HSCT
Number transplanted in 2013*	17 654	6455	1946	2554	109	Not reported
Common indications	ESRD due to: • Diabetes • Glomerulonephritis • Renovascular • PKD	Hepatitis B/C Alcoholic cirrhosis PBC/PSC Autoimmune hepatitis Malignancy Amyloidosis Acute liver failure	COPD including α-1 antitrypsin def. Cystic fibrosis Interstitial lung disease Primary PH Sarcoidosis	Idiopathic CM Ischemic CM Congenital HD HOCM (often after bridging with LVAD)	TPN-dependent short-bowel syndrome due to: • Mesenteric thrombosis • Crohn's disease • Radiation • Neoplasm	Leukemia, MDS Lymphoma Aplastic anemia Myeloma Sickle cell Malignancy SCID
Technical issues	Injury to the femoral nerve or lumbosacral plexus	Blood loss Fluid/electrolyte shifts Anastomotic leak or stenosis Air embolism	Phrenic nerve injury Reperfusion pulmonary edema Hypoxemia; PH May require bypass	Cardiac bypass Aortic cannulation Manipulation of the heart (air embolism) Arrhythmias	Bacterial translocation Malnutrition Hepatic cholestasis	Toxicity of conditioning regimens, including myeloablation and irradiation GVHD (allogeneic only)
Markers of posttransplant organ function	Making urine Serum creatinine	Making bile Serum bilirubin Synthetic function (e.g., INR)	Oxygenation Ability to extubate	Off circulatory support Cardiac index Blood pressure	Decreasing G tube returns Increasing ileostomy output	Engrafting, i.e., rising cell counts
Patient survival: 1-year vs. 5-year	95% 82%	88% 74%	83% 54%	88% 75%	89% 58%	Varies with disease (10–85%)
Specific neurologic complications	Hypertension-related stroke or ICH	CPM, HE Coagulopathy-related bleeding	HIE Phrenic nerve injury	Embolic stroke HIE	Wernicke's encephalopathy	Early infections Coagulopathy-related bleeding

Adapted from Dhar and Human (2011).
*Numbers from Israni et al. (2015).

CM, cardiomyopathy; CPM, central pontine myelinolysis; COPD, chronic obstructive pulmonary disease; ESRD, end-stage renal disease; GVHD, graft-versus-host disease; HD, heart disease; HE, hepatic encephalopathy; HIE, hypoxic-ischemic encephalopathy; HOCM, hypertrophic obstructive cardiomyopathy; HSCT, hematopoietic stem cell transplantation; ICH, intracranial hemorrhage; INR, international normalized ratio; LVAD, left-ventricular assist device; MDS, myelodysplastic syndrome; PBC, primary biliary cirrhosis; PKD, polycystic kidney disease; PH, pulmonary hypertension; PSC, primary sclerosing cholangitis; SCID, severe combined immunodeficiency; TPN, total parenteral nutrition.

bleeding in the period before engraftment of new stem cells occurs.

EPIDEMIOLOGY

A number of studies have attempted to ascertain the frequency and spectrum of neurologic complications after various types of organ transplantation (Table 30.2) (Graus et al., 1996; Antonini et al., 1998; Goldstein et al., 1998; Vecino et al., 1999; Bronster et al., 2000; Mayer et al., 2002; Lewis and Howdle, 2003; Sostak et al., 2003; Perez-Miralles et al., 2005; Denier et al., 2006; Saner et al., 2007; Siegal et al., 2007; Zierer et al., 2007; Dhar et al., 2008; van de Beek et al., 2008; Yardimci et al., 2008; Zivkovic et al., 2009, 2010; Mateen et al., 2010; Munoz et al., 2010; Vizzini ct al., 2011; Kim et al., 2015). However, many of these studies limit their evaluation to the acute period, use retrospective ascertainment which may miss subtle neurologic manifestations, and do not always have rigorous definitions of each complication and how underlying etiology was determined. Therefore, these are at best approximations of incidence, and most likely underestimate the true frequency. Older studies were omitted from this list, as most contemporary studies have demonstrated that complications have decreased in frequency over the past two decades. This may be attributed mainly to more careful initiation and titration of drug regimens that used to contribute to a high incidence of seizures and mental status alterations (Vizzini et al., 2011). A recent study of liver transplant recipients employed serial prospective evaluation of 134 adult recipients by

Table 30.2

Contemporary studies evaluating the incidence of neurologic complications (by type of transplant)

Organ	Studies	Numbers	Encephalopathy	Seizures	Cerebrovascular	Total rate
Heart	Mayer et al. (2002)	191	2%	7%	7.3%	36%[a]
	Perez-Miralles et al. (2005)	322	n/a	2%	3.5% ischemic; 0.6% ICH	14%
	Zierer et al. (2007)	200	5%	7%	3.5% ischemic; 2% ICH	23%
	Van de Beek et al. (2008)	313	9%	2.6%	2.2% stroke; 0.6% ICH	19%
	Munoz et al. (2010)	384	1%	2%	5% stroke; 1.6% ICH	20%
Lung	Goldstein et al. (1998)	100	3%	10%	5%	26%
	Zivkovic et al. (2009)	132	25%	8%	5% stroke; 1.5% ICH	68%
	Mateen et al. (2010)	120	24%	6%	10.8%	79%
Liver	Vecino et al. (1999)	43	16%	7%	9%	53%[b]
	Bronster et al. (2000)	463	12%	8%	0.6% ischemic; 1.5% ICH	20%
	Lewis and Howdle (2003)	711	10%	6%	1.7% ischemic; 2% ICH	26%
	Saner et al. (2007)	168	19%	5%	2.4%	27%
	Dhar et al. (2008)	101	28%	4%	None	31%
	Vizzini et al. (2011)	395	5%	1%	1.3% ischemia; 2% ICH	16%
	Kim et al. (2015)	791	3%	2%	2.1% ischemic; 1.3% ICH	8%
Intestinal	Zivkovic et al. (2010)	54	43%	17%	4%	85%[b]
Renal	Yardimici et al. (2008)	132	1%	2%	1.5%	14%[b]
HSCT	Graus et al. (1996)	425[c]	3%	4%	0.5% ischemic; 3.8% ICH[f]	11%
	Antonini et al. (1998)	115[c]	7%	7%	0.9% ICH	56%[a]
	Sostak et al. (2003)	71	3%	1%	4% ischemic; 2.8% ICH[f]	18%
	Denier et al. (2006)	361	3%	5%[e]	1.7% ICH	16%
	Siegal et al. (2007)	302	15%[d]	9%	0.3%	23%[g]

Adapted from Dhar and Human (2011).
[a]Includes polyneuropathy (in Antonini: rate of major complications was 24%).
[b]Includes headache.
[c]All had leukemia.
[d]Posterior reversible leukoencephalopathy syndrome in 7%.
[e]Mainly due to infections.
[f]Mainly subdural hematoma.
[g]9% at 30 days.
Note: all studies were retrospective except Vecino, Antonini, Sostak, and Siegal. Most studies had extended follow-up but the following focused only on the acute time period: Mayer, Zierer, Van de Beek, Lewis, Dhar, Antonini.
ICH, intracranial hemorrhage; HSCT, hematopoietic stem cell transplantation.

a neurologist (Bernhardt et al., 2015). Central nervous system (CNS) complications were seen in 44 (33%) cases, including a 6% incidence of seizures. Most complications (84%) were ascribed to toxic-metabolic etiologies, which were reversible and associated with low mortality. However, structural pathologies (primarily cerebrovascular in origin with rare fungal infections), although less common, were associated with high rates of death.

Incidence and distribution of complications may also vary in pediatric populations; most studies focus on adult recipients, while a few include both adult and pediatric patients. Rates of complications do appear to vary based on transplanted organ, with kidney transplants being the most common procedure, but harboring a relatively low rate of neurologic morbidity. Conversely, liver and lung transplants are extremely complex surgical procedures often performed in unstable recipients, and pose a high rate of acute neurologic complications. Overall, serious complications occur in 10–30% of transplant recipients, and are associated with (though not necessarily independently causative of) a higher mortality rate (Senzolo et al., 2009; Pustavoitau et al., 2011).

Timing of complications can inform the differential diagnosis. In the early postoperative period, encephalopathy is the most frequent syndrome, and drug toxicity is the major etiology. In contrast, opportunistic infections (OI) tend to occur 1 month or more after transplant, when the effects of immunosuppression have had time to set in. The spectrum of complications may be shifting, with fewer acute complications related to improved operative techniques and lower-intensity drug regimens. However, as transplant patients survive longer (e.g., 5-year survival of liver transplantation now exceeding 70%), the incidence of OI and posttransplant malignancies is likely to rise further. Beyond timing, morbidity can best be categorized by clinical syndrome (encephalopathy, seizures, focal deficits), and then by etiology. However, many events are multifactorial, with patient- and organ-specific factors contributing in tandem.

Morbidity after HSCT is common and harbors some distinct features. OI and bleeding complications are more likely than with organ transplantation. Allogeneic HSCT poses greater risk than autologous HSCT, especially given the need for longer-term immunosuppression and the ongoing risk of graft-versus-host disease (GVHD). Complication rates are also higher with unrelated HSCT donors and in those with underlying chronic myelogenous leukemia (Graus et al., 1996; de Brabander et al., 2000; Denier et al., 2006).

Risk factors for neurologic complications include premorbid and perioperative factors, as well as drug exposure and intensity. Some conditions leading to organ failure may predispose to postoperative complications; for example, those with alcoholic cirrhosis and preoperative hepatic encephalopathy are felt to be at higher risk for complications such as delirium and encephalopathy postoperatively (Buis et al., 2002; Dhar et al., 2008; Vizzini et al., 2011). Those undergoing repeat transplantation tend to be sicker and are also at higher risk (Bernhardt et al., 2015).

NEUROPATHOLOGY

Autopsy studies were excluded from this summary of incidence as they are, by their very nature, biased towards the most severe complications and likely find a higher incidence of abnormalities in this sick subpopulation than in a representative cohort. However, the variety of neuropathology seen in patients dying after organ transplantation is diverse and speaks to the serious complications that can occur. Many autopsy studies find a higher rate of osmotic demyelination, OI, and ischemic changes. Hypoxic-ischemic brain injury may be especially common in those dying of graft and multiorgan failure, or in those with terminal sepsis and shock (Blanco et al., 1995; Prayson and Estes, 1995; McCarron and Prayson, 1998; Idoate et al., 1999).

CLINICAL PRESENTATION

Acute encephalopathy

Alterations in consciousness and/or cognition remain the most common complications in most series. These may be termed delirium when primarily associated with fluctuating level of consciousness and impairment in attention. Such patients may have reduced level of consciousness or present with psychomotor agitation and hallucinations. Encephalopathy tends to occur mainly in the early postoperative period (i.e., prior to initial hospital and even intensive care unit (ICU) discharge) and may be significant in both extending hospital length of stay and, in some cases, as a risk factor for increased mortality (Bronster et al., 2000). Rates as high as 30–40% have been reported, especially in series involving liver, small bowel, and lung transplantation. These exhibit the greatest potential for perioperative metabolic derangements and complex postoperative courses. Encephalopathy is an important marker that a systemic or CNS derangement has developed, and is therefore important to recognize and evaluate. While findings are frequently nonfocal and related to diffuse brain dysfunction, multifocal CNS disorders such as strokes and infections can cause similar encephalopathy and mimic a toxic-metabolic process. Presentations may be divided into cases that present immediately postoperatively with failure to reawaken to normal mentation and acute confusional states that develop in the days after.

Failure to awaken

This is perhaps the most feared neurologic complication of any major surgery, but is fortunately relatively rare, even after transplant procedures. The CNS basis for such acute severe encephalopathy is usually either intraoperative global hypoxia-ischemia or multiple vascular events. The operative record should be reviewed for signs of hypotension (e.g., excess blood loss, cardiac arrhythmias/cardiac arrest, time on bypass); a particular postreperfusion syndrome can occur in liver transplantation after reanastomosis that can precipitate acute hypotension. Pulmonary dysfunction can lead to arterial desaturation and cerebral hypoxia. The resultant hypoxic-ischemic encephalopathy (HIE) with widespread damage to susceptible brain regions most often presents with failure to awaken, or severe encephalopathy after surgery, although immediate brain imaging with computed tomography (CT) may be unrevealing (Singh et al., 1994). Magnetic resonance imaging (MRI) may reveal diffusion-weighted signal restriction in a cortical laminar, patchy, or sometimes subcortical pattern. A subset of patients with failure to awaken after transplant are found to have serious graft dysfunction (i.e., primary nonfunction), resulting in hypoxemia (lung), hypotension (heart), or hepatic or uremic encephalopathy. This will usually be apparent and take precedence over the neurologic presentation, which may be missed amidst the need for stabilization, sedation, and even retransplantation (Pokorny et al., 2000; Varotti et al., 2005; Marasco et al., 2010). Not only does this second surgery have an even higher rate of neurologic complications, but cerebral damage from the initial HIE may only be detected many days later (and may have otherwise tempered the expediency of transplanting a second organ into a neurologically devastated recipient).

Vascular insults causing severe encephalopathy are almost always multifocal and therefore usually embolic in origin. Air embolism can occur during organ or vascular manipulation, and may be accompanied by seizures and even cardiovascular collapse (Starzl et al., 1978); brain imaging may reveal intracranial air. Multiple strokes from cardioembolic or aortic sources can occur with cardiac or pulmonary transplantation. As with HIE, immediate CT imaging may be negative or reveal multiple subtle hypodensities (Fig. 30.1), while MRI will usually reveal the underlying ischemic insults. Brainstem injury can occur in cases of CPM, primarily after liver transplantation and usually in association with preoperative hyponatremia and perioperative fluid and osmotic shifts. This can result in a locked-in or comatose state, but is fortunately relatively rare (estimated as < 1%).

While global and multifocal ischemic insults account for the core of serious and even irreversible causes of encephalopathy immediately after transplant, reversible systemic effects can mimic this syndrome. Patients with hepatic or renal dysfunction may exhibit delayed clearance and accumulation of anesthetic or sedative drugs (e.g., benzodiazepines, opioids, barbiturates) given intraoperatively or in the ICU, contributing to impaired awakening. Neuromuscular blocking agents can also have persistent effects if hepatic and renal clearance is impaired, resulting in prolonged paralysis lasting a day or more postoperatively (Watling and Dasta, 1994). Lack of motor response to pain should trigger train-of-four testing (applying peripheral nerve stimulation to the median, ulnar, or facial nerves). Neostigmine (given along with glycopyrrolate) can be administered to reverse the action of such agents. In recipients with evidence of hepatic dysfunction (e.g., liver allograft failure or hepatic dysfunction after HSCT), measuring ammonia

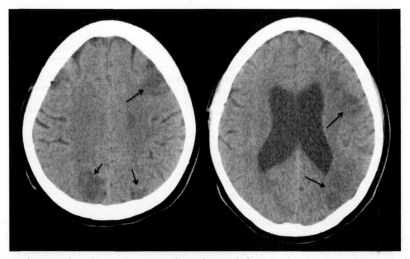

Fig. 30.1. Multiple hypodense regions (arrows) representing subacute infarcts on head computed tomography of a patient 4 days after cardiac transplant. Patient was confused and had mild aphasia. (Reproduced from Dhar and Human, 2011, with permission.)

levels may be helpful. Elevation in ammonia, while not predictive of severity or outcome, may point to hepatic encephalopathy as a contributing factor in cases of unexplained postoperative neurologic dysfunction (Ge and Runyon, 2014). Idiopathic hyperammonemia after HSCT is described later. Uremia, either from delayed graft function (occurring in 20–40% of renal transplants) or from acute kidney injury after cardiac, liver, or other transplants (often a sign of perioperative hypotension, medication effect, or sepsis), may be treated with dialysis, resulting in improvement in mental status. A distinct syndrome related to allograft rejection may occur, separate from the direct adverse consequences of graft dysfunction. The term rejection encephalopathy has been proposed to encapsulate this situation and has been described in patients with rejection after renal transplantation. It is hypothesized to occur secondary to cytokine release (Gross et al., 1982). Finally, nonconvulsive seizures can lead to failure to awaken and (unless a result of massive HIE or strokes) represents an important reversible etiology of coma.

Evaluation of transplant recipients with failure to awaken (Fig. 30.2) normally should begin with thorough assessment of clinical history, including: (1) premorbid status (e.g., pre-existing encephalopathy or neurologic disease); (2) intraoperative cardiovascular stability and notable surgical complications; (3) postoperative graft dysfunction (see Table 30.1 for signs in each organ setting) or rejection; (4) renal and hepatic function, and other metabolic derangements (e.g., hyponatremia); (5) sedative administration; and (6) signs of sepsis or multiorgan failure syndrome in the ICU (Young et al., 1990a). In particularly complex cases, this may best be accomplished by direct conversation and collaboration with the transplant surgeon, anesthesiologist, and ICU team (including pharmacist). Specific attention should be focused on any reversible etiologies and evaluation of catastrophic neurologic insults. If head CT is unrevealing

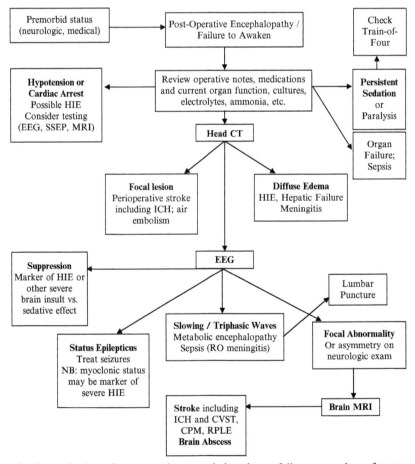

Fig. 30.2. Algorithm for the evaluation of postoperative encephalopathy or failure to awaken after transplant surgery. HIE, hypoxic-ischemic encephalopathy; EEG, electroencephalogram; SSEP, somatosensory evoked potential; MRI, magnetic resonance imaging; CT, computed tomography; ICH, intracranial hemorrhage; RO, rule out; CVST, cerebral venous sinus thrombosis; CPM, central pontine myelinolysis; RPLE, reversible posterior leukoencephalopathy. (Adapted from Dhar and Human, 2011, with permission.)

then an electroencephalogram (EEG) should be performed; this may uncover nonconvulsive seizures but can also point to focal abnormalities, triphasic waves as markers of metabolic encephalopathy, or severe suppression confirming diffuse brain injury or persistent sedative effect.

Postoperative encephalopathy and delirium

The spectrum of postoperative encephalopathy encompasses reduced levels of arousal (including stupor and coma) to alterations in awareness and content of cognition (i.e., delirium). Some of these presentations include a component of agitation and hallucinations, but others with encephalopathy are hypoactive and may not manifest as obviously (Beresford, 2001). Screening for delirium may be accomplished by the transplant team or bedside nurse using tools like the Confusion Assessment Method for the Intensive Care Unit (CAM-ICU) or Intensive Care Delirium Screening Checklist (ICDSC) (Bergeron et al., 2001; Ely et al., 2001).

A breakdown of common etiologies for encephalopathy is listed in Table 30.3; however, in many real-life cases, encephalopathy is multifactorial. Even preoperative factors may play a role: e.g., alcohol or drug withdrawal, including relating to cessation of prolonged sedation given to critically ill patients immediately preoperatively. Identification of likely etiologies is important, as both reversibility and prognosis are primarily determined by cause. Patients with remediable metabolic derangements or drug toxicity (each accounting for approximately one-third of cases) usually do well once the offending abnormality is removed or reversed. However, for those in whom encephalopathy heralds underlying CNS infection, HIE, osmotic demyelination, stroke, or other serious cerebral insult, prognosis is often guarded. No specific treatments for delirium have been proven effective beyond reversing or removing the offending etiology, if possible. Psychoactive medications that can exacerbate delirium (e.g., opiates, antihistamines, anticholinergics) should be avoided, while antipsychotic medications to control agitation should be reserved for severe cases where injury to self or others is a concern. Benzodiazepines should only be used in cases where encephalopathy is felt to result from alcohol or sedative withdrawal, or in the presence of concomitant seizures. There is also greater risk for medication interactions and impaired elimination in transplant patients with hepatic/renal dysfunction and polypharmacy.

Alterations in consciousness are particularly common after liver transplantation. Many such patients have a history of hepatic encephalopathy, which can present with confusion, inattention, apathy, and asterixis. Hepatic encephalopathy is related to not only excess ammonia,

Table 30.3

Differential diagnosis and evaluation of posttransplant encephalopathy

Cause of encephalopathy	Timing	Testing
Drug toxicity (especially cyclosporine and tacrolimus)	Usually acute (days to weeks)	Serum drug levels: may be elevated or normal MRI may show changes consistent with PRES
Metabolic: organ/graft failure, sepsis, electrolyte disturbance, medication effect (e.g., opiates, benzodiazepines)	Variable, often acute	Organ function panels (renal, hepatic, ABG) Electrolytes, glucose, calcium, ammonia level Cultures for infection
Stroke: ischemic or hemorrhagic, CVST, HIE	Perioperative (delayed in renal transplant)	Brain imaging (CT, MRI) MR venography if CVST suspected
Opportunistic infections and malignancies (especially PTLD)	Usually delayed	Brain imaging (CT, MRI) CSF analysis (cryptococcal antigen, fungal and AFB culture, PCR for viruses)
Osmotic demyelination syndrome (including central pontine myelinolysis)	Acute perioperative	MRI (but findings may be delayed up to 1 week)
Wernicke's encephalopathy	First few weeks postoperatively	Ataxia, ophthalmoplegia (absent caloric responses) Malnutrition; MRI has characteristic changes (mammillary bodies, periaqueductal)
Nonconvulsive seizures	Variable, usually acute	EEG

Adapted from Dhar and Human (2011).
ABG, arterial blood gas; AFB, acid-fast bacilli (e.g., Ziehl–Neelsen stain); CSF, cerebrospinal fluid; CT, computed tomography; CVST, cerebral venous sinus thrombosis; EEG, electroencephalogram; HIE, hypoxic-ischemic encephalopathy; MRI, magnetic resonance imaging; PCR, polymerase chain reaction; PRES, posterior reversible encephalopathy syndrome; PTLD, post-transplant lymphoproliferative disorder.

but a multifactorial failure to detoxify chemicals in the failing liver (Butterworth, 2003). This may alter the blood–brain barrier (BBB) and render these patients more susceptible to encephalopathy and drug neurotoxicity after liver transplantation. Interestingly, although acute liver failure (e.g., from acetaminophen overdose) can result in diffuse cerebral edema and coma, these patients often recover well neurologically if they receive transplantation before herniation has occurred (Dhar et al., 2008). Toxicity of myeloablative regimens in HSCT can lead to veno-occlusive disease of the liver, resulting in hepatomegaly, jaundice, and hepatic encephalopathy (Baglin et al., 1990; MacQuillan and Mutimer, 2004).

Akinetic mutism

Some transplant recipients develop a state of impaired verbal and motor responsiveness; they appear awake but do not speak or move spontaneously or to command (Starzl et al., 1978). While this syndrome has primarily been reported as a reversible complication of calcineurin inhibitors (i.e., cyclosporine and tacrolimus), it may also occur in cases of HIE or osmotic demyelination (Laureno and Karp, 1997). A similar picture was reported with amphotericin treatment of HSCT patients who received irradiation as part of their conditioning (Devinsky et al., 1987; Walker and Rosenblum, 1992). Radiation was postulated to open the BBB and facilitate drug toxicity in these cases. There was also a case of akinetic mutism in a heart transplant recipient after receiving the monoclonal antibody OKT3, which resolved once the drug was discontinued and CD3+ lymphocyte levels returned to normal (Pittock et al., 2003). Mutism without akinesia has been reported in 1% of liver transplant recipients within the first 10 days after surgery (often in association with seizures) and was responsive to discontinuation of calcineurin inhibitors (Bronster et al., 1995; Bianco et al., 2004). Finally, any patient with mutism should be evaluated for neuroleptic malignant syndrome, which can be rapidly fatal if not recognized. This syndrome is related to neuroleptic exposure and usually has associated fever, muscle rigidity, elevated creatine kinase, and dysautonomia (Garrido and Chauncey, 1998).

Seizures

Convulsive seizures are disturbing to patients, their families, and healthcare providers. They can lead to injury, hemodynamic instability, aspiration, and dislodgement of catheters and monitoring devices. They are also important warning signs of a serious CNS or metabolic derangement. Contemporary series suggest that seizures occur in 5–10% of transplant patients, with most clustered around the first weeks postoperatively (Bronster et al., 2000). The differential diagnosis of seizures overlaps that of encephalopathy and those developing seizures often exhibit prodromal or postictal encephalopathy in conjunction. As with encephalopathy, distinguishing benign reversible etiologies (e.g., metabolic/drug-related) from either primary CNS disorders or serious systemic disorders (e.g., organ/graft dysfunction) is imperative in determining prognosis. Systemic disorders typically present with generalized seizures, while focal-onset seizures signal an underlying CNS lesion (e.g., infection, stroke). However, focal features at onset of a seizure are often missed, so distinguishing based on semiology is unreliable. The most common cause of postoperative seizures is drug toxicity, which is often preceded by subtle behavioral and mental status changes. Nonetheless, drug levels are often not elevated in the face of probable drug-induced seizures. In these cases, the diagnosis rests on excluding other etiologies, but is supported by normalization of encephalopathy, and cessation of seizures, after drug discontinuation. A rapid rise in drug levels may contribute as much as raised absolute levels (Lane et al., 1988; Wijdicks et al., 1994, 1995, 1996b).

Nonconvulsive seizures can present primarily as unexplained and/or fluctuating encephalopathy and require EEG for diagnosis. Nonconvulsive status epilepticus (NCSE) should be considered in any patient with failure to awaken, even if a likely etiology has been revealed by initial evaluation. CNS or systemic dysfunction can cause encephalopathy, but presentation can be dramatically worsened if NCSE supervenes. Fortunately, NCSE is relatively rare, but if a patient remains unresponsive after apparent cessation of convulsive seizure activity, an urgent EEG should be obtained (Junna and Rabinstein, 2007). Status epilepticus can be the presenting mode of seizures and was reported as more common in pediatric series (Cordelli et al., 2014). In fact, 16 of 22 cases of seizures in this series of pediatric HSCT presented with status epilepticus; drug toxicity accounted for over half these seizures.

EEG can also be useful in differentiating epileptic from nonepileptic movements such as tremors or myoclonus, both of which can occur as a result of medication or metabolic toxicity. Multifocal myoclonus can mimic seizures in a comatose patient, but is a reliable marker of metabolic (i.e., nonepileptic) encephalopathy (Mahoney and Arieff, 1982). Conversely, actual myoclonic status epilepticus (where jerking is generalized rather than migratory, and corresponds to epileptic discharges on EEG) is a nearly uniform negative prognostic marker after HIE (especially if there is also a nonreactive EEG and poor neurologic exam) (Young et al., 1990b). Some patients develop myoclonic movements on emergence from anesthesia (especially with propofol), which can be mistaken for seizures (Walder et al., 2002).

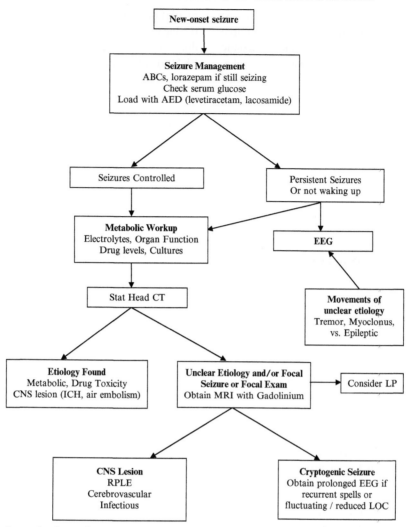

Fig. 30.3. Algorithm for evaluation and management of seizures after transplantation. ABCs, airway, breathing, circulation; AED, antiepileptic drug; EEG, electroencephalogram; CT, computed tomography; CNS, central nervous system; ICH, intracranial hemorrhage; MRI, magnetic resonance imaging; LP, lumbar puncture; RPLE, reversible posterior leukoencphalopathy; LOC, level of consciousness. (Adapted from Dhar and Human, 2011, with permission.)

An algorithm for the evaluation of seizures is shown in Figure 30.3.

TREATMENT

After stabilization of cardiopulmonary status and cessation of acute seizures, the priority of seizure management is rapid evaluation and reversal of underlying etiology. Hypoglycemia should be ruled out with bedside testing in all cases. First-line acute therapy for convulsions should be benzodiazepines. Many transplant patients will not require prolonged anticonvulsant therapy if a cause is found and removed. Administration of an antiepileptic drug (AED) can be considered while workup and general management are under way to prevent recurrent seizures. While phenytoin has traditionally been employed, as it is readily available in most hospital locations and can be given intravenously, it has significant limitations in this setting. It has a number of significant drug interactions, most seriously induction of cyclosporine metabolism (necessitating higher doses and active titration to maintain therapeutic levels) and high degree of protein binding (often altered in these sick patients, resulting in higher free and active drug levels than reflected in the commonly measured total phenytoin levels) (Keown et al., 1984; Glass et al., 2005). In patients with hepatic impairment, metabolism of phenytoin may be further impaired. Further, rapid administration of intravenous phenytoin can result in arrhythmias and hypotension, likely accentuated in hemodynamically susceptible transplant patients.

Given the limitations of phenytoin, newer AEDs are now preferred for seizures in transplant recipients. Both levetiracetam and lacosamide are attractive alternatives: they can be loaded rapidly by intravenous route, are free of significant drug interactions, do not undergo hepatic metabolism, and lack major relevant toxicities (Trinka, 2011). AED therapy can usually be safely discontinued after the period of risk has subsided; this may vary from days to weeks in cases of acute reversible derangements to longer/indefinitely in those with structural CNS lesions or epileptiform EEG findings. Prognosis of seizures is good if a reversible cause is found and corrected (with recurrence being rare), while it is poor if seizures herald a serious CNS process or a systemic illness, in whom seizures may simply be part of an agonal decline (Wijdicks et al., 1996b).

DRUG TOXICITY

Transplant patients are exposed to a number of potentially toxic medications as part of their management. These include immunosuppressive agents that have revolutionized graft survival by minimizing rejection (Calne et al., 1978). The first calcineurin inhibitor introduced was cyclosporine, although tacrolimus (formerly FK506) is now more frequently used due to lower rates of acute rejection, hirsutism, and hypertension (Webster et al., 2005; Hachem et al., 2007; Penninga et al., 2010). Induction agents may be added to the immunosuppressive regimen immediately after transplant to inhibit acute rejection: these include thymoglobulin, OKT3, basiliximab, or daclizumab. A calcineurin inhibitor, along with corticosteroids and often mycophenolate mofetil, then constitute maintenance therapy, at least to discharge, after which corticosteroids may be tapered. Most of these agents have narrow therapeutic windows and many have significant potential for neurotoxicity (as outlined in Table 30.4). For example, busulfan is used in some TBI-sparing induction regimens for HSCT, but can precipitate seizures in as many as 10% of patients (De La Camara et al., 1991). OKT3 has been associated with encephalopathy and seizures (Parizel et al., 1997;

Table 30.4

Drug toxicities: differential diagnosis classified by symptomatology

Symptom or syndrome	Medications	Management
Headache	CNI, MMF, sirolimus	Consider change to other CNI
	OKT3 and ATG (aseptic meningitis)	Try abortive agents (e.g., sumatriptan)
		Exclude underlying CNS lesion
Tremor	CNI, steroids, sirolimus	Observe/may resolve over time
Myoclonus	Cephalosporins, ketamine	
Psychiatric disturbances	CNI, steroids, MMF	Reduce dose if possible
Delirium/ encephalopathy	CNI, OKT3, steroids, ganciclovir	Hold CNI[a]
	May be multifactorial (consider concomitant benzodiazepines, opioids, etc.)	
Seizures	CNI, busulfan, OKT3 (rare), other medications (imipenem, beta-lactams, metronidazole, theophylline)	Hold CNI[a]
		See Fig. 30.3
Akinetic mutism	CNI, OKT3,	Hold CNI[a]
	amphotericin (with TBI)	Consider CPM, HIE, or other structural etiology if not resolve
PRES	CNI	Hold CNI*
		See Table 30.5
Paresthesias/ neuropathy	CNI, chemotherapy for HSCT, GVHD, thalidomide, triazole antifungals	Minimize drug exposure
		Consider CIP
Myopathy	Steroids,	Wean steroids if possible
	GVHD (polymyositis)	
Epidural lipomatosis	Steroids	May require surgical decompression

Adapted from Dhar and Human (2011).
*Consider restarting at lower dose or changing to another agent once symptoms resolve.
ATG, antithymocyte globulin; CIP, critical illness polyneuropathy; CNI, calcineurin inhibitors; CNS, central nervous system; CPM, central pontine myelinolysis; GVHD, graft-versus-host disease; HIE, hypoxic-ischemic encephalopathy; HSCT, hematopoietic stem cell transplantation; MMF, mycophenolate mofetil; PRES, posterior reversible leukoencephalopathy syndrome; TBI, total body irradiation.

Pittock et al., 2003). Newer induction agents like daclizumab may be associated with a lower rate of complications (Munoz et al., 2010). Ganciclovir for herpesvirus therapy or prophylaxis has been associated with encephalopathy that is reversible on discontinuation (Sharathkumar and Shaw, 1999; Sakamoto et al., 2013). In fact, a significant proportion of delirium and seizures after transplantation can be attributed to drug toxicity. Incidence peaks in the immediate postoperative period, when drugs are loaded and used in combination.

Calcineurin inhibitors are the leading cause of drug-related neurotoxicity in transplant recipients. These agents work by binding to proteins called immunophilins to form a complex that inhibits calcineurin (Ho et al., 1996). This results in inhibition of calcium-dependent signaling pathways that release interleukin-2 and activate T cells. Their neurotoxicity may also be mediated through their effects on calcineurin, a critical regulator of neuronal function and excitability (Gijtenbeek et al., 1999). Specifically, calcineurin inhibitors may affect regulation of the BBB and alter sympathetic activation and impair vasoconstriction of blood vessels in the brain (Dawson, 1996; Bechstein, 2000). Contributing factors such as pre-existing hepatic encephalopathy and perioperative hypotension may disrupt the BBB, allowing more drug to reach brain tissue and accentuate their effects. Low cholesterol facilitates higher unbound circulating drug levels to diffuse into the brain (de Groen et al., 1987). Drug effects include endothelial disruption and formation of vasogenic edema (Bunchman and Brookshire, 1991). Neuropathologic studies have confirmed such endothelial damage with vasogenic edema in the absence of infarction (Lavigne et al., 2006).

Calcineurin inhibitor toxicity may present with a myriad of symptoms ranging from minor complaints (e.g., headache, paresthesias, tremor – seen in up to half of treated patients) to neuropsychiatric disturbances (e.g., insomnia, anxiety, agitation). These symptoms may occur in isolation or precede the development of more overt delirium with hallucinations and delusions that may culminate in seizures and persistent encephalopathy if not detected. The radiographic correlate of calcineurin inhibitor toxicity in many cases is the posterior reversible encephalopathy syndrome (PRES), also known as reversible posterior leukoencephalopathy (RPLE).

PRES is a syndrome with many causes, but calcineurin inhibitor toxicity is perhaps second only to hypertension as a trigger in large observational studies (Hinchey et al., 1996). The acronym was developed based on a distinctive pattern of imaging findings: a white-matter predominant abnormality with a predilection for the posterior aspects of the cerebral hemispheres (Fig. 30.4). The pathologic hallmark is vasogenic

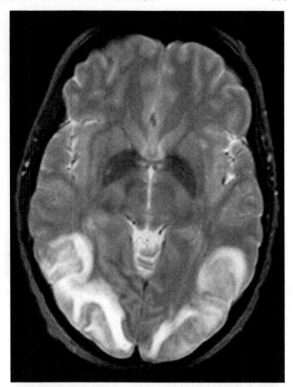

Fig. 30.4. Magnetic resonance imaging showing marked T2-weighted hyperintensities representing vasogenic edema associated with the use of calcineurin inhibitors (i.e., posterior reversible encephalopathy syndrome). (Reproduced from Wijdicks and Hocker, 2014, with permission.)

cerebral edema thought to result from disruption of the BBB with capillary leakage. White-matter involvement is usually bilateral, fairly symmetric, and, if caught early, reversible. However, as it is increasingly diagnosed with the advent of MRI, more variants and heterogeneity in its pattern are being recognized: for example, PRES may affect the frontal lobes and other regions of the cortex, may be asymmetric in some cases, and a brainstem and cerebellum-predominant variant may cause obstructive hydrocephalus and prominent mental status disturbances (Fig. 30.5) (Kumar et al., 2012). Typically, PRES presents with acute or subacute headache, confusion, and visual disturbances. In its severe form, it can lead to focal deficits, coma, and irreversible hemorrhage or stroke. Seizures are also common as the syndrome evolves; they can be a presenting symptom or later complication (Furukawa et al., 2001; Lee et al., 2008).

One large series found PRES in 0.5% of 4222 solid-organ transplant recipients, while another reported an incidence of 1.6% in HSCT (Wong et al., 2003; Bartynski et al., 2008). These numbers are likely underestimates, as there is no prospective study of transplant patients that evaluated each recipient with potential

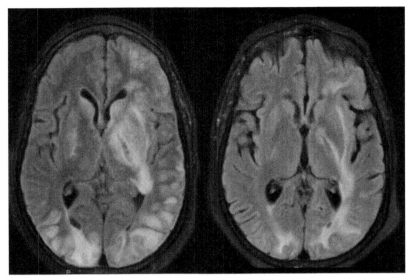

Fig. 30.5. Fluid-attenuated inversion recovery magnetic resonance imaging (MRI) (left) with extensive posterior but also asymmetric deep subcortical and basal ganglia edema in patient with seizures; follow-up MRI 6 days later (right) shows partial resolution.

symptoms using MRI. For example, a prospective series in HSCT found PRES in 7% by 1 year, with most cases being diagnosed within the first 30 days (Siegal et al., 2007). PRES likely accounts for many seizures and other neurologic abnormalities in the acute phase after transplantation. One older series found PRES in 4 of 17 patients with drug-induced seizures after liver transplant, but a contemporary study found MRI changes of PRES in all 4 patients with tacrolimus-induced seizures (Wijdicks et al., 1996b; Emiroglu et al., 2006). Half of the seizures seen in one series of heart transplant patients were attributed to PRES (Navarro et al., 2010). While the majority of cases of PRES occur within 30–90 days of transplant, late cases have been reported, especially in renal transplant recipients and in association with hypertension (Bartynski et al., 2008). Liver transplant recipients may be at particular risk, even in the absence of hypertension, since the BBB may become disrupted from portosystemic encephalopathy. PRES may also result from causes other than calcineurin toxicity. For example, sepsis, renal failure, and other medications (e.g., rituximab) may cause a similar syndrome (Jaiswal et al., 2015).

Prompt diagnosis of PRES is critical, as it is eminently reversible if caught early and calcineurin therapy is modified or discontinued. Diagnosis requires a high index of suspicion, even when manifestations are subtle. CT scan may be negative or only show subtle white-matter hypodensity. MRI is the preferred diagnostic tool, with fluid-attenuated inversion recovery sequences typically showing prominent signal abnormality (representing cerebral edema) in the white matter. These regions are typically diffusion weighted imaging-negative, showing increased, not decreased, diffusion, which differentiates these from acute infarction. They may sometimes involve the cortical ribbon, frontal lobes, or deeper white matter. Brainstem involvement in PRES may mimic central pontine demyelination, while hemispheric white-matter involvement can be confused with extrapontine demyelination in osmotic demyelination syndrome (ODS) or with demyelination from progressive multifocal leukoencephalopathy (PML). Human herpesvirus-6 (HHV-6) encephalitis (especially after HSCT) has also been reported to mimic PRES, with symmetric white-matter signal abnormality on MRI, but will also typically have systemic symptoms; cerebrospinal fluid (CSF) will harbor virus that can be detected by polymerase chain reaction (Gewurz et al., 2008). An approach to the differential diagnosis of white-matter diseases in transplant recipients is shown in Table 30.5.

Other particular neurologic deficits have been associated with calcineurin inhibitors. For example, internuclear ophthalmoplegia that was reversible on tacrolimus discontinuation has been reported (Oliverio et al., 2000; Lai et al., 2004). A rare pain syndrome involving the feet and legs has been described – the calcineurin inhibitor-induced pain syndrome (Grotz et al., 2001). Although the exact pathophysiology remains unclear, some feel it may represent a variant of reflex sympathetic dystrophy, as it can be accompanied by signs of limb (and marrow) edema, albeit with fewer trophic changes or vasomotor instability. Pain can be debilitating, but often resolves with reduction in drug dosage, and may be amenable to calcium channel-blocking medications.

Table 30.5

White-matter diseases affecting transplant recipients

	Posterior reversible encephalopathy syndrome	Osmotic demyelination syndrome	Progressive multifocal leukoencephalopathy
Timing after transplant	Majority occur acutely (<4 weeks)	Acute (<2 weeks), mainly after liver transplantation	Delayed (6 months–few years)
Location	Posterior predominant (although may also involve frontal lobes, brainstem, cerebellum)	Central pons (i.e., CPM) Basal ganglia External capsule	Subcortical white matter Cerebellar peduncles
Pattern	Usually symmetric	Trident-shaped lesion in pons	Asymmetric, multifocal
MRI signal	T2: high DWI: negative	T2: high DWI: variable	T2: high DWI: negative
Enhancement	Nonenhancing	Nonenhancing	Nonenhancing
Imaging vs. clinical	Imaging more sensitive	Imaging may lag behind clinical	
Etiology	Calcineurin inhibitor toxicity, hypertension	Shifts in serum sodium and osmolality, brain water shifts	Reactivation of JC virus
Pathology	Disruption of BBB, vasogenic edema	Myelin edema, demyelination	Oligodendrocyte infection resulting in demyelination
Clinical presentation	Headache, encephalopathy, seizures Cortical blindness	Quadriparesis, pseudobulbar state, locked-in syndrome, EPS	Confusion/behavioral change Focal and visual deficits
Onset	Acute	Acute	Subacute (weeks)
Diagnosis	High drug levels (variable) Reversible on stopping agent	Characteristic location, timing, and clinical context	PCR for JC virus in CSF (70%) Brain biopsy
Prognosis	Reversible, generally favorable unless hemorrhage has occurred	Some improvement possible in survivors	Progressive, often fatal
Treatment	Stop or reduce drug levels Control of hypertension	Supportive, reinstitution of hyponatremia (if acute)	Reduce immunosuppression Cytarabine?

Adapted from Dhar and Human (2011).
BBB, blood–brain barrier; CPM, central pontine myelinolysis; CSF, cerebrospinal fluid; DWI, diffusion-weighted imaging; EPS, extrapyramidal syndrome; MRI, magnetic resonance imaging; PCR, polymerase chain reaction.

RISK FACTORS FOR TOXICITY

Uncontrolled comparisons of cyclosporine vs. tacrolimus have not found higher rates of neurotoxicity with one drug compared to the other (Saner et al., 2007). One study found a neurotoxicity rate of 25% with both agents after liver transplant (Lewis and Howdle, 2003). However, there is significant variability in drug kinetics between individuals, as well as significant potential for drug–drug interactions; for example, tacrolimus bioavailability is increased in cases of diarrhea, since normal secretion of drug into the intestinal lumen is impaired by colonic epithelial dysfunction (Asano et al., 2004). Concurrent metoclopramide administration may increase motility and drug bioavailability, thereby precipitating toxicity (Prescott et al., 2004). Systemic factors such as hypocholesterolemia and hypomagnesemia, as well as hypertension, have been proposed as predisposing to toxicity despite normal drug levels. The incidence of toxicity appears to be decreasing in recent years, with delayed loading and more cautious drug dosing. A recent series evaluating liver transplant patients found an incidence of PRES at less than 1% (Vizzini et al., 2011).

MANAGEMENT OF SUSPECTED TOXICITY

Once neurotoxicity is suspected, investigations (such as MRI) to confirm the diagnosis should be planned, but dose reductions (or, in severe cases, drug discontinuation) should be considered immediately (Bechstein, 2000). However, this places the allograft at risk for rejection while therapy is interrupted, such that there should be a plan for reinstitution, rechallenge, or conversion to a new regimen (Guarino et al., 2006). Most commonly, one calcineurin inhibitor is substituted for another, or the same agent is restarted at a lower dose once symptoms have resolved; recurrence rates are low with either paradigm (Wijdicks et al., 1995). In cases of severe unequivocal toxicity, switching to other regimens may be preferred. Lack of neurotoxicity associated with newer agents (e.g., mTor inhibitors such as sirolimus, everolimus) makes them appealing alternatives to calcineurin inhibitor therapy, either in cases with toxicity or as first-line agents (Maramattom and Wijdicks, 2004; van de Beek et al., 2009; Bilbao et al., 2014). Sirolimus may be used either as replacement therapy, or in combination with low-dose calcineurin inhibitors (Sevmis et al., 2007; Vivarelli et al., 2010). Isolated cases of sirolimus-associated PRES have been reported, but only in association with hypertension (Bodkin and Eidelman, 2007; Moskowitz et al., 2007). Neurologic symptoms are generally reversible after discontinuation of calcineurin inhibitors, except in cases where permanent brain injury (i.e., stroke or hemorrhage) has occurred due to late detection. Symptoms and neurologic deficits usually resolve within a few days, but radiographic abnormalities may take weeks to fully disappear (Singh et al., 2000). Diagnostic criteria for calcineurin inhibitor neurotoxicity are shown below.

DIAGNOSTIC CRITERIA FOR CALCINEURIN INHIBITOR NEUROTOXICITY

The development of seizures and/or encephalopathy in a patient receiving cyclosporine or tacrolimus, associated with:

1. High serum trough levels of the drug at time of symptom onset
2. Rapid rise in drug levels prior to symptoms
3. Imaging features of PRES
4. Reversibility with discontinuation or dose reduction of offending agents

Stroke syndrome

The acute development of focal neurologic deficits (e.g., hemiparesis, aphasia) in a transplant recipient may signal a cerebrovascular or infectious CNS process. However, focal deficits may also be seen in the context of Todd's paresis after seizures, with drug neurotoxicity (e.g., asymmetric PRES), and even due to peripheral nerve dysfunction (e.g., perioperative compressive neuropathy). Conversely, focal CNS processes in transplant patients are actually often multifocal and present with nonspecific mental status changes and/or seizures (Kim et al., 2015). Brain imaging is critical to evaluate for focal lesions in most cases of neurologic symptoms after transplantation.

Transplant patients may be at higher risk for cerebrovascular complications, both ischemic and hemorrhagic. Thrombocytopenia is common with HSCT and may result in spontaneous intracerebral or subdural bleeding. Coagulopathy may also occur with hepatic dysfunction, or in association with anticoagulation (Fig. 30.6). Conversely, a hypercoagulable state can occur in the context of chemotherapy, which may trigger cerebral venous sinus thrombosis; this should be suspected in the presence of multiple simultaneous hemorrhagic lesions. Embolic strokes are particularly common after heart and lung transplantation (Fig. 30.1), as is global or watershed ischemic injury from hypotension, hypoxemia, or cardiac arrest (Sila, 1989). Hemorrhagic stroke can also occur after cardiac transplant relating to reperfusion of brain that was adapted to a low cardiac output state. One series of 384 heart transplant patients found stroke in 7% (mostly ischemic, with a few hemorrhagic strokes and HIE) (Munoz et al., 2010). The majority of these were detected immediately postoperatively. Infectious processes like varicella-zoster virus can cause cerebral vasculitis, and a similar picture can occur with GVHD (discussed subsequently). Transplant patients also often harbor significant vascular risk factors, like diabetes mellitus and hypertension, and may develop stroke related to these traditional triggers (e.g., after renal transplant).

Neuromuscular syndromes

Weakness due to neuromuscular disorders is less common than deficits due to CNS processes. However, certain nerve and muscle disorders may be unique or more common in transplant recipients. Postoperative mononeuropathies are usually related to nerve traction or compression during surgery (e.g., femoral neuropathy in 2% after kidney transplantation) (Sharma et al., 2002). Brachial or lumbosacral plexopathies may also result from operative positioning (Katirji, 1989; Dhillon and Sarac, 2000). Phrenic or recurrent laryngeal nerve injury can occur after heart or lung transplants and may impair respiratory/cough mechanisms and ventilator weaning (Maziak et al., 1996; Mateen et al., 2009).

More diffuse weakness may be attributed to critical illness myopathy or polyneuropathy in those recipients with complicated postoperative courses; this was described in 7% of one series after liver transplantation (Campellone et al., 1998). Cases of Guillain–Barré syndrome have been described after

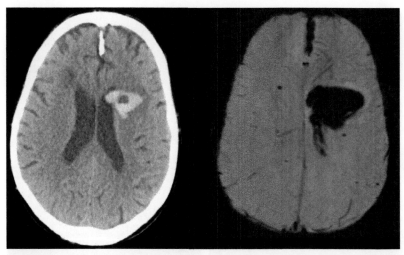

Fig. 30.6. Computed tomography scan (left) with acute intraparenchymal hemorrhage in a patient who presented with confusion and facial droop in the first month after bilateral lung transplant; he was on anticoagulation for venous thromboembolism and had negative cerebral angiogram. Susceptibility-weighted magnetic resonance imaging (right) confirms hematoma. Other infarcts were seen and this likely represents hemorrhagic transformation of embolic strokes.

transplantation, possibly in association with cytomegalovirus (CMV) infection or rejection (Amato et al., 1993; El-Sabrout et al., 2001). Severe electrolyte disorders such as hypophosphatemia, hypokalemia, or hypo/hypermagnesemia can also lead to generalized weakness. Inflammatory myopathies may occur with an increased incidence after HSCT, either in conjunction with GVHD or in isolation (Ruzhansky and Brannagan, 2015). Polymyositis has also been described in association with tacrolimus therapy (Vattemi et al., 2014). This must be differentiated from steroid myopathy, usually by elevated creatine kinase levels and electromyogram findings.

NEURODIAGNOSTICS AND IMAGING

Neuroimaging should be obtained for almost all acute neurologic symptoms in transplant recipients, even if associated with clear metabolic precipitant; this is especially true for focal deficits or seizures with focal semiology. CT of the head is a reasonable and rapidly available first-line investigation, which will reveal major lesions such as intracranial hemorrhages, brain abscesses, or cerebral edema, but will often miss acute ischemia and other subtle or evolving lesions (e.g., CPM, HIE). MRI is much more sensitive to ischemia and smaller lesions, and should be performed in the presence of persistent focal deficits or unexplained mental status changes. Contrast administration is preferable, as this will enhance detection of infectious and inflammatory disorders, but should be avoided in the presence of renal insufficiency (for both CT and MR contrast). MRI may also not be possible immediately after cardiac procedures if pacing wires are still in place. Notably, inflammatory/infectious lesions may not enhance as avidly in transplant recipients as in normal patients.

Neuroimaging may be diagnostic in many disorders relevant to this population. For example, changes of PRES (as discussed above) are usually characteristically posterior-predominant, affect white matter, and do not enhance. Osmotic demyelination has characteristic locations and patterns (i.e., central pons with CPM, with classic trident or bat-shaped lesion – Fig. 30.7A). PRES and osmotic demyelination are toxic-metabolic disorders that highlight the usually symmetric nature of systemic disorders affecting the brain (Fig. 30.7B), while focal infectious or vascular lesions are more likely asymmetric (e.g., PML, lymphoma). Pre-existing abnormalities may also be seen: for example, lesions due to organ failure (e.g., bilateral globus pallidus lesions in liver failure) (Fernandez-Rodriguez et al., 2010).

Electrophysiologic evaluation for nonconvulsive seizures (i.e., EEG) should be performed in those with unexplained encephalopathy after imaging and systemic studies have been unrevealing, or if subtle signs are seen (e.g., nystagmus, eye deviation, twitching). However, myoclonus, especially when multifocal, is more often a sign of metabolic derangement such as uremia than a marker of seizures. Similarly, urgent EEG should be obtained in any patient with convulsive seizures who has not awoken, or is not returning at least toward baseline mentation within a few hours. It should also be performed if no explanation is found for failure to awaken after transplant surgery.

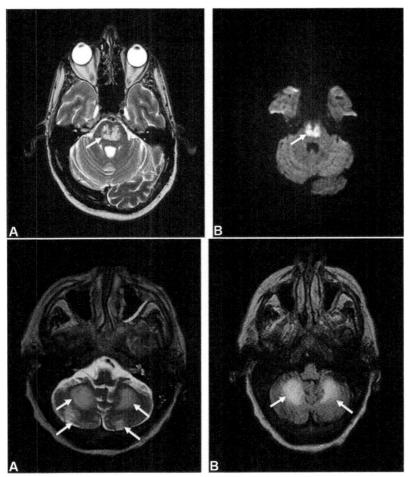

Fig. 30.7. (A) Central pontine myelinolysis (left is T2-weighted and right is diffusion-weighted image). (B) Extrapontine myelinolysis (left is T2-weighted and right is fluid-attenuated inversion recovery image) with symmetric involvement of cerebellar hemispheres. (Reproduced from Huq et al., 2007, with permission.)

Nerve conduction studies and electromyography may be useful for unexplained or unclear neuromuscular symptoms or deficits. It can differentiate diffuse weakness due to demyelinating disorders like Guillain–Barré syndrome from critical illness polyneuropathy, which has an axonal pattern. CSF testing should be performed in any cases with suspected CNS infections (Table 30.6).

OUTCOME

CNS complications may be a byproduct of serious systemic illness such as sepsis or a consequence of drug toxicity. They may also result from brain disorders such as strokes, tumors, or infections. The underlying etiology for the symptoms is critical in determining prognosis. Many neurologic complications in transplant recipients are transient and reversible, especially those related to drugs and metabolic disturbances (Bernhardt et al., 2015). However, graft failure may result in brain dysfunction and be associated with poor, or even fatal, outcomes. Similarly, catastrophic neurologic injury (e.g., HIE, osmotic demyelination, large intracranial hemorrhage) may similarly be associated with long-term morbidity and death.

CENTRAL PONTINE MYELINOLYSIS

CPM is a pathologic syndrome of pontine demyelination, initially described in malnourished alcoholics who experienced rapid rises in serum sodium and osmolality (Adams et al., 1959). It is now categorized under the broader rubric, ODS (King and Rosner, 2010) as brain imaging has now demonstrated that it can involve other brain regions, including the basal ganglia, external capsule, thalamus, as well as cerebellum (i.e., extrapontine

Table 30.6

Opportunistic infections affecting transplant recipients: overview of presentation, diagnosis, and treatment

Disease	Organism	Source	Primary site	Presentation	Diagnosis	Treatment
Listeriosis	*Listeria monocytogenes* (Gram-positive bacillus)	Contaminated milk or cheese	GI tract	Meningitis Rhombencephalitis	Cultures: CSF, blood	Ampicillin ± gentamicin
Nocardiosis	*Nocardia asteroides* (Gram-positive branching rod)	Inhalation	Lung	Brain abscess	Culture/acid-fast stain of sputum, skin or brain abscess	TMP-SMX ± third-generation cephalosporin
Aspergillosis	*Aspergillus fumigatus/flavus* (angioinvasive fungus)	Inhalation	Lung	Abscesses, multiple infarcts/bleeds	Galactomannan test (serum), histology/culture of lung or brain, culture	Voriconazole
Candidiasis	*Candida albicans*/other species (pseudohyphae)	Nosocomial/skin	Disseminated (lungs, skin)	Abscesses, meningitis	Blood culture, histology/culture of lungs, skin, brain	Amphotericin
Cryptococcus	*Cryptococcus neoformans* (encapsulated yeast)	Inhalation	Lung (skin, soft tissue)	Meningitis, abscess Hydrocephalus	Antigen – CSF, blood	Amphotericin and flucytosine
CMV, EBV, VZV, HHV-6	Herpesviruses	Reactivation or primary infection	Lungs, liver, skin (retinitis with CMV)	Meningitis + encephalitis PTLD with EBV	Viral PCR in CSF (CSF pleocytosis may be minimal/absent)	Acyclovir Ganciclovir or foscarnet (for CMV and HHV-6)
West Nile	WNV (flavivirus)	Donor-derived or outbreak	Systemic/lung	Meningitis/encephalitis, flaccid paralysis (poliomyelitis)	Serology (WNV-specific antibodies) in serum or CSF, PCR	Polyclonal immunoglobulin, ribivarin, interferon*
PML	JC virus (polyomavirus)	Reactivation	Brain	White-matter lesions, subacute	PCR for JC virus Brain biopsy	Reduce immune suppression Cytarabine*
Toxoplasma	*Toxoplasma gondii* (parasite)	Reactivation Donor-derived	Heart (donor)	Multiple small abscesses, meningitis	Response to empiric therapy vs. biopsy Serology in blood should be positive PCR for viral DNA	Pyrimethamine + sulfadiazine + folinic acid
TB	*Mycobacterium tuberculosis*	Primary or reactivation	Lung	Basilar meningitis, tuberculoma(s)	AFB stain of sputum or CSF, PCR, and culture (delayed), low CSF glucose	INH + rifampin + pyrazinamide + ethambutol; consider steroids

Adapted from Dhar and Human (2011).
*All experimental without solid evidence for benefit.
AFB, acid-fast bacilli; CMV, cytomegalovirus; CSF, cerebrospinal fluid; EBV, Epstein–Barr virus; GI, gastrointestinal; HHV-6, human herpesvirus 6; INH, isoniazid; PCR, polymerase chain reaction; PML, progressive multifocal leukoencephalopathy; PTLD, posttransplant lymphoproliferative disorder; TB, tuberculosis; TMP-SMX, trimethoprim-sulfamethoxazole; VZV, varicella-zoster virus; WNV, West Nile virus.

demyelination) (Huq et al., 2007). Shifts in sodium and brain water are theorized to result in a noninflammatory oligodendrocyte injury resulting in myelin loss in vulnerable brain regions (Wright et al., 1979). Due to predominant involvement of the basis pontis, CPM classically presents with quadriparesis, pseudobulbar palsy, and variable mental status alterations. If horizontal eye movements are abolished from involvement of the pontine tegmentum, then it may result in a locked-in syndrome where the patient is fully awake, but cannot move or respond to stimuli (except with vertical eye movements or blinking), so can be mistaken for coma (Messert et al., 1979). Extrapyramidal involvement in ODS can result in tremor, myoclonus, or an akinetic-rigid state.

CPM has been reported in the transplant population almost exclusively after liver transplantation, although a single case in a child after intestinal transplant has been reported (Starzl et al., 1978; Estol et al., 1989; Idoate et al., 1999). Large series of liver transplant recipients have estimated its incidence at 1–2% (Bronster et al., 2000; Lewis and Howdle, 2003). Two recent studies found 3 of 791 patients (0.4%) and 2 of 395 (0.5%) were diagnosed with CPM (presenting with encephalopathy) after liver transplant (Vizzini et al., 2011; Kim et al., 2015). Another retrospective review found 11 cases among 997 transplants (1.1%), including a majority with extrapontine involvement (Crivellin et al., 2015). As in other situations, it is felt to occur when there are large perioperative osmotic shifts (e.g., the rapid correction of hyponatremia that may occur in cirrhotic patients after transplantation) (Wszolek et al., 1989; Abbasoglu et al., 1998). Larger perioperative sodium shifts were found in one case-control study (17 vs. 10 mEq/L) (Crivellin et al., 2015). Such brain water shifts may be accentuated by the depletion of myo-inositol that occurs with pre-existing hepatic encephalopathy, as this is a known mediator of cerebral osmotic homeostasis (Haussinger et al., 1994).

However, not all patients who develop CPM have experienced large fluctuations in serum sodium. Similarly, the spectrum and severity of ODS after transplant are likely broader than severe quadriparetic forms of CPM. Some may instead present with nonfocal postoperative alterations in level of consciousness (Wijdicks et al., 1996a). Such encephalopathy can progress over a few days after transplant, or can present with failure to awaken immediately after surgery. Head CT is an insensitive test for detecting subtle demyelination, especially early on, while MRI may reveal the classic trident or bat-shaped increase in T2 signal in the central pons within a few days of symptom onset (Fig. 30.8) (Miller et al., 1988; Bronster et al., 2000).

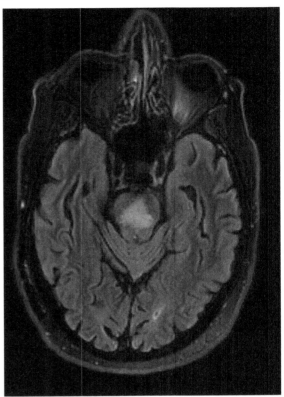

Fig. 30.8. Trident-shaped signal abnormality on fluid-attenuated inversion recovery magnetic resonance imaging characteristic for central pontine myelinolysis. (Reproduced from Dhar and Human, 2011, with permission.)

CPM was initially thought to be a catastrophic complication when it was only detected in autopsy series or in those with locked-in state. However, with detection of the full spectrum of ODS now possible with MRI, newer series have demonstrated potential for significant recovery, although some patients may be left with residual motor and/or cognitive deficits (Menger and Jorg, 1999). There are no established treatments for CPM, although reintroduction of hyponatremia in the acute phase has been proposed in cases where it is associated with a sudden rapid rise in osmolality/sodium (Soupart et al., 1999). Whether this can prevent myelin injury is unclear. Other series have proposed the use of plasmapharesis and/or immunoglobulin (Zhang et al., 2009; Ludwig et al., 2011). Prevention of CPM is eminently preferable, primarily through close monitoring of sodium and osmolality, and minimization of fluid shifts in the perioperative period.

GRAFT-VERSUS-HOST DISEASE

Allogeneic HSCT can be helpful in eradicating refractory neoplasms, as allogeneic stem cells may exhibit potent graft-versus-tumor effects. However, allogeneic exposure may also result in GVHD (Perreault et al., 1983). This multiorgan disorder results from immune-mediated attack by donor leukocytes on recipient tissues, and most commonly involves the skin, oral mucosa, lungs, intestinal tract, and liver (elevated bilirubin) (Deeg and Storb, 1984). Acute GVHD occurs in 20–40% of allogeneic HSCT, and forms the basis for prophylaxis with potent immunosuppression, typically methotrexate or mycophenolate early after HSCT, and maintenance therapy with a calcineurin inhibitor, plus antithymocyte globulin in particularly high-risk cases (Chao and Chen, 2006). If it develops, acute GVHD can be treated with high doses of corticosteroids, much like solid-organ rejection. Bowel involvement can induce thiamine deficiency, precipitating Wernicke's encephalopathy with subacute mental status changes and ophthalmoplegia (Bleggi-Torres et al., 2000; Choi et al., 2010). This may also be seen in intestinal recipients, who often have significant nutritional deficiencies pretransplant. GVHD is not seen after autologous HSCT, and these patients are generally at lower risk of complications once engraftment occurs and cell counts normalize.

The peripheral nervous system may be involved in chronic GVHD, with inflammatory myopathies being the most frequent manifestation (Stevens et al., 2003; Ruzhansky and Brannagan, 2015). CNS involvement is more controversial and should only be diagnosed after exclusion of other potential etiologies for neurologic symptoms, such as OI (Grauer et al., 2010). A few cases presenting with features of CNS vasculitis or immune-mediated encephalitis have been described (Sostak et al., 2010). Patients may develop cognitive and/or focal deficits months to years after HSCT, often following reduction in immunosuppression. To further substantiate the diagnosis of CNS GVHD, there should be evidence of involvement in other organs and a clear response to immunosuppressive therapy (e.g., pulse corticosteroids). CSF may show elevated protein concentration, oligoclonal bands, and variable pleocytosis. MRI may show white-matter lesions, not dissimilar to those seen in multiple sclerosis (Matsuo et al., 2009). Brain biopsy may be required if infectious etiologies cannot be excluded by imaging, serology, or CSF testing.

IDIOPATHIC HYPERAMMONEMIA

Idiopathic hyperammonemia is a rare but highly fatal syndrome that is seen in less than 1% of HSCT recipients during the period of severe neutropenia (Davies et al., 1996). It may begin with lethargy, confusion, and tachypnea (with respiratory alkalosis) progressing to seizures and coma. Ammonia levels are characteristically elevated (often > 200 μmol/L), while liver enzymes are normal or only mildly elevated. Brain imaging may reveal marked cerebral edema, similar to Reye syndrome (Metzeler et al., 2009). There may also be signal change in the insular and cingulate cortices, with sparing of the occipital cortex (in contrast to posterior and white-matter-predominant PRES pattern on MRI) (Bindu et al., 2009).

Urea cycle defects including ornithine transcarbamylase deficiency should be excluded by measurement of urinary amino acids and orotic acid, as these can present in a similar manner. High ammonia levels may also be seen with liver failure, multiple myeloma, or valproate drug therapy (Clay and Hainline, 2007). Effective treatments are lacking (hemodialysis or sodium benzoate to trap/remove ammonia have been attempted) but mortality remains high (del Rosario et al., 1997).

CNS INFECTIONS

The use of potent immunosuppressive regimens has reduced the risk of graft rejection and prolonged survival for transplant recipients. However, this success now exposes patients to long-term immunosuppression and longer survival, both of which result in higher risk of OI (Fishman, 2007). Recipients of heart and intestinal transplants are generally most heavily immunosuppressed, and may be more prone to these complications. In addition, as the host's immune response is blunted, and many offending organisms are less pathogenic, these potentially life-threatening infections can present in subtle, nonspecific ways, without acute or fulminant symptoms. For example, headache and low-grade fever (usually in the absence of neck stiffness) may be the only signs of CNS infection, while mental status changes comprise the most frequent presentation of even those with focal or multifocal brain abscesses (van de Beek et al., 2007). Any patient with such features more than a month after transplantation should be evaluated for CNS infection. Any CNS unexplained lesion should be assumed to be potentially infectious.

CNS infections may be broadly categorized into those transmitted from the organ donor, those representing reactivation of a pre-existing quiescent infection, those with CNS involvement as part of a new systemic infection, and those with direct CNS involvement. OIs are uncommon in the early posttransplant period, when nosocomial infections like pneumonia and wound infections predominate. The period of highest risk is typically between 1 and 6 months, when maximal immunosuppression has been achieved (except with HSCT, where it is earlier). Intensity of immunosuppression may be reduced after 6 months if rejection or GVHD has not occurred, minimizing subsequent risk of infection. Those requiring persistent high-dose immunosuppression and repeated courses of corticosteroids for rejection or GVHD remain at risk for infections. Transplant recipients are also at higher risk of contracting infections during local outbreaks, or when in endemic areas (Yango et al., 2014). Therefore, minimizing exposure is an important preventive measure for such patients. Common infections, modes of presentation, and means of diagnosis/treatment are outlined in Table 30.6.

Donor-derived infections

A number of serious infections can be acquired inadvertently from the organ donor. These include cases where the donor died of cryptogenic encephalitis or hemorrhage, which has been reported with rabies, West Nile virus, and lymphocytic choriomeningitis virus (Iwamoto et al., 2003; Srinivasan et al., 2005; Fischer et al., 2006; Basavaraju et al., 2014; Winston et al., 2014). Transmission of amebic (*Balamuthia mandrillaris*) encephalitis has also been recently reported (Gupte et al., 2014). For this reason, undiagnosed or suspicious cases of encephalitis usually represent exclusionary factors for organ donation. Such infections most commonly present in the first month after transplant, such that a careful history of the donor may be essential when recipients (especially clusters in the same region) present with unusual CNS symptoms. Other infections that can occur in this early period include those related to pretransplant colonization, or nosocomial infections acquired postoperatively while in the hospital.

Reactivation

Herpesviruses and the JC virus are commonly latent in the brain or cell ganglia. These can reactivate in periods of immune deficiency, causing herpes zoster (sometimes without skin lesions), encephalitis, or, in the case of JC virus, PML. Reactivation may also trigger Guillain–Barré syndrome (El-Sabrout et al., 2001). HHV-6 is increasingly recognized as a cause of limbic encephalitis in transplant recipients (especially after HSCT), and may present with cognitive dysfunction and seizures (Singh et al., 2000; Bollen et al., 2001). In fact, the acronym PALE (posttransplant acute limbic encephalitis) was coined to describe the syndrome of amnesia with limbic abnormalities on MRI associated with HHV-6 (Fig. 30.9) (Seeley et al., 2007; Vinnard et al., 2009). This study found an incidence of 1.5% among 584 patients undergoing allogeneic HSCT, with a median onset of 29 days posttransplant. GVHD and hyponatremia due to syndrome of inappropriate antidiuretic hormone production were common, and many patients also had seizures (with temporal-lobe EEG abnormalities). CSF pleocytosis was mild, and polymerase chain reaction (PCR) testing for HHV-6 was diagnostic. Treatment can be with valganciclovir or foscarnet.

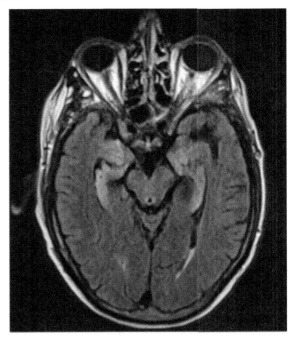

Fig. 30.9. Fluid-attenuated inversion recovery magnetic resonance imaging with bilateral hyperintensity in medial temporal lobes consistent with limbic encephalitis related to human herpesvirus-6 infection. (With permission from John Wiley and Sons.)

Toxoplasmosis is the most common protozoal infection in transplant recipients. It is usually related to primary infection, and occurs more often in heart transplant recipients, where cysts can be found in the myocardium (Munoz et al., 2010; Fernandez-Sabe et al., 2012). It usually presents with multiple enhancing brain abscesses in the first year after transplant. Toxoplasmosis is treatable and carries a relatively low mortality rate once identified. Amebic meningoencephalitis, in contrast, is a less common protozoal infection, but is almost uniformly fatal (Satlin et al., 2013).

PML is a subacute demyelinating disorder caused by reactivation of the JC polyomavirus in immunocompromised hosts (Fig. 30.10). The virus infects astrocytes and oligodendrocytes, resulting in loss of myelin. White-matter lesions are typically multifocal, asymmetric, and nonenhancing. PML remains fairly rare, with an estimated incidence of 1.2 per 1000 posttransplantation years, and is seen after both solid-organ and HSCT (Mateen et al., 2011). Symptoms (e.g., cognitive decline, aphasia, and motor deficits) usually develop gradually over weeks to months, and occur in a delayed fashion (on average 17 months) after transplant. The gold standard for diagnosis is brain biopsy, but positive PCR for JC virus in CSF can now confirm PML without need

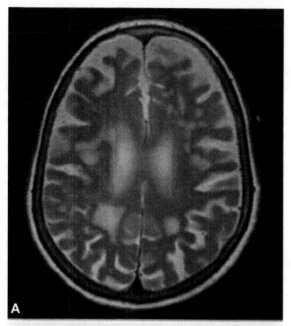

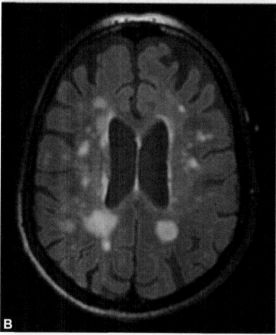

Fig. 30.10. Fluid-attenuated inversion recovery (**A**) and T2-weighted magnetic resonance imaging (**B**) showing multifocal bilateral asymmetric hyperintense white-matter lesions in the periventricular and subcortical regions corresponding to progressive multifocal leukoencephalopathy. These lesions did not exhibit mass effect or contrast enhancement (not shown). (With permission from John Wiley and Sons.)

for biopsy. While there is no specific therapy for PML, reducing immunosuppression is often attempted, and cytosine arabinoside or cidofovir may be administered. However, mortality is extremely high (approximately 80%), with a median survival of 6 months.

The spectrum of infections after transplant has been shifting with increased use of prophylaxis for specific common pathogens. For example, sulfa prophylaxis has reduced rates of *Listeria*, toxoplasmosis, and *Nocardia*, while CMV screening and pre-emptive treatment have reduced the consequences of CMV reactivation.

Lungs are constantly exposed to environmental pathogens and comprise a common portal for new infections that can access and involve the CNS; lung transplant recipients may be particularly at risk. For example, systemic *Nocardia* infection has a predilection for lung transplant recipients but now occurs rarely (less than 1%) with sulfa prophylaxis (Clark et al., 2013). CNS infection occurs as a complication of pulmonary involvement, and typically presents with solitary or multifocal brain abscesses, which may be diagnosed either by culture of brain or sputum.

Listeria monocytogenes meningitis is more likely in immunocompromised hosts and is transmitted from unpasteurized dairy products or undercooked meats. It may produce only low degree of CSF cell pleocytosis (without typical neutrophil predominance) and is slow to culture (in both blood and CSF). Listeriosis tends to involve the brainstem (i.e., rhombencephalitis), presenting with fever, headache, cranial nerve palsies, and pyramidal/cerebellar findings. Coverage of suspected meningitis in a transplant patient should include intravenous ampicillin for *Listeria* until cultures are all negative. Another cause of basal meningitis to be considered is tuberculosis, which can also present with multiple cranial nerve palsies and often hydrocephalus.

Fungal infections may be catastrophic in transplant recipients and occur at a higher rate than in nonimmunosuppressed populations. CNS infection by *Cryptococcus neoformans* is the most common cause of fungal meningitis, and may present subacutely with headache, neck rigidity, or signs of hydrocephalus and raised intracranial pressure (e.g., sixth cranial nerve palsy, papilledema). One series found that 0.5% of over 5000 transplant recipients developed cryptococcal meningitis, with a higher incidence in heart recipients (2%). Mortality was high, especially in those with liver transplant (Wu et al., 2002). Immune reconstitution syndrome (IRIS; see later section) may occur when cryptococcosis is treated (Sun et al., 2015). *Candida* usually involves the CNS, usually as a result of disseminated infection (with fungemia), and may present with chronic meningitis or with abscesses. Cerebral aspergillosis usually occurs in combination with invasive pulmonary infection, has earlier onset posttransplant, and is more likely after lung transplantation (Singh et al., 2009). It most often presents with multiple brain abscesses (which may be hemorrhagic due to its angioinvasive nature) and such advanced infection harbors an extremely high mortality. Survivors have been reported with prompt voriconazole therapy. Travel history may aid in identifying risk for other specific fungal pathogens (Kauffman et al., 2014).

CSF evaluation is important, but not always diagnostic. CSF protein may be elevated but cell counts may only be minimally elevated or even normal with encephalitis. Imaging findings may be less dramatic for brain abscesses, with less enhancement and edema than typical (especially with HSCT and total immune failure).

Immune reconstitution inflammatory syndrome

IRIS occurs when the immune response becomes dysregulated and exaggerated as a shift occurs from an immunosuppressed to proinflammatory state. This has primarily been described with antiretroviral treatment of human immunodeficiency virus (HIV) patients, but is increasingly recognized in immunocompromised transplant recipients (Sun and Singh, 2011). This occurs when pathogen-induced and iatrogenic immunosuppression is reversed during treatment of OI, with concomitant reduction in immunosuppressive medications. IRIS has been observed with effective treatment of fungal, mycobacterial, and CMV infections after transplant. Neurologic signs of IRIS may include clinical or radiologic evidence of inflammation (e.g., contrast enhancement of leptomeninges and CSF pleocytosis with negative cultures) and can otherwise mimic recurrent infection or meningitis. A recent series of 89 patients with posttransplant cryptococcosis found that 13 (14%) developed IRIS (Sun et al., 2015). Predictors of IRIS were CNS involvement by OI and discontinuation of calcineurin inhibitors. Half of patients with both these risk factors developed the syndrome. Neuroimaging findings of cryptococcal disease were also associated with IRIS. Falling titers of cryptococcal antigen suggest IRIS, rather than recurrence of initial infection. This syndrome may also be associated with higher risk of graft rejection, and is usually treated with corticosteroids (Singh et al., 2005).

POSTTRANSPLANTATION LYMPHOPROLIFERATIVE DISORDER (PLTD)

Systemic PTLD is typically associated with B-cell lymphoma and, even though CNS involvement is not frequent, represents the most common brain tumor seen in transplant recipients. The CNS may also rarely be the primary site of involvement (Buell et al., 2005). Most cases of PTLD are associated with Epstein–Barr virus infection and proliferation of donor B lymphocytes in the face of T-cell suppression. Incidence ranges from 1 to 8% (higher in heart, lung, and intestinal recipients) and peaks a few years after transplantation (Cavaliere et al., 2010; Evens et al., 2013). Tumor involvement is typically parenchymal, a pattern similar to acquired immunodeficiency syndrome (AIDS) patients, but different from nonimmunocompromised hosts, who often have leptomeningeal involvement. Multiple periventricular enhancing lesions are typical and may mimic toxoplasmosis (Fig. 30.11). Increased uptake on single-photon emission CT (SPECT) may suggest PTLD over infectious etiologies. Presentation is more often subacute in onset with headache, confusion, and sometimes focal deficits and seizures. CSF cytology is often negative and Epstein–Barr virus PCR can also be negative in some cases (Hamadani et al., 2007). Corticosteroids should be held, if possible, prior to biopsy, as their administration may induce tumor cell lysis and compromise histopathologic diagnosis (and lead to radiographic "vanishing tumor" transiently). They can be started after diagnosis to reduce edema pending irradiation/chemotherapy. If possible, reduction in immunosuppression should also be instituted concurrently. Primary glial neoplasms are much less common, although transmission of high-grade gliomas from organ donors has been rarely described (Schiff et al., 2001; Fatt et al., 2008). This could be accentuated in the setting of systemic spread due to CSF shunting or recent neurosurgical procedures.

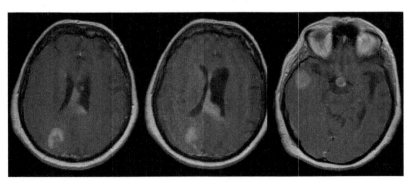

Fig. 30.11. Contrast-enhanced T1-weighted magnetic resonance imaging shows multifocal periventricular and subcortical enhancing masses representing primary central nervous system lymphoma. An incidental pituitary adenoma is also seen.

GUIDELINES

The European Federation of Neurological Societies prepared guidelines addressing the management of neurologic complications after liver transplantation (Guarino et al., 2006). They recommended minimizing drug neurotoxicity by cautious dosing of calcineurin inhibitors and strict monitoring of drug levels, as well as correction of systemic factors such as magnesium levels and blood pressure. MRI was the diagnostic tool recommended for evaluation of toxicity. It was also recommended for evaluation of seizures and possible osmotic demyelination (with serial or delayed imaging recommended to capture the latter). No other systematic guidelines were found in the literature.

REFERENCES

Abbasoglu O, Goldstein RM, Vodapally MS et al. (1998). Liver transplantation in hyponatremic patients with emphasis on central pontine myelinolysis. Clin Transplant 12: 263–269.

Adams RD, Victor M, Mancall EL (1959). Central pontine myelinolysis: a hitherto undescribed disease occurring in alcoholic and malnourished patients. AMA Arch Neurol Psychiatry 81: 154–172.

Amato AA, Barohn RJ, Sahenk Z et al. (1993). Polyneuropathy complicating bone marrow and solid organ transplantation. Neurology 43: 1513–1518.

Antonini G, Ceschin V, Morino S et al. (1998). Early neurologic complications following allogeneic bone marrow transplant for leukemia: a prospective study. Neurology 50: 1441–1445.

Asano T, Nishimoto K, Hayakawa M (2004). Increased tacrolimus trough levels in association with severe diarrhea, a case report. Transplant Proc 36: 2096–2097.

Baglin TP, Harper P, Marcus RE (1990). Veno-occlusive disease of the liver complicating ABMT successfully treated with recombinant tissue plasminogen activator (rt-PA). Bone Marrow Transplant 5: 439–441.

Bartynski WS, Tan HP, Boardman JF et al. (2008). Posterior reversible encephalopathy syndrome after solid organ transplantation. AJNR Am J Neuroradiol 29: 924–930.

Basavaraju SV, Kuehnert MJ, Zaki SR et al. (2014). Encephalitis caused by pathogens transmitted through organ transplants, United States, 2002–2013. Emerg Infect Dis 20: 1443–1451.

Bechstein WO (2000). Neurotoxicity of calcineurin inhibitors: impact and clinical management. Transpl Int 13: 313–326.

Beresford TP (2001). Neuropsychiatric complications of liver and other solid organ transplantation. Liver Transpl 7: S36–S45.

Bernhardt M, Pflugrad H, Goldbecker A et al. (2015). Central nervous system complications after liver transplantation: common but mostly transient phenomena. Liver Transpl 21: 224–232.

Bianco F, Fattapposta F, Locuratolo N et al. (2004). Reversible diffusion MRI abnormalities and transient mutism after liver transplantation. Neurology 62: 981–983.

Bilbao I, Dopazo C, Castells L et al. (2014). Immunosuppression based on everolimus in liver transplant recipients with severe early post-transplantation neurotoxicity. Transplant Proc 46: 3104–3107.

Bindu PS, Sinha S, Taly AB et al. (2009). Cranial MRI in acute hyperammonemic encephalopathy. Pediatr Neurol 41: 139–142.

Blanco R, De Girolami U, Jenkins RL et al. (1995). Neuropathology of liver transplantation. Clin Neuropathol 14: 109–117.

Bleggi-Torres LF, de Medeiros BC, Werner B et al. (2000). Neuropathological findings after bone marrow transplantation: an autopsy study of 180 cases. Bone Marrow Transplant 25: 301–307.

Bodkin CL, Eidelman BH (2007). Sirolimus-induced posterior reversible encephalopathy. Neurology 68: 2039–2040.

Bollen AE, Wartan AN, Krikke AP et al. (2001). Amnestic syndrome after lung transplantation by human herpes virus-6 encephalitis. J Neurol 248: 619–620.

Bronster DJ, Boccagni P, O'Rourke M et al. (1995). Loss of speech after orthotopic liver transplantation. Transpl Int 8: 234–237.

Bronster DJ, Emre S, Boccagni P et al. (2000). Central nervous system complications in liver transplant recipients – incidence, timing, and long-term follow-up. Clin Transplant 14: 1–7.

Buell JF, Gross TG, Hanaway MJ et al. (2005). Posttransplant lymphoproliferative disorder: significance of central nervous system involvement. Transplant Proc 37: 954–955.

Buis CI, Wiesner RH, Krom RAF et al. (2002). Acute confusional state following liver transplantation for alcoholic liver disease. Neurology 59: 601–605.

Bunchman TE, Brookshire CA (1991). Smooth muscle cell proliferation by conditioned media from cyclosporine-treated endothelial cells: a role of endothelin. Transplant Proc 23: 967–968.

Butterworth RF (2003). Pathogenesis of hepatic encephalopathy: new insights from neuroimaging and molecular studies. J Hepatol 39: 278–285.

Calne RY, White DJ, Thiru S et al. (1978). Cyclosporin A in patients receiving renal allografts from cadaver donors. Lancet 2: 1323–1327.

Campellone JV, Lacomis D, Kramer DJ et al. (1998). Acute myopathy after liver transplantation. Neurology 50: 46–53.

Cavaliere R, Petroni G, Lopes MB et al. (2010). Primary central nervous system post-transplantation lymphoproliferative disorder: an International Primary Central Nervous System Lymphoma Collaborative Group Report. Cancer 116: 863–870.

Chao NJ, Chen BJ (2006). Prophylaxis and treatment of acute graft-versus-host disease. Semin Hematol 43: 32–41.

Choi YJ, Park SJ, Kim JS et al. (2010). Wernicke's encephalopathy following allogeneic hematopoietic stem cell transplantation. Korean J Hematol 45: 279–281.

Clark NM, Reid GE, AST Infectious Diseases Community of Practice (2013). *Nocardia* infections in solid organ transplantation. Am J Transplant 13 (Suppl 4): 83–92.

Clay AS, Hainline BE (2007). Hyperammonemia in the ICU. Chest 132: 1368–1378.

Copelan EA (2006). Hematopoietic stem-cell transplantation. N Engl J Med 354: 1813–1826.

Cordelli DM, Masetti R, Zama D et al. (2014). Etiology, characteristics and outcome of seizures after pediatric hematopoietic stem cell transplantation. Seizure 23: 140–145.

Crivellin C, Cagnin A, Manara R et al. (2015). Risk factors for central pontine and extrapontine myelinolysis after liver transplantation: a single-center study. Transplantation 99: 1257–1264.

Davies SM, Szabo E, Wagner JE et al. (1996). Idiopathic hyperammonemia: a frequently lethal complication of bone marrow transplantation. Bone Marrow Transplant 17: 1119–1125.

Dawson TM (1996). Immunosuppressants, immunophilins, and the nervous system. Ann Neurol 40: 559–560.

de Brabander C, Cornelissen J, Smitt PA et al. (2000). Increased incidence of neurological complications in patients receiving an allogenic bone marrow transplantation from alternative donors. J Neurol Neurosurg Psychiatry 68: 36–40.

de Groen PC, Aksamit AJ, Rakela J et al. (1987). Central nervous system toxicity after liver transplantation. The role of cyclosporine and cholesterol. N Engl J Med 317: 861–866.

De La Camara R, Tomas JF, Figuera A et al. (1991). High dose busulfan and seizures. Bone Marrow Transplant 7: 363–364.

Deeg HJ, Storb R (1984). Graft-versus-host disease: pathophysiological and clinical aspects. Annu Rev Med 35: 11–24.

del Rosario M, Werlin SL, Lauer SJ (1997). Hyperammonemic encephalopathy after chemotherapy. Survival after treatment with sodium benzoate and sodium phenylacetate. J Clin Gastroenterol 25: 682–684.

Denier C, Bourhis JH, Lacroix C et al. (2006). Spectrum and prognosis of neurologic complications after hematopoietic transplantation. Neurology 67: 1990–1997.

Devinsky O, Lemann W, Evans AC et al. (1987). Akinetic mutism in a bone marrow transplant recipient following total-body irradiation and amphotericin B chemoprophylaxis. A positron emission tomographic and neuropathologic study. Arch Neurol 44: 414–417.

Dhar R, Human T (2011). Central nervous system complications of transplantation. Neurology Clinics 29 (4): 943–972.

Dhar R, Young G, Marotta P (2008). Perioperative neurological complications after liver transplantation are best predicted by pre-transplant hepatic encephalopathy. Neurocrit Care 8: 253–258.

Dhillon SS, Sarac E (2000). Lumbosacral plexopathy after dual kidney transplantation. Am J Kidney Dis 36: 1045–1048.

El-Sabrout RA, Radovancevic B, Ankoma-Sey V et al. (2001). Guillain–Barré syndrome after solid organ transplantation. Transplantation 71: 1311–1316.

Ely EW, Inouye SK, Bernard GR et al. (2001). Delirium in mechanically ventilated patients: validity and reliability of the Confusion Assessment Method for the Intensive Care Unit (CAM-ICU). JAMA 286: 2703–2710.

Emiroglu R, Ayvaz I, Moray G et al. (2006). Tacrolimus-related neurologic and renal complications in liver transplantation: a single-center experience. Transplant Proc 38: 619–621.

Estol CJ, Faris AA, Martinez AJ et al. (1989). Central pontine myelinolysis after liver transplantation. Neurology 39: 493–498.

Evens AM, Choquet S, Kroll-Desrosiers AR et al. (2013). Primary CNS posttransplant lymphoproliferative disease (PTLD): an international report of 84 cases in the modern era. Am J Transplant 13: 1512–1522.

Fatt MA, Horton KM, Fishman EK (2008). Transmission of metastatic glioblastoma multiforme from donor to lung transplant recipient. J Comput Assist Tomogr 32: 407–409.

Fernandez-Rodriguez R, Contreras A, De Villoria JG et al. (2010). Acquired hepatocerebral degeneration: clinical characteristics and MRI findings. Eur J Neurol 17: 1463–1470.

Fernandez-Sabe N, Cervera C, Farinas MC et al. (2012). Risk factors, clinical features, and outcomes of toxoplasmosis in solid-organ transplant recipients: a matched case-control study. Clin Infect Dis 54: 355–361.

Fischer SA, Graham MB, Kuehnert MJ et al. (2006). Transmission of lymphocytic choriomeningitis virus by organ transplantation. N Engl J Med 354: 2235–2249.

Fishman JA (2007). Infection in solid-organ transplant recipients. N Engl J Med 357: 2601–2614.

Furukawa M, Terae S, Chu BC et al. (2001). MRI in seven cases of tacrolimus (FK-506) encephalopathy: utility of FLAIR and diffusion-weighted imaging. Neuroradiology 43: 615–621.

Garrido SM, Chauncey TR (1998). Neuroleptic malignant syndrome following autologous peripheral blood stem cell transplantation. Bone Marrow Transplant 21: 427–428.

Ge PS, Runyon BA (2014). Serum ammonia level for the evaluation of hepatic encephalopathy. JAMA 312: 643–644.

Gewurz BE, Marty FM, Baden LR et al. (2008). Human herpesvirus 6 encephalitis. Curr Infect Dis Rep 10: 292–299.

Gijtenbeek JM, van den Bent MJ, Vecht CJ (1999). Cyclosporine neurotoxicity: a review. J Neurol 246: 339–346.

Glass GA, Stankiewicz J, Mithoefer A et al. (2005). Levetiracetam for seizures after liver transplantation. Neurology 64: 1084–1085.

Goldstein LS, Haug 3rd MT, Perl 2nd J et al. (1998). Central nervous system complications after lung transplantation. J Heart Lung Transplant 17: 185–191.

Grauer O, Wolff D, Bertz H et al. (2010). Neurological manifestations of chronic graft-versus-host disease after allogeneic haematopoietic stem cell transplantation: report from the Consensus Conference on Clinical Practice in chronic graft-versus-host disease. Brain 133: 2852–2865.

Graus F, Saiz A, Sierra J et al. (1996). Neurologic complications of autologous and allogeneic bone marrow transplantation in patients with leukemia: a comparative study. Neurology 46: 1004–1009.

Gross ML, Pearson RM, Kennedy J et al. (1982). Rejection encephalopathy. Lancet 2: 1217.

Grotz WH, Breitenfeldt MK, Braune SW et al. (2001). Calcineurin-inhibitor induced pain syndrome (CIPS): a severe disabling complication after organ transplantation. Transpl Int 14: 16–23.

Guarino M, Benito-Leon J, Decruyenaere J et al. (2006). EFNS guidelines on management of neurological problems in liver transplantation. Eur J Neurol 13: 2–9.

Gupte AA, Hocevar SN, Lea AS et al. (2014). Transmission of Balamuthia mandrillaris through solid organ transplantation: utility of organ recipient serology to guide clinical management. Am J Transplant 14: 1417–1424.

Hachem RR, Yusen RD, Chakinala MM et al. (2007). A randomized controlled trial of tacrolimus versus cyclosporine after lung transplantation. J Heart Lung Transplant 26: 1012–1018.

Hamadani M, Martin LK, Benson DM et al. (2007). Central nervous system post-transplant lymphoproliferative disorder despite negative serum and spinal fluid Epstein–Barr virus DNA PCR. Bone Marrow Transplant 39: 249–251.

Haussinger D, Laubenberger J, vom Dahl S et al. (1994). Proton magnetic resonance spectroscopy studies on human brain myo-inositol in hypo-osmolarity and hepatic encephalopathy. Gastroenterology 107: 1475–1480.

Hinchey J, Chaves C, Appignani B et al. (1996). A reversible posterior leukoencephalopathy syndrome. N Engl J Med 334: 494–500.

Ho S, Clipstone N, Timmermann L et al. (1996). The mechanism of action of cyclosporin A and FK506. Clin Immunol Immunopathol 80: S40–S45.

Huq S, Wong M, Chan H et al. (2007). Osmotic demyelination syndromes: central and extrapontine myelinolysis. J Clin Neurosci 14: 684–688.

Idoate MA, Martinez AJ, Bueno J et al. (1999). The neuropathology of intestinal failure and small bowel transplantation. Acta Neuropathol 97: 502–508.

Israni AK, Zaun DA, Rosendale JD et al. (2015). OPTN/SRTR 2013 annual data report: deceased organ donation. Am J Transplant 15 (Suppl 2): 1–13.

Iwamoto M, Jernigan DB, Guasch A et al. (2003). Transmission of West Nile virus from an organ donor to four transplant recipients. N Engl J Med 348: 2196–2203.

Jaiswal A, Sabnani I, Baran DA et al. (2015). A unique case of rituximab-related posterior reversible encephalopathy syndrome in a heart transplant recipient with posttransplant lymphoproliferative disorder. Am J Transplant 15: 823–826.

Junna MR, Rabinstein AA (2007). Tacrolimus induced leukoencephalopathy presenting with status epilepticus and prolonged coma. J Neurol Neurosurg Psychiatry 78: 1410–1411.

Katirji MB (1989). Brachial plexus injury following liver transplantation. Neurology 39: 736–738.

Kauffman CA, Freifeld AG, Andes DR et al. (2014). Endemic fungal infections in solid organ and hematopoietic cell transplant recipients enrolled in the Transplant-Associated Infection Surveillance Network (TRANSNET). Transpl Infect Dis 16: 213–224.

Keown PA, Laupacis A, Carruthers G et al. (1984). Interaction between phenytoin and cyclosporine following organ transplantation. Transplantation 38: 304–306.

Kim JM, Jung KH, Lee ST et al. (2015). Central nervous system complications after liver transplantation. J Clin Neurosci 22: 1355–1359.

King JD, Rosner MH (2010). Osmotic demyelination syndrome. Am J Med Sci 339: 561–567.

Kumar A, Keyrouz SG, Willie JT et al. (2012). Reversible obstructive hydrocephalus from hypertensive encephalopathy. Neurocrit Care 16: 433–439.

Lai MM, Kerrison JB, Miller NR (2004). Reversible bilateral internuclear ophthalmoplegia associated with FK506. J Neurol Neurosurg Psychiatry 75: 776–778.

Lane RJ, Roche SW, Leung AA et al. (1988). Cyclosporin neurotoxicity in cardiac transplant recipients. J Neurol Neurosurg Psychiatry 51: 1434–1437.

Laureno R, Karp BP (1997). Cyclosporine mutism. Neurology 48: 296–297.

Lavigne CM, Shrier DA, Ketkar M et al. (2006). Tacrolimus leukoencephalopathy: a neuropathologic confirmation. Neurology 63: 1132–1133.

Lee VH, Wijdicks EF, Manno EM et al. (2008). Clinical spectrum of reversible posterior leukoencephalopathy syndrome. Arch Neurol 65: 205–210.

Lewis MB, Howdle PD (2003). Neurologic complications of liver transplantation in adults. Neurology 61: 1174–1178.

Ludwig KP, Thiesset HF, Gayowski TJ et al. (2011). Plasmapheresis and intravenous immune globulin improve neurologic outcome of central pontine myelinolysis occurring post orthotopic liver transplant. Ann Pharmacother 45. e10.

MacQuillan GC, Mutimer D (2004). Fulminant liver failure due to severe veno-occlusive disease after haematopoietic cell transplantation: a depressing experience. QJM 97: 581–589.

Mahoney CA, Arieff AI (1982). Uremic encephalopathies: clinical, biochemical, and experimental features. Am J Kidney Dis 2: 324–336.

Maramattom BV, Wijdicks EF (2004). Sirolimus may not cause neurotoxicity in kidney and liver transplant recipients. Neurology 63: 1958–1959.

Marasco SF, Vale M, Pellegrino V et al. (2010). Extracorporeal membrane oxygenation in primary graft failure after heart transplantation. Ann Thorac Surg 90: 1541–1546.

Mateen FJ, van de Beek D, Kremers WK et al. (2009). Neuromuscular diseases after cardiac transplantation. J Heart Lung Transplant 28: 226–230.

Mateen FJ, Dierkhising RA, Rabinstein AA et al. (2010). Neurological complications following adult lung transplantation. Am J Transplant 10: 908–914.

Mateen FJ, Muralidharan R, Carone M et al. (2011). Progressive multifocal leukoencephalopathy in transplant recipients. Ann Neurol 70: 305–322.

Matsuo Y, Kamezaki K, Takeishi S et al. (2009). Encephalomyelitis mimicking multiple sclerosis associated with chronic graft-versus-host disease after allogeneic bone marrow transplantation. Intern Med 48: 1453–1456.

Mayer TO, Biller J, O'Donnell J et al. (2002). Contrasting the neurologic complications of cardiac transplantation in adults and children. J Child Neurol 17: 195–199.

Maziak DE, Maurer JR, Kesten S (1996). Diaphragmatic paralysis: a complication of lung transplantation. Ann Thorac Surg 61: 170–173.

McCarron KF, Prayson RA (1998). The neuropathology of orthotopic liver transplantation: an autopsy series of 16 patients. Arch Pathol Lab Med 122: 726–731.

Menger H, Jorg J (1999). Outcome of central pontine and extrapontine myelinolysis ($n = 44$). J Neurol 246: 700–705.

Messert B, Orrison WW, Hawkins MJ et al. (1979). Central pontine myelinolysis. Considerations on etiology, diagnosis, and treatment. Neurology 29: 147–160.

Metzeler KH, Boeck S, Christ B et al. (2009). Idiopathic hyperammonemia (IHA) after dose-dense induction chemotherapy for acute myeloid leukemia: case report and review of the literature. Leuk Res 33: e69–e72.

Miller GM, Baker Jr HL, Okazaki H et al. (1988). Central pontine myelinolysis and its imitators: MR findings. Radiology 168: 795–802.

Moskowitz A, Nolan C, Lis E et al. (2007). Posterior reversible encephalopathy syndrome due to sirolimus. Bone Marrow Transplant 39: 653–654.

Munoz P, Valerio M, Palomo J et al. (2010). Infectious and non-infectious neurologic complications in heart transplant recipients. Medicine (Baltimore) 89: 166–175.

Navarro V, Varnous S, Galanaud D et al. (2010). Incidence and risk factors for seizures after heart transplantation. J Neurol 257: 563–568.

Oliverio PJ, Restrepo L, Mitchell SA et al. (2000). Reversible tacrolimus-induced neurotoxicity isolated to the brain stem. AJNR Am J Neuroradiol 21: 1251–1254.

Parizel PM, Snoeck HW, van den Hauwe L et al. (1997). Cerebral complications of murine monoclonal CD3 antibody (OKT3): CT and MR findings. AJNR Am J Neuroradiol 18: 1935–1938.

Penninga L, Moller CH, Gustafsson F et al. (2010). Tacrolimus versus cyclosporine as primary immunosuppression after heart transplantation: systematic review with meta-analyses and trial sequential analyses of randomised trials. Eur J Clin Pharmacol 66: 1177–1187.

Perez-Miralles F, Sanchez-Manso JC, Almenar-Bonet L et al. (2005). Incidence of and risk factors for neurologic complications after heart transplantation. Transplant Proc 37: 4067–4070.

Perreault C, Gyger M, Boileau J et al. (1983). Acute graft-versus-host disease after allogeneic bone marrow transplantation. Can Med Assoc J 129: 969–974.

Pittock SJ, Rabinstein AA, Edwards BS et al. (2003). OKT3 neurotoxicity presenting as akinetic mutism. Transplantation 75: 1058–1060.

Pokorny H, Gruenberger T, Soliman T et al. (2000). Organ survival after primary dysfunction of liver grafts in clinical orthotopic liver transplantation. Transpl Int 13 (Suppl 1): S154–S157.

Prayson RA, Estes ML (1995). The neuropathology of cardiac allograft transplantation. An autopsy series of 18 patients. Arch Pathol Lab Med 119: 59–63.

Prescott Jr WA, Callahan BL, Park JM (2004). Tacrolimus toxicity associated with concomitant metoclopramide therapy. Pharmacotherapy 24: 532–537.

Pustavoitau A, Bhardwaj A, Stevens R (2011). Neurological complications of transplantation. J Intensive Care Med 26: 209–222.

Ruzhansky KM, Brannagan 3rd TH (2015). Neuromuscular complications of hematopoietic stem cell transplantation. Muscle Nerve 52: 480–487.

Sakamoto H, Hirano M, Nose K et al. (2013). A case of severe ganciclovir-induced encephalopathy. Case Rep Neurol 5: 183–186.

Saner FH, Sotiropoulos GC, Gu Y et al. (2007). Severe neurological events following liver transplantation. Arch Med Res 38: 75–79.

Satlin MJ, Graham JK, Visvesvara GS et al. (2013). Fulminant and fatal encephalitis caused by *Acanthamoeba* in a kidney transplant recipient: case report and literature review. Transpl Infect Dis 15: 619–626.

Schiff D, O'Neill B, Wijdicks E et al. (2001). Gliomas arising in organ transplant recipients: an unrecognized complication of transplantation? Neurology 57: 1486–1488.

Seeley WW, Marty FM, Holmes TM et al. (2007). Post-transplant acute limbic encephalitis: clinical features and relationship to HHV6. Neurology 69: 156–165.

Senzolo M, Marco S, Ferronato C et al. (2009). Neurologic complications after solid organ transplantation. Transpl Int 22: 269–278.

Sevmis S, Karakayali H, Emiroglu R et al. (2007). Tacrolimus-related seizure in the early postoperative period after liver transplantation. Transplant Proc 39: 1211–1213.

Sharathkumar A, Shaw P (1999). Ganciclovir-induced encephalopathy in a bone marrow transplant recipient. Bone Marrow Transplant 24: 421–423.

Sharma KR, Cross J, Santiago F et al. (2002). Incidence of acute femoral neuropathy following renal transplantation. Arch Neurol 59: 541–545.

Siegal D, Keller A, Xu W et al. (2007). Central nervous system complications after allogeneic hematopoietic stem cell transplantation: incidence, manifestations, and clinical significance. Biol Blood Marrow Transplant 13: 1369–1379.

Sila CA (1989). Spectrum of neurologic events following cardiac transplantation. Stroke 20: 1586–1589.

Singh N, Yu VL, Gayowski T (1994). Central nervous system lesions in adult liver transplant recipients: clinical review with implications for management. Medicine (Baltimore) 73: 110–118.

Singh N, Bonham A, Fukui M (2000). Immunosuppressive-associated leukoencephalopathy in organ transplant recipients. Transplantation 69: 467–472.

Singh N, Lortholary O, Alexander BD et al. (2005). Allograft loss in renal transplant recipients with *Cryptococcus neoformans* associated immune reconstitution syndrome. Transplantation 80: 1131–1133.

Singh N, Husain S, AST Infectious Diseases Community of Practice (2009). Invasive aspergillosis in solid organ transplant recipients. Am J Transplant 9 (Suppl 4): S180–S191.

Sostak P, Padovan CS, Yousry TA et al. (2003). Prospective evaluation of neurological complications after allogeneic bone marrow transplantation. Neurology 60: 842–848.

Sostak P, Padovan CS, Eigenbrod S et al. (2010). Cerebral angiitis in four patients with chronic GVHD. Bone Marrow Transplant 45: 1181–1188.

Soupart A, Ngassa M, Decaux G (1999). Therapeutic relowering of the serum sodium in a patient after excessive correction of hyponatremia. Clin Nephrol 51: 383–386.

Srinivasan A, Burton EC, Kuehnert MJ et al. (2005). Transmission of rabies virus from an organ donor to four transplant recipients. N Engl J Med 352: 1103–1111.

Starzl TE, Schneck SA, Mazzoni G et al. (1978). Acute neurological complications after liver transplantation with particular reference to intraoperative cerebral air embolus. Ann Surg 187: 236–240.

Stevens AM, Sullivan KM, Nelson JL (2003). Polymyositis as a manifestation of chronic graft-versus-host disease. Rheumatology (Oxford) 42: 34–39.

Sun HY, Singh N (2011). Opportunistic infection-associated immune reconstitution syndrome in transplant recipients. Clin Infect Dis 53: 168–176.

Sun HY, Alexander BD, Huprikar S et al. (2015). Predictors of immune reconstitution syndrome in organ transplant recipients with cryptococcosis: implications for the management of immunosuppression. Clin Infect Dis 60: 36–44.

Tome S, Wells JT, Said A et al. (2008). Quality of life after liver transplantation. A systematic review J Hepatol 48: 567–577.

Trinka E (2011). What is the evidence to use new intravenous AEDs in status epilepticus? Epilepsia 52 (Suppl 8): 35–38.

van de Beek D, Patel R, Daly RC et al. (2007). Central nervous system infections in heart transplant recipients. Arch Neurol 64: 1715–1720.

van de Beek D, Kremers W, Daly RC et al. (2008). Effect of neurologic complications on outcome after heart transplant. Arch Neurol 65: 226–231.

van de Beek D, Kremers WK, Kushwaha SS et al. (2009). No major neurologic complications with sirolimus use in heart transplant recipients. Mayo Clin Proc 84: 330–332.

Varotti G, Grazi GL, Vetrone G et al. (2005). Causes of early acute graft failure after liver transplantation: analysis of a 17-year single-centre experience. Clin Transplant 19: 492–500.

Vattemi G, Marini M, Di Chio M et al. (2014). Polymyositis in solid organ transplant recipients receiving tacrolimus. J Neurol Sci 345: 239–243.

Vecino MC, Cantisani G, Zanotelli ML et al. (1999). Neurological complications in liver transplantation. Transplant Proc 31: 3048–3049.

Vinnard C, Barton T, Jerud E et al. (2009). A report of human herpesvirus 6-associated encephalitis in a solid organ transplant recipient and a review of previously published cases. Liver Transpl 15: 1242–1246.

Vivarelli M, Dazzi A, Cucchetti A et al. (2010). Sirolimus in liver transplant recipients: a large single-center experience. Transplant Proc 42: 2579–2584.

Vizzini G, Asaro M, Miraglia R et al. (2011). Changing picture of central nervous system complications in liver transplant recipients. Liver Transpl 17: 1279–1285.

Walder B, Tramer MR, Seeck M (2002). Seizure-like phenomena and propofol: a systematic review. Neurology 58: 1327–1332.

Walker RW, Rosenblum MK (1992). Amphotericin B-associated leukoencephalopathy. Neurology 42: 2005–2010.

Watling SM, Dasta JF (1994). Prolonged paralysis in intensive care unit patients after the use of neuromuscular blocking agents: a review of the literature. Crit Care Med 22: 884–893.

Webster A, Woodroffe RC, Taylor RS et al. (2005). Tacrolimus versus cyclosporin as primary immunosuppression for kidney transplant recipients. Cochrane Database Syst Rev. CD003961.

White SL, Hirth R, Mahillo B et al. (2014). The global diffusion of organ transplantation: trends, drivers and policy implications. Bull World Health Organ 92: 826–835.

Wijdicks EF, Hocker SE (2014). Neurologic complications of liver transplantation. Handb Clin Neurol 121: 1257–1266.

Wijdicks EF, Wiesner RH, Dahlke LJ et al. (1994). FK506-induced neurotoxicity in liver transplantation. Ann Neurol 35: 498–501.

Wijdicks EF, Wiesner RH, Krom RA (1995). Neurotoxicity in liver transplant recipients with cyclosporine immunosuppression. Neurology 45: 1962–1964.

Wijdicks EF, Blue PR, Steers JL et al. (1996a). Central pontine myelinolysis with stupor alone after orthotopic liver transplantation. Liver Transpl Surg 2: 14–16.

Wijdicks EF, Plevak DJ, Wiesner RH et al. (1996b). Causes and outcome of seizures in liver transplant recipients. Neurology 47: 1523–1525.

Winston DJ, Vikram HR, Rabe IB et al. (2014). Donor-derived West Nile virus infection in solid organ transplant recipients: report of four additional cases and review of clinical, diagnostic, and therapeutic features. Transplantation 97: 881–889.

Wong R, Beguelin GZ, de Lima M et al. (2003). Tacrolimus-associated posterior reversible encephalopathy syndrome after allogeneic haematopoietic stem cell transplantation. Br J Haematol 122: 128–134.

Wright DG, Laureno R, Victor M (1979). Pontine and extrapontine myelinolysis. Brain 102: 361–385.

Wszolek ZK, McComb RD, Pfeiffer RF et al. (1989). Pontine and extrapontine myelinolysis following liver transplantation. Relationship to serum sodium. Transplantation 48: 1006–1012.

Wu G, Vilchez RA, Eidelman B et al. (2002). Cryptococcal meningitis: an analysis among 5,521 consecutive organ transplant recipients. Transpl Infect Dis 4: 183–188.

Yango AF, Fischbach BV, Levy M et al. (2014). West Nile virus infection in kidney and pancreas transplant recipients in the Dallas-Fort Worth Metroplex during the 2012 Texas epidemic. Transplantation 97: 953–957.

Yardimci N, Colak T, Sevmis S et al. (2008). Neurologic complications after renal transplant. Exp Clin Transplant 6: 224–228.

Young GB, Bolton CF, Austin TW et al. (1990a). The encephalopathy associated with septic illness. Clin Invest Med 13: 297–304.

Young GB, Gilbert JJ, Zochodne DW (1990b). The significance of myoclonic status epilepticus in postanoxic coma. Neurology 40: 1843–1848.

Zhang ZW, Kang Y, Deng LJ et al. (2009). Therapy of central pontine myelinolysis following living donor liver transplantation: report of three cases. World J Gastroenterol 15: 3960–3963.

Zierer A, Melby SJ, Voeller RK et al. (2007). Significance of neurologic complications in the modern era of cardiac transplantation. Ann Thorac Surg 83: 1684–1690.

Zivkovic SA, Jumaa M, Barisic N et al. (2009). Neurologic complications following lung transplantation. J Neurol Sci 280: 90–93.

Zivkovic SA, Eidelman BH, Bond G et al. (2010). The clinical spectrum of neurologic disorders after intestinal and multivisceral transplantation. Clin Transplant 24: 164–168.

Chapter 31

Neurologic complications of cardiac and vascular surgery

K.N. SHETH* AND E. NOUROLLAHZADEH

Division of Neurocritical Care and Emergency Neurology, Department of Neurology, Yale New Haven Hospital, New Haven, CT, USA

Abstract

This chapter will provide an overview of the major neurologic complications of common cardiac and vascular surgeries, such as coronary artery bypass grafting and carotid endarterectomy. Neurologic complications after cardiac and vascular surgeries can cause significant morbidity and mortality, which can negate the beneficial effects of the intervention. Some of the complications to be discussed include ischemic and hemorrhagic stroke, seizures, delirium, cognitive dysfunction, cerebral hyperperfusion syndrome, cranial nerve injuries, and peripheral neuropathies. The severity of these complications can range from mild to lethal. The etiology of complications can include a variety of mechanisms, which can differ based on the type of cardiac or vascular surgery that is performed. Our knowledge about neuropathology, prevention, and management of surgical complications is growing and will be discussed in this chapter. It is imperative for clinicians to be familiar with these complications in order to narrow the differential diagnosis, start early management, anticipate the natural history, and improve outcomes.

INTRODUCTION

Patients undergoing cardiac or vascular surgeries are considered to be at high risk for various perioperative complications, in part due to coexisting comorbidities. Neurologic complications have long been associated with these types of surgery. Management of cardiac and vascular patients should be initiated preoperatively by risk stratification based on factors such as age, atherosclerotic burden, and other comorbidities. This should be followed with careful intraoperative management by the surgeon and anesthesiologist targeted toward minimizing neurologic complications with specific attention to the patient's hemodynamic parameters. Postoperative management requires familiarity and a high level of suspicion by the clinician for the early diagnosis and management of common complications for each specific surgery. The combination of optimal management in pre-, intra-, and postoperative phases maximizes the benefits of surgery. This chapter will review various neurologic complications that are associated with cardiac and vascular surgeries, with specific focus on coronary artery bypass grafting (CABG), carotid endarterectomy (CEA), extracorporeal membrane oxygenation (ECMO), and aortic surgery.

CABG is a revascularization procedure that can improve survival in patients with presence of triple-vessel coronary artery disease, left main coronary artery disease ($\geq 50\%$ stenosis), and two-vessel disease with proximal involvement of the left anterior descending artery (Hillis et al., 2011). CABG is also preferred in patients with severely depressed left ventricular systolic function, and for patients who need surgery for cardiac conditions in addition to coronary artery disease, such as replacement of valves (Shekar, 2006). Access for the procedure is through a midline sternotomy, which is performed either with on-pump or off-pump techniques. During on-pump surgery, the heart is stopped and perfusion is maintained through a cardiopulmonary bypass machine. During off-pump procedures, the heart

---

*Correspondence to: Kevin N. Sheth, MD, Division of Neurocritical Care and Emergency Neurology, Department of Neurology, Yale New Haven Hospital, 15 York St, LCI 1003, New Haven, CT 06510, USA. Tel: +1-203-737-8051, E-mail: kevin.sheth@yale.edu

continues to beat and graft suturing can be performed using stabilizing devices. Studies have shown that patients with multiple comorbidities and severe cardiac disease have better outcomes with surgery compared to maximal medical management (Ferguson et al., 2002), but their outcome is significantly affected if perioperative neurologic complications occur (Hannan et al., 2005; Malenka et al., 2005). Major neurologic complications associated with CABG include ischemic stroke, delirium, and cognitive dysfunction (Table 31.1).

CEA or carotid artery stenting (CAS) should be considered for patients with symptomatic carotid stenosis (50–99%), and for those with asymptomatic stenosis of 70–99% (Brott et al., 2011). CEA can be performed under general or regional anesthesia, and involves a longitudinal neck incision along the anterior border of the sternocleidomastoid muscle, with exposure of the common and internal carotid arteries. The carotid artery is cross-clamped, and a shunt may be placed to bypass the clamped portion to preserve blood flow to the brain. After endarterectomy is completed, patch angioplasty may be performed to prevent restenosis. Significant improvement in outcomes after CEA, as well as carotid stenting, is highly dependent on low perioperative complication rates, especially stroke (Eckstein et al., 2008). Major neurologic complications associated with CEA include ischemic and hemorrhagic

Table 31.1

Epidemiology of major neurologic complications after vascular and cardiac surgeries

Complication	CABG		CEA	
	Incidence	Etiology	Incidence	Etiology
Ischemic stroke	1.6%	Embolization (atheroma, air, thrombus) Hypotension Atrial fibrillation	1.6% (0.3–7%)	Embolization Local thrombus or reocclusion Prolonged carotid cross-clamping
ICH	<0.1%	Hemorrhagic transformation Spontaneous (e.g., with antithrombotics)	0.3%	Hemorrhagic transformation Spontaneous CHS-related
Seizures	0.5–3.5%	Subtherapeutic AEDs New structural lesion (stroke, ICH) Acute metabolic derangement Medication-induced	0.8%	Same
Delirium	10%	Inflammatory Metabolic/infectious Neurotransmitter imbalance, etc.	19–39%	Same
Cognitive deficit	24% at 6 months 42% at 5 years	Hypotension Stroke/microemboli Exacerbating prior dementia		
Cerebral hyperperfusion syndrome			1–3%	Increased CBF with cerebral dysautoregulation
Cranial Nerve deficits			8.6% (2–15%)	Prolonged retraction Stretching Compression

Peripheral neuropathies:
  Brachial plexus
  Phrenic
  Recurrent laryngeal
  Saphenous nerve
	13%	Retraction
		Stretching
		Compression
		Hypothermia (phrenic nerve)		

AED, antiepileptic drug; CABG, coronary artery bypass grafting; CBF, cerebral blood flow; CEA, carotid endarterectomy; ICH, intracerebral hemorrhage.

stroke, cerebral hyperperfusion syndrome (CHS), cranial nerve (CN) deficits, seizures, and delirium (Table 31.1) (Greenstein et al., 2007).

ECMO or extracorporeal life support is a rescue therapy that provides temporary mechanical respiratory and/or cardiac support in patients with severe and potentially reversible cardiopulmonary failure. The use of ECMO in adults has continued to increase over the past decade, with improvement in survival rate (Sauer et al., 2015). ECMO works by drainage of venous blood, which then circulates in an external system where hemoglobin oxygenation and carbon dioxide removal occur. The filtered blood returns to the patient either through venous (venovenous) or arterial (venoarterial) cannulation. Venovenous ECMO provides only respiratory support, whereas venoarterial ECMO also provides hemodynamic support in patients with cardiac failure. Neurologic complications occur in about 15% (range of 10–50%) of patients undergoing ECMO. The main complications include stroke, intracerebral hemorrhage (ICH), seizures, and brain death. Other possible complications are encephalopathy (typically hypoxic-ischemic) and subarachnoid hemorrhage (Mateen et al., 2011; Nasr and Rabinstein, 2015; Lorusso et al., 2016).

EPIDEMIOLOGY

Ischemic stroke

The stroke rate after CABG has decreased steadily since the early 1980s to less than 2% (Tarakji et al., 2011). However, the rate of ischemic stroke may be underestimated, since clinically silent strokes can often only be detected using magnetic resonance imaging (MRI) (Selnes et al., 2012). Intraoperative stroke accounts for up to 40% of cases. Randomized trials demonstrate that the postoperative stroke rate does not vary significantly between the off- and on-pump techniques (Selnes et al., 2012). In comparison to CABG, percutaneous coronary intervention has been associated with significantly lower stroke rate within 30 days (Korn-Lubetzki et al., 2013; Palmerini et al., 2013).

With valve replacement surgery, there is a higher risk for an ischemic stroke compared to CABG. The most common procedures involve the mitral and aortic valves (Hogue et al., 2001). In a study using Society of Thoracic Surgeons data from 2002 to 2006, the risk for ischemic stroke after mitral or aortic valve replacement and mitral valve repair was about 3–4% within 30 days of surgery (O'Brien et al., 2009).

Periprocedural stroke can occur in about 0.3–7% of patients following CEA, when defined as a stroke within 30 days of surgery. In the Carotid Revascularization Endarterectomy versus Stenting Trial (CREST), the periprocedural incidence of any ischemic stroke within 30 days after CEA and CAS was 1.6% and 4.1%, respectively. When assessed for severity, the incidence of a major stroke after CEA and CAS was 0.3% and 0.9%, respectively, with major stroke being defined as having a National Institutes of Health stroke scale (NIHSS) ≥ 9 (Hill et al., 2012).

Periprocedural stroke rate in ECMO can vary due to factors such as the underlying etiology, duration of treatment, age, and other risk factors. Overall, ECMO-associated stroke has been reported to occur in 5–8% of patients (Lan et al., 2010; Mateen et al., 2011; Cheng et al., 2014; Nasr and Rabinstein, 2015).

Spinal cord infarction (SCI) can occur perioperatively in both cardiac and aortic surgeries, typically leading to paraparesis or paraplegia. Aortic surgery is required for treatment of aortic aneurysms or dissection. The risk for SCI is highly associated with operations involving the thoracic and thoracoabdominal regions (from the left subclavian branch to the aortic bifurication). Overall, there is a trend toward higher rates of SCI with open surgery compared to thoracic endovascular aortic repair (TEVAR), although data are inconclusive (Messe et al., 2008; Hiratzka et al., 2010). Patients undergoing type I or II aortic aneurysm treatment (Crawford classification) are particularly susceptible to SCI (Greenberg et al., 2008).

SCI is also a rare complication of CABG. There have been only 31 cases of paraplegia related to CABG in the literature, of which almost all were due to either intraaortic balloon pump or aortic dissection (Sevuk et al., 2016). There is a paucity of data on management of SCI, but patients may benefit from keeping mean arterial pressure (MAP) above 90 mmHg and using cerebrospinal fluid drainage to improve spinal cord perfusion (Romagnoli et al., 2012).

Cardiac surgery may also cause ophthalmologic complications. One prospective study from almost 30 years ago reported that a large proportion of patients (25%) developed ophthalmologic concerns that could be detected with careful bedside evaluation (Shaw et al., 1987). The most common complication was retinal infarction (more than 65%), but many patients were asymptomatic and did not have any lasting symptoms. There have been other reports of ophthalmic conditions (e.g., impaired saccadic and smooth-pursuit eye movements) after CABG and valvular heart surgery, but they are relatively rare (Devere et al., 1997; Solomon et al., 2008).

Intracerebral hemorrhage

ICH is an uncommon complication of cardiac surgery, accounting for only 1% of strokes (Likosky et al., 2003). The incidence of ICH is higher in patients undergoing urgent percutaneous coronary intervention due to the need for antiplatelet and anticoagulation therapy, and accounts for 8–46% of per-procedural strokes (Brown and Topol, 1993; Fuchs et al., 2002; Dukkipati et al., 2004). The ICH occurrence within 30 days of CEA has been reported to be at about 0.3–1.8%, and can be due to hemorrhagic transformation of an infarct or development of CHS (Rockman et al., 2000; Henderson et al., 2001). In the CREST trial, the incidence of ICH after CEA and CAS was 0.3% and 0.4%, respectively (Hill et al., 2012). The risk factors for development of ICH after CEA are recent ipsilateral stroke, greater than 90% stenosis of the ipsilateral ICA, and postoperative hypertension (Russell and Gough, 2004).

ICH occurs in about 1.8–3.6% of patients undergoing ECMO (Nasr and Rabinstein, 2015; Lorusso et al., 2016). Similar to other ECMO-related neurologic complications, ICH rate has declined over the past decade, most likely due to improved supportive medical management (Kasirajan et al., 1999; Lorusso et al., 2016).

Seizures

The incidence of postoperative seizures in patients undergoing CABG is about 0.5–3.5% (Roach et al., 1996). In patients undergoing CEA, the incidence of seizures has been reported to be approximately 0.8% within 30 days from the procedure. The majority of these patients developed seizures in the setting of high blood pressure due to the CHS (Naylor et al., 2003). Seizures can also occur during ECMO in about 1.8–4.1% of patients (Lorusso et al., 2016).

Delirium

Delirium occurs in about 10% of patients after CABG surgery in most large series. The incidence has been reported to be as high as 47% in smaller studies due to variability in assessment and diagnosis (Bucerius et al., 2004; Gottesman et al., 2010). Risk factors for the development of delirium after both cardiac and vascular surgeries include advanced age, previous cognitive impairment or depression, history of previous stroke, regular alcohol consumption, peripheral vascular disease, and renal failure (Veliz-Reissmuller et al., 2007; Groen et al., 2012). Post-CABG delirium can last for weeks and sometimes overlaps with cognitive impairment, which is discussed below.

The incidence of postoperative delirium in patients undergoing vascular surgery (including CEA, aortic and peripheral artery surgeries) is 19–39%, and these patients typically have a longer intensive care unit stay (Bohner et al., 2000; Ellard et al., 2014).

Cognitive impairment

Postoperative cognitive impairment is common, although the specific incidence varies depending on the criteria used for diagnosis (Selnes et al., 2006). Many of the studies investigating cognitive decline after CABG did not use any control group and may have overestimated the actual incidence. In one large prospective study, cognitive decline occurred in 53% of patients at discharge, 36% at 6 weeks, 24% at 6 months, and 42% at 5 years (Newman et al., 2001; Stygall et al., 2003). More recently, in two prospective studies, with inclusion of cardiovascular patients with or without intervention, there was similar cognitive decline when patients were assed with neuropsychologic tests (Sweet et al., 2008; Selnes et al., 2009).

Cerebral hyperperfusion syndrome

CHS occurs after carotid revascularization, and is characterized by ipsilateral headaches, seizures, and focal neurologic deficits with an increased cerebral blood flow (CBF) of more than 100% compared to baseline (van Mook et al., 2005). The first case of CHS following CEA was described in 1975 (Sundt et al., 1975). CHS occurs in 1–3% of patients following CEA, and usually presents 3–7 days after the surgery, but can occur up to 28 days postoperatively (Moulakakis et al., 2009). It should be noted that most patients develop increased CBF immediately after CEA, but it only lasts for a few hours and patients remain asymptomatic (Jorgensen and Schroeder, 1993).

Cranial nerve deficits

CN injuries are the most common neurologic complication of CEA, occurring in 2–15% of cases. CN VII, IX, X, and XII are most commonly affected. The incidence of CN deficits in the North American Symptomatic Carotid Endarterectomy (NASCET) and European Carotid Surgery Trial (ECST) was 8.6% and 5.1%, respectively (Ferguson et al., 1999; Cunningham et al., 2004). In NASCET, the most common CN injury was the hypoglossal (XII) (3.7%), followed by vagus (X) (2.5%), and facial (VII) nerves (2.2%). In ECST, operation time longer than 2 hours was the only factor independently associated with increased risk of CN injury (Cunningham et al., 2004). In other studies, aside from the duration of the procedure, general anesthesia, intraoperative hematoma, and repeat CEA were also factors associated with increased risk of CN injury (Hye et al., 2015). The risk of cranial or peripheral nerve injury in CAS is negligible.

Peripheral neuropathies

Peripheral neuropathies can occur in about 10–15% of patients undergoing CABG. The most common neuropathies involve the brachial plexus, with incidence ranging from 1.5% to 24%, phrenic nerve (26%, with range of 10–60%), and, less frequently, the recurrent laryngeal and saphenous nerves (Vander Salm, 1984; Efthimiou et al., 1991; O'Brien et al., 1991; DeVita et al., 1993; Sharma et al., 2000). In one study, the incidence of saphenous neuropathy was as high as 90% immediately after surgery (Nair et al., 1988). The wide range of peripheral neuropathy incidence could be due to different diagnostic criteria, and difficulty identifying patients with mild symptoms. For instance, phrenic injury in some studies was diagnosed based on diaphragmatic elevation on chest radiography, thus overestimating the incidence since other etiologies (e.g., atelectasis, pneumonia, or pleural effusion) may cause the same radiographic finding.

NEUROPATHOLOGY

Ischemic stroke

The mechanism of ischemic stroke post-CABG is usually due to embolization, cerebral hypoperfusion (e.g., watershed stroke), and an increased inflammatory response postoperatively (Gottesman et al., 2006; Raja and Berg, 2007). Most of the surgical technique modifications (e.g., off-pump CABG) have focused on minimizing intraoperative embolization. However, this has not led to significant reduction in the incidence of ischemic stroke or cognitive decline in randomized trials (Hernandez et al., 2007; Selnes et al., 2012). Arterial embolization is by far the predominant cause of intraoperative stroke, accounting for more than 60% of strokes, followed by hypoperfusion-related strokes at about 9% (Likosky et al., 2003). The thrombus or atheroma can be embolized due to clamping or unclamping of the ascending aorta, construction of coronary artery anastomoses, or during high turbulence within the diseased aortic segment. In the case of valvular surgeries, the source may also be from diseased and calcified aortic and mitral valves. Air embolization may also occur during CABG, but is relatively rare. Another important cause of delayed postoperative stroke is new-onset atrial fibrillation, which occurs in 20–50% of patients after cardiac surgery (Echahidi et al., 2008). The mechanism is most likely multifactorial and includes perioperative pericardial inflammation, catecholamine surge, and autonomic dysfunction (Echahidi et al., 2008). Global hypoperfusion can result from hypotension, typically from reduced cardiac output. This can directly lead to an ischemic stroke within the watershed areas along major vascular territories between middle cerebral artery and either posterior cerebral artery or anterior cerebral artery. There is some evidence that impaired clearance of emboli also contributes to the burden of watershed stroke (Caplan and Hennerici, 1998). The predisposing risk factors for perioperative ischemic stroke include aortic atherosclerotic burden, older age, diabetes mellitus, renal failure, peripheral artery disease, hypertension, recent myocardial infarction, history of previous ischemic stroke, and severe left ventricular dysfunction (Hogue et al., 1999).

Stroke etiology after CEA can be due to embolization, local thrombosis, or reocclusion within the surgical site, and to a lesser extent, prolonged carotid artery cross-clamping. As expected, patients with baseline extra- and intracranial vascular lesions and poor cortical collaterals are at higher risk for stroke. In patients with poor cortical collaterals or low carotid stump pressure, a temporary intra-arterial shunt can be used during the cross-clamping of the carotid artery to minimize stroke occurrence (Aburahma et al., 2010). Delayed strokes are typically due to embolization from the surgical site, or new-onset atrial fibrillation.

Stroke etiology during ECMO is typically embolic (clot or air) or due to hemodynamic instability (watershed stroke). The embolic stroke could be due to formation of thrombus or air within the nonbiologic surface of ECMO circuit or due to cardiac failure (i.e., blood stasis). There are no clear data to indicate a difference in stroke rate in relation to route of cannulation (venoarterial versus venovenous) in adults (Mateen et al., 2011; Luyt et al., 2016).

Spinal cord ischemia during aortic surgery can be due to mechanisms such as aortic cross-clamping, hypotension, stent-related blockage of aortic branches (collateral segmental arteries supplying the spinal cord), dislodgement of atherosclerotic emboli due to catheter manipulation, and injury to the hypogastric collateral network related to complicated groin puncture access (Ullery et al., 2011). The cervical and upper thoracic spinal cord is usually spared from complications since it is predominantly perfused through the vertebral arteries that supply the anterior and posterior spinal arteries. However, the spinal cord caudal to the upper thoracic region becomes increasingly more reliant on collateral perfusion from the aorta due to attenuation or discontinuation of anterior (and, to a lesser degree, posterior) spinal arteries (Shamji et al., 2003). More specifically, the aorta gives rise to the segmental arteries that ultimately branch off into anterior and posterior radicular arteries, which in turn form a collateral network with the attenuated anterior and posterior spinal arteries. The artery of Adamkiewicz (an anterior radicular artery, arising from intercostal and lumbar segmental arteries) is the predominant supplier of blood within the thoracolumbar spine, providing about 68% of blood flow, and thus patients undergoing aortic surgeries can be susceptible to SCI if this artery is compromised. Generally, the watershed areas within the thoracolumbar spine include T1, T5, T8, and T9 levels (Chakravorty, 1971; Ullery et al., 2011).

Intracerebral hemorrhage

Postsurgical ICH is typically secondary to hemorrhagic transformation of an initially ischemic stroke (Likosky et al., 2003). Hemorrhagic transformation occurs once there is restoration of blood flow to the stroke territory, which has abnormal vasculature due to disruption of the blood–brain barrier. Risk factors may include large-volume ischemic stroke, use of anticoagulation within 12 hours of stroke onset, and older age (Yatsu et al., 1988). ICH can also be spontaneous, possibly due to systemic anticoagulation during cardiopulmonary bypass or CHS.

In patients receiving ECMO, the mechanism of spontaneous ICH is typically primarily due to anticoagulation, with renal failure, low fibrinogen level, and thrombocytopenia also being potential factors (Kasirajan et al., 1999).

Seizures

Causes for seizure can be variable, but in the perioperative setting, seizures are most likely due to prior history of epilepsy with subtherapeutic antiepileptic medications, new or old structural brain lesions (e.g., stroke or brain tumor), hypoxic-ischemic encephalopathy, metabolic derangements such as hyponatremia or hypoglycemia, or certain medications, such as aminocaproic acid or tranexamic acid. With CEA, seizures may also be due to CHS.

Delirium

Delirium is likely attributable to multiple mechanisms, such as increased inflammation, neurotransmitter imbalance, metabolic dysfunction, hemodynamic instability, and underlying genetic factors (Brown, 2014). The aforementioned mechanisms are further affected by patients' baseline cognitive reserve, choice of sedatives, poor cerebral perfusion, microembolic strokes, and infections (Siepe et al., 2011).

Cognitive impairment

Short-term cognitive decline has been attributed to various factors, such as general anesthesia, medications, sleep disturbances, and the inflammatory response. Short-term cognitive decline is not unique to cardiac and vascular surgery, and may also occur with general anesthesia for other types of surgery (Monk et al., 2008). The most important predisposing factor of long-term cognitive decline is patients' underlying cerebrovascular disease, rather than the initial cardiac surgery, which was previously assumed to be the culprit. This has been demonstrated in recent studies that included control groups, showing similar cognitive decline among patients with or without cardiac surgery that had similar cerebrovascular risk factors (Selnes et al., 2012). Important intraoperative risk factors include severe hypotension or microembolic burden during aorta and heart manipulation, as shown with transcranial Doppler (Liu et al., 2009).

Cerebral hyperperfusion syndrome

CHS occurrence is related to chronic cerebral hypoperfusion prior to revascularization, and can be explained by a few contributing mechanisms. First, the sudden increase in CBF after recanalization is not compensated due to impaired CBF autoregulation. The autoregulatory curve is shifted, such that cerebral vasoconstriction cannot maintain constant flow. The impairment is due to endothelial injury from a combination of chronic small-vessel disease as well as oxygen free radicals produced during CEA (Skydell et al., 1987; Ogasawara et al., 2004). Rapid changes in systemic blood pressure are also not buffered owing to dysfunction and denervation of baroreceptor reflex incurred during CEA. These factors can lead to hydrostatic cerebral edema, especially in the vertebrobasilar territory, due to transudation of fluid into the insterstitium and astrocytes (van Mook et al., 2005). Some of the risk factors for CHS development include high-grade ipsilateral stenosis, bilateral carotid disease, and uncontrolled perioperative hypertension. Furthermore, the type of anesthesia may be a predisposing factor. For instance, isoflurane has been implicated as contributing to impaired autoregulation (Skydell et al., 1987).

Cranial nerve deficits

CN injuries that occur during CEA are due to prolonged retraction and transection, as well as inadvertent stretching and clamping of the CNs that are in the vicinity of the surgical field. The hypoglossal nerve (CN XII) is particularly prone to injury during retraction, as the exposure of the internal carotid artery may require inadvertent manipulation or mobilization. This nerve exits the skull through the hypoglossal canal, runs behind the carotid artery, and crosses the external and internal carotid arteries distal to the bifurcation. The risk for CN XII injury is increased in patients with a high carotid bifurcation (Bademci et al., 2005). The facial nerve (CN VII) does not lie within the surgical field, but can be injured due to prolonged retraction at the angle of the jaw. This is because the marginal mandibular branch of the facial nerve (the common site of injury) emerges from the parotid gland and is anterior to the masseter muscle, which is in close proximity to retraction instruments. The vagus nerve (CN X) exits the skull via the jugular foramen and courses between the internal carotid artery and the jugular vein. Vagus nerve injury occurs during the dissection of the carotid artery from the internal jugular vein, thus leading to stretching, inadvertent clamping, or transection. The glossopharyngeal nerve (CN IX) follows a similar course to the vagus nerve. CN IX injury is uncommon, since it is typically outside the surgical field, but this nerve can be mechanically injured when the posterior belly of the digastric muscle is mobilized for additional field exposure.

Sympathetic nerve injury (Horner's syndrome) is another rare complication of CEA. It occurs due to either ischemic or direct injury of the postganglionic sympathetic fibers within the carotid plexus (Perry et al., 2001). Direct injury of the superior cervical ganglion (or its postganglionic fibers) occurs during cases requiring a wide surgical field, with the exposure of the digastric muscle (similar to the CN IX injury). An ischemic injury causing Horner syndrome is due to compromised vascular supply of the superior cervical ganglion, due to compromised perforators from ascending pharyngeal and carotid arteries.

Peripheral neuropathy

In cardiac surgeries requiring sternotomy, sternal retraction, internal mammary artery (IMA) dissection, or hypothermia-related injuries are the common mechanisms leading to peripheral neuropathies. The brachial plexus anatomy involves joining of C5–T1 spinal nerves to become brachial plexus roots, which then join to become trunks (superior, middle, inferior) as they pass over the first rib and under the clavicle. The brachial plexus is susceptible to injury because it is superficial, its roots are fixed proximally and distally (making it susceptible to stretch injury), and its trunks pass through the narrow space between the first rib and clavicle (Kirsh et al., 1971).

The most common cause of brachial plexopathy is superior rotation of the first rib and subsequent downward displacement of the clavicle during the sternotomy procedure, thereby stretching and injuring the brachial plexus, most often the lower roots, C8–T1). Posterior fractures of the first rib can also lead to direct brachial plexopathy. Patients who undergo IMA grafting are at higher risk of developing brachial plexopathy (10.6% vs. 1% without) (Roy et al., 1988). This is because IMA grafting requires a larger sternotomy for better visualization. The phrenic nerve is formed by the ventral roots of the C3–C5 spinal cord segment. The left phrenic nerve runs lateral to the aortic arch and pericardium. The right phrenic nerve is lateral to the superior vena cava, right atrium, and inferior vena cava as it courses towards the diaphragm. Phrenic nerve injury is mainly due to topical hypothermia commonly used during CABG for myocardial protection. At temperatures below 5°C, there can be complete nerve conduction block, and the hypothermia-induced neuropathy can range from focal demyelination to axonal degeneration, depending on the temperature and duration (Denny-Brown et al., 1945; Basbaum, 1973). The left phrenic nerve is the most commonly affected due to its close proximity to the cooled myocardium. Another risk factor for phrenic nerve injury is the use of an IMA graft, which is due to either direct surgical damage or ischemia, since IMA branches supply the phrenic nerves. Bilateral phrenic nerve injury is rare, and is typically diagnosed in patients who are difficult to wean from mechanical ventilation.

Recurrent laryngeal nerve injury during CABG can occur either due to hypothermic injury or during IMA dissection, similar to phrenic nerve injury (Tewari and Aggarwal, 1996). This is because left recurrent laryngeal nerve is within proximity of the parietal pleura and pericardium as it circles around the aortic arch.

Saphenous nerve injury may occur during saphenous vein harvesting for CABG. The saphenous nerve is a sensory branch of the femoral nerve (L2–L4) that supplies the anteromedial region of the leg.

CLINICAL PRESENTATION

Patients undergoing vascular or cardiac surgeries require a pre- and postoperative focused neurologic examination to screen for any acute neurologic deterioration. Early detection and management of time-sensitive diagnoses such as stroke (ischemic or hemorrhagic) and seizure can significantly minimize morbidity and mortality. In a noncomatose patient, there should be an assessment of mental status (orientation, attention, and concentration), language, visual field, CNs, and lateralizing motor signs. In a comatose patient, an assessment of level of consciousness and motor examination must be done with loud verbal or noxious stimuli, as well as a CN/brainstem examination. Patients should be inspected for facial, oral, or limb twitching, as well as gaze deviation.

Typically, clinical presentation of stupor and coma is due to processes affecting bilateral cerebral hemispheres, thalami, or the brainstem. Unilateral lesions can also cause coma if there is sufficient mass effect to produce significant midline shift or compress the brainstem.

Metabolic disorders can impair consciousness by affecting both the reticular formation and the cerebral cortex. The various clinical presentations of stroke (ischemic or hemorrhagic) and seizures are discussed in relevant chapters.

Stroke

Ischemic and hemorrhagic stroke following cardiac and vascular surgery present in a similar fashion to strokes occurring in other settings, with focal motor, sensory, or CN deficits, as well as aphasia, hemianopia, and hemispatial neglect. Patients with embolic showers may develop infarction in multiple vascular territories and manifest with failure to awaken postoperatively. Watershed territory infarcts are usually bilateral, and may manifest with impaired consciousness, bilateral visual loss, and predominantly proximal upper-limb weakness ("man in the barrel" syndrome).

In SCI related to aortic surgery, clinical deficits become apparent within minutes to hours. Thoracic and lumbar spinal infarction can lead to paraplegia/paraparesis or distal weakness of legs, respectively. When the predominant blood supply of the thoracolumbar spine is compromised (i.e., aortic collaterals), the most common clinical presentation is anterior spinal artery syndrome (ASAS). ASAS at the level of the thoracic spine can cause partial or complete weakness of the lower extremities, impaired sensation of pain and temperature because of involvement of the spinothalamic tract, and preserved sensation of position and vibration due to sparing of the dorsal columns. Other symptoms include hypotension and impaired function of bowel and bladder, abnormal thermoregulation, sexual dysfunction, and eventual spasticity (except with concurrent lower motor neuron injury, as with infarcted anterior horn cells).

Seizures

Seizure manifestations can range from psychosis to focal or generalized tonic and/or clonic activity of limbs (Table 31.2). In patients undergoing CEA, seizures may present as one of the features of CHS, together with headache and elevated blood pressure. CHS-related seizures typically start as focal motor convulsions, originating in the cortex ipsilateral to the CEA, with possible secondary generalization into tonic-clonic seizures (Kieburtz et al., 1990; Naylor et al., 2003).

Delirium

As per the fifth edition of *Diagnostic and Statistical Manual of Mental Disorders* (DSM-5: American Psychiatric Association, 2013), delirium is defined as an impairment of attention and awareness, which develops over a short period of time (hours to days), and fluctuates in severity. There must also be an

Table 31.2

Clinical presentation and differential diagnosis of seizure

Clinical manifestations of seizure	
Cognitive	Memory loss
	Altered mental status:
	Fluctuating or persistent
	Severity: confusion to coma
	Echolalia, aphasia, and perseveration
	Psychosis, hallucinations, and catatonia
	Crying and laughter
Motor	Tonic and/or clonic activity
	Posturing
	Eye deviation, blinking, facial twitching, and nystagmus
Autonomic	Tachycardia, bradycardia, nausea, vomiting, miosis, mydriasis,
Seizure mimics	
Movement disorders	Chorea, dystonia, tics, myoclonus, and asterixis
Stroke	Ischemic or hemorrhagic:
	"Limb shaking" due to severe carotid stenosis
	Posterior-circulation stroke
	Brainstem hemorrhage
Herniation syndromes	Posturing (extensor and/or flexor)
Syncope	Cardiogenic or cataplexy (narcolepsy-related)
Other	Psychogenic, delirium, shivering, and tremor

additional cognitive disturbance (i.e., memory, orientation, language, visuospatial ability, or perception) (Lawlor and Bush, 2014).

Cognitive impairment

Cognitive impairment presents as dysfunction in one or more cognitive domains that are typically tested during neuropsychologic evaluation. These domains include memory, language, executive functions, attention, and motor functions. Thorough evaluation of cognitive impairment is most helpful if it can be performed and compared pre- and postoperatively.

Cerebral hyperperfusion syndrome

CHS may present with an ipsilateral headache, hypertension, seizures, focal neurologic deficits, or encephalopathy. In severe cases, patients may develop cerebral edema. Failure to treat CHS in a timely fashion may lead to ICH.

Cranial nerve deficits

The hypoglossal nerve (CN XII) supplies motor function to the tongue, with injury manifesting as deviation towards the site of injury, as well as dysarthria and dysphagia due to tongue weakness.

Facial nerve (CN VII) injury results in an ipsilateral facial droop, due to weakness of the orbicularis oris muscle, typically without major clinical consequences. Vagus nerve (CN X) injury can lead to ipsilateral paralysis of the soft palate, pharynx, and larynx, in turn causing hoarseness, dysphagia, and ipsilateral vocal cord paresis. The two at-risk branches of the vagus nerve are the superior laryngeal nerve (leading to changes in voice quality) and the recurrent laryngeal nerve, which supplies all the intrinsic muscles of the larynx, except the cricothyroid muscles, leading to hoarseness and vocal cord paresis. Glossopharyngeal nerve (CN IX) injury, an uncommon complication, can lead to loss of gag reflex, dysphagia, as well as bradycardia and hypotension due to its innervation of the carotid sinus. Damage to the sympathetic chain can also occur during CEA, which will lead to either Horner's syndrome (miosis, ptosis, and anhydrosis) or, rarely, the "first-bite syndrome" (unilateral pain in the parotid region after the first bite of each meal). This latter diagnosis is exceedingly rare, and may potentially be treated with local injection of botulinum toxin (Wang et al., 2013).

Peripheral neuropathy

Diagnosis of brachial plexopathy requires high clinical suspicion followed by a thorough motor, sensory, and reflex examination of the upper extremities. The brachial plexus innervates almost all muscles of the upper extremities except the trapezius and scapulae (Leffert, 1974). Typically, post-CABG brachial plexopathy involves the C8–T1 nerve roots (Ben-David and Stahl, 1997), with clinical presentation of weakness and/or sensory deficits in the affected myotome/dermatome of the upper extremities. Phrenic nerve injury, especially if unilateral, is relatively asymptomatic, as other respiratory muscles compensate for the ipsilateral diaphragmatic paresis (i.e., accessory, abdominal, and intercostal muscles). Phrenic nerve injury may cause exertional dyspnea or nocturnal orthopnea (Dimopoulou et al., 1998). More severe injury can lead to atelectasis and prolonged mechanical ventilation. Phrenic nerve injury occurs more commonly on the left, due to proximity of the left phrenic nerve to the topically cooled pericardium. Saphenous neuropathy usually presents with paresthesias and pain along the medial aspect of the calf and foot.

NEURODIAGNOSTICS AND IMAGING

All patients with acute neurologic deficits postoperatively should undergo head computed tomography (CT) and laboratory evaluation. Subsequent neurodiagnostic testing should be tailored based on patients' history, medications, type of surgery, and, most importantly, the neurologic examination (Fig. 31.1). Important diagnoses not to be missed are ischemic stroke, ICH, seizure, or CHS, due to the time-sensitive management of these conditions. In a comatose patient with an unremarkable head CT, if there are CN or brainstem deficits, the next neurodiagnostic study should assess for an arterial occlusion (e.g., CT angiography of head and neck) to rule out a posterior-circulation stroke. On the other hand, if a patient has rhythmic tonic/clonic activities of extremities, then continuous electroencephalogram (EEG) would be the next appropriate test after initial administration of antiepileptic drugs. If the diagnosis remains elusive, MRI of the brain is usually the next modality of choice, which should include, as a minimum, the following sequences: diffusion-weighted imaging, apparent diffusion coefficient, susceptibility-weighted imaging, and fluid-attenuated inversion recovery.

In patients with a high clinical suspicion for seizures, continuous EEG should be performed for at least 12–24 hours in noncomatose patients and 24–48 hours in comatose patients (Claassen et al., 2004). It should be noted that scalp EEG could be falsely negative, as it detects seizures only when a relatively large area of cortex (>10 cm^2) is affected (Tao et al., 2005). Patients with CHS may have abnormal CT or MRI findings, such as diffuse or patchy vasogenic edema in the white matter (typically in the parieto-occipital region) and ipsilateral petechial hemorrhage or ischemic strokes. Transcranial Doppler or a perfusion study (CT perfusion or perfusion-weighted MRI) may be helpful to confirm the diagnosis (Hingorani et al., 2002).

Delirium and cognitive dysfunction can be diagnosed with evaluation of attention and cognitive domains by neuropsychologic testing before and after surgery. Neuroimaging is often necessary to screen for contributory structural lesions. The laboratory workup may include evaluation of electrolytes, endocrine, hepatic and renal dysfunction, and vitamin deficiencies.

In order to minimize CN injury in complex CEA cases, intraoperative monitoring can help with localization of CNs within the surgical field. This has been successfully performed with nerve conduction studies (NCS) of both the vagus and hypoglossal nerves (Driscoll and Chalmer, 2002; Tomonori et al., 2012).

As for peripheral neuropathies, intraoperative monitoring with somatosensory evoked potentials of the

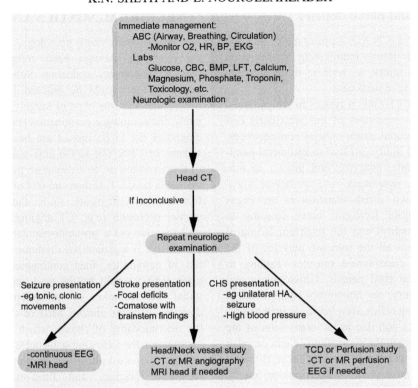

Fig. 31.1. Proposed neurodiagnostic studies for acute neurologic deficits after vascular or cardiac surgeries. BMP, basic metabolic profile; BP, blood pressure; CBC, complete blood count; CHS, cerebral hyperperfusion syndrome; CT, computed tomography; EEG, electroencephalogram; EKG, electrocardiogram; HA, headache; HR, heart rate; LFT, liver function test; O2, oxygen; MRI, magnetic resonance imaging; TCD, transcranial Doppler.

median and ulnar nerves has been shown to be sensitive in the diagnosis of brachial plexopathy (Hickey et al., 1993). Intraoperative monitoring of phrenic nerve function can also be performed. Phrenic nerve injury may be suspected based on chest radiography, and confirmed with real-time fluoroscopy ("sniff test"), diaphragmatic sonography, and NCS of the phrenic nerve. Chest radiographs alone have a high false-positive rate, since a finding of an elevated hemidiaphragm can also be due to other etiologies, such as pleural disease, atelectasis, and ileus. Fluoroscopy and sonography both enable direct visualization of diaphragmatic hypokinesis or paradoxic movement. However, these tests require the patient to be breathing spontaneously (DeVita et al., 1993). The gold-standard neurodiagnostic study is phrenic NCS, which typically shows an increased latency period (>9 ms) and decreased amplitude of compound muscle action potentials (Dimopoulou et al., 1998). Electromyography can also be helpful in assessing the severity of axonal injury in the phrenic nerve through evaluation of diaphragmatic spontaneous activity (e.g., fibrillation or positive sharp waves), motor unit recruitment, amplitude, duration, and morphology (DeVita et al., 1993).

HOSPITAL COURSE AND MANAGEMENT

In order to reduce the risk of periprocedural neurologic complications, pre-, intra-, and postoperative neuroprotective strategies are required (McKhann et al., 2006). Preoperatively, high-risk patients should be identified, especially those with significant cerebrovascular disease. In some cases, alternative or modified surgical procedures should be considered, although it remains unclear whether off-pump procedures reduce the risk of stroke. Neurologic concerns should be discussed with the anesthesiologist, such that appropriate intraoperative monitoring and care can be planned in advance. Postoperative care should begin with a focused neurologic examination, and include optimal blood pressure targets tailored for the individual patient, arrhythmia control, and efforts to minimize the chance of delirium.

Ischemic stroke

In patients undergoing CABG, recent American Heart Association (AHA) guidelines recommend that aspirin (100–325 mg/day) should be started preoperatively, or at the latest within 6 hours postoperatively to reduce saphenous vein graft closure, and for secondary stroke prevention. Moreover, all patients should be on statins, unless contraindicated, to reduce the low-density lipoprotein concentration by at least 30% and to a level below 100 mg/dL (<70 mg/dL in high-risk patients) (Hillis et al., 2011), although it is unclear whether this reduces the risk of early postoperative stroke. Similarly, patients undergoing CEA are recommended to be started on an antiplatelet agent and statin preoperatively (Engelter and Lyrer, 2003). Beta-blockers started 24 hours before CABG may reduce the incidence of new-onset atrial fibrillation. In patients with symptomatic carotid artery stenosis, or with high-grade bilateral stenosis, it may be appropriate to perform CEA or CAS prior to CABG.

Intraoperatively, caution should be exercised to avoid hypotension. In one study, patients with a drop of MAP by at least 10 mmHg were four times more likely to sustain watershed strokes (Gottesman et al., 2006). In another study, patients with an MAP of 80–100 mmHg had fewer neurologic complications, including ischemic stroke, compared with patients in whom blood pressure was maintained at a lower level (Gold et al., 1995). Intraoperative transesophageal echocardiography aids in optimal placement of aortic cross-clamps, to avoid disrupting atheromatous plaques. Transcranial Doppler of the middle cerebral artery can detect microemboli, although the specific relationship to stroke risk remains unclear.

Patients who are found to have an acute ischemic stroke postoperatively should be evaluated immediately by a neurologist and may be considered for intra-arterial thrombolysis or thrombectomy, since intravenous tissue plasminogen activator (tPA) is not recommended within 14 days of major surgery (Powers et al., 2015). Other neuroprotective management includes maintenance of euglycemia and euthermia (Hogue et al., 2006). Fever, defined as temperature greater than 38°C, can contribute to secondary neurologic injury through various mechanisms, including increasing levels of excitatory neurotransmitters, free radicals, and lactic acid, as well as increased ischemic depolarization and blood–brain barrier breakdown (Greer et al., 2008). Similarly, hyperglycemia can lead to local brain tissue acidosis, free-radical generation, and blood–brain barrier injury (Bruno et al., 2002; Parsons et al., 2002). The recommended target glucose level is in the range of 140–180 mg/dL (Powers et al., 2015).

In aortic surgeries, identifying high-risk patients (e.g., treatment requiring large stent coverage and prior history of aortic aneurysms) preoperatively may help reduce the risk of spinal cord ischemia. In such cases, clinicians can plan for lumbar drain placement, intraoperative neuromonitoring, minimization of sedation, and frequent neurologic examination postoperatively (Cheung et al., 2002; Khoynezhad et al., 2007; Schlosser et al., 2009; Hiratzka et al., 2010; Erbel et al., 2014). Intraoperative neuromonitoring with modalities such as somatosensory evoked potential and motor evoked potential has been shown to help in identifying SCI earlier. However, management of SCI continues to be predominantly supportive, with appropriate volume repletion, goal MAP ≥90 mmHg, mild hypothermia, and maintenance of spinal cerebrospinal fluid pressure ≤ 10 mmHg to improve spinal cord perfusion pressure (Hiratzka et al., 2010). Unfortunately, patients are typically not candidates for intravenous tPA. However, those undergoing TEVAR may be considered for intra-arterial thrombectomy if SCI is diagnosed early (within 6 hours from onset), especially with a femoral sheath already in place for arterial access.

Intracerebral hemorrhage

Patients with postoperative ICH should be treated as per recent AHA guidelines. The role for reversal of antiplatelet agents is uncertain. The blood pressure can be targeted to less than 140 mmHg, as this has shown to be safe and may improve functional outcome (Hemphill et al., 2015). As with ischemic stroke, euglycemia and euthermia should be maintained.

Seizures

As in other settings, timely detection of seizures and appropriate treatment are critical (Kapur and Macdonald, 1997). First-line agents are benzodiazepines. If the patient does not return to baseline, second-line drugs should be tailored to the patient according to their side-effect profile. Phenytoin and lacosamide are not preferred in patients with significant cardiac issues, as they may cause arrhythmias. Alternatives include valproate and levetiracetam, especially because these can be given intravenously.

Delirium

Therapy for delirium includes both pharmacologic and nonpharmacologic interventions. Preoperatively, some studies suggest that statins may reduce the incidence of delirium, which could be due to its effect on inflammatory response (Rothenhausler et al., 2005). One study suggested that use of resperidone may reduce the incidence of delirium after cardiac surgery (11% vs. 32%) (Prakanrattana and Prapaitrakool, 2007). However, there are currently insufficient data to support the routine use of antipsychotics to prevent delirium. Use of dexmedetomidine, which is an alpha-2 agonist, has been postulated to prevent delirium, due to improvement of sleep structure and analgesic effects that help minimize opioid requirements (Hsu et al., 2004). In multiple studies, dexmedetomidine was shown to improve the delirium rate in relation to other sedation strategies; however, this was not supported by a large meta-analysis (Maldonado et al., 2009; Shehabi et al., 2009). Other pharamacologic agents that have been assessed for the prevention of delirium include ivastigmine, ketamine, propofol, and clonidine (Mu et al., 2015). Nonpharmacologic strategies include minimization of medications that may contribute to delirium, such as benzodiazepines, opioids, and anticholinergic drugs. Other techniques include cognitive interventions (e.g., orientation protocol with schedule, name board, and discussion of current events), sleep–wake cycle maintenance (e.g., warm drink, relaxation tape, and noise reduction to promote sleep), early mobilization, visual and hearing interventions (i.e., use of adaptive devices or communication techniques to aid with vision/hearing impairment), and adequate hydration (Bogardus et al., 2003). In a meta-analysis of 14 large studies, nonpharmacologic interventions were associated with a reduction in delirium incidence, falls, and length of stay in a heterogeneous population of patients (Hshieh et al., 2015). Four of those studies were randomized trials, in which the delirium incidence was reduced by 44% with the use of nonpharmacologic interventions.

Cognitive impairment

Minimization of the risk of perioperative ischemic stroke also helps prevent cognitive impairment. A variety of pharmacologic strategies for preventing cognitive impairment have been assessed in small, single-center randomized controlled trials.

In one study, the intraoperative choice of anesthesia with sevoflurane was associated with improved cognitive outcome in four different cognitive tests when compared to propofol. This may be due to neuroprotective properties of sevoflurane, as shown previously *in vitro* in animal studies (Schoen et al., 2011).

Intraoperative use of lidocaine and piracetam has also been associated with lower rates of cognitive decline after CABG (Szalma et al., 2006; Ghafari et al., 2012). Lidocaine is a sodium channel blocker, which may potentially work by reducing the cerebral metabolic rate and decreasing excitotoxicity in the setting of cerebral ischemia. Piracetam may work by reducing microemboli, as it decreases platelet aggregation and blood viscosity.

Another intraoperative measure is the monitoring of regional cerebral oximetry in the frontal lobes using near infrared spectroscopy, which helps to detect and respond to low brain tissue oxygenation (Slater et al., 2009). In two randomized trials, post-CABG cognitive decline was significantly lower among patients with intraoperative cerebral oximetry monitoring within a 7-day follow-up period (Slater et al., 2009; Colak et al., 2015).

Postoperatively, acetylcholinesterase inhibitors (e.g., donepezil) have shown promise in reducing cognitive dysfunction. In a small trial, patients with cognitive dysfunction a year after CABG were randomized to 12 weeks of either donepezil or placebo (Doraiswamy et al., 2007). Patients in the donepezil group had significantly greater memory improvement; however, composite cognitive performance was not significantly different between groups.

Another medication, memantine (N-methyl-D-aspartate antagonist), has shown small benefits in cognition of patients with vascular dementia; however, there is paucity of evidence for its efficacy in patients with post-CABG cognitive decline. In a meta-analysis of five randomized trials, efficacy of memantine, galantamine, and rivastigmine was evaluated in patients with mild to moderate vascular dementia over a 6-month period. In all of the trials, patients on the study drug scored significantly better on the Alzheimer Disease Assessment Scale (cognitive) than the control group. However, cholinesterase inhibitors (galantamine and rivastigmine) had a significantly increased risk of gastrointestinal adverse events (i.e., anorexia, nausea, vomiting, and diarrhea) (Kavirajan and Schneider, 2007).

Cognitive decline post-CABG may also be improved through nonpharmacologic approaches, such as intensive physical therapy. In a small trial, patients randomized to three physical therapy session per day had a significantly smaller decline in their cognitive function, as assessed by four different cognitive assessment scales (Cavalcante et al., 2014). This may be due to neuroplasticity, with physical activity stimulating not only the motor and sensory pathways, but also the neurocognitive system. In another randomized study, patients undergoing 6 weeks of attention and memory training post-CABG had a significant improvement in the trained cognitive domains (de Tournay-Jette et al., 2012).

Cerebral hyperperfusion syndrome

Since most patients are discharged early after the CEA surgery, they should be instructed to look out for certain symptoms. Preoperatively, treatment with edaravone immediately prior to internal carotid artery clamping appeared to reduce the incidence of CHS in one single-center, matched-cohort study. This drug is a free-radical scavenger that is thought to work through inhibition of lipid peroxidation and reduced endothelial injury. However, further trials are needed to support the routine use of this medication (Ogasawara et al., 2004).

The most important method of CHS prevention is through strict postoperative blood pressure control.

Labetalol is a mixed alpha and beta antagonist, which can be administered intravenously, and has no direct effect on CBF (Jorgensen and Schroeder, 1993). Esmolol is an alternative intravenous beta-blocker which can be considered in an intensive care unit setting, with the advantage that it is very short acting and can be administered as an intravenous infusion. Clonidine is a central-acting sympatholytic drug, which may be useful since CHS is associated with catecholamine surges, although it is not available as an intravenous preparation in North America (Tietjen et al., 1996). Nitroprusside is a very effective, short-acting, intravenous antihypertensive, but may produce cerebral vasodilatation. Nicardipine is an alternative that may achieve more stable blood pressure control (Halpern et al., 1992; Dorman et al., 2001).

The duration of strict blood pressure control depends on the rate of normalization of autoregulation, which can be variable among patients. Some clinicians continue with tight blood pressure control up to 6 months, whereas others use transcranial Doppler and continue with strict blood pressure control until there is normalization of CBF velocity within the middle cerebral artery (Dalman et al., 1999). In patients with severe cerebral edema, osmotherapy with mannitol or hypertonic saline may be necessary.

Cranial nerve deficits and peripheral neuropathy

Most periprocedural CN deficits and peripheral neuropathies are transient and management is conservative. Intraoperative measures, such as use of NCS to further delineate the pertinent CNs during CEA, or use of pericardial insulation pads to minimize hypothermic injury to the phrenic nerve, are some of the possible preventive approaches. Patients with failed conservative management of vagus nerve injury can be assessed for surgical options. For instance, patients with vocal cord paralysis can undergo Teflon injection to increase tension and further support the vocal fold. Dysphagia can sometimes be treated with cricopharyngeal myotomy, as it can relieve the persistent spasm from the cricopharyngeus muscle (Buchholz and Neumann, 1997). The recovery of unilateral vocal cord dysfunction due to recurrent laryngeal neuropathy can take up to 1 year (Shafei et al., 1997). Patients with severe phrenic nerve injury may be candidates for diaphragm plication. This procedure can be performed with an open or minimally invasive approach, in which the diaphragm dome is flattened to provide larger expansion volume for the lungs. If plication fails to improve conditions, the diaphragm can be repaired or reinforced with synthetic or bioprosthetic mesh (Bowen et al., 1982; Tsakiridis et al., 2012).

CLINICAL TRIALS AND GUIDELINES

Relevant guidelines from the AHA include management recommendations relevant to acute ischemic stroke, ICH, CABG, CEA, and thoracic aortic disease (Hiratzka et al., 2010; Hillis et al., 2011; Hemphill et al., 2015; Powers et al., 2015).

COMPLEX CLINICAL DECISIONS

One of the complex decisions may involve the optimal timing of CABG in patients with recent ischemic stroke. Unfortunately, there is a paucity of data to guide clinical decisions. Considerations in the timing of CABG should be based on the urgency of CABG, stroke etiology, and infarct volume, along with other comorbidities. A related topic is the role of preoperative carotid artery screening and revascularization prior to CABG. As per the most recent AHA guidelines, a multidisciplinary approach is recommended in patients with significant carotid artery stenosis. Those with a previous history of transient ischemic attack or stroke are reasonable candidates for revascularization. In addition, revascularization may be considered in the presence of bilateral severe carotid stenosis (70–99%) or unilateral severe stenosis with a contralateral occlusion (Hillis et al., 2011). Lastly, patients with new postoperative acute ischemic stroke (including spinal cord ischemia) should be evaluated immediately by a vascular neurologist and may be considered for intra-arterial thrombolysis or thrombectomy, as intravenous tPA is not recommended within 14 days of major surgery (Powers et al., 2015).

OUTCOME PREDICTION

Ischemic and hemorrhagic stroke

Patient with post-CABG stroke have higher morbidity and mortality in comparison to those without (Braxton et al., 2000). The mortality risk at 10-year follow-up can be three times higher in those with stroke, with the greatest risk of death being in the first year (Dacey et al., 2005). However, overall mortality after CABG has continued to decrease since the early 1990s. In one study, in-hospital mortality during the 1990s was 7% compared to 3% in early 2000s (Maganti et al., 2009). Risk factors associated with higher stroke rates include older age, previous history of stroke, hypertension, diabetes mellitus, and smoking. The risk of a major ischemic stroke (NIHSS > 9) after carotid revascularization is relatively low, at 0.9% and 0.3% after CAS and CEA, respectively. However, these strokes were associated with significant morbidity and a threefold increase in mortality during 4 years of follow-up of patients in the CREST study (Hill et al., 2012).

The mortality rate in adult patients undergoing ECMO is dependent on selection criteria. In one cohort study, it was 44% at 1 month compared to approximately 100% mortality for comparable patients when ECMO was not offered (Mateen et al., 2011). Patients who develop neurologic complications (specifically ischemic stroke and ICH) during ECMO have significantly higher rates of morbidity and mortality (Lan et al., 2010; Nasr and Rabinstein, 2015; Luyt et al., 2016). In one study, ischemic stroke and ICH were determined to be one of the six independent predictors of mortality at discharge, with others being age, dialysis, infection, hypoglycemia, and alkalosis (Lan et al., 2010). In another study, neurologic complications were most common when ECMO was used subsequent to initiation of cardiopulmonary resuscitation. They also found that those with neurologic complications (stroke, ICH, seizure, and brain death) experienced significantly higher in-hospital mortality compared to those without them, at 89% and 57%, respectively (Lorusso et al., 2016). Lastly, there is a paucity of data regarding cognitive dysfunction in adult patients receiving ECMO. In the neonatal/pediatric population, rates of moderate to severe cognitive delay can be as high as 50% (Hamrick et al., 2003; Lequier et al., 2008).

In spinal cord ischemia, morbidity and mortality are influenced by the severity of initial impairment, older age, peripheral vascular disease, and underlying etiology. In one study, patients with SCI due to various etiologies (70% due to aortic surgeries) continued to have improved motor function when followed over an average of 3 years. The percentage of wheelchair-bound patients at onset compared to the follow-up was 81% and 35%, respectively (Robertson et al., 2012). In another study, aortic surgery-related SCI was significantly associated with mortality when compared to patients without that complication (39% vs. 14%) (Messe et al., 2008).

Seizures

Patients who develop post-CABG seizures, without a stroke, have similar outcomes compared with patients who do not develop seizures. However, in patients with seizures and concomitant acute stroke, morbidity and mortality rates are significantly increased (Ivascu et al., 2015). This suggests that seizure may not be a significant independent prognostic determinant. Alternatively, the lack of association with poor outcomes could also be due to small sample size or the self-limiting nature of post-CABG seizures. As for CEA-related seizures, there is a paucity of literature regarding the implications on outcome. Prompt detection and management of seizures are generally associated with improved outcome in critically ill patients (Treiman et al., 1998). In a retrospective study, patients who developed seizures after CEA were more likely to have a poor outcome. However, the majority of patients had CHS, rather than an isolated seizure, making the results difficult to interpret (Naylor et al., 2003).

Delirium

Delirium has been associated with worse functional outcome and cognitive decline at 1 month, as well as increased risk of mortality at 10-year follow-up (Hudetz et al., 2011). The death rate in patients with and without delirium was 16 and 7 per 100 person-years, respectively ($p<0.0001$) (Gottesman et al., 2010; Rudolph et al., 2010). In one study, postoperative delirium was significantly associated with a higher likelihood of developing cognitive impairment at 6 months relative to patients without delirium (40% and 24%, respectively) (Saczynski et al., 2012). Delirium is also associated with longer length of stay (McKhann et al., 2002).

Cognitive impairment

It is important to note that a significant number of candidates for CABG (ranging from 20% to 46%) already have some degree of cognitive impairment (Jensen et al., 2006). These patients typically have an increased burden of lacunar infarcts and evidence of small-vessel disease on MRI, which puts them at a higher perioperative stroke risk (Maekawa et al., 2008). Cognitive impairment after CABG has a pattern of early improvement followed by progressive decline. Perioperative cognitive impairment typically resolves within 3 months. Over the long term, the rate of cognitive decline appears to be similar between CABG and nonsurgical patients, when adjusted for cardiovascular risk factors (Selnes et al., 2008, 2009). Elderly patients with multiple cardiovascular risk factors are at higher risk of cognitive decline, and early cognitive decline at discharge is a significant predictor of long-term function (Hammon et al., 1997; Newman et al., 2001). Another factor that can be associated with cognitive decline is the perioperative stroke burden that can be visualized on MRI, especially using diffusion-weighted imaging (Barber et al., 2008; Hogan et al., 2013). Other predisposing factors include perioperative hypotension, exposure to general anesthesia, hypoxemia, and hyperthermia (Grocott et al., 2002; Gottesman et al., 2007; Hogan et al., 2013). As mentioned earlier, aggressive control of modifiable risk factors may be the key to minimizing long-term cognitive deficits in high-risk patients.

Cerebral hyperperfusion syndrome

Most patients with CHS, especially if diagnosed and treated early, will recover completely. In patients with severe CHS and delayed treatment, mortality can be as high as 50%, usually due to ICH (Piepgras et al., 1988; Meyers et al., 2000).

Cranial nerve deficits and peripheral neuropathy

Most CEA-related CN injuries resolve spontaneously, with fewer than 1% of patients having long-term deficits (Fokkema et al., 2014). In the CREST trial, 34% of CN injuries resolved within 30 days and more than 80% within the first year (Hye et al., 2015). Outcome from brachial plexopathies post cardiac surgery is typically favorable. In two prospective studies, the incidence of brachial plexopathy was about 5%, with only 1% having persistent symptoms beyond 3 months (Vahl et al., 1991). Patients with persistent symptoms should undergo further testing with NCS to delineate axonal or myelin involvement, as axonal injury is associated with longer recovery. Phrenic nerve injuries are usually asymptomatic, and the majority of patients recover within 3–6 months (DeVita et al., 1993). The required time for healing depends on the type of injury (e.g., demyelination versus axonal injury) and the regeneration distance.

NEUROREHABILITATION

Previously, the general belief was that regeneration in an adult central nervous system is very limited after brain injury. However, studies have shown the neuronal network capacity for adaptation in response to repeated stimulation; this has been termed neuroplasticity (Johnston, 2009; Mueller et al., 2009). This principle is the foundation of neurorehabilitation, which most likely leads to an actual reorganization of neuronal networks, thus promoting recovery. Patients and their caregivers are educated and trained for better adaptive and compensatory practices, such as assisted ambulation, transfer, feeding, and language skills. Stroke is one of the major complications of both vascular and cardiac surgeries, with patients commonly developing motor deficits (>70%), aphasia (>20%), and cognitive dysfunction (Hendricks et al., 2002; Dobkin, 2004). Most stroke recovery occurs within the first 3–6 months, with disability at 1 month being a reliable indicator of final recovery (Jorgensen et al., 1999; Ovbiagele et al., 2010). Intensive stroke rehabilitation should ideally start within the first 24 hours, and once the patient is able to participate in physical therapy for at least 3 hours per day, he or she is transferred to a rehabilitation unit, with a usual stay of about 2 weeks. Patients unable to participate in intensive programs are transferred to subacute rehabilitation settings or skilled nursing facilities. Following discharge home, patients should participate in home rehabilitation or outpatient programs.

CONCLUSION

Despite best effort by surgeons, anesthesiologists, and intensivists, postprocedural neurologic complications are still relatively common following cardiac and vascular surgery, due to patients' multiple comorbidities and the complexity of the procedure. These complications, especially hemorrhagic and ischemic strokes, have significant associated morbidity and mortality. Clinicians' familiarity with various complications is essential for early diagnosis and management. Continued investigation into mechanisms, preventive measures, and management of postsurgical complications is warranted to further improve clinical outcomes.

REFERENCES

Aburahma AF, Stone PA, Hass SM et al. (2010). Prospective randomized trial of routine versus selective shunting in carotid endarterectomy based on stump pressure. J Vasc Surg 51: 1133–1138.

American Psychiatric Association (2013). Diagnostic and Statistical Manual of Mental Disorders, 5th ed. American Psychiatric Publishing, Arlington, VA.

Bademci G, Batay F, Vural E et al. (2005). Microsurgical anatomical landmarks associated with high bifurcation carotid artery surgery and related to hypoglossal nerve. Cerebrovasc Dis 19: 404–406.

Barber PA, Hach S, Tippett LJ et al. (2008). Cerebral ischemic lesions on diffusion-weighted imaging are associated with neurocognitive decline after cardiac surgery. Stroke 39: 1427–1433.

Basbaum CB (1973). Induced hypothermia in peripheral nerve: electron microscopic and electrophysiological observations. J Neurocytol 2: 171–187.

Ben-David B, Stahl S (1997). Prognosis of intraoperative brachial plexus injury: a review of 22 cases. Br J Anaesth 79: 440–445.

Bogardus Jr ST, Desai MM, Williams CS et al. (2003). The effects of a targeted multicomponent delirium intervention on postdischarge outcomes for hospitalized older adults. Am J Med 114: 383–390.

Bohner H, Schneider F, Stierstorfer A et al. (2000). Delirium after vascular surgery interventions. Intermediate-term results of a prospective study. Chirurg 71: 215–221.

Bowen TE, Zajtchuk R, Albus RA (1982). Diaphragmatic paralysis managed by diaphragmatic replacement. Ann Thorac Surg 33: 184–188.

Braxton JH, Marrin CA, McGrath PD et al. (2000). Mediastinitis and long-term survival after coronary artery bypass graft surgery. Ann Thorac Surg 70: 2004–2007.

Brott TG, Halperin JL, Abbara S et al. (2011). 2011 ASA/ACCF/AHA/AANN/AANS/ACR/ASNR/CNS/SAIP/SCAI/SIR/SNIS/SVM/SVS guideline on the management of patients with extracranial carotid and vertebral artery disease. Circulation 124: e54–e130.

Brown CH (2014). Delirium in the cardiac surgical ICU. Curr Opin Anaesthesiol 27: 117–122.

Brown DL, Topol EJ (1993). Stroke complicating percutaneous coronary revascularization. Am J Cardiol 72: 1207–1209.

Bruno A, Levine SR, Frankel MR et al. (2002). Admission glucose level and clinical outcomes in the NINDS rt-PA Stroke Trial. Neurology 59: 669–674.

Bucerius J, Gummert JF, Borger MA et al. (2004). Predictors of delirium after cardiac surgery delirium: effect of beating-heart (off-pump) surgery. J Thorac Cardiovasc Surg 127: 57–64.

Buchholz DW, Neumann S (1997). Management of vagus nerve injury after carotid endarterectomy. Dysphagia 12: 58–59.

Caplan LR, Hennerici M (1998). Impaired clearance of emboli (washout) is an important link between hypoperfusion, embolism, and ischemic stroke. Arch Neurol 55: 1475–1482.

Cavalcante ED, Magario R, Conforti CA et al. (2014). Impact of intensive physiotherapy on cognitive function after coronary artery bypass graft surgery. Arq Bras Cardiol 103: 391–397.

Chakravorty BG (1971). Arterial supply of the cervical spinal cord (with special reference to the radicular arteries). Anat Rec 170: 311–329.

Cheng R, Hachamovitch R, Kittleson M et al. (2014). Complications of extracorporeal membrane oxygenation for treatment of cardiogenic shock and cardiac arrest: a meta-analysis of 1,866 adult patients. Ann Thorac Surg 97: 610–616.

Cheung AT, Weiss SJ, McGarvey ML et al. (2002). Interventions for reversing delayed-onset postoperative paraplegia after thoracic aortic reconstruction. Ann Thorac Surg 74: 413–419. discussion 420–411.

Claassen J, Mayer SA, Kowalski RG et al. (2004). Detection of electrographic seizures with continuous EEG monitoring in critically ill patients. Neurology 62: 1743–1748.

Colak Z, Borojevic M, Bogovic A et al. (2015). Influence of intraoperative cerebral oximetry monitoring on neurocognitive function after coronary artery bypass surgery: a randomized, prospective study. Eur J Cardiothorac Surg 47: 447–454.

Cunningham EJ, Bond R, Mayberg MR et al. (2004). Risk of persistent cranial nerve injury after carotid endarterectomy. J Neurosurg 101: 445–448.

Dacey LJ, Likosky DS, Leavitt BJ et al. (2005). Perioperative stroke and long-term survival after coronary bypass graft surgery. Ann Thorac Surg 79: 532–536. discussion 537.

Dalman JE, Beenakkers IC, Moll FL et al. (1999). Transcranial Doppler monitoring during carotid endarterectomy helps to identify patients at risk of postoperative hyperperfusion. Eur J Vasc Endovasc Surg 18: 222–227.

de Tournay-Jette E, Dupuis G, Denault A et al. (2012). The benefits of cognitive training after a coronary artery bypass graft surgery. J Behav Med 35: 557–568.

Denny-Brown D, Raymond MB, Adams RD et al. (1945). The pathology of injury to nerve induced by cold. J Neuropathol Exp Neurol 4: 305–323.

Devere TR, Lee AG, Hamill MB, Denny-Brown MB, Raymond D, Adams MD, Charles Brenner MD, Margaret M, Doherty AB (1997). Acquired supranuclear ocular motor paresis following cardiovascular surgery. J Neuroophthalmol 17: 189–193.

DeVita MA, Robinson LR, Rehder J et al. (1993). Incidence and natural history of phrenic neuropathy occurring during open heart surgery. Chest 103: 850–856.

Dimopoulou I, Daganou M, Dafni U et al. (1998). Phrenic nerve dysfunction after cardiac operations: electrophysiologic evaluation of risk factors. Chest 113: 8–14.

Dobkin BH (2004). Strategies for stroke rehabilitation. Lancet Neurol 3: 528–536.

Doraiswamy PM, Babyak MA, Hennig T et al. (2007). Donepezil for cognitive decline following coronary artery bypass surgery: a pilot randomized controlled trial. Psychopharmacol Bull 40: 54–62.

Dorman T, Thompson DA, Breslow MJ et al. (2001). Nicardipine versus nitroprusside for breakthrough hypertension following carotid endarterectomy. Journal of clinical anesthesia 13: 16–19.

Driscoll PJ, Chalmer RT (2002). The use of a nerve stimulator in difficult carotid surgery. J Vasc Surg 35: 627.

Dukkipati S, O'Neill WW, Harjai KJ et al. (2004). Characteristics of cerebrovascular accidents after percutaneous coronary interventions. J Am Coll Cardiol 43: 1161–1167.

Echahidi N, Pibarot P, O'Hara G et al. (2008). Mechanisms, prevention, and treatment of atrial fibrillation after cardiac surgery. J Am Coll Cardiol 51: 793–801.

Eckstein HH, Ringleb P, Allenberg JR et al. (2008). Results of the Stent-Protected Angioplasty versus Carotid Endarterectomy (SPACE) study to treat symptomatic stenoses at 2 years: a multinational, prospective, randomised trial. Lancet Neurol 7: 893–902.

Efthimiou J, Butler J, Woodham C et al. (1991). Diaphragm paralysis following cardiac surgery: role of phrenic nerve cold injury. Ann Thorac Surg 52: 1005–1008.

Ellard L, Katznelson R, Wasowicz M et al. (2014). Type of anesthesia and postoperative delirium after vascular surgery. J Cardiothorac Vasc Anesth 28: 458–461.

Engelter S, Lyrer P (2003). Antiplatelet therapy for preventing stroke and other vascular events after carotid endarterectomy. Cochrane Database Syst Rev. CD001458.

Erbel R, Aboyans V, Boileau C et al. (2014). 2014 ESC Guidelines on the diagnosis and treatment of aortic diseases: Document covering acute and chronic aortic diseases of the thoracic and abdominal aorta of the adult. The Task Force for the Diagnosis and Treatment of Aortic Diseases of the European Society of Cardiology (ESC). Eur Heart J 35: 2873–2926.

Ferguson GG, Eliasziw M, Barr HW et al. (1999). The North American Symptomatic Carotid Endarterectomy Trial: surgical results in 1415 patients. Stroke 30: 1751–1758.

Ferguson Jr TB, Hammill BG, Peterson ED et al. (2002). A decade of change – risk profiles and outcomes for isolated coronary artery bypass grafting procedures, 1990–1999: a report from the STS National Database Committee and the Duke Clinical Research Institute. Society of Thoracic Surgeons. Ann Thorac Surg 73: 480–489. discussion 489–490.

Fokkema M, de Borst GJ, Nolan BW et al. (2014). Clinical relevance of cranial nerve injury following carotid endarterectomy. Eur J Vasc Endovasc Surg 47: 2–7.

Fuchs S, Stabile E, Kinnaird TD et al. (2002). Stroke complicating percutaneous coronary interventions: incidence, predictors, and prognostic implications. Circulation 106: 86–91.

Ghafari R, Baradari AG, Firouzian A et al. (2012). Cognitive deficit in first-time coronary artery bypass graft patients: a randomized clinical trial of lidocaine versus procaine hydrochloride. Perfusion 27: 320–325.

Gold JP, Charlson ME, Williams-Russo P et al. (1995). Improvement of outcomes after coronary artery bypass. A randomized trial comparing intraoperative high versus low mean arterial pressure. J Thorac Cardiovasc Surg 110: 1302–1311. discussion 1311–1314.

Gottesman RF, Sherman PM, Grega MA et al. (2006). Watershed strokes after cardiac surgery: diagnosis, etiology, and outcome. Stroke 37: 2306–2311.

Gottesman RF, Hillis AE, Grega MA et al. (2007). Early postoperative cognitive dysfunction and blood pressure during coronary artery bypass graft operation. Arch Neurol 64: 1111–1114.

Gottesman RF, Grega MA, Bailey MM et al. (2010). Delirium after coronary artery bypass graft surgery and late mortality. Ann Neurol 67: 338–344.

Greenberg RK, Lu Q, Roselli EE et al. (2008). Contemporary analysis of descending thoracic and thoracoabdominal aneurysm repair: a comparison of endovascular and open techniques. Circulation 118: 808–817.

Greenstein AJ, Chassin MR, Wang J et al. (2007). Association between minor and major surgical complications after carotid endarterectomy: results of the New York Carotid Artery Surgery study. J Vasc Surg 46: 1138–1144. discussion 1145–1146.

Greer DM, Funk SE, Reaven NL et al. (2008). Impact of fever on outcome in patients with stroke and neurologic injury: a comprehensive meta-analysis. Stroke 39: 3029–3035.

Grocott HP, Mackensen GB, Grigore AM et al. (2002). Postoperative hyperthermia is associated with cognitive dysfunction after coronary artery bypass graft surgery. Stroke 33: 537–541.

Groen JA, Banayan D, Gupta S et al. (2012). Treatment of delirium following cardiac surgery. J Card Surg 27: 589–593.

Halpern NA, Goldberg M, Neely C et al. (1992). Postoperative hypertension: a multicenter, prospective, randomized comparison between intravenous nicardipine and sodium nitroprusside. Critical care medicine 20: 1637–1643.

Hammon Jr JW, Stump DA, Kon ND et al. (1997). Risk factors and solutions for the development of neurobehavioral changes after coronary artery bypass grafting. Ann Thorac Surg 63: 1613–1618.

Hamrick SE, Gremmels DB, Keet CA et al. (2003). Neurodevelopmental outcome of infants supported with extracorporeal membrane oxygenation after cardiac surgery. Pediatrics 111: e671–e675.

Hannan EL, Racz MJ, Walford G et al. (2005). Long-term outcomes of coronary-artery bypass grafting versus stent implantation. N Engl J Med 352: 2174–2183.

Hemphill 3rd JC, Greenberg SM, Anderson CS et al. (2015). Guidelines for the management of spontaneous intracerebral hemorrhage: a guideline for healthcare professionals from the American Heart Association/American Stroke Association. Stroke 46: 2032–2060.

Henderson RD, Phan TG, Piepgras DG et al. (2001). Mechanisms of intracerebral hemorrhage after carotid endarterectomy. J Neurosurg 95: 964–969.

Hendricks HT, van Limbeek J, Geurts AC et al. (2002). Motor recovery after stroke: a systematic review of the literature. Arch Phys Med Rehabil 83: 1629–1637.

Hernandez Jr F, Brown JR, Likosky DS et al. (2007). Neurocognitive outcomes of off-pump versus on-pump coronary artery bypass: a prospective randomized controlled trial. Ann Thorac Surg 84: 1897–1903.

Hickey C, Gugino LD, Aglio LS et al. (1993). Intraoperative somatosensory evoked potential monitoring predicts peripheral nerve injury during cardiac surgery. Anesthesiology 78: 29–35.

Hill MD, Brooks W, Mackey A et al. (2012). Stroke after carotid stenting and endarterectomy in the Carotid Revascularization Endarterectomy versus Stenting Trial (CREST). Circulation 126: 3054–3061.

Hillis LD, Smith PK, Anderson JL et al. (2011). 2011 ACCF/AHA guideline for coronary artery bypass graft surgery: a report of the American College of Cardiology Foundation/American Heart Association Task Force on Practice Guidelines. Circulation 124: e652–e735.

Hingorani A, Ascher E, Tsemekhim B et al. (2002). Causes of early post carotid endarterectomy stroke in a recent series: the increasing importance of hyperperfusion syndrome. Acta Chir Belg 102: 435–438.

Hiratzka LF, Bakris GL, Beckman JA et al. (2010). 2010 ACCF/AHA/AATS/ACR/ASA/SCA/SCAI/SIR/STS/SVM guidelines for the diagnosis and management of patients with Thoracic Aortic Disease. Circulation 121: e266–e369.

Hogan AM, Shipolini A, Brown MM et al. (2013). Fixing hearts and protecting minds: a review of the multiple, interacting factors influencing cognitive function after coronary artery bypass graft surgery. Circulation 128: 162–171.

Hogue Jr CW, Murphy SF, Schechtman KB et al. (1999). Risk factors for early or delayed stroke after cardiac surgery. Circulation 100: 642–647.

Hogue Jr CW, Barzilai B, Pieper KS et al. (2001). Sex differences in neurological outcomes and mortality after cardiac surgery: a Society of Thoracic Surgery national database report. Circulation 103: 2133–2137.

Hogue Jr CW, Palin CA, Arrowsmith JE (2006). Cardiopulmonary bypass management and neurologic outcomes: an evidence-based appraisal of current practices. Anesth Analg 103: 21–37.

Hshieh TT, Yue J, Oh E et al. (2015). Effectiveness of multicomponent nonpharmacological delirium interventions: a meta-analysis. JAMA Intern Med 175: 512–520.

Hsu YW, Cortinez LI, Robertson KM et al. (2004). Dexmedetomidine pharmacodynamics: part I: crossover comparison of the respiratory effects of dexmedetomidine and remifentanil in healthy volunteers. Anesthesiology 101: 1066–1076.

Hudetz JA, Iqbal Z, Gandhi SD et al. (2011). Postoperative delirium and short-term cognitive dysfunction occur more frequently in patients undergoing valve surgery with or without coronary artery bypass graft surgery compared with coronary artery bypass graft surgery alone: results of a pilot study. J Cardiothorac Vasc Anesth 25: 811–816.

Hye RJ, Mackey A, Hill MD et al. (2015). Incidence, outcomes, and effect on quality of life of cranial nerve injury in the Carotid Revascularization Endarterectomy versus Stenting Trial. J Vasc Surg 61: 1208–1214.

Ivascu NS, Gaudino M, Lau C et al. (2015). Nonischemic postoperative seizure does not increase mortality after cardiac surgery. Ann Thorac Surg 100: 101–106.

Jensen BO, Hughes P, Rasmussen LS et al. (2006). Cognitive outcomes in elderly high-risk patients after off-pump versus conventional coronary artery bypass grafting: a randomized trial. Circulation 113: 2790–2795.

Johnston MV (2009). Plasticity in the developing brain: implications for rehabilitation. Dev Disabil Res Rev 15: 94–101.

Jorgensen LG, Schroeder TV (1993). Defective cerebrovascular autoregulation after carotid endarterectomy. Eur J Vasc Surg 7: 370–379.

Jorgensen HS, Nakayama H, Raaschou HO et al. (1999). Stroke. Neurologic and functional recovery: the Copenhagen Stroke Study. Phys Med Rehabil Clin N Am 10: 887–906.

Kapur J, Macdonald RL (1997). Rapid seizure-induced reduction of benzodiazepine and Zn2+ sensitivity of hippocampal dentate granule cell GABAA receptors. J Neurosci 17: 7532–7540.

Kasirajan V, Smedira NG, McCarthy JF et al. (1999). Risk factors for intracranial hemorrhage in adults on extracorporeal membrane oxygenation. Eur J Cardiothorac Surg 15: 508–514.

Kavirajan H, Schneider LS (2007). Efficacy and adverse effects of cholinesterase inhibitors and memantine in vascular dementia: a meta-analysis of randomised controlled trials. Lancet Neurol 6: 782–792.

Khoynezhad A, Donayre CE, Bui H et al. (2007). Risk factors of neurologic deficit after thoracic aortic endografting. Ann Thorac Surg 83: S882–S889. discussion S890-882.

Kieburtz K, Ricotta JJ, Moxley 3rd RT (1990). Seizures following carotid endarterectomy. Arch Neurol 47: 568–570.

Kirsh MM, Magee KR, Gago O et al. (1971). Brachial plexus injury following median sternotomy incision. Ann Thorac Surg 11: 315–319.

Korn-Lubetzki I, Farkash R, Pachino RM et al. (2013). Incidence and risk factors of cerebrovascular events following cardiac catheterization. J Am Heart Assoc 2. e000413.

Lan C, Tsai PR, Chen YS et al. (2010). Prognostic factors for adult patients receiving extracorporeal membrane oxygenation as mechanical circulatory support – a 14-year experience at a medical center. Artif Organs 34: E59–E64.

Lawlor PG, Bush SH (2014). Delirium diagnosis, screening and management. Curr Opin Support Palliat Care 8: 286–295.

Leffert RD (1974). Brachial-plexus injuries. N Engl J Med 291: 1059–1067.

Lequier L, Joffe AR, Robertson CM et al. (2008). Two-year survival, mental, and motor outcomes after cardiac extracorporeal life support at less than five years of age. J Thorac Cardiovasc Surg 136 (976–983): e973.

Likosky DS, Marrin CA, Caplan LR et al. (2003). Determination of etiologic mechanisms of strokes secondary to coronary artery bypass graft surgery. Stroke 34: 2830–2834.

Liu YH, Wang DX, Li LH et al. (2009). The effects of cardiopulmonary bypass on the number of cerebral microemboli and the incidence of cognitive dysfunction after coronary artery bypass graft surgery. Anesth Analg 109: 1013–1022.

Lorusso R, Barili F, Mauro MD et al. (2016). In-hospital neurologic complications in adult patients undergoing venoarterial extracorporeal membrane oxygenation: results from the extracorporeal life support organization registry. Crit Care Med 44: e964–e972.

Luyt CE, Brechot N, Demondion P et al. (2016). Brain injury during venovenous extracorporeal membrane oxygenation. Intensive Care Med 42: 897–907.

Maekawa K, Goto T, Baba T et al. (2008). Abnormalities in the brain before elective cardiac surgery detected by diffusion-weighted magnetic resonance imaging. Ann Thorac Surg 86: 1563–1569.

Maganti M, Rao V, Brister S et al. (2009). Decreasing mortality for coronary artery bypass surgery in octogenarians. Can J Cardiol 25: e32–e35.

Maldonado JR, Wysong A, van der Starre PJ et al. (2009). Dexmedetomidine and the reduction of postoperative delirium after cardiac surgery. Psychosomatics 50: 206–217.

Malenka DJ, Leavitt BJ, Hearne MJ et al. (2005). Comparing long-term survival of patients with multivessel coronary disease after CABG or PCI: analysis of BARI-like patients in northern New England. Circulation 112: I371–I376.

Mateen FJ, Muralidharan R, Shinohara RT et al. (2011). Neurological injury in adults treated with extracorporeal membrane oxygenation. Arch Neurol 68: 1543–1549.

McKhann GM, Grega MA, Borowicz Jr LM et al. (2002). Encephalopathy and stroke after coronary artery bypass grafting: incidence, consequences, and prediction. Arch Neurol 59: 1422–1428.

McKhann GM, Grega MA, Borowicz Jr LM et al. (2006). Stroke and encephalopathy after cardiac surgery: an update. Stroke 37: 562–571.

Messe SR, Bavaria JE, Mullen M et al. (2008). Neurologic outcomes from high risk descending thoracic and thoracoabdominal aortic operations in the era of endovascular repair. Neurocrit Care 9: 344–351.

Meyers PM, Higashida RT, Phatouros CC et al. (2000). Cerebral hyperperfusion syndrome after percutaneous transluminal stenting of the craniocervical arteries. Neurosurgery 47: 335–343. discussion 343–335.

Monk TG, Weldon BC, Garvan CW et al. (2008). Predictors of cognitive dysfunction after major noncardiac surgery. Anesthesiology 108: 18–30.

Moulakakis KG, Mylonas SN, Sfyroeras GS et al. (2009). Hyperperfusion syndrome after carotid revascularization. J Vasc Surg 49: 1060–1068.

Mu JL, Lee A, Joynt GM (2015). Pharmacologic agents for the prevention and treatment of delirium in patients undergoing cardiac surgery: systematic review and metaanalysis. Crit Care Med 43: 194–204.

Mueller BK, Mueller R, Schoemaker H (2009). Stimulating neuroregeneration as a therapeutic drug approach for traumatic brain injury. Br J Pharmacol 157: 675–685.

Nair UR, Griffiths G, Lawson RA (1988). Postoperative neuralgia in the leg after saphenous vein coronary artery bypass graft: a prospective study. Thorax 43: 41–43.

Nasr DM, Rabinstein AA (2015). Neurologic complications of extracorporeal membrane oxygenation. Journal of Clinical Neurology 11: 383–389.

Naylor AR, Evans J, Thompson MM et al. (2003). Seizures after carotid endarterectomy: hyperfusion, dysautoregulation or hypertensive encephalopathy? Eur J Vasc Endovasc Surg 26: 39–44.

Newman MF, Kirchner JL, Phillips-Bute B et al. (2001). Longitudinal assessment of neurocognitive function after coronary-artery bypass surgery. N Engl J Med 344: 395–402.

O'Brien JW, Johnson SH, VanSteyn SJ et al. (1991). Effects of internal mammary artery dissection on phrenic nerve perfusion and function. Ann Thorac Surg 52: 182–188.

O'Brien SM, Shahian DM, Filardo G et al. (2009). The Society of Thoracic Surgeons 2008 cardiac surgery risk models: part 2 – isolated valve surgery. Ann Thorac Surg 88: S23–S42.

Ogasawara K, Inoue T, Kobayashi M et al. (2004). Pretreatment with the free radical scavenger edaravone prevents cerebral hyperperfusion after carotid endarterectomy. Neurosurgery 55: 1060–1067.

Ovbiagele B, Lyden PD, Saver JL et al. (2010). Disability status at 1 month is a reliable proxy for final ischemic stroke outcome. Neurology 75: 688–692.

Palmerini T, Biondi-Zoccai G, Riva DD et al. (2013). Risk of stroke with percutaneous coronary intervention compared with on-pump and off-pump coronary artery bypass graft surgery: evidence from a comprehensive network meta-analysis. Am Heart J 165 (910–917): e914.

Parsons MW, Barber PA, Desmond PM et al. (2002). Acute hyperglycemia adversely affects stroke outcome: a magnetic resonance imaging and spectroscopy study. Ann Neurol 52: 20–28.

Perry C, James D, Wixon C et al. (2001). Horner's syndrome after carotid endarterectomy – a case report. Vasc Surg 35: 325–327.

Piepgras DG, Morgan MK, Sundt Jr TM et al. (1988). Intracerebral hemorrhage after carotid endarterectomy. J Neurosurg 68: 532–536.

Powers WJ, Derdeyn CP, Biller J et al. (2015). 2015 American Heart Association/American Stroke Association focused update of the 2013 guidelines for the early management of patients with acute ischemic stroke regarding endovascular treatment: a guideline for healthcare professionals from the American Heart Association/American Stroke Association. Stroke 46: 3020–3035.

Prakanrattana U, Prapaitrakool S (2007). Efficacy of risperidone for prevention of postoperative delirium in cardiac surgery. Anaesth Intensive Care 35: 714–719.

Raja SG, Berg GA (2007). Impact of off-pump coronary artery bypass surgery on systemic inflammation: current best available evidence. J Card Surg 22: 445–455.

Roach GW, Kanchuger M, Mangano CM et al. (1996). Adverse cerebral outcomes after coronary bypass surgery. Multicenter Study of Perioperative Ischemia Research Group and the Ischemia Research and Education Foundation Investigators. N Engl J Med 335: 1857–1863.

Robertson CE, Brown Jr RD, Wijdicks EF et al. (2012). Recovery after spinal cord infarcts: long-term outcome in 115 patients. Neurology 78: 114–121.

Rockman CB, Jacobowitz GR, Lamparello PJ et al. (2000). Immediate reexploration for the perioperative neurologic event after carotid endarterectomy: is it worthwhile? J Vasc Surg 32: 1062–1070.

Romagnoli S, Ricci Z, Pinelli F et al. (2012). Spinal cord injury after ascending aorta and aortic arch replacement combined with antegrade stent grafting: role of postoperative cerebrospinal fluid drainage. J Card Surg 27: 224–227.

Rothenhausler HB, Grieser B, Nollert G et al. (2005). Psychiatric and psychosocial outcome of cardiac surgery with cardiopulmonary bypass: a prospective 12-month follow-up study. Gen Hosp Psychiatry 27: 18–28.

Roy RC, Stafford MA, Charlton JE (1988). Nerve injury and musculoskeletal complaints after cardiac surgery: influence of internal mammary artery dissection and left arm position. Anesth Analg 67: 277–279.

Rudolph JL, Inouye SK, Jones RN et al. (2010). Delirium: an independent predictor of functional decline after cardiac surgery. J Am Geriatr Soc 58: 643–649.

Russell DA, Gough MJ (2004). Intracerebral haemorrhage following carotid endarterectomy. Eur J Vasc Endovasc Surg 28: 115–123.

Saczynski JS, Marcantonio ER, Quach L et al. (2012). Cognitive trajectories after postoperative delirium. N Engl J Med 367: 30–39.

Sauer CM, Yuh DD, Bonde P (2015). Extracorporeal membrane oxygenation use has increased by 433% in adults in the United States from 2006 to 2011. ASAIO J 61: 31–36.

Schlosser FJ, Verhagen HJ, Lin PH et al. (2009). TEVAR following prior abdominal aortic aneurysm surgery: increased risk of neurological deficit. J Vasc Surg 49: 308–314. discussion 314.

Schoen J, Husemann L, Tiemeyer C et al. (2011). Cognitive function after sevoflurane- vs propofol-based anaesthesia for on-pump cardiac surgery: a randomized controlled trial. Br J Anaesth 106: 840–850.

Selnes OA, Pham L, Zeger S et al. (2006). Defining cognitive change after CABG: decline versus normal variability. Ann Thorac Surg 82: 388–390.

Selnes OA, Grega MA, Bailey MM et al. (2008). Cognition 6 years after surgical or medical therapy for coronary artery disease. Ann Neurol 63: 581–590.

Selnes OA, Grega MA, Bailey MM et al. (2009). Do management strategies for coronary artery disease influence 6-year cognitive outcomes? Ann Thorac Surg 88: 445–454.

Selnes OA, Gottesman RF, Grega MA et al. (2012). Cognitive and neurologic outcomes after coronary-artery bypass surgery. N Engl J Med 366: 250–257.

Sevuk U, Kaya S, Ayaz F et al. (2016). Paraplegia due to spinal cord infarction after coronary artery bypass graft surgery. J Card Surg 31: 51–56.

Shafei H, el-Kholy A, Azmy S et al. (1997). Vocal cord dysfunction after cardiac surgery: an overlooked complication. Eur J Cardiothorac Surg 11: 564–566.

Shamji MF, Maziak DE, Shamji FM et al. (2003). Circulation of the spinal cord: an important consideration for thoracic surgeons. Ann Thorac Surg 76: 315–321.

Sharma AD, Parmley CL, Sreeram G et al. (2000). Peripheral nerve injuries during cardiac surgery: risk factors, diagnosis, prognosis, and prevention. Anesth Analg 91: 1358–1369.

Shaw PJ, Bates D, Cartlidge NE et al. (1987). Neuro-ophthalmological complications of coronary artery bypass graft surgery. Acta Neurol Scand 76: 1–7.

Shehabi Y, Grant P, Wolfenden H et al. (2009). Prevalence of delirium with dexmedetomidine compared with morphine based therapy after cardiac surgery: a randomized controlled trial (DEXmedetomidine COmpared to Morphine-DEXCOM study). Anesthesiology 111: 1075–1084.

Shekar PS (2006). Cardiology patient page. On-pump and off-pump coronary artery bypass grafting. Circulation 113: e51–e52.

Siepe M, Pfeiffer T, Gieringer A et al. (2011). Increased systemic perfusion pressure during cardiopulmonary bypass is associated with less early postoperative cognitive dysfunction and delirium. Eur J Cardiothorac Surg 40: 200–207.

Skydell JL, Machleder HI, Baker JD et al. (1987). Incidence and mechanism of post-carotid endarterectomy hypertension. Arch Surg 122: 1153–1155.

Slater JP, Guarino T, Stack J et al. (2009). Cerebral oxygen desaturation predicts cognitive decline and longer hospital stay after cardiac surgery. Ann Thorac Surg 87: 36–44. discussion 44–45.

Solomon D, Ramat S, Tomsak RL et al. (2008). Saccadic palsy after cardiac surgery: characteristics and pathogenesis. Ann Neurol 63: 355–365.

Stygall J, Newman SP, Fitzgerald G et al. (2003). Cognitive change 5 years after coronary artery bypass surgery. Health Psychol 22: 579–586.

Sundt TM, Sandok BA, Whisnant JP (1975). Carotid endarterectomy. Complications and preoperative assessment of risk. Mayo Clin Proc 50: 301–306.

Sweet JJ, Finnin E, Wolfe PL et al. (2008). Absence of cognitive decline one year after coronary bypass surgery: comparison to nonsurgical and healthy controls. Ann Thorac Surg 85: 1571–1578.

Szalma I, Kiss A, Kardos L et al. (2006). Piracetam prevents cognitive decline in coronary artery bypass: a randomized trial versus placebo. Ann Thorac Surg 82: 1430–1435.

Tao JX, Ray A, Hawes-Ebersole S et al. (2005). Intracranial EEG substrates of scalp EEG interictal spikes. Epilepsia 46: 669–676.

Tarakji KG, Sabik 3rd JF, Bhudia SK et al. (2011). Temporal onset, risk factors, and outcomes associated with stroke after coronary artery bypass grafting. JAMA 305: 381–390.

Tewari P, Aggarwal SK (1996). Combined left-sided recurrent laryngeal and phrenic nerve palsy after coronary artery operation. Ann Thorac Surg 61: 1721–1722. discussion 1722–1723.

Tietjen CS, Hurn PD, Ulatowski JA et al. (1996). Treatment modalities for hypertensive patients with intracranial pathology: options and risks. Crit Care Med 24: 311–322.

Tomonori T, Minoru K, Norihiro S et al. (2012). Vagus nerve neuromonitoring during carotid endarterectomy. Perspect Vasc Surg Endovasc Ther 24: 137–140.

Treiman DM, Meyers PD, Walton NY et al. (1998). A comparison of four treatments for generalized convulsive status epilepticus. Veterans Affairs Status Epilepticus Cooperative Study Group. N Engl J Med 339: 792–798.

Tsakiridis K, Visouli AN, Zarogoulidis P et al. (2012). Early hemi-diaphragmatic plication through a video assisted mini-thoracotomy in postcardiotomy phrenic nerve paresis. Journal of Thoracic Disease 4 (Suppl 1): 56–68.

Ullery BW, Wang GJ, Low D et al. (2011). Neurological complications of thoracic endovascular aortic repair. Semin Cardiothorac Vasc Anesth 15: 123–140.

Vahl CF, Carl I, Muller-Vahl H et al. (1991). Brachial plexus injury after cardiac surgery. The role of internal mammary artery preparation: a prospective study on 1000 consecutive patients. J Thorac Cardiovasc Surg 102: 724–729.

van Mook WN, Rennenberg RJ, Schurink GW et al. (2005). Cerebral hyperperfusion syndrome. Lancet Neurol 4: 877–888.

Vander Salm T (1984). Brachial plexus injury after open-heart surgery. Ann Thorac Surg 38: 660–661.

Veliz-Reissmuller G, Aguero Torres H, van der Linden J et al. (2007). Pre-operative mild cognitive dysfunction predicts risk for post-operative delirium after elective cardiac surgery. Aging Clin Exp Res 19: 172–177.

Wang TK, Bhamidipaty V, MacCormick M (2013). First bite syndrome following ipsilateral carotid endarterectomy. Vasc Endovascular Surg 47: 148–150.

Yatsu FM, Hart RG, Mohr JP et al. (1988). Anticoagulation of embolic strokes of cardiac origin: an update. Neurology 38: 314–316.

Chapter 32

Neurology of cardiopulmonary resuscitation

M. MULDER[1] AND R.G. GEOCADIN[2]*

[1]Department of Critical Care and the John Nasseff Neuroscience Institute, Abbott Northwestern Hospital, Allina Health, Minneapolis, MN, USA

[2]Neurosciences Critical Care Division, Department of Anesthesiology and Critical Care Medicine and Departments of Neurology and Neurosurgery, Johns Hopkins University School of Medicine, Baltimore, MD, USA

Abstract

This chapter aims to provide an up-to-date review of the science and clinical practice pertaining to neurologic injury after successful cardiopulmonary resuscitation. The past two decades have seen a major shift in the science and practice of cardiopulmonary resuscitation, with a major emphasis on postresuscitation neurologic care. This chapter provides a nuanced and thoughtful historic and bench-to-bedside overview of the neurologic aspects of cardiopulmonary resuscitation. A particular emphasis is made on the anatomy and pathophysiology of hypoxic-ischemic encephalopathy, up-to-date management of survivors of cardiopulmonary resuscitation, and a careful discussion on neurologic outcome prediction. Guidance to practice evidence-based clinical care when able and thoughtful, pragmatic suggestions for care where evidence is lacking are also provided. This chapter serves as both a useful clinical guide and an updated, thorough, and state-of-the-art reference on the topic for advanced students and experienced practitioners in the field.

EPIDEMIOLOGY

Cardiac arrest is defined as a sudden pulseless state due to cessation of effective cardiac mechanical activity. This can be further classified as primary cardiac arrest due to an intrinsic cardiac condition and secondary due to noncardiac causes such as respiratory, neurologic, metabolic, toxic, asphyxia, drowning, trauma, and resulting from environmental exposure. Based on the most recent statistics available in 2016, each year there are over half a million cardiac arrests in the USA. Approximately 356 000 of these occur in the community and are assessed and treated by emergency medical personnel, and another approximately 209 000 cases occur in healthcare facilities in the USA every year (Mozaffarian et al., 2016). North American, European, and Asian data estimate the incidence of cardiac arrest to vary widely between 50 and 110 per 100 000 in the general population (Fishman et al., 2010). Current data from the USA estimate the yearly incidence of out-of-hospital cardiac arrest (OHCA) at 110.8 individuals per 100 000 population (Mozaffarian et al., 2016). Early North American studies in OHCA showed that overall survival ranged from 2.1% to 7.3% (Hillis et al., 1993; Westfal et al., 1996), while ventricular tachycardia (VT)/ventricular fibrillation (VF) patients who were successfully resuscitated in the field and brought to the hospital showed survival rates ranging from around 8% to 40% (Nichol et al., 2008). Data from another large North American registry of OHCA presumed to be of cardiac etiology have shown a survival rate to hospital admission of 26.3%, and an overall survival rate to hospital discharge of 9.6% (McNally et al., 2009). In the USA current overall survival rates of OHCA are approximately 10.6% (Mozaffarian et al., 2016).

In-hospital cardiac arrest (IHCA) also displays a significant variability in survival and neurologic outcomes. This is due to a number of not fully understood factors

---

*Correspondence to: Romergryko G. Geocadin, Neurosciences Critical Care Division, Department of Anesthesiology and Critical Care Medicine; Johns Hopkins University School of Medicine, Baltimore, MD, USA. E-mail: rgeocad1@jhmi.edu

and mechanisms, such as the underlying cause, initial rhythm on electrocardiogram (ECG) (Nadkarni et al., 2006; Holmgren et al., 2010; Terman et al., 2014), time to return of spontaneous circulation (ROSC) (Meaney et al., 2010), location (Gwinnutt et al., 2000), and time of day (Peberdy et al., 2008), being witnessed or not (Brady et al., 2011), bystander cardiopulmonary resuscitation (CPR) (Wissenberg et al., 2013; Malta Hansen et al., 2015) and many of the same factors affecting OHCA (Huang et al., 2002; Brady et al., 2011; Merchant et al., 2011, 2012; Wallmuller et al., 2012; Nürnberger et al., 2013). Survival from IHCA historically ranges from 37% to 42% in VT/VF arrests to as low as 6.2–12% for asystolic and pulseless electric activity (PEA) arrests respectively (Gwinnutt et al., 2000; Meaney et al., 2010). A recent study on outcomes of IHCA showed that clinically significant neurologic disability among survivors has decreased over time, with a risk-adjusted rate of 32.9% in 2000 and 28.1% in 2009 (Girotra et al., 2012). This improvement over time is thought to be due to improvements in "the chain of survival," as well as postresuscitation care, including targeted temperature management (TTM), whereas the overall improvements in outcomes from IHCA compared to OHCA are usually attributed to shorter times to initiation of CPR (including bystander CPR), early defibrillation, and postresuscitation care. Current IHCA survival to hospital discharge rates in the USA are about 19% (Merchant et al., 2012; Kazaure et al., 2013). A recent study in the USA based on data from the 2003–2011 Nationwide Inpatient Sample databases of adult inpatients who underwent CPR analyzed 838 465 patients with IHCA (Kolte et al., 2015). Previous data estimated the incidence of IHCA in the USA to be 0.92 per 1000 hospital bed days (interquartile range 0.58–1.2) (Merchant et al., 2011). The geographic distribution showed that 19.4% occurred in the Northeast, 19.0% in the Midwest, 37.7% in the South, and 23.9% were in the West (Kolte et al., 2015). Based on data from this same study, IHCA incidence in the USA was 2.85 per 1000 hospital admissions, with significant regional variation in IHCA incidence; it was lowest in the Midwest and highest in the West (2.33 and 3.73 per 1000 hospital admissions, respectively) (Kolte et al., 2015). Additionally, risk-adjusted survival to discharge was also highest in the Midwest (odds ratio, 1.33; 95% confidence interval, 1.31–1.36).

Data from the UK, which included both OHCA and IHCA, demonstrated that only 28.6% of patients who achieved ROSC survived to discharge (Nolan et al., 2007). Study of an Irish registry showed that overall survival to hospital discharge improved significantly from 2.6% to 11.3% over the time period between 2003 and 2008 (Margey et al., 2011); and results from a Norwegian trial showed survival to hospital discharge rates to be between 9.2% and 10.5% (Olasveengen et al., 2009).

Cardiac arrests are believed to account for over half of all deaths from ischemic heart disease and for approximately 15–20% of all deaths (Deo and Albert, 2012). The economic burden imposed by cardiac arrest worldwide is not known, but likely to be staggering considering the cost of critical care, rehabilitation, loss of productivity, and toll on families (Suchard et al., 1999; Hamel et al., 2002; Burke et al., 2005; Søholm et al., 2014).

NEUROPATHOLOGY

Cellular injury during cardiac arrest is principally mediated by oxygen deprivation, but additional neurologic injury is incurred depending on associated factors such as concomitant toxic exposures such as carbon monoxide, metabolic stressors such as hypoglycemia, and free-radical production after restoration of blood flow in the so-called reperfusion injury. Based on the study of animal models, the magnitude of neurologic injury and the extent of possible recovery are directly related to the duration of the arrest, with up to 95% of brain tissue affected at 15 minutes of arrest in a rabbit model (Ames et al., 1968). The effect of duration of arrest in humans is intuitively also a significant predictor of the degree of neurologic injury and potential for recovery; however there are too many unknowns and potential variables, especially in OHCA, to predict outcomes solely based on arrest time.

Cellular hypoxia in the brain results in a decrease of adenosine triphosphate (ATP) production with resulting cellular energy starvation as well as a breakdown of cellular integrity from dysfunction of membrane ATP-dependent Na-K pumps. This in turn leads to the uncontrolled release of glutamate, an excitatory neurotransmitter which leads to injury from excitotoxicity (Rothman and Olney, 1986; Vaagenes et al., 1996; Belousov, 2012) mediated mainly through N-methyl-D-aspartate (NMDA) receptors (Lipton and Rosenberg, 1994). In this chain reaction of cellular injury, inhibitory neurotransmitters that normally dampen glutamate excitotoxicity, such as γ-aminobutyric acid (GABA) and glycine (Gundersen et al., 2005), are also downregulated (Globus et al., 1991). The glutamate-NMDA excitotoxic process described results in an intracellular calcium influx that activates a number of second messengers that amplify cellular injury by increasing the calcium permeability and increasing glutamate release, leading to a vicious cycle (Choi, 1994; Zipfel et al., 2000; Kaindl et al., 2012). This complex chain of events, triggered by circulatory arrest and leading to intracellular calcium overload, is almost immediate, and results in the activation of neuronal nitric oxide synthase (nNOS). The production of oxygen free-radical species by nNOS causes direct cellular injury by direct DNA fragmentation, protein oxidation, lipid peroxidation (Moore and Traystman, 1994), and disruption of the mitochondrial respiratory chain. These reactive oxygen species are released by calcium-mediated mitochondrial disruption as well as during reperfusion (Traystman et al., 1991), when renewed oxygen supplies act as a substrate for enzymatic oxidative reactions (Rodrigo et al., 2005). Oxidative stress mediated by the aforementioned mechanisms transcends cellular injury to tissue-level injury through complement activation and subsequent degradation (van Beek et al., 2003; Yang et al., 2013), cytokine production (interleukin-1 (IL-1), IL-6, IL-8, and tumor necrosis factor-α (TNF-α)), expression of leukocyte adhesion molecules, and microvascular dysfunction (Donadello et al., 2011). A delayed brain injury results from cerebral edema with resulting elevations in intracranial pressure, nonconvulsive seizures and blood–brain barrier disruption, again causing tissue-level damage. Cytotoxic edema is a result of excitotoxicity and ionic pump failure, as well as cellular water shift across cell membranes with impaired aquaporin function (Lo Pizzo et al., 2013).

Ischemia and reperfusion injury result in a breakdown of the blood–brain barrier, resulting in an influx of proteins into the brain (Wiklund et al., 2012). Cerebral edema plays a role in the degradation of the blood–brain barrier, and has been shown to be regulated by aquaporin-4, a membrane protein on glia that regulates water transport, and matrix metalloprotease-9 (Amiry-Moghaddam et al., 2003). Cardiac arrest leads to upregulation of aquaporin-4, a process mitigated by hypothermia (Suehiro et al., 2004; Xiao et al., 2004). Oxidative stress results in both cellular necrosis and apoptosis, DNA fragmentation, nicotinamide adenine dinucleotide depletion, p53 activation, and mitochondrial disruption (Kataoka and Yanase, 1998; Greer, 2006; Harukuni and Bhardwaj, 2006; Mongardon et al., 2011).

The areas of the brain responsible for arousal are most commonly and severely affected, while areas responsible for cranial nerves (CNs) and sensory motor reflexes seem to be more resistant and are affected only in more severe injuries (Fujioka et al., 1994). The areas of the brain most affected by a hypoxic-ischemic injury tend to be those newest in phylogenetic terms and responsible for consciousness and higher function, whereas those more primitive and responsive for vegetative functions appear to be more resistant. To illustrate this concept, cortical projection neurons, the posterior cingulate cortex/precuneus, medial prefrontal cortex, and bilateral temporoparietal junctions (Hoesch et al., 2008; Norton et al., 2012), the cerebellar Purkinje cells (Paine et al., 2012) and the CA-1 area of the hippocampus (Wijdicks et al., 2001; Sekeljic et al., 2012) are critical to consciousness and arousal but are some of the most affected. By contrast, the more phylogenetically "primitive" brainstem appears to be somewhat more resilient and resistant to hypoxic-ischemic injury (Brierley et al., 1971), which is probably why not all arrests compromise the midbrain and lead to brain death. Animal evidence also points to hypothalamic activation with increased levels of cerebrospinal fluid, norepinephrine, and acetylcholinesterase activity (Wortsman et al., 1987). Injury to basal ganglia structures and to the cerebellum can also be incurred and often results in various movement disorders and ataxia (Barrett et al., 2007).

On gross pathologic examination, Wijdicks and Pfeifer (2008) noted that specimens from cardiac arrest victims display extensive neocortical, diencephalic, and brainstem necrosis consistent with global hypoxic-ischemic encephalopathy (HIE). In that study, patients who died with post cardiac arrest syndrome (PCAS) displayed the most severe diffuse neuronal damage in the series, compared to patients with trauma, intracranial hemorrhage, and stroke with malignant edema. This leads to the conclusion that post cardiac arrest hypoxic-ischemic brain injury is likely a different histopathologic entity than brain injury from other causes, as those findings are consistent with a more recent series (Dragancea et al., 2012). More recent studies using diffusion tensor imaging (DTI) also show a difference between hemispheric axonal injury in cardiac arrest as opposed to a predominance of central myelin damage in traumatic brain injury (van der Eerden et al., 2014), as well as good correlation of DTI with histopathology (Gerdes et al., 2014). DTI showed well-localized evidence of white-matter injury, macrophage presence, and astrogliosis of the corpus callosum and bilateral white-matter tracts of the frontal, parietal, and occipital lobes. Injury to the temporal lobes, hippocampus, basal ganglia, and thalamus also correlated well to the histopathologic examination.

CLINICAL PRESENTATION

The neurologic and systemic presentations and sequelae of the multisystem injury incurred after a cardiac arrest have led to the recognition of a distinct syndrome. The PCAS concept has its roots in the 1972 publication by Negovsky, who initially coined the phrase "post-resuscitation disease" for the pathophysiologic process that followed circulatory arrest and the subsequent multi-organ ischemia and reperfusion states that ensued. The modern term of PCAS was defined in 2008 in the International Liaison Committee on Resuscitation (ILCOR) consensus statement on cardiac arrest (Neumar et al., 2008), and it defined the following four key components: (1) post cardiac arrest brain injury; (2) post cardiac arrest myocardial dysfunction; (3) systemic ischemia/reperfusion response; and (4) persistent precipitating pathology.

Post cardiac arrest brain injury manifests as a host of neurologic conditions which during the acute phase center mostly on disorders of consciousness. This clinical spectrum of neurologic disorders correlates well with the selective vulnerability of the cortex, the arousal systems, thalamus, and cerebellum, as described above. Coma and the minimally conscious states are typically observed. Coma is defined as unresponsiveness to internal and external stimuli in the absence of arousal, whereas the minimally conscious states (vegetative states) may include intermittent arousal and responsiveness to stimuli, but lack other features of consciousness and cognition (Giacino et al., 2002; Huff et al., 2012). With the selective injury of other neural areas such as the basal ganglia and cerebellum, movement disorders and ataxia can be among the neurologic sequelae of cardiac arrest. Rarely, cardiac arrest survivors can also suffer from other dyskinesias, including parkinsonism, dystonias, chorea, athetosis, and various tremors, post hypoxic myoclonus (PHM), and the Lance–Adams syndrome, in which a cardiac arrest survivor who has regained consciousness develops myoclonus days to weeks after the cardiac arrest (Shin et al., 2012). In some patients focal deficits (paresis) may be noted and are related to watershed infarcts caused by the abrupt reduction in perfusion, and can occur anteriorly at the junction of the anterior and middle cerebral arteries and posteriorly at the junction of middle and posterior cerebral arteries.

As the care of patients with the PCAS has improved, and with the presence of more survivors, there is a growing realization of the lack in the long-term follow-up of cardiac arrest survivors. The extensive cortical injury has been shown to cause cognitive deficits mainly focused on memory, attention, and executive function (Boyce-van der Wal et al., 2015; Sabedra et al., 2015). According to some studies, the quality of life of PCAS survivors is significantly decreased, with only 20% being able to resume their prior employment (de Vos et al., 1999). Due to cortical injury, the incidence of seizures in PCAS survivors can be as high as 40% (Khot and Tirschwell, 2006).

Post cardiac arrest myocardial dysfunction is the second component in the PCAS definition; this may be due to a variety of factors which may be primary contributors to the arrest, such as a myocardial infarction (Wolfrum et al., 2008; McManus et al., 2012), a primary arrhythmia (Roy et al., 1983; Brady et al., 1995, 1999), or secondary myocardial dysfunction (Laurent et al., 2002) following ischemia and reperfusion of the heart. This dysfunction is characterized by decreased left ventricular function, which may even manifest as cardiogenic shock or stress-induced cardiomyopathy (Kurisu et al., 2010). Respiratory problems encountered in resuscitated cardiac arrest victims range from pulmonary contusions, as well as costal and sternal fractures resulting from CPR (Kim et al., 2011a), leading to hypoxia and alterations in chest wall compliance, aspiration pneumonia/pneumonitis due to loss of airway reflexes or emergent intubation (Virkkunen et al., 2007) and pneumothoraces, all of which may lead to respiratory failure with or without acute respiratory distress syndrome (ARDS). The PCAS has been described as a "sepsis-like" state (Adrie et al., 2002), with vasodilatation and cardiodepression mediated by cytokine release and chemotactic factors, as noted in the previous section; this results in sequestration of activated neutrophils in the lungs that may be at least partially responsible. There is some indication that this "sepsis-like" state may be related to intestinal bacterial translocation secondary to ischemic injury to the intestines, leading to increased serum endotoxin levels (Korth et al., 2003; Grimaldi et al., 2013). Intestinal ischemia can also complicate the postresuscitation course by resulting in refractory lactic acidosis or, in some cases, lead to perforation (Katsoulis et al., 2012) or even withdrawal of care when the entire bowel is deemed nonviable.

The kidneys and the liver are two organs commonly thought to be injured by ischemia-reperfusion or low-flow states; however, few objective data exist on these entities in the setting of PCAS. Hypoxic hepatitis or "shock liver" is not infrequently seen in the PCAS, and carries a high mortality in survivors of cardiac arrest (Raurich et al., 2011); on the other hand acute kidney injury following cardiac arrest seems to have been overestimated, particularly if cardiogenic shock is not present (Chua et al., 2012). Patients suffer multiple endocrine derangements, including changes in ADH, cortisol, adrenocorticotropic hormone, insulin, and glucagon levels (de Jong et al., 2008; Oshima et al., 2010; Kim et al., 2011c) as part of their postresuscitation course, and little is yet known or understood about these conditions and the role of therapeutic hypothermia (TH) in their course. Hematologic and coagulation disorders are also recognized; however, they too are not well described or fully understood. There is activation of coagulation factors as well as hyperfibrinolysis (Mehta et al., 1972; Adrie et al., 2005; Schöchl et al., 2012). A recent study (Sutherasan et al., 2015) has looked at the incidence of various forms of organ failure and complications of care in PCAS patients and found the following: ARDS 4–7%, hospital-acquired pneumonias 4–13%, sepsis 3–19%, cardiovascular failure 19–48%, renal failure 20–30%, hepatic failure 2–13%.

NEURODIAGNOSTICS AND IMAGING

The diagnosis of cardiac arrest is clinical based on absent pulse and unresponsiveness, with electrocardiography and echocardiography as useful adjuncts in determining initial management and etiology. For the purposes of this chapter we will focus on the diagnosis of HIE as part of the PCAS. After the patient has recovered spontaneous circulation and respiratory effort, if the patient remains unresponsive without other clear causes (medications, toxins, and gross metabolic derangements such as profound hypoxemia or hypercapnea), the clinical exam suffices for the determination of HIE and the initiation post resuscitative cares targeting HIE.

Clinical examination

The clinical evaluation of a patient successfully resuscitated from cardiac arrest calls for a careful appraisal of neurologic function in the setting of multiple confounders. The vast majority of patients will have some degree of disorder of consciousness and at the same time will be subjected to physiologic conditions and drugs that will obscure the neurologic assessment. Sedatives, paralytics, hypothermia, and hypoperfusion have to be considered and corrected for in order to make a valid neurologic assessment.

The overall goal is to have a full neurologic assessment of the patient using the conventional neurologic approach. Assessment of level of arousal requires the patient to be provided with the appropriate stimuli to elicit a response. If the patient is unresponsive, the examination focuses on the patient's "best response." It is important to establish whether the patient provided a meaningful response and if the patient is able to "follow commands" (i.e., open or close eyes, show two fingers, give a thumbs up). If this is noted it signifies cortical and arousal system function and a more detailed cognitive examination can be undertaken. The absence of any meaningful response leads to the assessment of the unresponsive patient by "best response," which is based mostly on automatic responses or reflexive actions. These responses can range from "localizing and avoiding noxious stimulus," to reflexive responses such as flexor or extensor posturing.

The assessment of CNs is undertaken as in a routine neurologic assessment. It is critical to note these because they have important implications in prognostication. Ophthalmoscopy to assess CN II (optic nerve) can be undertaken, but the evaluation of CN III, the oculomotor nerve, may be one of the most important. This is tested by assessing the pupillary light reflex, which assesses the constriction, shape, and size of the pupils. Ocular movements (CN III, IV, and VI) in the unresponsive may be assessed by eliciting the doll's-eyes reflex. The corneal reflex assesses the trigeminal nerve (CN V) as afferent and the facial nerve (CN VII) as efferent components. The corneal reflex is among the critical parameters assessed in prognostication. CN VIII may be assessed using cold caloric testing. CN IX and CN X are tested by gag reflex. CN XI is assessed by the shrug response and CN XII is assessed by examining tongue movements. The motor assessment in the unresponsive patient can be incorporated in the sensory assessment, typically in response to noxious stimulation, such as suctioning or deep sternal pressure. Assessment of deep tendon reflexes is done in the usual manner.

Clinical examination tools such as the Glasgow Coma Scale (GCS) (Teasdale and Jennett, 1974; Teasdale et al., 2014) were developed to rapidly assess the unresponsive patient. While the GCS has been successful as a tool for coma, it does not provide the critical details needed to fully assess the extent of the neurologic injury. The recently developed FOUR (Full Outline of UnResponsiveness) score was introduced to overcome many of the limitations in the GCS, such as inability to assess verbal function in intubated patients and lack of brainstem testing (Wijdicks et al., 2005; Iyer et al., 2009). This scale evaluates ocular, motor, and brainstem responses as well as respiratory patterns. Despite its slow adoption and increased complexity, it is more suited to the task. Though neither of these scales is a substitute for a careful neurologic evaluation, they certainly have the advantage of easily and reproducibly conveying a systematic evaluation. Currently the most widely accepted measure to determine HIE and initiate specific post resuscitative care in the setting of a resuscitated cardiac arrest patient is a motor GCS (mGCS) score of less than 6 (Callaway et al., 2015) (Table 32.1).

Blood-based biomarkers

Biochemical markers of brain injury derived from systemic blood in the setting of HIE are not necessary to make the diagnosis, but have mostly been evaluated for prognostic performance. As a group of assays they mostly are markers of brain injury and their values are generally considered to be proportional to the extent of the injury. None of these makers are specific to HIE and can be elevated and of some utility in other neurologic conditions. The most commonly used biomarker is neuron-specific enolase (NSE), which can be measured in peripheral blood in the presence of neuronal damage and disruption of the blood–brain barrier. As such, it can also be measured in cerebrospinal fluid, though this is seldom practical or indicated in the setting of cardiac arrest. Serum S-100B protein can also be measured after various forms of brain injury, including HIE,

Table 32.1

Comparison of commonly used coma examination scales

The Glasgow Coma Scale (GCS)	The Full Outline of UnResponsiveness (FOUR score)
Eye response 4 = Eyes open spontaneously 3 = Eyes closed but open to voice 2 = Eyes closed but open to pain 1 = Eyes closed and do not open **Motor response** 6 = Follows commands 5 = Localizes to pain 4 = Withdraws from pain 3 = Flexor response to pain 2 = Extensor response to pain 1 = No response to pain **Verbal response** 5 = Conversant and oriented 4 = Conversant and confused 3 = Conversant with inappropriate words 2 = Incomprehensible sounds 1 = No verbal response	**Eye response** 4 = Eyes open or opened; tracks or blinks to command 3 = Eyes open but not tracking 2 = Eyes closed but open to loud voice 1 = Eyes closed but open to pain 0 = Eyes closed and do not open to pain **Motor response** 4 = Thumbs up, fist, or peace sign to command 3 = Localizes to pain 2 = Flexor response to pain 1 = Extensor response to pain 0 = No response to pain or generalizes status myoclonus **Brainstem reflexes** 4 = Pupillary and corneal reflexes intact 3 = One pupil is fixed and dilated 2 = Absent pupillary or corneal reflexes 1 = Absent pupillary and corneal reflexes 0 = Absent pupillary, corneal, and cough reflexes **Respiration** 4 = Not intubated, regular respiratory pattern 3 = Not intubated, Cheyne–Stokes respirations 2 = Not intubated, irregular respiratory pattern 1 = Intubated, breathes over set ventilator rate 0 = Apneic or breathes at set ventilator rate

Adapted from Teasdale and Jennett (1974), and Wijdicks et al. (2005), with permission from John Wiley.

and it is released from astroglial cells. S-100B is much less studied and limited in use than NSE. In the modern postresuscitation era, the first major study investigating their clinical utility was in 2003 (Tiainen et al., 2003). While many studies have been undertaken using biomarkers, many concerns remain regarding their ability to reliably prognosticate outcomes. These biomarkers reflect injury to neurons and astroglia in general, and provide no specific information related to specific functions of the cell-type injury. The utility of biomarkers for neuroprognostication will be discussed later in this chapter.

Neurophysiologic testing

Much like biomarkers, neurophysiologic testing has little role in the initial diagnosis of HIE, and most of their clinical utility and use are based on the detection and diagnosis of potential complications like nonconvulsive status epilepticus (NCSE) and in neuroprognostication. Electroencephalography (EEG), somatosensory evoked potentials (SSEP), and, most recently, functional magnetic resonance imaging (MRI) have been evaluated in the setting of HIE, but again with a focus on neurologic outcome prediction.

Neuroimaging

Neuroimaging plays a vital role in the early evaluation of comatose cardiac arrest survivors as part of the diagnostic workup to rule out a neurologic cause for the arrest (such as a subarachnoid hemorrhage) or to rule out associated trauma. Once again, the intent is not to diagnose HIE *per se*. Computed tomography (CT) imaging is ideally suited to this task, being rapid and readily available in emergency departments. Because of this there has been much interest in the use of this modality, to attempt to use it to quantify the amount of brain injury and therefore as an early predictor of outcome (Kjos et al., 1983; Yanagawa et al., 2005). MRI is more sensitive to early cerebral ischemic injury; however it is significantly more time- and labor-intensive as well as less readily available in the emergency setting (Wijdicks et al., 2001; Wu et al., 2009). Like MRI, but with even higher resolution of the degree of neural damage, DTI promises many advantages; however, at this time it is clinically impractical in the vast majority of time-critical clinical scenarios (Gerdes et al., 2014; van der Eerden et al., 2014).

HOSPITAL COURSE AND MANAGEMENT

After ROSC, if not already done, the airway should be definitively secured with endotracheal intubation and mechanical ventilation instituted. All patients should have a 12-lead ECG performed immediately after ROSC; this is vital to establish the diagnosis of an ST-elevation myocardial infarct, as this has an immediate bearing on acute patient management. Acute myocardial infarction is an important and reversible cause of cardiac arrest, and early diagnosis and treatment via revascularization play a crucial role in survival and outcome (Wolfrum et al., 2008; Callaway et al., 2015). After the airway is secured in definitive fashion, immediate portable chest radiography should be obtained to confirm endotracheal tube placement and placement of central venous lines and to gain an approximation of cardiopulmonary status by evaluating for pulmonary infiltrates or edema, cardiac size, and vascular markings. Enteral access should be obtained via orogastric tube, gastric contents emptied, and the tube placed on low intermittent suction.

After confirmation of adequate peripheral intravenous access, central venous access should be obtained. The indications for central access are for administration of vasopressors, additional access, hemodynamic monitoring, and endovascular temperature management, when available and applicable. Arterial access should also be obtained expeditiously to enable hemodynamic monitoring and serial arterial blood gas measurements. Immediate echocardiography should be performed as well to evaluate for left ventricular size, shape, and function; to rule out pericardial tamponade, evaluate for right ventricular strain, and evidence of massive pulmonary embolus; and to assess for regional wall motion abnormalities or stress-induced cardiomyopathy (Volpicelli, 2011). In cases of ongoing resuscitation emergent transthoracic echocardiography may aid in the decision to proceed with resuscitative efforts or not (Tarmey et al., 2011; Cureton et al., 2012; Oren-Grinberg et al., 2012).

Transthoracic echocardiography should suffice for the initial emergent evaluation and should be combined with an extended focused assessment with sonography for trauma (EFAST) exam to rule out pneumothoraces missed on radiography, pleural effusions, and intra-abdominal fluid. Of note, once stabilized and in the intensive care unit, all patients should have a complete formal transthoracic echocardiogram. Finally patients should have an initial emergency cranial CT, as this serves to rule out complicating conditions with potential to alter immediate management, such as subarachnoid or intra-parenchymal hemorrhage, ischemic stroke, and cerebral herniation or conditions that may affect the use of antiplatelet agents or anticoagulants, if there is a concomitant myocardial infarction (Kürkciyan et al., 2001; Inamasu et al., 2009; Naples et al., 2009; Prout and Nolan, 2009; Cocchi et al., 2010; Skrifvars and Parr, 2012).

Once life support devices are in place and basic diagnostic workup, including laboratory evaluation, is under way, the decision to initiate TH – nowadays more often referred to as TTM – must be made. The decision-making process will be discussed in detail later in this chapter in the section on complex clinical decisions; however, most comatose cardiac arrest survivors generally undergo TTM when in experienced centers.

Assuming the patient is to undergo TTM, the next step is to induce TH to goal temperature, which can range from 32 to 36°C. This variation will also be discussed later on in the chapter. The induction of TH may have already begun during transport in the form of ice packs or from environmental exposure. With regard to the timing of induction of TH, it is generally accepted that prompt initiation of hypothermia and achievement of target temperature are optimal (Sendelbach et al., 2012). Data from laboratory studies (Abella et al., 2004; Nozari et al., 2006) indicate that intra-arrest TH (IATH) improves survival and neurologic recovery compared to postarrest TH (PATH) or normothermic controls. Laboratory animal data have also shown improved ROSC rates (Riter et al., 2009; Yannopoulos et al., 2009), limitation of myocardial infarction size, and improvement in cardiac function when compared to PATH (Tsai et al., 2008; Yannopoulos et al., 2009). Experimental evidence also seems to indicate that IATH improves cerebral perfusion and decreases cerebral metabolism when compared to normothermia (Nordmark et al., 2009).

Clinical evidence in humans undergoing IATH is scarce, but had initially appeared rather promising (Deasy et al., 2011), and prehospital induction of hypothermia seemed like a good natural extension of this concept (Callaway et al., 2002; Bruel et al., 2008; Kämäräinen et al., 2008; Castrén et al., 2010; Garrett et al., 2011). It is important to consider once again the discrepancies between preclinical data and small studies to the results of large randomized clinical trials. A large trial of prehospital induction of TH failed to demonstrate improved clinical outcomes despite earlier achievement of target temperature as well as increased pulmonary edema (Kim et al., 2013). It is interesting to note that animal data indicate that initiating hypothermia after 12 hours has no benefit (Kuboyama et al., 1993).

There are a multitude of commercially available devices, ranging from surface-cooling systems that include headgear (Hachimi-Idrissi et al., 2001; Storm et al., 2008), pads for the torso and extremities (Haugk et al., 2007; Heard et al., 2010), garments (Laish-Farkash et al., 2011), mattresses (Hypothermia after Cardiac Arrest Study Group, 2002) and nasal (Castrén et al., 2010) cooling systems, to invasive cooling via endovascular catheters (Flemming et al., 2006; Arrich and European Resuscitation Council Hypothermia After Cardiac Arrest Registry Study Group, 2007; Pichon et al., 2007) and automated peritoneal lavage systems (de Waard et al., 2013; Polderman et al., 2015), to simpler methods such as applying ice packs and administering chilled intravenous fluids. Experiments on local brain cooling with intracranial cooling devices are still in the animal experiment stage, but appear to show some promise (Moomiaie et al., 2012). All these systems may be roughly classified as whole-body cooling or selective brain cooling; whole-body cooling remains the standard, though selective brain cooling has been shown to be feasible in adult PCAS (Hachimi-Idrissi et al., 2001; Storm et al., 2008; Castrén et al., 2010). Evidence from the pediatric literature (Sarkar et al., 2012) seems to suggest that selective brain cooling may not provide benefits comparable to whole-body cooling, especially for deeper brain structures. One could postulate that this difference may become accentuated in adult brains. Regardless, much research remains to be learned with regard to the ideal system to induce and maintain TTM. Given the importance of avoiding overcooling and rebound post-TH fevers, the automated closed-loop systems may offer some advantages for TTM (Finley Caulfield et al., 2011; Bro-Jeppesen et al., 2013; Nielsen et al., 2013).

Once the patient has been rewarmed, the therapeutic temperature management system should remain in place for a further 48–72 hours to ensure normothermia, protecting the brain from the detrimental effects of hyperthermia (Bro-Jeppesen et al., 2013). Rebound pyrexia is a common phenomenon, occurring in about 40% of patients post-TTM, with temperatures >38.7°C being associated with worse neurologic outcomes in patients who survive to discharge (Leary et al., 2012). In general, the main determinants of the ability to cool patients, regardless of the method, are the initial ambient and patient temperatures, body surface area, age, and the level of impairment of the endogenous thermoregulatory mechanisms (Lyden et al., 2012). This should by no means discourage clinicians without access to these systems from providing TTM to their patients, as it has been clearly shown that TTM can effectively and safely be applied with simple measures like chilled fluid infusions and ice packs (Bernard et al., 2002, 2010, 2012; Arrich and European Resuscitation Council Hypothermia After Cardiac Arrest Registry Study Group, 2007; Carlson et al., 2012).

In order to facilitate cooling, sedation and neuromuscular blockade (NMB) are decisions as important as the choice of the method to achieve and maintain target temperature. With regard to the choice of paralytic agent, the use of a nondepolarizing NMB with no histamine release potential and short to intermediate duration, such as rocuronium, vecuronium, or cisatracurium, is preferable. The timely and judicious use of these agents as well as careful pharmacologic consideration are crucial. The rationale for the use of sedation is to optimize ventilator synchrony (Patel and Kress, 2012), prevent shivering, thereby facilitating induction of hypothermia (Badjatia et al., 2008; Polderman, 2009), and to minimize endogenous stress-induced catecholamine surges. There is great variability in the practice and methods for both sedation and analgesia in TH for the PCAS (Chamorro et al., 2010), and this is due to the lack of clear evidence pointing to the choice of ideal agents.

Some practical considerations should be kept in mind when selecting drugs for sedation and analgesia in the setting of the PCAS. Multiple factors, such as age, weight, and obesity, hemodynamic status, renal and hepatic function, pharmacokinetics and dynamics, hypothermia, and the need to accurately and quickly assess neurologic function complicate the choice of an agent. There is no clear guidance on the best agents to use; however the use of sedoanalgesia seems to be theoretically preferable. Narcotic analgesic agents such as fentanyl and remifentanyl seem to be reasonable choices that need to be paired with a hypnotic sedative such as propofol, dexmedetomidine, or benzodiazepines (Sato et al., 2010; Hoy and Keating, 2011; Spies et al., 2011; Bjelland et al., 2012; Futier et al., 2012; Schoeler et al., 2012).

The management of shivering is a significant concern, as shivering is a centrally mediated thermoregulatory response that normally sets in at 35.5°C, and is usually overcome below 34°C. However, these reference temperatures apply to healthy individuals, and may not be the same in all PCAS patients. The absence of shivering after induction of hypothermia, or spontaneous hypothermia prior to induction of hypothermia, has been associated with worse outcomes (Benz-Woerner et al., 2012). It is possible that damage to the hypothalamus impairing thermoregulation may be a marker for more severe injury. A clinical scale to quantify and assess shivering has been developed (Badjatia et al., 2008) and can be used to suppress shivering in a stepwise fashion. The management of shivering is therefore most important during the induction and rewarming phases of TTM. The use of NMB and sedoanalgesia are the cornerstones of shivering management. Additionally an initial 4–6-gram load of intravenous magnesium sulfate to achieve serum magnesium levels of 3–4 mg/dL is recommended to correct potential hypomagnesemia and to facilitate hypothermia by decreasing shivering as part of the initial stabilization and induction of hypothermia (Wadhwa et al., 2005). Meperidine and buspirone have a role in the control and management of hypothermia, as well as the use of skin counterwarming measures (Mokhtarani et al., 2001; Kimberger et al., 2007; Choi et al., 2011).

The head of the bed should remain elevated to 30° to decrease potential for ventilator-associated pneumonia, and in the midline position to improve cerebral venous drainage. Nursing staff should record vital signs and core temperature hourly during cooling and rewarming phases until target temperature is reached, every 2 hours during the maintenance phase, and every 4–6 hours during the normothermia phase. Urinary output should be measured and recorded every 2 hours throughout the intensive care unit stay. Electrolyte replacement protocols may be employed during all but the rewarming phase, as extracellular shifting of potassium results in increased serum concentration. Though there is no question that patients undergoing TH post cardiac arrest should receive stress ulcer prophylaxis, as they are at high risk for developing significant gastrointestinal bleeding due to ischemia, hypoperfusion, physiologic stress, and coagulopathy, it is currently unclear as to which agents are to be used. There has been some evidence that pantoprazole may have the least likelihood of interfering with clopidogrel activity; there is much debate and contradicting evidence on this point (Ogilvie et al., 2011; Douglas et al., 2012; Goodman et al., 2012). Of note, H_2 receptor blockers in general, such as cimetidine and ranitidine, should be avoided as they may potentiate coagulopathy by platelet inhibition (Nakamura et al., 1999), although no clinical evidence supports this claim definitively. Deep-vein thrombosis prophylaxis is indicated with pneumatic compression devices and appropriately weight-based prophylactic heparin infusions. Glycemic control should likewise be maintained with an insulin infusion if necessary; current guidelines suggest maintaining blood glucose levels between 144 and 180 mg/dL as well as aggressively treating levels below 80 mg/dL (Peberdy et al., 2010). The subcutaneous administration of medications can result in erratic and unpredictable absorption given changes in subcutaneous tissue perfusion during TTM.

Mechanical ventilation should be aimed at maintaining tissue normoxia and normocapnea. Arterial oxygen concentration or saturation goals have not been studied rigorously, but there seems to be some indication that supranormal values may in fact be counterproductive (Kuisma et al., 2006; Kilgannon et al., 2010, 2011). Current guidelines suggest discontinuation of 100% Fio_2 once ROSC is achieved, and that oxygen delivery be titrated to maintain an arterial oxygen saturation of 94–98% as soon as possible in the postresuscitation phase (Peberdy et al., 2010), as hyperoxia seems to have detrimental effects (Janz et al., 2012). Normocapnea with P_{CO_2} values between 40 and 45 mmHg and end-tidal CO_2 values of 35–40 mmHg should be targeted. One must pay attention to the temperature and corrections used by the laboratory to ensure correct interpretation of arterial blood gases in view of alpha-stat and pH-stat analysis, which could have significant implications (Sakamoto et al., 2004; Kollmar et al., 2009; Huff et al., 2012). Hyperventilation can decrease cardiac output via increased intrathoracic pressure, potential decreases in cerebral blood flow (Czosnyka et al., 2009; Willie et al., 2012), and a decrease in seizure threshold (Bergsholm et al., 1984; Loo et al., 2010). Though low tidal volume lung-protective ventilation strategies are standard of care for patients with ARDS and acute lung injury (The Acute Respiratory Distress Syndrome Network, 2000; Needham et al., 2012), we also recommend ventilation with tidal volumes of 6 mL/kg of ideal body weight in patients without evidence of ARDS (Serpa Neto et al., 2012). It is recommended that continuous pulse oximetry and end-tidal carbon dioxide be monitored, with frequent correlation to measured arterial blood gas measurements. Both techniques have been used extensively in the setting of resuscitation; however, transcutaneous pulse oximetry can be potentially confounded in a linear fashion by increasing hypothermia, and both oximetry and capnography can be further affected by poor perfusion states.

TH induces important changes in metabolism and homeostasis, resulting in changes to urine output (cold diuresis) with alterations in potassium, magnesium, and phosphate; changes in the electrolyte composition of intra- and extracellular compartments have significant impact on potassium levels and crystalloid infusions further complicate the picture (Polderman, 2009). The importance of control over potassium and magnesium levels is related to their importance in cardiac conduction and role in arrhythmogenesis; potassium levels of >4 mEq/L and magnesium levels >2.0 mEq/L should be maintained, with levels in the 3–4 mEq/L range to reduce shivering being preferred. Serum sodium concentrations should be maintained in the normal range whenever possible, and any corrections should be made with careful consideration of potential chronic sodium disorders that should not be corrected too rapidly. In cases of cerebral edema, elevated intracranial pressure and herniation and increased sodium goals must be individualized. Hypotonic intravenous fluids should be avoided. Current guidelines highlight the lack of quality data regarding hemodynamic goals in the setting of resuscitated cardiac arrest, with the only clear guidance being to avoid and immediately correct hypotension, keeping systolic blood pressures greater than 90 mmHg and mean arterial blood pressures above 65 mmHg; beyond that clinicians must individualize and tailor hemodynamics to the patient's specific circumstances (Callaway et al., 2015).

It is recommended that patients remain at target temperature (32–36°C) for 24 hours, after which the rewarming process should begin. Prior to initiation of the rewarming, care must be taken to ensure that the patient is intravascularly replete and euvolemic, as increasing core temperature will result in reversal of peripheral vasoconstriction and patients may become hypotensive. Depending on individual patient volume and hemodynamic profiles judicious volume loading should precede and be continued during the rewarming phase. It is also important to keep in mind that, during rewarming, extracellular shifting of potassium results in increased serum concentration of potassium. Hypothermia decreases insulin secretion and sensitivity and great care must be taken to avoid hypoglycemia during rewarming, as insulin sensitivity increases to baseline. It is recommended that point-of-care blood glucose measurements be obtained hourly. Care must also be taken to avoid the occurrence of shivering during rewarming, for the reasons already mentioned earlier in this chapter. Careful down-titration of sedoanalgesia is warranted as the patient approaches normothermia with close attention to patient comfort and shivering. Once the patient is normothermic, sedoanalgesia should be further minimized, balancing patient safety, comfort, and the ability to assess the neurologic and cognitive examination.

Active, controlled rewarming at a rate of 0.25°C per hour is recommended until a core temperature of 36–37°C is achieved (Peberdy et al., 2010). Once the patient has been rewarmed, the therapeutic temperature management system should remain in place for a further 48–72 hours to ensure normothermia, protecting the brain from the detrimental effects of hyperthermia (Bro-Jeppesen et al., 2013). The recommendations for the speed of rewarming and avoidance of hyperthermia after rewarming are still somewhat controversial. They are based on assumptions and observations from both laboratory research and clinical studies indicating a correlation between worse outcomes, markers of neuronal damage, or dysfunction and damage in the setting of pyrexia (Takasu et al., 2001; Zeiner et al., 2001; Hata et al., 2008; Jia et al., 2008a, b; Badjatia, 2009; Polderman and Herold, 2009; Suffoletto et al., 2009; Gordan et al., 2010).

In HIE secondary to cardiac arrest, the incidence of NCSE is estimated to be as high as 24% (Mani et al., 2012; Rittenberger et al., 2012) and is associated with worse outcomes (Rossetti et al., 2010a; Nielsen et al., 2011). Therefore, neuromonitoring with continuous surface EEG should be strongly considered (Callaway et al., 2015). There is insufficient evidence to recommend prophylactic use of antiepileptic drugs at this time (Callaway et al., 2015). Also, the optimal antiepileptic medication for the treatment of NCSE in the setting of the PCAS is not clear. It is reasonable to use less sedating agents or those with a short half-life (midazolam, levetiracetam, fosphenytoin, or valproic acid) to avoid clouding the neurologic evaluation for purposes of neurologic prognostication once the patient has been rewarmed and adequate time for observation has been provided. In HIE patients with status epilepticus, existing management guidelines for the treatment of status epilepticus should be followed (Callaway et al., 2015). Acute PHM occurs in about 30% of cardiac arrest patients with HIE, and can be divided into status myoclonus and multifocal myoclonus. Though little evidence exists regarding the optimal therapy for PHM, subcortical myoclonus usually responds best to clonazepam. Propofol, though not an antiepileptic drug *per se*, does suppress myoclonic as well as electrographic activity, but longer-term control after weaning of propofol is problematic; alternatively, valproate, phenytoin, phenobarbital, or other benzodiazepines can be used, but seem to be less effective (Bouwes et al., 2012a).

Cerebral edema is thought to play a major role in the pathophysiology of brain injury following cardiac arrest. Cerebral edema is apparent on the initial cranial CT in approximately 30% of patients following cardiac arrest (Naples et al., 2009). TH has an effect in decreasing cerebral edema; however, additional measures such as osmotherapy and barbiturate coma have not been well studied and at this time are not recommended for routine use. In cases of cerebral herniation, the use of hypertonic saline and osmotherapy in an attempt to reverse herniation could be considered (Koenig et al., 2008). However, if a patient develops cerebral herniation due to diffuse brain swelling in the setting of HIE, this is an ominous sign and a re-evaluation of expected outcomes and goals is warranted.

CLINICAL TRIALS AND GUIDELINES

TH for comatose survivors of cardiac arrest did not receive widespread notice or clinical consideration until the simultaneous publication of two seminal clinical trials of TH for resuscitated VT/VF arrests in 2002. The European multicenter trial by Holzer et al. (Hypothermia after Cardiac Arrest Study Group, 2002) was the largest of the two landmark trials, with 137 patients randomized to the experimental group and 138 to the normothermic control group. In this protocol patients were cooled to 32–34°C for 24 hours with a cooling mattress. The primary endpoint was 6-month neurologic outcome, and the secondary endpoints were

6-month mortality and complication rates at 1 week. In this study patients treated with TH had a favorable neurologic outcome in 55% of cases, while only 39% of controls had a favorable neurologic outcome; the mortality in the TH group was 41% and 55% in the normothermia group.

The second study, by Bernard and colleagues (2002), was carried out in four hospitals in Melbourne, Australia. This protocol randomized 77 patients: 34 to standard care and 43 to hypothermia induced with ice packs to a goal temperature of 33°C for 12 hours. The primary outcome measure was survival to hospital discharge with a neurologic status sufficient to allow discharge to home or to a rehabilitation facility. The TH group had a favorable neurologic outcome in 49% of cases while only 26% did in the normothermic controls.

These two studies showed that a relatively simple and cost-effective intervention at the time had a significant impact in survival, with a number need to treat of six (Holzer et al., 2005), and neurologic outcomes leading to rapid endorsement by international resuscitation guidelines (Nolan et al., 2003; International Liaison Committee on Resuscitation, 2005) and worldwide adoption. Following the initial landmark trials, ILCOR and the American Heart Association (AHA) published an interim scientific statement in 2003 recommending the use of TH in comatose survivors of cardiac arrest (Nolan et al., 2003). This was followed in 2005 by an update to the AHA guidelines for cardiopulmonary resuscitation and emergency cardiovascular care (American Heart Association, 2005).

In the following decade, both AHA and ILCOR recommendations for postresuscitation care included the use of TH, based mainly on these two studies and a number of smaller, mostly nonrandomized, clinical trials. Notable exceptions include a single-center randomized trial comparing target cooling temperatures (Lopez-de-Sa et al., 2012), which randomized 36 patients: 18 to undergo TH at the upper range of mild TH (34°C) and another 18 to the lower range, at 32°C. This study was also the first major study to include patients who presented in asystole; however, PEA arrests were excluded. The primary outcome was survival with functional independence at 6 months. In the 32°C group, 44.4% met the primary endpoint, compared with 11.1% in the 34°C group. The incidence of complications was similar in both groups, with bradycardia being more common in the 32°C group. It is interesting to note that there was a marked reduction in the incidence of clinical seizures, which was lower (1 versus 11) in patients assigned to 32°C. This study had significant clinical implications as it compared both ends of the temperature spectrum employed in TH at the time, and showed a difference in outcomes, favoring cooling to 32°C. This study came a year after the 2011 single-center randomized trial by Kim et al. (2011b), where 62 patients were randomized to TH of 34, 33, and 32°C and no difference in mortality or neurologic outcomes was noted. However, there was more hypotension in the 32°C group compared to the other two.

The most recent, and certainly the largest prospective randomized clinical trial of TH (TTM 33–36) was published in late 2013 by Nielsen et al., comparing TH of 33°C versus 36°C regardless of initial rhythm. This multicenter study in 36 intensive care units in Europe and Australia randomized 939 comatose OHCA survivors to cooling to 33 or 36°C, with protocolized sedation, rewarming, and prognostic evaluation, and found no significant difference in outcomes or complications between the two temperatures. This study sparked a fierce debate regarding the continued use of TH/TTM and spawned a number of *post hoc* analyses attempting to address a number of questions and criticisms leveled at the initial trial, including the surprisingly high proportion of bystander CPR and short durations of arrest. The most current joint AHA-ILCOR advisory statement from 2015 on temperature management after cardiac arrest (Donnino et al., 2015) provides six key recommendations to guide the use of TTM in the PCAS (Table 32.2).

COMPLEX CLINICAL DECISIONS

Despite considerable international interest and effort in resuscitation science, many important clinical questions remain unclear. The most fundamental questions pertain to the particulars of the provision of TTM and the prognostication of neurologic outcomes with an emphasis on early prediction. The most hotly debated question in the resuscitation community and one that occurs at the bedside of each comatose cardiac arrest victim is what temperature to cool to. Current guidelines only recommend the provision of TTM, with no clear guidance on what temperature to select, ranging from 32 to 36°C (Callaway et al., 2015; Donnino et al., 2015). These same guidelines recommend treating survivors of nonshockable arrests with TTM, albeit with a weak recommendation with low-quality evidence. The authors of this chapter are of the opinion that, pending further evidence and based on similar safety, patients be maintained at a temperature of 32–34°C, especially in settings of prolonged arrest and no bystander CPR. Additionally we recommend cooling both shockable and nonshockable arrests if families are interested in pursuing aggressive care. A comparison of TTM in nonshockable rhythms seems to favor survival and improved neurologic outcomes in those cooled versus those not cooled (Perman et al., 2015; Sung et al., 2016).

Table 32.2

Summary of the 2015 American Heart Association–International Liaison Committee on Resuscitation (AHA-ILCOR) consensus advisory on the use of targeted temperature management in comatose survivors of cardiac arrest

2015 AHA-ILCOR advisory	Strength of recommendation	Quality of evidence
Recommend treatment with TTM as opposed to no TTM for comatose adult survivors of OHCA with VT/VF arrest	Strong	Low
Suggest TTM for comatose adult survivors of OHCA with asystole/PEA arrest	Weak	Low
Suggest TTM for comatose adult survivors of IHCA regardless of initial rhythm	Weak	Very low
Recommend constant target temperature between 32°C and 36°C when using TTM	Strong	Moderate
Recommend against routine prehospital initiation of cooling with large-volume infusions of cooled intravenous fluids	Strong	Moderate
Suggest that when using TTM target temperature be maintained at least 24 hours	–	–

Adapted from Donnino et al. (2015).
TTM, targeted temperature management; OHCA, out-of-hospital cardiac arrest; VT, ventricular tachycardia; VF, ventricular fibrillation; PEA, pulseless electric activity; IHCA, in-hospital cardiac arrest.

Clinically the most challenging decisions in the care of PCAS survivors, particularly those who have undergone TTM, are neurologic outcome prognostication and the methods and timing involved to reach these decisions. This is critical, as accurate early prediction of poor outcomes can result in less anguish for patient families, less suffering and more dignity for patients, and significant cost savings in both the acute care and posthospital care for societies. At this time, there is consensus on neurologic outcome prognostication for comatose survivors of cardiac arrest who have not undergone TTM (Wijdicks et al., 2006); however, there is much debate as to how and when to prognosticate in the setting of TTM. There is conceptual agreement in the resuscitation community that a multimodal predictive algorithm including clinical examination, neurophysiologic testing, imaging, and biomarkers should provide the best results, but beyond that there is little agreement or solid evidence. The merits of various modalities for neurologic outcome prognostication will be discussed in the following section.

OUTCOME PREDICTION

Clinical examination

The clinical examination remains a fundamental part of any neurologic prognostication algorithm. As noted above, the 2006 prediction of outcome in comatose survivors after cardiopulmonary resuscitation practice parameters of the American Academy of Neurology (AAN) provide specific recommendations for the prognostication of neurologic outcomes for cardiac arrest survivors (Wijdicks et al., 2006); however, these recommendations are based mainly on dated observations from the pre-TH era. The 2015 AHA guidelines make a clear distinction between the use of the physical exam in patients who have been treated with TTM and those who have not (Callaway et al., 2015). The timing of the exam in patients who have not undergone TTM is 72 hours from ROSC, provided that there is no concern for residual drug effects (Sandroni et al., 2013). In patients treated with TTM, the recommended timing for prediction of neurologic outcome based on clinical examination is 72 hours after rewarming is complete and provided residual sedation and paralysis are excluded (Sandroni et al., 2013; Mulder et al., 2014).

The traditional clinical examination findings for prediction of poor outcome (absent pupillary light reflex, absent bilateral corneal reflexes, and extensor posturing or no motor response to pain) have been re-evaluated in the setting of TTM (Al Thenayan et al., 2008; Fugate et al., 2010; Rossetti et al., 2010b; Bisschops et al., 2011; Samaniego et al., 2011; Bouwes et al., 2012c). Of the three, only the absence of pupillary light reflex at greater than 72 hours after arrest achieves an acceptable false-positive ratio (FPR 0%, 95% confidence interval (CI), 0–3%) (Callaway et al., 2015). Additionally, the presence of myoclonus, which traditionally had been considered a universally ominous finding, has been found to be less certain, particularly with a clear distinction being made between the presence of myoclonus and status myoclonus. The presence of status myoclonus, defined as continuous myoclonic jerks lasting more than 30 minutes, noted in the first 72 hours postarrest does still carry predictive weight for a poor outcome (FPR 0%, 95% CI, 0–4%), but this condition has to be carefully differentiated from other similar conditions (e.g., simple myoclonus, Lance–Adams myoclonic syndrome) (Callaway et al., 2015).

Blood-based biomarkers

As discussed earlier in the chapter, the laboratory tests for neurologic prognostication in PCAS are mainly NSE and S-100B. In the AAN guidelines, an NSE value >33 μg/L obtained within the first 72 hours is assigned an FPR of 0, with a 95% CI of 0–3% in patients treated with TH. A subsequent study (Steffen et al., 2010) has questioned the cutoff value in patients who have undergone TTM, where in order to have 100% specificity the cutoff needed to be raised to 78.9 μg/L. This study find a similar cutoff value to that stated by the AAN guidelines for patients who were not treated with TH.

Two other studies raise important concerns regarding the applications of NSE in neuroprognostication post-TH (Fugate et al., 2010; Bouwes et al., 2012c). Other studies of S-100B in serum and cerebrospinal fluid alone and in combination with other biomarkers have also been explored (Song et al., 2010; Oda et al., 2012; Rana et al., 2012), but have had little impact. The current AHA guidelines have stated that neither laboratory test nor combination thereof should be made to make a prediction of poor neurologic outcome; however, persistently elevated levels of NSE can be used to support a prediction of poor outcome made on the grounds of other modalities 72 hours or more postarrest (Callaway et al., 2015).

Neurophysiologic testing

Neurophysiologic testing refers primarily to the use of EEG and SSEPs to predict neurologic outcome in the PCAS. EEG has been extensively studied for the prognostication of neurologic outcomes in cardiac arrest (Legriel et al., 2009; Stammet et al., 2009; Wennervirta et al., 2009; Leary et al., 2010; Rossetti et al., 2010b; Rundgren et al., 2010; Cloostermans et al., 2012; Oh et al., 2012; Crepeau et al., 2013). The 2006 AAN practice parameters assign EEG an FPR of 3% with a 95% CI of 0.9–11%, making it the least predictive method to predict neurologic outcomes in its review. A pooled analysis of four existing studies (Abend et al., 2012) on EEG in PCAS patients who had undergone TH found that 29% of these patients had acute electrographic NCSE. This has important clinical repercussions and illustrates the need for continuous EEG monitoring of these patients until they recover consciousness, as aggressive antiepileptic treatment should be instated to avoid falling into self-fulfilling prophecies, equating NCSE to a poor outcome (Geocadin and Ritzl, 2012). This is important, as 6% of patients in the pooled sample recovered consciousness, including several with minimal residual neurologic deficits (Abend et al., 2012).

This study is in contrast to the findings of another study of continuous EEG in PCAS patients who underwent TH and in whom treatment of seizures did not seem to impact outcomes (Crepeau et al., 2013). However, this was a small retrospective study whose main goal was to validate the prognostic value of a 3-point EEG rating scale. At this time guidelines consider it reasonable to predict a poor outcome in comatose PCAS survivors treated with TTM based on the persistent absence of EEG reactivity to physical stimuli and on persistent burst suppression on EEG once the patient has been rewarmed (FPR 0%, 95% CI 0–3%). Additionally, intractable status epilepticus lasting greater than 72 hours accompanied by nonreactivity to stimuli may be reasonable to predict a poor outcome; though no FPR is given, it carries the same grade recommendation and level of evidence as the aforementioned recommendation (class IIb, B nonrandomized) (Callaway et al., 2015). Of note, it is important that clinicians make every possible effort in treating seizures and proving them to be truly intractable. In patients not treated with TTM, the same guidelines deem EEG burst suppression at 72 hours postarrest as a reasonable predictor (FPR 0%, 95% CI 0–11%), but to be used in combination with other predictors to prognosticate a poor neurologic outcome given the wide confidence interval.

The use of SSEPs, which had been previously considered one of the best predictors of poor outcome, now is considered to have a higher FPR and many caveats to consider in its application in patients treated with TTM (Tiainen et al., 2005; Bouwes et al., 2012b, c). A recent study comparing SSEP and continuous EEG found EEG to be superior in terms of its sensitivity to predict poor neurologic outcomes in PCAS treated with TH (Cloostermans et al., 2012). Another publication demonstrated that neurologic recovery is possible despite absent or minimally present median nerve N20 responses greater than 24 hours after cardiac arrest (Leithner et al., 2010). As is unfortunately the case in many studies of prognosticators in cardiac arrest victims, there is often a concern for self-fulfilling prophecies when the study is not appropriately blinded, and this is often the case in clinical practice as well. Guidelines have not changed with regard to the use of SSEPs in patients who have not undergone TTM, where bilaterally absent N20 responses at 24, 48, or 72 hours postarrest predict a poor outcome (FPR 0%, 95% CI 0–12%). With regard to patients treated with TTM, the AHA guidelines have been updated with bilateral absence of N20 SSEPs 24–72 hours postarrest, preferably after rewarming, to state that they are predictors of poor outcomes with FPR 1%, 95% CI 0–3%.

Neuroimaging

The use of imaging for the prediction of poor outcomes in the PCAS has seen a major update in the 2015 AHA guidelines, with imaging studies now being considered reasonable predictors in certain situations and in combination with other established predictors (Callaway et al., 2015). Imaging employed for prognostication of poor neurologic outcome in PCAS mainly takes the shape of brain CT, where loss of gray/white-matter differentiation, diffuse cerebral edema, and obvious infarction have been used to augment clinical prediction. More recently, quantitative measurements of signal change on both CT and MRI have attempted to improve the predictive abilities of imaging studies in the PCAS. Early studies in the 1980s characterized the "classic" findings of hypoxic-ischemic brain damage on CT in order of incidence (Kjos et al., 1983): diffuse cerebral edema with cortical sulcal and brainstem cistern effacement, decreased intensity of the cortical gray matter, loss of gray/white-matter differentiation, hypodensities in the basal ganglia, and appearance of watershed infarcts. Later studies have found correlation between time to ROSC and the incidence of sulcal effacement and loss of gray/white-matter differentiation (Inamasu et al., 2010). Subsequent studies have looked at signal changes in various regions of interest that include the putamen, posterior limb of the internal capsule, caudate, hippocampus, cerebellum, as well as midbrain and cortical structures with both CT and MRI (Wijman et al., 2009; Mlynash et al., 2010; Wu et al., 2011). The updated guidelines state that it may be reasonable for centers with expertise to use marked reductions in gray/white-matter ratios on CT scans obtained at least 2 hours after time of arrest in patients not treated with TTM to predict poor outcomes (Callaway et al., 2015) (Fig. 32.1). With respect to MRI, the same guidelines consider it is reasonable to consider marked diffusion restriction on MRI obtained 2–6 days postarrest (Fig. 32.2). It is the authors' belief that at this stage these tools simply can provide supporting information in an overall multimodal prognostication strategy; no decisions should be made based on imaging alone.

Multimodal prognostication

As emphasized by various guidelines, neurologic prognostication should not rely on a single predictor based on what we currently know (Cronberg et al., 2013; Callaway et al., 2015). The question that remains is

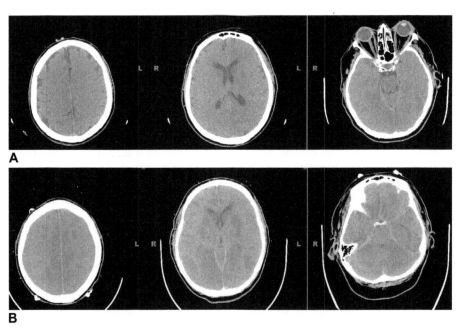

Fig. 32.1. Computed tomography (CT) imaging of hypoxic-ischemic encephalopathy. (**A**) CT scan obtained shortly after resuscitation from an out-of-hospital cardiac arrest. There is no evidence of diffuse cerebral edema; there are normal ventricles and cisterns as well as preservation of differentiation between gray and white matter. (**B**) CT scan obtained 3 days after resuscitation. There is significant diffuse cerebral edema with effacement of sulci and cisterns and narrowing of the lateral ventricles and effacement of the occipital horns. Widespread loss of gray–white differentiation and low attenuation of the basal ganglia are also noted. This CT scan was obtained due to the presence of status myoclonus refractory to treatment. This patient died after withdrawal of life-sustaining treatment based on family decisions and newly found advance directives from the patient.

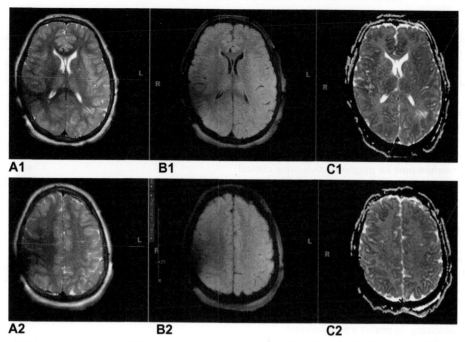

Fig. 32.2. Magnetic resonance imaging (MRI) of hypoxic-ischemic encephalopathy. (**A1, A2**) MRI T2; (**B1, B2**) MRI T2 fluid attenuation inversion recovery (FLAIR); (**C1, C2**) MRI apparent diffusion coefficient. MRI sequences on postarrest day number 5 showing diffuse cortical diffusion restriction, as well as diffuse cerebral edema with effacement of the overlying sulci. This MRI was obtained due to persistent coma with present corneal reflexes and cough and an otherwise poor exam, despite appropriate sedation washout period. Continuous electroencephalogram showed diffuse slowing; generalized epileptiform discharges and neuron-specific enolase levels were persistently elevated at 48 and 72 hours postarrest. Patient died after withdrawal of life-sustaining treatment on postarrest day 6 after a declaration of poor neurologic outcome.

which combination of tests can provide an early and unerring determination of poor prognosis. It would seem logical at this time to use as many predictors as available, to ensure that medication effects are accounted for and to wait at least 72 hours postrewarming before making a conclusive statement of neurologic prediction for patients in whom aggressive cares are warranted and this is in keeping with the patient's wishes (Table 32.3).

NEUROREHABILITATION

Relative to the amount of publications on the resuscitation and postresuscitation care of cardiac arrests, there is scarce literature pertaining to the neurorehabilitation of PCAS survivors. Early studies seem to echo clinical experience in the postresuscitation arena, in that the survivors of primary cardiac arrests progress more rapidly in their rehabilitation course than those who survive cardiac arrests not of a primary cardiac etiology, and the costs involved in the rehabilitation of the latter are also higher (Burke et al., 2005). Posthospital discharge of PCAS survivors evaluations show that approximately 25% of survivors suffer from anxiety and depression 6 months after discharge (Lilja et al., 2015), and a similar number demonstrated cognitive impairment. There are conflicting data regarding the impact of cognitive impairment and autonomy and perceived quality of life (Cronberg et al., 2009; Boyce-van der Wal et al., 2015).

There is a wide spectrum of outcomes from PCAS, where patients essentially return to their normal lives shortly after discharge, to patients surviving with profound disorders of consciousness, cognitive or physical impairment, psychosocial morbidity, and any combination thereof. It is interesting to note that, in a German study of patients referred to a rehabilitation facility with profound HIE, 6.2% eventually achieved good functional outcomes and many required up to 3 months to begin making significant progress (Howell et al., 2013). A recent randomized controlled trial in the Netherlands of neurologically focused follow-up showed promising improvements in quality of life, emotional, mental, and physical health as well as significant improvements in early return to work (Moulaert et al., 2015). Much remains to be learned, and programs such as those described need to be widely implemented to provide comprehensive and holistic care to PCAS survivors going forward.

Table 32.3

Considerations for neuroprognostication after cardiac arrest according to use of targeted temperature management

	Intervention	False-positive rate (FPR)	Recommendation and level of evidence (LOE)
TTM	The earliest time for prognostication using clinical examination in patients treated with TTM, where sedation or paralysis could be a confounder, is 72 hours after normothermia		Class IIb LOE C-EO
	The earliest time for prognostication using clinical examination in patients treated with TTM, where sedation or paralysis could be a confounder, is 72 hours after normothermia		Class IIb LOE C-EO
	In comatose patients who are treated with TTM, the absence of pupillary reflex to light at 72 hours or greater after arrest is a useful predictor of poor neurologic outcome	FPR, 1% 95% CI: 0–3%	Class I LOE B-NR
	In comatose postcardiac-arrest patients treated with TTM, it may be reasonable to consider persistent absence of EEG reactivity to external stimuli at 72 hours after cardiac arrest, and persistent burst suppression on EEG after rewarming, to predict a poor outcome	FPR, 0% 95% CI: 0–3%	Class IIb LOE B-NR
No TTM	The earliest time to prognosticate a poor neurologic outcome using clinical examination in patients not treated with TTM is 72 hours after cardiac arrest		Class I, LOE B-NR
	In patients who are comatose after resuscitation not treated with TTM, it may be reasonable to use the presence of a marked reduction of the GWR on brain CT obtained within 2 hours after cardiac arrest to predict poor outcome		Class IIb, LOE B-NR
	In resuscitated comatose patients who are not treated with TTM, the absence of pupillary reflex to light at 72 hours or more after cardiac arrest is a reasonable exam finding with which to predict a poor neurologic outcome	FPR, 0%; 95% CI: 0–8%	Class IIa, LOE B-NR
	In comatose postarrest patients not treated with TTM, it may be reasonable to consider the presence of burst suppression on EEG at 72 hours or more after cardiac arrest, in combination with other predictors, to predict a poor neurologic outcome	FPR, 0%; 95% CI: 0–11%	Class IIb, LOE B-NR
Both	These times until prognostication can be even longer than 72 hours after cardiac arrest if the residual effect of sedation or paralysis confounds the clinical examination		Class IIa LOE C-LD
	The motor examination may be a reasonable means to identify patients who need further prognostic testing to predict poor outcomes		Class IIb LOE B-NR
	Intractable and persistent (>72 hours) status epilepticus in the absence of EEG reactivity to external stimuli may be reasonable to predict poor outcome		Class IIb LOE B-NR
	It may be reasonable to consider extensive restriction of diffusion on brain MRI at 2–6 days after cardiac arrest in combination with other established predictors to predict a poor neurologic outcome		Class IIb LOE B-NR
	Given the possibility of high FPRs, blood levels of NSE and S-100B should not be used as single predictors of poor neurologic outcomes		Class III: Harm LOE C-LD
	When performed with other prognostic tests at >72 hours postarrest, it may be reasonable to consider high serum values of NSE at 48–72 hours after cardiac arrest to		* Class IIb LOE B-NR

Table 3.2
Continued

Intervention	False-positive rate (FPR)	Recommendation and level of evidence (LOE)
support the prognosis of a poor neurologic outcome,* especially if repeated sampling reveals persistently high values†		†Class IIb LOE C-LD
In comatose survivors of cardiac arrest regardless of treatment with TTM, it is reasonable to consider bilateral absence of N20 SSEPs 24–72 hours after arrest or rewarming a predictor of poor outcome	FPR, 1%; 95% CI: 0–3%	Class IIa LOE B-NR
Based on unacceptable FPRs, the findings of either absent motor movements or extensor posturing should not be used alone for predicting a poor neurologic outcome	FPR, 10%; 95% CI: 7–15% to FPR, 15%; 95% CI: 5–31%	Class III: Harm LOE B-NR
The presence of myoclonus, which is distinct from status myoclonus, should not be used to predict poor neurologic outcomes because of high FPR	FPR, 5%; 95% CI: 3–8% to FPR, 11%; 95% CI: 3–26%	Class III: Harm LOE B-NR
In combination with other diagnostic tests at >72 hours after arrest, the presence of status myoclonus during the first 72–120 hours after arrest is a reasonable finding to help predict poor neurologic outcomes	FPR, 0%; 95% CI: 0–4%	Class IIa LOE B-NR

Adapted from Callaway et al. (2015).
TTM, targeted temperature management; C-EO, FPR, false-positive ratio; CI, confidence interval; B-NR, EEG, electroencephalogram; GWR: gray- to white-matter ratio; CT, computed tomography; NSE, neuron-specific enolase; C-LD, S-100B, protein S-100B; MRI, magnetic resonance imaging; SSEPs, somatosensory evoked potentials.

References

Abella BS, Zhao D, Alvarado J et al. (2004). Intra-arrest cooling improves outcomes in a murine cardiac arrest model. Circulation 109 (22): 2786–2791.

Abend NS, Mani R, Tschuda TN et al. (2012). EEG monitoring during therapeutic hypothermia in neonates, children, and adults. Am J Electroneurodiagnostic Technol 51 (3): 1–20.

Adrie C, Adib-Conguy M, Laurent I et al. (2002). Successful cardiopulmonary resuscitation after cardiac arrest as a "sepsis-like" syndrome. Circulation 106 (5): 562–568.

Adrie C, Monchi M, Laurent I et al. (2005). Coagulopathy after successful cardiopulmonary resuscitation following cardiac arrest: implication of the protein C anticoagulant pathway. J Am Coll Cardiol 46 (1): 21–28.

Al Thenayan E, Savard M, Sharpe M et al. (2008). Predictors of poor neurologic outcome after induced mild hypothermia following cardiac arrest. Neurology 71 (19): 1535–1537.

American Heart Association (2005). American Heart Association guidelines for cardiopulmonary resuscitation and emergency cardiovascular care part 7.5: postresuscitation support. Circulation 112 (24 suppl.): 84–89.

Ames 3rd A, Wright RL, Kowada M et al. (1968). Cerebral ischemia. II The no-reflow phenomenon. Am J Pathol 52 (2): 437–453.

Amiry-Moghaddam M, Otsuka T, Hurn PD et al. (2003). An alpha-syntrophin-dependent pool of AQP4 in astroglial end-feet confers bidirectional water flow between blood and brain. Proc Natl Acad Sci U S A 100 (4): 2106–2111.

Arrich J European Resuscitation Council Hypothermia After Cardiac Arrest Registry Study Group (2007). Clinical application of mild therapeutic hypothermia after cardiac arrest. Crit Care Med 35 (4): 1041–1047.

Badjatia N (2009). Hyperthermia and fever control in brain injury. Crit Care Med 37 (Suppl): S250–S257.

Badjatia N, Strongilis E, Gordon E et al. (2008). Metabolic impact of shivering during therapeutic temperature modulation: the Bedside Shivering Assessment Scale. Stroke 39 (12): 3242–3247.

Barrett KM, Freeman WD, Windling SM et al. (2007). Brain injury after cardiopulmonary arrest and its assessment with diffusion-weighted magnetic resonance imaging. Mayo Clin Proc 82 (7): 828–835.

Belousov AB (2012). Novel model for the mechanisms of glutamate-dependent excitotoxicity: role of neuronal gap junctions. Brain Res 1487: 123–130.

Benz-Woerner J, Delodder F, Benz R et al. (2012). Body temperature regulation and outcome after cardiac arrest and therapeutic hypothermia. Resuscitation 83 (3): 338–342.

Bergsholm P, Gran L, Bleie H (1984). Seizure duration in unilateral electroconvulsive therapy. The effect of hypocapnia induced by hyperventilation and the effect of ventilation with oxygen. Acta Psychiatr Scand 69 (2): 121–128.

Bernard SA, Gray TW, Buist MD et al. (2002). Treatment of comatose survivors of out-of-hospital cardiac arrest with induced hypothermia. New Engl J Med 346 (8): 557–563.

Bernard SA, Smith K, Cameron P et al. (2010). Induction of therapeutic hypothermia by paramedics after resuscitation from out-of-hospital ventricular fibrillation cardiac arrest: a randomized controlled trial. Circulation 122 (7): 737–742.

Bernard SA, Smith K, Cameron P et al. (2012). Induction of prehospital therapeutic hypothermia after resuscitation from nonventricular fibrillation cardiac arrest. Crit Care Med 40 (3): 747–753.

Bisschops LL, van Alfen N, Bons S et al. (2011). Predictors of poor neurologic outcome in patients after cardiac arrest treated with hypothermia: a retrospective study. Resuscitation 82 (6): 696–701.

Bjelland TW, Dale O, Kaisen K et al. (2012). Propofol and remifentanil versus midazolam and fentanyl for sedation during therapeutic hypothermia after cardiac arrest: a randomised trial. Intensive Care Med 38 (6): 959–967.

Bouwes A, van Poppelen D, Koelman JH et al. (2012a). Acute posthypoxic myoclonus after cardiopulmonary resuscitation. BMC Neurol 12 (63): 63.

Bouwes A, Binnekade JM, Verbaan BW et al. (2012b). Predictive value of neurological examination for early cortical responses to somatosensory evoked potentials in patients with postanoxic coma. J Neurol 259 (3): 537–541.

Bouwes A, Binnekade JM, Kuiper M et al. (2012c). Prognosis of coma after therapeutic hypothermia: a prospective cohort study. Ann Neurol 71 (2): 206–212.

Boyce-van der Wal LW, Volker WG, Vliet Vlieand TP et al. (2015). Cognitive problems in patients in a cardiac rehabilitation program after an out-of-hospital cardiac arrest. Resuscitation 93: 63–68.

Brady W, Meldon S, DeBehnke D (1995). Comparison of prehospital monomorphic and polymorphic ventricular tachycardia: prevalence, response to therapy, and outcome. Ann Emerg Med 25 (1): 64–70.

Brady WJ, DeBehnke DJ, Laundrie D (1999). Prevalence, therapeutic response, and outcome of ventricular tachycardia in the out-of-hospital setting: a comparison of ventricular tachycardia, polymorphic ventricular tachycardia, and torsades de pointes. Acad Emerg Med 6 (6): 609–617.

Brady WJ, Gurka KK, Mehring B et al. (2011). In-hospital cardiac arrest: impact of monitoring and witnessed event on patient survival and neurologic status at hospital discharge. Resuscitation 82 (7): 845–852.

Brierley JB, Graham DI, Adams JH et al. (1971). Neocortical death after cardiac arrest. The Lancet 298 (7724): 560–565.

Bro-Jeppesen J, Hassager C, Wanscher M et al. (2013). Post-hypothermia fever is associated with increased mortality after out-of-hospital cardiac arrest. Resuscitation 84 (12): 1734–1740.

Bruel C, Parienti JJ, Marie W et al. (2008). Mild hypothermia during advanced life support: a preliminary study in out-of-hospital cardiac arrest. Critical Care (London, England) 12 (1): R31.

Burke DT, Shah MK, Dorvlo AS et al. (2005). Rehabilitation outcomes of cardiac and non-cardiac anoxic brain injury: a single institution experience. Brain Inj 19 (9): 675–680.

Callaway CW, Tadler SC, Katz LM et al. (2002). Feasibility of external cranial cooling during out-of-hospital cardiac arrest. Resuscitation 52 (2): 159–165.

Callaway CW, Donnino MW, Fink EL et al. (2015). Part 8: Post-cardiac arrest care: 2015 American Heart Association guidelines update for cardiopulmonary resuscitation and emergency cardiovascular care. Circulation 132 (18): S465–S482.

Carlson DW, Pearson RD, Haggerty PF et al. (2012). Commotio cordis, therapeutic hypothermia, and evacuation from a United States military base in Iraq. J Emerg Med 1–5 (January).

Castrén M, Nordberg P, Svensson L et al. (2010). Intra-arrest transnasal evaporative cooling: a randomized, prehospital, multicenter study (PRINCE: Pre-ROSC IntraNasal Cooling Effectiveness). Circulation 122 (7): 729–736.

Chamorro C, Borralio JM, Romera MA et al. (2010). Anesthesia and analgesia protocol during therapeutic hypothermia after cardiac arrest: a systematic review. Anesth Analg 110 (5): 1328–1335.

Choi D (1994). Calcium and excitotoxic neuronal injury. Ann N Y Acad Sci 747: 162–171.

Choi HA, Ko SB, Presciutti M et al. (2011). Prevention of shivering during therapeutic temperature modulation: the Columbia anti-shivering protocol. Neurocrit Care 14 (3): 389–394.

Chua H-R, Glassford N, Bellomo R (2012). Acute kidney injury after cardiac arrest. Resuscitation 83 (6): 721–727.

Cloostermans MC, Van Meulen FB, Eertman CJ et al. (2012). Continuous electroencephalography monitoring for early prediction of neurological outcome in postanoxic patients after cardiac arrest: a prospective cohort study. Crit Care Med 40 (10): 2867–2875.

Cocchi MN, Lucas JM, Salciccioli J et al. (2010). The role of cranial computed tomography in the immediate post-cardiac arrest period. Intern Emerg Med 5 (6): 533–538.

Crepeau AZ, Rabinstein AA, Fugate JE et al. (2013). Continuous EEG in therapeutic hypothermia after cardiac arrest: prognostic and clinical value. Neurology 80 (4): 339–344.

Cronberg T, Lilia G, Rundgren M et al. (2009). Long-term neurological outcome after cardiac arrest and therapeutic hypothermia. Resuscitation 80 (10): 1119–1123.

Cronberg T, Brizzi M, Liedholm LJ et al. (2013). Neurological prognostication after cardiac arrest – recommendations from the Swedish Resuscitation Council. Resuscitation 84 (7): 867–872.

Cureton EL, Young LY, Kwan RO et al. (2012). The heart of the matter: utility of ultrasound of cardiac activity during traumatic arrest. J Trauma Acute Care Surg 73 (1): 102–110.

Czosnyka M, Brady K, Reinhard M et al. (2009). Monitoring of cerebrovascular autoregulation: facts, myths, and missing links. Neurocrit Care 10 (3): 373–386.

de Jong MF, Beishuizen A, de Jong MJ et al. (2008). The pituitary–adrenal axis is activated more in non-survivors than in survivors of cardiac arrest, irrespective of therapeutic hypothermia. Resuscitation 78 (3): 281–288.

de Vos R, de Haes HC, Koster RW et al. (1999). Quality of survival after cardiopulmonary resuscitation. Arch Intern Med 159 (3): 249–254.

de Waard MC, Biermann H, Brinckman SL et al. (2013). Automated peritoneal lavage: an extremely rapid and safe way to induce hypothermia in post-resuscitation patients. Crit Care 17 (1): R31.

Deasy C, Bernard S, Cameron P et al. (2011). Design of the RINSE trial: the rapid infusion of cold normal saline by paramedics during CPR. BMC Emerg Med 11 (1): 17.

Deo R, Albert CM (2012). Epidemiology and genetics of sudden cardiac death. Circulation 125 (4): 620–637.

Donadello K, Favory R, Salgado-Ribeiro D et al. (2011). Sublingual and muscular microcirculatory alterations after cardiac arrest: a pilot study. Resuscitation 82 (6): 690–695.

Donnino MW, Anderson LW, Berg KM et al. (2015). Temperature management after cardiac arrest. Circulation 132 (25): 2448–2456.

Douglas IJ, Evans SJ, Hingorani AD et al. (2012). Clopidogrel and interaction with proton pump inhibitors: comparison between cohort and within person study designs. BMJ 345: 1–14, e4388.

Dragancea I, Rundgren M, Englund E et al. (2012). The influence of induced hypothermia and delayed prognostication on the mode of death after cardiac arrest. Resuscitation: 6–11.

Finley Caulfield A, Rachabattula S, Eyngorn I et al. (2011). A comparison of cooling techniques to treat cardiac arrest patients with hypothermia. Stroke Research and Treatment 2011: 690506.

Fishman GI, Chugh SS, Dimarco JP et al. (2010). Sudden cardiac death prediction and prevention: report from a National Heart, Lung, and Blood Institute and Heart Rhythm Society Workshop. Circulation 122 (22): 2335–2348.

Flemming K, Simonis G, Ziegs E et al. (2006). Comparison of external and intravascular cooling to induce hypothermia in patients after CPR. Ger Med Sci 4: 31–34.

Fugate JE, Wijdicks EF, Mandrekar J et al. (2010). Predictors of neurologic outcome in hypothermia after cardiac arrest. Ann Neurol 68 (6): 907–914.

Fujioka M, Okuchi K, Sakaki T et al. (1994). Specific changes in human brain following reperfusion after cardiac arrest. Stroke 25 (10): 2091–2095.

Futier E, Chanques G, Cayot Constantin S et al. (2012). Influence of opioid choice on mechanical ventilation duration and ICU length of stay. Minerva Anestesiol 78 (1): 46–53.

Garrett JS, Studnek JR, Blackwell T et al. (2011). The association between intra-arrest therapeutic hypothermia and return of spontaneous circulation among individuals experiencing out of hospital cardiac arrest. Resuscitation 82 (1): 21–25.

Giacino JT, Ashwal S, Childs N et al. (2002). The minimally conscious state: definition and diagnostic criteria. Neurology 58 (3): 349–353; PMID: 11839831.

Geocadin RG, Ritzl EK (2012). Seizures and status epilepticus in post cardiac arrest syndrome: therapeutic opportunities to improve outcome or basis to withhold life sustaining therapies? Resuscitation 83 (7): 791–792.

Gerdes JS, Walther EU, Jaganjac S et al. (2014). Early detection of widespread progressive brain injury after cardiac arrest: a single case DTI and post-mortem histology study. PLoS One 9 (3): 1–5.

Girotra S, Nallamothu BK, Spertus JA et al. (2012). Trends in survival after in-hospital cardiac arrest. N Engl J Med 367 (20): 1912–1920.

Globus M, Ginsberg M, Busto R (1991). Excitotoxic index – a biochemical marker of selective vulnerability. Neurosci Lett 127 (1): 39–42.

Goodman SG, Clare R, Pieper KS et al. (2012). Association of proton pump inhibitor use on cardiovascular outcomes with clopidogrel and ticagrelor: insights from the platelet inhibition and patient outcomes trial. Circulation 125 (8): 978–986.

Gordan ML, Kellemann K, Blobner M et al. (2010). Fast rewarming after deep hypothermic circulatory arrest in rats impairs histologic outcome and increases NFκB expression in the brain. Perfusion 25 (5): 349–354.

Greer DM (2006). Mechanisms of injury in hypoxic-ischemic encephalopathy: implications to therapy. Semin Neurol 26 (4): 373–379.

Grimaldi D, Guivarch E, Neveux N et al. (2013). Markers of intestinal injury are associated with endotoxemia in successfully resuscitated patients. Resuscitation 84 (1): 60–65.

Gundersen RY, Vaagenes P, Breivik T et al. (2005). Glycine – an important neurotransmitter and cytoprotective agent. Acta Anaesthesiol Scand 49 (8): 1108–1116.

Gwinnutt CL, Columb M, Harris R (2000). Outcome after cardiac arrest in adults in UK hospitals: effect of the 1997 guidelines. Resuscitation 47 (2): 125–135.

Hachimi-Idrissi S, Corne L, Ebinger G et al. (2001). Mild hypothermia induced by a helmet device: a clinical feasibility study. Resuscitation 51 (3): 275–281.

Hamel MB, Phillips R, Teno J et al. (2002). Cost effectiveness of aggressive care for patients with nontraumatic coma. Crit Care Med 30 (6): 1191–1196.

Harukuni I, Bhardwaj A (2006). Mechanisms of brain injury after global cerebral ischemia. Neurol Clin 24 (1): 1–21.

Hata JS, Shelsky CR, Hindman BJ et al. (2008). A prospective, observational clinical trial of fever reduction to reduce systemic oxygen consumption in the setting of acute brain injury. Neurocrit Care 9 (1): 37–44.

Haugk M, Sterz F, Grassberger M et al. (2007). Feasibility and efficacy of a new non-invasive surface cooling device in post-resuscitation intensive care medicine. Resuscitation 75 (1): 76–81.

Heard KJ, Peberdy MA, Sayre MR et al. (2010). A randomized controlled trial comparing the arctic sun to standard cooling for induction of hypothermia after cardiac arrest. Resuscitation 81 (1): 9–14.

Hillis M, Sinclair D, Butler G et al. (1993). Prehospital cardiac arrest survival and neurologic recovery. J Emerg Med 11: 245–252.

Hoesch RE, Koenig MA, Geocadin RG (2008). Coma after global ischemic brain injury: pathophysiology and emerging therapies. Crit Care Clin 24 (1): 25–44.

Holmgren C, Bergfeldt L, Edvardsson N et al. (2010). Analysis of initial rhythm, witnessed status and delay to treatment among survivors of out-of-hospital cardiac arrest in Sweden. Heart 96 (22): 1826–1830.

Holzer M, Bernard SA, Hachimi-Idrissi S et al. (2005). Hypothermia for neuroprotection after cardiac arrest: systematic review and individual patient data meta-analysis. Crit Care Med 33 (2): 414–418.

Howell K, Grill E, Klein AM et al. (2013). Rehabilitation outcome of anoxic-ischaemic encephalopathy survivors with prolonged disorders of consciousness. Resuscitation 84 (10): 1409–1415.

Hoy SM, Keating GM (2011). Dexmedetomidine: a review of its use for sedation in mechanically ventilated patients in an intensive care setting and for procedural sedation. Drugs 71 (11): 1481–1501.

Huang C-H, Chen WJ, Ma MH et al. (2002). Factors influencing the outcomes after in-hospital resuscitation in Taiwan. Resuscitation 53 (3): 265–270.

Huff JS, Stevens RD, Weingart SD et al. (2012). Emergency neurological life support: approach to the patient with coma. Neurocrit Care 17 (Suppl 1): S54–S59.

Hypothermia after Cardiac Arrest Study Group (2002). Mild therapeutic hypothermia to improve the neurologic outcome after cardiac arrest. N Engl J Med 346 (8): 549–556.

Inamasu J, Miyatake S, Tomioka H et al. (2009). Subarachnoid haemorrhage as a cause of out-of-hospital cardiac arrest: a prospective computed tomography study. Resuscitation 80 (9): 977–980.

Inamasu J, Miyatake S, Suzuki M et al. (2010). Early CT signs in out-of-hospital cardiac arrest survivors: temporal profile and prognostic significance. Resuscitation 81 (5): 534–538.

International Liaison Committee on Resuscitation (2005). 2005 International Consensus on Cardiopulmonary Resuscitation and Emergency Cardiovascular Care Science with Treatment Recommendations. Part 2: Adult basic life support. Resuscitation 67 (2-3): 187–201.

Iyer VN, Mandrekar JN, Danielson RD et al. (2009). Validity of the FOUR score coma scale in the medical intensive care unit. Mayo Clin Proc 84 (8): 694–701.

Janz DR, Hollenbeck RD, Pollock JS et al. (2012). Hyperoxia is associated with increased mortality in patients treated with mild therapeutic hypothermia after sudden cardiac arrest. Crit Care Med 40 (12): 3135–3139.

Jia X, Koenig MA, Nickl R et al. (2008a). Early electrophysiologic markers predict functional outcome associated with temperature manipulation after cardiac arrest in rats. Crit Care Med 36 (6): 1909–1916.

Jia X, Koenig MA, Venkatraman A et al. (2008b). Post-cardiac arrest temperature manipulation alters early EEG bursting in rats. Resuscitation 78 (3): 367–373.

Kaindl AM, Degos V, Peineau S et al. (2012). Activation of microglial N-methyl-D-aspartate receptors triggers inflammation and neuronal cell death in the developing and mature brain. Ann Neurol 72 (4): 536–549.

Kämäräinen A, Virkkunen I, Tenhunen J et al. (2008). Induction of therapeutic hypothermia during prehospital CPR using ice-cold intravenous fluid. Resuscitation 79 (2): 205–211.

Kataoka K, Yanase H (1998). Mild hypothermia – a revived countermeasure against ischemic neuronal damages. Neurosci Res 32 (2): 103–117.

Katsoulis IE, Balanika A, Sakalidou M et al. (2012). Extensive colonic necrosis following cardiac arrest and successful cardiopulmonary resuscitation: report of a case and literature review. World Journal of Emergency Surgery : WJES 7 (1): 35.

Kazaure HS, Roman SA, Sosa JA (2013). Epidemiology and outcomes of in-hospital cardiopulmonary resuscitation in the United States, 2000–2009. Resuscitation 84 (9): 1255–1260.

Khot S, Tirschwell DL (2006). Long-term neurological complications after hypoxic-ischemic encephalopathy. Semin Neurol 4 (26): 422–431.

Kilgannon JH, Jones AE, Shapiro NI et al. (2010). Association between arterial hyperoxia following resuscitation from cardiac arrest and in-hospital mortality. JAMA 303 (21): 2165–2171.

Kilgannon JH, Jones AE, Parrillo JE et al. (2011). Relationship between supranormal oxygen tension and outcome after resuscitation from cardiac arrest. Circulation 123 (23): 2717–2722.

Kim EY, Yang HJ, Sung YM et al. (2011a). Multidetector CT findings of skeletal chest injuries secondary to cardiopulmonary resuscitation. Resuscitation 82 (10): 1285–1288.

Kim JJ, Yang HJ, Lim YS et al. (2011b). Effectiveness of each target body temperature during therapeutic hypothermia after cardiac arrest. Am J Emerg Med 29 (2): 148–154.

Kim JJ, Hyun SY, Hwang SY et al. (2011c). Hormonal responses upon return of spontaneous circulation after cardiac arrest: a retrospective cohort study. Crit Care 15 (1): R53.

Kim F, Nichol G, Maynard C et al. (2013). Effect of prehospital induction of mild hypothermia on survival and neurological status among adults with cardiac arrest: a randomized clinical trial. JAMA 98104: 1–8.

Kimberger O, Ali SZ, Markstaller M et al. (2007). Meperidine and skin surface warming additively reduce the shivering threshold: a volunteer study. Crit Care 11 (1): R29.

Kjos BO, Brant-Zawadzki M, Young RG (1983). Early CT findings of global central nervous system hypoperfusion. AJR Am J Roentgenol 141 (6): 1227–1232.

Koenig M, Bryan M, Lewin 3rd JL et al. (2008). Reversal of transtentorial herniation with hypertonic saline. Neurology 70 (13): 1023–1029.

Kollmar R, Georgiadis D, Schwab S (2009). Alpha-stat versus pH-stat guided ventilation in patients with large ischemic stroke treated by hypothermia. Neurocrit Care 10 (2): 173–180.

Kolte D, Khera S, Aronow WS et al. (2015). Regional variation in the incidence and outcomes of in-hospital cardiac arrest in the United States. Circulation 131 (16): 1415–1425.

Korth U, Krieter H, Denz C et al. (2003). Intestinal ischaemia during cardiac arrest and resuscitation: comparative analysis of extracellular metabolites by microdialysis. Resuscitation 58 (2): 209–217.

Kuboyama K, Safar P, Radovsky A et al. (1993). Delay in cooling negates the beneficial effect of mild resuscitative cerebral hypothermia after cardiac arrest in dogs: a prospective, randomized study. Crit Care Med 21 (9): 1348–1358.

Kuisma M, Boyd J, Voipio V et al. (2006). Comparison of 30 and the 100% inspired oxygen concentrations during early post-resuscitation period: a randomised controlled pilot study. Resuscitation 69 (2): 199–206.

Kurisu S, Inoue I, Kawagoe T et al. (2010). Tako-tsubo cardiomyopathy after successful resuscitation of out-of-hospital cardiac arrest. J Cardiovasc Med 11 (6): 465–468.

Kürkciyan I, Meron G, Sterz F et al. (2001). Spontaneous subarachnoid haemorrhage as a cause of out-of-hospital cardiac arrest. Resuscitation 51 (1): 27–32.

Laish-Farkash A, Matetzky S, Oieru D et al. (2011). Usefulness of mild therapeutic hypothermia for hospitalized comatose patients having out-of-hospital cardiac arrest. Am J Cardiol 108 (2): 173–178.

Laurent I, Monchi M, Chiche JD et al. (2002). Reversible myocardial dysfunction in survivors of out-of-hospital cardiac arrest. J Am Coll Cardiol 40 (12): 2110–2116.

Leary M, Fried DA, Gaieski DF et al. (2010). Neurologic prognostication and bispectral index monitoring after resuscitation from cardiac arrest. Resuscitation 81 (9): 1133–1137.

Leary M, Grossestreuer AV, Iannacone S et al. (2012). Pyrexia and neurologic outcomes after therapeutic hypothermia for cardiac arrest. Resuscitation: 7–9.

Legriel S, Bruneel F, Sediri H et al. (2009). Early EEG monitoring for detecting postanoxic status epilepticus during therapeutic hypothermia: a pilot study. Neurocrit Care 11 (3): 338–344.

Leithner C, Ploner CJ, Hasper D et al. (2010). Does hypothermia influence the predictive value of bilateral absent N20 after cardiac arrest? Neurology 74 (12): 965–969.

Lilja G, Nilsson G, Nielsen N et al. (2015). Anxiety and depression among out-of-hospital cardiac arrest survivors. Resuscitation 97: 68–75.

Lipton SA, Rosenberg PA (1994). Excitatory amino acids as a final common pathway for neurologic disorders. N Engl J Med 330 (9): 613–622.

Lo Pizzo M, Sciera G, Di Liegro I et al. (2013). Aquaporin-4 distribution in control and stressed astrocytes in culture and in the cerebrospinal fluid of patients with traumatic brain injuries. Neurol Sci 34: 1309–1314.

Loo CK, Kaill A, Paton P et al. (2010). The difficult-to-treat electroconvulsive therapy patient – strategies for augmenting outcomes. J Affect Disord 124 (3): 219–227.

Lopez-de-Sa E, Rey JR, Armada E et al. (2012). Hypothermia in comatose survivors from out-of-hospital cardiac arrest: pilot trial comparing 2 levels of target temperature. Circulation 126 (24): 2826–2833.

Lyden P, Ernstrom K, Cruz-Flores S et al. (2012). Determinants of effective cooling during endovascular hypothermia. Neurocrit Care 16 (3): 413–420.

Malta Hansen C, Kragholm K, Pearson DA et al. (2015). Association of bystander and first-responder intervention with survival after out-of-hospital cardiac arrest in North Carolina, 2010–2013. JAMA 314 (3): 255–264.

Mani R, Schmitt SE, Mazer M et al. (2012). The frequency and timing of epileptiform activity on continuous electroencephalogram in comatose post-cardiac arrest syndrome patients treated with therapeutic hypothermia. Resuscitation 83 (7): 840–847.

Margey R, Browne L, Murphy E et al. (2011). The Dublin cardiac arrest registry: temporal improvement in survival from out-of-hospital cardiac arrest reflects improved pre-hospital emergency care. Europace 13 (8): 1157–1165.

McManus DD, Aslam F, Goyal P et al. (2012). Incidence, prognosis, and factors associated with cardiac arrest in patients hospitalized with acute coronary syndromes (the Global Registry of Acute Coronary Events Registry). Coron Artery Dis 23 (2): 105–112.

McNally B, Robb R, Mehta M et al. (2009). CARES: Cardiac Arrest Registry to Enhance Survival. Ann Emerg Med 54 (5): 674–683.e2.

Meaney PA, Nadkarni VM, Kern KB et al. (2010). Rhythms and outcomes of adult in-hospital cardiac arrest. Crit Care Med 38 (1): 101–108.

Mehta B, Briggs DK, Sommers SC et al. (1972). Disseminated intravascular coagulation following cardiac arrest: a study of 15 patients. Am J Med Sci 264 (5): 353–363.

Merchant RM, Yang L, Becker LB et al. (2011). Incidence of treated cardiac arrest in hospitalized patients in the United States. Crit Care Med 39 (11): 2401–2406.

Merchant RM, Yang L, Becker LB et al. (2012). Variability in case-mix adjusted in-hospital cardiac arrest rates. Med Care 50 (2): 124–130.

Mlynash M, Campbell DM, Leproust EM et al. (2010). Temporal and spatial profile of brain diffusion-weighted MRI after cardiac arrest. Stroke 41 (8): 1665–1672.

Mokhtarani M, Mahgoub AN, Morioka N et al. (2001). Buspirone and meperidine synergistically reduce the shivering threshold. Anesth Analg 93 (5): 1233–1239.

Mongardon N, Dumas F, Ricome S et al. (2011). Postcardiac arrest syndrome: from immediate resuscitation to long-term outcome. Ann Intensive Care 1 (1): 45.

Moomiaie RM, Gould G, Solomon D et al. (2012). Novel intracranial brain cooling catheter to mitigate brain injuries. Journal of Neurointerventional Surgery 4 (2): 130–133.

Moore LE, Traystman RJ (1994). Role of oxygen free radicals and lipid peroxidation in cerebral reperfusion injury. Adv Pharmacol 31: 565–576.

Moulaert VRM, van Heugten CM, Winkens B et al. (2015). Early neurologically-focused follow-up after cardiac arrest improves quality of life at one year: a randomised controlled trial. Int J Cardiol 193: 8–16.

Mozaffarian D, Benjamin EJ, Go AS et al. (2016). Heart disease and stroke statistics – 2016 update: a report from the American Heart Association. Circulation http:/dx.doi.org/10.1161/CIR.0000000000000350.

Mulder M, Gibbs HG, Smith SW et al. (2014). Awakening and withdrawal of life-sustaining treatment in cardiac arrest survivors treated with therapeutic hypothermia. Crit Care Med 42 (12): 2493–2499.

Nadkarni VM, Larkin GL, Peberdy MA et al. (2006). First documented rhythm and clinical outcome from in-hospital cardiac arrest among children and adults. JAMA 295 (1): 50–57.

Nakamura K, Kariyazono H, Shinakawa T et al. (1999). Inhibitory effects of H2-receptor antagonists on platelet function in vitro. Hum Exp Toxicol 18 (8): 487–492.

Naples R, Ellison E, Brady WJ (2009). Cranial computed tomography in the resuscitated patient with cardiac arrest. Am J Emerg Med 27 (1): 63–67.

Needham D, Colantuoni E, Mendez-Tellez PA et al. (2012). Lung protective mechanical ventilation and two year survival in patients with acute lung injury: prospective cohort study. BMJ 344, e2124.

Negovsky VA (1972). The second step in resuscitation – the treatment of the "post-resuscitation disease". Resuscitation 1 (1): 1–7.

Neumar RW, Nolan JP, Adrie C et al. (2008). Post-cardiac arrest syndrome: epidemiology, pathophysiology, treatment, and prognostication. A consensus statement from the International Liaison Committee on Resuscitation. Circulation 118 (23): 2452–2483.

Nichol G, Thomas E, Callaway CW et al. (2008). Regional variation in out-of-hospital cardiac arrest incidence and outcome. JAMA 300 (12): 1423–1431.

Nielsen N, Sunde K, Hovdenes J et al. (2011). Adverse events and their relation to mortality in out-of-hospital cardiac arrest patients treated with therapeutic hypothermia. Crit Care Med 39 (1): 57–64.

Nielsen N, Wetterslev J, Cronberg T et al. (2013). Targeted temperature management at 33°C versus 36°C after cardiac arrest. N Engl J Med 369 (23): 2197–2206.

Nolan JP, Morley PT, Vanden Hoek TL et al. (2003). Therapeutic hypothermia after cardiac arrest: an advisory statement by the Advanced Life Support Task Force of the International Liaison Committee on Resuscitation. Circulation 108: 118–121.

Nolan JP, Laver SR, Welch CA et al. (2007). Outcome following admission to UK intensive care units after cardiac arrest: a secondary analysis of the ICNARC Case Mix Programme Database. Anaesthesia 62 (12): 1207–1216.

Nordmark J, Enblad P, Rubertsson S (2009). Cerebral energy failure following experimental cardiac arrest hypothermia treatment reduces secondary lactate/pyruvate-ratio increase. Resuscitation 80 (5): 573–579.

Norton L, Hutchison RM, Young GB et al. (2012). Disruptions of functional connectivity in the default mode network of comatose patients. Neurology 78 (3): 175–181.

Nozari A, Safar P, Stezoski SW et al. (2006). Critical time window for intra-arrest cooling with cold saline flush in a dog model of cardiopulmonary resuscitation. Circulation 113 (23): 2690–2696.

Nürnberger A, Sterz F, Malzer R et al. (2013). Out of hospital cardiac arrest in Vienna: incidence and outcome. Resuscitation 84: 42–47.

Oda Y, Tsuruta R, Fujita M et al. (2012). Prediction of the neurological outcome with intrathecal high mobility group box 1 and S100B in cardiac arrest victims: a pilot study. Resuscitation 83 (8): 1006–1012.

Ogilvie BW, Yerino P, Kazmi F et al. (2011). The proton pump inhibitor, omeprazole, but not lansoprazole or pantoprazole, is a metabolism-dependent inhibitor of CYP2C19 : implications for coadministration with clopidogrel. Drug Metabol Dipos 39 (11): 2020–2033.

Oh SH, Park KN, Kim YM et al. (2012). The prognostic value of continuous amplitude-integrated electroencephalogram applied immediately after return of spontaneous circulation in therapeutic hypothermia-treated cardiac arrest patients. Resuscitation 84 (2): 200–205.

Olasveengen TM, Sunde K, Brunborg C et al. (2009). Intravenous drug administration during out-of-hospital cardiac arrest. JAMA 302 (20): 2222–2229.

Oren-Grinberg A, Gulati G, Fuchs L et al. (2012). Echo rounds: hand-held echocardiography in the management of cardiac arrest. Anesth Analg 115 (5): 1038–1041.

Oshima C, Kaneko T, Tsuruta R et al. (2010). Increase in plasma glucagon, a factor in hyperglycemia, is related to neurological outcome in postcardiac-arrest patients. Resuscitation 81 (2): 187–192.

Paine MG, Che D, Li L et al. (2012). Cerebellar Purkinje cell neurodegeneration after cardiac arrest: effect of therapeutic hypothermia. Resuscitation 83 (12): 1511–1516.

Patel SB, Kress JP (2012). Sedation and analgesia in the mechanically ventilated patient. Am J Respir Crit Care Med 185 (5): 486–497.

Peberdy MA, Ornato JP, Larkin GL et al. (2008). Survival from in-hospital cardiac arrest during nights and weekends. JAMA 299 (7): 785–792.

Peberdy MA, Callaway CW, Neumar RW et al. (2010). Part 9: post-cardiac arrest care: 2010 American Heart Association guidelines for cardiopulmonary resuscitation and emergency cardiovascular care. Circulation 122 (18 Suppl 3): S768–S786.

Peberdy MA, Callaway CW, Neumar RW, Geocadin RG, Zimmerman JL, Donnino M et al. (2010 Nov 2). Part 9: post-cardiac arrest care: 2010 American Heart Association Guidelines for Cardiopulmonary Resuscitation and Emergency Cardiovascular Care. Circulation 122 (18 Suppl 3): S768–S786..

Perman SM, Grossestreuer AV, Wiebe DJ et al. (2015). The utility of therapeutic hypothermia for post-cardiac arrest syndrome patients with an initial nonshockable rhythm. Circulation 132 (22): 2146–2151.

Pichon N, Amiel JB, Francois B et al. (2007). Efficacy of and tolerance to mild induced hypothermia after out-of-hospital cardiac arrest using an endovascular cooling system. Crit Care 11 (3): R71.

Polderman KH (2009). Mechanisms of action, physiological effects, and complications of hypothermia. Crit Care Med 37 (7 Suppl): S186–S202.

Polderman KH, Herold I (2009). Therapeutic hypothermia and controlled normothermia in the intensive care unit:

practical considerations, side effects, and cooling methods. Crit Care Med 37 (3): 1101–1120.

Polderman KH, Noc M, Beishulzen A et al. (2015). Ultrarapid induction of hypothermia using continuous automated peritoneal lavage with ice-cold fluids. Crit Care Med 43: 2191–22201.

Prout R, Nolan J (2009). Out-of-hospital cardiac arrest: an indication for immediate computed tomography brain imaging? Resuscitation 80 (9): 969–970.

Rana OR, Schroder JW, Kuhnen JS et al. (2012). The Modified Glasgow Outcome Score for the prediction of outcome in patients after cardiac arrest: a prospective clinical proof of concept study. Clin Res Cardiol 101 (7): 533–543.

Raurich JM, Llompart-Pou JA, Ferrerueala M et al. (2011). Hypoxic hepatitis in critically ill patients: incidence, etiology and risk factors for mortality. J Anesth 25 (1): 50–56.

Riter HG, Brooks LA, Pretorius AM et al. (2009). Intra-arrest hypothermia: both cold liquid ventilation with perfluorocarbons and cold intravenous saline rapidly achieve hypothermia, but only cold liquid ventilation improves resumption of spontaneous circulation. Resuscitation 80 (5): 561–566.

Rittenberger JC, Popescu A, Brenner RP et al. (2012). Frequency and timing of nonconvulsive status epilepticus in comatose post-cardiac arrest subjects treated with hypothermia. Neurocrit Care 16 (1): 114–122.

Rodrigo J, Fernandez AP, Serrano J et al. (2005). The role of free radicals in cerebral hypoxia and ischemia. Free Radic Biol Med 39 (1): 26–50.

Rossetti AO, Urbano LA, Delodder F et al. (2010a). Prognostic value of continuous EEG monitoring during therapeutic hypothermia after cardiac arrest. Crit Care 14 (5): R173.

Rossetti AO, Oddo M, Logroscino G et al. (2010b). Prognostication after cardiac arrest and hypothermia: a prospective study. Ann Neurol 67 (3): 301–307.

Rothman SM, Olney JW (1986). Glutamate and the pathophysiology of hypoxic–ischemic brain damage. Ann Neurol 19 (2): 105–111.

Roy D, Waxman HL, Kienzle MG et al. (1983). Clinical characteristics and long-term follow-up in 119 survivors of cardiac arrest: relation to inducibility at electrophysiologic testing. Am J Cardiol 52 (8): 969–974.

Rundgren M, Westhall E, Cronberg T et al. (2010). Continuous amplitude-integrated electroencephalogram predicts outcome in hypothermia-treated cardiac arrest patients. Crit Care Med 38 (9): 1838–1844.

Sabedra AR, Kristan J, Raina K et al. (2015). Neurocognitive outcomes following successful resuscitation from cardiac arrest. Resuscitation 90: 67–72.

Sakamoto T, Kurosawa H, Shin'oka T et al. (2004). The influence of pH strategy on cerebral and collateral circulation during hypothermic cardiopulmonary bypass in cyanotic patients with heart disease: results of a randomized trial and real-time monitoring. J Thorac Cardiovasc Surg 127 (1): 12–19.

Samaniego EA, Mlynash M, Caulfield AF et al. (2011). Sedation confounds outcome prediction in cardiac arrest survivors treated with hypothermia. Neurocrit Care 15 (1): 113–119.

Sandroni C, Cavallaro F, Callaway CW et al. (2013). Predictors of poor neurological outcome in adult comatose survivors of cardiac arrest: a systematic review and meta-analysis. Part 2: Patients treated with therapeutic hypothermia. Resuscitation 84 (10): 1324–1338.

Sarkar S, Donn SM, Bapurai JR et al. (2012). Distribution and severity of hypoxic-ischaemic lesions on brain MRI following therapeutic cooling: selective head versus whole body cooling. Archives of Disease in Childhood. Fetal and Neonatal Ed 97 (5): F335–F339.

Sato K, Kimura T, Nishikawa T et al. (2010). Neuroprotective effects of a combination of dexmedetomidine and hypothermia after incomplete cerebral ischemia in rats. Acta Anaesthesiol Scand 54 (3): 377–382.

Schöchl H, Cadamuro J, Seidl S et al. (2012). Hyperfibrinolysis is common in out-of-hospital cardiac arrest: results from a prospective observational thromboelastometry study. Resuscitation: 6–11.

Schoeler M, Loetscher PD, Rossaint R et al. (2012). Dexmedetomidine is neuroprotective in an in vitro model for traumatic brain injury. BMC Neurol 12 (1): 20.

Sekeljic V, Bataveljic D, Stamenkovic S et al. (2012). Cellular markers of neuroinflammation and neurogenesis after ischemic brain injury in the long-term survival rat model. Brain Struct Funct 217 (2): 411–420.

Sendelbach S, Hearst MO, Johnson PJ et al. (2012). Effects of variation in temperature management on cerebral performance category scores in patients who received therapeutic hypothermia post cardiac arrest. Resuscitation 83 (7): 829–834.

Serpa Neto A, Cardoso SO, Manetta JA et al. (2012). Association between use of lung-protective ventilation with lower tidal volumes and clinical outcomes among patients without acute respiratory distress. JAMA 308 (16): 1651–1659.

Shin JH, Park JM, Kim AR et al. (2012). Lance-adams syndrome. Ann Rehabil Med 36 (4): 561–564. http://dx.doi.org/10.5535/arm.2012.36.4.561. Epub 2012 Aug 27.

Skrifvars MB, Parr MJ (2012). Incidence, predisposing factors, management and survival following cardiac arrest due to subarachnoid haemorrhage: a review of the literature. Scand J Trauma Resusc Emerg Med 20 (1): 75.

Søholm H, Hassager C, Lippert F et al. (2014). Factors associated with successful resuscitation after out-of-hospital cardiac arrest and temporal trends in survival and comorbidity. Ann Emerg Med 65 (5): 1–9.

Song KJ, Shin SD, Ong ME et al. (2010). Can early serum levels of S100B protein predict the prognosis of patients with out-of-hospital cardiac arrest? Resuscitation 81 (3): 337–342.

Spies C, Macguill M, Heymann A et al. (2011). A prospective, randomized, double-blind, multicenter study comparing remifentanil with fentanyl in mechanically ventilated patients. Intensive Care Med 37 (3): 469–476.

Stammet P, Werer C, Mertens L et al. (2009). Bispectral index (BIS) helps predicting bad neurological outcome in comatose survivors after cardiac arrest and induced therapeutic hypothermia. Resuscitation 80 (4): 437–442.

Steffen IG, Hasper D, Ploner CJ et al. (2010). Mild therapeutic hypothermia alters neuron specific enolase as an outcome predictor after resuscitation: 97 prospective hypothermia patients compared to 133 historical non-hypothermia patients. Crit Care 14 (2): R69.

Storm C, Schefold JC, Kerner T et al. (2008). Prehospital cooling with hypothermia caps (PreCoCa): a feasibility study. Clin Res Cardiol 97 (10): 768–772.

Suchard JR, Fenton FR, Powers RD (1999). Medicare expenditures on unsuccessful out-of-hospital resuscitations. J Emerg Med 17 (5): 801–805.

Suehiro E, Fujisawa H, Akimura T et al. (2004). Increased matrix metalloproteinase-9 in blood in association with activation of interleukin-6 after traumatic brain injury: influence of hypothermic therapy. J Neurotrauma 21 (12): 1706–1711.

Suffoletto B, Peberdy MA, van der Hoek T et al. (2009). Body temperature changes are associated with outcomes following in-hospital cardiac arrest and return of spontaneous circulation. Resuscitation 80 (12): 1365–1370.

Sung G, Bosson N, Kaji AH et al. (2016). Therapeutic hypothermia after resuscitation from a non-shockable rhythm improves outcomes in a regionalized system of cardiac arrest care. Neurocrit Care 24 (24): 90–96.

Sutherasan Y, Penueleas O, Muriel A et al. (2015). Management and outcome of mechanically ventilated patients after cardiac arrest. Crit Care 19: 215.

Takasu A, Saitoh D, Kaneko N et al. (2001). Hyperthermia: is it an ominous sign after cardiac arrest? Resuscitation 49 (3): 273–277.

Tarmey NT, Park CL, Bartels OJ et al. (2011). Outcomes following military traumatic cardiorespiratory arrest: a prospective observational study. Resuscitation 82 (9): 1194–1197.

Terman SW, Hume B, Meurer WJ et al. (2014). Impact of presenting rhythm on short- and long-term neurologic outcome in comatose survivors of cardiac arrest treated with therapeutic hypothermia. Crit Care Med (c): 1–10.

Teasdale G, Jennett B (1974). Assessment of coma and impaired consciousness. A practical scale. Lancet 2 (7872): 81–84. PMID: 4136544.

Teasdale G, Maas A, Lecky F et al. (2014). The Glasgow Coma Scale at 40 years: standing the test of time. Lancet Neurol 13 (8): 844–854. http://dx.doi.org/10.1016/S1474-4422(14)70120-6.

The Acute Respiratory Distress Syndrome Network (2000). Ventilation with lower tidal volumes as compared with traditional tidal volumes for acute lung injury and the acute respiratory distress syndrome. N Engl J Med 342 (18): 1301–1308.

Tiainen M, Roine RO, Pettila V et al. (2003). Serum neuron-specific enolase and S-100B protein in cardiac arrest patients treated with hypothermia. Stroke 34 (12): 2881–2886.

Tiainen M, Kovala TT, Takkunen OS et al. (2005). Somatosensory and brainstem auditory evoked potentials in cardiac arrest patients treated with hypothermia. Crit Care Med 33 (8): 1736–1740.

Traystman RJ, Kirsch JR, Koehler RC (1991). Oxygen radical mechanisms of brain injury following ischemia and reperfusion: Oxygen radical mechanisms of brain injury following ischemia and reperfusion. J Appl Physiol 71: 1185–1195.

Tsai M-S, Barbut D, Wang H et al. (2008). Intra-arrest rapid head cooling improves postresuscitation myocardial function in comparison with delayed postresuscitation surface cooling. Crit Care Med 36 (Suppl): S434–S439.

Vaagenes P, Ginsberg M, Ebmeyer U et al. (1996). Cerebral resuscitation from cardiac arrest: pathophysiologic mechanisms. Crit Care Med 24 (2): S57–S68.

van Beek J, Elward K, Gasque P (2003). Activation of complement in the central nervous system: roles in neurodegeneration and neuroprotection. Ann N Y Acad Sci 992 (992): 56–71.

van der Eerden AW, Khalilzadeh O, Perlbarg V et al. (2014). White matter changes in comatose survivors of anoxic ischemic encephalopathy and traumatic brain injury: comparative diffusion-tensor imaging study. Radiology 270 (2): 506–516.

Virkkunen I, Ryynanen S, Kujala S et al. (2007). Incidence of regurgitation and pulmonary aspiration of gastric contents in survivors from out-of-hospital cardiac arrest. Acta Anaesthesiol Scand 51 (2): 202–205.

Volpicelli G (2011). Usefulness of emergency ultrasound in nontraumatic cardiac arrest. Am J Emerg Med 29 (2): 216–223.

Wadhwa A, Senggupta P, Durrani J et al. (2005). Magnesium sulphate only slightly reduces the shivering threshold in humans. Br J Anaesth 94 (6): 756–762.

Wallmuller C, Meron G, Kurkciyan I et al. (2012). Causes of in-hospital cardiac arrest and influence on outcome. Resuscitation 83 (10): 1206–1211.

Wennervirta JE, Ermes MJ, Tiainen SM et al. (2009). Hypothermia-treated cardiac arrest patients with good neurological outcome differ early in quantitative variables of EEG suppression and epileptiform activity. Crit Care Med 37 (8): 2427–2435.

Westfal RE, Reissman S, Doering G (1996). Out-of-hospital cardiac arrests: an 8-year New York City experience. Am J Emerg Med 14 (4): 364–368.

Wijdicks EFM, Pfeifer EA (2008). Neuropathology of brain death in the modern transplant era. Neurology 70 (15): 1234–1237.

Wijdicks EF, Campeau NG, Miller GM (2001). MR imaging in comatose survivors of cardiac resuscitation. AJNR 22 (8): 1561–1565.

Wijdicks EFM, Bamlet WR, Maramattom BV et al. (2005). Validation of a new coma scale: the FOUR score. Ann Neurol 58 (4): 585–593.

Wijdicks EFM, Hildar A, Young GB et al. (2006). Practice parameter: prediction of outcome in comatose survivors after cardiopulmonary resuscitation (an evidence-based review): report of the Quality Standards Subcommittee of the American Academy of Neurology. Neurology 67 (2): 203–210.

Wijman CA, Miynash M, Caulfield AF et al. (2009). Prognostic value of brain diffusion-weighted imaging after cardiac arrest. Ann Neurol 65 (4): 394–402.

Wiklund L, Martin C, Miclescu A et al. (2012). Central nervous tissue damage after hypoxia and reperfusion in conjunction with cardiac arrest and cardiopulmonary resuscitation: mechanisms of action and possibilities for mitigation. Int Rev Neurobiol 102: 173–187.

Willie C, Macleod DB, Shaw AD et al. (2012). Regional brain blood flow in man during acute changes in arterial blood gases. J Physiol 590: 3261–3275.

Wissenberg M, Lippert FK, Folke F et al. (2013). Association of national initiatives to improve cardiac arrest management with rates of bystander intervention and patient survival after out-of-hospital cardiac arrest. JAMA 310 (13): 1377–1384.

Wolfrum S, Pierau C, Radke PW et al. (2008). Mild therapeutic hypothermia in patients after out-of-hospital cardiac arrest due to acute ST-segment elevation myocardial infarction undergoing immediate percutaneous coronary intervention. Crit Care Med 36 (6): 1780–1786.

Wortsman J, Foley PJ, Tacker WA et al. (1987). Cerebrospinal fluid changes in experimental cardiac arrest (maximal stress). American Journal of Phisiology - Endocrinology and Metabolism 252: E756–E761.

Wu O, Sorensen AG, Benner T et al. (2009). Comatose patients with cardiac arrest: predicting clinical outcome with diffusion-weighted MR imaging. Radiology 252 (1): 173–181.

Wu O, Batista LM, Lima FO et al. (2011). Predicting clinical outcome in comatose cardiac arrest patients using early noncontrast computed tomography. Stroke 42 (4): 985–992.

Xiao F, Arnold TC, Zhang S et al. (2004). Cerebral cortical aquaporin-4 expression in brain edema following cardiac arrest in rats. Acad Emerg Med 11 (10): 1001–1007.

Yanagawa Y, Un-no Y, Sakamoto T et al. (2005). Cerebral density on CT immediately after a successful resuscitation of cardiopulmonary arrest correlates with outcome. Resuscitation 64 (1): 97–101.

Yang J, Ahn HN, Chang M et al. (2013). Complement component 3 inhibition by an antioxidant is neuroprotective after cerebral ischemia and reperfusion in mice. J Neurochem 124 (4): 523–535.

Yannopoulos D, Zviman M, Castro V et al. (2009). Intracardiopulmonary resuscitation hypothermia with and without volume loading in an ischemic model of cardiac arrest. Circulation 120 (120): 1426–1435.

Zeiner A, Holzer M, Sterz F et al. (2001). Hyperthermia after cardiac arrest is associated with an unfavorable neurologic outcome. Arch Intern Med 161 (16): 2007–2012.

Zipfel G, Babcock DJ, Lee JM et al. (2000). Neuronal apoptosis after CNS injury: the roles of glutamate and calcium. J Neurotrauma 17 (10): 857–869.

Chapter 33

Therapeutic hypothermia protocols

N. BADJATIA*

Department of Neurology, University of Maryland School of Medicine, Baltimore, MD, USA

Abstract

The application of targeted temperature management has become common practice in the neurocritical care setting. It is important to recognize the pathophysiologic mechanisms by which temperature control impacts acute neurologic injury, as well as the clinical limitations to its application. Nonetheless, when utilizing temperature modulation, an organized approach is required in order to avoid complications and minimize side-effects. The most common clinically relevant complications are related to the impact of cooling on hemodynamics and electrolytes. In both instances, the rate of complications is often related to the depth and rate of cooling or rewarming. Shivering is the most common side-effect of hypothermia and is best managed by adequate monitoring and stepwise administration of medications specifically targeting the shivering response. Due to the impact cooling can have upon pharmacokinetics of commonly used sedatives and analgesics, there can be significant delays in the return of the neurologic examination. As a result, early prognostication posthypothermia should be avoided.

INTRODUCTION

Therapeutic hypothermia (TH) is an effective treatment in models of experimental brain injury, including cardiac arrest, ischemic stroke, and traumatic brain injury (TBI). Current literature does not support the routine utilization of hypothermia for conditions other than cardiac arrest (Callaway et al., 2015). Hypothermia for raised intracranial pressure remains a consideration, but only after other measures such as osmotic therapy have failed. Other considerations for hypothermia remain experimental.

NEUROPATHOLOGY

As a neuroprotectant, the efficacy of hypothermia varies with depth, delay, and duration of treatment (MacLellan et al., 2009). In focal ischemia, blood flow in the core and penumbra is reduced to approximately 20% and 50% of baseline (Morikawa et al., 1992; Kawai et al., 2000; Yanamoto et al., 2001). Hypothermia does not potentiate the decrease in cerebral blood flow (CBF) during ischemia (Busto et al., 1987), but reduces metabolism more so than decreasing CBF. Increased blood pressure, a sympathetic response observed during cooling and shivering, a common clinical problem during cooling, may improve collateral reperfusion (Kawamata et al., 1997; Hayashi et al., 1998; Lin et al., 2002), though the metabolic increases observed may negate the overall benefit of cooling.

One of the primary mechanisms through which hypothermia confers neuroprotection is by a dramatic reduction in cerebral metabolism. Decline in core temperature to 32°C can lower the cerebral metabolic rate for oxygen ($CMRO_2$) by 15–30% (Rosomoff and Holaday, 1954; Hagerdal et al., 1975; Michenfelder and Milde, 1991; Nakashima et al., 1995; Okubo et al., 2001; Erecinska et al., 2003). Likewise, there are many studies indicating that intraischemic TH delays adenosine triphosphate (ATP), lactate, and pyruvate expenditure, but does not prevent it (Nilsson et al., 1975; Welsh et al., 1990; Sutton et al., 1991; Ibayashi et al., 2000; Kimura et al., 2002; Erecinska et al., 2003). Nonetheless, TH improves recovery of high-energy phosphate metabolites and reverses acidosis produced by lactate accumulation during reperfusion (Chopp et al., 1989; Sutton et al., 1991;

---

*Correspondence to: Neeraj Badjatia, MD MS FCCM, R Adam Cowley Shock Trauma Center, University of Maryland School of Medicine, 22 South Greene Street, Baltimore MD 21201, USA. Tel: +1-410-328-4515, E-mail: nbadjatia@umm.edu

Kimura et al., 2002; Erecinska et al., 2003). Even though there is a significant decrease in metabolism caused by TH, this cannot fully account for potent neuroprotection afforded by intraischemic cooling. Cooling also alters metabolism in ways that confer protection (Kaibara et al., 1999). The decrease in oxygen availability during ischemia disrupts aerobic glycolysis and the ensuing anaerobic metabolism leads to lactic acidosis.

Intraischemic hypothermia can partially diminish Ca^{2+} influx, thereby ameliorating cell death (Kristian et al., 1992; Moyer et al., 1992). Delayed and prolonged hypothermia also presumably attenuates delayed excitotoxicity that occurs through the ischemia-induced downregulation of the GluR2 subunit of AMPA channels, which normally prevents Ca^{2+} and Zn^{2+} influx, in the CA1 region after global ischemia (Bennett et al., 1996; Colbourne et al., 2003). Cooling also lessens protein kinase C and Ca^{2+}/calmodulin-dependent protein kinase II translocation and activation, possibly by decreasing Ca^{2+} (Cardell and Wieloch, 1993; Busto et al., 1994; Chapman et al., 1995; Hu and Wieloch, 1995; Tohyama et al., 1998; Harada et al., 2002).

Many neurotransmitters are released after ischemia due to spreading depression and the inability to regulate their release. Cooling decreases the propagation of spreading depolarization (Takaoka et al., 1996), thereby ameliorating total neurotransmitter release and the possibility of further excitotoxic injury. Cooling decreases the immediate postischemic hyperperfusion, as well as glutamate and intracellular Ca^{2+} toxicity, thereby reducing injury and the inflammatory response (Karibe et al., 1994; Huang et al., 1998). By affecting these processes, hypothermia indirectly affects $\cdot O_2^-$ production. TH also decreases ·nitric oxide (NO) production (Kader et al., 1994; Han et al., 2002), possibly due to reduced inflammation. NO synthase (NOS) has been used as a marker of ·NO production in several studies. Upregulation of both neuronal NOS and microglia-inducible NOS after focal ischemia can be prevented by intra- and postischemic TH (Han et al., 2002; Karabiyikoglu et al., 2003; Van Hemelrijck et al., 2005; Mueller-Burke et al., 2008).

Inflammation

The inflammatory response that follows ischemia is a complex process, mediated by many factors, that is predominantly regulated through activated microglia, resulting in a phagocytic response to neuronal stress and injury (Ceulemans et al., 2010). Hypothermia reduces the total number of microglia, as well as reactive microglia (Inamasu et al., 2000; Deng et al., 2003; Han et al., 2003; Fukui et al., 2006; Florian et al., 2008; Webster et al., 2009; Ceulemans et al., 2011; Drabek et al., 2012; Fries et al., 2012), which may be the result of a decrease in expression of transcription factor NF-κB under lower temperatures (Han et al., 2003; Yenari and Han, 2006; Florian et al., 2008; Webster et al., 2009). An immunomodulatory benefit of hypothermia may be a decrease in the secretion of tumor necrosis factor-α and interleukin-6, without affecting the release of beneficial neurotrophic factors (Xiong et al., 2009; Ceulemans et al., 2011).

Edema and increased intracranial pressure (ICP) after ischemia is partially due to dysregulation of water flow, as well as disruption of the blood–brain barrier (BBB). TH has a favorable effect on edema after experimental ischemia (Karibe et al., 1994; Preston and Webster, 2004; Xiao et al., 2004; Kallmunzer et al., 2012). Both intra- and postischemic cooling can lessen the amount of BBB disruption (Karibe et al., 1994; Wagner et al., 2003; Hamann et al., 2004; Preston and Webster, 2004; Lee et al., 2005; Nagel et al., 2008; Baumann et al., 2009), possibly by preserving the shape of endothelial cells and avoiding pericyte disassociation (Duz et al., 2007). Cooling attenuates the increase in metallomatrix proteins in experimental (Wagner et al., 2003; Hamann et al., 2004; Lee et al., 2005; Nagel et al., 2008; Kallmunzer et al., 2012) and clinical stroke (Horstmann et al., 2003), thereby diminishing BBB disruption.

Cardiac arrest

During no-flow states, such as observed in cardiac arrest, there is membrane depolarization, calcium influx, glutamate release, acidosis, and activation of lipases, proteases, and nucleases. Upon reflow, there is reoxygenation injury involving iron, free radicals, nitric oxide, catecholamines, excitatory amino acid release, and renewed calcium shifts (Polderman, 2009). During postischemic reperfusion, even after prolonged ischemic periods, the high-energy ATP load recovers rapidly and approaches normal levels quickly after return of spontaneous circulation (ROSC); however, tissue injury continues after reperfusion. The observation of morphologic changes (cytosolic microvacuolation) seen in hippocampal hilar, CA1 pyramidal neurons, and cortical pyramidal neurons of layers 3 and 5 after reperfusion has led to recognition of the concepts of reperfusion injury and selective neuronal vulnerability (Polderman, 2009). As a result, much of the brain injury after even brief periods of anoxia is due to the reperfusion injury occurring after ROSC. Experimental studies have shown that these mechanisms can be minimized or prevented with the application of hypothermia.

Traumatic brain injury

The application of hypothermia after TBI can have very different effects based upon patient selection, as well as the timing, duration, and depth of cooling. Cooling within minutes to hours after the injury is intended as a neuroprotective strategy, mitigating many of the cellular mechanisms that eventually result in further damage. As hours and days postinjury continue, the cumulative effect of these mechanisms is observed clinically, where TH can also be applied as an effective adjunctive therapy for processes leading to raised ICP.

Subarachnoid and intracerebral hemorrhage

The focus of TH in the acute phase of subarachnoid hemorrhage (SAH) is on mitigating the effects of the initial hemorrhage ("early brain injury"). Experimental studies have demonstrated that mild to moderate hypothermia reverses hypoperfusion that is unrelated to cerebral perfusion pressure, enhances recovery of posthemorrhagic CBF, and attenuates edema formation. The vascular effects may be attributed to hypothermia-induced vasodilatation, or to the prevention of autoregulatory impairment, whereas prevention of lactate accumulation may help reverse post-SAH cerebral edema. The application of hypothermia after parenchymal hemorrhage is understudied. Similar to SAH, clinical studies in intracerebral hemorrhage (ICH) have not been adequately designed to understand the impact of TH on outcome, though there has been a consistent finding that cooling reduces hemorrhage-related cerebral edema (Howell et al., 1956; Fingas et al., 2007).

Spinal cord injury

Approximately 11 000–12 000 individuals sustain a spinal cord injury from motor vehicle accidents, sport-related injuries, and direct trauma each year in the USA (Dietrich et al., 2011). Recent surgical advancements have reduced mortality and morbidity, but long-term disability remains a major concern (Fehlings et al., 2012). Currently, there are no proven medical treatments that protect against the consequences of spinal cord injury. Experimental models have reliably demonstrated a strong benefit of TH (Dietrich et al., 2011), but clinical studies are lacking to better inform the timing, depth, and duration for hypothermia in this clinical setting.

NEURODIAGNOSTICS AND IMAGING

Electrophysiologic testing

The effects of hypothermia on electroencephalography (EEG) have been well studied experimentally, with notable decreases in voltage, and slower frequencies with progressive hypothermia (ten Cate et al., 1949; Chatfield et al., 1951; Lyman and Chatfield, 1953; Callaghan et al., 1954; Owens, 1958; Woodhall et al., 1958; Massopust et al., 1970). Similar effects of EEG suppression during hypothermia have been shown in the clinical setting, although there is considerable variation between individual patients (Quasha et al., 1981). The temperature necessary to induce significant suppression or electrocerebral silence has been reported to range anywhere from 12°C and 33°C (Stecker et al., 2001), which may have important implications when considering specific temperature goals for the treatment of status epilepticus, or in the interpretation of EEG findings.

Somatosensory evoked potentials (SSEPs) can be used to assess the integrity of the arousal system, specifically assessing the somatosensory pathway and maintenance or restoration of normal thalamocortical coupling (Madhok et al., 2012). Hypothermia can have a profound impact on the amplitude and latency of the signals obtained with median nerve stimulation. Nonetheless, the absence of cortical SSEP components (N_2O) after stimulation of the median nerve during hypothermia remains a reliable electrophysiologic indicator of an unfavorable prognosis in the cardiac arrest population.

HOSPITAL COURSE AND MANAGEMENT

Temperature modulation techniques

When considering TH in the neurocritical care setting, after deciding the target temperature, the next management decision centers on the modality of cooling. Recent technologic advances have brought about many new devices targeting core body temperature that allow for tight regulation of temperature to within ±0.2°C. However, it is important to recognize that induction of hypothermia can also be achieved efficiently without devices.

Infusion of cold fluids is effective at initiating the cooling process (Polderman et al., 2005). There is variability as to the type of fluid and volume necessary; however, the most reliable method to utilize in neurologically injured patients is to infuse 2 liters of isotonic fluid cooled to 4°C. Initiating shivering control prior to infusion is important, as cold crystalloid infusion without effective shivering control is associated with a much slower rate of cooling. The most commonly observed adverse effect associated with cold fluid induction is pulmonary congestion, which has been noted to occur with higher volumes of infusion (>2 liters).

Induced hypothermia can be separated into three phases: induction, maintenance, and rewarming. Since the development of automated cooling systems, the clinical application of targeted temperature management (TTM) has become safer and simpler. Each cooling device works by promoting conductive heat loss, either through surface or intravascular cooling.

Surface cooling systems consist of pads containing circulating forced cold air or fluid which are applied to the skin (Lay and Badjatia, 2010), whereas intravascular systems consist of endovascular heat exchange catheters that are placed into a central vein to cool the blood (Owens, 1958). Both methods can be used effectively to induce and maintain hypothermia (Hoedemaekers et al., 2007). Some clinicians speculate that intravascular devices cause less shivering than surface cooling methods; however, no direct comparisons between the two methods have been performed. All the devices work via a feedback loop that adjusts the temperature of the water circulating through the cooling system to maintain a constant target core body temperature, which should be continuously monitored via a probe in the bladder, esophagus, or rectum. Each of these advanced temperature modulation devices has shown superior temperature regulation as compared to more traditional methods such as ice packets, cooling blankets, or fans (Diringer and Neurocritical Care Fever Reduction Trial Group, 2004; Mayer et al., 2004; Steinberg et al., 2004; Broessner et al., 2009). However, there are few data to support that these temperature-modulating devices improve outcome, with a recent trial in cardiac arrest patients demonstrating no benefit in long-term outcome of intravascular cooling vs. traditional cooling methods (Deye et al., 2015).

Induction

Application of ice packs and infusion of cold intravenous fluids (e.g., 4°C normal saline or lactated Ringer's solution at 30–40 mL/kg over 1 hour) is the simplest and least expensive method of inducing hypothermia (Polderman et al., 2005). Body temperature typically falls by 2°C after the 1-hour infusion. Large volumes of refrigerated saline should not be given to patients with active clinical signs of symptoms of congestive heart failure, as this treatment could cause exacerbation of pulmonary edema.

Maintenance

Once induction has begun, a device is applied to complete the cooling and begin the maintenance phase. During the maintenance period, advanced cooling technology can maintain core body temperature with only minor fluctuations ($\pm 0.2°C$). Fever, the most frequent clinical sign of infection, no longer occurs when hypothermia is induced. The temperature of the circulating water in the cooling device can be used as a surrogate marker for increased heat production by the patient: water temperature lower than 10°C is indicative of a febrile response.

Rewarming

Rewarming is the most dangerous phase of hypothermia, particularly in patients with cerebral edema and intracranial mass effect, who are at risk of elevated ICP. A rapid increase in body temperature can cause systemic vasodilation and hypotension, which can in turn trigger cerebral vasodilation and ICP plateau waves. Rates of rewarming vary depending on the indication for TTM. In general, rewarming can be performed at rates as fast as 0.5°C/hour if patients have no ICP-related issues, and as slow as 0.1°C/hour if elevated ICP is a concern. If ICP elevation is suspected or observed, rewarming should be slowed or even halted. In most cases, a rewarming rate of 0.25°C/hour is recommended, and rewarming should always be performed in a controlled manner to avoid overshooting and developing hyperthermia.

CLINICAL TRIALS AND GUIDELINES

Cardiac arrest

Despite knowledge of the benefits of TH after experimental cardiac arrest for several decades (Stub et al., 2011), it has only recently been studied extensively in humans. A decade has passed since the results of two randomized controlled trials provided evidence that TH (32–34°C) for 12–24 hours is an effective treatment for patients who remain comatose after resuscitation from out-of-hospital cardiac arrest when the initial cardiac rhythm is ventricular fibrillation or pulseless ventricular tachycardia (Bernard et al., 2002; The Hypothermia After Cardiac Arrest Study Group, 2002). As with other therapeutic interventions after brain injury, time to treatment is important and this therapy should only be initiated within 6 hours of injury and without delay. In 2010, the American Heart Asssociation recommended as part of routine post cardiac arrest care that comatose adult patients surviving out-of-hospital ventricular fibrillation or pulseless ventricular tachycardia cardiac arrest should be cooled to 32–34°C for 12–24 hours. Further, less robust recommendations were made for TH for comatose adult patients after in-hospital cardiac arrest or after out-of-hospital cardiac arrest with an initial rhythm of pulseless electric activity or asystole (Peberdy et al., 2010).

A subsequent, larger, well-conducted randomized controlled trial found that neurologic outcomes and survival at 6 months after out-of-hospital cardiac arrest were not inferior when temperature was controlled at 36°C versus 33°C (Nielsen et al., 2013). There are no direct comparisons of different durations of TTM in post cardiac arrest patients. The largest trials and studies of TTM maintained temperatures for 24 hours (Hypothermia After Cardiac Arrest Study Group, 2002) or 28 hours (Nielsen et al., 2013), followed by a gradual (approximately 0.25°C/hour) return to normothermia.

The American Heart Association has recently changed its guideline for cooling after hypothermia to recommend selecting and maintaining a constant temperature between 32°C and 36°C during TTM (class I, Level Of Evidence - B Randomized study (LOE B-R)).

Initiating cooling in the prehospital setting has failed to demonstrate a benefit versus initiating once a patient has arrived in the Emergency Department (Bernard et al., 2010, 2012; Debaty et al., 2014; Kim et al., 2014). In neonatal hypoxic encephalopathy, trials have shown that cooling comatose newborn patients for 72 hours improves the proportion of patients with an intact neurologic recovery (Gluckman et al., 2005; Shankaran et al., 2005). The results of clinical trials in the setting of pediatric cardiac care are, however, difficult to interpret. A recent trial of 260 pediatric patients with out-of-hospital cardiac arrest reported neurocognitive outcomes in 20% of patients cooled to 33°C as compared to 12% of those maintained at 36.8°C, but this difference was not statistically significant (Moler et al., 2015). It is important to note that normothermia was actively controlled, further indicating the importance of TTM after cardiac arrest.

Traumatic brain injury

Thirteen controlled single-center studies conducted in adult TBI patients demonstrated significantly better outcomes associated with TH (Polderman et al., 2004). In contrast, three multicenter randomized controlled trials that tested early short-term (maximum 48 hours) TH (Clifton et al., 2001; Shiozaki et al., 2001; Clifton et al., 2011) found no benefit with regard to survival and neurologic outcome.

The most two recent studies published are the National Acute Brain Injury Study: Hypothermia II (NABIS:HII) and EuroTherm trial. NABIS:HII was a multicenter trial including patients who were 16–45 years old after severe, nonpenetrating TBI, treated with very early TH, initiated in the prehospital setting (Clifton et al., 2011). The trial was stopped after inclusion of 108 patients, and no effect on outcome was observed. Subgroup analysis found that patients with surgically evacuated hematomas treated with TH had better outcome, while those with diffuse brain injury treated with hypothermia had a trend towards worsened outcome. One of the reasons for improvement in such a subpopulation may be due to the impact of temperature control on reperfusion injury-related spreading depolarizations, as recently reported by the Co-operative Study of Brain Injury Depolarizations (COSBID) study group (Hartings et al., 2009, 2011).

The EuroTherm trial assessed functional outcomes after TBI when hypothermia was used as a second-line treatment for elevated ICP, prior to use of osmotic therapy (Polderman et al., 2004; Andrews et al., 2015). The results indicated that the hypothermia group was less likely to require the use of stage 3 treatments, but this was not associated with an improvement in outcome. In fact, a favorable recovery occurred less often in the hypothermia group (26%) than in the control group (37%; $p=0.03$). Serious adverse events were also reported more often with hypothermia (33 vs. 10 events).

In pediatric TBI populations, there have been four randomized controlled trials comparing hypothermia to normothermia (Biswas et al., 2002; Adelson et al., 2005, 2013; Hutchison et al., 2008). In each of these studies, hypothermia therapy was associated with a lower ICP, but failed to have an impact on outcome, with rebound intracranial hypertension or systemic hypotension noted to be associated with treatment. Most recently, the Cool Kids trial, a phase III multinational 15-center trial, randomized 77 children to hypothermia or normothermia for 48–72 hours followed by a slow rewarming rate of 0.5–1°C over 12–24 hours. The study was terminated for futility on interim analysis without a difference in outcome or adverse events between treatment groups (Adelson et al., 2013).

There are 17 controlled trials investigating the impact of TH on outcome in adult patients with severe TBI patients and refractory intracranial hypertension and most of these studies demonstrate that hypothermia is an effective method for lowering ICP, though the data on outcome are inconsistent (Polderman et al., 2004). The magnitude of the effect of TH on ICP reduction is estimated to be approximately 10 mmHg (range 5–23 mmHg). Across the studies analyzed, the effect of TH on ICP reduction was superior to that achieved with moderate hyperventilation, barbiturates, and mannitol, but less effective than hemicraniectomy and hypertonic saline (Schreckinger and Marion, 2009).

The optimal target temperature of TH when used for ICP control is not well defined. There is experimental evidence that decreasing body temperature to 35–35.5°C effectively treats intracranial hypertension, while maintaining sufficient cerebral perfusion pressure without cardiac dysfunction or oxygen debt (Tokutomi et al., 2007). Resting energy expenditure and cardiac output decrease progressively with hypothermia, reaching very low levels at temperatures below 35°C (Tokutomi et al., 2007). At core temperatures below 35°C there is a concomitant significant decrease in brain tissue oxygenation (Gupta et al., 2002). Thus, 35–35.5°C may be the optimal temperature at which to treat patients with intracranial hypertension following severe TBI. However, instead of applying fixed temperature targets, TH may be better applied by titrating temperature to maintain ICP below 20 mmHg. On the basis of a meta-analysis, some have advocated that duration of TH longer than 48 hours may be beneficial (McIntyre et al., 2003), but the optimal duration of cooling is not known. Rather than focusing on optimal timing, a better target for cooling is ongoing efficacy for reduction of ICP weighed against the risks associated with deep sedation and impaired immune function that accompany prolonged cooling.

Rewarming remains the most dangerous aspect of hypothermia management. Large fluctuations in temperature can reverse the protective effects of cooling, and aggravate secondary brain injury (Suehiro and Povlishock, 2001; Ueda et al., 2003). This is evidenced by impaired cerebrovascular vasoreactivity, hyperemia, and rebound intracranial hypertension (Lavinio et al., 2007). Studies have documented rapid rewarming to be associated with more episodes of rebound intracranial hypertension, and worse outcomes (Thompson et al., 2010; Clifton et al., 2011). A very slow, controlled rewarming (0.1–0.2°C/hour) was utilized in both recent randomized controlled trials of hypothermia to reduce the risk of rebound cerebral edema and intracranial hypertension.

Intracerebral hemorrhage

TH has been less rigorously studied in patients with ICH. Initial studies reported on the potential benefits of TH on clinical signs and symptoms related to raised ICP after ICH (Howell et al., 1956). The most recent study dealing with hypothermia in ICH prospectively included 12 patients with large supratentorial spontaneous ICH, who were compared to a historical control group of patients matched for ICH volume (Kollmar et al., 2010). All patients were cooled to 35°C using an endovascular approach for 10 days. In the TH group, perihemorrhagic edema volume remained stable during 14 days of computed tomography follow-up, whereas it significantly increased to double its initial volume in the control group within the same time span. Beyond high rates of pneumonia, no other major complications of hypothermia were observed. This study was not powered to assess differences in outcome (Staykov et al., 2011).

Subarachnoid hemorrhage

The Intraoperative Hypothermia for Aneurysm Surgery Trial (IHAST), a phase III multicenter randomized controlled trial, compared the use of mild hypothermia at a target temperature of 33°C during aneurysm surgery to normothermia (target temperature of 36.5°C) (Todd et al., 2005). Treatment was performed within 14 days of symptom onset and hypothermia was induced with a forced-air blanket. The study involved predominantly low-grade SAH patients, ranging from World Federation of Neurological Societies (WFNS) I to III. The study did not demonstrate a significant difference in clinical outcome between the two groups 90 days after enrollment.

There was a slightly higher rate of bacteremia in the hypothermia group (5% vs. 3%; $p=0.05$). One possible explanation for the lack of improvement in mortality and functional outcome may be the relatively good clinical condition of the majority of study participants, with the consequent lack of a temperature-modifiable brain lesion (Linares and Mayer, 2009). Additionally, the therapy was limited to the operative setting and did not extend for subsequent days, during which there may have been greater benefit for hypothermia.

The use of prolonged TH for treatment of intractable increases in ICP or vasospasm, mainly in patients with poor-grade SAH, has been reported in several case series (Kawamura et al., 2000; Nagao et al., 2000, 2003; Gasser et al., 2003; Seule et al., 2009). The largest of these reported on poor-grade SAH patients who developed elevated ICP and/or symptomatic vasospasm refractory to conventional treatment (Seule et al., 2009). Hypothermia was achieved with an endovascular approach and maintained at 33–34°C for up to 1 week. The acuity of this population made it unclear as to whether the high rate of complications was related to disease severity or hypothermia. Nonetheless, there was a higher than expected proportion of patients with good outcomes reported at 3 months.

Ischemic stroke

The perceived need for a secure airway, mechanical ventilation, and control of shivering has limited the use of hypothermia as a therapeutic approach in stroke patients. However, some studies have shown that it is possible to cool nonintubated stroke patients, albeit with variable success (Kammersgaard et al., 2000; Zweifler et al., 2003; Guluma et al., 2008). Schwab et al. (2001) reported on two uncontrolled trials of induced hypothermia as salvage therapy for patients with established middle cerebral artery (MCA) territory infarction. Patients were admitted to an intensive care unit and hypothermia was achieved with surface cooling. ICP was monitored with parenchymal sensors placed ipsilateral to the infarct. In the first of these studies, hypothermia was induced in 25 malignant MCA infarct patients at an average of 14 hours after stroke onset, and temperature was maintained at 33°C for 48–72 hours (Schwab et al., 2001). There was significant morbidity associated with cerebral edema due to uncontrolled rewarming. Further data in MCA infarct patients suggest that controlled rewarming rates of ≤0.1°C/hour allow for improved control of ICP when compared with patients in whom rewarming is achieved in a passive, uncontrolled fashion (Schwab et al., 2001; Steiner et al., 2001).

In a prospective randomized study, Els et al. (2006) enrolled 25 consecutive patients with an ischemic infarction of more than two-thirds of one hemisphere to either hemicraniectomy alone, or in combination with hypothermia. Safety parameters were compared between treatment groups and the clinical outcome assessed at 6 months. Overall mortality was 12% (2/13 vs. 1/12 in the two groups), but none of the 3 patients died due to treatment-related complications. There were no severe adverse effects of hypothermia. The clinical results showed a tendency for a better outcome in the hemicraniectomy plus moderate hypothermia group after 6 months. Delayed cooling for the treatment of cytotoxic brain edema does not provide definitive treatment for malignant cerebral edema, and should not be used as an alternative to the proven therapy of hemicraniectomy (Juttler et al., 2007; Hofmeijer et al., 2009). However, these results suggest that hypothermia may still be of benefit, even in patients who have undergone hemicraniectomy.

There are results from two recent phase II studies that utilize advanced cooling techniques in ischemic stroke patients. In a small study of 36 patients randomized 1:1 to mild hypothermia (35°C) or to standard stroke unit care within 6 hours of symptom onset, hypothermia was induced with a surface-cooling device and cold saline infusions, and maintained for 12 hours, with gradual rewarming to normothermia (Pirronen et al., 2014). The majority (83%) of hypothermia-treated patients reached target temperature within the prespecified time period without a higher rate of adverse effects. There was a trend towards higher rates of poor outcome in the normothermia-treated patients.

The multicenter randomized controlled ICTuS-L trial investigated the feasibility and safety of the combination of endovascular hypothermia and intravenous thrombolysis with rt-PA (Hemmen et al., 2010). Of the 59 patients included, 44 were enrolled within 3 hours of symptom onset, and 48 were treated with thrombolysis. Twenty-eight patients were randomized to hypothermia at 33°C for 24 hours. Pneumonia occurred at a significantly higher rate in the hypothermia group, as compared to controls (50% vs. 10%). Outcome assessed with the modified Rankin scale 3 months after symptom onset did not differ between the two groups. A pivotal phase III study is under way to determine the overall efficacy of hypothermia in this population.

Other investigators are studying the effects of mild hypothermia combined with additional neuroprotective agents, such as caffeine and ethanol; however, until tested in a prospective controlled study, TH, either as a standalone therapy or as an adjunct, remains experimental in this setting (Abou-Chebl et al., 2004; Hemmen et al., 2010).

Spinal cord injury

The only human evidence thus far in the literature is a single-center study from the University of Miami involving 14 patients with an average age of 39 years (range, 16–62 years), with acute, complete (American Spinal Injury Association (ASIA) A) cervical spinal cord injuries, that were treated using an intravascular cooling catheter to achieve modest (33°C) systemic hypothermia for 48 hours (Levi et al., 2009). In this small series, the investigators found the cooling approach to be feasible, with no increase in the rate of complications. Even though they noted 6 of 14 patients converted from ASIA A status, large prospective studies are needed before TH can be considered as part of standard care in this population.

COMPLEX CLINICAL DECISIONS

Shivering

Shivering, coupled with vasoconstriction, is a thermoregulatory defense to maintain body temperature at the hypothalamic set point. In healthy humans, peripheral vasoconstriction is triggered at approximately 36.5°C and shivering at 35.5°C (Sessler, 2009). Temperature thresholds for vasoconstriction and shivering are sometimes higher than normal in brain-injured patients; therefore, these thermoregulatory defenses may occur at higher temperatures (Badjatia, 2009).

Control of shivering is essential for effective hypothermia. If left uncontrolled, shivering will interfere with the cooling process, and can result in increments in systemic and cerebral energy consumption and metabolic demand (Badjatia et al., 2008, 2009). The first step in managing shivering is to have an effective tool for measurement. The Bedside Shivering Assessment Scale is a simple, validated four-point scale that enables repeated quantification of shivering at the bedside. It has been utilized in conjunction with continuous bispectral index monitoring to provide detection of shivering in moderately or deeply sedated critically ill patients.

Therapy for shivering should ideally suppress the central thermoregulatory reflex rather than just uncoupling this response from skeletal muscle contraction, as this does not mitigate the ongoing cerebral and systemic stress response. Initial measures have been focused on nonsedating or minimally sedating pharmacologic interventions, so as not to impair the ability to track neurologic exam changes and increase the risk for complications related to prolonged mechanical ventilation (Choi et al., 2011).

The first step utilizes acetaminophen, buspirone, and magnesium infusion (Mokhtarani et al., 2001; Kasner et al., 2002; Zweifler et al., 2004). Additionally, patients should be treated with forced warm-air skin counter-warming. An increment in mean skin temperature by 4°C, without affecting core body temperature, can increase the sensation of warmth and blunt the shivering reflex by about 1°C (Lennon et al., 1990; Badjatia et al., 2008). Approximately half of the patients who shiver in response to TTM require additional pharmacologic therapy to prevent this response. Dexmeditomidine is a central-acting α_2-receptor agonist that has been shown to decrease the shivering threshold (Lennon et al., 1990). Propofol and the opioid meperidine are also effective at reducing shivering, but can cause oversedation and prolong the need for mechanical ventilation when given at high doses (Lennon et al., 1990; Matsukawa et al., 1995). If all other options to prevent shivering are exhausted, pharmacologic paralysis may be needed (Choi et al., 2011).

Reduced electrolyte levels

In addition to decreased systemic and cerebral metabolism, other physiologic changes commonly occur in patients treated with hypothermia. Cooling drives electrolytes into the intracellular compartment and results in decreased levels of serum potassium, magnesium, and phosphate (Polderman et al., 2001). However, during rewarming these electrolytes are released from intracellular stores and move to extracellular spaces. Care should be taken, therefore, to avoid excessive potassium replacement during the maintenance phase to avoid rebound hyperkalemia during rewarming (Polderman, 2004). The rate of rewarming is likely closely associated with rebound hyperkalemia, with faster rates resulting in a higher incidence of hyperkalemia.

Acid–base status

As patients are cooled, carbon dioxide becomes more soluble, the partial pressure of carbon dioxide ($p\text{CO}_2$) falls, and the pH rises. There are two management strategies for this change in solubility during cooling: alpha-stat management refers to the practice of interpreting blood gas values at 37°C regardless of the patient's actual body temperature, and pH-stat management is when blood gas values are corrected to account for the colder body temperature. To maintain normal $p\text{CO}_2$ and pH levels with pH-stat management, a state of hypoventilation and hypercarbia is maintained, which results in cerebral vasodilation and could, in theory, lead to an increase in CBF and ICP. Substantial controversy exists over which method of acid–base management, if either, is preferable (Bacher, 2005; Polderman, 2009; Lay and Badjatia, 2010). In general, only one method should be utilized, with respiratory management performed accordingly.

Insulin resistance and kidney dysfunction

Insulin resistance occurs during hypothermia, but often this is difficult to differentiate from insulin resistance known to occur with critical illness. Regardless, the rate of rewarming may increase insulin sensitivity rapidly, and may lead to hypoglycemia (Polderman, 2004). Peripheral vasoconstriction during hypothermia can cause a diversion of blood to the kidneys, which results in mild renal tubular dysfunction. The combination of cooling and renal dysfunction produces a "cold diuresis" effect (Knight and Horvath, 1985; Guluma et al., 2010), which can make fluid management during hypothermia challenging.

Finally, cooling has an unmeasurable impact on renal clearance of common pharmaceuticals, especially analgesics and sedatives. Hypothermia is particularly impactful on the cytochrome P450 enzymes, which play an essential role in metabolizing medications commonly used in critical care medicine, including benzodiazepines, calcium channel blockers, anesthetics, and opioids (Empey et al., 2012). As a result, there is often a sustained impact of analgesia and sedation long after cooling has been discontinued. This can affect the reliability of the neurologic examination, and may impact on the timing for prognostication.

Cardiac function

Core body temperatures between 33°C and 35°C are generally well tolerated by the heart. As long as shivering is well controlled, cooling results in bradycardia and reduced myocardial contractility, which causes reduced cardiac output and blood pressure. Temperatures below 32°C can lead to serious cardiac arrhythmias such as atrial and ventricular tachycardia and fibrillation (Polderman, 2004; Bergman et al., 2010).

Impaired immune function

Cooling impairs leukocyte phagocytic function and causes immunosuppression, which explains the increased risk for bacterial infections during hypothermia (Todd et al., 2005; Seule et al., 2009; Hemmen et al., 2010). This risk appears to increase especially with prolonged hypothermia, though it is not clear at which time point the risk is unavoidable. Tracking the development of infections can also be difficult in the absence of temperature elevations and raised white blood cell counts. There are currently no reliable methods to overcome these limitations and reliably track the development of infections during hypothermia. A high index of suspicion is required. A recent retrospective cohort study of cardiac arrest patients who underwent hypothermia and treated prophylactically with antibiotics had a fourfold reduction in pneumonia rates (Gagnon et al., 2015). However, there was limited information about microbiologic organisms, the type of antibiotic prophylaxis given, and timing and duration of therapy, limiting the ability of generalizing this practice without further prospective study results.

Hematologic effects

Platelet dysfunction, increased fibrinolysis, and decreased activity of coagulation cascade enzymes all contribute to bleeding during hypothermia. However, clinically relevant coagulopathy and thrombocytopenia seem to occur more frequently in spontaneous hypothermia after trauma than after medically induced hypothermia. Mild coagulopathy and platelet dysfunction also occur at temperatures above 35°C, but most trials have not shown an increased risk of serious bleeding, even in patients with pre-existing intracranial hemorrhage (Schefold et al., 2009).

OUTCOME PREDICTION

Assessing the ability for recovery after a neurologic injury is often a multimodality determination. Electrophysiologic tests such as EEG and evoked potentials may have dampened signals or latencies due to cooling, however, their prognostic value has not been shown to be impacted by hypothermia. Likewise, neuroimaging and biomarker assessments, such as neuron-specific enolase, have not been shown to be impacted by core body temperature (Blondin and Greer, 2011). However, these modalities are usually ancillary to the neurologic examination, which remains the cornerstone of prognostication for all types of brain injury. Hypothermia can have a profound impact on the ability to track the neurologic examination due to the lingering effects of sedatives and analgesics. A recent study in cardiac arrest patients found that treatment with hypothermia was associated with a variable delay in awakening for up to 5 days postarrest (Grossestreuer et al., 2013). Therefore, it is important to account for an adequate time period after hypothermia before determining prognosis on the basis of the neurologic examination.

REFERENCES

Abou-Chebl A, DeGeorgia MA, Andrefsky JC et al. (2004). Technical refinements and drawbacks of a surface cooling technique for the treatment of severe acute ischemic stroke. Neurocrit Care 1: 131–143.

Adelson PD, Ragheb J, Kanev P et al. (2005). Phase II clinical trial of moderate hypothermia after severe traumatic brain injury in children. Neurosurgery 56: 740–754.

Adelson PD, Wisniewski SR, Beca J et al. (2013). Comparison of hypothermia and normothermia after severe traumatic brain injury in children (Cool Kids): a phase 3, randomised controlled trial. Lancet Neurol 12: 546–553.

Andrews PJ, Sinclair HL, Rodriguez A et al. (2015). Hypothermia for intracranial hypertension after traumatic brain injury. N Engl J Med 373: 2403–2412.

Bacher A (2005). Effects of body temperature on blood gases. Intensive Care Med 31: 24–27.

Badjatia N (2009). Hyperthermia and fever control in brain injury. Crit Care Med 37: S250–S257.

Badjatia N, Strongilis E, Gordon E et al. (2008). Metabolic impact of shivering during therapeutic temperature modulation: the Bedside Shivering Assessment Scale. Stroke 39: 3242–3247.

Badjatia N, Strongilis E, Prescutti M et al. (2009). Metabolic benefits of surface counter warming during therapeutic temperature modulation. Crit Care Med 37: 1893–1897.

Baumann E, Preston E, Slinn J et al. (2009). Post-ischemic hypothermia attenuates loss of the vascular basement membrane proteins, agrin and SPARC, and the blood–brain barrier disruption after global cerebral ischemia. Brain Res 1269: 185–197.

Bennett M, Pellegrini-Giampietro D, Gorter J et al. (1996). The GluR2 hypothesis: Ca++-permeable AMPA receptors in delayed neurodegeneration. Cold Spring Harbor Symposia on 61: 373–384.

Bergman R, Braber A, Adriaanse MA et al. (2010). Haemodynamic consequences of mild therapeutic hypothermia after cardiac arrest. Eur J Anaesthesiol 27: 383–387.

Bernard SA, Gray TW, Buist MD et al. (2002). Treatment of comatose survivors of out-of-hospital cardiac arrest with induced hypothermia. N Engl J Med 346: 557–563.

Bernard SA, Smith K, Cameron P et al. (2010). Induction of therapeutic hypothermia by paramedics after resuscitation from out-of-hospital ventricular fibrillation cardiac arrest: a randomized controlled trial. Circulation 122: 737–742.

Bernard SA, Smith K, Cameron P et al. (2012). Induction of prehospital therapeutic hypothermia after resuscitation from nonventricular fibrillation cardiac arrest. Crit Care Med 40: 747–753.

Biswas AK, Bruce DA, Sklar FH et al. (2002). Treatment of acute traumatic brain injury in children with moderate hypothermia improves intracranial hypertension. Crit Care Med 30: 2742–2751.

Blondin NA, Greer DM (2011). Neurologic prognosis in cardiac arrest patients treated with therapeutic hypothermia. Neurologist 17: 241–248.

Broessner G, Beer R, Lackner P et al. (2009). Prophylactic, endovascularly based, long-term normothermia in ICU patients with severe cerebrovascular disease: bicenter prospective, randomized trial. Stroke 40: e657–e665.

Busto R, Dietrich W, Globus M-T et al. (1987). Small differences in intraischemic brain temperature critically determine the extent of ischemic neuronal injury. J Cereb Blood Flow Metab 7: 729–738.

Busto R, Globus MY, Neary JT et al. (1994). Regional alterations of protein kinase C activity following transient cerebral ischemia: effects of intraischemic brain temperature modulation. J Neurochem 63: 1095–1103.

Callaghan JC, McQueen DA, Scott JW et al. (1954). Cerebral effects of experimental hypothermia. Arch Surg 68: 208–215.

Callaway CW, Donnino MW, Fink EL et al. (2015). Part 8: Post-Cardiac Arrest Care: 2015 American Heart Association guidelines update for cardiopulmonary resuscitation and emergency cardiovascular care. Circulation 132: S465–S482.

Cardell M, Wieloch T (1993). Time course of the translocation and inhibition of protein kinase C during complete cerebral ischemia in the rat. J Neurochem 61: 1308–1314.

Ceulemans AG, Zgavc T, Kooijman R et al. (2010). The dual role of the neuroinflammatory response after ischemic stroke: modulatory effects of hypothermia. J Neuroinflammation 7: 74.

Ceulemans AG, Zgavc T, Kooijman R et al. (2011). Mild hypothermia causes differential, time-dependent changes in cytokine expression and gliosis following endothelin-1-induced transient focal cerebral ischemia. J Neuroinflammation 8: 60.

Chapman PF, Frenguelli BG, Smith A et al. (1995). The alpha-Ca2+/calmodulin kinase II: a bidirectional modulator of presynaptic plasticity. Neuron 14: 591–597.

Chatfield PO, Lyman CP, Purpura DP (1951). The effects of temperature on the spontaneous and induced electrical activity in the cerebral cortex of the golden hamster. Electroencephalogr Clin Neurophysiol 3: 225–230.

Choi HA, Ko SB, Presciutti M et al. (2011). Prevention of shivering during therapeutic temperature modulation: the Columbia Anti-Shivering Protocol. Neurocrit Care 14: 389–394.

Chopp M, Knight R, Tidwell C et al. (1989). The metabolic effects of mild hypothermia on global cerebral ischemia and recirculation in the cat: comparison to normothermia and hyperthermia. J Cereb Blood Flow Metab 9: 141–148.

Clifton GL, Miller ER, Choi SC et al. (2001). Lack of effect of induction of hypothermia after acute brain injury. N Engl J Med 344: 556–563.

Clifton GL, Valadka A, Zygun D et al. (2011). Very early hypothermia induction in patients with severe brain injury (the National Acute Brain Injury Study: Hypothermia II): a randomised trial. Lancet Neurol 10: 131–139.

Colbourne F, Grooms SY, Zukin RS et al. (2003). Hypothermia rescues hippocampal CA1 neurons and attenuates down-regulation of the AMPA receptor GluR2 subunit after forebrain ischemia. Proc Natl Acad Sci U S A 100: 2906–2910.

Debaty G, Maignan M, Savary D et al. (2014). Impact of intra-arrest therapeutic hypothermia in outcomes of prehospital cardiac arrest: a randomized controlled trial. Intensive Care Med 40: 1832–1842.

Deng H, Han HS, Cheng D et al. (2003). Mild hypothermia inhibits inflammation after experimental stroke and brain inflammation. Stroke 34: 2495–2501.

Deye N, Cariou A, Girardie P et al. (2015). Endovascular versus external targeted temperature management for patients with out-of-hospital cardiac arrest: a randomized, controlled study. Circulation 132: 182–193.

Dietrich WD, Levi AD, Wang M et al. (2011). Hypothermic treatment for acute spinal cord injury. Neurotherapeutics 8: 229–239.

Diringer MN. Neurocritical Care Fever Reduction Trial Group (2004). Treatment of fever in the neurologic intensive care unit with a catheter-based heat exchange system. Crit Care Med 32: 559–564.

Drabek T, Janata A, Jackson EK et al. (2012). Microglial depletion using intrahippocampal injection of liposome-encapsulated clodronate in prolonged hypothermic cardiac arrest in rats. Resuscitation 83: 517–526.

Duz B, Oztas E, Erginay T et al. (2007). The effect of moderate hypothermia in acute ischemic stroke on pericyte migration: an ultrastructural study. Cryobiology 55: 279–284.

Els T, Oehm E, Voigt S et al. (2006). Safety and therapeutical benefit of hemicraniectomy combined with mild hypothermia in comparison with hemicraniectomy alone in patients with malignant ischemic stroke. Cerebrovasc Dis 21: 79–85.

Empey PE, Miller TM, Philbrick AH et al. (2012). Mild hypothermia decreases fentanyl and midazolam steady-state clearance in a rat model of cardiac arrest. Crit Care Med 40: 1221–1228.

Erecinska M, Thoresen M, Silver IA (2003). Effects of hypothermia on energy metabolism in mammalian central nervous system. J Cereb Blood Flow Metab 23: 513–530.

Fehlings MG, Vaccaro A, Wilson JR et al. (2012). Early versus delayed decompression for traumatic cervical spinal cord injury: results of the Surgical Timing in Acute Spinal Cord Injury Study (STASCIS). PLoS One 7. e32037.

Fingas M, Clark DL, Colbourne F (2007). The effects of selective brain hypothermia on intracerebral hemorrhage in rats. Exp Neurol 208: 277–284.

Florian B, Vintilescu R, Balseanu AT et al. (2008). Long-term hypothermia reduces infarct volume in aged rats after focal ischemia. Neurosci Lett 438: 180–185.

Fries M, Brucken A, Cizen A et al. (2012). Combining xenon and mild therapeutic hypothermia preserves neurological function after prolonged cardiac arrest in pigs. Crit Care Med 40: 1297–1303.

Fukui O, Kinugasa Y, Fukuda A et al. (2006). Post-ischemic hypothermia reduced IL-18 expression and suppressed microglial activation in the immature brain. Brain Res 1121: 35–45.

Gagnon DJ, Nielsen N, Fraser GL et al. (2015). Prophylactic antibiotics are associated with a lower incidence of pneumonia in cardiac arrest survivors treated with targeted temperature management. Resuscitation 92: 154–159.

Gasser S, Khan N, Yonekawa Y et al. (2003). Long-term hypothermia in patients with severe brain edema after poor-grade subarachnoid hemorrhage: feasibility and intensive care complications. J Neurosurg Anesthesiol 15: 240–248.

Gluckman PD, Wyatt JS, Azzopardi D et al. (2005). Selective head cooling with mild systemic hypothermia after neonatal encephalopathy: multicentre randomised trial. Lancet 365: 663–670.

Grossestreuer AV, Abella BS, Leary M et al. (2013). Time to awakening and neurologic outcome in therapeutic hypothermia-treated cardiac arrest patients. Resuscitation 84: 1741–1746.

Guluma KZ, Oh H, Yu SW et al. (2008). Effect of endovascular hypothermia on acute ischemic edema: morphometric analysis of the ICTuS trial. Neurocrit Care 8: 42–47.

Guluma KZ, Liu L, Hemmen TM et al. (2010). Therapeutic hypothermia is associated with a decrease in urine output in acute stroke patients. Resuscitation 81: 1642–1647.

Gupta AK, Al-Rawi PG, Hutchinson PJ et al. (2002). Effect of hypothermia on brain tissue oxygenation in patients with severe head injury. Br J Anaesth 88: 188–192.

Hagerdal M, Harp J, Siesjo BK (1975). Effect of hypothermia upon organic phosphates, glycolytic metabolites, citric acid cycle intermediates and associated amino acids in rat cerebral cortex. J Neurochem 24: 743–748.

Hamann GF, Burggraf D, Martens HK et al. (2004). Mild to moderate hypothermia prevents microvascular basal lamina antigen loss in experimental focal cerebral ischemia. Stroke 35: 764–769.

Han HS, Qiao Y, Karabiyikoglu M et al. (2002). Influence of mild hypothermia on inducible nitric oxide synthase expression and reactive nitrogen production in experimental stroke and inflammation. J Neurosci 22: 3921–3928.

Han HS, Karabiyikoglu M, Kelly S et al. (2003). Mild hypothermia inhibits nuclear factor-kappaB translocation in experimental stroke. J Cereb Blood Flow Metab 23: 589–598.

Harada K, Maekawa T, Tsuruta R et al. (2002). Hypothermia inhibits translocation of CaM kinase II and PKC-alpha, beta, gamma isoforms and fodrin proteolysis in rat brain synaptosome during ischemia-reperfusion. J Neurosci Res 67: 664–669.

Hartings JA, Strong AJ, Fabricius M et al. (2009). Spreading depolarizations and late secondary insults after traumatic brain injury. J Neurotrauma 26: 1857–1866.

Hartings JA, Bullock MR, Okonkwo DO et al. (2011). Spreading depolarisations and outcome after traumatic brain injury: a prospective observational study. Lancet Neurol 10: 1058–1064.

Hayashi T, Abe K, Itoyama Y (1998). Reduction of ischemic damage by application of vascular endothelial growth factor in rat brain after transient ischemia. J Cereb Blood Flow Metab 18: 887–895.

Hemmen TM, Raman R, Guluma KZ et al. (2010). Intravenous thrombolysis plus hypothermia for acute treatment of ischemic stroke (ICTuS-L): final results. Stroke 41: 2265–2270.

Hoedemaekers CW, Ezzahti M, Gerritsen A et al. (2007). Comparison of cooling methods to induce and maintain normo- and hypothermia in intensive care unit patients: a prospective intervention study. Crit Care 11: R91.

Hofmeijer J, Kappelle LJ, Algra A et al. (2009). Surgical decompression for space-occupying cerebral infarction (the Hemicraniectomy After Middle Cerebral Artery infarction with Life-threatening Edema Trial [HAMLET]): a multicentre, open, randomised trial. Lancet Neurol 8: 326–333.

Horstmann S, Kalb P, Koziol J et al. (2003). Profiles of matrix metalloproteinases, their inhibitors, and laminin in stroke patients: influence of different therapies. Stroke 34: 2165–2170.

Howell DA, Posnikoff J, Stratford JG (1956). Prolonged hypothermia in treatment of massive cerebral haemorrhage; a preliminary report. Can Med Assoc J 75: 388–394.

Hu BR, Wieloch T (1995). Persistent translocation of Ca2+/calmodulin-dependent protein kinase II to synaptic junctions in the vulnerable hippocampal CA1 region following transient ischemia. J Neurochem 64: 277–284.

Huang F-P, Zhou L-F, Yang G-Y (1998). The effect of extending mild hypothermia on focal cerebral ischemia and reperfusion in the rat. Neurol Res 20: 57–62.

Hutchison JS, Ward RE, Lacroix J et al. (2008). Hypothermia therapy after traumatic brain injury in children. N Engl J Med 358: 2447–2456.

Hypothermia After Cardiac Arrest Study Group (2002). Mild therapeutic hypothermia to improve the neurologic outcome after cardiac arrest. N Engl J Med 346: 549–556.

Ibayashi S, Takano K, Ooboshi H et al. (2000). Effect of selective brain hypothermia on regional cerebral blood flow and tissue metabolism using brain thermo-regulator in spontaneously hypertensive rats. Neurochem Res 25: 369–375.

Inamasu J, Suga S, Sato S et al. (2000). Post-ischemic hypothermia delayed neutrophil accumulation and microglial activation following transient focal ischemia in rats. J Neuroimmunol 109: 66–74.

Juttler E, Schwab S, Schmiedek P et al. (2007). Decompressive Surgery for the Treatment of Malignant Infarction of the Middle Cerebral Artery (DESTINY): a randomized, controlled trial. Stroke; a journal of cerebral circulation 38: 2518–2525.

Kader A, Frazzini VI, Baker CJ et al. (1994). Effect of mild hypothermia on nitric oxide synthesis during focal cerebral ischemia. Neurosurgery 35: 272–277. discussion 277.

Kaibara T, Sutherland GR, Colbourne F et al. (1999). Hypothermia: depression of tricarboxylic acid cycle flux and evidence for pentose phosphate shunt upregulation. J Neurosurg 90: 339–347.

Kallmunzer B, Schwab S, Kollmar R (2012). Mild hypothermia of 34 degrees C reduces side effects of rt-PA treatment after thromboembolic stroke in rats. Exp Transl Stroke Med 4: 3.

Kammersgaard LP, Rasmussen BH, Jorgensen HS et al. (2000). Feasibility and safety of inducing modest hypothermia in awake patients with acute stroke through surface cooling: a case-control study; the Copenhagen Stroke Study. Stroke; a journal of cerebral circulation 31: 2251–2256.

Karabiyikoglu M, Han HS, Yenari MA et al. (2003). Attenuation of nitric oxide synthase isoform expression by mild hypothermia after focal cerebral ischemia: variations depending on timing of cooling. J Neurosurg 98: 1271–1276.

Karibe H, Zarow GJ, Graham SH et al. (1994). Mild intraischemic hypothermia reduces postischemic hyperperfusion, delayed postischemic hypoperfusion, blood–brain barrier disruption, brain edema, and neuronal damage volume after temporary focal cerebral ischemia in rats. J Cereb Blood Flow Metab 14: 620–627.

Kasner SE, Wein T, Piriyawat P et al. (2002). Acetaminophen for altering body temperature in acute stroke: a randomized clinical trial. Stroke 33: 130–134.

Kawai N, Okauchi M, Morisaki K et al. (2000). Effects of delayed intraischemic and postischemic hypothermia on a focal model of transient cerebral ischemia in rats. Stroke 31: 1982–1989.

Kawamata T, Dietrich WD, Schallert T et al. (1997). Intracisternal basic fibroblast growth factor enhances functional recovery and up-regulates the expression of a molecular marker of neuronal sprouting following focal cerebral infarction. Proc Natl Acad Sci U S A 94: 8179–8184.

Kawamura S, Suzuki A, Hadeishi H et al. (2000). Cerebral blood flow and oxygen metabolism during mild hypothermia in patients with subarachnoid haemorrhage. Acta Neurochir (Wien) 142: 1117–1121. discussion 1121–1122.

Kim F, Nichol G, Maynard C et al. (2014). Effect of prehospital induction of mild hypothermia on survival and neurological status among adults with cardiac arrest: a randomized clinical trial. JAMA 311: 45–52.

Kimura T, Sako K, Tanaka K et al. (2002). Effect of mild hypothermia on energy state recovery following transient forebrain ischemia in the gerbil. Exp Brain Res 145: 83–90.

Knight DR, Horvath SM (1985). Urinary responses to cold temperature during water immersion. Am J Physiol 248: R560–R566.

Kollmar R, Staykov D, Dorfler A et al. (2010). Hypothermia reduces perihemorrhagic edema after intracerebral hemorrhage. Stroke 41: 1684–1689.

Kristian T, Katsura K, Siesjo BK (1992). The influence of moderate hypothermia on cellular calcium uptake in complete ischaemia: implications for the excitotoxic hypothesis. Acta Physiol Scand 146: 531–532.

Lavinio A, Timofeev I, Nortje J et al. (2007). Cerebrovascular reactivity during hypothermia and rewarming. Br J Anaesth 99: 237–244.

Lay C, Badjatia N (2010). Therapeutic hypothermia after cardiac arrest. Curr Atheroscler Rep 12: 336–342.

Lee JE, Yoon YJ, Moseley ME (2005). Reduction in levels of matrix metalloproteinases and increased expression of tissue inhibitor of metalloproteinase-2 in response to mild hypothermia therapy in experimental stroke. J Neurosurg 103: 289–297.

Lennon RL, Hosking MP, Conover MA et al. (1990). Evaluation of a forced-air system for warming hypothermic postoperative patients. Anesth Analg 70: 424–427.

Levi AD, Green BA, Wang MY et al. (2009). Clinical application of modest hypothermia after spinal cord injury. J Neurotrauma 26: 407–415.

Lin CS, Ho HC, Chen KC et al. (2002). Intracavernosal injection of vascular endothelial growth factor induces nitric oxide synthase isoforms. BJU Int 89: 955–960.

Linares G, Mayer SA (2009). Hypothermia for the treatment of ischemic and hemorrhagic stroke. Crit Care Med 37: S243–S249.

Lyman CP, Chatfield PO (1953). Hibernation and cortical electrical activity in the woodchuck (Marmota monax). Science (New York, NY) 117: 533–534.

MacLellan CL, Clark DL, Silasi G et al. (2009). Use of prolonged hypothermia to treat ischemic and hemorrhagic stroke. J Neurotrauma 26: 313–323.

Madhok J, Wu D, Xiong W et al. (2012). Hypothermia amplifies somatosensory-evoked potentials in uninjured rats. J Neurosurg Anesthesiol 24: 197–202.

Massopust Jr LC, Wolin LR, White RJ et al. (1970). Electroencephalographic characteristics of brain cooling and rewarming in monkey. Exp Neurol 26: 518–526.

Matsukawa T, Kurz A, Sessler DI et al. (1995). Propofol linearly reduces the vasoconstriction and shivering thresholds. Anesthesiology 82: 1169–1180.

Mayer SA, Kowalski RG, Presciutti M et al. (2004). Clinical trial of a novel surface cooling system for fever control in neurocritical care patients. Crit Care Med 32: 2508–2515.

McIntyre LA, Fergusson DA, Hebert PC et al. (2003). Prolonged therapeutic hypothermia after traumatic brain injury in adults: a systematic review. JAMA 289: 2992–2999.

Michenfelder JD, Milde JH (1991). The relationship among canine brain temperature, metabolism, and function during hypothermia. Anesthesiology 75: 130–136.

Mokhtarani M, Mahgoub AN, Morioka N et al. (2001). Buspirone and meperidine synergistically reduce the shivering threshold. Anesth Analg 93: 1233–1239.

Moler FW, Silverstein FS, Holubkov R et al. (2015). Therapeutic hypothermia after out-of-hospital cardiac arrest in children. N Engl J Med 372: 1898–1908.

Morikawa E, Ginsberg MD, Dietrich WD et al. (1992). The significance of brain temperature in focal cerebral ischemia: histopathological consequences of middle cerebral artery occlusion in the rat. J Cereb Blood Flow Metab 12: 380–389.

Moyer DJ, Welsh FA, Zager EL (1992). Spontaneous cerebral hypothermia diminishes focal infarction in rat brain. Stroke 23: 1812–1816.

Mueller-Burke D, Koehler RC, Martin LJ (2008). Rapid NMDA receptor phosphorylation and oxidative stress precede striatal neurodegeneration after hypoxic ischemia in newborn piglets and are attenuated with hypothermia. Int J Dev Neurosci 26: 67–76.

Nagao S, Irie K, Kawai N et al. (2000). Protective effect of mild hypothermia on symptomatic vasospasm: a preliminary report. Acta Neurochir Suppl 76: 547–550.

Nagao S, Irie K, Kawai N et al. (2003). The use of mild hypothermia for patients with severe vasospasm: a preliminary report. J Clin Neurosci 10: 208–212.

Nagel S, Su Y, Horstmann S et al. (2008). Minocycline and hypothermia for reperfusion injury after focal cerebral ischemia in the rat-Effects on BBB breakdown and MMP expression in the acute and subacute phase. Brain Res 1188: 198–206.

Nakashima K, Todd MM, Warner DS (1995). The relation between cerebral metabolic rate and ischemic depolarization. A comparison of the effects of hypothermia, pentobarbital, and isoflurane. Anesthesiology 82: 1199–1208.

Nielsen N, Wetterslev J, Cronberg T et al. (2013). Targeted temperature management at 33 degrees C versus 36 degrees C after cardiac arrest. N Engl J Med 369: 2197–2206.

Nilsson L, Kogure K, Busto R (1975). Effects of hypothermia and hyperthermia on brain energy metabolism. Acta Anaesthesiol Scand 19: 199–205.

Okubo K, Itoh S, Isobe K et al. (2001). Cerebral metabolism and regional cerebral blood flow during moderate systemic cooling in newborn piglets. Pediatr Int 43: 496–501.

Owens G (1958). Effect of hypothermia on seizures induced by physical and chemical means. Am J Physiol 193: 560–562.

Peberdy MA, Callaway CW, Neumar RW et al. (2010). Part 9: post-cardiac arrest care: 2010 American Heart Association guidelines for cardiopulmonary resuscitation and emergency cardiovascular care. Circulation 122: S768–S786.

Polderman KH (2004). Application of therapeutic hypothermia in the intensive care unit. Opportunities and pitfalls of a promising treatment modality – Part 2: practical aspects and side effects. Intensive Care Med 30: 757–769.

Polderman KH (2009). Mechanisms of action, physiological effects, and complications of hypothermia. Crit Care Med 37: S186–S202.

Polderman KH, Peerdeman SM, Girbes AR (2001). Hypophosphatemia and hypomagnesemia induced by cooling in patients with severe head injury. J Neurosurg 94: 697–705.

Polderman KH, Ely EW, Badr AE et al. (2004). Induced hypothermia in traumatic brain injury: considering the conflicting results of meta-analyses and moving forward. Intensive Care Med 30: 1860–1864.

Polderman KH, Rijnsburger ER, Peerdeman SM et al. (2005). Induction of hypothermia in patients with various types of neurologic injury with use of large volumes of ice-cold intravenous fluid. Crit Care Med 33: 2744–2751.

Preston E, Webster J (2004). A two-hour window for hypothermic modulation of early events that impact delayed opening of the rat blood–brain barrier after ischemia. Acta Neuropathol 108: 406–412.

Quasha AL, Tinker JH, Sharbrough FW (1981). Hypothermia plus thiopental: prolonged electroencephalographic suppression. Anesthesiology 55: 636–640.

Rosomoff HL, Holaday DA (1954). Cerebral blood flow and cerebral oxygen consumption during hypothermia. Am J Physiol 179: 85–88.

Schefold JC, Storm C, Joerres A et al. (2009). Mild therapeutic hypothermia after cardiac arrest and the risk of bleeding in patients with acute myocardial infarction. Int J Cardiol 132: 387–391.

Schreckinger M, Marion DW (2009). Contemporary management of traumatic intracranial hypertension: is there a role for therapeutic hypothermia? Neurocrit Care 11: 427–436.

Schwab S, Georgiadis D, Berrouschot J et al. (2001). Feasibility and safety of moderate hypothermia after massive hemispheric infarction. Stroke; a journal of cerebral circulation 32: 2033–2035.

Sessler DI (2009). Defeating normal thermoregulatory defenses: induction of therapeutic hypothermia. Stroke 40: e614–e621.

Seule MA, Muroi C, Mink S et al. (2009). Therapeutic hypothermia in patients with aneurysmal subarachnoid hemorrhage, refractory intracranial hypertension, or cerebral vasospasm. Neurosurgery 64: 86–92. discussion 92–93.

Shankaran S, Laptook AR, Ehrenkranz RA et al. (2005). Whole-body hypothermia for neonates with hypoxic-ischemic encephalopathy. N Engl J Med 353: 1574–1584.

Shiozaki T, Hayakata T, Taneda M et al. (2001). A multicenter prospective randomized controlled trial of the efficacy of mild hypothermia for severely head injured patients with low intracranial pressure. Mild Hypothermia Study Group in Japan. J Neurosurg 94: 50–54.

Staykov D, Schwab S, Doerfler A et al. (2011). Hypothermia reduces perihemorrhagic edema after intracerebral hemorrhage: but does it influence functional outcome and mortality? Therapeutic Hypothermia and Temperature Management 1: 105–106.

Stecker MM, Cheung AT, Pochettino A et al. (2001). Deep hypothermic circulatory arrest: I. Effects of cooling on electroencephalogram and evoked potentials. Ann Thorac Surg 71: 14–21.

Steinberg GK, Ogilvy CS, Shuer LM et al. (2004). Comparison of endovascular and surface cooling during unruptured cerebral aneurysm repair. Neurosurgery 55: 307–314. discussion 314–315.

Steiner T, Friede T, Aschoff A et al. (2001). Effect and feasibility of controlled rewarming after moderate hypothermia in stroke patients with malignant infarction of the middle cerebral artery. Stroke; a journal of cerebral circulation 32: 2833–2835.

Stub D, Bernard S, Duffy SJ et al. (2011). Post cardiac arrest syndrome: a review of therapeutic strategies. Circulation 123: 1428–1435.

Suehiro E, Povlishock JT (2001). Exacerbation of traumatically induced axonal injury by rapid posthypothermic rewarming and attenuation of axonal change by cyclosporin A. J Neurosurg 94: 493–498.

Sutton LN, Clark BJ, Norwood CR et al. (1991). Global cerebral ischemia in piglets under conditions of mild and deep hypothermia. Stroke 22: 1567–1573.

Takaoka S, Pearlstein RD, Warner DS (1996). Hypothermia reduces the propensity of cortical tissue to propagate direct current depolarizations in the rat. Neurosci Lett 218: 25–28.

ten Cate J, Horsten GPM, Koopman LJ (1949). The influence of the body temperature on the EEG of the rat. Electroencephalogr Clin Neurophysiol 1: 231–235.

Thompson HJ, Kirkness CJ, Mitchell PH (2010). Hypothermia and rapid rewarming is associated with worse outcome following traumatic brain injury. J Trauma Nurs 17: 173–177.

Todd MM, Hindman BJ, Clarke WR et al. (2005). Mild intraoperative hypothermia during surgery for intracranial aneurysm. N Engl J Med 352: 135–145.

Tohyama Y, Sako K, Yonemasu Y (1998). Hypothermia attenuates the activation of protein kinase C in focal ischemic rat brain: dual autoradiographic study of [3H]phorbol 12,13-dibutyrate and iodo[14C]antipyrine. Brain Res 782: 348–351.

Tokutomi T, Morimoto K, Miyagi T et al. (2007). Optimal temperature for the management of severe traumatic brain injury: effect of hypothermia on intracranial pressure, systemic and intracranial hemodynamics, and metabolism. Neurosurgery 61: 256–265. discussion 265–266.

Ueda Y, Wei EP, Kontos HA et al. (2003). Effects of delayed, prolonged hypothermia on the pial vascular response after traumatic brain injury in rats. J Neurosurg 99: 899–906.

Van Hemelrijck A, Hachimi-Idrissi S, Sarre S et al. (2005). Post-ischaemic mild hypothermia inhibits apoptosis in the penumbral region by reducing neuronal nitric oxide synthase activity and thereby preventing endothelin-1-induced hydroxyl radical formation. Eur J Neurosci 22: 1327–1337.

Wagner S, Nagel S, Kluge B et al. (2003). Topographically graded postischemic presence of metalloproteinases is inhibited by hypothermia. Brain Res 984: 63–75.

Webster CM, Kelly S, Koike MA et al. (2009). Inflammation and NFkappaB activation is decreased by hypothermia following global cerebral ischemia. Neurobiol Dis 33: 301–312.

Welsh FA, Sims RE, Harris VA (1990). Mild hypothermia prevents ischemic injury in gerbil hippocampus. J Cereb Blood Flow Metab 10: 557–563.

Woodhall B, Reynolds DH, Mahaley Jr S et al. (1958). The physiologic and pathologic effects of localized cerebral hypothermia. Ann Surg 147: 673–683.

Xiao F, Arnold TC, Zhang S et al. (2004). Cerebral cortical aquaporin-4 expression in brain edema following cardiac arrest in rats. Acad Emerg Med 11: 1001–1007.

Xiong M, Yang Y, Chen GQ et al. (2009). Post-ischemic hypothermia for 24 h in P7 rats rescues hippocampal neuron: association with decreased astrocyte activation and inflammatory cytokine expression. Brain Res Bull 79: 351–357.

Yanamoto H, Nagata I, Niitsu Y et al. (2001). Prolonged mild hypothermia therapy protects the brain against permanent focal ischemia. Stroke 32: 232–239.

Yenari MA, Han HS (2006). Influence of hypothermia on post-ischemic inflammation: role of nuclear factor kappa B (NFkappaB). Neurochem Int 49: 164–169.

Zweifler RM, Voorhees ME, Mahmood MA et al. (2003). Induction and maintenance of mild hypothermia by surface cooling in non-intubated subjects. J Stroke Cerebrovasc Dis 12: 237–243.

Zweifler RM, Voorhees ME, Mahmood MA et al. (2004). Magnesium sulfate increases the rate of hypothermia via surface cooling and improves comfort. Stroke 35: 2331–2334.

Chapter 34

Neurologic complications of polytrauma

R.M. JHA AND L. SHUTTER*
Department of Critical Care Medicine, University of Pittsburgh, Pittsburgh, PA, USA

Abstract

Neurologic complications in polytrauma can be classified by etiology and clinical manifestations: neurovascular, delirium, and spinal or neuromuscular problems. Neurovascular complications include ischemic strokes, intracranial hemorrhage, or the development of traumatic arteriovenous fistulae. Delirium and encephalopathy have a reported incidence of 67–92% in mechanically ventilated polytrauma patients. Causes include sedation, analgesia/pain, medications, sleep deprivation, postoperative state, toxic ingestions, withdrawal syndromes, organ system dysfunction, electrolyte/metabolic abnormalities, and infections. Rapid identification and treatment of the underlying cause are imperative. Benzodiazepines increase the risk of delirium, and alternative agents are preferred sedatives. Pharmacologic treatment of agitated delirium can be achieved with antipsychotics. Nonconvulsive seizures and status epilepticus are not uncommon in surgical/trauma intensive care unit (ICU) patients, require electroencephalography for diagnosis, and need timely management. Spinal cord ischemia is a known complication in patients with traumatic aortic dissections or blunt aortic injury requiring surgery. Thoracic endovascular aortic repair has reduced the paralysis rate. Neuromuscular complications include nerve and plexus injuries, and ICU-acquired weakness. In polytrauma, the neurologic examination is often confounded by pain, sedation, mechanical ventilation, and distracting injuries. Regular sedation pauses for examination and maintaining a high index of suspicion for neurologic complications are warranted, particularly because early diagnosis and management can improve outcomes.

INTRODUCTION

Polytrauma typically refers to severely injured patients with two or more significant traumatic injuries, with a total injury severity score of greater than 15, or an abbreviated injury scale > 2 in at least two injury severity score regions (Butcher and Balogh, 2009; Pape, 2012; Butcher et al., 2014). (For an explanation of severity scoring, see www.surgicalcriticalcare.net/Resources/injury_severity_scoring.pdf.) Neurologic injury is frequently part of the presentation, and other than musculoskeletal injury, traumatic brain injury (TBI) has been cited as the most common constituent of polytrauma (Sumann et al., 2002; Gross et al., 2012; Andruszkow et al., 2013). Moderate TBI in addition to extracranial injury doubles the predicted mortality rate (McMahon et al., 1999; Sumann et al., 2002). Spinal cord injury occurs in approximately 6–10% of polytrauma patients, and the residual disability in survivors can consume a significant proportion of healthcare resources (estimated to be $9.7 billion per year in 1998) (Oliver et al., 2012; Stephan et al., 2015). Even in the absence of primary central or peripheral nervous system injury, polytrauma patients frequently have neurologic complications such as debilitating ischemic stroke, delirium, and critical illness neuropathy and myopathy. This chapter focuses on presentation, diagnosis, and management of these complications, and categorizes them into different etiologies, including neurovascular, seizures, encephalopathies, and spinal or neuromuscular (Table 34.1).

---

*Correspondence to: Lori Shutter, MD, Critical Care Medicine, Scaife Hall, 6th floor, 3550 Terrace Street, Pittsburgh PA 15261, USA. Tel: +1-412-647-8410, Fax: +1-412-647-8060, E-mail: shutterla@upmc.edu

Table 34.1

Categorization of polytrauma-related neurologic complications

Neurovascular	Dissections	Blunt carotid-vertebral injury (BCVI)
		Blunt carotid injury (BCI)
		Blunt vertebral injury (BVI)
		Aortic dissections extending into cervical vessels
	Coagulopathy	Ischemic (posttraumatic cerebral infarction, e.g., from treatment with recombinant factor VIIa or disseminated intravascular coagulation (DIC)
		Hemorrhagic (e.g., from thrombocytopenia, DIC)
	Embolic strokes	Fat emboli (long-bone fractures)
		Air emboli
		Cardiac emboli (pre-existing or new-onset arrhythmias)
	Hypoperfusion	e.g., from hemorrhagic shock
		Bilateral watershed infarcts (prolonged hypotension)
		Focal infarcts (particularly with pre-existing cervical/intracranial stenosis)
	Other	Arteriovenous fistulae (carotid-cavernous, dural arteriovenous fistula)
		Venous sinus thrombosis
		Cortical vein thrombosis
		Posterior reversible encephalopathy syndrome
Altered mental status/ delirium	Encephalopathy	Infectious
		Metabolic (e.g., renal failure, uremia, liver failure, hyperammonemia)
		Fluid/electrolyte disturbances
		Organ system failure (hypoxic, hypercarbic)
		Nutritional deficiency (e.g., Wernicke's encephalopathy)
	Iatrogenic	Sedatives/analgesics
		Psychoactive medications
		Polypharmacy
		Drug interactions
	Intoxication/withdrawal	Common substances include alcohol, benzodiazepines, opiates, tricyclic antidepressants, acetaminophen
	ICU delirium	
Seizures	Clinical seizures	Generalized
	Nonconvulsive seizures	Lateralized
	Periodic epileptiform discharges	Bilateral independent
		Stimulus-induced
	Cortical spreading depolarization	
Spine	Blunt aortic injury	
Neuromuscular	Nerve and plexus injuries	Peripheral nerve entrapment
		Brachial plexus injuries
		Lumbosacral plexus injuries
	Critical illness weakness/ ICU-acquired weakness	Critical illness polyneuropathy
		Critical illness myopathy

ICU, intensive care unit.

NEUROVASCULAR

Neurovascular complications in patients with polytrauma include arterial ischemic strokes (from vertebral or carotid dissection, hypoperfusion, disseminated intravascular coagulation (DIC) or other coagulopathy-related phenomena, air and fat emboli, or cardiac emboli), venous sinus thromboses, and development of traumatic arteriovenous fistulae (e.g., dural arteriovenous fistula and carotid cavernous fistula) (Ho, 1998; Blacker and Wijdicks, 2004).

Arterial dissection

EPIDEMIOLOGY

Blunt carotid-vertebral injury (BCVI) is relatively rare overall, with a total reported incidence ranging from 0.1% to 2.5% of all traumatic injuries (Prall et al., 1998; Hwang et al., 2010), with isolated carotid injury incidence of 0.1–0.4% (Hwang et al., 2010). These frequencies are likely underreported given the lack of standardized screening protocols (Biffl et al., 2009). Rapid diagnosis and management of this entity are imperative, given the associated high morbidity (37–80%) and mortality (5–43%) (Hwang et al., 2010).

Motor vehicle collisions (MVC) are the most frequent cause of BCVI, responsible for 45–96% of cases; other etiologies include assault, crush injuries, falls, pedestrian trauma, and strangulation (Fabian et al., 1996; Biffl et al., 1999a; Miller et al., 2001). BCVI is associated especially with direct blows to the neck, hyperextension, contralateral rotation, facial/intraoral trauma, and basal skull fractures (Biffl et al., 2001). Patients with associated cervical spine fractures are at particularly high risk (Cothren et al., 2007; Hwang et al., 2010), and up to 71% of patients with BCVI also have cervical spine injuries (Biffl et al., 2000). There is a 30% rate of BCVI in patients where cervical spine fractures involve the foramen transversarium (McKinney et al., 2007).

NEUROPATHOLOGY

The pathophysiology of either carotid or vertebral artery injury involves intimal disruption, which subsequently becomes a nidus for platelet adhesion and aggregation. The formed clot could result in emboli or could potentially occlude the vessel (Burlew and Biffl, 2010).

CLINICAL PRESENTATION

Most patients with BCVI are initially asymptomatic, making early recognition of this process challenging (Biffl et al., 1999b, 2001, 2002; Miller et al., 2001). Approximately 60% of these patients are found to have focal deficits more than 12 hours after the injury (Cogbill et al., 1994; Fabian et al., 1996; Parikh et al., 1997; Hwang et al., 2010). Further complicating diagnosis are issues such as the inability to perform a good neurologic examination due to sedation and absence of screening protocols (Hwang et al., 2010). Clinical symptoms can present within an hour to years later, but typically between 10 and 72 hours after injury (Cogbill et al., 1994; Biffl et al., 1999b, 2002; Cothren et al., 2004a, 2009). Patients with BCVI may develop stroke symptoms if not treated appropriately with antiplatelets or anticoagulation (Burlew and Biffl, 2010). Symptoms may include Horner's syndrome or cranial nerve deficits caused by dissection or related hematomas. Deficits from cerebral ischemia are related to the location of infarct and may include sensorimotor deficits, aphasia, ataxia, visual field deficits, or hemineglect (Burlew and Biffl, 2010).

NEURODIAGNOSTICS AND IMAGING

The Denver classification system was established in 1999 as a radiographic scale to grade BCVI injuries (Table 34.2a) (Biffl et al., 1999a). This scale classifies the injury based on the arteriographic appearance of vessels, as determined by conventional angiography, which remains the gold standard for diagnosis (Biffl et al., 1999a).

CT angiography (CTA) is a viable alternative to catheter-based imaging, as it is noninvasive, easily available, and has a lower complication rate (Biffl et al., 2002; Cothren et al., 2004a; Burlew and Biffl, 2010). Some studies using multidetector CTAs are now reporting sensitivity and specificity rates approaching 100% for detection of BCVI, when interpreted by experienced radiologists (Berne et al., 2006; Biffl et al., 2006; Utter et al., 2006; Malhotra et al. 2007). However, it is important to be aware of conflicting data suggesting much lower sensitivity (54%), despite utilization of newer technologies such as 64-slice CTA (Goodwin et al., 2009). The most challenging areas for detection of BCVI with CTA are at the skull base.

Magnetic resonance imaging (MRI) and duplex ultrasonography are not considered reliable imaging tools for

Table 34.2a

Denver criteria for blunt carotid injury grading scale

I	<25% narrowing of the vessel lumen with irregularity or dissection
II	≥25% narrowing of the vessel lumen with intramural hematoma or dissection
III	Pseudoaneurysm
IV	Complete occlusion
V	Vessel transection

BCVI (Burlew and Biffl, 2010). Diffusion-weighted MRI is useful to detect infarcts as a result of emboli or occlusions.

CLINICAL TRIALS AND GUIDELINES

There are no double-blind randomized controlled trials evaluating the appropriate screening criteria or management of BCVI.

Given the high morbidity and mortality associated with BCVI, along with an initial clinically silent period making diagnosis challenging, the development of appropriate screening protocols is imperative for early recognition and treatment. Early diagnosis and therapy have been associated with improved neurologic outcomes. Protocols typically suggest screening for high-risk injury patterns (Table 34.2b) (Biffl et al., 1999b; Cothren et al., 2003). Notwithstanding proposed protocols based on high-risk patterns of injury and presentation, there remains significant controversy in the literature on the advantages of screening, with some groups arguing for more aggressive and broader screening and others questioning its utility (Biffl et al., 2009; Berne et al., 2010; Bromberg et al., 2010; Burlew and Biffl, 2010). Based on publications from the Eastern Association for the Surgery of Trauma and Western Trauma Association, screening recommendations at many institutions classify patients based on mechanism of injury, as well as injury patterns, and perform screening imaging on high-risk patients (Biffl et al., 2009; Bromberg et al., 2010).

Focal neurologic deficits and poor outcomes in polytrauma patients were often attributed to TBI prior to recognition of BCVI (Burlew and Biffl, 2010). The use of screening protocols and increasing recognition of this entity, combined with prompt initiation of treatment, has reduced the incidence and morbidity of BCVI-related stroke (Miller et al., 2001, 2002; Biffl et al., 2002, 2009; Bromberg et al., 2010). Some groups have reported a 10- and 20-fold reduction in stroke risk from carotid and vertebral artery injuries, respectively, after either anticoagulation or administration of antiplatelet agents. Other groups have shown a stroke rate as low as 0.3% in patients with BCVI who were treated with antiplatelets, compared with 21% in untreated patients (Cothren et al., 2009; Burlew and Biffl, 2010). The optimal treatment regimen has not been established.

In early studies, unfractionated heparin was used as the treatment of choice, based on the assumption that it would promote clot stabilization and prevent further clot formation (Burlew and Biffl, 2010). Studies suggested improved outcome in BCVI patients treated with anticoagulation (Cothren et al., 2004b). Given the concern of bleeding risk, goal partial thromboplastin time (PTT) levels have been reduced, thereby reducing this complication risk to as low as 1–4% (Biffl et al., 2002; Cothren et al., 2009). Heparin infusions are often titrated to achieve PTT levels between 40 and 50 seconds, and a loading dose is generally not used (Biffl et al., 2009). Several small case series and retrospective studies have reported the use of antiplatelet therapies as an alternative to anticoagulation (Biffl et al., 2009). Unfortunately, there are no multicenter randomized trials comparing the efficacy and complication rates of these two therapies. In retrospective analyses, antiplatelet therapy and anticoagulation have been shown to be equivalent in the prevention of stroke, as well as the resolution of the vascular injury (Miller et al., 2002; Cothren et al., 2009; Burlew and Biffl, 2010). The largest comparison of the two treatment modalities was a retrospective review performed over 10 years in 422 patients with BCVI. There was no significant difference in stroke risk or injury healing rates with heparin versus antiplatelet agents (Cothren et al., 2009). The optimal duration of therapy is unknown, but it is most often continued for 3–6 months.

More invasive treatment strategies such as endovascular stenting or surgery have also been considered for BCVI. Most BCVI locations are surgically inaccessible (Burlew and Biffl, 2010). The data for endovascular stenting consist primarily of case reports and series (Seth et al., 2013; Brzezicki et al., 2016). It is important to remember that endovascular stenting often requires use of dual antiplatelet therapy postprocedurally. This can be problematic in patients who are at high risk for bleeding complications.

HOSPITAL COURSE AND COMPLEX CLINICAL DECISIONS

Two primary clinical decisions are necessary in the polytrauma patient: when to screen for BCVI and what treatment should be initiated if screening is positive. It is

Table 34.2b

Blunt carotid vertebral injury screening indications

Carotid injuries	Vertebral injuries
Cervical hyperextension/rotation/ hyperflexion injuries	Cervical spine subluxations
Midface or mandibular fractures	Foramen transversarium fractures
Diffuse axonal injury	
Cervical region seat belt abrasions	
Soft-tissue injury of the neck	Fractures involving C1–C3
Altered mental status	
Basal skull fractures	
Cervical vertebral fractures	
Glasgow Coma Scale (GCS) score < 6	

reasonable to initially screen using CTA if there is the presence of a cervical fracture, complex facial fractures, a seatbelt sign, or neurologic deficits (Harrigan et al., 2011; Crawford et al., 2015). In light of the significant morbidity and mortality in untreated patients, a high degree of suspicion should be maintained in high-risk patients.

Based on the Denver criteria, the risk of stroke after BCVI appears to increase with higher grade of carotid, but not vertebral artery, injury (Biffl et al., 2002). Despite the reduced risk of stroke, timing of treatment initiation needs to be balanced with other considerations, including the potential for bleeding complications from other injuries and surgical planning. The Denver group recommends initial treatment of BCVI with intravenous heparin, given the ease with which it can be discontinued and reversed in the case of a bleeding complication. For grade 1 injuries, follow-up imaging can be obtained in 7–10 days, because more than 50% completely heal and it is possible that treatment can be discontinued. For those that do not heal, and most other injuries (grades 2–5), treatment can be transitioned to antiplatelet regimen at discharge, with repeat imaging at 6 months to re-evaluate the injury and healing (Burlew and Biffl, 2010). However, this strategy has not been shown to be superior to the use of aspirin only, with which some clinicians are more comfortable.

If patients develop strokes secondary to BCVI, they are usually not candidates for systemic thrombolysis in light of their recent trauma, although endovascular therapy could be considered. Additionally, diagnosis is frequently delayed, since neurologic deficits are often not easily detected in patients with distracting injuries or those requiring sedation. In delayed diagnosis, poststroke management may be limited to secondary measures, such as control of cerebral edema.

Coagulopathy-related neurovascular phenomena

Epidemiology

Acute coagulopathies in polytrauma patients are one of the most common causes of mortality (Lippi and Cervellin, 2010). Most patients with severe injuries are coagulopathic on admission, and this is particularly pronounced in those requiring massive transfusion (Tieu et al., 2007). During resuscitation of polytrauma patients with bleeding, one of the major challenges is to prevent coagulation disturbances and the development of DIC, which can result in either excessive bleeding or clotting. The incidence of DIC has been noted to be as high as 70% in patients with severe trauma (Levi and Cate, 1999).

Posttraumatic coagulopathies in trauma have been associated with ischemic and hemorrhagic strokes and unfavorable outcomes (Saggar et al., 2009; Lustenberger et al., 2010; Chen et al., 2013). Posttraumatic cerebral infarction (PTCI) is a well-documented entity in patients with TBI and polytrauma (Server et al., 2001; Chen et al., 2013). Underlying etiologies include intracranial mass effect causing vascular impingement, vasospasm, and posttraumatic coagulopathy, including DIC (Chen et al., 2013). One retrospective study found that more than 25% of patients with moderate to severe head injury who develop DIC will also develop early PTCI (Chen et al., 2013). Another risk factor for PTCI is treatment with recombinant factor VIIa, which is occasionally necessary during massive resuscitation (Tawil et al., 2008). On the other hand, posttraumatic thrombocytopenia, platelet dysfunction, or fibrinolytic coagulopathy can increase the risk of hemorrhagic complications and death (Nekludov et al., 2007b; Chen et al., 2013).

Neuropathology

After moderate or severe TBI, there is an imbalance between the natural hemostatic versus anticoagulation cascades, predisposing towards DIC, thrombosis, and ischemia (Nekludov et al., 2007a; Halpern et al., 2008; Alexiou et al., 2011; Chen et al., 2013). Thromboplastin is upregulated and released from damaged brain parenchyma, and subsequently activates procoagulant factors, leading to derangements in the prothrombin time (PT) and PTT, as well as intravascular deposition of fibrin, which in turn may result in an increased risk of cerebral infarction (Carrick et al., 2005; Chen et al., 2013).

Clinical presentation

PTCI usually occurs within the first week posttrauma and can be seen in multiple vascular territories (Chen et al., 2013). DIC can also result in cerebral microhemorrhages. Clinical presentation of these secondary neurologic complications will vary from undetectable to major deficits, including coma, depending on the extent and location. Unexplained alterations in mental status or focal neurologic deficits should initiate additional diagnostic workup, including neuroimaging.

Neurodiagnostics and imaging

Diagnostic workup includes standard coagulation studies such as fibrinogen level, D-dimer, PT, and PTT. Noncontrast head CT scans can diagnose both hemorrhagic and subacute ischemic lesions. Patients with concurrent TBI and Glasgow Coma Scale (GCS) scores < 9 often have intracranial pressure (ICP) monitors that can detect changes from cerebral edema that may occur either in the setting of blossoming contusions or developing infarctions. MRI is more sensitive than CT at detecting ischemic lesions using diffusion-weighted imaging (DWI), as well as microhemorrhages with gradient echo or susceptibility-weighted imaging. Thus, MRI is the diagnostic test of choice if DIC-related microvascular phenomena are suspected.

Clinical trials and guidelines

No large clinical trials or guidelines are currently available regarding the ideal diagnostic workup and management for secondary neurovascular complications related to traumatic coagulopathy. Based on retrospective data, development of PTCI is associated with greater severity of DIC ("DIC score" > 5), D-dimer > 2 mg/L, abnormal fibrinogen and PTT values, and thrombocytopenia (Chen et al., 2013). Development of PTCI is, in turn, associated with unfavorable outcomes (Saggar et al., 2009; Lustenberger et al., 2010; Chen et al., 2013).

Hospital course and complex clinical decisions

The hospital course of polytrauma patients with coagulopathies may also include bleeding from mucosal lesions, vascular access sites, or excessive clotting and multiorgan system failure.

Treatment is primarily supportive, requiring adequate resuscitation and correction of the underlying coagulopathy. In patients with ongoing hemorrhage, it is also important to correct any underlying acidosis and hypothermia (Tieu et al., 2007). Resuscitation with lactated Ringer's may be preferable to normal saline to avoid hyperchloremic metabolic acidosis, but should be used with caution in the presence of TBI. Artificial colloids (e.g., starches) have been associated with development of coagulopathy (Tieu et al., 2007). Although systemic anticoagulation is occasionally considered for treatment of DIC with associated systemic microthrombi (Levi and Cate, 1999), there is no evidence suggesting this is beneficial in managing PTCI. Moreover, since PTCI commonly occurs in patients with coexisting TBI, any theoretic benefit of anticoagulation must be weighed against the risk of worsening intracranial hemorrhage. If strokes or hemorrhages do occur, management must include addressing secondary processes, such as cerebral edema.

When patients present with hemorrhagic shock requiring massive transfusion, they may require treatment with procoagulant agents such as tranexamic acid or (rarely) recombinant factor VIIa, despite the subsequently increased risk of thrombosis and stroke (Tawil et al., 2008). These risk–benefit analyses are standard considerations for polytrauma patients and warrant routine neurologic examinations with sedation pauses. Given the underlying systemic pathophysiology, ischemic events secondary to DIC can be diffuse and span multiple vascular territories, leading to devastating neurologic injuries, particularly if not recognized and managed early.

It is important to remember that there are multiple potential etiologies for ischemic events in trauma patients. Although coagulopathies are present in many trauma patients, they may not necessarily be causatively related to central nervous system (CNS) injury. Attributing posttraumatic strokes to underlying systemic coagulopathies is not straightforward. Vascular imaging and echocardiograms may be warranted to identify other etiologies. Patients with coagulopathy-related neurologic injury are likely to have other systemic manifestations of problems with bleeding or clotting, and evidence of significant coagulopathy by laboratory values.

Embolic strokes

Epidemiology

Polytrauma patients are at increased risk for cerebral embolism due to fat, air, or cardiac sources (Blacker and Wijdicks, 2004). The exact incidence is unknown. One large study reported that 231 (0.6%) of 40 846 polytrauma patients had "thromboembolic" stroke (Lichte et al., 2015). In a 10-year single-center retrospective review, fat emboli were reported in 0.9% of patients with long-bone fractures, with over half experiencing mental status changes (Bulger et al., 1997). The most common cause of fat embolism syndrome (FES) is femoral and pelvic fractures after an MVC (Caplan, 2009). FES is unusual in children or patients with isolated upper-extremity fractures (Parisi et al., 2002). Cerebral air emboli are rare, but a few cases have been reported as complications of blunt and penetrating chest trauma (Reith et al., 2015).

Neuropathology

Posttraumatic cardioembolic strokes can occur as a direct result of cardiac thrombus formation after blunt or penetrating injuries (Blacker and Wijdicks, 2004; Neidlinger et al., 2004). These can also occur in the setting of preexisting cardiac arrhythmias such as atrial fibrillation or new-onset cardiac arrhythmias that develop in the setting of critical illness.

Like any cerebral embolus, air emboli can occlude blood flow in the cerebral vasculature. Typically the occlusion is immediate yet transient, since air can rapidly move through capillary beds into venules and dissipate (Menkin and Schwartzman, 1977; Caplan, 2009). The pathology is thought to be secondary to gas bubble-induced arterial vasoconstriction, followed by dilation and stasis of blood flow (Menkin and Schwartzman, 1977; Caplan, 2009).

The pathology of fat emboli is theorized to have two main components: mechanical obstruction triggering platelet aggregation, fibrin generation, and mechanical obstruction, and biochemical destruction due to a proinflammatory state created by the glycerol and toxic free fatty acids activating cytokine cascades (Kosova et al., 2015). Brain autopsies of patients with FES have shown ring-shaped, perivascular hemorrhages, cerebral edema, and microinfarction (Kamenar and Burger, 1980). Fat staining reveals multiple fat globules within hemorrhagic lesions, as well as in CNS microvasculature. A high quantity of fat globules is required to produce neurologic symptoms, and an autopsy study documented as many as 100 fat globules per mm^2 of brain (Kamenar and Burger, 1980).

Clinical presentation

Most posttrauma strokes have a delayed clinical presentation, with the assessment initially confounded by sedation, physiologic instability, or head trauma (Blacker and Wijdicks, 2004). Cardioembolic strokes present with variable neurologic symptoms, depending on the location of the infarct. Those with embolic showers may have nonspecific symptoms, including global alteration in consciousness.

The clinical presentation for cerebral air emboli has primarily been studied in patients with diving-related incidents and consists of sudden loss of consciousness, paresthesias, weakness, blurred vision, headaches, seizures, and focal neurologic signs related to lesions in the brainstem and cerebellum (Cantais et al., 2003; Demaerel et al., 2003; Valentino et al., 2003).

The classic triad of FES includes respiratory distress with hypoxemia, petechiae, and neurologic abnormalities (Kosova et al., 2015). This syndrome typically occurs 24–72 hours after an injury, but can occur up to a few days later. Respiratory complications are the most common, and have been reported to affect up to 96% of patients (Kosova et al., 2015). Neurologic manifestations from cerebral fat emboli are also common, occurring in more than 80% of patients (Caplan, 2009) and consisting of confusion, delirium, encephalopathy, agitation, and decreased arousal. The alteration in mental status often progresses to stupor or coma. Focal and generalized seizures frequently occur early in FES. Focal neurologic signs are noted in up to one-third of patients (Caplan, 2009).

Neurodiagnostics and imaging

Imaging characteristics of posttrauma-related cardioembolic strokes on MRI are typical for embolic infarcts, and include lesions in specific vascular territories, or embolic showers in multiple distributions. Acute infarcts are best seen on DWI. Echocardiography is useful to evaluate for cardiac source of thrombus, and cardiac monitoring should be continued to evaluate for predisposing arrhythmias. Transthoracic echocardiograms visualize the body of the left atrium well, but not the atrial appendage, which is better visualized using transesophageal echocardiography (Caplan, 2009). Transthoracic echocardiogram is generally preferred as the initial modality, given that it is noninvasive and readily available (Caplan, 2009). However, based on the clinical scenario and degree of suspicion, if there is high suspicion for cardioembolic stroke from cardiac thrombus or aortic plaques, transesophageal echocardiography should be considered (Caplan, 2009).

Air emboli can be seen on CT scans (Fig. 34.1A) and further microemboli can theoretically be detected by

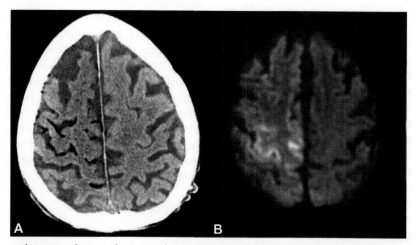

Fig. 34.1. (A) Computed tomography scan demonstrating scattered air emboli in the right hemisphere. (B) Subsequent diffusion-weighted sequence on magnetic resonance imaging demonstrating gyriform pattern of infarction at 17 hours. (Courtesy of Dr. Ashutosh Jadhav.)

transcranial Doppler. DWI on MRI can show brain infarction, usually in multiple vascular territories (Fig. 34.1B).

Fat emboli are best seen on DWI, fluid-attenuated inversion recovery, and contrast-enhanced MRI images that can detect both microinfarcts and microhemorrhages (Parizel et al., 2001; Caplan, 2009). The MRI appearance of cerebral fat emboli classically demonstrates multiple, diffuse, scattered hyperintense lesions on T2-weighted sequences. These are more prevalent in white-matter and borderzone regions, but can affect both gray and white matter, and are usually associated with edema (Parizel et al., 2001; Caplan, 2009). DWI sequencing can demonstrate a "starfield" pattern that is typical for fat emboli (Parizel et al., 2001; Caplan, 2009). Most often, CT scans are normal, but may show hypo- or hyperdensities based on infarct characteristics. Transcranial Doppler may be able to detect fat microemboli (Forteza et al., 2002). MR spectroscopy has been used to identify fat molecules, specifically free lipids, and can provide supportive diagnostic evidence (Guillevin et al., 2005).

CLINICAL TRIALS AND GUIDELINES

There are no clinical trials or formal guidelines regarding the diagnostic evaluation or management of air, fat, or cardiac emboli in the context of trauma.

HOSPITAL COURSE AND COMPLEX CLINICAL DECISIONS

Management of thrombotic cardioembolic strokes in polytrauma patients requires careful and dynamic assessment of the risks versus benefits of treatment. The presence of a left ventricular thrombus would significantly increase risk of stroke, but this needs to be weighed against the hemorrhagic risk of other concurrent injuries. Additionally, if a large stroke has already occurred with a known cardiac thrombus, anticoagulation may need to be delayed due to risk of hemorrhagic transformation. As with any cardioembolic stroke, recovery is slow, requires significant neurorehabilitation, and may be prolonged in the setting of critical illness and other concurrent injuries.

Although there are no formal guidelines, cerebral air emboli are often treated acutely with 100% oxygen and, if feasible, hyperbaric oxygen (Caplan, 2009). However, not all centers have hyperbaric chambers available for treatment in a timely fashion.

Clinical manifestations of FES may be present for anywhere from 7 to 30 days depending on the severity. Clinical resolution typically occurs prior to radiographic improvement, which may take up to 3 months (Buskens et al., 2008; Kellogg et al., 2013). It remains unclear whether clinical severity correlates with MRI findings. ICP monitoring may be considered in selected comatose trauma patients with suspected FES given its inflammatory and edematous nature, especially if the patient has signs of elevated ICP or imaging suggests diffuse edema (Kellogg et al., 2013). Electroencephalographic (EEG) monitoring may also be of utility since these patients may have subclinical seizures. Treatment of FES is primarily supportive; in rare cases, the degree of cerebral edema may be severe. Medications such as corticosteroids, heparin in patients with concomitant DIC, and intravenous 5% alcohol (for its theorized lipolytic capacity) have been suggested, but their effectiveness is unknown and these strategies are not routinely recommended (Caplan, 2009). The mortality rate of FES is as high as 50%, although this has decreased with time and improvement in supportive care (Parisi et al., 2002; Caplan, 2009). Patients with FES may develop a constellation of other systemic findings, including anemia, thrombocytopenia, coagulopathy progressing to DIC, urine fat globules, retinal infarcts, fever, and jaundice. Mortality is significantly increased in the setting of multisystem involvement.

Hypoperfusion-related cerebral ischemia

EPIDEMIOLOGY

Despite the frequency of hemorrhagic shock, borderzone infarctions after polytrauma are rare and the literature is limited to case reports and series (Blacker and Wijdicks, 2004; Takaoka et al., 2004; Liu et al., 2011).

NEUROPATHOLOGY

Animal studies demonstrate that ischemic injury from hypotension occurs in watershed areas of both the cerebrum and cerebellum, typically when mean arterial pressures (MAPs) are 25–35 mmHg for prolonged periods of time and autoregulation fails (Graham et al., 1979). The borderzone region includes territories of both the anterior and posterior circulation that are furthest from the parent arterial trunks, making them most vulnerable to low perfusion pressure. This pattern is different than cardiac arrest, where blood supply to the entire brain is simultaneously halted and diffuse ischemic changes occur throughout the cortex in areas vulnerable to ischemia, rather than just borderzone regions (Fujioka et al., 1994). Although not proven, there has also been some suggestion of microembolism and intravascular coagulopathy playing a role in borderzone infarctions (Takaoka et al., 2004).

CLINICAL PRESENTATION

Multiple different mechanisms of injury can produce cerebral hypoperfusion injuries. The common factor is significant hypotension requiring prolonged resuscitation. In approximately half of cases there was no report of pre-existing craniocervical stenosis or injury (Takaoka et al., 2004). In other cases, documented stenosis of the carotid artery resulted in ipsilateral borderzone strokes (Blacker and Wijdicks, 2004; Liu et al., 2011).

Clinical presentations prompting further radiographic imaging varied from minimal deficits to seizures and decreased consciousness. In one case, the diagnosis was made at autopsy (Blacker and Wijdicks, 2004).

Neurodiagnostics and imaging

Initial imaging obtained in most cases is a noncontrast head CT, which may show hypodensities in borderzone regions. Diagnosis of borderzone infarction is confirmed using DWI MRI. CT or MR angiography and echocardiography should generally be obtained in order to rule out other causes of stroke.

Hospital course and complex clinical decisions

The hospital course will vary depending on the degree and duration of hypotension, baseline neurologic status, pre-existing CNS vascular lesions, severity of other injuries, and development of other intensive care unit (ICU) complications, such as infections. The reported cases illustrate this diversity, and many patients experience a relatively benign hospital course with minimal neurologic deficits (Takaoka et al., 2004; Liu et al., 2011). Others are left with significant residual deficits, including quadriparesis, persistent vegetative state, or death secondary to massive cerebral edema (Blacker and Wijdicks, 2004; Takaoka et al., 2004).

In most of the reported cases of borderzone infarction, diagnosis was delayed by several hours to days. One potential reason is that the neurologic exam may be nonfocal, with only depressed consciousness that may have a number of alternative explanations.

Clinical trials and guidelines

No clinical trials or guidelines are currently available. Prompt diagnosis is often challenging given the need for sedation and presence of distracting injuries. Cerebral hypoperfusion must be considered in at-risk patients who develop focal neurologic deficits or have persistent depressed consciousness when sedation is paused. It is not practical or necessary to obtain routine head CT scans on all polytrauma patients, particularly if they do not have suggestion of CNS compromise on clinical examination. For the majority of cases, daily sedation pauses are crucial since neurologic sequelae may only be suspected or diagnosed based on subtle changes in the neurologic exam or level of arousal.

Other vascular complications

Less common vascular complications of trauma include the formation of arteriovenous fistulas, cortical venous or venous sinus thrombosis, and posterior reversible encephalopathy syndrome related to medications, transfusions, or autonomic dysregulation.

ALTERED MENTAL STATUS

This topic is covered in more detail in Chapters 25 and 30 of this volume. This section will focus on altered mental status specifically in trauma patients.

Epidemiology

Alteration in mental status is a frequent and challenging finding in trauma units. Aside from vascular events or seizures, critically ill patients with polytrauma are at high risk for delirium or encephalopathy. These terms are frequently used interchangeably in the literature and in clinical practice. Delirium is an acute disorder of attention that results in fluctuating global cognitive dysfunction. Encephalopathy more specifically refers to an underlying pathology causing the global disturbance of cerebral function, resulting from derangements of other organ systems (e.g., septic encephalopathy) (Fig. 34.2). The DSM-IV-TR diagnostic criteria for the diagnosis of delirium due to a general medical condition have been previously defined (American Psychiatric Association, 2000).

Delirium in trauma patients can result from many etiologies, including prolonged sedation and analgesia, pain, sleep deprivation, medications, encephalopathies, nutritional deficiencies, toxic ingestions, as well as withdrawal from substances such as ethanol, opiates, or benzodiazepines. Indeed, the incidence of delirium in mechanically ventilated trauma patients is as high as 67–92% (Pandharipande et al., 2007a; Bryczkowski et al., 2014). Types of encephalopathies seen in trauma patients include hepatic, septic, hypoxic-ischemic, hypo- or hyperglycemic, uremic, and hypercalcemic. Most encephalopathies are reversible with correction of the underlying problem, and do not typically require treatment with antipsychotic agents.

Previously identified risk factors for delirium in critically ill patients are numerous (Table 34.3a) (Bryczkowski et al., 2014). In the trauma population, use of benzodiazepines, increasing age, blood transfusions, low GCS score, and multiple-organ failure are the strongest predictors of developing delirium (Pandharipande et al., 2006; Angles et al., 2008). For every year older than 50, the relative odds of delirium has been reported to increase by 10% (Bryczkowski et al., 2014).

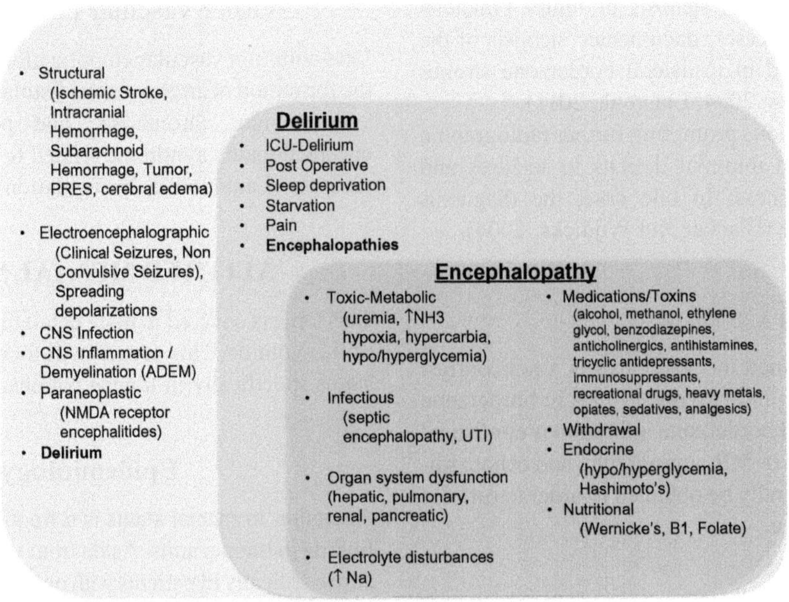

Fig. 34.2. Basic differential diagnosis for altered mental status in an intensive care unit (ICU). PRES, posterior reversible encephalopathy syndrome; CNS, central nervous system; ADEM, acute disseminated encephalomyelitis; NMDA, N-methyl-D-aspartate; UTI, urinary tract infection.

Table 34.3a

Risk factors for delirium in the intensive care unit

Medical
Coma
Mechanical ventilation
Indwelling bladder catheters
Fever
Hospital-acquired infection
Metabolic disarray
    Fluid/electrolyte disorders
    Malnutrition
Medications
    Greater than three new medications per day
    Psychoactive agents
    Sedatives (benzodiazepines)
    Analgesics (opiates)
Transfusions
Patient characteristics/behavioral
Age
Smoking
Alcohol consumption
Drug abuse
Pre-existing cognitive or psychiatric impairment
Environmental
Restraints
Absence of visible daylight
Lack of visitors
Disruption of sleep–wake cycle

Delirium secondary to alcohol withdrawal syndrome (AWS) is particularly common, and its symptomatology and management may differ from other causes. Ethanol abuse has been implicated in almost one-third of motor vehicle fatalities and 40% of violent crimes. Approximately 50% of patients admitted to trauma centers in the USA report alcohol usage, and elevated blood ethanol levels have been documented in almost half of patients. However, incidence of AWS in trauma patients has a variable reported range from <1% to 33%. The lower reported incidence might be related to recognition failure. Postulated reasons may include physician familiarity with signs and symptoms of AWS, confounding head trauma, and use of pain and sedative medications. Wernicke's encephalopathy is rare, but given the high prevalence of alcoholics in the polytrauma population, it should be considered when there is a history of chronic alcoholism or malnutrition (Ma et al., 2013).

Neuropathology

The underlying pathophysiology of delirium associated with trauma or critical illness is unknown. It is hypothesized that the GABAergic and cholinergic neurotransmitter pathways play a crucial role, in part due to an increased association of delirium with GABA agonists and anticholinergic medications. Other theories implicate increased dopaminergic activity (supporting use of antipsychotic medications), and neurotoxic effects of inflammatory cytokines (Reade and Finfer, 2014). Given the multiple different etiologies, there may be a final common pathway that has yet to be determined. Interestingly, MRI demonstrates a correlation between duration of delirium and cerebral atrophy, as well as white-matter disruption, suggesting that delirium can result in structural CNS alterations and may have long-term cognitive consequences (Gunther et al., 2012; Morandi et al., 2012).

Clinical presentation

Unfortunately, a sizable proportion of delirium cases in ICUs remain undiagnosed. This has been attributed to two primary reasons: (1) healthcare teams are more aware of hyperactive than hypoactive delirium; and (2) confounding factors such as medications or severity of illness interfere with the diagnosis. The ability to diagnose delirium has improved with the introduction of two standardized, validated scales: the Confusion Assessment Method for the ICU (CAM-ICU) and the Intensive Care Delirium Screening Checklist (ICDSC) (Bergeron et al., 2001; Ely et al., 2004).

In polytrauma patients, delirium or encephalopathy can occur at any point depending on the underlying cause. For example, trauma patients who develop sepsis may simultaneously become encephalopathic, and the degree of encephalopathy can mirror the severity of their systemic illness. Postoperative delirium may occur after surgery. A prospective study in trauma patients found that delirium occurred most often on postinjury day 2, and lasted for approximately 4.7 ± 4.5 days (Angles et al., 2008). The most common form of delirium was hypoactive, seen in 46% of patients, followed in frequency by mixed delirium in 39%, and hyperactive delirium in only 15% of patients. Hyperactive delirium typically manifests as agitation, whereas hypoactive delirium is characterized by decreased level of consciousness, disordered thinking, and inattention without agitation. Delirium occurs frequently, is associated with morbidity and mortality, and can be difficult to recognize in trauma patients.

AWS may be difficult to diagnose due to sedation, analgesia, confounding issues, and brain injury (Lukan et al., 2002). Additionally, manifestations of AWS such as agitation and tachycardia may be interpreted as responses to pain or traumatic injury (Lukan et al., 2002). AWS is a neurologic emergency and can occasionally manifest as early as 6–8 hours after last ethanol consumption. However, the more typical time course is 3–5 days after cessation of alcohol. Duration is variable, but commonly about 72 hours. Patients will become restless and develop hallucinations, confusion, tremors, heightened autonomic nervous system activity (diaphoresis, tachycardia, hypertension, fever), and seizures. In trauma patients, a normal mean corpuscular volume and serum aspartate aminotransferase despite a positive toxicology screen for ethanol has been associated with a low ($\sim$4%) risk of developing AWS (Findley et al., 2010).

Wernicke's encephalopathy has traditionally been reported to present with the triad of confusion, ataxia, and oculomotor abnormalities, but fewer than 20% of patients present with all three symptoms. Confusion and delirium are the most common initial presentations.

Neurodiagnostics and imaging

Delirium and encephalopathy are clinical diagnoses and there are usually no specific imaging characteristics that are useful in the diagnostic evaluation. Wernicke's encephalopathy is one of the few causes that does have characteristic findings on brain MRI, which consist of T2 hyperintensity in the mammillary bodies, periaqueductal region, splenium of the corpus callosum, and occasionally thalamus and hypothalamus. There is usually also corresponding restricted diffusion.

It is important to rule out structural brain injury as the cause of an alteration in mental status; a screening CT scan is generally warranted. MRI is more sensitive to assess for diffuse axonal injury. EEGs can be useful to distinguish between seizures or other patterns characteristic of underlying encephalopathy. Lumbar puncture may be performed to evaluate for CNS infection or inflammation, but must be avoided if there is significant intracranial mass effect.

Clinical trials and guidelines

Guidelines for the management of delirium in critically ill patients have been published and can be applied in the setting of trauma (Barr et al., 2013). Use of sedation is frequently implicated in development of altered mental status and delirium in ICUs. There are abundant data from randomized controlled trials supporting the benefits of minimizing sedation levels, which include fewer ICU days, reduced length of mechanical ventilation, and lower mortality (Girard et al., 2008; Reade and Finfer, 2014). It has also been noted that patients in dedicated neurologic ICUs receive less intravenous sedation than those in other ICUs (Kurtz et al., 2011). However, minimizing sedation is particularly challenging in polytrauma patients, where adequate treatment of injury-related pain and anxiety may result in higher levels of sedation. Sedation interruptions allow for routine neurologic examinations, which in turn allow detection of secondary CNS insults and delirium.

Although there are numerous trials evaluating different sedative agents, no drug has demonstrated superiority (Roberts et al., 2012; Reade and Finfer, 2014). Only a handful of these trials have been carried out specifically in trauma patients (Pandharipande et al., 2008; Lat et al., 2009). Use of benzodiazepines is common in many ICUs, but these agents should be minimized given their documented increased risk for delirium (Pandharipande et al., 2008). The association of opiates with delirium in this population is controversial (Pandharipande et al., 2008; Lat et al., 2009). Propofol and dexmedetomidine remain reasonable alternatives. The advantages of dexmedetomidine include less respiratory depression and some degree of analgesia. Compared with benzodiazepines, dexmedetomidine produces less delirium, but there was no definite advantage over propofol in clinical trials (Pandharipande et al., 2007b; Riker et al., 2009; Jakob et al., 2012). Delirium prevention is an important concept, but the key placebo-controlled trials advocating for use of pharmacologic prophylaxis for delirium with haloperidol, risperidone or ketamine were conducted in patients undergoing elective surgeries, such that the results may not be directly applicable to polytrauma patients (Wang et al., 2012).

Information regarding management of established delirium in ICUs is limited, and most efforts are directed towards rapid identification and treatment of the underlying etiology. Available studies comparing treatments are small, not specific to trauma patients, and do not distinguish hypo- versus hyperactive delirium. Haloperidol has traditionally been the agent of choice in trauma patients because it has few side-effects, does not cause hypotension, and has multiple formulations. In a prospective trial of 73 patients in medical and surgical ICUs, haloperidol showed equivalent efficacy to olanzapine, although the latter was associated with fewer extrapyramidal side-effects (Skrobik et al., 2004). One prospective randomized placebo study showed faster resolution of delirium with quetiapine, but another one comparing haloperidol, ziprasidone, and placebo showed no difference in outcomes (Devlin et al., 2010; Girard et al., 2010). Dexmedetomidine has recently gained traction as a potential treatment for delirium. A pilot study found that, compared to haloperidol, dexmedetomidine use resulted in shorter time to extubation and ICU length of stay (Reade et al., 2009). These findings were subsequently confirmed in a multicenter randomized controlled trial demonstrating that, in patients with agitated delirium, use of a dexmedetomidine infusion (starting at 0.5 μg/kg/hour, to a maximum of 1.5 μg/kg/hour) is associated with earlier extubation readiness (Reade et al., 2016). There were relatively few trauma patients in this study.

Early mobilization of medical ICU patients during sedation holidays has also been shown to decrease duration of delirium (Schweickert et al., 2009).

Hospital course and complex clinical decisions

In polytrauma patients, there is a delicate balance between adequate analgesia, sedation, and delirium management. Untreated pain and agitation can contribute towards delirium. Although it is important to eliminate underlying primary CNS pathology, alterations in mental status from delirium and encephalopathy are common problems and carry high rates of morbidity and mortality (Ely et al., 2004; Bryczkowski et al., 2014).

In the trauma population, delirium has been noted to occur as early as the second day after injury, and can persist for more than 2 weeks depending on the underlying etiology. Identification and treatment of the causal factors are key to resolution. The diagnostic evaluation should be tailored to individual clinical scenarios. Based on common etiologies of encephalopathy and delirium, the workup should focus on metabolic, infectious and pharmacologic causes (Table 34.3b). More often than not, encephalopathy or delirium will be multifactorial from multiple causes.

While some risk factors for delirium are not modifiable, others can be addressed (Zaal et al., 2015). Mechanical ventilation should be weaned as rapidly as possible. Polypharmacy and benzodiazepine use should be minimized. Propofol and dexmedetomidine are preferred agents due to their short half-life and lower risk of inducing delirium, although benzodiazepines are appropriate for treatment of alcohol withdrawal. Other measures include regulation of sleep–wake cycles, frequent reorientation, noise reduction, and early mobilization. Correcting any underlying abnormality may take time, and improvement of the mental status typically lags, particularly in patients with renal or hepatic dysfunction or prolonged deep sedation. If antipsychotics are required, haloperidol is effective with minimal adverse effects,

Table 34.3b

Delirium and encephalopathy diagnostic workup

Basic electrolyte panel
Glucose levels
Renal function
Liver function, including ammonia and lipase
Arterial blood gas for hypoxia, hypercarbia, or acid–base disturbances
Complete blood counts for anemia and leukocytosis
Infectious workup
Medication review

and atypical antipsychotics are equally effective with fewer extrapyramidal manifestations and potentially better sleep patterns.

Treatment for AWS in polytrauma patients may be challenging. Traditionally, benzodiazepines have been the drug of choice. However, given their risks for inducing delirium, contributing to oversedation and respiratory depression, other agents such as low-dose phenobarbital are increasingly being considered. Some data suggest equal efficacy of phenobarbital with benzodiazepines, and potentially a benefit from combined usage, with decreased need for mechanical ventilation, less nosocomial pneumonia, and shorter ICU stays (Rosenson et al., 2013; Duby et al., 2014). Mechanistically, phenobarbital increases the frequency of GABA channel opening. However, unlike benzodiazepines, it also affects α-amino-3-hydroxy-5-methyl-4-isoxazolepropionic acid (AMPA) and kainite N-methyl-D-aspartate (NMDA) receptors, thereby modulating glutamatergic activity and treating AWS with a potentially more comprehensive approach. Dosing regimens vary, but generally are determined by ideal body weight-based bolus dosing, followed by gradually decreasing maintenance therapy. The doses of phenobarbital used for treatment of AWS are sufficiently low that they do not cause respiratory depression or somnolence, and levels are often similar to those achieved for outpatient seizure control.

Neurorehabilitation/outcomes

Development of delirium in critically ill patients is a predictor of increased mortality (Ely et al., 2004). Patients with delirium have a longer time on the ventilator, ICU and hospital length of stay, and higher cost of care (Milbrandt et al., 2004). Moreover, some of these patients have significant cognitive impairment even after discharge (Jackson et al., 2004).

SEIZURES AND ELECTROENCEPHALOGRAPHY

This topic is discussed in more detail in Chapters 9 and 28 of this volume.

Epidemiology

The incidence of seizures is unclear in polytrauma patients without concomitant head injury. Data from surgical ICU patients (some of whom had sustained polytrauma) with altered mental status suggest that the incidence of nonconvulsive seizures (NCSz) may be as high as 11–16%, and that of nonconvulsive status epilepticus (NCSE) 5–8% (Kamel et al., 2012; Kurtz et al., 2013). However, these were nonconsecutive patients, and there is some possibility of selection bias. In one study, periodic epileptiform discharges were detected in 29% of patients (Kurtz et al., 2013). Infections, electrolyte disturbances, and substance abuse or withdrawal increase the risk for seizures (Claassen et al., 2013; The Participants in the International Multi-disciplinary Consensus Conference on Multimodality Monitoring et al., 2014).

The incidence of seizures in TBI ranges from 22% to 33% within the first week after injury despite seizure prophylaxis (Vespa et al., 1999; Ronne-Engstrom and Winkler, 2006), and most are nonconvulsive. Known risk factors for NCSz include depressed skull fracture, penetrating injury, and cortical contusion/hematomas. In addition, it is estimated that traumatic injuries occur in 0.5% of patients with seizures (Souverein et al., 2005; Gill et al., 2015), due to the violent nature of the seizure itself, or from ensuing accidents.

Clinical presentation

It has been estimated that 90% of ICU patients with seizures have nonconvulsive episodes that can only be identified with continuous EEG (Claassen et al., 2004). This poses a diagnostic challenge for patients in the trauma ICU, who are frequently sedated, delirious, agitated, or have other explanations for altered mental status. Convulsive seizures should be easier to identify, but nonepileptic movements such as rigors or myoclonus can lead to misdiagnosis.

Neurodiagnostics and imaging

The development of seizures should prompt an urgent CT scan to rule out acute intracranial pathology. Trauma patients can harbor occult vascular injuries that result in delayed ischemic strokes, which may in turn manifest as seizures. In addition, patients may have diffuse axonal injury that was not detected on admission CT, and requires MRI for identification.

EEG is the only method to definitively diagnose nonconvulsive events. EEG monitoring can also detect other electrographic patterns that do not meet the definition for seizure, but may still represent ictal or interictal phenomena (Ng et al., 2014). Lateralized periodic or rhythmic patterns have been associated with increased risk of true seizures and worse outcomes, but the need for treatment is controversial (Pedersen et al., 2013; Ng et al., 2014).

A high index of suspicion for NCSz is necessary during the workup of unexplained or fluctuating mental status. When identified, workup of seizures should include an evaluation of potential underlying etiologies such as metabolic or electrolyte derangements, toxins or withdrawal, infections, medication effects, FES, hypoxic-ischemic encephalopathy, undiagnosed TBI, hyperthermia, and secondary strokes. Medications that are commonly used in trauma patients and are associated with NCSE include third- and fourth-generation cephalosporins, quinolones, carbapenems (primarily imipenem), and antipsychotics (Maganti et al., 2008). Consulting a neurologist or neurointensivist may be helpful (Mittal et al., 2015).

Hospital course and complex clinical decisions

No one agent has been determined as ideal for treatment of seizures in critically ill polytrauma patients. Levetiracetam is commonly used because of minimal adverse effects and drug interactions and the possibility of intravenous administration. It can be activating, so agitated patients may benefit from a sedating antiseizure agent. Convulsive and nonconvulsive status epilepticus are medical emergencies, and treatment algorithms have been well defined (Brophy et al., 2012; Betjemann and Lowenstein, 2015). If status epilepticus is not controlled within 30–60 minutes after initial drug therapy, then it is classified as refractory, and treatment should rapidly progress to use of anesthetic agents (Table 34.4).

Clinical trials and guidelines

The 2013 European Society for Intensive Care Medicine guidelines recommend continuous EEG in comatose ICU patients without acute brain injury who have unexplained altered mental status or neurologic deficits, especially those with sepsis, renal failure, or hepatic failure (Claassen et al., 2013). The American Clinical Neurophysiology Society guideline recommendations regarding indications for continuous EEG to identify NCSz and NCSE in critically ill adults are listed in Table 34.5 (Herman et al., 2015).

Neurorehabilitation/outcomes

Delay in the diagnosis and treatment of NCSz has been associated with increased mortality (Vespa et al., 1999; Friedman et al., 2009). NCSE-related mortality at hospital discharge ranges from 18% to 52% (Brophy et al., 2012). Critically ill patients greater than 75 years old are particularly susceptible to unfavorable outcomes (Bottaro et al., 2007).

SPINAL COMPLICATIONS

Epidemiology

Spinal cord injuries are common in polytrauma patients, and may be caused by direct primary injury or secondary

Table 34.4

Management of status epilepticus (convulsive or nonconvulsive)

Initial steps
Assess need for airway protection
Establish intravenous access

Pharmacologic interventions	Dosing
Empiric medications	
Thiamine	100 mg
Dextrose	D50, 50 mL
Emergent seizure control	
Lorazepam (Ativan)	0.1 mg/kg IV, give in 4-mg increments
Midazolam (Versed)	0.2 mg/kg IM, give up to 10 mg per dose
Urgent seizure control	
Phenytoin (Dilantin) or	Load: 20 mg/kg IV; maintenance: 4–6 mg/kg/day divided in 2–3 doses
Fosphenytoin (Cerebryx)	
Valproate sodium (Depacon)	Load: 20–40 mg/kg IV; maintenance: 10–15 mg/kg/day divided into 2–4 doses
Levetiracetam (Keppra)	1000–3000 mg/day IV in two divided doses
Lacosamide (Vimpat)	Load: 200–400 mg/day IV; maintenance: 200 mg every 12 hours
Refractory status epilepticus	
Midazolam (Versed)	Bolus: 0.2 mg/kg IV; infusion: 0.05–2 mg/kg/hour
Propofol (Diprovan)	Bolus: 1–2 mg/kg IV; infusion: 30–250 µg/kg/min
Pentobarbital (Nembutal)	Bolus: 10–15 mg/kg IV; infusion: 0.5–5 mg/kg/hour

complications. Paraplegia due to spinal cord ischemia is a known complication in patients with traumatic aortic injuries requiring operative repair. The incidence of paralysis after aortic surgery in the 1990s was 9% with use of motor evoked responses for monitoring (Svensson, 2005). Changes in operative techniques and the addition of thoracic endovascular aortic repair have lowered the rate of paralysis to less than 3% (Demetriades et al., 2008; DuBose et al., 2015).

Neuropathology

The mechanism of spinal cord injury after aortic repair is multifactorial. The primary cause is reduction of blood flow distal to the aortic clamp. Additional contributors include changes in cerebrospinal fluid (CSF) pressures

Table 34.5

American Clinical Neurophysiology Society indications for continuous electroencephalogram (EEG) (Herman et al., 2015)

- Persistently altered mental status after clinically evident seizures (i.e., no improvement after 10 minutes, or altered consciousness after 30 minutes)
- Evidence of acute supratentorial pathology with altered mental status
- Fluctuating mental status or altered mental status of unknown etiology without primary brain injury
- Presence of generalized periodic epileptiform discharges, lateralized periodic discharges, or bilateral independent periodic discharges on routine or emergent spot EEG
- Use of neuromuscular blockade for nonneurologic reasons (hypothermia, acute respiratory distress syndrome, etc.) in patients with high seizure risk
- Clarification is required regarding paroxysmal movements (e.g., chewing, yawning, apnea, autonomic changes) that have potential to be seizures

and variability in vascular anatomy and collateralization. Increased risk for paraplegia postoperatively is primarily related to cross-clamp times exceeding 30 minutes (Hershberger and Cho, 2014). Additional risk factors include the extent of aorta repaired, rupture, age, proximal aneurysm, and baseline renal dysfunction (Hershberger and Cho, 2014). The fact that paraplegia can also occur in patients with short cross-clamp times is likely related to the blood supply and collateral flow to the spinal cord.

Clinical presentation

Most patients present with an anterior spinal artery syndrome, with weakness of the lower extremities. Motor deficits are initially flaccid, but progressively become spastic with corticospinal signs. Sphincter dysfunction is common, with initial urinary retention followed by spasmodic bladder. Pain and temperature deficits are noted, since the spinothalamic tracts are supplied by the anterior spinal artery. Depending on the extent of infarction and edema, the sensory level is typically in the thoracic region (Robertson et al., 2012). Neurogenic shock with autonomic dysfunction may ensue due to decreased vasomotor tone (Etz et al., 2015). The cervical cord is well collateralized, and is therefore more resilient to ischemic lesions from hypoperfusion. The artery of Adamkiewicz is the largest of the radicular arteries that contributes to the anterior spinal artery, and is the most susceptible to injury because of tenuous collateralization at this level. This artery originates around T9–T12 in approximately 75% of cases, but can originate anywhere from T8 to L3. Although symptoms are typically acute in onset, delayed paraplegia may also occur with transient postoperative hypotension (Hershberger and Cho, 2014).

Neurodiagnostics and imaging

Most patients with traumatic aortic injury do not survive to reach a healthcare facility. In those who do, aortic injury can be detected on vascular imaging, most often CTA. Preoperative identification of the artery of Adamkiewicz with CT or MR angiography is increasingly utilized if the clinical situation allows (Panthee and Ono, 2015). Use of intraoperative motor and somatosensory evoked potentials for monitoring has contributed to reducing spinal cord injury (Panthee and Ono, 2015). Spinal cord infarction is detected with MRI.

Clinical trials and guidelines

A randomized clinical trial that aimed to decrease CSF pressure to less than 10 mmHg and continue CSF drainage for 48 hours after surgery showed an 80% reduction in the relative risk of postoperative paraplegia (Coselli et al., 2002). This has been confirmed in subsequent studies (Hnath et al., 2008). There are no large randomized controlled trials to support postoperative augmentation of MAP; however, given the underlying pathogenesis of spinal cord ischemia, it is recommended to maintain a MAP of 80–100 mmHg to ensure a spinal cord perfusion pressure of at least 70 mmHg (Etz et al., 2015).

Hospital course and complex clinical decisions

Surgical strategies to minimize occurrence of postoperative paraplegia include techniques to limit cross-clamp time, use of endovascular methods whenever possible, induction of moderate hypothermia, and intraoperative monitoring of evoked potentials (Etz et al., 2015). CSF drainage is used intra- and postoperatively to improve net spinal cord perfusion pressure. Discussions between the ICU and surgical teams are imperative to convey relevant intraoperative events such as blood loss, hypotension, and risk for subsequent spinal cord ischemia. Postoperatively, the critical care team is crucial to reduce the risk of paralysis. Hourly neurologic examinations should be performed. Blood pressure should be closely monitored, preferably with an arterial line, and vasopressors used as needed to maintain a goal MAP > 80 mmHg or spinal perfusion pressure ≥70 mmHg. Lumbar drainage should be continued for approximately 48 hours, with a goal CSF pressure of less than 10 mmHg. Caution regarding overdrainage should be exercised, since this can result in intracranial hypotension, tonsillar herniation, or subdural hematomas. If neurogenic shock occurs, supportive care with crystalloid infusion and vasopressor agents may be required. If hemodynamically significant bradycardia does not resolve with atropine and chronotropic drugs, a pacemaker may be indicated.

Neurorehabilitation and outcomes

Most studies evaluating outcomes after spinal cord infarct are limited and have very small sample sizes. One of the largest reports is a retrospective review of 115 patients, of which 69% were post aortic surgery (Robertson et al., 2012), with a 3-year mortality of 23%. At hospital discharge, 81% required a wheelchair and 86% required catheterization. This improved in the 3-year survivors to 42% and 54%, respectively.

NEUROMUSCULAR

Nerve and plexus injuries

EPIDEMIOLOGY

Injuries to peripheral nerves and brachial or lumbosacral plexuses can be complications of bone fractures, direct injury, or compression. The incidence of brachial plexus injury (BPI) is difficult to estimate, but has increased recently (Sakellariou et al., 2014). Approximately 70% of BPIs are due to traffic accidents, predominantly involving motorcycles (Sakellariou et al., 2014). Avulsion injury is the most common, followed by compression or crush injuries, anterior shoulder dislocation, and occasionally iatrogenic surgical positioning (Limthongthang et al., 2013).

Traumatic lumbosacral plexus injury (LPI) is less common given its limited mobility within the retroperitoneal space (Wilbourn, 2007). LPI is usually associated with sacral pelvic fractures or sacroiliac joint separation. Incidence is estimated at 0.7–0.8% in patients with acetabular or pelvic fractures (Kutsy et al., 2000). Etiologies include high-speed MVC, gunshot wounds, and retroperitoneal hematomas (Kutsy et al., 2000). Isolated nerve injuries are much more common than plexus injuries. The incidence of nerve injury with hip displacement ranges from 10% to 20%, most commonly involving the sciatic nerve (Cornwall and Radomisli, 2000).

NEUROPATHOLOGY

Nerve injuries have been classified as neurapraxia, axonotmesis, or neurotmesis, depending on the extent of injury (Seddon, 1943). Neurapraxia is dysfunction without macroscopic involvement of the nerve, and deficits may last anywhere from a few hours to months, with nerve conduction studies (NCS) showing lack of conductivity at the point of injury, but normal distal conduction. Axonotmesis occurs when continuity of individual nerve fibers is disrupted, with preservation of perineurium and epineurium. Neurotmesis refers to complete axonal disruption, including the perineurium and epineurium.

CLINICAL PRESENTATION

Clinical deficits from BPI and LPI localize to the innervation provided by the affected nerve(s), as shown in Figure 34.3. Specific localization may be difficult in this patient population, due to confounding injuries, sedation, and altered mental status. About 75% of BPIs are closed injuries affecting roots and trunks in the supraclavicular zone. Most are complete C5–T1 plexus injuries, resulting in total paralysis and sensory loss in the extremity. Partial injuries are less common, with 20–25% involving C5–C6 and less than 5% involving C8–T1. Avulsion injuries occur most frequently with forceful rotation of the head/neck away from the ipsilateral shoulder, or high-energy traction to the arm, causing disruption of the C5–C7 root or upper trunk. Lower-trunk injuries occur with forceful abduction of the arm above the head. Infra- and retroclavicular injuries are less common. Open BPI occurs with penetrating wounds, or open shoulder fractures. Gunshot wounds produce extensive initial neurologic deficit, but transected nerve lesions are uncommon, ranging from 12% to 15% (Kline, 2009; Limthongthang et al., 2013).

Associated vascular injuries to the subclavian or axillary artery and subclavian vein occur in as many as 23% of BPI patients (Narakas, 1985). Infraclavicular injuries are associated with axillary artery rupture, and trauma to a cervical rib, cervical vertebral fractures, or T1 root injuries can produce a preganglionic Horner's syndrome.

Traumatic LPI from hip dislocation is a common clinical presentation, and predominantly involves major branches of the sciatic nerve. Symptoms include pain, decreased mobility, weakness, and sensory disturbances in the involved nerve distributions. Most frequently affected is the common peroneal nerve, followed by the gluteal, tibial, and obturator nerves (Fig. 34.3B).

NEURODIAGNOSTICS AND IMAGING

Plain radiographs or CT scans of the cervical spine, shoulder, and chest may reveal vertebral and clavicular fractures that raise suspicion for a BPI. Similarly, high-energy MVC resulting in pelvic trauma should raise concern for LPI.

CT and CT myelography are particularly helpful for evaluating nerve injury. Blood clots at the point of avulsion injuries appear as overshadowing on CT myelography. Pseudomeningocele formation is a common but delayed finding after root avulsion. MRI T2 sequences demonstrate contrast of the spinal cord roots relative to CSF. Use of overlapping coronal oblique MRI slices has a reported 93% sensitivity and 81% specificity for root avulsion detection, although movement artifact

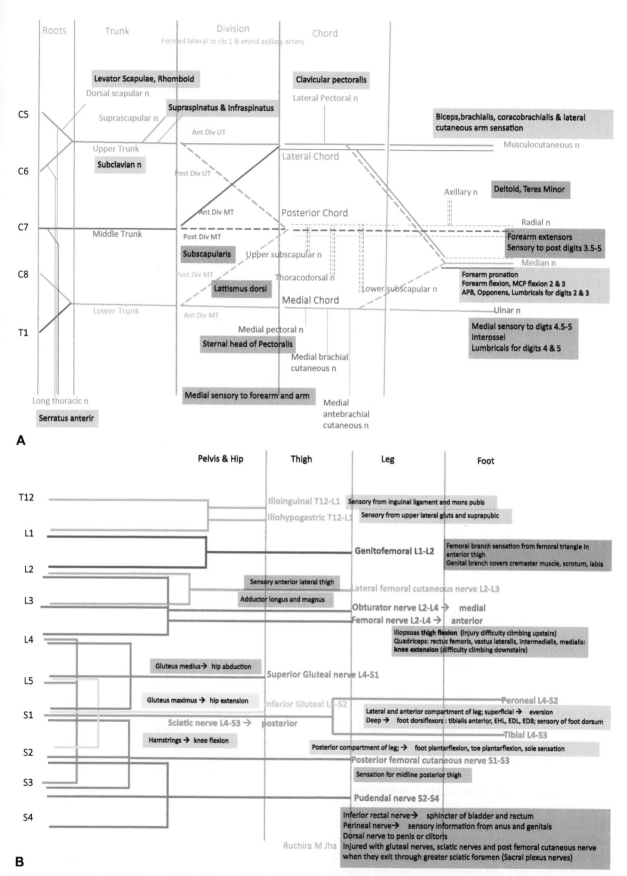

Fig. 34.3. (A) Brachial plexus; (B) lumbosacral plexus. UT, Upper Trunk; MT, Middle Trunk; MCP, Metacarpophalangeal; ABP, Abductor Pollicis Brevis; EDL, Extensor Digitorum Longus; EDB, Extensor Digitorum Brevis.

may limit image quality. Given the high incidence of associated vascular injuries, dedicated vessel imaging with conventional CT or MR angiography should be considered (Limthongthang et al., 2013).

Electromyography (EMG) and NCS are useful diagnostically, prognostically, and for localization. NCS typically show decreased amplitudes in both motor and sensory action potentials, with only mild reductions in conduction velocity. For BPI, preoperative sensory nerve action potentials can diagnose preganglionic nerve root avulsion and the level of injury. Intraoperative action potentials may guide surgical decisions. EMG demonstrates muscle denervation 2–3 weeks postinjury (Harper, 2005). In lumbosacral plexopathies, fibrillations on EMG spare paraspinal muscles, which can distinguish LPI from radiculopathies and radiculoplexopathies.

CLINICAL TRIALS AND GUIDELINES

There are no clinical trials or formal guidelines regarding diagnostic evaluation or management of polytrauma-related plexus injuries.

HOSPITAL COURSE AND COMPLEX CLINICAL DECISIONS

High clinical suspicion for plexus injuries should be maintained in patients with trauma to neighboring structures. Radiographic imaging is the first diagnostic step. EMG and NCS provide additional diagnostic, localizing, and prognostic information.

BPI often requires operative repair, although controversy exists regarding the type and timing of surgery. Surgical delays result in denervation atrophy and may limit reinnervation. However, immediate or early intervention does not allow for spontaneous recovery, which occurs within 3–4 months in 40% of C5–C6 injuries and 15% of C5–C7 injuries (Kline, 2009). Typically, multiple staged reconstructive procedures are required (Limthongthang et al., 2013). Immediate repair, or exploration and reconstruction within 3–4 weeks are recommended for penetrating injuries and preganglionic total-arm BPI. In crush injury or rupture, nerves are tagged and then re-explored within 3–4 weeks. Otherwise, reconstructive surgery within 6–9 months is typical, which allows time for axonal regeneration before irreversible denervation atrophy (Limthongthang et al., 2013). However, others have supported reconstruction within 14 days or exploration within 3 months for BPI to avoid fibrotic scar tissue formation (Birch, 2009). The treatment therefore depends in large part on practical feasibility and surgeon preference. Reconstructive surgery includes nerve transfers or functioning free muscle transfers in total BPI, and concomitant vascular injury is associated with higher failure rates (Limthongthang et al., 2013).

LPI management is primarily pain control and physical therapy, but surgery is considered in certain cases (Dyck and Thaisetthawatkul, 2014). Posterior hip dislocation, with compression of the sciatic nerve, is an orthopedic emergency requiring rapid reduction to prevent irreversible nerve necrosis (Cornwall and Radomisli, 2000). Management of femoral dislocations and fractures is more controversial, as some urge immediate open reduction and surgical exploration, but others recommend a more conservative approach, with initial closed reduction (Cornwall and Radomisli, 2000).

NEUROREHABILITATION

The outcome of traumatic BPI depends on multiple variables, including age, injury mechanism, level of injury, timing of surgical intervention, and other comorbidities. Young age and upper-trunk lesions are associated with favorable outcomes. High-energy injuries and those with vascular damage are associated with less favorable prognosis. Avulsion injuries, supraclavicular lesions, mixed-nerve lesions, comorbidities, evidence of fibrosis, and degeneration of target muscles at the time of surgery, and persistent pain after 6 months have worse prognosis (Sakellariou et al., 2014). Traumatic LPI generally has a poor prognosis, although a neuropraxic pattern has better recovery potential. Outcomes of nerve injuries are highly variable in traumatic hip dislocations and fracture-dislocations, with partial recovery occurring in 60–70% of patients, depending on injury type and fracture severity. Tibial injuries may recover more rapidly than those involving the peroneal division (Fassler et al., 1993). Early rehabilitation efforts while in the ICU are crucial to prevent contractures, pressure ulcers, and permanent extremity deformities (Cornwall and Radomisli, 2000).

REFERENCES

Alexiou GA, Pahatouridis D, Vulgaris S (2011). Coagulopathy in traumatic brain injury. Injury 42 (1): 113. author reply 114.

American Psychiatric Association (2000). Diagnostic and Statistical Manual of Mental Disorders, 4th Edition, Text Revision. Washington, DC.

Andruszkow H, Probst C, Grun O et al. (2013). Does additional head trauma affect the long-term outcome after upper extremity trauma in multiple traumatized patients: is there an additional effect of traumatic brain injury? Clin Orthop Relat Res 471 (9): 2899–2905.

Angles EM, Robinson TN, Biffl WL et al. (2008). Risk factors for delirium after major trauma. AJS 196 (6): 864–870.

Barr J et al. (2013). American College of Critical Care Medicine. Clinical practice guidelines for the management of pain, agitation, and delirium in adult patients in the intensive care unit. Crit Care Med 41 (1): 263–306.

Bergeron N, Dubois MJ, Dumont M et al. (2001). Intensive Care Delirium Screening Checklist: evaluation of a new screening tool. Intensive Care Med 27 (5): 859–864.

Berne JD, Reuland KS, Villarreal DH et al. (2006). Sixteen-slice multi-detector computed tomographic angiography improves the accuracy of screening for blunt cerebrovascular injury. J Trauma 60 (6): 1204–1209. discussion 1209–1210.

Berne JD, Cook A, Rowe SA et al. (2010). A multivariate logistic regression analysis of risk factors for blunt cerebrovascular injury. J Vasc Surg 51 (1): 57–64.

Betjemann JP, Lowenstein DH (2015). Status epilepticus in adults. Lancet Neurol 14 (6): 615–624.

Biffl WL, Moore EE, Offner PJ et al. (1999a). Blunt carotid arterial injuries: implications of a new grading scale. J Trauma 47 (5): 845–853.

Biffl WL, Moore EE, Offner PJ et al. (1999b). Optimizing screening for blunt cerebrovascular injuries. Am J Surg 178 (6): 517–521.

Biffl WL, Moore EE, Elliott JP et al. (2000). The devastating potential of blunt vertebral arterial injuries. Ann Surg 231 (5): 672–681.

Biffl WL, Moore EE, Offner PJ et al. (2001). Blunt carotid and vertebral arterial injuries. World J Surg 25 (8): 1036–1043.

Biffl WL, Ray Jr CE, Moore EE et al. (2002). Treatment-related outcomes from blunt cerebrovascular injuries: importance of routine follow-up arteriography. Ann Surg 235 (5): 699–706. discussion 706–707.

Biffl WL, Egglin T, Benedetto B et al. (2006). Sixteen-slice computed tomographic angiography is a reliable noninvasive screening test for clinically significant blunt cerebrovascular injuries. J Trauma 60 (4): 745–751. discussion 751–752.

Biffl WL, Cothren CC, Moore EE et al. (2009). Western Trauma Association critical decisions in trauma: screening for and treatment of blunt cerebrovascular injuries. J Trauma 67 (6): 1150–1153.

Birch R (2009). Brachial plexus injury: The London experience with supraclavicular traction lesions. Neurosurg Clin N Am 20 (1): 15–23.

Blacker DJ, Wijdicks EFM (2004). Clinical characteristics and mechanisms of stroke after polytrauma. Mayo Clin Proc 79 (5): 630–635.

Bottaro FJ, Martinez OA, Pardal MM et al. (2007). Nonconvulsive status epilepticus in the elderly: a case-control study. Epilepsia 48 (5): 966–972.

Bromberg WJ, Collier BC, Diebel LN et al. (2010). Blunt cerebrovascular injury practice management guidelines: the Eastern Association for the Surgery of Trauma. J Trauma 68 (2): 471–477.

Brophy GM, Bell R, Claassen J et al. (2012). Guidelines for the evaluation and management of status epilepticus. Neurocrit Care 17 (1): 3–23.

Bryczkowski SB, Lopreiato MC, Yonclas PP et al. (2014). Risk factors for delirium in older trauma patients admitted to the surgical intensive care unit. J Trauma Acute Care Surg 77 (6): 944–951.

Brzezicki G, Rivet DJ, Reavey-Cantwell J (2016). Pipeline embolization device for treatment of high cervical and skull base carotid artery dissections: clinical case series. Journal of neurointerventional surgery 8: 722–728.

Bulger EM, Smith DG, Maier EV et al. (1997). Fat embolism syndrome. A 10-year review. Arch surg (Chicago, Ill : 1960) 132 (4): 435–439.

Burlew CC, Biffl WL (2010). Blunt cerebrovascular trauma. Curr Opin Crit Care 16 (6): 587–595.

Buskens CJ, Gratama JW, Hogervorst M et al. (2008). Encephalopathy and MRI abnormalities in fat embolism syndrome: a case report. Med Sci Monit : Int Med J Exp Clin Res 14 (11): CS125–CS129.

Butcher N, Balogh ZJ (2009). The definition of polytrauma: the need for international consensus. Injury 40 (Suppl 4): S12–S22.

Butcher NE, D'Este C, Balogh ZJ (2014). The quest for a universal definition of polytrauma. J Trauma Acute Care Surg 77 (4): 620–623.

Cantais E, Louge P, Suppini A et al. (2003). Right-to-left shunt and risk of decompression illness with cochleovestibular and cerebral symptoms in divers: case control study in 101 consecutive dive accidents. Crit Care Med 31 (1): 84–88.

Caplan LR (2009). Caplan's Stroke, 4th edn. Elsevier Health Sciences, Philadelphia.

Carrick MM, Tyroch AH, Youens CA et al. (2005). Subsequent development of thrombocytopenia and coagulopathy in moderate and severe head injury: support for serial laboratory examination. J Trauma 58 (4): 725–729. discussion 729–730.

Chen H, Xue LX, Guo Y et al. (2013). The influence of hemocoagulation disorders on the development of posttraumatic cerebral infarction and outcome in patients with moderate or severe head trauma. Biomed Res Int 2013: 685174.

Claassen J, Mayer SA, Kowalski RG et al. (2004). Detection of electrographic seizures with continuous EEG monitoring in critically ill patients. Neurology 62: 1743–1748.

Claassen J, Taccone FS, Horn P et al. (2013). Recommendations on the use of EEG monitoring in critically ill patients: consensus statement from the neurointensive care section of the ESICM. Intensive Care Med 39 (8): 1337–1351.

Cogbill TH, Moore EE, Meissner M et al. (1994). The spectrum of blunt injury to the carotid artery: a multicenter perspective. J Trauma 37 (3): 473–479.

Cornwall R, Radomisli TE (2000). Nerve injury in traumatic dislocation of the hip. Clin Orthop Relat Res® (377): 84–91.

Coselli JS, LeMaire SA, Koksoy C et al. (2002). Cerebrospinal fluid drainage reduces paraplegia after thoracoabdominal aortic aneurysm repair: results of a randomized clinical trial. YMVA 35 (4): 631–639.

Cothren CC, Moore EE, Biffle WL et al. (2003). Cervical spine fracture patterns predictive of blunt vertebral artery injury. J Trauma 55 (5): 811–813.

Cothren CC, Moore EE, Biffl WL et al. (2004a). Anticoagulation is the gold standard therapy for blunt carotid injuries to reduce stroke rate. Arch surg (Chicago, Ill : 1960) 139 (5): 540–545. discussion 545–546.

Cothren CC, Moore EE, Biffl WL et al. (2004b). Anticoagulation is the gold standard therapy for blunt carotid injuries to reduce stroke rate. Arch Surg (Chicago, Ill : 1960) 139 (5): 540–545. discussion 545–546.

Cothren CC, Moore EE, Ray Jr CE et al. (2007). Cervical spine fracture patterns mandating screening to rule out blunt cerebrovascular injury. Surgery 141 (1): 76–82.

Cothren CC, Biffl WL, Moore EE et al. (2009). Treatment for blunt cerebrovascular injuries: equivalence of anticoagulation and antiplatelet agents. Arch Surg (Chicago, Ill : 1960) 144 (7): 685–690.

Crawford JD, Allan KM, Patel KU et al. (2015). The natural history of indeterminate blunt cerebrovascular injury. JAMA Surg 150: 841–847.

Demaerel P, Gevers AM, De Bruecker Y et al. (2003). Stroke caused by cerebral air embolism during endoscopy. Gastrointest Endosc 57 (1): 134–135.

Demetriades D, Velmahos GC, Scalea TM et al. (2008). Operative repair or endovascular stent graft in blunt traumatic thoracic aortic injuries: results of an American Association for the Surgery of Trauma multicenter study. J Trauma 64 (3): 561–571.

Devlin JW, Roberts RJ, Fong JJ et al. (2010). Efficacy and safety of quetiapine in critically ill patients with delirium: a prospective, multicenter, randomized, double-blind, placebo-controlled pilot study. Crit Care Med 38 (2): 419–427.

DuBose JJ, Leake SS, Brenner M et al. (2015). Contemporary management and outcomes of blunt thoracic aortic injury. J Trauma Acute Care Surg 78 (2): 360–369.

Duby JJ, Berry AJ, Ghayyem P et al. (2014). Alcohol withdrawal syndrome in critically ill patients. J Trauma Acute Care Surg 77 (6): 938–943.

Dyck PJB, Thaisetthawatkul P (2014). Lumbosacral plexopathy. Continuum (Minneapolis, Minn.). 20 (5 Peripheral Nervous System Disorders): 1343–1358.

Ely EW, Shintani A, Truman B et al. (2004). Delirium as a predictor of mortality in mechanically ventilated patients in the intensive care unit. JAMA 291 (14): 1753–1762.

Etz CD, Weigang E, Hartert M et al. (2015). Contemporary spinal cord protection during thoracic and thoracoabdominal aortic surgery and endovascular aortic repair: a position paper of the vascular domain of the European Association for Cardio-Thoracic Surgery. Eur J Cardio Thorac Surg : official journal of the European Association for Cardio-thoracic Surgery 47 (6): 943–957.

Fabian TC, Patton Jr JH, Croce MA et al. (1996). Blunt carotid injury. Importance of early diagnosis and anticoagulant therapy. Ann Surg 223 (5): 513–522. discussion 522–525.

Fassler PR, Swiontkowski MF, Kilroy AW et al. (1993). Injury of the sciatic nerve associated with acetabular fracture. J Bone Joint Surg Am 75 (8): 1157–1166.

Findley JK, Park LT, Siefert CJ et al. (2010). Two routine blood tests – mean corpuscular volume and aspartate aminotransferase – as predictors of delirium tremens in trauma patients. J Trauma 69 (1): 199–201.

Forteza AM, Rabinstein A, Koch S et al. (2002). Endovascular closure of a patent foramen ovale in the fat embolism syndrome: changes in the embolic patterns as detected by transcranial Doppler. Arch Neurol 59 (3): 455–459.

Friedman D, Claassen J, Hirsch LJ (2009). Continuous electroencephalogram monitoring in the intensive care unit. Anesth Analg 109 (2): 506–523.

Fujioka M, Okuchi K, Sakaki T et al. (1994). Specific changes in human brain following reperfusion after cardiac arrest. Stroke; a journal of cerebral circulation 25 (10): 2091–2095.

Gill JR, Murphy CG, Quansah B et al. (2015). Seizure induced polytrauma; not just posterior dislocation of the shoulder. BMJ Case Reports. http://dx.doi.org/10.1136/bcr-2015-211445.

Girard TD, Kress JP, Fuchs BD et al. (2008). Efficacy and safety of a paired sedation and ventilator weaning protocol for mechanically ventilated patients in intensive care (Awakening and Breathing Controlled trial): a randomised controlled trial. Lancet 371 (9607): 126–134.

Girard TD, Pandharipande PP, Carson SS et al. (2010). Feasibility, efficacy, and safety of antipsychotics for intensive care unit delirium: the MIND randomized, placebo-controlled trial. Crit Care Med 38 (2): 428–437.

Goodwin RB, Beery 2nd PR, Dorbish RJ et al. (2009). Computed tomographic angiography versus conventional angiography for the diagnosis of blunt cerebrovascular injury in trauma patients. J Trauma 67 (5): 1046–1050.

Graham DI, Fitch W, MacKenzie ET et al. (1979). Effects of hemorrhagic hypotension on the cerebral circulation. III. Neuropathology. Stroke; a journal of cerebral circulation 10 (6): 724–727.

Gross T, Schuepp M, Attenberger C et al. (2012). Outcome in polytraumatized patients with and without brain injury. Acta Anaesthesiol Scand 56 (9): 1163–1174.

Guillevin R, Vallée JN, Demeret S et al. (2005). Cerebral fat embolism: usefulness of magnetic resonance spectroscopy. Ann Neurol 57 (3): 434–439.

Gunther ML, Morandi A, Krauskopf E et al. (2012). The association between brain volumes, delirium duration, and cognitive outcomes in intensive care unit survivors: the VISIONS cohort magnetic resonance imaging study. Crit Care Med 40 (7): 2022–2032.

Halpern CH, Reilly PM, Turtz AR et al. (2008). Traumatic coagulopathy: the effect of brain injury. J Neurotrauma 25 (8): 997–1001.

Harper CM (2005). Preoperative and intraoperative electrophysiologic assessment of brachial plexus injuries. Hand Clin 21 (1): 39–46, vi.

Harrigan MR, Weinberg JA, Peaks YS et al. (2011). Management of blunt extracranial traumatic cerebrovascular injury: a multidisciplinary survey of current practice. World journal of emergency surgery : WJES 6: 11.

Herman ST, Abend NS, Bleck TP et al. (2015). Consensus statement on continuous EEG in critically ill adults and children, part I: indications. Journal of clinical neurophysiology : official publication of the American Electroencephalographic Society 32 (2): 87–95.

Hershberger R, Cho JS (2014). Neurologic complications of aortic diseases and aortic surgery. Handb Clin Neurol 119: 223–238.

Hnath JC, Mehta M, Taggert JB et al. (2008). Strategies to improve spinal cord ischemia in endovascular thoracic aortic repair: outcomes of a prospective cerebrospinal fluid drainage protocol. J Vasc Surg 48 (4): 836–840.

Ho AM (1998). Acute causes of circulatory collapse and neurologic dysfunction after trauma. Canadian journal of anaesthesia = Journal canadien d'anesthésie 45 (12): 1223.

Hwang PYK, Lewis PM, Balasubramani YV et al. (2010). The epidemiology of BCVI at a single state trauma centre. Injury 41 (9): 929–934.

Jackson JC, Gordon SM, Hart RP et al. (2004). The association between delirium and cognitive decline: a review of the empirical literature. Neuropsychol Rev 14 (2): 87–98.

Jakob SM, Ruokonen E, Grounds RM et al. (2012). Dexmedetomidine vs midazolam or propofol for sedation during prolonged mechanical ventilation: two randomized controlled trials. JAMA 307 (11): 1151–1160.

Kamel H, Betjemann JP, Navi BB et al. (2012). Diagnostic yield of electroencephalography in the medical and surgical intensive care unit. Neurocrit Care 19 (3): 336–341.

Kamenar E, Burger PC (1980). Cerebral fat embolism: a neuropathological study of a microembolic state. Stroke; a journal of cerebral circulation 11 (5): 477–484.

Kellogg RG, Fontes RBV, Lopes DK (2013). Massive cerebral involvement in fat embolism syndrome and intracranial pressure management. J Neurosurg 119 (5): 1263–1270.

Kline DG (2009). Timing for brachial plexus injury: a personal experience. Neurosurg Clin N Am 20 (1): 24–26.

Kosova E, Bergmark B, Piazza G (2015). Fat embolism syndrome. Circulation 131 (3): 317–320.

Kurtz P, Fitts V, Sumer Z et al. (2011). How does care differ for neurological patients admitted to a neurocritical care unit versus a general ICU? Neurocrit Care 15 (3): 477–480.

Kurtz P, Gaspard N, Wahl AS et al. (2013). Continuous electroencephalography in a surgical intensive care unit. Intensive Care Med 40 (2): 228–234.

Kutsy RL, Robinson LR, Routt ML (2000). Lumbosacral plexopathy in pelvic trauma. Muscle Nerve 23 (11): 1757–1760.

Lat I, McMillian W, Taylor S et al. (2009). The impact of delirium on clinical outcomes in mechanically ventilated surgical and trauma patients. Crit Care Med 37 (6): 1898–1905.

Levi M, Cate Ten H (1999). Disseminated intravascular coagulation. N Engl J Med 341 (8): 586–592.

Lichte P, Kobbe P, Almahmoud K et al. (2015). Post-traumatic thrombo-embolic complications in polytrauma patients. Int Orthop 39 (5): 947–954.

Limthongthang R, Bachoura A, Songcharoen P et al. (2013). Adult brachial plexus injury: evaluation and management. Orthop Clin North Am 44 (4): 591–603.

Lippi G, Cervellin G (2010). Disseminated intravascular coagulation in trauma injuries. Semin Thromb Hemost 36 (04): 378–387.

Liu Y-H, Lin CK, Chen CW et al. (2011). Unilateral borderzone infarction in a young polytrauma patient. International journal of surgery case reports 2 (8): 235–238.

Lukan JK, Reed Jr DN, Looney SW et al. (2002). Risk factors for delirium tremens in trauma patients. J Trauma 53 (5): 901–906.

Lustenberger T, Talving P, Kobayashi L et al. (2010). Time course of coagulopathy in isolated severe traumatic brain injury. Injury 41 (9): 924–928.

Ma L, Lin Z, Chen D (2013). Wernicke encephalopathy following nutritional deficiency in a patient with multiple trauma. Anaesth Intensive Care 41 (6): 816–817.

Maganti R, Gerber P, Drees C et al. (2008). Nonconvulsive status epilepticus. Epilepsy Behav 12 (4): 572–586.

Malhotra AK, Camacho M, Ivatury RR et al. (2007). Computed tomographic angiography for the diagnosis of blunt carotid/vertebral artery injury: a note of caution. Ann Surg 246 (4): 632–642. discussion 642–643.

McKinney A, Ott F, Short J et al. (2007). Angiographic frequency of blunt cerebrovascular injury in patients with carotid canal or vertebral foramen fractures on multidetector CT. Eur J Radiol 62 (3): 385–393.

McMahon CG, Yates DW, Campbell FM et al. (1999). Unexpected contribution of moderate traumatic brain injury to death after major trauma. J Trauma 47 (5): 891–895.

Menkin M, Schwartzman RJ (1977). Cerebral air embolism. Report of five cases and review of the literature. Arch Neurol 34 (3): 168–170.

Milbrandt EB, Deppen S, Harrison PL et al. (2004). Costs associated with delirium in mechanically ventilated patients. Crit Care Med 32 (4): 955–962.

Miller PR, Fabian TC, Bee TK et al. (2001). Blunt cerebrovascular injuries: diagnosis and treatment. J Trauma 51 (2): 279–285. discussion 285–286.

Miller PR, Fabian TC, Croce MA et al. (2002). Prospective screening for blunt cerebrovascular injuries: analysis of diagnostic modalities and outcomes. Ann Surg 236 (3): 386–393. discussion 393–395.

Mittal MK, Kashyap R, Herasevich V et al. (2015). Do patients in a medical or surgical ICU benefit from a neurologic consultation? Int J Neurosci 125 (7): 512–520.

Morandi A, Rodgers BP, Gunther ML et al. (2012). The relationship between delirium duration, white matter integrity, and cognitive impairment in intensive care unit survivors as determined by diffusion tensor imaging: the VISIONS prospective cohort magnetic resonance imaging study. Crit Care Med 40 (7): 2182–2189.

Narakas AO (1985). The treatment of brachial plexus injuries. Int Orthop 9 (1): 29–36.

Neidlinger NA, Puzziferri N, Victorino GP et al. (2004). Cardiac thromboemboli complicating a stab wound to the heart. Cardiovasc Pathol 13 (1): 56–58.

Nekludov M, Antovic J, Bredbacka S et al. (2007a). Coagulation abnormalities associated with severe isolated traumatic brain injury: cerebral arterio-venous differences in coagulation and inflammatory markers. J Neurotrauma 24 (1): 174–180.

Nekludov M, Bellander B-M, Blomback M et al. (2007b). Platelet dysfunction in patients with severe traumatic brain injury. J Neurotrauma 24 (11): 1699–1706.

Ng KWP, Wong HC, Rathakrishnan R (2014). Should we treat patients with impaired consciousness and periodic patterns on EEG? Seizure: European Journal of Epilepsy 23 (8): 622–628.

Oliver M, Inaba K, Tang A et al. (2012). The changing epidemiology of spinal trauma: a 13-year review from a level I trauma centre. Injury 43 (8): 1296–1300.

Pandharipande P, Morandi A, Adams JR et al. (2006). Lorazepam is an independent risk factor for transitioning to delirium in intensive care unit patients. Anesthesiology 104 (1): 21–26.

Pandharipande P, Cotton BA, Shintane A et al. (2007a). Motoric subtypes of delirium in mechanically ventilated surgical and trauma intensive care unit patients. Intensive Care Med 33 (10): 1726–1731.

Pandharipande PP, Pun BT, Herr DL et al. (2007b). Effect of sedation with dexmedetomidine vs lorazepam on acute brain dysfunction in mechanically ventilated patients: the MENDS randomized controlled trial. JAMA 298 (22): 2644–2653.

Pandharipande P, Cotton BA, Shintani A et al. (2008). Prevalence and risk factors for development of delirium in surgical and trauma intensive care unit patients. J Trauma 65 (1): 34–41.

Panthee N, Ono M (2015). Spinal cord injury following thoracic and thoracoabdominal aortic repairs. Asian Cardiovascular and Thoracic Annals 23 (2): 235–246.

Pape HC (2012). Classification of patients with multiple injuries – is the polytrauma patient defined adequately in 2012? Injury 43 (2): 127–128.

Parikh AA, Luchette FA, Valente JF et al. (1997). Blunt carotid artery injuries. J Am Coll Surg 185 (1): 80–86.

Parisi DM, Koval K, Egol K (2002). Fat embolism syndrome. Am J Orthop (Belle Mead NJ) 31 (9): 507–512.

Parizel PM, Demey HE, Veeckmans G et al. (2001). Early diagnosis of cerebral fat embolism syndrome by diffusion-weighted MRI (starfield pattern). Stroke; a journal of cerebral circulation 32 (12): 2942–2944.

Pedersen GL, Rasmussen SB, Gyllenborg J et al. (2013). Prognostic value of periodic electroencephalographic discharges for neurological patients with profound disturbances of consciousness. Clin Neurophysiol 124 (1): 44–51.

Prall JA, Brega KE, Coldwell DM et al. (1998). Incidence of unsuspected blunt carotid artery injury. Neurosurgery 42 (3): 495–498. discussion 498–499.

Reade MC et al. (2016). Australian and New Zealand intensive care society clinical trials group. Effect of dexmedetomidine added to standard care on ventilator-free time in patients with agitated delirium: A Randomized Clinical Trial. JAMA 315 (14): 1460–1468. Erratum in: JAMA. 2016 Aug 16; 316 (7): 775.

Reade MC, Finfer S (2014). Sedation and delirium in the intensive care unit. N Engl J Med 370 (5): 444–454.

Reade MC, O'Sullivan K, Bates S et al. (2009). Dexmedetomidine vs. haloperidol in delirious, agitated, intubated patients: a randomised open-label trial. Crit Care 13 (3): R75.

Reith G, Bouillon B, Sakka SG et al. (2015). Massive cerebral air embolism after blunt chest trauma with full neurological recovery. CJEM 1–4.

Riker RR, Shehabi Y, Bokesch PM et al. (2009). Dexmedetomidine vs midazolam for sedation of critically ill patients: a randomized trial. JAMA 301 (5): 489–499.

Roberts DJ, Haroon B, Hall RI (2012). Sedation for critically ill or injured adults in the intensive care unit: a shifting paradigm. Drugs 72 (14): 1881–1916.

Robertson CE, Brown Jr RD, Wijdicks EF et al. (2012). Recovery after spinal cord infarcts: long-term outcome in 115 patients. Neurology 78 (2): 114–121.

Ronne-Engstrom E, Winkler T (2006). Continuous EEG monitoring in patients with traumatic brain injury reveals a high incidence of epileptiform activity. Acta Neurol Scand 114 (1): 47–53.

Rosenson J, Clements C, Simon B et al. (2013). Phenobarbital for acute alcohol withdrawal: a prospective randomized double-blind placebo-controlled study. J Emerg Med 44 (3): 592–598. e2.

Saggar V, Mittal RS, Vyas MC (2009). Hemostatic abnormalities in patients with closed head injuries and their role in predicting early mortality. J Neurotrauma 26 (10): 1665–1668.

Sakellariou VI, Badilas NK, Mazis GA et al. (2014). Brachial plexus injuries in adults: evaluation and diagnostic approach. ISRN orthopedics 2014: 726103.

Schweickert WD, Pohlman MC, Pohlman AS et al. (2009). Early physical and occupational therapy in mechanically ventilated, critically ill patients: a randomised controlled trial. Lancet 373 (9678): 1874–1882.

Seddon HJ (1943). Three types of nerve injury. Brain 66: 237.

Server A, Dullerud R, Haakonsen M et al. (2001). Posttraumatic cerebral infarction. Neuroimaging findings, etiology and outcome. Acta radiologica (Stockholm, Sweden : 1987) 42 (3): 254–260.

Seth R, Obuchowski AM, Zoarski GH (2013). Endovascular repair of traumatic cervical internal carotid artery injuries: a safe and effective treatment option. AJNR Am J Neuroradiol 34 (6): 1219–1226.

Skrobik YK, Bergeron N, Dumont M et al. (2004). Olanzapine vs haloperidol: treating delirium in a critical care setting. Intensive Care Med 30 (3): 444–449.

Souverein PC, Webb DJ, Petri H et al. (2005). Incidence of fractures among epilepsy patients: a population-based retrospective cohort study in the General Practice Research Database. Epilepsia 46 (2): 304–310.

Stephan K, Huber S, Haberle S et al. (2015). Spinal cord injury—incidence, prognosis, and outcome: an analysis of the TraumaRegister DGU. Spine J 15 (9): 1994–2001.

Sumann G, Kampfl A, Wenzel V et al. (2002). Early intensive care unit intervention for trauma care: what alters the outcome? Curr Opin Crit Care 8 (6): 587–592.

Svensson LG (2005). Paralysis after aortic surgery: in search of lost cord function. Surgeon 3 (6): 396–405.

Takaoka M, Matsusaka M, Ishikawa K et al. (2004). Multiple border-zone infarcts after hemorrhagic shock in trauma victims: three case reports. J Trauma 56 (5): 1152–1155.

Tawil I, Stein DM, Mirvis SE et al. (2008). Posttraumatic cerebral infarction: incidence, outcome, and risk factors. J Trauma 64 (4): 849–853.

The Participants in the International Multi-disciplinary Consensus Conference on Multimodality Monitoring, Claassen J, Vespa P (2014). Electrophysiologic monitoring in acute brain injury. Neurocritical Care 21 (S2): 129–147.

Tieu BH, Holcomb JB, Schreiber MA (2007). Coagulopathy: its pathophysiology and treatment in the injured patient. World J Surg 31 (5): 1055–1064.

Utter GH, Hollingworth W, Hallam DK et al. (2006). Sixteen-slice CT angiography in patients with suspected blunt carotid and vertebral artery injuries. J Am Coll Surg 203 (6): 838–848.

Valentino R, Hilbert G, Vargas F et al. (2003). Computed tomographic scan of massive cerebral air embolism. Lancet 361 (9372): 1848.

Vespa PM, Nuwer MR, Nenov V et al. (1999). Increased incidence and impact of nonconvulsive and convulsive seizures after traumatic brain injury as detected by continuous electroencephalographic monitoring. J Neurosurg 91 (5): 750–760.

Wang W, Li HL, Wang DX et al. (2012). Haloperidol prophylaxis decreases delirium incidence in elderly patients after noncardiac surgery: a randomized controlled trial. Crit Care Med 40 (3): 731–739.

Wilbourn AJ (2007). Plexopathies. Neurol Clin 25 (1): 139–171.

Zaal IJ, Devlin JW, Peelen LM et al. (2015). A systematic review of risk factors for delirium in t ICU. Crit Care Med 43 (1): 40–47.

The incidence of ICU admissions for pregnant patients (including up to 6 weeks postpartum) ranged from 0.7 to 13.5 per 1000 deliveries worldwide. The median was 2.6 ICU admissions per 1000, with no difference in mean between developing and developed countries (Pollock et al., 2010). Obstetric ICU admissions accounted for 0.4–16% of total ICU admissions (mean = 2.2%). The highest rates of admission were in the UK and the USA (Gaffney, 2014). ICU admissions occurred more frequently in the postpartum period rather than the antepartum period (Gaffney, 2014). One study reported a rate of ICU admission in developed countries of 2–4 per 1000 obstetric patients (Zeeman, 2006). Differences in rates may be attributable to variations in patient population, ICU admission criteria, nursing policies, and available treatment options. Most women were admitted for less than 48 hours (Zeeman, 2006). A study completed in the USA showed that 0.4% of obstetric patients were admitted to the ICU (Kilpatrick and Matthay, 1992). Of these patients, 34% were antepartum and 66% were postpartum (Kilpatrick and Matthay, 1992). Four patients (12% of those admitted to the ICU) in this study died, including the only 2 with cerebral hemorrhages and 2 of the 8 patients with acute lung injury (Kilpatrick and Matthay, 1992).

Neurologic complications in critically ill obstetric patients range from 18 to 50% (Kilpatrick and Matthay, 1992; Platteau et al., 1997; Karnad et al., 2004; Munnur et al., 2005). Neurologic symptoms can result from pregnancy-related conditions, including eclampsia, acute fatty liver of pregnancy, or amniotic fluid embolism (Karnad and Guntupalli, 2004). Neurologic symptoms can also occur if pre-existing medical disorders such as epilepsy (Kaymar and Varner, 2013), hypertension, or intracranial neoplasms worsen during pregnancy (Lapinksy et al., 1997; Qhah et al., 2001). Pregnancy may predispose to certain illnesses, some of which may present in nonpregnant patients but are more common or worse in pregnant patients, such as cerebral venous sinus thrombosis (CVST) (Karnad and Guntupalli, 2004; Karnad et al., 2004; Munnur et al., 2005).

Though data are conflicting (Sharshar et al., 1995; Kittner et al., 1996), the balance of the evidence indicates that stroke risk is increased during pregnancy and the puerperium (Jeng et al., 2004). An older study argued that the incidence of nonhemorrhagic stroke was not increased during pregnancy or early puerperium, but that eclampsia is the cause of the neurologic findings attributed to stroke (Sharshar et al., 1995). A classic study by Kittner et al. (1996) found a relative increase in ischemic stroke and particularly in hemorrhagic stroke in the postpartum period. In a study looking at ischemic stroke, hemorrhagic stroke, and cerebral venous sinus thrombosis (CVST), Skidmore et al. (2001) found that strokes are more likely to occur in the first week postpartum, but that the risk is generally increased in the third trimester and postpartum period. A more recent study supports that when all stroke types are considered (ischemic, hemorrhagic, subarachnoid hemorrhage, and CVST), the incidence is 25–34 per 100 000 deliveries, and that this represents an increase in incidence since the 1990s (Kuklina et al., 2011). This increase has been largely attributed to concurrent hypertensive disorders or heart disease (Kuklina et al., 2011). Even though the incidence of stroke in pregnant and postpartum women is still relatively low, stroke in this patient population causes significant long-term disability (Feske and Singhal, 2014). The reported death rate following stroke in pregnancy varies, possibly due to geographic setting (Skidmore et al., 2001). Stroke of all types accounts for 12% of maternal death (Donaldson and Lee, 1994; Witlin et al., 1997). Reported percentage of death due to intracerebral hemorrhage (ICH) in one study was 7.2% (Gaffney, 2014). The incidence of hemorrhagic stroke is higher in Japanese women when compared with other western countries (Yoshida et al., 2016). Amniotic fluid embolism (Goldman et al., 1964) and choriorcarcinoma (Weir et al., 1978) can cause stroke, but stroke is an uncommon complication of these rare conditions.

NEUROPATHOLOGY

Neurologic symptoms can be the initial presentation of a disease that is not pregnancy related. Neurologic symptoms may also occur in the setting of an exacerbation of pre-existing neurologic disorders; some, such as epilepsy, are known to worsen during pregnancy (Neligan and Laffey, 2011). Pregnancy itself can put patients at risk for acute neurologic symptoms because of factors such as hypertension, hypercoaguability, and the effects of hormonal fluctuations (Edlow et al., 2013).

The risks of pregnancy include changes that affect the nervous system, such as vascular changes, edema, and alterations in the immune system (Edlow et al., 2013). Hemodynamic changes in the postpartum period include an increase in cardiac output, acute blood loss at delivery, and a decrease in plasma oncotic pressure (Kilpatrick and Matthay, 1992). Retention of water causes decrease of serum sodium and osmolality (Skidmore et al., 2001). Pregnancy is a prothrombotic state, and the risk of venous and arterial thrombotic events is increased (Lanska and Kryscio, 1998; Edlow et al., 2013). Near the end of the third trimester, increased progesterone levels lead to increased venous distensiblity, allowing for venous stasis and increasing the risk of venous thromboembolism (Tettenborn, 2012).

Pre-eclampsia, formerly known as toxemia, is a disease characterized by vasospasm and coagulation system

Takaoka M, Matsusaka M, Ishikawa K et al. (2004). Multiple border-zone infarcts after hemorrhagic shock in trauma victims: three case reports. J Trauma 56 (5): 1152–1155.

Tawil I, Stein DM, Mirvis SE et al. (2008). Posttraumatic cerebral infarction: incidence, outcome, and risk factors. J Trauma 64 (4): 849–853.

The Participants in the International Multi-disciplinary Consensus Conference on Multimodality Monitoring, Claassen J, Vespa P (2014). Electrophysiologic monitoring in acute brain injury. Neurocritical Care 21 (S2): 129–147.

Tieu BH, Holcomb JB, Schreiber MA (2007). Coagulopathy: its pathophysiology and treatment in the injured patient. World J Surg 31 (5): 1055–1064.

Utter GH, Hollingworth W, Hallam DK et al. (2006). Sixteen-slice CT angiography in patients with suspected blunt carotid and vertebral artery injuries. J Am Coll Surg 203 (6): 838–848.

Valentino R, Hilbert G, Vargas F et al. (2003). Computed tomographic scan of massive cerebral air embolism. Lancet 361 (9372): 1848.

Vespa PM, Nuwer MR, Nenov V et al. (1999). Increased incidence and impact of nonconvulsive and convulsive seizures after traumatic brain injury as detected by continuous electroencephalographic monitoring. J Neurosurg 91 (5): 750–760.

Wang W, Li HL, Wang DX et al. (2012). Haloperidol prophylaxis decreases delirium incidence in elderly patients after noncardiac surgery: a randomized controlled trial. Crit Care Med 40 (3): 731–739.

Wilbourn AJ (2007). Plexopathies. Neurol Clin 25 (1): 139–171.

Zaal IJ, Devlin JW, Peelen LM et al. (2015). A systematic review of risk factors for delirium in t ICU. Crit Care Med 43 (1): 40–47.

Chapter 35

Neurologic complications in critically ill pregnant patients

W.L. WRIGHT*
Neuroscience Intensive Care Unit, Emory University Hospital Midtown, Atlanta, GA, USA

Abstract

Neurologic complications in a critically ill pregnant woman are uncommon but some of the complications (such as eclampsia) are unique to pregnancy and the puerperal period. Other neurologic complications (such as seizures in the setting of epilepsy) may worsen during pregnancy. Clinical signs and symptoms such as seizure, headache, weakness, focal neurologic deficits, and decreased level of consciousness require careful consideration of potential causes to ensure prompt treatment measures are instituted to prevent ongoing neurologic injury. Clinicians should be familiar with syndromes such as pre-eclampsia, eclampsia, stroke, posterior reversible encephalopathy syndrome, and reversible cerebral vasoconstriction syndrome. Necessary imaging studies can usually be performed safely in pregnancy. Scoring systems for predicting maternal mortality are inadequate, as are recommendations for neurorehabilitation. Tensions can arise when there is conflict between the interests of the mother and the interests of the fetus, but in general maternal health is prioritized. The complexity of care requires a multidisciplinary and multiprofessional approach to achieve best outcome in an often unexpected situation.

INTRODUCTION

Critically ill pregnant women are at risk for a broad spectrum of neurologic complications, only a few of which are unique to pregnancy. Some of the risks arise from hypertension, hypercoagulability, and fragility of blood vessels due to hormonal and hemodynamic changes (Edlow et al., 2013). Despite neurologic complications, maternal mortality is rare in developed countries (Gaffney, 2014). Assessing the incidence of neurologic complications in critically ill obstetric patients is difficult, partially because there is neither a standard definition of a critical care bed (Wunsch et al., 2008) nor standard critical care unit admission criteria (Gaffney, 2014). Therefore, admission to an intensive care unit (ICU) is often based on local referral patterns (Gaffney, 2014). There are several serious disorders which require immediate attention and are thus included in this volume. Because pregnant women are generally young and otherwise in good health, care should be optimal and aggressive. The complexity of care requires a multidisciplinary and multiprofessional approach to achieve best outcome. Clinicians should consider early transfer to a medical center that can provide multidisciplinary care, including expertise in obstetrics, maternal-fetal medicine, neurology, critical care, neurosurgery, and radiology (Edlow et al., 2013).

EPIDEMIOLOGY

The incidence of pre-eclampsia is reported to be 2–8%, depending on the population studied (American College of Obstetricians and Gynecologists, 2002; Neligan and Laffey, 2011). The incidence of eclampsia in one study was 0.71%; 31.14% of these patients had neurologic abnormalities (Tank et al., 2004). In developed countries, morbidity and mortality from pre-eclampsia and eclampsia are low, but still remain an issue in developing areas (Villar et al., 2003). One study done in the USA estimated severe maternal morbidity from all causes to occur in 12.9 in every 1000 delivery hospitalizations, and 2.9 in every 1000 postpartum hospitalizations (Callaghan et al., 2012). Total mortality data are sparse (Gaffney, 2014).

---

*Correspondence to: Wendy L. Wright, MD, FNCS, FCCM, Emory University Hospital Midtown, 550 Peachtree Street NE, Atlanta GA 30309, USA. Tel: +1-404-778-3752, E-mail: wwrigh2@emory.edu

The incidence of ICU admissions for pregnant patients (including up to 6 weeks postpartum) ranged from 0.7 to 13.5 per 1000 deliveries worldwide. The median was 2.6 ICU admissions per 1000, with no difference in mean between developing and developed countries (Pollock et al., 2010). Obstetric ICU admissions accounted for 0.4–16% of total ICU admissions (mean = 2.2%). The highest rates of admission were in the UK and the USA (Gaffney, 2014). ICU admissions occurred more frequently in the postpartum period rather than the antepartum period (Gaffney, 2014). One study reported a rate of ICU admission in developed countries of 2–4 per 1000 obstetric patients (Zeeman, 2006). Differences in rates may be attributable to variations in patient population, ICU admission criteria, nursing policies, and available treatment options. Most women were admitted for less than 48 hours (Zeeman, 2006). A study completed in the USA showed that 0.4% of obstetric patients were admitted to the ICU (Kilpatrick and Matthay, 1992). Of these patients, 34% were antepartum and 66% were postpartum (Kilpatrick and Matthay, 1992). Four patients (12% of those admitted to the ICU) in this study died, including the only 2 with cerebral hemorrhages and 2 of the 8 patients with acute lung injury (Kilpatrick and Matthay, 1992).

Neurologic complications in critically ill obstetric patients range from 18 to 50% (Kilpatrick and Matthay, 1992; Platteau et al., 1997; Karnad et al., 2004; Munnur et al., 2005). Neurologic symptoms can result from pregnancy-related conditions, including eclampsia, acute fatty liver of pregnancy, or amniotic fluid embolism (Karnad and Guntupalli, 2004). Neurologic symptoms can also occur if pre-existing medical disorders such as epilepsy (Kaymar and Varner, 2013), hypertension, or intracranial neoplasms worsen during pregnancy (Lapinksy et al., 1997; Qhah et al., 2001). Pregnancy may predispose to certain illnesses, some of which may present in nonpregnant patients but are more common or worse in pregnant patients, such as cerebral venous sinus thrombosis (CVST) (Karnad and Guntupalli, 2004; Karnad et al., 2004; Munnur et al., 2005).

Though data are conflicting (Sharshar et al., 1995; Kittner et al., 1996), the balance of the evidence indicates that stroke risk is increased during pregnancy and the puerperium (Jeng et al., 2004). An older study argued that the incidence of nonhemorrhagic stroke was not increased during pregnancy or early puerperium, but that eclampsia is the cause of the neurologic findings attributed to stroke (Sharshar et al., 1995). A classic study by Kittner et al. (1996) found a relative increase in ischemic stroke and particularly in hemorrhagic stroke in the postpartum period. In a study looking at ischemic stroke, hemorrhagic stroke, and cerebral venous sinus thrombosis (CVST), Skidmore et al. (2001) found that strokes are more likely to occur in the first week postpartum, but that the risk is generally increased in the third trimester and postpartum period. A more recent study supports that when all stroke types are considered (ischemic, hemorrhagic, subarachnoid hemorrhage, and CVST), the incidence is 25–34 per 100 000 deliveries, and that this represents an increase in incidence since the 1990s (Kuklina et al., 2011). This increase has been largely attributed to concurrent hypertensive disorders or heart disease (Kuklina et al., 2011). Even though the incidence of stroke in pregnant and postpartum women is still relatively low, stroke in this patient population causes significant long-term disability (Feske and Singhal, 2014). The reported death rate following stroke in pregnancy varies, possibly due to geographic setting (Skidmore et al., 2001). Stroke of all types accounts for 12% of maternal death (Donaldson and Lee, 1994; Witlin et al., 1997). Reported percentage of death due to intracerebral hemorrhage (ICH) in one study was 7.2% (Gaffney, 2014). The incidence of hemorrhagic stroke is higher in Japanese women when compared with other western countries (Yoshida et al., 2016). Amniotic fluid embolism (Goldman et al., 1964) and choriorcarcinoma (Weir et al., 1978) can cause stroke, but stroke is an uncommon complication of these rare conditions.

NEUROPATHOLOGY

Neurologic symptoms can be the initial presentation of a disease that is not pregnancy related. Neurologic symptoms may also occur in the setting of an exacerbation of pre-existing neurologic disorders; some, such as epilepsy, are known to worsen during pregnancy (Neligan and Laffey, 2011). Pregnancy itself can put patients at risk for acute neurologic symptoms because of factors such as hypertension, hypercoaguability, and the effects of hormonal fluctuations (Edlow et al., 2013).

The risks of pregnancy include changes that affect the nervous system, such as vascular changes, edema, and alterations in the immune system (Edlow et al., 2013). Hemodynamic changes in the postpartum period include an increase in cardiac output, acute blood loss at delivery, and a decrease in plasma oncotic pressure (Kilpatrick and Matthay, 1992). Retention of water causes decrease of serum sodium and osmolality (Skidmore et al., 2001). Pregnancy is a prothrombotic state, and the risk of venous and arterial thrombotic events is increased (Lanska and Kryscio, 1998; Edlow et al., 2013). Near the end of the third trimester, increased progesterone levels lead to increased venous distensiblity, allowing for venous stasis and increasing the risk of venous thromboembolism (Tettenborn, 2012).

Pre-eclampsia, formerly known as toxemia, is a disease characterized by vasospasm and coagulation system activation that leads to impaired organ perfusion. It is a multisystem disorder characterized by hypertension, edema, and proteinuria (Feske and Singhal, 2014). Pre-eclampsia may be due to a maladapted immune response to antigens expressed on the invading trophoblast (Feske and Singhal, 2014), resulting in a premature halting of trophoblastic invasion into the maternal arteries (Hanna et al., 2006). In order to mitigate the resulting poor placental perfusion, the placenta will produce angiogenic factors, which cause symptoms when they alter maternal vascular endothelial function (Baumwell and Karumanchi, 2007; Karumanchi and Epstein, 2007). The seizures in eclampsia probably result from intracranial vasospasm, local ischemia, and endothelial dysfunction, leading to vasogenic and cytotoxic edema (Neligan and Laffey, 2011).

Hypertensive encephalopathy is a pathophysiologic model for the cerebrovascular abnormalities in eclampsia (Schwartz et al., 2000). Decreased cerebrovascular resistance causes increased pressure on the microvascular circulation, resulting in vasogenic edema. This mechanism is likely similar or perhaps even identical to the processes involved in posterior reversible encephalopathy syndrome (PRES) (Hinchey et al., 1996).

The causes of stroke in pregnant and postpartum women are diverse. Estrogen levels stimulate the hepatic synthesis of clotting factors (Tettenborn, 2012; Wabnitz and Bushnell, 2014), increasing the risk of thromboembolic events in both the venous and the arterial systems (Lanska and Kryscio, 1998; Wabnitz and Bushnell, 2014). Estrogen also increases cholesterol levels, increasing the risk of hyperlipidemia and vascular disease (Tettenborn, 2012). Craniocervical dissection from labor-related Valsalva or neck extension can lead to ischemic stroke (Edlow et al., 2013).

CLINICAL PRESENTATION

As with most patients experiencing neurologic symptoms, the clinical presentation of pregnant and postpartum women will help identify neurologic signs, symptoms, and syndromes, which can guide neurodiagnostic and imaging studies. The timing of onset of signs and symptoms may help establish a differential diagnosis. Sudden-onset symptoms require urgent and thorough diagnostic investigation to help identify potential causes (Table 35.1). In critically ill patients especially, the etiology of the symptoms needs to be identified expeditiously in order to prevent ongoing neurologic damage.

Seizures

Preventing and treating seizures helps avoid morbidity to an obstetric patient and her fetus (Sheth and Sheth,

Table 35.1

Causes of sudden onset of neurologic symptoms

Ischemic stroke
Intracerebral hemorrhage
Subarachnoid hemorrhage
Cerebral venous sinus thrombosis
Epidural hematoma
Brain mass lesions
Meningitis or other central nervous system infection
Demyelination
Migraine headache
Seizures with postictal symptoms
Toxic and metabolic encephalopathies
Posterior reversible encephalopathy syndrome
Reversible cerebral vasoconstriction syndrome
Psychogenic
Transient global amnesia
Cardiac syncope
Eclampsia or severe pre-eclampsia
Thrombotic thrombocytopenic purpura
Nervous system trauma

Information from Jamieson (2009); Sheth and Sheth (2012); Edlow et al. (2013).

2012). When a critically ill pregnant woman has a seizure, immediate attention turns to airway, breathing, and circulation (Sheth and Sheth, 2012). While stabilizing the patient, initial measures to stop an ongoing seizure or prevent another seizure should also be implemented. Quick consideration of the differential diagnosis may reveal additional diagnostic and treatment priorities (Sheth and Sheth, 2012).

The differential diagnosis for seizure in a pregnant or postpartum woman is broad, including pre-existing epilepsy (Karnad and Guntupalli, 2005; Edlow et al., 2013), eclampsia, intracranial hemorrhage, CVST, reversible cerebral vasoconstriction syndrome (RCVS), posterior reversible leukoencephalopathy, thrombotic thrombocytopenic purpura (TTP), intracranial tumor (Edlow et al., 2013; Bove and Klein, 2014), meningitis, encephalitis, neuroinflammation (Bove and Klein, 2014), metabolic derangements (Karnad and Guntupalli, 2005) such as hypoglycemia (Edlow et al., 2013), and other intracranial lesions (Karnad and Guntupalli, 2005).

Seizure is a defining feature of eclampsia (Edlow et al., 2013). Up to 40% of seizures in eclampsia occur after delivery (Neligan and Laffey, 2011). In a critically ill woman whose pregnancy has passed 20 weeks' gestation presenting with a seizure, eclampsia should always be considered as a potential etiology, even if the woman has a known seizure disorder (Edlow et al., 2013). However, a clinician should not assume eclampsia is the cause of the seizure because this may result in other serious diagnoses being overlooked, such as ICH or CVST (Edlow et al., 2013).

Patients should undergo a diagnostic workup that is similar to those who are not pregnant. (Edlow et al., 2013). This generally means that a patient will need intracranial imaging to assess for causes of seizure (Edlow et al., 2013). If a patient is in status epilepticus, it is imperative that the usual evidence-based treatment protocols for status epilepticus be followed, with the goal of terminating seizures quickly. Preventing seizure recurrence is in the best interest of the patient and the fetus, so antiepileptic drugs (AEDs) should not be withheld solely because of pregnancy (Sheth and Sheth, 2012). While epileptic seizures may require treatment with lorazepam or even a specific antiepileptic medication, seizures due to eclampsia should be treated with magnesium sulfate. A commonly recommended dose of magnesium sulfate is 4–6 grams in 100 mL of fluid infused over 15–20 minutes (Sheth and Sheth, 2012), followed by a continuous infusion at a rate of 1–2 g/hour (Zeeman, 2006) until 24 hours after delivery (Sheth and Sheth, 2012). The goal magnesium level is generally 4–7 mEq/L. The most serious side-effect of magnesium toxicity is respiratory depression, often preceded by the loss of patellar reflexes. The definitive treatment of eclampsia is delivery of the fetus (Sheth and Sheth, 2012), which should proceed immediately after stabilization of the patient (Frontera and Ahmed, 2014).

Headaches

Headache is a common neurologic symptom and also has an extensive differential diagnosis. Tension-type headache and migraine headache are the commonest causes of headache in pregnant and nonpregnant women (Edlow et al., 2013). Migraine headaches usually improve during pregnancy, but can worsen when estrogen levels fall postpartum (Sances et al., 2003; Goadsby et al., 2008). Migraines can first develop during pregnancy (Chancellor et al., 1990), but be cautious making this diagnosis since other serious causes need to be excluded (Edlow et al., 2013). Acephalgic migraine may present with only focal neurologic deficits and no headache (Edlow et al., 2013).

An obstetric patient with new-onset headache past 20 weeks' gestation must be screened for pre-eclampsia (Edlow et al., 2013). Abrupt-onset or severe headache should prompt a workup for subarachnoid hemorrhage, RCVS, craniocervical dissection, and PRES (Edlow et al., 2013). Since computed tomography (CT) scan or lumbar puncture (LP) may be nondiagnostic for some of these disorders, magnetic resonance imaging (MRI) may be needed or, in some cases, preferred (Edlow et al., 2013). Causes of postpartum headache include pre-eclampsia, eclampsia, CVST, pituitary apoplexy (Stella and Jodicke, 2007), or postepidural headache (Klein and Loder, 2010). Postepidural headache is the result of low cerebrospinal fluid (CSF) pressure (Edlow et al., 2013). Though uncommon, postpartum patients can get a dural tear from pushing during labor, causing low CSF pressure headache in the absence of spinal anesthesia (Edlow et al., 2013). Headaches from low CSF pressure are often self-limited, but serious complications such as subdural hematoma (Zeidan et al., 2006), CVST (Lockhart and Baysinger, 2007), and PRES (Ho and Chan, 2007) can result. In general, secondary causes of headache should be investigated if the patient has new-onset headache, worsening headache, headache that has changed in character, visual disturbance, seizure, altered mental status, or focal neurologic signs (Edlow et al., 2013). Emergency and pathologic causes of headache in pregnant and postpartum woman are listed in Table 35.2.

New focal neurologic signs

Focal signs are hemiparesis, monoparesis, a language or speech disturbance, ataxia, cranial nerve dysfunction, numbness, neglect, or sudden changes in vision or sensation; more complex behavior syndromes (e.g., alexia) also require prompt and thorough investigation (Edlow et al., 2013). Stroke in pregnant women is uncommon, but the risk is increased when compared to nonpregnant age-matched controls (Kittner et al., 1996; Salonen Ros et al., 2001). Thrombotic thrombocytopenic purpura (TTP) can present with stroke-like symptoms (Edlow et al., 2013). Focal neurologic symptoms can also be the result of severe pre-eclampsia, eclampsia, intracranial hemorrhage, subarachnoid hemorrhage, PRES (Edlow et al., 2013), intracranial mass or postictal state. Migraine with aura can cause focal symptoms, even without accompanying headache (Edlow et al., 2013).

Table 35.2

Diagnoses that require emergent neurosurgical evaluation

Subarachnoid hemorrhage
Intracerebral hemorrhage
Ischemic stroke or hemorrhage in the cerebellum
Large hemispheric ischemic stroke
Epidural hematoma in the brain or spinal cord
Subdural hematoma
Acute spinal cord compression
Hydrocephalus
Intracranial mass lesion
Brain or spinal cord abscess
Pituitary apoplexy
Increased intracranial pressure
Penetrating brain, spinal cord, or nerve injury

Abnormal level of consciousness

Abnormal level of consciousness is a common neurologic presentation in critically ill obstetric patients (Karnad et al., 2004; Munnur et al., 2005). The differential of stupor or coma includes ischemic stroke, ICH, subarachnoid hemorrhage, CVST, intracranial tumor, metabolic derangement (Karnad and Guntupalli, 2005; Price et al., 2008), traumatic head injury, hypotension, hypoxia, demyelination, pre-eclampsia (Price et al., 2008), eclampsia, drugs, toxins, and seizures (Karnad and Guntupalli, 2005). Hyperemesis gravidarum can lead to Wernicke's encephalopathy or osmotic demyelination syndrome (Zara et al., 2012). Focal neurologic signs can offer a clue about underlying etiology, helping to guide further workup (Zeigler, 1985). Coma with focal deficits may indicate underlying ischemic stroke, ICH, CVST, brain abscess, or intracranial tumor (Karnad and Guntupalli, 2005). As with nonpregnant patients, the first priority in patients with altered mental status or coma is to ensure adequate airway, breathing, and circulation.

Increased intracranial pressure

Increased intracranial pressure (ICP) can be rapidly life-threatening if the cause is not rapidly detected and reversed. Workup and management strategies for elevated ICP should generally be similar to nonpregnant patients. When considering osmotherapy, hypertonic saline should be used preferentially over mannitol due to the potential for fetal hypoxia and acid–base disturbances with mannitol administration (Frontera and Ahmed, 2014). An additional consideration is that during labor, ICP can increase as high as 70 cm H_2O due to bearing down and forceful uterine contractions (Stevenson and Thompson, 2005). Some experts consider cesarean section in patients who are at risk for or known to have increased ICP (Stevenson and Thompson, 2005). Others recommend shortening the second stage of labor by assisting with forceps or vacuum extraction (Carhuapoma et al., 1999).

Generalized weakness

The clinician must ensure that a woman presenting with generalized weakness has sufficient ororpharyngeal function and respiratory effort to maintain a safe airway and adequate breathing. Additionally, clinicians should promptly assess for a sensory level or other signs of spinal cord compression, such as loss of bowel or bladder control, as emergent neurosurgical intervention may be indicated. Causes of weakness in critically ill pregnant or postpartum women include demyelination (such as from multiple sclerosis or transverse myelitis), trauma, myopathy, Guillain–Barré syndrome (GBS), myasthenia gravis, or other neuromuscular disorders (Karnad and Guntupalli, 2005).

GBS rarely complicates pregnancy, but has been associated with maternal mortality. In general, the incidence of GBS is not different during pregnancy than in the general population (Sax and Rosenbaum, 2006), but there may be an increased risk within the first 2 weeks after delivery (Yamada et al., 2001). In one study, one-third of pregnant patients with GBS required mechanical ventilation (Chan and Tsui Leung, 2004). There should be a low threshold for elective intubation to support respiratory function (Sax and Rosenbaum, 2006). Treatment includes immunomodulating therapy such as intravenous immunoglobulin (IVIG) or plasma exchange (PLEX) (Kuller et al., 1995; Yamada et al., 2001; Chan and Tsui Leung, 2004). When choosing between the two therapies, one should consider that IVIG can precipitate renal failure and thrombotic complications. The main risk of PLEX during pregnancy is hypotension from the alteration of blood volume (Parry and Heiman-Patterson, 1988). Termination of pregnancy does not seem to alter disease course (Chan and Tsui Leung, 2004). GBS does not affect uterine contractility, so vaginal delivery is a reasonable expectation (Sax and Rosenbaum, 2006). Babies delivered to mothers with GBS are not usually weak (Nelson and McLean, 1985). There is a risk of maternal relapse in the postpartum period (Meenakshi-Sundaram et al., 2014). Supportive care is essential, including deep-vein thrombosis prophylaxis and close monitoring of cardiac and respiratory function (Parry and Heiman-Patterson, 1988; Sax and Rosenbaum, 2006).

Pregnancy has a variable effect on the course of myasthenia gravis. Maternal mortality risk is 4% (Plauché, 1979), with risk factors including respiratory failure, cholinergic crisis, and magnesium administration (Plauché, 1991). Anticholinesterase medications, which are pregnancy category C, may need to be adjusted due to changes in intestinal absorption and renal clearance. Prednisone and methylprednisolone are also pregnancy category C (Ciafaloni and Massey, 2004). Severe exacerbations can be treated with IVIG or PLEX. Myasthenia does not affect uterine smooth muscle, but may cause fatigue in the striated muscles used during labor. Magnesium sulfate is considered contraindicated in myasthenia gravis as it can cause profound weakness (Sax and Rosenbaum, 2006). Azathioprine and cyclosporine should be avoided due to significant risks to the fetus (Ciafaloni and Massey, 2004).

PREGNANCY-RELATED CLINICAL SYNDROMES

Pre-eclampsia is a multisystem disorder characterized by hypertension in a previously normotensive pregnant woman (Zeeman, 2006). The clinical criteria for pre-eclampsia are classically described as newly diagnosed hypertension and proteinuria after the 20th week of gestation, and resolving within 6–12 weeks of delivery (Neligan and Laffey, 2011). Aggressive blood pressure control with intravenous antihypertensive medications such as labetalol and hydralazine is an important treatment goal and can help prevent ICH and hypertensive encephalopathy (Zeeman, 2006). Pregnant women who present with new-onset hypertension (systolic blood pressure > 140 mmHg or diastolic blood pressure >90 mmHg) should be assumed to have pre-eclampsia (Williams et al., 2008). Recent guidelines from the American College of Obstetricians and Gynecologists (2013) do not require the presence of proteinuria to meet the definition of pre-eclampsia if other signs of severe pre-eclampsia are present. Severe pre-eclampsia is defined as severe hypertension, proteinuria of greater than 5 grams in 24 hours, oliguria of less than 400 mL in 24 hours, cerebral irritability, epigastric (or right upper quadrant) pain, or pulmonary edema (Neligan and Laffey, 2011). Patients with severe eclampsia should be prepared for delivery. Magnesium sulfate should be started by loading with 4–6 g IV in 100 mL of fluid over 20 minutes, then infusing 2 g/hour (Williams et al., 2008) to maintain therapeutic range of 4–7 mEq/mL (Zeeman, 2006; Sheth and Sheth, 2012).

Eclampsia is defined as seizure in the setting of pre-eclampsia (Zeeman, 2006; Williams et al., 2008) and in the absence of another condition to which the seizure can be attributed (Neligan and Laffey, 2011; Edlow et al., 2013). Other neurologic symptoms of eclampsia and pre-eclampsia can include visual disturbance, altered mental status, coma (Zeeman, 2006), and headache (Williams et al., 2008; Edlow et al., 2013). Neurologic symptoms of eclampsia do not commonly persist and epilepsy is not usually a long-term complication, but ischemic stroke and ICH are well-known complications of eclampsia that can lead to persistent neurologic disability or death (Zeeman, 2006; Williams and Fletcher, 2010). Seizures should be treated with magnesium sulfate, as described previously. Mortality from eclampsia in the developed world has become uncommon (Neligan and Laffey, 2011). The definitive treatment of eclampsia is to deliver the fetus (American College of Obstetricians and Gynecologists, 2002).

Amniotic fluid embolism is a rare but catastrophic obstetric emergency (Williams et al., 2008) that usually occurs within 24 hours of delivery (Neligan and Laffey, 2011). Previously, symptoms of amniotic fluid embolism were thought to be due to embolization of amniotic fluid into the pulmonary circulation. However, more recent theories suggest that an anaphylactic response or hypersensitivity reaction to one or more components of the amniotic fluid is to blame (Clark, 1990; Williams et al., 2008). Amniotic fluid embolism can present with seizures or mental status change (Williams et al., 2008), as well as shock, acute hypoxic respiratory failure, and disseminated intravascular coagulation (Neligan and Laffey, 2011). The incidence is not well delineated, with reports in the literature varying between 1 in 8000 and 1 in 80 000 (Clark et al., 1995). Mortality can be up to one-quarter to one-third of patients, and long-term neurologic deficits can persist in as many as 10–20% (Williams et al., 2008). Patients are managed with aggressive life support measures, including ICU management, mechanical ventilation, high levels of supplemental oxygen, and aggressive measures to maintain circulatory perfusion (such as fluids, pressors, and blood products as needed) (Williams et al., 2008). Hypothermia (Ocegueda-Pacheco et al., 2014) and extracorporeal membrane oxygenation (Conde-Agudelo and Romero, 2009) have been used in the setting of cardiovascular collapse. Emergency delivery of the fetus may be required (Williams et al., 2008).

Choriocarcinoma is a rare cancer of trophoblastic tissue (Edlow et al., 2013). It causes highly vascular and aggressive tumors that are prone to hemorrhage (Bove and Klein, 2014). Besides local invasive effects, choriocarcinoma can metastasize to liver, lungs, spine, and brain (Smith et al., 2005). Neurologic symptoms occur when central nervous system tumors cause mass effect, bleeding, and/or invasion of local blood vessels (Edlow et al., 2013). Chemotherapy should be initiated, as this is a highly chemosensitive tumor (Frontera and Ahmed, 2014).

TTP usually presents in the late second or early third trimester of pregnancy (Martin et al., 2008; McCrae, 2010). Clinical symptoms are classically a pentad of thrombocytopenia, microangiopathic hemolytic anemia, fever, renal dysfunction, and neurologic dysfunction (Edlow et al., 2013). More than half of patients with TTP have neurologic symptoms, which include headache, seizures, altered mental status, and focal neurologic deficits (Edlow et al., 2013). Treatment with PLEX should be started as soon as the diagnosis of TTP is clinically suspected (Frontera and Ahmed, 2014).

Stroke in pregnant women is rare, but increased when compared to nonpregnant age-matched controls (Kittner et al., 1996; Salonen Ros et al., 2001). Causes of stroke can often be determined with rigorous laboratory and imaging studies (Skidmore et al., 2001). Pre-eclampsia and eclampsia are considered causative in up to 25–50% of cerebrovascular events in this patient population (Sharshar et al., 1995; Jaigobin and Silver, 2000).

The evaluation for ischemic stroke is similar to that of a nonpregnant young woman, including an echocardiogram, MRI, and assessment of the intracranial and extracranial vasculature. If those studies are unrevealing, a hypercoagulable workup should be sent (Skidmore et al., 2001). Some rare causes to consider are peripartum cardiomyopathy, choriocarcinoma, and amniotic fluid embolism (Feske and Singhal, 2014). The efficacy and safety of intravenous tissue plasminogen activator (IV tPA) in pregnancy are unknown (Sheth and Sheth, 2012).

Pregnancy increases the risk of hemorrhagic stroke more so than ischemic stroke (Feske and Singhal, 2014), especially in the early postpartum period. The increased risk of ICH is attributed largely to preeclampsia and eclampsia. Any coagulopathies due to systemic disease or anticoagulant use should be corrected in the setting of ICH (Jauch et al., 2013). Other common etiologies of ICH in this patient population include ruptured arteriovenous malformations (AVMs) and cerebral aneurysms (Feske and Singhal, 2014). Therefore, patients with ICH require workup for underlying structural lesions (Sharshar et al., 1995; Jaigobin and Silver, 2000). Data are conflicting about the influence of pregnancy on AVMs (Feske and Singhal, 2014). There does seem to be a higher risk of rupture on the day of delivery (Parkinson and Bachers, 1980). Expert recommendations are to treat a known AVM beforehand if a woman anticipates pregnancy. If an AVM is discovered during pregnancy but has not bled, expert recommendations are to avoid treatment during pregnancy. If an AVM bleeds during pregnancy, then treatment before delivery should be considered, taking into account the grade of the lesion and the length of time it will take to achieve therapeutic benefit from the treatment strategy (Ogilvy et al., 2001).

The risk of subarachnoid hemorrhage is not thought to be increased during pregnancy (Bateman et al., 2012). Underlying structural lesions, most commonly cerebral aneurysms, are usually the cause (Sharshar et al., 1995; Jaigobin and Silver, 2000). The rupture of a cerebral aneurysm, however, is considered coincident to pregnancy (Gaffney, 2014). Mortality for pregnant women is about 35% (Dias and Sekhar, 1990), and this is the third leading cause of maternal death that is not related to obstetric complications (Bateman et al., 2012). Fetal mortality is 17% (Dias and Sekhar, 1990).

CVST often presents with a headache that progresses in severity (Edlow et al., 2013). Other signs and symptoms can include dizziness, nausea, seizures, lethargy, coma, or focal neurologic signs (Masuhr et al., 2004). Risk of CVST is increased during pregnancy due to hypercoagulability (Feske and Singhal, 2014). The greatest risk period is during the third trimester and up to 4 weeks postpartum (Saposnik et al., 2011). Approximately three-quarters of the cases are postpartum (Coutinho et al., 2009). Neurologic symptoms arise due to edema surrounding the thrombosed cerebral venous sinuses (Feske and Singhal, 2014), and venous infarct can result (Skidmore et al., 2001). Additionally, hemorrhage may occur into the parenchyma, or even extend into the subarachnoid, subdural, and intraventricular spaces (Feske and Singhal, 2014). Clinicians need to maintain a high degree of clinical suspicion in pregnant and postpartum women presenting with new or worsening headache, as this diagnosis is often made in a delayed fashion. The diagnostic study of choice in pregnant women is an MR venogram (Frontera and Ahmed, 2014). The treatment of CVST is anticoagulation.

PRES commonly presents with encephalopathy, visual changes, seizures, and headache plus vasogenic edema on imaging (Staykov and Schwab, 2011). Focal signs such as weakness or sensory deficits may develop (Bartynski, 2008). There is a potential to reverse the radiographic findings and clinical symptoms with adequate treatment (Staykov and Schwab, 2011), including the removal of causative factors and treatment of blood pressure (Servillo et al., 2007). In pregnant or postpartum women, PRES is thought to be the radiographic correlate of preeclampsia and eclampsia (Staykov and Schwab, 2011). Other diseases and drugs can cause PRES, such as sepsis, hypertensive emergency, or immunosuppressive agents (Raroque et al., 1990; Bartynski, 2008; Bartynski and Boardmand, 2007).

RCVS is a spectrum of disorders that present with the abrupt onset of severe headaches plus multifocal areas of (potentially) reversible cerebral vasoconstriction (Ducros and Brousser, 2009). CT scans are usually negative, unless there is ICH. Rather, vascular imaging such as MR angiography, CT angiography, or conventional angiogram is needed to make the diagnosis (Ducros and Brousser, 2009). Recurrent daily thunderclap headaches are highly suggestive of the diagnosis (Chen et al., 2006; Calabrese et al., 2007; Ducros and Brousser, 2009). Complications include seizures, brain edema, lobar hemorrhage, subarachnoid hemorrhage, and ischemic stroke (Chen et al., 2006; Calabrese et al., 2007; Ducros and Brousser, 2009; Feske and Singhal, 2014). RCVS can be a manifestation of pre-eclampsia or eclampsia, but it can also be caused by craniocervical dissection, blood transfusion, cocaine, serotonin reuptake inhibitors, immunosuppressive drugs, and other medications (Ducros, 2012). Postpartum angiopathy falls within the spectrum of RCVS, and usually occurs within 1 week of delivery and after a normal pregnancy (Ducros, 2012; Fugate et al., 2012). Treatment goals include blood pressure control and seizure management (Feske and Singhal, 2014). Induced hypertension and balloon angioplasty have been used in cases of refractory or severe vasoconstriction (Frontera and Ahmed, 2014). Some authors promote the use of magnesium as part of the treatment strategy (Feske and Singhal, 2014). Symptoms usually subside over 2–3 months (Calabrese et al., 2007; Ducros and Brousser, 2009).

Brain tumors are rare in pregnancy. Patients with intracranial tumors can present with seizures, headache, focal neurologic deficits, or signs of increased ICP (such as nausea and vomiting, visual changes, or altered mental status). Brain tumors are not more common in pregnant and postpartum women, but symptoms may present or worsen during pregnancy as a result of an increase in extracellular fluid volume in the late second and third trimesters (Antonelli et al., 1996; Stevenson and Thompson, 2005), and due to hormonal changes that promote the growth of certain tumors (Antonelli et al., 1996; To and Cheung, 1997; Stevenson and Thompson, 2005). The most frequent brain tumors in pregnant women are gliomas (38%), meningiomas (28%), and acoustic neuromas (14%) (Stevenson and Thompson, 2005). MRI is the preferred imaging study to detect suspected brain tumors (Stevenson and Thompson, 2005).

Acute infarction or hemorrhage of the pituitary gland can present with headache, visual loss, ophthalmoplegia, and alteration of consciousness (Edlow et al., 2013). Pituitary apoplexy usually occurs in the setting of a pituitary adenoma (Edlow et al., 2013), but in pregnant patients it can occur when no adenoma is present (Ranabir and Baruah, 2011). The pituitary gland enlarges during pregnancy but this rarely leads to apoplexy (Edlow et al., 2013). Pituitary apoplexy associated with altered mental status, visual changes, or other clinical instability is a neurosurgical emergency (Ranabir and Baruah, 2011). Delay in treatment can lead to permanent visual loss and long-term endocrinopathies (Ranabir and Baruah, 2011). Additionally, corticosteroids should be given to avoid an addisonian crisis (Sheth and Sheth, 2012).

NEURODIAGNOSTICS AND IMAGING

The use of diagnostic studies should be guided by results of the history and physical (Zeigler, 1985). Evaluation for suspected cerebrovascular disease or other neurologic emergencies should not be delayed purely due to perceived risks of radiologic studies to the fetus (Sheth and Sheth, 2012). In pregnant woman it is especially prudent to coordinate needed imaging studies with radiology, in order to minimize duplicative or unnecessary tests (Edlow et al., 2013).

LP to obtain CSF for study is an important test to send whenever meningitis or encephalitis is suspected. LP is also an important diagnostic step in patients whose radiographic imaging is negative but suspicion remains for subarachnoid hemorrhage. Once a differential diagnosis is carefully considered, LP can help exclude or confirm inflammatory conditions and can allow for cytologic examination of the CSF. Risks of LP include infection, bleeding, headache, CSF leak and, in some cases, cerebral herniation. Cerebral imaging is usually done before LP for any patient suspected of having cerebral edema, obstructive hydrocephalus, or intracranial mass lesion that may lead to cerebral herniation. In cases where herniation risk is high, defer LP until the risk decreases and treat the patient empirically (for example, with antibiotics after blood cultures are sent in cases of possible central nervous system infection) (Roos and Greenlee, 2011).

Electroencephalogram (EEG) should be used whenever new-onset seizure has occurred or ongoing seizure activity is suspected. This includes patients who have coma or altered mental status with no known etiology (Towne et al., 2000). EEG may help narrow down potential etiologies of encephalopathy, such as hepatic encephalopathy or that caused by herpes encephalitis. EEG can help individually guide the use of therapy for status epilepticus, and should be performed in any patient who is not regaining consciousness 2 hours after being treated for status epilepticus; those requiring pharmacologic coma for refractory status epilepticus in order to evaluate the possibility of ongoing seizure without overt clinical signs, also called subclinical status (Manno, 2003); and patients whose clinical features are suggestive of nonepileptic seizures (Lawn and Wijdicks, 2002).

Electromyogram (EMG) and nerve conduction studies (NCS) can help diagnose peripheral nerve, neuromuscular junction, and muscle disease (Patten, 1996). In some diseases, the extent of damage as revealed on EMG/NCS can help predict prognosis for recovery (Bella and Chad, 1998). Pregnancy is not a contraindication to either EMG or NCS.

In general, elective neuroimaging should be deferred until after pregnancy (Bove and Klein, 2014). However, in the case of neurologic emergency, imaging studies should be selected with the goals of establishing diagnosis while minimizing potential risks (Bove and Klein, 2014). When possible, risks and benefits should be discussed with the pregnant patient (Bove and Klein, 2014) or her surrogate decision maker, and informed consent should be obtained (American College of Obstetricians and Gynecologists, 2004; Klein and Hsu, 2011). Collaborative decision making with radiology can help ensure efficient selection of the proper imaging study in order to save time and money while minimizing risks (Edlow et al., 2013). In general, most necessary tests can be performed safely (Honiden et al., 2013).

Ionizing radiation associated with conventional X-ray and CT can have teratogenic, carcinogenic, and mutagenic effects (American College of Obstetricians and

Gynecologists, 2004). The risk of teratogenesis is probably highest between 8 and 15 weeks of gestation (American College of Obstetricians and Gynecologists, 2004). Fetal radiation exposure during a noncontrasted head CT is minimal (American College of Obstetricians and Gynecologists, 2004; Williams and Fletcher, 2010; Klein and Hsu, 2011), but any exposure may increase the likelihood of childhood malignancy, especially leukemia (Centers for Disease Control and Prevention, 2014). However, there is no absolute contraindication to any radiologic test that is medically necessary during pregnancy (Gaffney, 2014). In nonpregnant women, noncontrasted head CT scan is usually the radiographic study of choice to assess for emergency conditions such as subarachnoid hemorrhage, ICH, and obstructive hydrocephalus. CT is useful in identifying emergencies because it requires very little time to perform, is sensitive for blood, and can reveal preliminary information about changes in brain structure (Jauch et al., 2013). Iodinated contrast is pregnancy category B, and carries a possible risk of neonatal hypothyroidism (Bove and Klein, 2014). Ideally, informed consent would be obtained before administration of iodinated contrast during pregnancy (Honiden et al., 2013). In practice, iodinated contrast is usually avoided in this patient population (Bove and Klein, 2014) unless essential for proper diagnosis (American College of Obstetricians and Gynecologists, 2004).

Ultrasound and noncontrasted MRI are considered safe during pregnancy (Gaffney, 2014). MRI provides more detailed brain imaging and is the study of choice to confirm acute ischemic stroke, and to look for brain tumors, focal areas of infection, inflammatory conditions, and demyelinating diseases. MRI is recommended preferentially over CT for imaging spinal cord pathology, but a CT of the spine to assess bony anatomy is often obtained before MRI in the setting of trauma (Burns and Selzer, 2009). MRI is useful in diagnosing PRES, as it will show transient fluid-attenuated inversion recovery and T2 abnormalities in the subcortical regions of the parietal occipital lobes (Zeeman, 2006). Early cerebral imaging in suspected stroke is important for prompt diagnosis and treatment to improve maternal morbidity and mortality. MRI is not usually done prior to consideration of t-PA for stroke because of the time it takes to complete the study. MRI of the brain with MR angiography of the head and neck should be performed in patients with transient ischemic attack to maximize stroke risk factor management, provided there is no contraindication to MRI such as pacemaker (DeLaPaz et al., 2011). MRI is preferred if CVST is suspected for several reasons: it can provide vessel imaging without contrast, it can demonstrate infarcts, and it spares the fetus the risks of ionizing radiation (Bove and Klein, 2014). In nonpregnant patients, MRI is usually part of the workup for new-onset seizures, if head CT is unrevealing of etiology (Bleck, 2012). In pregnant women with new-onset seizures, MRI should probably be the initial imaging study of choice over CT if time permits, since MRI is more sensitive for common seizure etiologies such as PRES (Edlow et al., 2013). Gadolinium enhancement for MRI is pregnancy category C, and therefore best avoided if possible (De Santis et al., 2007) unless the information gained would directly benefit the patient or fetus (Webb et al., 2005). Gadolinium may theoretically introduce a risk of nephrogenic systemic fibrosis of fetal kidneys and therefore the risks and benefits of administration should be carefully considered (Chen et al., 2008).

HOSPITAL COURSE AND MANAGEMENT

An initial treatment priority in critically ill pregnant and postpartum women is to ensure adequate airway, breathing, and circulation. In addition to issues of gas exchange causing respiratory failure, airway compromise may occur due to altered mental status or cranial nerve dysfunction from neurologic complications. Weakness due to spinal cord or neuromuscular disease may cause hypoventilation, as might irregular respiratory patterns from brain pathology (Roppolo and Walters, 2004). Some possible neurologic triggers for intubation include a Glasgow Coma Scale of less than 8 (Dunham et al., 2005), impaired gag reflex, inability to handle oral secretions, failure to oxygenate, failure to ventilate, and anticipated clinical decline (Roppolo and Walters, 2004).

Noninvasive positive-pressure ventilation is not commonly used in pregnant patients due to the increased risk of aspiration (Al-Ansari et al., 2007). If it is considered in patients with rapidly reversible causes of respiratory insufficiency, careful monitoring is required (Mallampalli et al., 2010). Noninvasive positive-pressure ventilation should not be used in patients with impaired mental status, hemodynamic instability, or difficulty handling oral secretions due to cranial nerve abnormalities (Roppolo and Walters, 2004; Mallampalli et al., 2010).

Besides the common preventive measures implemented to prevent thromboembolic disease, constipation, peptic ulcer disease, and hospital-acquired infections, patients with central nervous system injury require efforts to prevent secondary brain injury. In general, this requires promoting the balance between oxygen supply and demand, while ameliorating factors that contribute to inflammatory or excitotoxic damage. Oxygen supply can be improved by avoiding hypoxia, controlling ICP to allow for arterial delivery of blood, and determining appropriate blood pressure goals (Wright and Geogadin, 2006). Fever and seizure can also cause increased brain oxygen demand out of proportion to supply (Wright and Geogadin, 2006) and therefore should be avoided or aggressively treated. Therapeutic hypothermia has not been studied in pregnancy, but it is used in the treatment of severely increased or refractory intracranial hypertension (Frontera and Ahmed, 2014), and has been reported as a supportive measure in the setting of hemodynamic collapse from amniotic fluid embolism (Ocegueda-Pacheco et al., 2014). Hyponatremia is the most commonly encountered electrolyte disorder in pregnancy (Pazhayattil et al., 2015), but it is not well tolerated in patients with acute intracranial processes, because it can cause cerebral edema and is a risk factor for seizures (Wright, 2012). In pregnant and postpartum women, relieving pain and anxiety is an important treatment goal (Price et al., 2008) and can help maintain the balance between cerebral metabolic supply and cerebral metabolic demand (Wright and Geogadin, 2006).

Treatment of specific conditions

SEIZURES

It is unclear why seizure rate goes up in pregnant women (Meador, 2014). More studies are needed to figure out which AEDs have the lowest teratogenicity rates and yet control the seizures, especially when used in combination (Meador, 2014). Based on a recent update on the AEDs introduced since 1990, levetiracetam seems to be a good choice in pregnant women, whereas valproic acid seems to confer higher risk than other AEDs (Vadja et al., 2014).

Magnesium sulfate is the medication of choice to prevent recurrent eclamptic seizures (Neligan and Laffey, 2011). Respiratory depression or arrest can result from magnesium toxicity, but can be reversed with calcium. Magnesium toxicity is rare in the absence of renal failure (Neligan and Laffey, 2011).

ISCHEMIC STROKE

Pregnant women were excluded from tPA trials, so many questions about safety and efficacy remain unanswered (Feske and Singhal, 2014). Pregnancy is a relative contraindication to intravenous tPA, but not an absolute contraindication (Jauch et al., 2013). Cases of successful IV tPA and intra-arterial therapy have been reported (Johnson et al., 2005; Murugappan et al., 2006; Del Zotto et al., 2011). Case series indicate that adverse effects are uncommon, therefore IV tPA should be considered during pregnancy (Frontera and Ahmed, 2014). IV tPA should probably not be given in patients whose stroke is due to hypertensive encephalopathy associated with pre-eclampsia or eclampsia due to risk of cerebral hemorrhage (Feske and Singhal, 2014). Patients with contraindications to IV tPA or those who have large-vessel strokes and are not improving after IV tPA should be evaluated for intra-arterial thrombolysis or mechanical clot retrieval (Frontera and Ahmed, 2014).

ANEURYSMAL SUBARACHNOID HEMORRHAGE

Early treatment of a ruptured aneurysm reduces mortality for mother and fetus (Dias and Sekhar, 1990). Therefore, treatment should proceed as in nonpregnant patients (Feske and Singhal, 2014). The choice to clip or coil an aneurysm remains controversial. Some authors suggest that endovascular coiling in this patient population is preferable to the operative risks of clipping, even though it comes with greater radiation exposure (Frontera and Ahmed, 2014). Others preferentially support aneurysm clipping, but suggest that endovascular coiling may be considered as a treatment option with proper shielding to minimize fetal radiation (Meyers et al., 2000). As with nonpregnant patients, aneurysmal subarachnoid hemorrhage patients should be treated at high-volume centers, and by a multidisciplinary team that can evaluate each patient for the proper aneurysm obliteration technique and treat the serious complications that can result from subarachnoid hemorrhage, including cerebral vasospasm and hydrocephalus (Diringer et al., 2011). If there are urgent obstetric issues, an emergent caesarean section may need to take place, followed by surgical or endovascular control of the aneurysm (Feske and Singhal, 2014).

BRAIN TUMORS

It has been hypothesized that hormonal changes during pregnancy can influence tumor growth and trigger the development of neurologic symptoms (Stevenson and Thompson, 2005; Cowppli-Bony et al., 2011). In order to safeguard maternal outcome, the same options that are appropriate for nonpregnant women should be considered for pregnant women, and administered without delay if possible (Verheecke et al., 2014). Neurosurgical resection is the definitive therapy for most malignant and growing benign intracranial tumors (Verheecke et al., 2014). Optimal timing is a matter of debate, since surgery in the first trimester can increase the risk of miscarriage, yet early intervention can prevent progressive maternal neurologic decline (Verheecke et al., 2014). For example, high-grade glioma should be resected as soon as possible (Antonelli et al., 1996; Stevenson and Thompson, 2005). Steroids can help reduce cerebral edema associated with brain tumors (Stevenson and Thompson, 2005). The choice for adjuvant chemotherapy in pregnant women is a more complex matter that is beyond the scope of this chapter. However, in general, the different treatment options should be coordinated in a multidisciplinary setting based on available evidence regarding the risks and benefits (Stevenson and Thompson, 2005). For the most part, treatment with standard protocols during pregnancy with a goal of a full-term vaginal delivery is feasible (Verheecke et al., 2014).

CLINICAL TRIALS AND GUIDELINES

Clinical trials and guidelines specifically addressing the issue of neurologic complications in critically ill pregnant women are lacking. Many of the guidelines published that help guide the care of pregnant women do not address the intersection of critical illness and neurologic complications, other than eclampsia. The Maternal Critical Care Working Group published guidelines in 2011 that recommended care settings based on level of illness. For example, some indications for critical care unit placement could include the need for ICP monitoring or magnesium infusion to control seizures. The Food and Drug Administration assigns risk categories to drugs depending on evidence of risk of harm to the fetus (Doering et al., 2014). While this can be helpful when choosing between medications, some neurologic emergencies present limited options for treatment, and the "safest" option for the fetus is the one that will best treat the mother.

It is increasingly common that hospitals have protocols to address neurologic emergencies such as stroke and status epilepticus. However, these guidelines do not commonly address issues specific to pregnancy. One can seek guidance from guidelines based on specific diagnoses such as acute ischemic stroke (Lansberg et al., 2012; Jauch et al., 2013), subarachnoid hemorrhage (Diringer et al., 2011), ICH (Morgenstern et al., 2010), transient ischemic attack (Easton et al., 2009), nontraumatic headache (Cortelli et al., 2004; Edlow et al., 2008), neuromuscular disease (Gronseth and Barohn, 2000; Hughes et al., 2003; Skeie and Apostolski, 2010; Patwa et al., 2012), traumatic brain injury (The Brain Trauma Foundation, 2007), spinal cord injury (Consortium for Spinal Cord Medicine, 2008), status epilepticus (Brophy et al., 2012), and bacterial meningitis (Chaudhuri et al., 2008).

Status epilepticus

When a pregnant woman presents with status epilepticus, she should be treated with standard protocols because stopping any ongoing seizures will minimize harm to the patient and fetus (Kass, 2014). Magnesium should be given if eclampsia is suspected or confirmed, but additional AEDs may be needed (Brophy et al., 2012). Recently published guidelines for the treatment of status epilepticus recommend lorazepam for initial benzodiazepine administration, followed by either fosphenytoin or levetiracetam infusion as emergent initial therapy (Brophy et al., 2012). Fosphenytoin is pregnancy class D, but it is within the standard of care to use it when treating this life-threatening emergency (Kass, 2014). Data from recent registries suggest that newer AEDs are associated with fewer risks to the fetus (Molgaard-Nielsen, 2011), therefore levetiracetam for the treatment of status epilepticus in pregnant women can be considered instead of fosphenytoin (Brophy et al., 2012).

Severe hypertension, severe pre-eclampsia, and eclampsia

The American College of Obstetricians and Gynecologists recommends treating severe systolic blood pressure elevations (defined as greater than 160 mmHg) or diastolic blood pressure elevations (defined as greater than 110 mmHg) that are sustained for at least 15 minutes. Labetalol and hydralazine are generally considered first-line agents, but nifedipine was added as an alternative for first-line therapy in recent guidelines (American College of Obstetricians and Gynecologists, 2015). Magnesium sulfate administration is indicated to prevent seizures in patients with pre-eclampsia and to prevent seizure recurrence in eclampsia (American College of Obstetricians and Gynecologists, 2015). Delivery of the fetus is recommended after the patient is stabilized (American College of Obstetricians and Gynecologists, 2002).

Acute ischemic stroke

When a pregnant or postpartum woman presents with signs and symptoms consistent with stroke, first ensure adequate airway, breathing, and circulation. Perform a National Institutes of Health Stroke Scale (NIHSS) and document the score. One must exclude ICH as a cause of symptoms. Pregnancy is a relative contraindication to IV tPA, but not an absolute contraindication (Jauch et al., 2013). The time window for delivery of IV tPA was traditionally 3 hours, but it can be extended from 3 to 4.5 hours if the patient is not on warfarin (regardless of international normalized ratio), if the NIHSS is less than 25, and provided that the patient does not have a history of both diabetes and previous stroke. In order to receive IV tPA, the patient or surrogate should provide informed consent, and blood pressure must be controlled such that systolic blood pressureP is less than 185 mmHg and diastolic blood pressure is less than 110 mmHg (Jauch et al., 2013). The guideline recommendations for aspirin administration, statin therapy, and glucose control are applicable regardless of pregnancy status (Jauch et al., 2013).

Cerebral venous thrombosis

The treatment for CVST is anticoagulation. Unfractionated heparin has been associated with developmental abnormalities and bleeding in the fetus. Therefore, low-molecular-weight heparin (LMWH) is preferred over unfractionated heparin. Vitamin K antagonists such as warfarin are contraindicated during pregnancy. Full anticoagulant doses of LMWH should be continued throughout pregnancy, and LMWH or vitamin K antagonists should be continued for at least 6 weeks postpartum, for a minimum treatment time of 6 months (Saposnik et al., 2011). Fibrinolytic therapy can be considered in patients who deteriorate despite full systemic anticoagulation, or in patients who have absolute contraindications to anticoagulation. Anticoagulation therapy should not be withheld in the presence of hemorrhagic infarction that occurred as a consequence of the CVST. In the setting of rapid deterioration due to cerebral edema, decompressive hemicraniectomy can be life-saving (Saposnik et al., 2011).

COMPLEX CLINICAL DECISIONS

When an ethical debate occurs in the context of the maternal–fetal relationship, it is often centered on the idea that the fetus is considered an individual patient (van Bogaert and Dhai, 2008). Clinicians should use respect for the pregnant woman's autonomy as the guiding principle through such ethical dilemmas (van Bogaert and Dhai, 2008). If neurologic complications prevent the woman from making her own decisions, surrogate decisions makers and substituted judgment will need to be relied upon (van Bogaert and Dhai, 2008). Maternal critical illness poses significant risk to the fetus (Price et al., 2008). It is usually the case that attending to the health of the mother supports the well-being of the fetus (Honiden et al., 2013). Rarely, however, the best interests of the mother and the best interests of the fetus conflict. American professional organizations have emphasized that the professional's obligation to save the life of the mother supersedes the duty to the fetus. The principle of autonomy dictates that no action should proceed that is against the mother's expressed wishes (Honiden et al., 2013). At times, improving maternal pathophysiology is dependent on delivering the fetus. Clinicians must balance the risks of premature delivery versus the maternal benefit from delivery, but the mother's health should ultimately be the priority (Price et al., 2008).

Since stabilizing the mother is essential to good fetal care, one should consider early transfer to a critical care unit (Neligan and Laffey, 2011). Neurocritical care units offer multidisciplinary expertise for critically ill patients with neurologic illness or neurologic complications (Sheth and Sheth, 2012). Ideally, such patients should be cared for in a setting that allows access to expertise in obstetrics, neurology, neurosurgery, radiology, and critical care (Edlow et al., 2013) and in a way that applies critical care principles based on the physiologic changes unique to pregnancy (Gaffney, 2014).

Advances in treatment for acutely ill neurologic patients include the proliferation of neurocritical care units. Neurointensive care units provide multidisciplinary, multiprofessional care in a setting rich with advanced neurologic monitoring devices (Wright, 2007). Neurocritical care providers are skilled at providing brain-oriented intensive care, including aggressive ICP management with modalities such as osmotherapy and hypothermia. (Wright and Geogadin, 2006). Patients with nonsurgical neurologic emergencies that could benefit from a neurointensive care setting include those being evaluated for eclampsia, status epilepticus, central nervous system infection, increased ICP, large ischemic stroke, CVST, rapidly progressive weakness, coma, and, of course, any patient with an acute neurologic condition that would meet traditional hemodynamic or respiratory triggers for ICU care.

Diagnoses that prompt urgent or emergent evaluation by a neurosurgical service (Table 35.3) should similarly

Table 35.3

Emergency and pathologic causes of headache during pregnancy and the puerperium

Emergency causes
Severe pre-eclampsia or eclampsia
Subarachnoid hemorrhage
Acute expansion of unruptured intracranial aneurysm, intracranial tumor, or other intracranial mass lesion
Intracerebral hemorrhage
Cerebral venous sinus thrombosis
Ischemic stroke
Internal carotid or vertebral artery dissection
Posterior reversible leukoencephalopathy
Reversible cerebral vasoconstriction syndrome/postpartum angiopathy
Epidural hematoma
Subdural hematoma
Pituitary apoplexy
Obstructive hydrocephalus
Meningitis
Herpes simplex encephalitis
Brain abscess
Other pathologic causes
Arteriovenous malformation
Brain tumor
Communicating hydrocephalus
Reversible cerebral vasoconstriction syndrome
Low-pressure headache
Idiopathic intracranial hypertension
Central nervous system infection
Idiopathic intracranial hypertension
Traumatic brain injury
Viral encephalitis and meningitis
Central nervous system vasculitis

Information from Cortelli et al. (2004); Edlow et al. (2008); Jamieson (2009); Green (2012); Lester and Liu (2013).

prompt admission to a critical care unit. Subarachnoid hemorrhage patients need emergent evaluation for ruptured aneurysm. If an aneurysm is found, early intervention to secure it can present the catastrophic consequences of re-rupture (Diringer et al., 2011). ICH does not usually require surgical intervention, but it may be caused by vascular malformation. Additionally, patients with ICH or subarachnoid hemorrhage may need CSF diversion via ventriculostomy. Neurosurgeons should be made aware of patients with stroke or hemorrhage of the cerebellum, as these patients may need suboccipital craniectomy for cerebral edema. Similarly, large ischemic cortical stroke (which usually means a stroke greater than one-third to one-half of the middle cerebral artery territory) may require hemicraniectomy if swelling becomes significant. Acute obstructive hydrocephalus requires emergency CSF diversion, and further management will depend on the cause.

Maternal brain death and somatic support

When a pregnant woman suffers cardiac arrest or if maternal death is considered imminent, the welfare of a potentially viable fetus should take a more prominent role in the treatment paradigm. Brain death is defined as the irreversible loss of all brain function, including that of the brainstem. Once a formal declaration of death is made, it is futile and unethical to continue to provide support for vital organs (also called somatic support). However, maternal brain death is a potential exception since a fetus is present (Farragher and Laffey, 2005), especially if that fetus is near or has reached viability (Mallampalli and Guy, 2005).

Providing somatic support after brain death is rare (Powner and Bernstein, 2003) and it should not be done if it is against the express wishes of the mother. Providing somatic support may seem expensive, but Neligan and Laffey (2011) point out that the cost may actually be lower than the long-term cost of supporting a severely premature neonate. A brain-dead mother can make fetal support difficult due to issues such as hypothermia, panhypopituitarism, and hypotension. If cardiac arrest occurs, cesarean delivery performed within 4 minutes of cardiac arrest may optimize fetal and maternal outcomes. In a nonbrain-dead pregnant woman who suffers a cardiac arrest, cardiopulmonary resuscitation should continue uninterrupted during and following delivery (Honiden et al., 2013). However, in a woman who has already been declared brain-dead, cardiopulmonary resuscitation would not continue after delivery of the fetus. As with somatic support, delivery in the setting of cardiac arrest should not be performed if it is against the mother's stated wishes (Honiden et al., 2013).

OUTCOME PREDICTION

Data regarding outcome prediction in critically ill obstetric patients are lacking. Speculating on the prognosis of the most critically ill patients is difficult due to lack of accurate prognostic models. Clinical nihilism can lead to a self-fulfilling prophecy, meaning that a falsely skewed perception early in treatment that the patient will not do well will lead to early termination of life-sustaining therapies, which forces the poor-prognosis prediction to become true (Hemphill and White, 2009). Scoring systems designed for critically ill patients tend to be inaccurate when applied to pregnant women for a variety of reasons (Lapinsky, 2014). First, obstetric patients only account for a small percentage of ICU patients, so they tend to be underrepresented in critical care studies (Pollock et al., 2010). Second, normal physiologic changes in pregnancy are scored as "abnormal" in many of the traditional ICU predictive scoring models (Honiden et al., 2013). Additionally, some obstetric-specific disorders have unique treatments not available otherwise (Honiden et al., 2013). Ideally, one would use a scale with diagnosis-specific weighting, but such weighting is not usually available for obstetric diagnoses (Lapinsky, 2014).

Advances in critical care have resulted in reduced mortality (Adhikari et al., 2010), so more patients are surviving critical illness. However, questions remain about the effect of critical illness on brain structure and function (Hopkins and Jackson, 2012). Survivors of critical illness are prone to post-ICU cognitive impairment, though etiology and mechanisms are unclear (Hopkins and Jackson, 2012). Some of the identified mechanisms include hypoxemia, hypotension, glucose dysregulation, inflammation, and cytokine-activated immune system dysregulation (Elenkov et al., 2005; Hopkins et al., 2010). Other possible risk factors probably include an interaction of factors like genetic markers, age, and comorbid chronic medical disorders (Hopkins and Jackson, 2012). It is unclear if pregnancy confers any additional risks for long-term cognitive deficits. It is known that, in general, the overall prognosis in pregnant women with ICH is worse than pregnant women with ischemic stroke (Sharshar et al., 1995). However, specialized neurocritical care units have been shown to improve outcomes in intracranial hemorrhage (Diringer and Edwards, 2001; Damian et al., 2013), myasthenia gravis, and GBS (Damian et al., 2013). A registry of pregnant and postpartum patients with acute neurologic emergencies might improve the understanding of outcomes in this patient population (Edlow et al., 2013). In general, however, a clinician should not be nihilistic when caring for critically ill obstetric patients (Sheth and Sheth, 2012). Aggressive care can allow a patient to leave the ICU and potentially participate in a rehabilitation program (Sadowsky et al., 2011).

NEUROREHABILITATION

Critically ill patients with neurologic complications should be assessed by physical therapy, occupational therapy, and speech-language pathology as indicated, to assist in proper transition out of the hospital. The patient might need additional multiprofessional, multidisciplinary rehabilitation services after a hospital stay if her illness or injury has caused functional limitation in order to help her improve or recover lost function (Sadowsky et al., 2011).

Studies on the rehabilitation needs of critically ill obstetric women are lacking. In one study, two-thirds of obstetric patients with ischemic and hemorrhagic stroke left the hospital with neurologic deficits (Skidmore et al., 2001). Most (63%) patients with hemorrhagic stroke were discharged to a nursing home, but most (73%) with ischemic stroke were able to be discharged home (Skidmore et al., 2001). One case report in a pregnant woman with GBS indicates that pregnant women have some additional considerations for rehabilitation, including the need to perform childcare activities after childbirth (Wada et al., 2010). Many patients leaving the ICU may need cognitive rehabilitation (Hopkins and Jackson, 2012). Understanding the post-ICU neurologic deficits and cognitive morbidities will help more firmly formulate rehabilitation needs and could help better quantify the benefits of rehabilitation (Hopkins and Jackson, 2012).

REFERENCES

Adhikari NK, Fowler RA, Bhagwanjee S et al. (2010). Critical care and the global burden of illness in adults. Lancet 376 (9794): 1339–1346.

Al-Ansari MA, Hameed AA, Al-jawder SE et al. (2007). Use of noninvasive positive pressure ventilation during pregnancy: Case series. Ann Thorac Med 2 (1): 23–25.

American College of Obstetricians and Gynecologists (2002). Practice bulletin: diagnosis and management of preeclampsia and eclampsia. Clinical management guidelines for obstetrician-gynecologists. Obstet Gynecol 99: 159–167.

American College of Obstetricians and Gynecologists (2004). Committee opinion: guidelines for diagnostic imaging during pregnancy. Obstet Gynecol 104 (3): 647–651.

American College of Obstetricians and Gynecologists (2013). Hypertension in pregnancy: executive summary. Obstet Gynecol 122: 1122–1131.

American College of Obstetricians and Gynecologists (2015). Emergency therapy for acute-onset, severe hypertension during pregnancy and the postpartum period. Obstet Gynecol 125: 521–525.

Antonelli NM, Dotter DJ, Katz VL et al. (1996). Cancer in pregnancy: a review of the literature: part I. Clin Obstet Gynecol 48: 24–37.

Bartynski WS (2008). Posterior reversible encephalopathy syndrome, part 1: fundamental imaging and clinical features. Am J Neuroradiol 29 (6): 1036–1042.

Bartynski WS, Boardmand JF (2007). Distinct imaging patterns and lesion distribution in posterior reversible encephalopathy syndrome. Am J Neuroradiol 28 (7): 1320–1327.

Bateman BT, Olbrecht VA, Berman MF et al. (2012). Peripartum subarachnoid hemorrhage: nationwide data and institutional experience. Anesthesiology 116 (2): 324–333.

Baumwell S, Karumanchi SA (2007). Pre-eclampsia: clinical manifestations and molecular mechanisms. Nephron 106: c72–c81.

Bella I, Chad DA (1998). Neuromuscular disorders and acute respiratory failure. Neuro Clin 16 (2): 391–417.

Bleck TP (2012). Status epilepticus and the use of continuous EEG monitoring in the intensive care unit. Continuum 18 (3): 560–578.

Bove RM, Klein JP (2014). Neuroradiology in women of childbearing age. Continuum 20 (1): 23–41.

Brophy GM, Bell R, Claassen J et al. (2012). Guidelines for the evaluation and management of status epilepticus. Neurocrit Care 17 (1): 3–23.

Burns SA, Selzer ME (2009). Acute spinal cord injury. In: RT Johnson, JW Griffin, JC McArthur (Eds.), Current Therapy in Neurologic Disease, 7th edn. Mosby, Philadelphia, pp. 241–244.

Calabrese LH, Doddick DW, Schwedt TJ et al. (2007). Ann Intern Med 146: 34–44.

Callaghan WM, Creanga AA, Kuklina EV (2012). Severe maternal morbidity among delivery and postpartum hospitalizations in the United States. Obstet Gynecol 120: 1029–1036.

Carhuapoma JR, Tomlinson MW, Levine SR (1999). Neurologic diseases. In: DK James, PJ Steer, CP Weiner et al. (Eds.), High Risk Pregnancy: Management Options, 2nd edn. WB Saunders, London, pp. 803–806.

Centers for Disease Control and Prevention (2014). Radiation and pregnancy: a fact sheet for clinicians, Available at: http://emergency.cdc.gov/radiation/prenatalphysician.asp. Accessed June 11, 2015.

Chan LY-S, Tsui Leung TN (2004). Guillain–Barré syndrome in pregnancy. Acta Obstet Gynecol Scand 83: 319–325.

Chancellor AM, Wroe SJ, Cull RE (1990). Migraine occurring for the first time in pregnancy. Headache 30: 224–227.

Chaudhuri A, Martinez-Martin P, Kennedy PG et al. (2008). EFMS guidelines on the management of community-required bacterial meningitis: a report of an EFNS Task Force on acute bacterial meningitis in older children and adults. Eur J Neurol 15 (7): 649–659.

Chen SP, Fuh JH, Lirng JF et al. (2006). Recurrent primary thunderclap headache and benign CNS angiopathy: spectra of the same disorder? Neurology 67: 2164–2169.

Chen MM, Coakly FV, Kaimal A et al. (2008). Guidelines for computed tomography and magnetic resonance imaging use during pregnancy and lactation. Obstet Gynecol 112: 1170–1175.

Ciafaloni E, Massey JM (2004). The management of myasthenia gravis in pregnancy. Neurol Clin 24: 95–100.

Clark SL (1990). New concepts of amniotic fluid embolism. Am J Obstet Gynecol 45: 360–368.

Clark SL, Hankins GD, Dudley DA et al. (1995). Amniotic fluid embolism: analysis of the national registry. Am J Obstet Gynecol 172: 1158–1167.

Conde-Agudelo A, Romero R (2009). Amniotic fluid embolism: an evidence-based review. Am J Obstet Gynecol 201 (5): 445.e1–445.e13.

Consortium for Spinal Cord Medicine (2008). Early acute management in adults with spinal cord injury: a clinical practice guideline for health care professionals. J Spinal Cord Med 31 (4): 403–479.

Cortelli P, Cevoli S, Nonino F et al. (2004). Evidence-based diagnosis of nontraumatic headache in the Emergency Department: a consensus statement on four clinical scenarios. Headache 44 (6): 587–595.

Coutinho JM, Ferro JM, Canhao P et al. (2009). Stroke in pregnancy and the puerperium. J Neurol 245: 305–313.

Cowppli-Bony A, Bouvier G, Rue M et al. (2011). Brain tumors and hormonal factors: a review of the epidemiological literature. Cancer Causes Control 22: 697–714.

Damian MS, Ben-Shlomo Y, Howard R et al. (2013). The effect of secular trends and specialist neurocritical care on mortality for patients with intracerebral haemorrhage, myasthenia gravis and Guillain-Barré syndrome admitted to critical care. Intensive Care Med 39: 1405–1412.

De Santis M, Straface G, Cavaliere AF et al. (2007). Gadolinium periconceptional exposure: pregnancy and neonatal outcome. Acta Obstet Gynecol Scand 86: 99–101.

Del Zotto E, Giossi A, Volonghi I et al. (2011). Ischemic stroke during pregnancy and puerperium. Stroke Res Treat 2011: 606–780.

DeLaPaz RL, Wippold FJ, Cornelious RS et al. (2011). ACR appropriateness criteria on cerebrovascular disease. J Am Coll Radiol 8: 532–538.

Dias MS, Sekhar LN (1990). Intracranial hemorrhage from aneurysms and arteriovenous malformations during pregnancy and the puerperium. Neurosurgery 27 (6): 855–865.

Diringer MN, Edwards DF (2001). Admission to a neurologic/neurosurgical intensive care unit is associated with reduced mortality rate after intracerebral hemorrhage. Crit Care Med 29: 635–640.

Diringer MN, Bleck TP, Hemphill HC (2011). Critical care management of patients following aneurysmal subarachnoid hemorrhage: recommendations from the Neurocritical Care Society's Multidisciplinary Consensus Conference. Neurocrit Care 15: 211–240.

Doering PL, Boothby LA, Cheok M (2014). Review of pregnancy labeling of prescription drugs: is the current system adequate to inform risks? Am J Obstet Gynecol 187: 333–339.

Donaldson JE, Lee NS (1994). Arterial and venous stroke associated with pregnancy. Neurol Clin 12: 583–599.

Ducros A (2012). Reversible cerebral vasoconstriction syndrome. Lancet Neurol 11 (10): 906–917.

Ducros A, Brousser MG (2009). Reversible cerebral vasoconstriction syndrome. Pract Neurol 9: 256–267.

Dunham CM, Barraco RD, Clark DE et al. (2005). Guidelines for emergency tracheal intubation immediately after traumatic injury. J Trauma 55: 162–179.

Easton JD, Saver JL, Albers GW et al. (2009). Definition and evaluation of transient ischemic attack: a scientific statement for healthcare professionals. Stroke 40: 2276–2293.

Edlow JA, Panagos PD, Goodwin SA et al. (2008). Clinical policy: critical issues in the evaluation and management of adults presenting to the emergency department with acute headache. Ann Emerg Med 52 (4): 407–436.

Edlow J, Caplan L, O'Brien K et al. (2013). Diagnosis of acute neurological emergencies in pregnant and post-partum women. Lancet Neurol 12: 175–185.

Elenkov IJ, Iezzoni DG, Daly A et al. (2005). Cytokine dysregulation, inflammation and well-being. Neuroimmunomodulation 12 (5): 255–269.

Farragher RA, Laffey JG (2005). Maternal brain death and somatic support. Neurocrit Care 3: 66–106.

Feske SK, Singhal AB (2014). Cerebrovascular disorders complicating pregnancy. Continuum 20 (1): 80–99.

Frontera J, Ahmed A (2014). Neurocritical care complications of pregnancy and peurperium. J Crit Care 29: 1069–1081.

Fugate JE, Ameriso SF, Ortiz G et al. (2012). Variable presentations of postpartum angiopathy. Stroke 43 (3): 670–676.

Gaffney A (2014). Critical care in pregnancy – is it different? Semin Perinatol 38: 329–340.

Goadsby PJ, Godlberg J, Silberstein SD (2008). Migraine in pregnancy. BMJ 336: 1502–1504.

Goldman JA, Eckerling B, Gans B (1964). Intracranial venous sinus thrombosis in pregnancy and the puerperium: report of fifteen cases. J Obstet Gynaecol Br Commonw 71: 791–796.

Green MW (2012). Secondary headaches. Continuum 181 (4): 783–795.

Gronseth G, Barohn RJ (2000). Practice parameter: thymectomy for autoimmune myasthenia gravis (an evidence-based review): report of the Quality Standards Subcommittee of the American Academy of Neurology. Neurology 55: 7–15.

Hanna J, Goldman-Wohl D, Hamani Y et al. (2006). Decidual NK cells regulate key developmental processes at the human fetal-maternal interface. Nat Med 12: 1065–1074.

Hemphill JC, White DB (2009). Clinical nihilism in neuroemergencies. Emerg Med Clini N Am 27: 27–37.

Hinchey J, Chaves C, Appignani B et al. (1996). A reversible posterior leukoencephalopathy syndrome. N Engl J Med 334: 494–500.

Ho CM, Chan KH (2007). Posterior reversible leukoencephalopathy syndrome with vasospasm in a postpartum woman after postdural puncher headache following spinal anesthesia. Anesth Analg 105: 770–772.

Honiden S, Abdel-Razeq SS, Siegel MD (2013). The management of the critically ill obstetric patient. J Intensive Care Med 28 (2): 93–106.

Hopkins RO, Jackson JC (2012). Neuroimaging findings in survivors of critical illness. NeuroRehabilitation 31 (3): 311–318.

Hopkins RO, Suchyta MR, Snow GL et al. (2010). Blood glucose regulation and cognitive outcomes in ARDS survivors. Brain Inj 24 (12): 1478–1484.

Hughes RAC, Wijdicks EFM, Barohn R et al. (2003). Practice parameter: immunotherapy for Guillain-Barré syndrome: report of the Quality Standards Subcommittee of the American Academy of Neurology. Neurology 61: 736–740.

Jaigobin C, Silver FL (2000). Stroke and pregnancy. Stroke 31: 2948–2951.

Jamieson DG (2009). Diagnosis of ischemic stroke. Am J Med 122: S14–S20.

Jauch ED, Saver JL, Adams HP et al. (2013). Guidelines for the early management of patients with acute ischemic stroke: a guideline for healthcare professionals from the American Heart Association/American Stroke Association. Stroke 44 (3): 870–947.

Jeng JS, Tang SC, Yip PK (2004). Stroke in women of reproductive age: comparison between stroke related and unrelated to pregnancy. J Neurol Sci 221: 25–29.

Johnson DM, Kramer DC, Cohen E et al. (2005). Thrombolytic therapy for acute stroke in late pregnancy with intra-arterial recombinant tissue plasminogen activator. Stroke 37 (8): 2168–2169.

Karnad DR, Guntupalli KK (2004). Critical illness and pregnancy: review of a global problem. Crit Care Clin 20: 555–576.

Karnad DR, Guntupalli KK (2005). Neurologic disorders in pregnancy. Crit Care Med 33: S362–S371.

Karnad DR, Lapsia V, Krishnan A et al. (2004). Prognostic factors in obstetric patients admitted to an Indian intensive care unit. Crit Care Med 31: 1294–1299.

Karumanchi SA, Epstein FH (2007). Placental ischemia and soluble fms-like tyrosine kinase I: cause or consequence of preeclampsia? Kidney Int 71: 959–961.

Kass JS (2014). Epilepsy and pregnancy: a practical approach to mitigating legal risk. Continuum 20 (1): 181–185.

Kaymar M, Varner M (2013). Epilepsy in pregnancy. Clin Obstet Gynecol 56: 330–341.

Kilpatrick SJ, Matthay MA (1992). Obstetric patients requiring critical care: a five-year review. Chest 101: 1407–1412.

Kittner SJ, Stern BJ, Feeser BR et al. (1996). Pregnancy and the risk of stroke. N Engl J Med 335: 768–774.

Klein JP, Hsu L (2011). Neuroimaging during pregnancy. Semin Neurol 31: 361–373.

Klein AM, Loder E (2010). Postpartum headache. Int J Obstet Anesth 19: 422–430.

Kuklina EV, Tong C, Bansil P et al. (2011). Trends in pregnancy hospitalizations that include a stroke in the United States from 1994 to 2007: reasons for concern? Stroke 42 (9): 2564–2570.

Kuller JA, Katz VL, McCoy MC et al. (1995). Pregnancy complicated by Guillian-Barre syndrome. South Med J 88 (9): 987–989.

Lansberg MG, O'Donnell MJ, Khatri P et al. (2012). Antithrombotic and thrombolytic therapy for ischemic stroke: antithrombotic therapy and prevention of thrombosis, 9th ed. American College of Chest Physicians evidence-based clinical practice guidelines. Chest 141: e601S–e636S.

Lanska DJ, Kryscio RJ (1998). Stroke and intracranial venous thrombosis during pregnancy and puerperium. Neurology 51: 1622–1628.

Lapinsky SE, Kruczynski K, Seaward GR et al. (1997). Critical care management of obstetric patients. Can J Anaesth 44: 325–329.

Lapinsky SE (2014). Severity of illness in pregnancy. Crit Care Med 42 (5): 1284–1285.

Lawn ND, Wijdicks EFM (2002). Status epilepticus: a critical review of management options. Can J Neurol Sci 29: 206–215.

Lester MS, Liu BP (2013). Imaging evaluation of headache. Med Clin N Am 97: 243–265.

Lockhart EM, Baysinger CL (2007). Intracranial venous thrombosis in the parturient. Anesthesia 107: 652–658.

Mallampalli A, Guy E (2005). Cardiac arrest in pregnancy and somatic support after brain death. Crit Care Med 33: S325–S331.

Mallampalli A, Hanania NA, Guntupalli KK (2010). Acute lung injury and acute respiratory distress syndrome (ARDS) during pregnancy. In: M Belfort, GR Saade, M Foley et al. (Eds.), Critical Care Obstetrics, 5th edn. Wiley-Blackwell, New York, pp. 3338–3348.

Manno EM (2003). New management strategies in the treatment of status epilepticus. Mayo Clin Proc 78 (4): 508–518.

Martin JN, Bailey AP, Rehbergh JF et al. (2008). Thrombotic thrombocytopenic purpura in 166 pregnancies: 1995–2006. Am J Obstet Gynecol 199: 98–104.

Masuhr F, Mehraein S, Einhaupl K (2004). Cerebral venous sinus thrombosis. J Neurol 251: 11–23.

Maternal Critical Care Working Group (2011). Providing equity of critical and maternity care for the critically ill pregnant or recently pregnant woman, Available at: https://www.rcog.org.uk/globalassets/documents/guidelines/prov_eq_matandcritcare.pdf. Accessed May 8, 2015.

McCrae KR (2010). Thrombocytopenia in pregnancy. Hematology Am Soc Hematol Educ Program 2010: 397–402.

Meador K (2014). Pregnancy in women with epilepsy – risks and management. Nat Rev Neurol 10: 614–616.

Meenakshi-Sundaram S, Swaminathan K, Karthik SN et al. (2014). Replapsing Guillian-Barré syndrome in pregnancy and postpartum. Ann Indian Acad Neurol 17 (3): 352–354.

Meyers PM, Halbach VV, Malek AM et al. (2000). Endovascular treatment of cerebral artery aneurysms during pregnancy: report of three cases. Am J Neuroradiol 21 (7): 1306–1311.

Molgaard-Nielsen D (2011). Newer-generation antiepileptic drugs and the risk of major birth defects. J Am Med Assoc 305 (19): 1996–2002.

Morgenstern LB, Hemphill JC, Anderson C et al. (2010). Guidelines for the management of spontaneous intracerebral hemorrhage. Stroke 41: 2109–2129.

Munnur U, Karnad DR, Bandi VAP et al. (2005). Critically ill obstetric patients in an American and an Indian public hospital: comparison of case-mix, organ dysfunction, intensive care requirements and outcomes. Intensive Care Med 31: 1087–1094.

Murugappan A, Coplin WM, Al-Sadat AN et al. (2006). Thrombolytic therapy of acute ischemic stroke during pregnancy. Neurology 66 (5): 768–770.

Neligan PJ, Laffey JG (2011). Clinical review: special populations – critical illness and pregnancy. Crit Care 15: 227–236.

Nelson LH, McLean WT (1985). Management of Landry-Guillian-Barré syndrome in pregnancy. Obstet Gynecol 106: 25S–29S.

Ocegueda-Pacheco C, Garcia JC, Varon J et al. (2014). Therapeutic hypothermia for cardiovascular collapse and severe respiratory distress after amniotic fluid embolism. Ther Hypothermia Tem Manag 4 (2): 96–98.

Ogilvy CS, Steig PE, Awac I et al. (2001). Recommendations for the management of intracranial arteriovenous malformations: a statement for health care professionals for a special writing group of the Stroke Council, American Stroke Association. Stroke 32 (6): 1458–1471.

Parkinson D, Bachers G (1980). Arteriovenous malformations: summary of 100 consecutive supratentorial cases. J Neurosurg 53 (3): 285–299.

Parry GJ, Heiman-Patterson TD (1988). Pregnancy and autoimmune neuromuscular disease. Semin Neurol 8: 197–204.

Patten J (1996). Peripheral neuropathy and diseases of the lower motor neuron. In: Differential Diagnosis of Neurologic Disease, 2nd edn. Springer, New York, pp. 333–356.

Patwa HS, Chaudry V, Katzberg H et al. (2012). Evidence-based guideline: Intravenous immunoglobulin in the treatment of neuromuscular disorders: report of the Therapeutics and Technology Assessment Subcommittee of the American Academy of Neurology. Neurology 78: 1009-1–1009-15.

Pazhayattil GS, Rastegar A, Brewster UC (2015). Approach to the diagnosis and treatment of hyponatremia in pregnancy. Am J Kidney Dis 65 (4): 623–627.

Platteau P, Engelhardt T, Moodley J et al. (1997). Obstetric and gynaecological patients in an intensive care unit: a one year review. Trop Doctor 27: 202–206.

Plauché WC (1979). Myasthenia gravis in pregnancy: an update. Am J Obstet Gynecol 135: 691–697.

Plauché WC (1991). Myasthenia gravis in mothers and their newborns. Clin Obstet Gynaecol 34: 82–99.

Pollock W, Rose L, Dennis CL (2010). Pregnant and postpartum admissions to the intensive care unit: a systematic review. Intensive Care Med 36: 1465–1474.

Powner DJ, Bernstein IM (2003). Extended somatic support for pregnant women after brain death. Crit Care Med 31: 1241–1249.

Price LC, Slack A, Nelson-Piercy C (2008). Aims of obstetric critical care management. Best Pract Res Clin Obstet Gynaecol 22 (5): 775–799.

Qhah TCC, Chiu JW, Tan KH et al. (2001). Obstetric admissions to the intensive care unit of a tertiary care institution. Ann Acad Med Singapore 30: 250–253.

Ranabir S, Baruah MP (2011). Pituitary apoplexy. Indian J Endocrin Metab 15 (Suppl 3): S188–S196.

Raroque HG, Orrison WW, Rosenberg GA (1990). Neurologic involvement in toxemia of pregnancy: reversible MRI lesions. Neurology 40: 167–169.

Roos KL, Greenlee KE (2011). Meningitis and encephalitis. Continuum 17 (5): 101–1023.

Roppolo LP, Walters K (2004). Airway management in neurological emergencies. Neurocrit Care 1: 405–414.

Sadowsky CL, Becker D, Bosques G et al. (2011). (2011). Rehabilitation in transverse myelitis. Continuum 17 (4): 816–830.

Salonen Ros H, Lichtenstein P, Bellocco R et al. (2001). Increased risks of circulatory disease in late pregnancy and puerperium. Epidemiology 12: 456–460.

Sances G, Granella F, Nappi RE et al. (2003). Course of migraine during pregnancy and postpartum: a prospective study. Cephalalgia 23: 197–205.

Saposnik G, Barinagarrementeria F, Brown R et al. (2011). Diagnosis and management of cerebral venous thrombosis: a statement for health care professionals from the American Heart Association/American Stroke Association. Stroke 42: 1158–1192.

Sax TW, Rosenbaum RB (2006). Neuromuscular disorders in pregnancy. Muscle Nerve 34: 559–571.

Schwartz RB, Feske SK, Polak JF et al. (2000). Pre-eclampsia and eclampsia: clinical and neuroradiographic correlates and insights into the pathogenesis of hypertensive encephalopathy. Radiology 217: 371–376.

Servillo G, Bifulco F, De Robertis E et al. (2007). Posterior reversible encephalopathy syndrome in intensive care medicine. Intensive Care Med 33 (2): 230–236.

Sharshar T, Lamy C, Mas JL (1995). Incidence and causes of strokes associated with pregnancy and puerperium. Stroke 26: 930–936.

Sheth S, Sheth K (2012). Treatment of neurocritical care emergencies in pregnancy. Current Treatment Opinions in Neurology 14: 197–210.

Skeie GO, Apostolski S (2010). Guidelines for the treatment of autoimmune neuromuscular transmission disorders. Eur J Neurol 17 (7): 893–902.

Skidmore FM, Williams LS, Fradkin KD et al. (2001). Presentation, etiology, and outcome of stroke in pregnancy and puerperium. J Stroke Cerebrovasc Dis 10 (1): 1–10.

Smith HO, Kohorn E, Cole LA (2005). Choriocarcinoma and gestational trophoblastic disease. Obstet Gynecol Clin North Am 32 (4): 661–684.

Staykov D, Schwab S (2011). Posterior reversible encephalopathy syndrome. J Intensive Care Med 27: 11–24.

Stella CL, Jodicke CD (2007). Postpartum headache: is your work-up complete? Am J Obstet Gynecol 196: 318.

Stevenson CB, Thompson RB (2005). The clinical management of intracranial neoplasms in pregnancy. Clin Obstet Gynecol 48: 24–37.

Tank PD, Chauhan AR, Bhattacharya MS et al. (2004). Neurological complications in eclampsia: a case series. Int J Fertil Womens Med 49 (2): 61–69.

Tettenborn B (2012). Stroke and pregnancy. Neurol Clin 30: 913–924.

The Brain Trauma Foundation (2007). Guidelines for management of severe traumatic brain injury. 3rd ed. J Neurotrauma 24: S1–S106.

To WK, Cheung RTF (1997). Neurological disorders in pregnancy. Hong Kong Med 3: 400–408.

Towne AR, Waterhouse EJ, Boggs JG et al. (2000). Prevalence of nonconvulsive status epilepticus in comatose patients. Neurology 54 (2): 1421–1423.

Vadja FJ, O'Brien TJ, Lander CM et al. (2014). The tetragenicity of the newer antiepileptic drugs- and update. Acta Neuorl Scand 130 (4): 234–238.

Van Bogaert JL, Dhai A (2008). Ethical challenges of treating the critically ill pregnant patient. Best Pract Res Clin Obstet Gynaecol 22: 983–996.

Verheecke M, Halaska MJ, Lok CA et al. (2014). Primary brain tumours, meningiomas and brain metastases in pregnancy: report on 27 cases and review of the literature. Eur J Cancer 50: 1462–1471.

Villar J, Say J, Gulmezoglu et al. (2003). Eclampsia and pre-eclampsia: a world health problem for 2000 years. In: H Critchley, A Maclead, L Poston, Walker (Eds.), Preeclapmsia, Royal College of Obstetricians and Gynaecologists, London, pp. 210–225.

Wabnitz A, Bushnell C (2014). Migraine, cardiovascular disease and stroke during pregnancy. Cephalgia 1–8.

Wada S, Kawate N, Morotomi N et al. (2010). Experience of rehabilitation for Guillain-Barré syndrome during and after pregnancy: a case study. Disabil Rehabil 32 (34): 2056–2059.

Webb JA, Thomsen HS, Morcos SK (2005). The use of iodinated and gadolinium contrast media during pregnancy and lactation. Eur Radiol 15 (6): 1234–1240.

Weir B, MacDonald N, Mielke B (1978). Intracranial vascular complications of choriocarcinoma. Neurosurgery 2: 138–142.

Williams PM, Fletcher S (2010). Health effects of prenatal radiation exposure. Am Fam Physician 82 (5): 488–493.

Williams J, Mozurkewich E, Chilimigras J et al. (2008). Critical care in obstetrics: pregnancy-specific conditions. Best Pract Res Clin Obstet Gynaecol 22 (5): 825–846.

Witlin AG, Friedman SA, Egerman RS et al. (1997). Cerebrovascular disorders complicating pregnancy beyond eclampsia. Am J Obstet Gynecol 176: 1139–1148.

Wright WL (2007). Multimodal monitoring in the ICU: when could it be useful? Neurol Sci 261: 10–15.

Wright WL (2012). Sodium and fluid management in acute brain injury. Cur Neuro Neurosci Rep 12 (4): 466–473.

Wright WL, Geogadin RG (2006). Postresuscitative intensive care: neuroprotective strategies after cardiac arrest. Semin Neurol 26 (4): 396–402.

Wunsch H, Angus DC, Harrison DA et al. (2008). Variation in critical care services across North America and Western Europe. Crit Care Med 36: 2787–2793.

Yamada H, Noro N, Kato EH et al. (2001). Massive intravenous immunoglobulin treatment in pregnancy complicated by Guillain-Barré syndrome. Eur J Obstet Gyn Reprod Bio 97: 101–104.

Yoshida K, Takahashi JC, Takenobu Y et al. (2016). Strokes associated with pregnancy and puerperium: a nationwide study by the Japan Stroke society. Stroke. Epub ahead of print.

Zara G, Codemo V, Palmieri A et al. (2012). Neurological complications in hyperemesis gravidarum. Neurol Sci 33: 133–135.

Zeeman GG (2006). Obstetric critical care: a blueprint for improved outcomes. Crit Care Med 34 (9): S208–S214.

Zeidan A, Farhat O, Maaliki H et al. (2006). Does postdural puncture headache left untreated lead to subdural hematoma? Case report and review of the literature. Int J Obstet Anesth 15: 50–58.

Zeigler DK (1985). Is the neurologic examination becoming obsolete? Neurology 35: 599.

Chapter 36

Neurologic complications of sepsis

E. SCHMUTZHARD* AND B. PFAUSLER

Neurocritical Care Unit, Department of Neurology, Medical University Innsbruck, Innsbruck, Austria

Abstract

Over the past decades, the incidence of sepsis and resultant neurologic sequelae has increased, both in industrialized and low- or middle-income countries, by approximately 5% per year. Up to 300 patients per 100 000 population per year are reported to suffer from sepsis, severe sepsis, and septic shock. Mortality is up to 30%, depending on the precision of diagnostic criteria. The increasing incidence of sepsis is partially explained by demographic changes in society, with aging, increasing numbers of immunocompromised patients, dissemination of multiresistant pathogens, and greater availability of supportive medical care in both industrialized and middle-income countries. This results in more septic patients being admitted to intensive care units. Septic encephalopathy is a manifestation especially of severe sepsis and septic shock where the neurologist plays a crucial role in diagnosis and management. It is well known that timely treatment of sepsis improves outcome and that septic encephalopathy may precede other signs and symptoms. Particularly in the elderly and immunocompromised patient, the brain may be the first organ to show signs of failure. The neurologist diagnosing early septic encephalopathy may therefore contribute to the optimal management of septic patients. The brain is not only an organ failing in sepsis (a "sepsis victim" – as with other organs), but it also overwhelmingly influences all inflammatory processes on a variety of pathophysiologic levels, thus contributing to the initiation and propagation of septic processes. Therefore, the best possible pathophysiologic understanding of septic encephalopathy is essential for its management, and the earliest possible therapy is crucial to prevent the evolution of septic encephalopathy, brain failure, and poor prognosis.

EPIDEMIOLOGY

Sepsis is an increasing major health problem, affecting millions of people annually, on a worldwide basis. Up to 750 000 persons in the USA and more than one million in the European Union are diagnosed with sepsis each year (Engel et al., 2007; Kumar et al., 2011; Angus and van der Poll, 2013; Chaudhary et al., 2014; Cohen et al., 2015; Cross et al., 2015; Tiru et al., 2015; Yealy et al., 2015).

Up to a quarter of patients in general intensive care units (ICUs) have severe sepsis (Engel et al., 2007). Besides the aging of society, it is progress in all fields of medicine, ranging from emergency medicine to surgery, from dermatology to transplantation medicine, from neurology to internal medicine and immunology and, in particular, advances in immunomodulating therapies, which has contributed to the increase of sepsis syndrome, severe sepsis, and septic shock and, in turn, septic encephalopathy (Gaieski and Goyal, 2013; Chaudhary et al., 2014; Kaukonen et al., 2015; Yealy et al., 2015).

Within less than a decade, the hospital admission rate for severe sepsis in the USA increased from 143 per 100 000 persons to 343 per 100 000 (Kumar et al., 2011). There was a similar increase in European countries and an even greater one in middle-income countries (Engel et al., 2007). On a worldwide basis, up to 1400 patients die from sepsis each day (Gaieski and Goyal, 2013). Mortality due to sepsis is more frequent than mortality due to myocardial infarction or ischemic stroke. There is a strong relationship between the subtype of

---

*Correspondence to: Erich Schmutzhard, Department of Neurology, Neurocritical Care Unit, Medical University Innsbruck, Austria. E-mail: erich.schmutzhard@i-med.ac.at

sepsis and mortality. Around 70% of all patients with sepsis present only with sepsis syndrome, with a mortality of 5–15%. Another 20% have severe sepsis, with an expected mortality of 25–35%, while 10% are admitted to the ICU with septic shock, with a mortality ranging from 40 to 60% (Winters et al., 2010). Septic shock involves failure of the cardiocirculatory system, with vasodilatation and therapy-refractory arterial hypotension, as well as myocardial dysfunction ("septic cardiomyopathy"), which in turn leads to reduced myocardial output (Kumar et al., 2000, 2006; Janssens and Graf, 2008; Yealy et al., 2015). If septic shock involves a failing brain (failing due to multiple factors, such as increased sympathetic stress response, cerebral edema, impaired blood–brain barrier, disrupted mitochondrial function, apoptosis, impaired astrocytic function and activated microglia), all these pathophysiologic processes, aggravated by impaired microcirculation and secondary hypoxia, contribute to severe neurologic morbidity and even mortality (Bolton et al., 1993; Eidelman et al., 1996; Ebersoldt et al., 2007; Terborg, 2012; Sonneville et al., 2013; Hocker and Wijdicks, 2014). Keeping these aspects in mind, and considering the description of pathophysiologic processes involved in organ failure, it is evident that timeliness is crucial in the management of sepsis syndrome.

Sepsis is a medical and neurologic emergency. The time elapsing between the first signs and symptoms and the initiation of appropriate therapy is important for prognosis, morbidity, and mortality (Schmidbauer et al., 2013). In one large study, each hour of delay in specific antimicrobial chemotherapy was associated with a 7% increase in the probability of death (Kumar et al., 2006). Thus, in the critical care of sepsis, the first hour is termed the "golden hour" of management. Treatment during this "golden hour" should consist of rapid antibiotic therapy, fluid resuscitation, and adequate oxygen delivery, together termed early goal-directed therapy (Dellinger et al., 2013). Although the management of a patient with severe sepsis or septic shock with early and late bundled therapy has been correctly criticized (Rivers et al., 2012; ARISE Investigators et al., 2014; Marik, 2015), rapid emergency management is still essential, particularly in the early phases of sepsis management (Rivers et al., 2001; Dellinger et al., 2013; Marik, 2014).

NEUROPATHOLOGY

Over decades, a range of synonymous terms has been used for what we call today septic encephalopathy. The term sepsis-associated delirium (Ely et al., 2001a, b; Rosengarten et al., 2011) is used for milder cases of functional impairment of consciousness in septic patients, whereas in more severe cases, most likely with structural damage in the brain, the term sepsis-associated encephalopathy or septic encephalopathy is in use (Bolton et al., 1993; Eidelman et al., 1996; Ebersoldt et al., 2007; Terborg, 2012; Sonneville et al., 2013). However, it needs to be noted that both terms are not easy to differentiate and are more likely entities of an evolutionary process in sepsis, developing into severe sepsis or septic shock (Sonneville et al., 2013).

This chapter does not deal with severe infections of the brain or its surrounding tissues, although sepsis syndrome might accompany and aggravate bacterial meningitis, and bacterial meningitis might accompany and aggravate a sepsis syndrome. Rather, it deals with the influence of sepsis on the function and morphology of the brain, a highly important and frequently overlooked interplay, since the brain is also one of the first (frequently the first) organs failing in sepsis.

Septic encephalopathy is defined as a diffuse, possibly also multifocal, disturbance of cerebral function as a consequence of the systemic inflammatory response triggered by the contact of the host with potent pathogenic microbes (Wilson and Young, 2003).

The brain itself – at least in the early phase – is not directly "attacked" by the pathogenic microbes. Instead, it is the overwhelming inflammatory response that is the major contributor to the disturbance of cerebral function seen in sepsis encephalopathy. Therefore, noninfectious processes, which can elicit a similar overwhelming systemic inflammatory response, might also lead to the signs and symptoms of sepsis, including neurologic dysfunction, not to be differentiated from sepsis encephalopathy originating from systemic bacterial or fungal infection. Up to 70% of patients suffering from sepsis show at least some kind of cerebral dysfunction, thus qualifying as having septic encephalopathy (Bone et al., 2009; Terborg, 2012). Septic encephalopathy is diagnosed both clinically and electrophysiologically in patients with sepsis as severe sepsis (Levy et al., 2003).

The pathophysiologic processes hypothesized or known to contribute to cerebral dysfunction, thus forming at least part of the basis of septic encephalopathy, are summarized here.

Endotoxins

Lipopolysaccharides are known to lead to blood–brain barrier impairment and activation of intracranial cytokines and free radicals, thereby influencing thermoregulatory or vegetative centers (Töllner et al., 2000).

Inflammatory mediators

Release of proinflammatory cytokines (e.g., tumor necrosis factor-alpha, interferon-gamma, and interleukin-6) (Thibeault et al., 2001) is provoked early in disease. This hyperinflammatory response is followed by anti-inflammatory counterregulation, with increased levels of interleukin-10, leading to a state of immune paralysis (Semmler et al., 2008). Interleukin-6 induces cyclooxygenase II in glial cells, provoking (via prostaglandins) the activation of the hypothalamus–pituitary axis, thereby triggering fever and upregulating inducible nitric oxide synthase, with consequent excessive formation of nitric oxide (Rosengarten et al., 2009). This highly diffusible radical easily penetrates the blood–brain barrier and contributes to the formation of oxygen radicals and consequent oxidative stress within the brain tissue, thereby severely impeding brain function (Fialkow et al., 2007). These reactive oxygen and nitrogen species are produced by granulocytes and macrophages. The associated inflammatory irritation depletes the antioxidative capacity of the brain and endothelial cells (Huet et al., 2008), thereby contributing to disturbance of the microcirculation.

Microcirculation

Within a few hours of the systemic inflammatory response, the regulatory mechanisms of the microcirculation become severely impaired (Rosengarten et al., 2009). Proinflammatory cytokines (e.g., tumor necrosis factor-alpha, interleukin-1B) provoke the expression of adhesion molecules (e.g., intercellular adhesion molecule (ICAM) or vascular cell adhesion molecule (VCAM)) and lead, within a short time period, to adhesion of leukocytes, erythrocytes, and thrombocytes on to the endothelial lining (so-called "rolling" or "sticking"), in turn leading to cellular infiltration into blood vessel walls and provoking the extravasation of inflammatory mediators and fluid (Hoffmann et al., 1999). Such early disturbance of the microcirculation is difficult to recognize. If it has already led to multiple organ failure, the reversal is extremely difficult, and overall mortality is high.

Disturbance of the microcirculation may be responsible for both functional and structural impairment and damage to the brain, but may also contribute to disturbances of circulation, leading to hypotension or impairment of perfusion in other organs, provoking multiorgan failure (Ragaller, 2008; Pottecher et al., 2010). Each failing organ system may contribute further to brain dysfunction, aggravating septic encephalopathy via metabolic mechanisms, including hepatic encephalopathy, coagulation disturbances, and secondary bacteremia (Gram-negative bacteria easily migrate through the hypoperfused intestinal wall). In patients with septic encephalopathy, the vasoreactivity (Terborg et al., 2001; Rosengarten et al., 2008) of intracranial blood vessels (employing acetazolamide testing) has been shown to be impaired (Szatmári et al., 2010; Teboul and Duranteau, 2012; De Backer and Durand, 2014; Vaskó et al., 2014). In areas of the brain suffering from reduced perfusion, the inadequate supply of oxygen and energy is paralleled by an increase in biomarkers for hypoxia (Mihaylova et al., 2012). This inadequate supply of substrate increases the susceptibility of neuronal cells to even minor secondary or tertiary impacts, such as mildly increased intracranial pressure or reduced cerebral perfusion pressure (Rosengarten et al., 2009).

Blood–brain barrier breakdown

Early in the course of sepsis, the blood–brain barrier may be disrupted by inhibition of complement, endothelial inflammation, and nitric oxide-mediated disturbances of perfusion. Vasogenic edema is quickly paralleled by cytotoxic edema, and a cascade of excitotoxicity is provoked, thereby contributing to the supply and demand imbalance within neuronal cells. Blood–brain barrier breakdown allows catecholamines to aggravate vasoconstriction, initiate microthrombus formation, and further reduce brain perfusion (Davies, 2002).

Mitochondrial dysfunction

Cytokines, reactive oxygen and nitrogen species, and nitric oxide quickly interfere with mitochondrial function, eventually depleting the neuronal cells of adenosine triphosphate and, again, tilting the energy balance of neurons towards insufficient supply (Messaris et al., 2004; d'Avila et al., 2008; Harrois et al., 2009; Bozza et al., 2013).

Apoptosis

Caspase 8 and 9 are triggered early during the course of sepsis and jointly stimulate caspase 3, which eventually induces apoptosis of neuronal cells (Sharshar et al., 2003; Messaris et al., 2004, 2008; Semmler et al., 2005). This apoptosis of neuronal cells is seen predominantly in highly active brain areas, such as the hippocampus, basal ganglia, or cerebellum (Hotchkiss and Nicholson, 2006; Polito et al., 2011; Götz et al., 2014).

Astrocytic function and microglia activation

Microglial cells are activated and astrocytes malfunction during the acute phase of inflammation. This activation may persist, and prolonged inflammatory processes contribute to faster aging of the brain (Semmler et al., 2005; Widmann and Heneka, 2014). Sepsis surveillance studies show continued cognitive impairment over time, paralleled by hippocampal atrophy and electroencephalogram (EEG) changes (Semmler et al., 2005, 2013; Jackson et al., 2009; Iwashyna et al., 2010).

Fever

Hyperpyrexia has been shown to be a negative factor, adding damage to otherwise stressed neuronal cells and eventually impairing their function, again via an imbalance of energy supply and demand (Sharshar et al., 2005; Adam et al., 2013).

The inflammatory response activates various hormonal regulatory circles, including the hypothalamus–pituitary–adrenal axis and the renin–angiotensin–alddosterone system. Cortisol, known to suppress the inflammatory cascade and immune function, activates the gluconeogenesis in the liver and depletes glucose from muscles and perhaps also from the brain, further contributing to energy imbalance. Angiotensin II causes vasoconstriction, possibly contributing to deterioration of microcirculation and organ perfusion (Sharshar et al., 2010).

In sepsis, the autonomic nervous system is known to be clearly impaired. The highly complex interaction and interplay of the sympathetic and parasympathetic nervous system (Pavlov and Tracey, 2012) are typically disturbed, with dysregulation of both inotropic and chronotropic effects in the heart and dysmotility of the gastrointestinal tract (Borovikova et al., 2000a, b; Tracey, 2009). Activation of the sympathetic nervous system leads to vasoconstriction (Rosas-Ballina and Tracey, 2009; Gamboa et al., 2013).

CLINICAL PRESENTATION AND NEURODIAGNOSTICS

The diagnosis of septic encephalopathy requires full recording of laboratory parameters relevant to sepsis and organ dysfunction, bearing in mind that kidney, liver, coagulation, and respiratory failure may all contribute to impairment of brain function (Bone et al., 1992, 2009; Adam et al., 2013; Hocker and Wijdicks, 2014).

Clinically, the first sign and symptom of a failing brain in a septic patient is qualitative disturbance of consciousness, with reduced memory, impairment of higher cortical functions, and disorientation, potentially consistent with delirium (Ely et al., 2001a). Only later in the course of disease does quantitative impairment of consciousness, with the development of stupor and coma, occur. Delirium, assessed using standardized tests (such as the Confusion Assessment Method for the Intensive Care Unit (CAM-ICU) or Intensive Care Delirium Screening Checklist (ICDSC)) have high validity and interrater reliability, such that even early signs of delirium can be detected (Bergeron et al., 2001; Ely et al., 2001a, b). The neurologic signs and symptoms are usually diffuse rather than focal, or perhaps multifocal (Adam et al., 2013). Typically, the cranial nerves do not show any involvement. Multifocal myoclonus or "flapping tremor" is usually not seen in sepsis encephalopathy, but may be a consequence of hypoxia or hepatic encephalopathy. Overt convulsive seizures are rare, whereas nonconvulsive status epilepticus may be detected in patients who undergo continuous EEG monitoring (Chong and Hirsch, 2005; Kurtz et al., 2014; Gilmore et al., 2015; Oddo and Taccone, 2015).

The severity of septic encephalopathy can be described by means of quantitative scales (e.g., Glasgow Coma Scale), neuropsychologic bedside testing, and EEG (Gilmore et al., 2015).

Table 36.1 lists typical neurologic and neuropsychiatric signs and symptoms of sepsis encephalopathy.

The use of CAM-ICU or ICDSC testing is highly recommended, although analgesics, sedatives, and other medications may severely influence this test (Bergeron et al., 2001; Ely et al., 2001a). Neither meningismus nor focal neurologic signs and symptoms are seen in septic encephalopathy. If focality is present, further active diagnostic workup is warranted to exclude venous sinus thrombosis, meningitis, brain abscess, endocarditis with embolization, or other causes of focal signs and symptoms.

Increased levels of S100β (marker of glial damage) and neuron-specific enolase (a marker for neuronal damage) are sometimes found in septic encephalopathy (Hamed et al., 2009), but are highly nonspecific and should not be used for the quantification of the severity of septic encephalopathy (Piazza et al., 2007; Hsu et al., 2008; Zenaide and Gusmao-Flores, 2013; Honoré et al., 2014). Metabolic encephalopathies need to be excluded by appropriate laboratory testing (Casserly et al., 2015).

Cerebrospinal fluid may show increased protein, but not pleocytosis or increments in glucose or lactate. The patient may undergo lumbar puncture to exclude a direct infectious cause of the brain dysfunction, such as meningitis or encephalitis.

Convulsive seizures, nonconvulsive status epilepticus, periodic discharges (generalized, lateralized, or bilaterally independent) and triphasic waves have all been described in patients with septic encephalopathy (Young et al., 1996, 1999; Chong and Hirsch, 2005; Kamel et al., 2013; Gilmore et al., 2015; Oddo and Taccone, 2015). The presence and severity of EEG background abnormalities are associated with altered clinical, laboratory, and neurologic findings, as well as the severity of septic encephalopathy and increased mortality (Young et al., 1999; Gilmore et al., 2015).

Table 36.1

Signs and symptoms of septic encephalopathy (Iacobone et al., 2009; Lamar et al., 2011; Adam et al., 2013; Hosokawa et al., 2014)

Fluctuating impairment of concentration
Disorientation
Impairment of attention
Psychomotor slowing
Agitation
Quantitative impairment of consciousness

In very severe septic encephalopathy, suppression patterns, continuous attenuation, or even-burst suppression may be found. Lack of EEG reactivity has been reported to be associated with worse outcome. Triphasic waves, periodic epileptic discharges, frontal intermittent rhythmic delta activity, or electrographic seizures have been shown to be associated with altered mental status and severity of sepsis encephalopathy (Gilmore et al., 2015). Status epilepticus is an independent predictor of poor outcome.

Somatosensory evoked potentials show prolonged interpeak latencies in patients with sepsis compared to controls (Zauner et al., 2002; Rinaldi et al., 2008; Hosokawa et al., 2014). In particular, the late-response N70 generated by the somatosensory cortex ipsilaterally to the stimulated median nerve and the N20–N70 latency are more frequently prolonged. This delay is related to the severity of sepsis, but unaffected by sedation. Auditory evoked potentials have also been found to be impaired in septic patients (Rinaldi et al., 2008), with I–V interwave latency being prolonged in the majority of patients. Visual evoked potentials do not appear to be affected by sepsis encephalopathy (Hosokawa et al., 2014).

Brain computed tomography (CT) or magnetic resonance imaging (MRI) should be performed in all patients who show focal neurologic signs and symptoms, epileptic seizures, or meningismus to exclude other causes of the brain dysfunction, especially cerebral ischemia, intracranial hemorrhage, meningitis, cerebritis, or brain abscess, as well as cerebral venous thrombosis. MRI may reveal posterior reversible encephalopathy syndrome (PRES) in some patient, with vasogenic edema that is usually occipitally accentuated and may involve both cortical and subcortical structures. In animal experiments, vasogenic and cytotoxic brain edema have been found both in the basal brain regions, as well as in the thalamus and cortex (Piazza et al., 2009). MR spectroscopy yields a reduced N-acetylaspartate/choline ratio as an indicator of neuronal damage (Davies et al., 2006; Sharshar et al., 2007).

HOSPITAL COURSE AND MANAGEMENT

Management of sepsis requires appropriate resuscitation, specific antimicrobial, source control, and adjunctive therapy (Marik, 2014). These therapies should be initiated as quickly as possible (i.e., within the golden hours) (Gaieski et al., 2010; Levy et al., 2012; Chaudhary et al., 2014; Huet and Chin-Dusting, 2014).

The Surviving Sepsis Campaign has proposed for the first hour to administer the antimicrobial drug(s) thought to be the most appropriate; initiate volume resuscitation; measure lactate; and draw blood cultures.

Organ failure and septic encephalopathy (Siami et al., 2008) will benefit from rapid administration of specific antimicrobial therapy and surgical source control. Other potential adjunctive therapeutic strategies are given in Tables 36.2 and 36.3.

Table 36.2

Adjunctive therapeutic strategies in severe sepsis/septic shock (Rivers et al., 2012; Cohen et al., 2015; Mouncey et al., 2015; Wisdom et al., 2015)

Blood product	Antithrombin
	Reduction of duration of disseminated intravascular coagulation, reduction of mortality
Nutrition	Immunonutrition is not supported by literature; may lead to increased mortality
Glucose control	Avoid glucose variability
Intravenous immunoglobulins	Do not improve outcome
Selenium	Contradictory results
Statins	No effect

Table 36.3

Further adjunctive therapeutic measures according to the guidelines of Surviving Sepsis Campagn (Gaieski et al., 2010, 2014; Levy et al., 2012; Dellinger et al., 2013; Schmidbauer et al., 2013; ARISE Investigators et al., 2014; Wisdom et al., 2015; Yealy et al., 2015)

Measure	Recommendation
Mechanical ventilation	Tidal volume 6 mL/kg kg, PEEP (8 mmHg or higher), elevated head of bed (30–45°)
Analgesia and sedation	Depending on the underlying disease, as little as possible during weaning and awakening
Relaxation	If possible, no muscle relaxants If (for pulmonary reason) unavoidable, maximum 48 hours (duration)
Acute renal failure	Intermittent dialysis or continuous hemofiltration
Bicarbonate	If pH >7.15: no bicarbonate
Deep venous thrombosis prophylaxis	Low-molecular-weight heparin, mechanical prophylaxis
Stress ulcer prophylaxis	H_2-blocking agents and proton pump inhibitors in case of increased risk of hemorrhage If no elevated risk, no prophylaxis necessary

PEEP, positive end-expiratory pressure.

Whether the application of cholinesterase inhibiting agents or recombinant activated protein C are helpful for septic encephalopathy patients has never been evaluated in prospective studies (Spapen et al., 2010). Long-term sequelae of septic encephalopathy need to be identified when present, and symptomatic therapy offered (Annane and Sharshar, 2015).

Severe bacterial meningitis, especially pneumococcal and meningococcal, may present with sepsis syndrome, purpura fulminans, or Waterhouse–Friderichsen syndrome. Manifestations may be particularly overwhelming in splenectomized patients. These conditions are discussed further in Chapter 13. Sepsis may also be complicated by ICU-acquired weakness, which is addressed in Chapter 29.

REFERENCES

Adam N, Kandelman S, Mantz J et al. (2013). Sepsis-induced brain dysfunction. Expert Rev Anti Infect Ther 11: 211–221.

Angus DC, van der Poll T (2013). Severe sepsis and septic shock. N Engl J Med 369: 840–851.

Annane D, Sharshar T (2015). Cognitive decline after sepsis. Lancet Respir Med 3: 61–69.

ARISE Investigators, ANZICS Clinical Trials Group, Peake SL et al. (2014). Goal-directed resuscitation for patients with early septic shock. N Engl J Med 371: 1496–1506.

Bergeron N, Dubois MJ, Dumont M et al. (2001). Intensive Care Delirium Screening Checklist: evaluation of a new screening tool. Intensive Care Med 27: 859–864.

Bolton CF, Young GB, Zochodne DW (1993). The neurological complications of sepsis. Ann Neurol 33: 94–100.

Bone RC, Balk RA, Cerra FB et al. (1992). Definitions for sepsis and organ failure and guidelines for the use of innovative therapies in sepsis. The ACCP/SCCM Consensus Conference Committee. American College of Chest Physicians/Society of Critical Care Medicine, Chest 101: 1644–1655.

Bone RC, Balk RA, Cerra FB et al. (2009). Definitions for sepsis and organ failure and guidelines for the use of innovative therapies in sepsis. The ACCP/SCCM Consensus Conference Committee. American College of Chest Physicians/Society of Critical Care Medicine 1992, Chest 136: e28.

Borovikova LV, Ivanova S, Nardi D et al. (2000a). Role of vagus nerve signaling in CNI-1493-mediated suppression of acute inflammation. Auton Neurosci 85: 141–147.

Borovikova LV, Ivanova S, Zhang M et al. (2000b). Vagus nerve stimulation attenuates the systemic inflammatory response to endotoxin. Nature 405: 458–462.

Bozza FA, D'Avila JC, Ritter C et al. (2013). Bioenergetics, mitochondrial dysfunction, and oxidative stress in the pathophysiology of septic encephalopathy. Shock 1: 10–16.

Casserly B, Phillips GS, Schorr C et al. (2015). Lactate measurements in sepsis-induced tissue hypoperfusion: results from the Surviving Sepsis Campaign database. Crit Care Med 43: 567–573.

Chaudhary T, Hohenstein C, Bayer O (2014). The golden hour of sepsis: initial therapy should start in the prehospital setting. Med Klin Intensivmed Notfmed 109: 104–108.

Chong DJ, Hirsch LJ (2005). Which EEG patterns warrant treatment in the critically ill? Reviewing the evidence for treatment of periodic epileptiform discharges and related patterns. J Clin Neurophysiol 22: 79–91.

Cohen J, Vincent JL, Adhikari NK et al. (2015). Sepsis: a roadmap for future research. Lancet Infect Dis 15: 581–614.

Cross G, Bilgrami I, Eastwood G et al. (2015). The epidemiology of sepsis during rapid response team reviews in a teaching hospital. Anaesth Intensive Care 43: 193–198.

Davies DC (2002). Blood–brain barrier breakdown in septic encephalopathy and brain tumours. J Anat 200: 639–646.

Davies NW, Sharief MK, Howard RS (2006). Infection-associated encephalopathies: their investigation, diagnosis, and treatment. J Neurol 253: 833–845.

d'Avila JC, Santiago AP, Amâncio RT et al. (2008). Sepsis induces brain mitochondrial dysfunction. Crit Care Med 36: 1925–1932.

De Backer D, Durand A (2014). Monitoring the microcirculation in critically ill patients. Best Pract Res Clin Anaesthesiol 28: 441–451.

Dellinger RP, Levy MM, Rhodes A et al. (2013). Surviving Sepsis campaign: international guidelines for management of severe sepsis and septic shock, 2012. Intensive Care Med 39: 165–228.

Ebersoldt M, Sharshar T, Annane D (2007). Sepsis-associated delirium. Intensive Care Med 33: 941–950.

Eidelman LA, Putterman D, Putterman C et al. (1996). The spectrum of septic encephalopathy. Definitions, etiologies, and mortalities. JAMA 275: 470–473.

Ely EW, Gautam S, Margolin R et al. (2001a). The impact of delirium in the intensive care unit on hospital length of stay. Intensive Care Med 27: 1892–1900.

Ely EW, Inouye SK, Bernard GR et al. (2001b). Delirium in mechanically ventilated patients: validity and reliability of the confusion assessment method for the intensive care unit (CAM-ICU). JAMA 286: 2703–2710.

Engel C, Brunkhorst FM, Bone HG et al. (2007). Epidemiology of sepsis in Germany: results from a national prospective multicenter study. Intensive Care Med 33: 606–618.

Fialkow L, Wang Y, Downey GP (2007). Reactive oxygen and nitrogen species as signaling molecules regulating neutrophil function. Free Radic Biol Med 42: 153–164.

Gaieski DF, Goyal M (2013). What is sepsis? What is severe sepsis? What is septic shock? Searching for objective definitions among the winds of doctrines and wild theories. Expert Rev Anti Infect Ther 11: 867–871.

Gaieski DF, Mikkelsen ME, Band RA et al. (2010). Impact of time to antibiotics on survival in patients with severe sepsis or septic shock in whom early goal-directed therapy was initiated in the emergency department. Crit Care Med 38: 1045–1053.

Gaieski DF, Edwards JM, Kallan MJ et al. (2014). The relationship between hospital volume and mortality in severe sepsis. Am J Respir Crit Care Med 190: 665–674.

Gamboa A, Okamoto LE, Raj SR et al. (2013). Nitric oxide and regulation of heart rate in patients with postural tachycardia syndrome and healthy subjects. Hypertension 61: 376–381.

Gilmore EJ, Gaspard N, Choi HA et al. (2015). Acute brain failure in severe sepsis: a prospective study in the medical intensive care unit utilizing continuous EEG monitoring. Intensive Care Med 41: 686–694.

Götz T, Günther A, Witte OW et al. (2014). Long-term sequelae of severe sepsis: cognitive impairment and structural brain alterations – an MRI study (LossCog MRI). BMC Neurol 14: 145.

Hamed SA, Hamed EA, Abdella MM (2009). Septic encephalopathy: relationship to serum and cerebrospinal fluid levels of adhesion molecules, lipid peroxides and S-100B protein. Neuropediatrics 40: 66–72.

Harrois A, Huet O, Duranteau J (2009). Alterations of mitochondrial function in sepsis and critical illness. Curr Opin Anaesthesiol 22: 143–149.

Hoffmann JN, Vollmar B, Inthorn D et al. (1999). A chronic model for intravital microscopic study of microcirculatory disorders and leukocyte/endothelial cell interaction during normotensive endotoxemia. Shock 12: 355–364.

Hocker SE, Wijdicks EF (2014). Neurologic complications of sepsis. Continuum (Minneap Minn) Neurology of Systemic Disease 20: 598–613.

Honoré PM, Jacobs R, De Waele E et al. (2014). Biomarkers to detect sepsis: a "burning issue" but still a long way to go. Crit Care Med 42: 2137–2138.

Hosokawa K, Gaspard N, Su F et al. (2014). Clinical neurophysiological assessment of sepsis-associated brain dysfunction: a systematic review. Crit Care 18: 674.

Hotchkiss RS, Nicholson DW (2006). Apoptosis and caspases regulate death and inflammation in sepsis. Nat Rev Immunol 6: 813–822.

Hsu AA, Fenton K, Weinstein S et al. (2008). Neurological injury markers in children with septic shock. Pediatr Crit Care Med 9: 245–251.

Huet O, Chin-Dusting JP (2014). Septic shock: desperately seeking treatment. Clin Sci (Lond) 126: 31–39.

Huet O, Cherreau C, Nicco C et al. (2008). Pivotal role of glutathione depletion in plasma-induced endothelial oxidative stress during sepsis. Crit Care Med 36: 2328–2334.

Iacobone E, Bailly-Salin J, Polito A et al. (2009). Sepsis-associated encephalopathy and its differential diagnosis. Crit Care Med 37: 331–336.

Iwashyna TJ, Ely EW, Smith DM et al. (2010). Long-term cognitive impairment and functional disability among survivors of severe sepsis. JAMA 304: 1787–1794.

Jackson JC, Hopkins RO, Miller RR et al. (2009). Acute respiratory distress syndrome, sepsis, and cognitive decline: a review and case study. South Med J 102: 1150–1157.

Janssens U, Graf J (2008). Cardiovascular dysfunction in sepsis. Anasthesiol Intensivmed Notfallmed Schmerzther 43: 56–63.

Kamel H, Betjemann JP, Navi BB et al. (2013). Diagnostic yield of electroencephalography in the medical and surgical intensive care unit. Neurocrit Care 19: 336–341.

Kaukonen KM, Bailey M, Pilcher D et al. (2015). Systemic inflammatory response syndrome criteria in defining severe sepsis. N Engl J Med 372: 1629–1638.

Kumar A, Haery C, Parrillo JE (2000). Myocardial dysfunction in septic shock. Crit Care Clin 16: 251–287.

Kumar A, Haery C, Paladugu B et al. (2006). The duration of hypotension before the initiation of antibiotic treatment is a critical determinant of survival in a murine model of *Escherichia coli* septic shock: association with serum lactate and inflammatory cytokine levels. J Infect Dis 193: 251–258.

Kumar G, Kumar N, Taneja A et al. (2011). Nationwide trends of severe sepsis in the 21st century (2000–2007). Chest 140: 1223–1231.

Kurtz P, Gaspard N, Wahl AS et al. (2014). Continuous electroencephalography in a surgical intensive care unit. Intensive Care Med 40: 228–234.

Lamar CD, Hurley RA, Taber KH (2011). Sepsis-associated encephalopathy: review of the neuropsychiatric manifestations and cognitive outcome. J Neuropsychiatry Clin Neurosci 23: 237–241.

Levy MM, Fink MP, Marshall JC et al. (2003). 2001 SCCM/ESICM/ACCP/ATS/SIS International Sepsis Definitions Conference. Crit Care Med 31: 1250–1256.

Levy MM, Artigas A, Phillips GS et al. (2012). Outcomes of the Surviving Sepsis Campaign in intensive care units in the USA and Europe: a prospective cohort study. Lancet Infect Dis 12: 919–924.

Marik PE (2014). Don't miss the diagnosis of sepsis. Crit Care 18: 529.

Marik PE (2015). The demise of early goal-directed therapy for severe sepsis and septic shock. Acta Anaesthesiol Scand 59: 561–567.

Messaris E, Memos N, Chatzigianni E et al. (2004). Time-dependent mitochondrial-mediated programmed neuronal cell death prolongs survival in sepsis. Crit Care Med 32: 1764–1770.

Messaris E, Memos N, Chatzigianni E et al. (2008). Apoptotic death of renal tubular cells in experimental sepsis. Surg Infect 9: 377–388.

Mihaylova S, Killian A, Mayer K et al. (2012). Effects of anti-inflammatory vagus nerve stimulation on the cerebral microcirculation in endotoxinemic rats. J Neuroinflammation 9: 183.

Mouncey PR, Osborn TM, Power GS et al. (2015). Trial of early, goal-directed resuscitation for septic shock. N Engl J Med 372: 1301–1311.

Oddo M, Taccone FS (2015). How to monitor the brain in septic patients? Minerva Anestesiol 81: 776–788.

Pavlov VA, Tracey KJ (2012). The vagus nerve and the inflammatory reflex – linking immunity and metabolism. Nat Rev Endocrinol 8: 743–754.

Piazza O, Russo E, Cotena S et al. (2007). Elevated S100B levels do not correlate with the severity of encephalopathy during sepsis. Br J Anaesth 99: 518–521.

Piazza O, Cotena S, De Robertis E et al. (2009). Sepsis associated encephalopathy studied by MRI and cerebral spinal fluid S100B measurement. Neurochem Res 34: 1289–1292.

Polito A, Brouland JP, Porcher R et al. (2011). Hyperglycaemia and apoptosis of microglial cells in human septic shock. Crit Care 15: R131.

Pottecher J1, Deruddre S, Teboul JL (2010). Both passive leg raising and intravascular volume expansion improve sublingual microcirculatory perfusion in severe sepsis and septic shock patients. Intensive Care Med 36: 1867–1874.

Ragaller M (2008). Microcirculation in sepsis and septic schock – therapeutic options? Anaesthesiol Intensivmed Notfallmed Schmerzther 43: 48–53.

Rinaldi S, Consales G, De Gaudio AR (2008). Changes in auditory evoked potentials induced by postsurgical sepsis. Minerva Anestesiol 74: 245–250.

Rivers E, Nguyen B, Havstad S et al. (2001). Early goal-directed therapy in the treatment of severe sepsis and septic shock. N Engl J Med 345: 1368–1377.

Rivers EP, Katranji M, Jaehne KA et al. (2012). Early interventions in severe sepsis and septic shock: a review of the evidence one decade later. Minerva Anestesiol 78: 712–724.

Rosas-Ballina M, Tracey KJ (2009). The neurology of the immune system: neural reflexes regulate immunity. Neuron 64: 28–32.

Rosengarten B, Hecht M, Wolff S et al. (2008). Autoregulative function in the brain in an endotoxic rat shock model. Inflamm Res 57: 542–546.

Rosengarten B, Wolff S, Klatt S et al. (2009). Effects of inducible nitric oxide synthase inhibition or norepinephrine on the neurovascular coupling in an endotoxic rat shock model. Crit Care 13: R139.

Rosengarten B, Mayer K, Weigand MA (2011). Clinical neurological diagnosis of sepsis-associated delirium. Nervenarzt 82: 1578–1583.

Schmidbauer W, Stuhr M, Veit C et al. (2013). Sepsis in emergency medicine-pre-hospital and early in-hospital emergency treatment. Anasthesiol Intensivmed Notfallmed Schmerzther 48: 524–530.

Semmler A, Okulla T, Sastre M et al. (2005). Systemic inflammation induces apoptosis with variable vulnerability of different brain regions. Chem Neuroanat 30: 144–157.

Semmler A, Hermann S, Mormann F et al. (2008). Sepsis causes neuroinflammation and concomitant decrease of cerebral metabolism. J Neuroinflammation 5: 38.

Semmler A, Widmann CN, Okulla T et al. (2013). Persistent cognitive impairment, hippocampal atrophy and EEG changes in sepsis survivors. J Neurol Neurosurg Psychiatry 84: 62–69.

Sharshar T, Gray F, Lorin de la Grandmaison G et al. (2003). Apoptosis of neurons in cardiovascular autonomic centres triggered by inducible nitric oxide synthase after death from septic shock. Lancet 29 (362): 1799–1805.

Sharshar T, Hopkinson NS, Orlikowski D et al. (2005). Science review: the brain in sepsis – culprit and victim. Crit Care 9: 37–44.

Sharshar T, Carlier R, Bernard F et al. (2007). Brain lesions in septic shock: a magnetic resonance imaging study. Intensive Care Med 33: 798–806.

Sharshar T, Polito A, Checinski A et al. (2010). Septic-associated encephalopathy–everything starts at a microlevel. Crit Care 14: 199.

Siami S, Annane D, Sharshar T (2008). The encephalopathy in sepsis. Crit Care Clin 24: 67–82.

Spapen H, Nguyen DN, Troubleyn J et al. (2010). Drotrecogin alfa (activated) may attenuate severe sepsis-associated encephalopathy in clinical septic shock. Crit Care 14: R54.

Sonneville R, Verdonk F, Rauturier C et al. (2013). Understanding brain dysfunction in sepsis. Ann Intensive Care 3: 15.

Szatmári S, Végh T, Csomós A et al. (2010). Impaired cerebrovascular reactivity in sepsis-associated encephalopathy studied by acetazolamide test. Crit Care 14: R50.

Teboul JL, Duranteau J (2012). Alteration of microcirculation in sepsis: a reality but how to go further? Crit Care Med 40: 1653–1654.

Terborg C (2012). Septic encephalopathy. Med Klin intensivmed Notfmed 107: 629–633.

Terborg C, Schummer W, Albrecht M et al. (2001). Dysfunction of vasomotor reactivity in severe sepsis and septic shock. Intensive Care Med 27: 1231–1234.

Thibeault I, Laflamme N, Rivest S (2001). Regulation of the gene encoding the monocyte chemoattractant protein 1 (MCP-1) in the mouse and rat brain in response to circulating LPS and proinflammatory cytokines. J Comp Neurol 434: 461–477.

Tiru B, DiNino EK, Orenstein A et al. (2015). The economic and humanistic burden of severe sepsis. Pharmacoeconomics 33: 925–937.

Töllner B, Roth J, Störr B et al. (2000). The role of tumor necrosis factor (TNF) in the febrile and metabolic responses of rats to intraperitoneal injection of a high dose of lipopolysaccharide. Pflugers Arch 440: 925–932.

Tracey KJ (2009). Reflex control of immunity. Nat Rev Immunol 9: 418–428.

Vaskó A, Siró P, László I et al. (2014). Assessment of cerebral tissue oxygen saturation in septic patients during acetazolamide provocation – a near infrared spectroscopy study. Acta Physiol Hung 101: 32–39.

Widmann CN, Heneka MT (2014). Long-term cerebral consequences of sepsis. Lancet Neurol 13: 630–636.

Wilson JX, Young GB (2003). Progress in clinical neurosciences: sepsis-associated encephalopathy: evolving concepts. Can J Neurol Sci 30: 98–105.

Winters BD, Eberlein M, Leung J et al. (2010). Long-term mortality and quality of life in sepsis: a systematic review. Crit Care Med 38: 1276–1283.

Wisdom A, Eaton V, Gordon D et al. (2015). INITIAT-E.D.: Impact of timing of initiation of antibiotic therapy on mortality of patients presenting to an emergency department with sepsis. Emerg Med Australas 27: 196–201.

Yealy DM, Huang DT, Delaney A et al. (2015). Recognizing and managing sepsis: what needs to be done? BMC Med 13: 98.

Young GB, Jordan KG, Doig GS (1996). An assessment of nonconvulsive seizures in the intensive care unit using continuous EEG monitoring: an investigation of variables associated with mortality. Neurology 47: 83–89.

Young GB, Kreeft JH, McLachlan RS et al. (1999). EEG and clinical associations with mortality in comatose patients in a general intensive care unit. J Clin Neurophysiol 16: 354–360.

Zauner C, Gendo A, Kramer L et al. (2002). Impaired subcortical and cortical sensory evoked potential pathways in septic patients. Crit Care Med 30: 1136–1139.

Zenaide PV, Gusmao-Flores D (2013). Biomarkers in septic encephalopathy: a systematic review of clinical studies. Rev Bras Ter Intensiva 25: 56–62.

Chapter 37

Neurologic complications of acute environmental injuries

I.R.F. DA SILVA[1] AND J.A. FRONTERA[2]*
[1]Neurocritical Care Unit, Americas Medical City, Rio de Janeiro, Brazil
[2]Neurological Institute, Cleveland Clinic, Cleveland, OH, USA

Abstract

Environmental injuries can result in serious neurologic morbidity. This chapter reviews neurologic complications of thermal burns, smoke inhalation, lightning strikes, electric injury, near drowning, decompression illness, as well as heat stroke and accidental hypothermia. Knowing the pathophysiology and clinical presentation of such injuries is essential to proper management of primary and secondary medical complications. This chapter highlights the most frequently encountered neurologic injuries secondary to common environmental hazards, divided into the topics: injuries related to fire, electricity, water, and the extremes of temperature.

INTRODUCTION

Environmental injuries are common and often preventable events that can be associated with multisystem organ failure. Management of these injuries often requires a multidisciplinary approach at centers which specialize in treating these complex patients - mostly burn centers with over 100 in the US. The prevalence of environmental injuries varies depending on geographic location, as well as the age of the victim. Knowing the pathophysiology and clinical presentation of such injuries is essential to proper management of primary and secondary medical complications. This chapter highlights the most frequently encountered neurologic injuries secondary to common environmental hazards, divided into the topics: injuries related to fire, electricity, water, and the extremes of temperature. We will discuss each injury separately.

INJURIES RELATED TO FIRE (THERMAL BURNS AND RESPIRATORY INTOXICATION)

Epidemiology

Burns are a major cause of accidental deaths and injuries worldwide, accounting for more than 2600 deaths and 13000 injuries in 2010 in the USA (Istre et al., 2014). Firefighters, military personnel, and industrial workers are particularly at risk, and burn injuries are associated with high morbidity and disability (Chapman et al., 2008; Mian et al., 2011; Mason et al., 2012; Matt et al., 2012). Moreover, a nation's income level is negatively correlated with burn mortality (Peck and Pressman, 2013).

Neuropathology

Survivors of large fires are exposed to different mechanisms of injury, including direct thermal injuries, carbon monoxide (CO) and cyanide poisoning. Thermal burns lead not only to extensive deep compromise of the skin layer, but also to severe systemic inflammation, a hypermetabolic, catabolic state, and multiple associated metabolic derangements (Farina et al., 2013). Patients with severe burns can also present with altered pharmacokinetics, due to protein loss, hypermetabolism, hepatic failure, shock, and renal failure (Blanchet et al., 2008; Steele et al., 2015).

An entity previously known as "burn encephalopathy" is now best described as burn-induced delirium. It is believed to be secondary to the multitude of metabolic derangements burn victims usually experience, such as infections, hypotension, renal failure, electrolyte

---

*Correspondence to: Jennifer Ann Frontera, MD, Neurological Institute, Cleveland Clinic, 9500 Euclid Ave. S80, Cleveland OH 44195, USA. Tel: +1-216-444-5536, E-mail: jenfrontera@hotmail.com

disturbances, altered pharmacokinetics, electrolyte disturbances, hypoxia, pain, and systemic inflammation, among many others. Psychiatry illnesses secondary to the traumatic experience can also be contributors. More recently, burn-induced cerebral inflammation has emerged as a putative mechanism for the less frequently encountered diffuse cerebral edema and neuronal dysfunction seen in patients with extensively burned body surface (Flierl et al., 2009).

Peripheral neuropathies are commonly seen in patients with extensive and deep burns. These neuropathies derive from neuronal destruction in deep burns as well as vasa nervorum ischemia, and/or compression due to tight dressing and incorrect splinting. Medication toxicity, myopathy/neuropathy of the critically ill patient, as well as vitamin and mineral deficiencies may cause secondary neuropathies in burn patients (Lee et al., 2009b; Tamam et al., 2013).

Survivors of fires are usually exposed to dangerous fumes leading to respiratory intoxication, including CO and cyanide. CO is an odorless, colorless gas that is a product of incomplete combustion of fuels from multiple sources. Other sources of CO include improperly functioning heaters or generators and motor vehicles operating in inadequately ventilated areas. Methylene chloride is a component of paint remover that is metabolized to CO by the liver and can cause toxicity. CO exerts its effects by interfering with systemic oxygenation by rapidly diffusing through the alveolar-capillary membrane and binding to hemoglobin more than 200 times faster than oxygen. The byproduct of this binding is called carboxyhemoglobin, a compound with much less capacity of carrying oxygen, which causes a left shift in the oxyhemoglobin dissociation curve. CO poisoning not only exerts its effects in highly metabolic organs through severe hypoxia, but also causes mitochondrial dysfunction, loss of cerebral autoregulation, production of reactive oxygen species, peroxidation of lipids, and deposition of peroxynitrite in blood vessel endothelium, among other inflammatory and immune responses (Betterman and Patel, 2014).

There are acute as well as delayed neurologic effects following CO intoxication. Acute symptoms are usually the consequence of hypoxia in highly susceptible areas, such as cerebellar Purkinje cells, globus pallidus, watershed areas, and the second and third cortical layers. Many of the delayed neurologic complications may be mediated by lipid peroxidation, radical oxygen species generated by xanthine oxidase, as well as inflammatory and immune responses, leading to subcortical and periventricular demyelination and axonal destruction. Some evidence suggests that a mechanism similar to ischemia-reperfusion injury may occur, such that exposure to hyperoxia may worsen the initial CO injury (Weaver, 1999; Tomaszewski, 1999).

Cyanide inhalation can occur with the combustion of material containing organic nitrogen, such as wood and plastics. Tobacco smoking also leads to elevated blood cyanide levels. Exposure can also occur with parenteral, enteric, and dermal exposure. Intravenous sodium nitroprusside use in doses of 5–10 μg/kg/min for as few as 3 hours can result in death. Limiting light exposure to the drug (wrapping tubing in foil) can prevent conversion of nitroprusside into cyanide moieties. Excessive consumption of bitter almonds, and the pits or seeds of apricots, cherries, plum, peach, pear, and apples, can cause cyanide toxicity. Cyanide and CO poisoning usually occur concomitantly. Cyanide disrupts normal mitochondrial function by inhibiting reoxidation of cytochrome oxidase and blocking oxidative phosphorylation (Lawson-Smith et al., 2011). It leads to cellular hypoxia and blockage of aerobic metabolism. Blood oxygen concentration is typically normal and cyanide poisoning is sometimes referred to as "asphyxia without peripheral cyanosis."

Clinical presentation

Burn encephalopathy can present early or late, sometimes weeks from the initial insult. It can occur in up to 80% of hospitalized ventilated burn patients (Agarwal et al., 2010), and there is a trend for higher prevalence in patients with extensive burns (Agarwal et al., 2010; Palmu et al., 2011). Multiple manifestations have been reported, including catatonia, seizures, confusion, myoclonus, rigidity, hallucinations, disorientation, aphasia, agitation, mania, depression, anxiety, or apathy (Quinn, 2014). In some cases, delirium can result from sensory deprivation due to blindness from occipital infarcts or anterior ischemic optic neuropathy (Zeng et al., 2010; Medina et al., 2015). Most patients will completely recover from burn encephalopathy, but some will experience some degree of long-term cognitive disturbances (Andreasen et al., 1974; Quinn, 2014).

Burn-related peripheral neuropathies have been reported in 2–84% of patients (Helm et al., 1985; Marquez et al., 1993; Khedr et al., 1997). Recent studies have shown that polyneuropathies and axonal neuropathy are more frequent than mononeuropathy and demyelination (Lee et al., 2009b; Tamam et al., 2013). Flame and electric injuries, as well third-degree injuries, were the most common type of burns in patients with peripheral neuropathy in these two studies.

CO poisoning can cause acute and delayed signs and symptoms. Acute symptoms depend on the concentration of carboxyhemoglobin in the blood (Table 37.1). Patients with underlying chronic heart and lung disease may present earlier at lower CO concentrations and with more severe symptoms (Betterman and Patel, 2014). Headache is one of the most common presenting features

Table 37.1

Acute symptoms of carbon monoxide poisoning

Carbon monoxide concentration	Symptoms	Time from exposure to symptoms	Carboxyhemoglobin level in the blood
35 ppm	Headache and dizziness	6–8 hours	<10%
100 ppm	Slight headache	2–3 hours	>10%
200 ppm	Slight headache and loss of judgment	2–3 hours	20%
400 ppm	Headache	1–2 hours	25%
800 ppm	Dizziness, nausea, and seizures	45 minutes	30%
1600 ppm	Headache, tachycardia, dizziness, and nausea	20 minutes (death <2 hours)	40%
3200 ppm	Headache, dizziness, and nausea	5–10 minutes (death in 30 minutes)	50%
6400 ppm	Headache and dizziness	1–2 minutes (death < 20 minutes)	60%
12 800 ppm	Death	<3 minutes	>70%

Modified with permission from Betterman and Patel (2014). ppm, parts per million.

of CO poisoning, occurring in over 80% of victims and mimicking tension headache or a migraine attack (Handa and Tai, 2005). Other neurologic symptoms can occur, such as generalized weakness, nausea, confusion, ataxia, and in severe cases seizures, coma, cerebral edema, and death can happen. The lips and skin may classically be "cherry red," though this is not a sensitive or specific sign. Chronic exposure to CO can mimic a flu-like state.

Survivors of CO intoxication can present with a delayed neuropsychiatric syndrome, usually the consequence of delayed inflammation and demyelination of the brain. It has been reported to occur 2–40 days after the initial exposure (Tibbles and Perrotta, 1994), in roughly 3% of exposed patients (Mimura et al., 1999). Severe dementia, psychosis, anxiety, personality changes, urinary and fecal incontinence, parkinsonism, and other neuropsychologic symptoms may occur after an initially full recovery following the primary exposure (Betterman and Patel, 2014). Roughly 60–75% of patients will recover within a year, but in 15% dementia and parkinsonism become long-term sequelae (Bhatia et al., 2007).

Cyanide poisoning can initially present with a brief period of hyperpnea, due to stimulation of chemoreceptors in the carotid bodies. Approximately 60% of patients are able to detect the odor of bitter almonds following cyanide inhalation. Because of decreased oxygen utilization by tissue, venous oxyhemoglobin levels remain high, giving venous blood and skin a bright cherry-red color. Milder cases can present with headache, nausea, vertigo, anxiety, altered mental status, tachypnea, hypertension, loss of consciousness, seizures, cardiovascular collapse, pulmonary edema, and death in severe exposures (Lawson-Smith et al., 2011). Mild exposures can lead to long-term neurologic sequelae, such as parkinsonism, dystonia, ataxia, optic neuropathies, and chorea, as well as permanent vegetative state (Uitti et al., 1985; Rosenberg et al., 1989; Messing, 1991). Tobacco amblyopia, which presents as progressive vision loss, occurs in cigarette smokers who may have an inherent inability to detoxify cyanide. Tropical ataxic neuropathy, which occurs with excessive cassava ingestion (which contains cyanogen), is a demyelinating disorder associated with ataxia, optic atrophy, paresthesias, and hearing loss. Both of these entities can be treated with vitamin B_{12} or hydroxocobalamin.

Neurodiagnostics and imaging

Burn encephalopathy is usually a state secondary to metabolic disturbances. Severe cases can lead to cerebral edema, and there are reports of posterior reversible encephalopathy in patients with extensive burns (Tan et al., 2014). Electroencephalogram (EEG) can disclose excessive slowing of brain waves (Andreasen et al., 1977).

The most common finding in brain magnetic resonance imaging (MRI) of patients with CO intoxication is T2 hyperintensity of the globus pallidus and subcortical white matter, and less frequently the involvement of the thalamus, lentiform nucleus, and caudate nucleus (Beppu, 2014). Some of the MRI changes (such as those seen on diffusion-weighted imaging) can be reversible (Beppu, 2014). Apparent diffusion coefficient values on MRI can help predict neurologic prognosis following CO exposure (Chen et al., 2013). Co-oximetry is necessary to quantify carboxyhemoglobin levels. Smokers may have carboxyhemoglobin levels as high as 10–15%, while nonsmokers typically have levels <3%. Carboxyhemoglobin levels do not precisely correlate with the degree of poisoning or the risk of delayed neurologic sequelae. This may be related to the timing of blood sampling.

Cyanide poisoning can lead to hemorrhagic necrosis in the striatum and globus pallidus on MRI, as well as pseudolaminar necrosis along the central cerebral cortex (Rosenow et al., 1995; Rachinger et al., 2002). ^{18}F-2-Fluoro-2-deoxyglucose positron emission tomography can show regional reduction of the glucose metabolism in the posterior putamen and temporoparieto-occipital and cerebellar cortex (Rosenow et al., 1995).

Hospital course and management

There is no specific treatment for burn encephalopathy. However, aggressive supportive management of metabolic disarray is warranted, as delirium has been shown to lead to worse outcomes in diverse clinical situations (Ely et al., 2004; Thomason et al., 2005; Pauley et al., 2015). A large study of delirium in burn care units has shown that exposure to benzodiazepines, organ failure, hyponatremia, and hypoglycemia were directly correlated with the incidence of delirium, whereas exposure to intravenous opiates and methadone lowered the risk of delirium (Agarwal et al., 2010). Other studies have also confirmed the correlation between benzodiazepine exposure and delirium in burn patients (Stanford and Pine, 1988), and haloperidol has been shown to be safe and effective in this setting (Brown et al., 1996).

Patients with CO poisoning should be treated with 100% oxygen, as this shortens the half-life of carboxyhemoglobin from 4–5 hours to 1 hour (Betterman and Patel, 2014). The use of hyperbaric oxygen (HBO) therapy remains controversial for the treatment of CO poisoning, as previous studies showed conflicting results and a Cochrane review of six clinical trials did not support its use (Buckley et al., 2011). HBO therapy consists of inhalation of 100% oxygen at higher than normal ambient pressures to significantly increase the amount of oxygen that is physically dissolved in blood, reducing the half-life of carboxyhemoglobin to roughly 20 minutes (Betterman and Patel, 2014). It is currently suggested in patients with severe intoxication with carboxyhemoglobin levels of >25%, presenting with syncope, coma, seizures, focal neurologic deficits, severe acidosis, or myocardial ischemia (Betterman and Patel, 2014). It is also highly recommended in pregnant women, as the fetus is especially vulnerable to intoxication, with high rates of fetal mortality and morbidity (Prockop and Chichkova, 2007). HBO may mitigate against late neurocognitive deficits. It is most beneficial if instituted within the first 6 hours of exposure and treatment beyond 12 hours may have negligible benefit.

Cyanide exposure is normally approached with the use of antidotes, combined with advanced life support. Intubated patients or those at low aspiration risk can receive activated charcoal within the first 2 hours of exposure. Antidote treatment consists of binding cyanide, induction of methemoglobinemia (methemoglobin acts as an alternative binding site for cyanide), and use of sulfur donors. Hydroxycobalamin is a precursor of vitamin B_{12} and directly binds cyanide. It is the preferred antidote for cyanide poisoning (dosed 70 mg/kg IV). Amyl nitrite, sodium nitrite, and 4-dimethylaminopyridine can induce low levels of methemoglobinemia to scavenge cyanide. Methylene blue, which is used to reverse methemoglobinemia, should be avoided in the context of cyanide poisoning as it can lead to the release of free cyanide. Additionally, induction of methemoglobinemia should be avoided in the context of smoke inhalation and/or CO poisoning since this could lead to lethal tissue hypoxia. Sodium thiosulfate is a sulfur donor which converts cyanide into thiocyanate, which is renally excreted. For many centers, the optimal antidotes to cyanide poisoning consist of hydroxocobalamin and sodium thiosulfate. In patients with coexposure to CO, HBO may be helpful (Lawson-Smith et al., 2011).

Complex clinical decisions

All of the forms of intoxication in this section require aggressive supportive care, and often admission to an intensive care unit. Patients exposed to CO should be closely followed for at least 2 months, as delayed symptoms can ensue weeks after the initial exposure. The use of HBO in patients with CO poisoning is controversial and should be decided on a case-by-case basis.

Outcome prediction

The clinical outcome of patients exposed to CO is highly variable, and the real incidence of delayed neuropsychologic deficits is not known (Betterman and Patel, 2014). Weaver and colleagues (2002) found 46% incidence of delayed neurologic deficits at 6 weeks after exposure, and another study showed that depression and anxiety were present in 45% of patients at 6 weeks, 44% at 6 months, and 43% at 12 months (Jasper et al., 2005). Patients with lesions on brain MRI may be at higher risk of developing delayed symptoms (Sohn et al., 2000; Parkinson et al., 2002). A large retrospective study showed that predictors for the development of delayed symptoms might be CO exposure duration >6 hours, a Glasgow Coma Scale (GCS) score

<9, seizures, systolic blood pressure <90 mmHg, elevated creatine phosphokinase concentration, and leukocytosis (Pepe et al., 2011).

Neurorehabilitation

There is no specific treatment for the sequelae of these types of neurologic injuries. Physical, occupational, speech, respiratory, and vocational therapy, as well as intensive rehabilitation can help ameliorate neurologic deficits. Levodopa and anticholinergics have generally had unclear benefit in post-CO poisoning parkinsonism (Lee and Marsden, 1994). Case reports have shown benefits of using donepezil for cognitive symptoms and amantadine for attention deficits (Price and Grimley Evans, 2002; Shprecher and Mehta, 2010).

INJURIES RELATED TO ELECTRICITY (LIGHTNING AND ACCIDENTS WITH POWER LINES)

Epidemiology

In the USA, lightning causes more deaths than do most other natural hazards, even though the incidence of lightning-related deaths has decreased since the 1950s (CDC, 1998). Data from the Centers for Disease Control have shown an average of 82 deaths per year from lightning strikes during the period from 1980 through 1995 (CDC, 1998), with electric trauma from other sources accounting for as many as 1000 additional deaths annually (Cooper, 1995). Many of the survivors will suffer from disability, derived from electric injury to the nervous system and other organ systems (Koumbourlis, 2002). Risk for occupational-related electric injury increases substantially depending on profession (linemen, construction workers, electricians) (Jumbelic, 1995). Lightning strikes are a hazard related to outdoor leisure (mountain climbing, golf, boating) and are more common in the southeast of the USA (Lopez and Holle, 1995). Fatal lightning strikes are much more common in men (85% of fatalities), 90% occur in the spring and summer, and 70% occur in the afternoon and evening (CDC, 1998). Stun guns and tasers deliver a similar type of injury because the high-voltage current can lead to fatal arrhythmias and cardiac arrest.

Neuropathology

Electric injuries are divided as either high-voltage (1000 V and higher) or low-voltage (<1000 V), and most accidents in the USA are caused by low-voltage (110–240 V), 60-Hz household alternating current electricity (Sanford and Gamelli, 2014). High-voltage accidents are usually seen in industrial accidents or in electric company line workers. The most striking difference between lightning and high-voltage electric accidents is the longer duration of exposure to electricity in the latter (Yarnell, 2005).

Even though the body of the survivor is exposed to extremely high voltage, most of the severe brain injuries are related to hypoxic-ischemic encephalopathy after cardiac arrest and severe traumatic brain injuries secondary to falls or strikes to the head (Yarnell, 2005).

In the acute setting, patients can present to the hospital unable to move their limbs, with physical findings of flaccid paralysis and variable sensory findings. The impairment commonly resolves over a 24-hour period and has been termed "keraunoparalysis," a transient form of loss of motor and sensory function, usually seen in lightning strike survivors (Lammertse, 2005). The pathogenesis is not completely understood, but it is believed that an acute dysfunction of the autonomic nervous system plays an important role (Rahmani et al., 2015). Signs and symptoms include pulselessness, pallor, or cyanosis and motor and sensory loss in the affected extremities, usually the lower limbs, which resolves in a matter of minutes or hours (Rahmani et al., 2015). Dysfunction of the autonomic nervous system can also add to long-term morbidity and mortality, as with the occurrence of complex regional pain syndrome and cardiovascular abnormalities (Rahmani et al., 2015).

Acute findings can sometimes be irreversible, if the patient receives direct injury to peripheral nerves and/or spinal cord, mediated either thermally or electrically. Electricity can cause electroconformational changes in membrane proteins, and the formation of pores in the cell membrane, a phenomenon called electroporation (Lee et al., 1995). Electric damage to a cell is particularly related to its length in the direction of the electric field (Thaventhiran et al., 2001). Acute vasospasm and endothelial damage may also play a role in tissue injury (Ko et al., 2004).

Delayed symptoms may occur days to weeks after the initial insult, resulting in permanent disability, with motor deficits more likely to occur than sensory findings (Lammertse, 2005). It is hypothesized that a delayed degeneration of very small-sized blood vessels could be the mechanism of insult (Ko et al., 2004). Electroporation, DNA breakdown, protein denaturation, free radical production, and glutaminergic hyperstimulation could also play a role in the pathogenesis of delayed neurologic deterioration (Reisner, 2013).

Finally, in most patients multiple mechanisms of injuries can coexist, particularly in spinal cord injuries, where the acute traumatic injury may be related to a fall in conjunction with acute and delayed electricity-induced injuries (Lammertse, 2005).

Clinical presentation

Neurologic symptoms secondary to electricity-related injuries can be divided into four categories:

- category I: signs and symptoms are temporary and usually benign
- category II: acute injury leads to prolonged or permanent neurological disability
- category III: delayed neurologic syndromes
- category IV: neurologic lesions resulting from trauma secondary to falls or blast effects (Cherington, 2005).

Table 37.2 summarizes the four categories. Acutely injured patients can present to the hospital with brief loss of consciousness (up to 75% of patients), amnesia, confusion, headache, paresthesia, and transient limb weakness or paresthesia (approximately 80% of patients) (Rahmani et al., 2015). Most of these symptoms are benign and transient. Keraunoparalysis, a transient form of loss of motor and sensory function, is usually seen in lightning strike survivors, and can present as pulselessness, pallor, or cyanosis and motor and sensory loss in the affected extremities, usually the lower limbs. It can mimic an acute spinal cord injury, but symptoms typically resolve in 1 day. Acute dysfunction of the autonomic nervous system can lead to hypertension and cardiac arrhythmias.

Acute permanent or prolonged neurologic injuries are usually the result of direct thermal and electric injuries to the central and peripheral nervous system. Symptoms include behavior and neuropsychologic disorders, encephalopathy, myelopathy, complex regional pain syndrome, peripheral neuropathy, cerebellar syndrome, and epilepsy (Cherington, 2005).

Delayed neurologic symptoms can occur days to years after the initial injury. The most commonly observed and described syndromes are delayed myelopathy, parkinsonism, dystonia, tics, motor neuron disease, and cognitive dysfunction (Ko et al., 2004; Cherington, 2005; Sanford and Gamelli, 2014).

Neurologic injuries secondary to traumatic brain injury, traumatic spinal cord injury, and hypoxic-ischemic encephalopathy following cardiac arrest are likely the most common sequelae observed in survivors of lightning strikes and electric injuries (Yarnell, 2005).

Finally, patients can suffer other consequences such as cataracts (usually star-shaped), corneal burns, intraocular hemorrhage or thrombosis, uveitis, retinal detachment, choroidoretinitis, iridocyclitis, hyphema, ruptured tympanic membrane, and disruption of auditory ossicles (Saffle et al., 1985; Cooper, 1995).

Neurodiagnostics and imaging

Generally, neuroimaging with MRI and computed tomography (CT) studies of lightning injury patients and electric injury patients usually reveal no abnormalities (Reisner, 2013). However, patients with secondary traumatic injuries or cardiac arrest may have imaging findings reflective of hypoxic-ischemic insults or trauma.

Patients with delayed spinal cord symptoms weeks to months from the initial insult may have MRI evidence of either cord hyperintensity on T2-weighted images or atrophy (Reisner, 2013). Reversible MRI changes can be seen in the spinal cord in T2-weighted and short tau inversion recovery sequences (Freeman et al., 2004). Bilateral basal ganglia hemorrhages have been described in lightning strikes of the head, and it is hypothesized that the pathways through the basal ganglia offer less resistance to the conduction of electricity (Ozgun and Castillo, 1995; Kint et al., 1999).

Peripheral neuropathies and plexopathies are common after high-voltage electric shocks, occurring in 30–50% of survivors (Haberal et al., 1996; Ferreiro et al., 1998).

Table 37.2

Categories of neurologic symptoms and syndromes secondary to electricity-related injuries

Category	Evolution	Symptoms and syndromes
I	Acute, temporary, and usually benign	Headache, loss of consciousness, numbness and weakness, including keraunoparalysis
II	Acute, prolonged, or permanent	Posthypoxic encephalopathy, intracerebral hemorrhages, cerebral infarction, cerebellar syndrome, myelopathy, peripheral nerve lesions, autonomic disorders, behavior and neuropsychological problems, epilepsy
III	Delayed neurologic syndromes (weeks to months after the insult)	Myelopathy, motor neuron syndrome, parkinsonism, dystonia, and tics
IV	Neurologic lesions resulting from trauma	Falls, blast effects, or trauma from flying debris

Hospital course and management

Supportive care is the mainstay for patients who sustain electric injury. Cardiac resuscitation, and therapeutic hypo- or normothermia should follow conventional post cardiac arrest guidelines. Additional management of other injuries, such as cardiac arrhythmias, rhabdomyolysis, acute renal failure, pulmonary edema, and respiratory arrest, typically require intensive care (Koumbourlis, 2002). Patients who also suffer severe thermal injuries in points of contact with electricity often require deep and extensive debridement.

Anecdotal use of steroids in the acute phase of spinal cord injury secondary to electricity and the use of prostaglandins in delayed myelopathy have been reported (Ko et al., 2004). Posttraumatic stress disorder and also depression are commonly seen and require psychiatric intervention (Primeau, 2005).

Clinical trials and guidelines

There are no current guidelines addressing the care of this population. Future studies should be designed to better delineate the prevention and management of delayed neurologic manifestations.

Complex clinical decisions

Survivors of accidents with electricity are often critically ill and have multisystem organ dysfunctions. Goals of care should be addressed with family members and other members of the multidisciplinary team (Gordon et al., 2009; Shanmugarajah et al., 2011).

Outcome prediction

Approximately 30% of patients struck by lightning die and 74% of survivors have long-term disability (Browne and Gaasch, 1992; Zafren et al., 2005; Ritenour et al., 2008). Most deaths related to lightning strike are due to arrhythmias resulting in cardiac arrest shortly after the injury.

The first 24 hours are critical in determining the extent of spinal cord injuries due to accidents with electricity, as patients with "temporary" paralysis will recover motor function during this time period (Lammertse, 2005). Accidents where electricity flows from a contact point in the head to an upper extremity or from one upper extremity to another are much more likely to cause permanent quadriplegia than accidents involving the exit point in the lower extremities (Ko et al., 2004).

Accidents with high voltage in industrial settings or power lines are related to more long-term sequelae, as patients are exposed to electricity for a longer time (Sanford and Gamelli, 2014). In one study of patients who sustained work-related electric injuries, 82% had long-term neurologic symptoms, and 71% had psychologic complaints (Singerman et al., 2008). Despite optimal medical management, only about 50% of low-voltage work-related electrical injury patients will return to a professionally productive life (Theman et al., 2008). In contrast, a longitudinal study of 10 survivors of lightning strikes over 12.3 years found that no patient had long-term neurologic or psychologic deficits (Muehlberger et al., 2001). The effectiveness of cardiac resuscitation in the field may be a larger predictor of outcome than any sequelae derived from the electricity discharge (Sanford and Gamelli, 2014). Finally, better recovery from electricity-related injuries is observed in acute neurologic insults as compared to delayed neurologic injury (Sanford and Gamelli, 2014).

Though many patients experience unfavorable outcomes, there are case reports of recovery several months to years after injury in severely affected patients with quadriplegia due to spinal cord injuries or polyneuropathy (Christensen et al., 1980; Breugem et al., 1999; Thaventhiran et al., 2001).

Neurorehabilitation

Rehabilitation planning should begin early after the incident and may continue for extended periods of time (Yarnell, 2005). Some of the most challenging issues encountered in patients admitted for rehabilitation after electric injuries are pain, memory and cognitive problems, neglect, bladder dysfunction, and spasticity (Yarnell, 2005).

INJURIES RELATED TO WATER (NONFATAL DROWNING AND DECOMPRESSION ILLNESS)

Epidemiology

Nonfatal drowning and decompression sickness (DCS) are the most dramatic accidents related to water submersion. According to the World Health Organization, more than 500 000 deaths occur annually worldwide as the consequence of drowning (WHO, 2003). Drowning is the second leading cause of injury-related death in the USA among children 1–4 years of age, with a death rate of 3 per 100 000 (Borse et al., 2008). The most relevant risk factors for drowning are low income, alcohol abuse, low educational status, male sex, age <14, lack of supervision, aquatic exposure, rural residency, risky behavior, and epilepsy (Szpilman et al., 2012).

Decompression illness is normally secondary to sports or work-related accidents. It is uncommon if appropriate decompression procedures are taken during ascent, with an incidence of 0.015% for scientific divers, 0.01–0.019% for recreational divers, 0.030% for US Navy divers, and 0.095% for commercial divers (Ladd et al., 2002; Vann, 2004).

Neuropathology

The World Health Organization defined drowning in 2005 as "the process of experiencing respiratory impairment from submersion/immersion in liquid" (van Beeck et al., 2005). The term nonfatal drowning is used if the person is rescued and the process of drowning is interrupted, and fatal drowning is the term applied if the person dies at any time as a result of drowning (Szpilman et al., 2012). Terms such as "near drowning," "dry or wet drowning," "secondary drowning," "active and passive drowning," and "delayed onset of respiratory distress" should be avoided (van Beeck et al., 2005).

During the initial minutes after submersion, small amounts of fluid are aspirated into the hypopharynx, triggering laryngospasm (Ibsen and Koch, 2002). In 85–90% of cases, the initial laryngospasm abates and the victim aspirates large volumes of water, sometimes followed by aspiration of gastric contents (Ibsen and Koch, 2002).

During the first minutes following submersion, the brain is severely deprived of oxygen, and when the cardiovascular system fails, cerebral blood flow decreases, leading to further ischemia. Selective vulnerability of the brain to hypoxic-ischemic injuries varies depending on the patient's age, and the region of the brain (Ibsen and Koch, 2002). The most affected areas are the hippocampus, insular cortex, basal ganglia, and in vascular "watershed" areas, and in cases of severe hypoxia/ischemia, extensive and global neocortical damage will occur (Ibsen and Koch, 2002).

DCS is a disorder resulting from supersaturation of inert gases in blood and tissues that result in free gas formation (Bove, 2014). Arterial gas embolism (AGE) is another disorder that results from diving and from other mechanisms that result in injection of air into the circulation. Decompression illness is the term used for the two disorders, as they can be hard to differentiate, and therapy for both relies on similar approaches. Although AGE can arise after ascent from shallow depths, DCS almost never occurs after a single dive to depths of less than 6 meters, even for an extended time (Vann et al., 2011).

Barotrauma can occur during descent or ascent. During descent, organs can be compressed due to increased pressure and corresponding decreased volume of gas (according to Boyle's law, which states that volume and pressure are inversely related in a fixed mass of an ideal gas at constant temperature). This can lead to pulmonary hemorrhages (Bove, 2014). In ascent, barotrauma can occur due to inadequate exhalation and overexpansion of the lungs. Elevated intra-alveolar pressures can lead to their rupture, with consequent migration of air to the pleural space, mediastinum, peritoneum, subcutaneous space, and in some circumstances air may enter the arterial circulation. The most frightening complication of pulmonary barotrauma is cerebral embolization of air.

DCS occurs when inert gas (normally nitrogen or helium) comes out of solution, as bubbles form secondary to a reduction in surrounding pressure during ascent. The amount of inert gas dissolved depends on the depth and the duration of the dive. Once gas bubbles form in tissues and in venous blood, they can trigger a cascade of events, leading to oxidative stress, excitatory amino acid release, inflammation, endothelial injury resulting in extravasation of plasma into the interstitium, neutrophil margination, and loss of vasoreactivity (Massey and Moon, 2014). Bubbles can have mechanical, embolic, and biochemical effects, leading to specific organ dysfunction. The severity of DCS is related to the extent of bubble formation.

Clinical presentation

The diagnosis of both DCS and AGE is based almost exclusively on the clinical and neurologic examination, and details of the dive history. It is difficult to differentiate neurologic symptoms due to DCS versus AGE, as they often occur together. Moreover, such differentiation is not of clinical importance, as the treatment for both conditions is essentially the same.

Classically, decompression illness is classified into type I and type II. The minor form (type I) includes musculoskeletal symptoms, skin rash, livedo reticularis, or swelling and pain in lymph nodes. The major form (type II) is characterized by neurologic, inner-ear, and cardiopulmonary symptoms (Table 37.3). Cerebral involvement occurs in 30% of cases of type II DCS (Hawes and Massey, 2008).

Table 37.3

Classification of decompression illness

Classification	Symptoms
Decompression sickness Type I (mild form)	Joint and muscle pain, skin rash, livedo reticularis, swelling and pain in lymph nodes, malaise, fatigue, headache
Type II (major form)	Paresthesia, hearing loss and vestibular dysfunction, confusion, lethargy, difficulty with concentration, visual disturbances, seizures, hemiparesis, quadriparesis, and coma
Acute gaseous emboli (AGE)	Subcutaneous emphysema, pneumothorax, pneumomediastinum, pneumoperitoneum, cerebral ischemia (hemiparesis, sensory deficits, cortical blindness, coma), shortness of breath, chest pain, cardiovascular collapse

Neurologic presentations of DCS can range from malaise, fatigue, headache, pain, paresthesias, hearing loss and vestibular dysfunction, confusion, lethargy, "mental cloudiness," difficulty with concentration, visual disturbances, seizures, hemiparesis, quadriparesis, and coma (Massey and Moon, 2014).

Injury to the spinal cord, usually at levels below T11 and T12 due to common anatomic factors leading to bubble migration, can cause paresthesias, weakness, or paralysis of the lower extremities, urinary retention, bowel or bladder incontinence, and sexual impotence (Hawes and Massey, 2008; Vann et al., 2011).

Nondermatomal hypoesthesia and truncal ataxia are common manifestations of DCS. Additionally, coordination can be disproportionately affected. The National Institutes of Health Stroke Scale has been used to help with severity stratification and has adequate predictive ability while providing a more standardized scale (Holck and Hunter, 2006).

Divers with patent foramen ovale (PFO) have an increased risk of developing decompressive illness, as venous bubbles crossing the atrial septum can cause AGE (Wilmshurst et al., 1994). The risk seems higher in patients with larger PFOs (Torti et al., 2004). Billinger and colleagues (2011) found a high incidence of MRI lesions in divers with PFOs and showed that PFO closure reduces the risk of recurrent symptomatic DCS and asymptomatic ischemic neurologic events in patients who continue to dive. Experts suggest that PFO closure should be considered in commercial divers who experience repeated DCS and have a large PFO (Moon and Bove, 2004; Bove, 2014).

Other factors can obscure the neurologic evaluation in divers immediately after resurfacing, such as CO intoxication due to faulty recirculation, nitrogen narcosis producing an anesthetic effect, and oxygen toxicity leading to auditory and visual hallucinations, and seizures, due to excess of oxygen and nitrogen in breathing mixtures (Bove, 2014).

Neurodiagnostics and imaging

The neurodiagnostics for near drowning are similar to those employed for other forms of hypoxic/anoxic brain injury. An initial head CT may disclose loss of differentiation of the gray–white junction, sulcal effacement or, in severe cases, global cerebral edema. However, a normal head CT does not necessarily imply a good neurologic outcome. EEG may be used to detect seizures or abnormal patterns reflective of global anoxia, such as burst suppression pattern. Factors that predict a poor neurologic outcome in survivors of cardiac arrest may also apply to patients with near drowning. Absent bilateral N_2O responses 24–72 hours following near drowning, myoclonic status epilepticus, and a burst suppression pattern on EEG (not induced by sedative medications) may all forebode a poor neurologic outcome. MRI may reveal evidence of anoxic brain injury with imaging patterns similar to cardiac arrest survivors. When a diving injury complicates a near-drowning episode, cervical spine CT and MRI are required.

A systematic review of several studies using MRI as the neuroimaging of choice after nonfatal drowning with neurologic symptoms identified eight studies with 68 cases (Nucci-da-Silva and Amaro, 2009). Of these, the main finding was the presence of brain edema in 78%. Cerebral edema was described as focal in 14% of patients and generalized in 86%. The second most prevalent finding was abnormal MRI signal in basal ganglia in 75% (lentiform nucleus, globus pallidus, caudate, putamen, subthalamic and thalamic nuclei). Additionally, diffuse cerebral atrophy has been described in 36% of patients and infarcts in 15%. The authors compared the findings to the ones normally seen in hypoxic-ischemic encephalopathy cardiac arrest (of other causes) and suggested that some of the differences in lesion distribution might be accounted for by other associated factors, such as hypothermia during drowning.

There are no imaging, serum, or neurophysiology tests to accurately detect DCS. Several MRI studies in patients with neurologic type 2 DCS have been reported, with variable and often controversial results regarding the sensitivity and utility of this modality (Kamtchum Tatuene et al., 2014). This could be likely explained by variable study designs, MRI techniques and imaging protocols, as well as variable time delays between symptom onset and imaging (Kamtchum Tatuene et al., 2014). It should be noted that normal MRI findings of the spinal cord do not exclude DCS as the cause of myelopathy (Kamtchum Tatuene et al., 2014) and the negative predictive value of spine MRI following DCS is only 77% (Kamtchum Tatuene et al., 2014). However, hemorrhagic cord injury is correlated with poor neurologic outcome and patients with lone spinal cord edema or normal spine MRI typically have good neurologic recovery.

AGE may lead to DWI restricted lesions in the brain (with low apparent diffusion coefficient values (Warren et al., 1988; Kamtchum Tatuene et al., 2014) and T2/fluid-attenuated inversion recovery hypersignal in the spinal cord, primarily in the lateral and posterior column white matter (Kim et al., 1984; Sparacia et al., 1997; Manabe et al., 1998; Vollmann et al., 2011). There are case reports of intracranial hemorrhages (Josefsen and Wester, 1999) and extensive gas inclusions in cerebrospinal fluid spaces (Ozdoba et al., 2005).

Hospital course and management

Survivors of severe drowning require intensive care during the first days to mitigate complications to the lungs and brain. The current treatment of survivors who have been rescued from drowning resembles that of patients with the acute respiratory distress syndrome, and the same ventilation-protective strategies should be followed (Szpilman et al., 2012). A systemic inflammatory response syndrome after resuscitation is sometimes observed in persons who have been rescued from drowning, and sepsis and disseminated intravascular coagulation can occur within the first 72 hours.

Recent reports on drowning have documented good outcomes with the use of therapeutic induction of hypothermia after resuscitation (Guenther et al., 2009; Vanden Hoek et al., 2010).

The mainstay of treatment for patients with DCS or AGE is recompression in an HBO chamber (Bennett et al., 2012; Bove, 2014). Once stabilization is achieved, the patient is decompressed slowly to surface pressure. Recompression therapy for decompressive illness decreases the risk of permanent injury, but the severity of the initial injury and the time to effective therapy affect outcome (Bove, 2014). There should be no delay in initiating therapy, but treatment even several days after injury has demonstrated efficacy (Bove, 2014).

Antiplatelet agents have been recommended to minimize ischemia following DCS and AGE, but clinical trials demonstrating their efficacy are lacking (Bennett et al., 2010, 2012). Bennett and colleagues (2003) used tenoxicam (a nonsteroidal anti-inflammatory medication) in a trial with patients with decompression illness, and found that its use was associated with fewer recompression treatments, but no significant change in clinical outcomes.

Clinical trials and guidelines

The Undersea and Hyperbaric Medical Society and the US Navy Department have published guidelines on the use of HBO therapy (US Navy, 2008; Weaver, 2014). Guidelines are also available for the rescue and initial resuscitation of near-drowning victims (Soar et al., 2010; Vanden Hoek et al., 2010).

Complex clinical decisions

In survivors of nonfatal drowning, immediate decisions about withdrawal of life-sustaining therapy based on neurologic examination on arrival should be avoided, as some patients can improve substantially in the following days. Accidental hypothermia may severely impair the initial neurologic exam and confound prognostication (Ibsen and Koch, 2002).

Providers should be vigilant for any subtle signs of decompression illness in divers, as most of the symptoms will be present soon after ascent. A thorough neurologic exam should be done, including cognition testing, as symptoms sometimes are vague and nonspecific. Appropriate therapy should be instituted promptly if neurologic findings are present.

Outcome prediction

Permanent neurologic sequelae are the most feared outcome in patients who have been resuscitated after a drowning incident. In cases where cardiopulmonary resuscitation is required, the risk of long-term neurologic damage is similar to that in other causes of cardiac arrest (Szpilman et al., 2012). However, hypothermia associated with drowning might provide a protective mechanism that allows patients (particularly children) to survive prolonged submersion episodes (Szpilman et al., 2012). Traumatic brain injury during submersion can also increase morbidity. Long-term sequelae of hypoxic-ischemic injury can manifest as cognitive and memory problems, myoclonus, dystonia, parkinsonism, epilepsy, myelopathy, dysfunction of arousal mechanisms, including minimally conscious states and permanent vegetative states (Lu-Emerson and Khot, 2010).

Prediction of functional neurologic outcome in patients after nonfatal drowning becomes more reliable with time. The neurologic examination in the first 24–72 hours of therapy is one of the best predictors of neurologic outcome (Ibsen and Koch, 2002). Historically, a number of investigators have noted that a GCS of <5 on presentation identified children at high risk for poor outcome (Allman et al., 1986; Bratton et al., 1994), but other case series have shown a significant number of survivors with excellent neurologic recovery, despite having a GCS <5 on arrival (Lavelle and Shaw, 1993). Bratton and colleagues (1994) found that all survivors with good neurologic outcome had spontaneous, purposeful movement within 24 hours of rescue.

A study using MRI in children after nonfatal drowning found that generalized or occipital edema correlated with poor outcome, and indistinct lentiform nuclei margins on T1-weighted images were a frequent finding (Dubowitz et al., 1998). A persistently attenuated record on EEG without medications is predictive of a poor neurologic outcome (Kruus et al., 1979).

Fortunately, most divers achieve a complete recovery from decompression illness. In a case series of divers with difficulty walking due to spinal cord DCS, at long-term follow-up 50% had no residual symptoms, and only one-third had manifestations that impacted daily activities (Vann et al., 2011). The most common residual symptoms were peripheral paresthesias, and a minority of patients experienced spasticity, urinary incontinence, impotence, or weakness (Massey and Moon, 2014). Similarly, the majority of divers who survive AGE typically make a full recovery. Roughly 60%

experience full recovery (van Hulst et al., 2003; Trytko and Bennett, 2008). Even divers who do not fully return to normal after AGE report only minor symptoms that do not affect quality of life (Trytko and Bennett, 2008). A study in Norway has shown that professional divers had more neurologic symptoms and neurologic findings than nondivers, primarily difficulties with concentration and short- and long-term memory, as well as abnormal findings such as distal spinal cord and nerve root dysfunction (Todnem et al., 1990).

Neurorehabilitation

Survivors of nonfatal drowning can experience neurologic symptoms of hypoxic-ischemic encephalopathy, including cognitive and memory problems, myoclonus, dystonia, parkinsonism, epilepsy, myelopathy, minimally conscious states, and permanent vegetative states. Recently, levodopa, amantadine, zolpidem, baclofen, dorsal column stimulation, and deep-brain stimulation have been used in some studies to improve rehabilitation in patients with chronic disorders of consciousness, with some success (Pistoia et al., 2010; Oliveira and Fregni, 2011; Giacino et al., 2012; Yamamoto et al., 2013; Formisano and Zasler, 2014).

Even though the vast majority of divers will recover without major sequelae, the minority will experience some degree of disability. Rehabilitation should focus on cognitive issues, such as memory and behavioral disturbances, as well as on symptoms of chronic myelopathy, including spasticity, urinary incontinence, impotence, and weakness.

INJURIES RELATED TO EXTREMES OF TEMPERATURE (HEAT STROKE AND ACCIDENTAL HYPOTHERMIA)

Epidemiology

Accidental hypothermia is usually defined as an involuntary drop of core body temperature to below 35°C (95°F) (Brown et al., 2012). Such accidents carry significant morbidity and mortality (Mair et al., 1994; Silfvast and Pettila, 2003), and approximately 1500 patients die each year in the USA due to accidental hypothermia (Baumgartner et al., 2008). Predisposing conditions include old age, malnutrition, alcohol and substance abuse, self-neglect, poverty, chronic debilitating conditions, dementia, hypothyroidism, hypopituitarism, hypocortisolism, drowning or immersion, neuroleptic drugs, severe dermatologic problems, and sepsis (Epstein and Anna, 2006).

Hyperthermia can be defined as a rise in body temperature above the hypothalamic set point when heat-dissipating mechanisms are impaired or overwhelmed

Table 37.4

Systemic manifestations of elevated body heat

Category	Body temperature	Systemic manifestations
Heat stress	<37°C	Mild discomfort, cramps, physiologic strain, possibly syncope (hot environment)
Heat exhaustion	37–40°C	Water and salt depletion. Mild neurologic symptoms (headache, dizziness, anxiety, fainting), but no organ dysfunction
Heat stroke	>40°C	Anxiety, confusion, bizarre behavior, loss of coordination, hallucinations, agitation, seizures, coma, acute renal failure, rhabdomyolysis, multiorgan failure

by external or internal heat (Bouchama and Knochel, 2002). Heat stroke is an entity defined as a rise in core body temperature above 40°C that is accompanied by hot, dry skin and central nervous system dysfunction, potentially leading to multiorgan failure (Bouchama and Knochel, 2002; Table 37.4). Heat stroke can result from exposure to high environmental temperature or strenuous exercise. Risk factors for heat stroke include extremes of age, genetic predispositions, medical comorbidities, poor access to climate-controlled environments, psychosocial problems, as well as occupational or recreational activities involving strenuous exercise in exceedingly hot environments (Atha, 2013). Each year in the USA, approximately 400 deaths are attributed to excessive natural heat (CDC, 2002). In an extreme event, the death of 14 800 individuals was attributed to a heat wave in France in 2003 (Argaud et al., 2007). If global temperatures continue to rise as predicted throughout the next century, it is hypothesized that the incidence of heat-related illnesses is likely to increase dramatically (Atha, 2013).

Neuropathology

Hypothermia can decrease cerebral metabolism by 6–10% for each 1°C reduction in body temperature. When the core body temperature drops to 32°C, the metabolic rate decreases to 50–65% of normal, and oxygen consumption and CO_2 production will decrease by the same percentage (Polderman, 2009). The decrease in cerebral metabolism is the explanation for the progressive blunting of the neurologic exam in the context of hypothermia. Cardiac output decreases due to hypothermia-induced bradycardia. Because the metabolic rate also decreases, the balance between supply and demand remains stable. Hypothermia can induce neuroprotective responses, such as mitigation of apoptosis and mitochondrial dysfunction, decreased secretion of neuroexcitatory transmitters, decreased production of proinflammatory cytokines and free radicals, improved brain glucose metabolism, protection of the blood–brain barrier, and also decreases vascular permeability (Polderman, 2009). This could explain why patients with accidental hypothermia and prolonged cardiac arrest might still experience surprisingly positive neurologic outcomes (Brown et al., 2012). Targeted temperature management has been shown to improve neurologic outcomes after cardiac arrest (Bernard et al., 2002; HACA, 2002; Nielsen et al., 2013) and in neonates after perinatal asphyxia (Azzopardi et al., 2009, 2014; Shankaran et al., 2012; Jacobs et al., 2013).

The human body relies on thermoregulatory mechanisms to avoid overheating. Heat is gained from the environment and produced by metabolism. Any elevation in body temperature about 37°C is sensed by the hypothalamus. Under normal conditions the hypothalamus signals the body to shunt blood to the periphery/subcutaneous tissue so heat can be exchanged with the environment. Sweat will vaporize and help cool the body surface. Table 37.5 describes other mechanisms of heat exchange between the human body and the surrounding environment.

Table 37.5

Mechanisms of heat exchange

Type	Heat loss percentage	Mechanism
Radiation	60%	Emanation of electromagnetic infrared heat rays
Evaporation	20%	Water or sweat evaporation; respiration
Conduction	15%	Transfer of heat through kinetic energy to an object of lower temperature
Convection	5%	Air currents on the surface

Adapted from Gomez (2014).

However, if the above mechanisms fail to stabilize core body temperature, a stress response is mounted and heat-related illness can develop. Excessive heat leads to secretion of several cytokines and proinflammatory cellular mediators, resulting in a response that resembles systemic inflammatory response syndrome (Atha, 2013). If not corrected, it can result in apoptosis and cell death, endothelial injury, protein denaturation, as well as direct activation of the coagulation cascade and progression to disseminated intravascular coagulation (Atha, 2013). Heat shock and acute-phase proteins are produced through gene transcription to help counterbalance the systemic inflammation and mitigate further damage (Bouchama and Knochel, 2002).

The brain (particularly the cerebellum) and the liver are the most vulnerable organs to damage by elevated temperatures (Atha, 2013). Bazille and colleagues (2005) reported almost total loss of Purkinje cells in 3 patients with heat stroke in a postmortem study. Increased expression of heat shock protein 70 was noted near remaining Purkinje cells and adjacent Bergmann glia, suggesting selective vulnerability of Purkinje cells to heat injury (Bazille et al., 2005).

Splanchnic ischemia, acute renal injury, rhabdomyolysis, acute respiratory distress syndrome, pancreatic injury, and hemorrhagic complications can occur, and more rarely, a direct cardiomyopathy without evidence of coronary artery disease has been described.

Clinical presentation

The neurologic exam varies depending on core temperature in the context of hypothermia. Initially, patients can present with shivering. In patients with lower temperatures, sluggishness, incoordination, worsening confusion, and blunted deep tendon reflexes can occur. Pupils tend to become dilated as temperature further drops, and in patients with severe hypothermia, coma, rigidity, and decreased or absent activity on EEG can be present (Gomez, 2014; Table 37.7). Paradoxic undressing can occur when peripheral vessels passively dilate after protracted constriction to maintain core body temperature. This passive vasodilation creates the sensation of being overheated and patients may undress to relieve this discomfort.

Heat stroke is defined by cerebral dysfunction in the context of body core temperature >40°C. The usually observed neurologic findings are anxiety, confusion, bizarre behavior, loss of coordination, hallucinations, agitation, seizures, and coma in severe cases (Yeo, 2004). The cerebellum is the part of the central nervous system most commonly affected (Kosgallana et al., 2013). Seizures might occur, especially during cooling

Table 37.6

Complications of hypothermia

Physiologic changes and complications by systems (hypothermia)	Comments
Cardiovascular	
Hypovolemia	Normally secondary to cold diuresis
Heart rate	Patients usually develop bradycardia, but matched with decreased body metabolism. Malignant bradycardia and decreased stroke volumes normally only seen with temperature < 30°C (patients should be rewarmed, as atropine is usually ineffective)
Electrocardiogram changes	Increase in PR, QT, and QRS intervals. Arrhythmias normally seen in temperature <30°C (atrial fibrillation and ventricular arrhythmias). Rewarming is the most effective treatment in this situation. Osborn waves normally noted only in cases of severe accidental hypothermia
Renal and electrolytes	Intracellular electrolyte shifts can occur. Acute tubular necrosis only observed in case of accidental hypothermia with temperature <28°C
Endocrine	Insulin resistance with hyperglycemia commonly seen
Infections	Hypothermia can increase the chances of developing infections, particularly pulmonary infections due to decreased airway ciliary clearance
Blood and coagulation	Mild leukopenia, as well as mild coagulation cascade and platelet dysfunction, but without significant clinical impact. Disseminated intravascular coagulation is mostly commonly observed in accidental hypothermia with temperature <30°C
Thermoregulation	Shivering markedly increases brain and body metabolism
Gastrointestinal	
Motility	Patients can develop ileus and delayed gastric emptying
Pancreas and liver	Liver metabolism is markedly decreased and mild increases in liver function enzymes can be observed, but usually of no significance. Amylase and lipase can also be mildly elevated, but without representing cell injury. Pancreatitis has been described in patients with severe accidental hypothermia
Drug metabolism	Due to liver function decrease, drug metabolism is usually compromised and half-lives of drugs primarily cleared by the enzymatic system may be prolonged
Skin	Patients are at high risk for bedsores, due to skin vasoconstriction and immune suppression
Respiratory and blood gases	O_2 and CO_2 can be overestimated and pH underestimated if blood gas analyses are not corrected to actual body temperature, due to changes in gas solubility with temperature. Patients with hypothermia tend to have low P_{CO_2}, due to decreased metabolism. It is controversial whether pH and P_{CO_2} value management should be guided by corrected ABG (alpha-stat versus pH-stat theories). The authors believe that a combination of corrected ABG aiming for P_{CO_2} levels mildly lower than normal (around 35 mmHg) could avoid cerebral ischemia secondary to extremely low P_{CO_2} or hyperemia (with increased ICP) due to high "actual" P_{CO_2} levels
Neurologic	
32–35°C	Shivering, confusion, agitation
28–32°C	Sluggishness, incoordination, worsening confusion
20–28°C	Amnesia, irrational behavior, severe incoordination, dilated pupils, hyporeflexia, decreased responsiveness
<20°C	Coma, rigidity, decreased or absent activity on electroencephalogram

ABG, arterial blood gas; ICP, intracranial pressure.

(Bouchama and Knochel, 2002). There are case reports of posterior reversible encephalopathy secondary to hyperthermia (Tan et al., 2014). Virtually all patients have tachycardia and hyperventilation (Bouchama and Knochel, 2002). Heat stroke can lead to splanchnic ischemia, acute renal injury, rhabdomyolysis, acute respiratory distress syndrome, pancreatic injury, hemorrhagic complications, and cardiomyopathy. Table 37.6 describes some of the physiologic and clinical changes secondary to hyperthermia.

Table 37.7

Complications of hyperthermia

Physiologic changes and complications by system (hyperthermia)	Comments
Cardiovascular	
Hemodynamics	The initial response is hyperdynamic, characterized by an elevated cardiac index, decreased systemic vascular resistance, and elevated central venous pressure. Later, cardiovascular system becomes compromised from the effects of dehydration and vasoconstriction, leading to low cardiac index and increased systemic vascular resistance
Arrhythmias	Sinus tachycardia, atrial fibrillation, and supraventricular tachycardia
Heart	Thermal myocardial dysfunction, myocardial ischemia
Electrocardiogram	Right bundle branch block, intraventricular conduction delays, prolongation of the Q-T interval, and nonspecific ST segment changes
Acid–base disorders	Lactic acidosis, respiratory alkalosis
Muscular	Rhabdomyolysis
Renal	Acute renal failure secondary to rhabdomyolysis, systemic inflammation, disseminated intravascular coagulation
Respiratory	Acute respiratory distress syndrome
Hematologic	
Coagulation	Disseminated intravascular coagulation
Platelets	Thrombocytopenia
Gastrointestinal	
Liver	Hepatocellular injury, liver failure
Bowel	Intestinal ischemia or infarction
Neurologic	Secondary to metabolic disturbances, metabolic encephalopathy, cerebral edema and ischemia, and possibly hypernatremic cerebral damage. Seizures, weakness and uncoordinated movements, encephalopathy, coma, anxiety, hallucination, agitation, cerebellar syndrome, infarcts

Neurodiagnostics and imaging

MRI of the brain in patients with heat stroke may show no acute abnormalities (Fushimi et al., 2012). Abnormal signal may appear in cerebellar efferent pathways over time, suggesting a possible mechanism of deafferentation (Fushimi et al., 2012). Cytotoxic edema (increased diffusion-weighted imaging signal and reduced apparent diffusion coefficient signal changes) has been described in the bilateral dentate nuclei (Lee et al., 2009a), bilateral superior cerebellar peduncles and thalami (Ookura et al., 2009), and decussation of the superior cerebellar peduncles (Bazille et al., 2005). In chronic cases, diffuse atrophy of the cerebellum on MRI can be observed (Albukrek et al., 1997).

Hospital course and management

Priorities for prehospital treatment in patients with accidental hypothermia include careful transport of the patient (to avoid cardiac arrhythmias), advanced life support, and passive/active external rewarming (Brown et al., 2012). In hemodynamically stable patients, active external and minimally invasive rewarming is indicated, to avoid further complications, as well as the absence of firm evidence that invasive treatments might change the outcome (Brown et al., 2012). In patients with cardiac instability or cardiac arrest, fast rewarming should be attempted, and extracorporeal membrane oxygenation (ECMO) can play a major role. ECMO has been shown to be safe and effective (Walpoth et al., 1997; Silfvast and Pettila, 2003), and among patients treated with it, the rate of survival with good neurologic outcome is 47–63% (Walpoth et al., 1997; Farstad et al., 2001; Silfvast and Pettila, 2003; Ruttmann et al., 2007). In facilities without the availability of ECMO, cardiopulmonary resuscitation should not be stopped until the core temperature has reached approximately 35°C, as there are reports of patients resuscitated for up to 390 minutes with good neurologic outcome (Lexow, 1991).

Heat stroke is a medical emergency, and evidence suggests that morbidity is significantly reduced if cooling measures are started within 30 minutes of recognition (Heled et al., 2004). Currently, there are no controlled studies comparing the effects of different cooling methods on cooling times and outcome in patients with heat-related illness. Traditionally, cold water and ice packages have been used to decrease core temperature, which might lead to overshooting and consequent hypothermia (Bouchama and Knochel, 2002). It might be reasonable to use newer surface and intravascular cooling devices to reach goal core temperature without overshooting. Cooling blankets are often ineffective, and alternative internal cooling measures, such as cold-water gastric, peritoneal, rectal, or bladder lavages, are insufficiently studied in humans and can lead to water intoxication (Atha, 2013).

Clinical trials and guidelines

There are no current clinical trials delineating management strategies for hypo- or hyperthermia. Trials analyzing different strategies to achieve goal core temperature (either in hypothermic or hyperthermic patients) using newer devices (surface and intravascular cooling methods) are necessary.

Complex clinical decisions

It can be extremely difficult to decide when to stop cardiopulmonary resuscitation in patients with accidental hypothermia, since rewarming is often necessary before return of spontaneous circulation can be achieved. As with other patients who suffer cardiac arrest, prognostication may be most reasonable after several days of treatment and observation of the neurologic exam.

Outcome prediction

The lowest reported core body temperature associated with full neurologic recovery was 13.7°C (57°F) in a patient with accidental hypothermia (Gilbert et al., 2000) and 9°C (48°F) in a case of induced hypothermia during surgery (Niazi and Lewis, 1958). Virtually all patients with cardiac stability treated with active external and minimally invasive rewarming in a study were neurologically intact at discharge (Kornberger et al., 1999). In patients treated with ECMO, the rate of survival with good neurologic outcome is 47–63% (Walpoth et al., 1997; Farstad et al., 2001; Silfvast and Pettila, 2003; Ruttmann et al., 2007). The presence of associated major trauma or avalanche burial for more than 35 minutes is a marker of unfavorable prognosis (Brown et al., 2012).

Dematte and colleagues (1998) studied 58 patients with classic heat stroke admitted to an intensive care unit and analyzed their long-term outcome. All patients experienced multiorgan dysfunction with neurologic impairment, with an in-hospital mortality of 21%. Most survivors recovered near-normal renal, hematologic, and respiratory status, but moderate to severe functional impairment was present in 33% of patients at hospital discharge. At discharge, 24% did not have neurologic impairment, 43% had minimal impairment, and 33% had moderate to severe impairment. At 1 year, no patient had improved functional status, and an additional 28% of patients had died. Recovery of cerebral function during cooling is a favorable prognostic sign and should be expected in the majority of patients who receive immediate and effective therapy (Bouchama and Knochel, 2002).

Neurorehabilitation

Survivors of severe heat stroke can experience significant incoordination and imbalance due to cerebellar dysfunction, and some might develop cognitive impairment and parkinsonism. Currently, no evidence exists for the care of this subset of patients, but traditional approaches to these issues derived from studies in other neurologic diseases might be entertained.

References

Agarwal V, O'Neill PJ, Cotton BA et al. (2010). Prevalence and risk factors for development of delirium in burn intensive care unit patients. J Burn Care Res 31: 706–715.

Albukrek D, Bakon M, Moran DS et al. (1997). Heat-stroke-induced cerebellar atrophy: clinical course, CT and MRI findings. Neuroradiology 39: 195–197.

Allman FD, Nelson WB, Pacentine GA et al. (1986). Outcome following cardiopulmonary resuscitation in severe pediatric near-drowning. Am J Dis Child 140: 571–575.

Andreasen NJ, Hartford CE, Knott JR et al. (1974). Letter: cerebral deficits after burn encephalopathy. N Engl J Med 290: 1487–1488.

Andreasen NJ, Hartford CE, Knott JR et al. (1977). EEG changes associated with burn delirium. Dis Nerv Syst 38: 27–31.

Argaud L, Ferry T, Le QH et al. (2007). Short- and long-term outcomes of heatstroke following the 2003 heat wave in Lyon, France. Arch Intern Med 167: 2177–2183.

Atha WF (2013). Heat-related illness. Emerg Med Clin North Am 31: 1097–1108.

Azzopardi DV, Strohm B, Edwards AD et al. (2009). Moderate hypothermia to treat perinatal asphyxial encephalopathy. N Engl J Med 361: 1349–1358.

Azzopardi D, Strohm B, Marlow N et al. (2014). Effects of hypothermia for perinatal asphyxia on childhood outcomes. N Engl J Med 371: 140–149.

Baumgartner EA, Belson M, Rubin C et al. (2008). Hypothermia and other cold-related morbidity emergency department visits: United States, 1995–2004. Wilderness Environ Med 19: 233–237.

Bazille C, Megarbane B, Bensimhon D et al. (2005). Brain damage after heat stroke. J Neuropathol Exp Neurol 64: 970–975.

Bennett M, Mitchell S, Dominguez A (2003). Adjunctive treatment of decompression illness with a non-steroidal anti-inflammatory drug (tenoxicam) reduces compression requirement. Undersea Hyperb Med 30: 195–205.

Bennett MH, Lehm JP, Mitchell SJ et al. (2010). Recompression and adjunctive therapy for decompression illness: a systematic review of randomized controlled trials. Anesth Analg 111: 757–762.

Bennett MH, Lehm JP, Mitchell SJ et al. (2012). Recompression and adjunctive therapy for decompression illness. Cochrane Database. Syst Rev 5. Cd005277.

Beppu T (2014). The role of MR imaging in assessment of brain damage from carbon monoxide poisoning: a review of the literature. AJNR Am J Neuroradiol 35: 625–631.

Bernard SA, Gray TW, Buist MD et al. (2002). Treatment of comatose survivors of out-of-hospital cardiac arrest with induced hypothermia. N Engl J Med 346: 557–563.

Betterman K, Patel S (2014). Neurologic complications of carbon monoxide intoxication. Handb Clin Neurol 120: 971–979.

Bhatia R, Chacko F, Lal V et al. (2007). Reversible delayed neuropsychiatric syndrome following acute carbon monoxide exposure. Indian J Occup Environ Med 11: 80–82.

Billinger M, Zbinden R, Mordasini R et al. (2011). Patent foramen ovale closure in recreational divers: effect on decompression illness and ischaemic brain lesions during long-term follow-up. Heart 97: 1932–1937.

Blanchet B, Jullien V, Vinsonneau C et al. (2008). Influence of burns on pharmacokinetics and pharmacodynamics of drugs used in the care of burn patients. Clin Pharmacokinet 47: 635–654.

Borse NN, Gilchrist J, Dellinger AM et al. (2008). CDC childhood injury report: patterns of unintentional injuries among 0–19 year olds in the United States, 2000–2006, U. S. Department of Health and Human Services Centers for Disease Control and Prevention, Atlanta, GA.

Bouchama A, Knochel JP (2002). Heat stroke. N Engl J Med 346: 1978–1988.

Bove AA (2014). Diving medicine. Am J Respir Crit Care Med 189: 1479–1486.

Bratton SL, Jardine DS, Morray JP (1994). Serial neurologic examinations after near drowning and outcome. Arch Pediatr Adolesc Med 148: 167–170.

Breugem CC, Van Hertum W, Groenevelt F (1999). High voltage electrical injury leading to a delayed onset tetraplegia, with recovery. Ann N Y Acad Sci 888: 131–136.

Brown RL, Henke A, Greenhalgh DG et al. (1996). The use of haloperidol in the agitated, critically ill pediatric patient with burns. J Burn Care Rehabil 17: 34–38.

Brown DJ, Brugger H, Boyd J et al. (2012). Accidental hypothermia. N Engl J Med 367: 1930–1938.

Browne JB, Gaasch WR (1992). Electrical injuries and lightning. Emerg Med Clin North Am 10 (2): 211–229.

Buckley NA, Juurlink DN, Isbister G et al. (2011). Hyperbaric oxygen for carbon monoxide poisoning. Cochrane Database. Syst Rev. Cd002041.

CDC (1998). Lightning-associated deaths – United States, 1980–1995. MMWR Morb Mortal Wkly Rep 47: 391–394.

CDC (2002). Heat-related deaths – four states, July–August 2001, and United States, 1979–1999. MMWR Morb Mortal Wkly Rep 51: 567–570.

Chapman TT, Richard RL, Hedman TL et al. (2008). Military return to duty and civilian return to work factors following burns with focus on the hand and literature review. J Burn Care Res 29: 756–762.

Chen NC, Huang CW, Lui CC et al. (2013). Diffusion-weighted imaging improves prediction in cognitive outcome and clinical phases in patients with carbon monoxide intoxication. Neuroradiology 55: 107–115.

Cherington M (2005). Spectrum of neurologic complications of lightning injuries. NeuroRehabilitation 20: 3–8.

Christensen JA, Sherman RT, Balis GA et al. (1980). Delayed neurologic injury secondary to high-voltage current, with recovery. J Trauma 20: 166–168.

Cooper MA (1995). Emergent care of lightning and electrical injuries. Semin Neurol 15: 268–278.

Dematte JE, O'Mara K, Buescher J et al. (1998). Near-fatal heat stroke during the 1995 heat wave in Chicago. Ann Intern Med 129: 173–181.

Dubowitz DJ, Bluml S, Arcinue E et al. (1998). MR of hypoxic encephalopathy in children after near drowning: correlation with quantitative proton MR spectroscopy and clinical outcome. AJNR Am J Neuroradiol 19: 1617–1627.

Ely EW, Shintani A, Truman B et al. (2004). Delirium as a predictor of mortality in mechanically ventilated patients in the intensive care unit. JAMA 291: 1753–1762.

Epstein E, Anna K (2006). Accidental hypothermia. BMJ 332: 706–709.

Farina Jr JA, Rosique MJ, Rosique RG (2013). Curbing inflammation in burn patients. Int J Inflam 2013: 715645.

Farstad M, Andersen KS, Koller ME et al. (2001). Rewarming from accidental hypothermia by extracorporeal circulation. A retrospective study. Eur J Cardiothorac Surg 20: 58–64.

Ferreiro I, Melendez J, Regalado J et al. (1998). Factors influencing the sequelae of high tension electrical injuries. Burns 24: 649–653.

Flierl MA, Stahel PF, Touban BM et al. (2009). Bench-to-bedside review: burn-induced cerebral inflammation – a neglected entity? Crit Care 13: 215.

Formisano R, Zasler ND (2014). Posttraumatic parkinsonism. J Head Trauma Rehabil 29: 387–390.

Freeman CB, Goyal M, Bourque PR (2004). MR imaging findings in delayed reversible myelopathy from lightning strike. AJNR Am J Neuroradiol 25: 851–853.

Fushimi Y, Taki H, Kawai H et al. (2012). Abnormal hyperintensity in cerebellar efferent pathways on diffusion-weighted imaging in a patient with heat stroke. Clin Radiol 67: 389–392.

Giacino JT, Whyte J, Bagiella E et al. (2012). Placebo-controlled trial of amantadine for severe traumatic brain injury. N Engl J Med 366: 819–826.

Gilbert M, Busund R, Skagseth A et al. (2000). Resuscitation from accidental hypothermia of 13.7 degrees C with circulatory arrest. Lancet 355: 375–376.

Gomez CR (2014). Disorders of body temperature. Handb Clin Neurol 120: 947–957.

Gordon CR, Siemionow M, Papay F et al. (2009). The world's experience with facial transplantation: what have we learned thus far? Ann Plast Surg 63: 572–578.

Guenther U, Varelmann D, Putensen C et al. (2009). Extended therapeutic hypothermia for several days during extracorporeal membrane-oxygenation after drowning and cardiac arrest: two cases of survival with no neurological sequelae. Resuscitation 80: 379–381.

Haberal MA, Gurer S, Akman N et al. (1996). Persistent peripheral nerve pathologies in patients with electric burns. J Burn Care Rehabil 17: 147–149.

HACA (2002). Mild therapeutic hypothermia to improve the neurologic outcome after cardiac arrest. N Engl J Med 346: 549–556.

Handa PK, Tai DY (2005). Carbon monoxide poisoning: a five year review at Tan Tock Seng Hospital, Singapore. Ann Acad Med Singapore 34: 611–614.

Hawes J, Massey EW (2008). Neurologic injuries from scuba diving. Neurol Clin 26: 297–308. xii.

Heled Y, Rav-Acha M, Shani Y et al. (2004). The "golden hour" for heatstroke treatment. Mil Med 169: 184–186.

Helm PA, Pandian G, Heck E (1985). Neuromuscular problems in the burn patient: cause and prevention. Arch Phys Med Rehabil 66: 451–453.

Holck P, Hunter RW (2006). NIHSS applied to cerebral neurological dive injuries as a tool for dive injury severity stratification. Undersea Hyperb Med 33: 271–280.

Ibsen LM, Koch T (2002). Submersion and asphyxial injury. Crit Care Med 30: S402–S408.

Istre GR, McCoy MA, Moore BJ et al. (2014). Preventing deaths and injuries from house fires: an outcome evaluation of a community-based smoke alarm installation programme. Inj Prev 20: 97–102.

Jacobs SE, Berg M, Hunt R et al. (2013). Cooling for newborns with hypoxic ischaemic encephalopathy. Cochrane Database Syst Rev 1. Cd003311.

Jasper BW, Hopkins RO, Duker HV et al. (2005). Affective outcome following carbon monoxide poisoning: a prospective longitudinal study. Cogn Behav Neurol 18: 127–134.

Josefsen R, Wester K (1999). Cerebellar hemorrhage – a rare, but serious complication in decompression disease. Tidsskr Nor Laegeforen 119: 3901–3902.

Jumbelic MI (1995). Forensic perspectives of electrical and lightning injuries. Semin Neurol 15: 342–350.

Kamtchum Tatuene J, Pignel R, Pollak P et al. (2014). Neuroimaging of diving-related decompression illness: current knowledge and perspectives. AJNR Am J Neuroradiol 35: 2039–2044.

Khedr EM, Khedr T, el-Oteify MA et al. (1997). Peripheral neuropathy in burn patients. Burns 23: 579–583.

Kim RC, Smith HR, Henbest ML et al. (1984). Nonhemorrhagic venous infarction of the spinal cord. Ann Neurol 15: 379–385.

Kint PA, Stroy JP, Parizel PM (1999). Basal ganglia hemorrhage secondary to lightning stroke. Jbr-btr 82: 113.

Ko SH, Chun W, Kim HC (2004). Delayed spinal cord injury following electrical burns: a 7-year experience. Burns 30: 691–695.

Kornberger E, Schwarz B, Lindner KH et al. (1999). Forced air surface rewarming in patients with severe accidental hypothermia. Resuscitation 41: 105–111.

Kosgallana AD, Mallik S, Patel V et al. (2013). Heat stroke induced cerebellar dysfunction: a "forgotten syndrome". World J Clin Cases 1: 260–261.

Koumbourlis AC (2002). Electrical injuries. Crit Care Med 30: S424–S430.

Kruus S, Bergstrom L, Suutarinen T et al. (1979). The prognosis of near-drowned children. Acta Paediatr Scand 68: 315–322.

Ladd G, Stepan V, Stevens L (2002). The Abacus project: establishing the risk of recreational scuba death and decompression illness. South Pacific Underwater Medicine Society Journal 32: 124–128.

Lammertse DP (2005). Neurorehabilitation of spinal cord injuries following lightning and electrical trauma. NeuroRehabilitation 20: 9–14.

Lavelle JM, Shaw KN (1993). Near drowning: is emergency department cardiopulmonary resuscitation or intensive care unit cerebral resuscitation indicated? Crit Care Med 21: 368–373.

Lawson-Smith P, Jansen EC, Hyldegaard O (2011). Cyanide intoxication as part of smoke inhalation – a review on diagnosis and treatment from the emergency perspective. Scand J Trauma Resusc Emerg Med 19: 14.

Lee MS, Marsden CD (1994). Neurological sequelae following carbon monoxide poisoning clinical course and outcome according to the clinical types and brain computed tomography scan findings. Mov Disord 9: 550–558.

Lee RC, Aarsvold JN, Chen W et al. (1995). Biophysical mechanisms of cell membrane damage in electrical shock. Semin Neurol 15: 367–374.

Lee JS, Choi JC, Kang SY et al. (2009a). Heat stroke: increased signal intensity in the bilateral cerebellar dentate nuclei and splenium on diffusion-weighted MR imaging. AJNR Am J Neuroradiol 30: E58.

Lee MY, Liu G, Kowlowitz V et al. (2009b). Causative factors affecting peripheral neuropathy in burn patients. Burns 35: 412–416.

Lexow K (1991). Severe accidental hypothermia: survival after 6 hours 30 minutes of cardiopulmonary resuscitation. Arctic Med Res 50 (Suppl 6): 112–114.

Lopez RE, Holle RL (1995). Demographics of lightning casualties. Semin Neurol 15: 286–295.

Lu-Emerson C, Khot S (2010). Neurological sequelae of hypoxic-ischemic brain injury. NeuroRehabilitation 26: 35–45.

Mair P, Kornberger E, Furtwaengler W et al. (1994). Prognostic markers in patients with severe accidental hypothermia and cardiocirculatory arrest. Resuscitation 27: 47–54.

Manabe Y, Sakai K, Kashihara K et al. (1998). Presumed venous infarction in spinal decompression sickness. AJNR Am J Neuroradiol 19: 1578–1580.

Marquez S, Turley JJ, Peters WJ (1993). Neuropathy in burn patients. Brain 116 (Pt 2): 471–483.

Mason ST, Esselman P, Fraser R et al. (2012). Return to work after burn injury: a systematic review. J Burn Care Res 33: 101–109.

Massey EW, Moon RE (2014). Neurology and diving. Handb Clin Neurol 120: 959–969.

Matt SE, Shupp JW, Carter EA et al. (2012). When a hero becomes a patient: firefighter burn injuries in the National Burn Repository. J Burn Care Res 33: 147–151.

Medina 3rd MA, Moore DA, Cairns BA (2015). A case series: bilateral ischemic optic neuropathy secondary to large volume fluid resuscitation in critically ill burn patients. Burns 41: e19–e23.

Messing B (1991). Extrapyramidal disturbances after cyanide poisoning (first MRT-investigation of the brain). J Neural Transm Suppl 33: 141–147.

Mian MA, Mullins RF, Alam B et al. (2011). Workplace-related burns. Ann Burns Fire Disasters 24: 89–93.

Mimura K, Harada M, Sumiyoshi S et al. (1999). Long-term follow-up study on sequelae of carbon monoxide poisoning; serial investigation 33 years after poisoning. Seishin Shinkeigaku Zasshi 101: 592–618.

Moon RE, Bove AA (2004). Transcatheter occlusion of patent foramen ovale: a prevention for decompression illness? Undersea Hyperb Med 31: 271–274.

Muehlberger T, Vogt PM, Munster AM (2001). The long-term consequences of lightning injuries. Burns 27: 829–833.

Navy US (2008). Diagnosis and treatment of decompression sickness and arterial gas embolism. US Navy Department, Washington, DC.

Niazi SA, Lewis FJ (1958). Profound hypothermia in man; report of a case. Ann Surg 147: 264–266.

Nielsen N, Wetterslev J, Cronberg T et al. (2013). Targeted temperature management at 33 degrees C versus 36 degrees C after cardiac arrest. N Engl J Med 369: 2197–2206.

Nucci-da-Silva MP, Amaro Jr E (2009). A systematic review of magnetic resonance imaging and spectroscopy in brain injury after drowning. Brain Inj 23: 707–714.

Oliveira L, Fregni F (2011). Pharmacological and electrical stimulation in chronic disorders of consciousness: new insights and future directions. Brain Inj 25: 315–327.

Ookura R, Shiro Y, Takai T et al. (2009). Diffusion-weighted magnetic resonance imaging of a severe heat stroke patient complicated with severe cerebellar ataxia. Intern Med 48: 1105–1108.

Ozdoba C, Weis J, Plattner T et al. (2005). Fatal scuba diving incident with massive gas embolism in cerebral and spinal arteries. Neuroradiology 47: 411–416.

Ozgun B, Castillo M (1995). Basal ganglia hemorrhage related to lightning strike. AJNR Am J Neuroradiol 16: 1370–1371.

Palmu R, Suominen K, Vuola J et al. (2011). Mental disorders after burn injury: a prospective study. Burns 37: 601–609.

Parkinson RB, Hopkins RO, Cleavinger HB et al. (2002). White matter hyperintensities and neuropsychological outcome following carbon monoxide poisoning. Neurology 58: 1525–1532.

Pauley E, Lishmanov A, Schumann S et al. (2015). Delirium is a robust predictor of morbidity and mortality among critically ill patients treated in the cardiac intensive care unit. Am Heart J 170: 79–86. 86.e71.

Peck M, Pressman MA (2013). The correlation between burn mortality rates from fire and flame and economic status of countries. Burns 39: 1054–1059.

Pepe G, Castelli M, Nazerian P et al. (2011). Delayed neuropsychological sequelae after carbon monoxide poisoning: predictive risk factors in the emergency department. A retrospective study. Scand J Trauma Resusc Emerg Med 19: 16.

Pistoia F, Mura E, Govoni S et al. (2010). Awakenings and awareness recovery in disorders of consciousness: is there a role for drugs? CNS Drugs 24: 625–638.

Polderman KH (2009). Mechanisms of action, physiological effects, and complications of hypothermia. Crit Care Med 37: S186–S202.

Price JD, Grimley Evans J (2002). An N-of-1 randomized controlled trial ('N-of-1 trial') of donepezil in the treatment of non-progressive amnestic syndrome. Age Ageing 31: 307–309.

Primeau M (2005). Neurorehabilitation of behavioral disorders following lightning and electrical trauma. NeuroRehabilitation 20: 25–33.

Prockop LD, Chichkova RI (2007). Carbon monoxide intoxication: an updated review. J Neurol Sci 262: 122–130.

Quinn DK (2014). "Burn catatonia": a case report and literature review. J Burn Care Res 35: e135–e142.

Rachinger J, Fellner FA, Stieglbauer K et al. (2002). MR changes after acute cyanide intoxication. AJNR Am J Neuroradiol 23: 1398–1401.

Rahmani SH, Faridaalaee G, Jahangard S (2015). Acute transient hemiparesis induced by lightning strike. Am J Emerg Med 33: 984.e981–984.e983.

Reisner AD (2013). Possible mechanisms for delayed neurological damage in lightning and electrical injury. Brain Inj 27: 565–569.

Ritenour AE, Morton MJ, McManus JG et al. (2008). Lightning injury: a review. Burns 34 (5): 585–594.

Rosenberg NL, Myers JA, Martin WR (1989). Cyanide-induced parkinsonism: clinical, MRI, and 6-fluorodopa PET studies. Neurology 39: 142–144.

Rosenow F, Herholz K, Lanfermann H et al. (1995). Neurological sequelae of cyanide intoxication – the patterns of clinical, magnetic resonance imaging, and positron emission tomography findings. Ann Neurol 38: 825–828.

Ruttmann E, Weissenbacher A, Ulmer H et al. (2007). Prolonged extracorporeal membrane oxygenation-assisted support provides improved survival in hypothermic patients with cardiocirculatory arrest. J Thorac Cardiovasc Surg 134: 594–600.

Saffle JR, Crandall A, Warden GD (1985). Cataracts: a long-term complication of electrical injury. J Trauma 25: 17–21.

Sanford A, Gamelli RL (2014). Lightning and thermal injuries. Handb Clin Neurol 120: 981–986.

Shankaran S, Pappas A, McDonald SA et al. (2012). Childhood outcomes after hypothermia for neonatal encephalopathy. N Engl J Med 366: 2085–2092.

Shanmugarajah K, Hettiaratchy S, Clarke A et al. (2011). Clinical outcomes of facial transplantation: a review. Int J Surg 9: 600–607.

Shprecher D, Mehta L (2010). The syndrome of delayed posthypoxic leukoencephalopathy. NeuroRehabilitation 26: 65–72.

Silfvast T, Pettila V (2003). Outcome from severe accidental hypothermia in Southern Finland – a 10-year review. Resuscitation 59: 285–290.

Singerman J, Gomez M, Fish JS (2008). Long-term sequelae of low-voltage electrical injury. J Burn Care Res 29: 773–777.

Soar J, Perkins GD, Abbas G et al. (2010). European Resuscitation Council guidelines for resuscitation 2010 Section 8. Cardiac arrest in special circumstances: electrolyte abnormalities, poisoning, drowning, accidental hypothermia, hyperthermia, asthma, anaphylaxis, cardiac surgery, trauma, pregnancy, electrocution. Resuscitation 81: 1400–1433.

Sohn YH, Jeong Y, Kim HS et al. (2000). The brain lesion responsible for parkinsonism after carbon monoxide poisoning. Arch Neurol 57: 1214–1218.

Sparacia G, Banco A, Sparacia B et al. (1997). Magnetic resonance findings in scuba diving-related spinal cord decompression sickness. MAGMA 5: 111–115.

Stanford GK, Pine RH (1988). Postburn delirium associated with use of intravenous lorazepam. J Burn Care Rehabil 9: 160–161.

Steele AN, Grimsrud KN, Sen S et al. (2015). Gap analysis of pharmacokinetics and pharmacodynamics in burn patients: a review. J Burn Care Res 36: e194–e211.

Szpilman D, Bierens JJ, Handley AJ et al. (2012). Drowning. N Engl J Med 366: 2102–2110.

Tamam Y, Tamam C, Tamam B et al. (2013). Peripheral neuropathy after burn injury. Eur Rev Med Pharmacol Sci 17 (Suppl 1): 107–111.

Tan JL, McClure J, Hennington L et al. (2014). The heat is on: a case of hyperthermia-induced posterior reversible encephalopathy syndrome (PRES). Neurol Sci 35: 127–130.

Thaventhiran J, O'Leary MJ, Coakley JH et al. (2001). Pathogenesis and recovery of tetraplegia after electrical injury. J Neurol Neurosurg Psychiatry 71: 535–537.

Theman K, Singerman J, Gomez M et al. (2008). Return to work after low voltage electrical injury. J Burn Care Res 29: 959–964.

Thomason JW, Shintani A, Peterson JF et al. (2005). Intensive care unit delirium is an independent predictor of longer hospital stay: a prospective analysis of 261 non-ventilated patients. Crit Care 9: R375–R381.

Tibbles PM, Perrotta PL (1994). Treatment of carbon monoxide poisoning: a critical review of human outcome studies comparing normobaric oxygen with hyperbaric oxygen. Ann Emerg Med 24: 269–276.

Todnem K, Nyland H, Kambestad BK et al. (1990). Influence of occupational diving upon the nervous system: an epidemiological study. Br J Ind Med 47: 708–714.

Tomaszewski C (1999). Carbon monoxide poisoning. Early awareness and intervention can save lives. Postgrad Med 105: 39.

Torti SR, Billinger M, Schwerzmann M et al. (2004). Risk of decompression illness among 230 divers in relation to the presence and size of patent foramen ovale. Eur Heart J 25: 1014–1020.

Trytko BE, Bennett MH (2008). Arterial gas embolism: a review of cases at Prince of Wales Hospital, Sydney, 1996 to 2006. Anaesth Intensive Care 36: 60–64.

Uitti RJ, Rajput AH, Ashenhurst EM et al. (1985). Cyanide-induced parkinsonism: a clinicopathologic report. Neurology 35: 921–925.

van Beeck EF, Branche CM, Szpilman D et al. (2005). A new definition of drowning: towards documentation and prevention of a global public health problem. Bull World Health Organ 83: 853–856.

van Hulst RA, Klein J, Lachmann B (2003). Gas embolism: pathophysiology and treatment. Clin Physiol Funct Imaging 23: 237–246.

Vanden Hoek TL, Morrison LJ, Shuster M et al. (2010). Part 12: cardiac arrest in special situations: 2010 American Heart Association guidelines for cardiopulmonary resuscitation and emergency cardiovascular care. Circulation 122: S829–S861.

Vann R (2004). Mechanisms and risks of decompression. In: A Bove (Ed.), Bove and Davis' Diving Medicine. Saunders, Philadelphia.

Vann RD, Butler FK, Mitchell SJ et al. (2011). Decompression illness. Lancet 377: 153–164.

Vollmann R, Lamperti M, Magyar M et al. (2011). Magnetic resonance imaging of the spine in a patient with decompression sickness. Clin Neuroradiol 21: 231–233.

Walpoth BH, Walpoth-Aslan BN, Mattle HP et al. (1997). Outcome of survivors of accidental deep hypothermia and circulatory arrest treated with extracorporeal blood warming. N Engl J Med 337: 1500–1505.

Warren Jr LP, Djang WT, Moon RE et al. (1988). Neuroimaging of scuba diving injuries to the CNS. AJR Am J Roentgenol 151: 1003–1008.

Weaver LK (1999). Carbon monoxide poisoning. Crit Care Clin 15: 297.

Weaver LK (2014). Hyperbaric oxygen therapy indications: the Hyperbaric Oxygen Therapy Committee report. 13th edition. Undersea and Hyperbaric Medical Society. Last accessed 10/17/2016. https://www.uhms.org.

Weaver LK, Hopkins RO, Chan KJ et al. (2002). Hyperbaric oxygen for acute carbon monoxide poisoning. N Engl J Med 347: 1057–1067.

WHO (2003). Injuries and violence prevention: noncommunicable diseases and mental health: fact sheet on drowning [Online]. Geneva. Available: http://www.who.int/violence_injury_prevention/other_injury/drowning/en/. (accessed November 18th 2015).

Wilmshurst PT, Treacher DF, Crowther A et al. (1994). Effects of a patent foramen ovale on arterial saturation during exercise and on cardiovascular responses to deep breathing, Valsalva manoeuvre, and passive tilt: relation to history of decompression illness in divers. Br Heart J 71: 229–231.

Yamamoto T, Katayama Y, Obuchi T et al. (2013). Deep brain stimulation and spinal cord stimulation for vegetative state and minimally conscious state. World Neurosurg 80: S30. e31–S30.e39.

Yarnell PR (2005). Neurorehabilitation of cerebral disorders following lightning and electrical trauma. NeuroRehabilitation 20: 15–18.

Yeo TP (2004). Heat stroke: a comprehensive review. AACN Clin Issues 15: 280–293.

Zafren K, Durrer B, Herry JP et al. (2005). Lightning injuries: prevention and on-site treament in mountains and remote areas. Official guidelines of the International Commission for Mountain Emergency Medicine and the Medical Commission of the International Mountaineering and Climbing Federation (ICAR and UIAA MEDCOM). Resuscitation 65 (3): 369–372.

Zeng Y, Zhang Y, Xin G et al. (2010). Cortical blindness – a rare complication of severe burns. A report of seven cases and review of the literature. Burns 36: e1–e3.

Chapter 38

Neurologic manifestations of major electrolyte abnormalities

M. DIRINGER*
Department of Neurology, Washington University, St. Louis, MO, USA

Abstract

The brain operates in an extraordinarily intricate environment which demands precise regulation of electrolytes. Tight control over their concentrations and gradients across cellular compartments is essential and when these relationships are disturbed neurologic manifestations may develop.

Perturbations of sodium are the electrolyte disturbances that most often lead to neurologic manifestations. Alterations in extracellular fluid sodium concentrations produce water shifts that lead to brain swelling or shrinkage. If marked or rapid they can result in profound changes in brain function which are proportional to the degree of cerebral edema or contraction. Adaptive mechanisms quickly respond to changes in cell size by either increasing or decreasing intracellular osmoles in order to restore size to normal. Unless cerebral edema has been severe or prolonged, correction of sodium disturbances usually restores function to normal. If the rate of correction is too rapid or overcorrection occurs, however, new neurologic manifestations may appear as a result of osmotic demyelination syndrome.

Disturbances of magnesium, phosphate and calcium all may contribute to alterations in sensorium. Hypomagnesemia and hypocalcemia can lead to weakness, muscle spasms, and tetany; the weakness from hypophosphatemia and hypomagnesemia can impair respiratory function. Seizures can be seen in cases with very low concentrations of sodium, magnesium, calcium, and phosphate.

The nervous system operates in an exquisitely complex environment which requires precise distribution and concentration of electrolytes in various compartments. Normal neurologic function requires tight control of their gradients across cellular compartments and when these relationships are disturbed neurologic manifestations may occur (Table 38.1). In addition electrolyte disturbances can have a profound effect on brain function when they lead to fluid shifts that cause cells to swell and shrink.

Perturbations of sodium are the electrolyte disturbance that most often leads to neurologic manifestations. They are particularly common in patients with central nervous system (CNS) disease due to the essential role the CNS plays in regulation of sodium and water homeostasis. Additionally, seriously ill neurologic patients are often administered treatments designed to perturb normal regulation of electrolytes and water.

PHYSIOLOGIC REGULATION OF SODIUM AND OSMOLALITY

The manifestations of sodium disturbances are best understood in the context of the physiologic regulation of extracellular osmolality and body water.

Water regulation

Changes in osmolality are detected by specialized neurons called osmoreceptors in the anterior hypothalamus (Leng et al., 1985). When an osmotic gradient develops between the intracellular and extracellular compartments, water moves passively to maintain osmolal equilibrium and the size changes. The osmoreceptors communicate this change to the magnocellular neurons of the supraoptic and paraventricular nuclei of the hypothalamus which synthesize antidiuretic hormone (ADH,

---

*Correspondence to: Michael Diringer, MD, Department of Neurology, Campus box 8111, 660 S. Euclid Ave., St. Louis MO 63110, USA. Tel: +1-314-362-2999, E-mail: diringerm@wustl.edu

Table 38.1

Electrolyte disturbances

Electrolyte disturbance	Causes	Signs/symptoms
Hyponatremia	Water retention SIADH Postoperative Heart failure Cirrhosis Sodium loss Gastrointestinal losses Renal losses Adrenal insufficiency	Muscle cramps Weakness Alter sensorium Coma Seizures
Hypernatremia	Diabetes insipidus Excess sweating or diarrhea	Confusion Altered sensorium Coma Seizures
Hypomagnesemia	Diuretics Osmotic agents Hypervolemic therapy Metabolic acidosis	Muscle spasms/weakness Paresthesia Disorientation/coma Seizures Arrhythmias Hypotension Heart failure
Hypomagnesemia	Renal failure Iatrogenic	Weakness
Hypophosphatemia	Diuretics Osmotic agents Acidemia Hypomagnesemia Hyperparathyroidism	Muscle weakness Tremor Altered sensorium Rhabdomyolysis Hemolysis
Hypocalcemia	Diuretics Osmotic agents Hypervolemic therapy Sepsis Chelators Hyperphosphatemia	Hypotension Cardiac insufficiency Bradycardia Muscle spasms Tetany

SIADH, syndrome of inappropriate secretion of antidiuretic hormone.

also known as arginine vasopressin). The peptide travels down their axons, passing through the median eminence, to the neurohypophysis or the posterior lobe of the pituitary, where they release ADH into the circulation. Stimulation of this system by as little as a 1% increase in extracellular fluid (ECF) osmolality causes secretion of ADH (Robertson et al., 1977). Maximal osmotic stimulation of ADH release occurs at approximately 310 mosmol/kg; ADH concentrations fall to almost zero when osmolality is below about 280 mosmol/kg. The system is sensitive to administration of hypertonic solutions – hyperosmolar solutions that do not easily cross the cell membrane such as sodium, glucose, or mannitol and produce water shifts. Hyperosmolar urea, on the other hand, is not hypertonic since it crosses the cell membrane and does not produce water shifts or change in cell size and does not stimulate ADH release.

The magnocellular neurons that synthesize and release ADH also respond to large changes in intravascular volume and blood pressure. Low-pressure baroreceptors in the left atrium of the heart and high-pressure baroreceptors in the aortic arch and carotid sinus monitor blood pressure and volume. This information is transmitted via cranial nerves IX and X to relay stations in the brainstem, the nucleus tractus solitarius, and ultimately to the hypothalamus (Share and Levy, 1962; Coenraad et al., 2001). Input from baroreceptors indicating moderate to severe hypotension or hypovolemia will overwhelm osmotic regulation and produce massive release of ADH (Dunn et al., 1973). Therefore, in the setting of hypovolemia, ADH levels are higher than would be seen for a given osmolality. This acts to preserve intravascular volume at the expense of osmolality. Additional nonosmotic stimuli to ADH release include physical pain, emotional stress, nausea, emesis, and hypoxia (Robertson, 1976). The presence of these factors often leads to high ADH levels in trauma and postoperative patients (Robertson et al., 1977).

Circulating ADH stimulates receptors in the kidneys and blood vessels. It increases water permeability in the final portion of the distal tubule and collecting duct of nephrons, resulting in water retention, a fall in ECF osmolality, which in turn leads to suppression of ADH secretion. ADH also acts to increase vascular resistance (Cowley, 1985), helping preserve blood pressure during hypovolemia or hypotension. In pharmacologic doses it can act as a vasopressor.

Even with maximal renal water conservation, water intake is essential to maintain homeostasis. Elevations of osmolality over approximately 300 mosmol/kg stimulate thirst (Vokes and Robertson, 1988; Ober, 1991), initiating behavioral responses directed toward increasing water intake. Under normal circumstances, the response to thirst is potent enough to maintain normal water balance even in the absence of ADH release.

Sodium regulation

Regulation of body sodium content is directed toward maintaining adequate blood pressure and intravascular volume. High- and low-pressure baroreceptors monitor these parameters and relay information to hypothalamic centers which integrate and coordinate a multifaceted response. Both neural and humeral messengers are used to produce renal, vascular, and behavioral changes directed toward restoring homeostasis.

Sodium regulation involves opposing systems. The renin–angiotensin–aldosterone system and the sympathetic nervous system promote sodium retention and vasoconstriction. They are opposed by a number of natriuretic factors that promote renal sodium loss and relaxation of vascular smooth muscle, producing vasodilation.

Blood pressure is monitored by renal baroreceptors located in the preglomerular arterioles. When blood pressure falls the baroreceptors stimulate the release of renin. The sodium concentration of the distal tubule of the kidney also acts to influence renin release. Another important modulator of renin release is sympathetic input via the renal nerve. Acting via angiotensinogen and angiotensin I, renin stimulates the production of angiotensin II which in turn induces aldosterone secretion from the adrenal medulla. Additional actions of angiotensin II include stimulating thirst (Lappe et al., 1986) and acting as a potent vasoconstrictor. Aldosterone stimulates the distal tubule of the nephron to retain sodium in exchange for potassium and hydrogen.

Under normal conditions about 98% of the sodium filtered by the glomerulus is reabsorbed in the proximal tubule. Minor changes in how much is reabsorbed can lead to large changes in sodium excretion. In addition to its role in influencing renin release, the sympathetic nervous system has an important modulatory influence on sodium reabsorption in the proximal tubule. (DiBona, 1977, 1985) Thus, neural mechanisms alter sodium excretion by acting on both the renin–angiotensin–aldosterone system and the kidney.

Opposing systems promote sodium excretion and vasodilation. The prototype, atrial natriuretic factor (ANF), acts both centrally and peripherally to promote sodium and volume loss. It is released from the heart by rises in atrial pressure (Lang et al., 1985; Laragh, 1985). Experimental infusions of ANF produce natriuresis and diuresis, relax vascular smooth muscle, and antagonize the release of ADH, renin, and aldosterone (Laragh, 1985). Central administration of ANF produces natriuresis and diuresis (Samson, 1987), inhibits angiotensin II-induced drinking, and inhibits ADH release. Brain natriuretic peptide, another centrally and peripherally acting peptide, plays a similar role in regulating sodium homeostasis.

PATHOPHYSIOLOGY AND ADAPTIVE RESPONSES TO SODIUM DISTURBANCES

Hyponatremia

Hyponatremia develops when there is an excess of extracellular water relative to sodium. In congestive heart failure and cirrhosis a reduction in effective arterial blood volume stimulates release of ADH and aldosterone and increases sympathetic nervous system activity (Schrier, 1988a, b). This state leads to hyponatremia by increasing the body's water content more than it increases sodium content.

Hyponatremia can also exist when there is an increase in water but no change in sodium content. Excessive water ingestion can lead to acute hyponatremia such as seen in psychiatric patients (psychogenic polydipsia), overhydrated runners (exercise-associated hyponatremia) (Urso et al., 2014), and consumption of substantial amounts of beer (beer potomania) (Sanghvi et al., 2007). Excessive water retention leading to dilutional hyponatremia also develops when there is nonosmotic release of ADH stimulated by pain, stress, and drugs, or in the syndrome of inappropriate secretion of ADH (SIADH) (Lester and Nelson, 1981).

Finally, hemodynamic factors, independent of osmolality, influence ADH secretion. Moderate or marked hypovolemia or hypotension is a potent stimulus to

ADH release (Schrier et al., 1979). In this setting ADH is acting to restore hemodynamic homeostasis at the expense of osmotic regulation. Gastrointestinal and renal losses of sodium in excess of water can also produce hyponatremia. Renal loss may be due to diuretic excess (especially thiazides), mineralocorticoid or glucocorticoid deficiency, hypothyroidism, or salt-losing nephritis.

In conditions where multiple nonosmotic stimuli to release ADH exist, intake of hypotonic fluids can rapidly lead to dilutional hyponatremia. Postoperative patients are at a relatively high risk for hyponatremia as stress, nausea, pain, volume contraction, and medications all promote nonosmotic ADH secretion (Chung et al., 1986). Administration of hypotonic fluids during a period of nonosmotic release of ADH can lead to severe symptomatic hyponatremia and even respiratory arrest (Arieff, 1986).

CEREBRAL ADAPTATION TO HYPONATREMIA

Large rapid falls in sodium concentration lead to marked cellular swelling and global cerebral edema (Holliday et al., 1968; Rymer and Fishman, 1973). Within hours, the swelling triggers the initiation of adaptive mechanisms to restore homeostasis which continue to act over the course of several days. Initially, interstitial fluid pressure rises, which drives brain ECF into the cerebrospinal fluid space (Pullen et al., 1987). Adaptive mechanisms allow electrolytes to exit brain cells within hours (rapid adaptation) and intracellular potassium is lost rapidly through activation of ion channels, limiting the adverse consequences of cellular swelling (Strange and Jackson, 1995). As brain cells begin to lose solutes an osmotic gradient is created which drives movement of water out of the cells (Strange, 1992; McManus et al., 1995). Later organic solutes are lost and substantial depletion of brain organic osmolytes occurs within 24 hours. Osmolytes, including amino acids (e.g., taurine, glutamine, glutamate, and aspartate), polyalcohols, and methylamine, are extruded (Massieu et al., 2004).

Hypernatremia

Hypernatremia results when there is a shift in the ratio of sodium to water that favors less water and/or more sodium. The primary causes of extrarenal water loss are sweating and diarrhea. Renal loss of water may occur through osmotic diuresis from hyperglycemia or administration of mannitol, nephrogenic diabetes insipidus when kidney fails to respond to ADH, or insufficient level of circulating ADH (central diabetes insipidus). Despite considerable water loss, hypernatremia will generally not develop when thirst mechanisms are intact and access to water is unrestricted. Primary hyperaldosteronism, Cushing's syndrome, and excessive sodium administration in hypertonic intravenous or oral solutions can result in hypernatremia through retention of sodium in excess of water.

CEREBRAL ADAPTATION TO HYPERNATREMIA

A rise in ECF osmolality leads to rapid movement of water out of the cellular compartment, leading to cell shrinkage. While this has little effect on other organs, CNS function can be severely affected. Within several hours the brain begins to adapt to this hyperosmolar stress. The number of intracellular osmotically active particles is increased in order to draw in water and return cell size to normal.

The first step involves the uptake of solutes by the cells. This acts to move water from the cerebrospinal fluid, through the interstitial space, to the intracellular compartment, restoring the cell size. The uptake of solutes begins with sodium, potassium, and chloride (Melton et al., 1987), followed by idiogenic osmoles (McDowell et al., 1955) which are, in part, made up of myo-inositol, glutamine, glutamate, and taurine.

This phenomenon has important therapeutic implications. When the brain has been exposed to hypertonic conditions sustained for several hours to days, compensatory mechanisms have been activated. The accumulation of osmolytes makes the brain very susceptible to cellular swelling from hypotonic solutions. Therefore, such patients should have their hyperosmolal state corrected very slowly, usually over about a week, to prevent rebound cerebral edema (Fig. 38.1).

Hypernatremia usually results from the combination of an insufficient release of ADH from the neurohypophysis and inadequate water intake. Disturbed water regulation in adults is usually characterized by polyuria, thirst, polydipsia, and occasionally a preference for ice water (Berl et al., 1976). Water intake may be insufficient to keep up with losses when thirst mechanisms are disturbed or access to water is limited due to paresis, disorientation, or obtundation (Hammond et al., 1986). In rare cases, severe hypernatremia can result in brain dehydration severe enough to tear the bridging veins, causing subdural hematomas.

Iatrogenic causes of hypernatremia in hospitalized patients may be inadvertent or deliberate. Failure to pay strict attention to water and electrolyte losses and adjust fluid appropriately can lead to hypernatremia in patients with unrecognized large free-water losses. In addition, the use of osmotic therapy with mannitol or hypertonic saline in neurologic patients is a frequent cause of hypernatremia; mannitol acts by increasing water loss and hypertonic saline through raising sodium content relative to water.

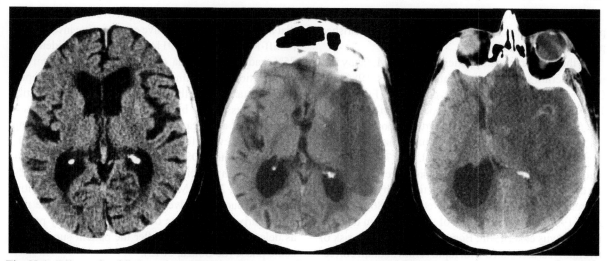

Fig. 38.1. Effects of rapidly lowering sodium in a hypernatremic patient. Following admission for a large hemispheric stroke (left panel), this 75-year-old patient was empirically treated with osmotic therapy and sodium rose to 168 mEq/L on day 2 (middle panel). On day 4 the family chose to limit care to comfort measures and fluids were changed to one-half normal saline. Thirty-six hours later the patient had become unresponsive and a computed tomography scan demonstrated a dramatic increase in edema and shift (right panel).

NEUROLOGIC MANIFESTATIONS

Acute hyponatremia

The symptoms of hyponatremia are primarily neurologic and their severity is determined by the degree of cerebral edema. In rapidly developing hyponatremia water enters the intracellular compartment of the brain and causes rapid cerebral swelling (Strange, 1992). Initially there is a reduction in cerebrospinal fluid and blood volume to compensate for the swelling and keep intracranial pressure from rising; however, given the limited space inside the cranial vault, those mechanisms are rapidly exhausted and intracranial pressure rises. As pressure rises cerebral perfusion is reduced and eventually, if the progression is not reversed, fails entirely, causing global ischemic infarction. While there is a correlation between symptoms and degree of hyponatremia, the rapidity with which the hyponatremia develops is probably a more important determinant of clinical symptoms (Holliday et al., 1968; Cowley, 1985; Rossi and Cadnapaphornchai, 1987).

Acute hyponatremia is most often seen in the setting of acute excessive water ingestion that may be self-induced, as in psychogenic polydipsia, beer potomania, and exercise-associated hyponatremia. Iatrogenic causes include administration of hypotonic fluids to postoperative or trauma patients. Finally an acute "relative hyponatremia" can occur when hyperosmolar patients, often as a result of aggressive osmotic therapy, have their osmolality rapidly lowered toward normal.

The earliest clinical manifestations of acute hyponatremia include nausea and malaise, which can be seen when sodium levels fall to < 125–130 mEq/L. With further declines, headache, lethargy, and obtundation follow. In severe cases seizures, coma, and respiratory arrest may occur when serum sodium concentration rapidly falls below 115–120 mEq/L (Ayus et al., 1992; Ellis, 1995; Sterns, 2015). Patients with pre-existing focal intracranial disease may be particularly sensitive to sudden drops in sodium concentration (regardless of starting sodium level), especially vasogenic lesions such as tumors, ischemic stroke, and inflammation. They will typically exhibit worsening of pre-existing signs followed by altered sensorium progressing to coma.

Chronic hyponatremia

When the onset of hyponatremia is slower and adaptive mechanisms have sufficient time to minimize edema formation, symptoms are limited to mild cognitive impairment or may be absent. This accounts for the asymptomatic presentation (Sterns et al., 1986) in cases of slowly developing, chronic hyponatremia. Symptoms that may appear are relatively nonspecific and include headache, nausea, vomiting, dizziness, gait disturbances, forgetfulness, confusion, muscle cramps, lethargy, and irritability (Sterns et al., 1994; Ellis, 1995).

Central pontine myelinolysis and osmotic demyelination syndrome

For decades the management of hyponatremia has been complicated by fear of inducing central pontine myelinolysis (CPM). CPM, also referred to as osmotic demyelination syndrome (ODS) (Sterns et al., 1986, 2010), is a demyelinating condition, sometimes restricted to the pons, originally described in alcoholic and malnourished patients (Adams et al., 1959).

In the initial reports of patients with CPM, many had normal serum sodium concentrations (Narins, 1986) and CPM was linked to hypoxia, liver disease (Estol et al., 1989), and hyperosmolality independent of hyponatremia (McKee et al., 1988). As numerous reports appeared associating the condition with hyponatremia and its treatment (Sterns et al., 1986; Ayus et al., 1987), two explanations were offered as to the mechanism responsible for the disorder: either too rapid correction of hyponatremia (Norenberg et al., 1982; Sterns et al., 1986; Brunner et al., 1990) or overcorrection of serum sodium concentration (Ayus et al., 1985; Boon et al., 1988). The risk of myelinolysis associated with rapid serum sodium correction increases when the brain's adaptive responses to cerebral edema have been activated for a period of time, as in chronic rather than acute hyponatremia. Clinically, ODS is more frequently associated with rapid correction or overcorrection of chronic but not acute hyponatremia. Correction of hyponatremia that has only been present for several hours has a very low risk of inducing ODS since cerebral adaptation is at an early stage.

The pathogenesis of ODS appears to be related to the brain's adaptive response to hyponatremia. The solutes that were lost from the intracellular compartment to adapt to the hyponatremic state require time to replace. If the rise of sodium concentration outpaces the ability to replace lost intracellular solutes the brain rapidly swings from a swollen to a shrunken state. This, along with possible injury to the blood–brain barrier, may damage astrocytes, lead to impaired myelin production by oligodendrocytes, release inflammatory cytokines, and activate microglia (Gankam Kengne et al., 2011).

The clinical signs of ODS generally do not appear until 2–6 days after correction of hyponatremia (Sterns et al., 1986, 1994, 2010). The clinical picture is that of initial neurologic improvement, with the rise in sodium concentration followed by deterioration days later.

The entity of CPM was first described by Adams and colleagues in 1959 and they state that:

> *The neurologic illness was characterized by flaccid quadriplegia, weakness of the face and tongue, and inability to speak and swallow. There was no response to painful stimuli; eventually the corneal reflexes were lost. All of these signs are readily explained by the lesion in the basis pontis.*

Of note, in these patients CPM occurred in the context of alcoholism and malnutrition.

The first cases were identified at autopsy. With the subsequent development of magnetic resonance imaging (MRI), a much broader spectrum of disease has been identified. A more recent retrospective review suggests that about half of patients have extrapontine involvement (Graff-Radford et al., 2011). In the 24 cases identified, 75% had hyponatremia and 75% had alcoholism and malnutrition. Presenting symptoms included encephalopathy, ataxia, dysarthria, eye movement abnormalities, and seizures. Initial brain MRI was normal in 20% of patients, but all became abnormal with serial imaging. A literature review found that the most common predisposing factor to ODS was hyponatremia (78%) and the most common presentation was encephalopathy (39%) (Singh et al., 2014). Movement disorders and cognitive problems due to involvement of the basal ganglia may occur early in the course of the illness, or may present as delayed manifestations after the patient survives the acute phase (de Souza, 2013).

Risk factors for developing ODS in the setting of hyponatremia are the serum sodium concentration at presentation, the duration of the hyponatremia, and the rate of correction. The vast majority of reported cases are in patients with a serum sodium of <120 mEq/L (Sterns et al., 1986, 1994; Barth et al., 2007; Sterns, 2015). Relowering of the serum sodium may be beneficial in patients where the correction has been too rapid (Oya et al., 2001; Perianayagam et al., 2008).

In a review of cases, favorable recovery occurred in about half and death in a quarter of patients. Outcome was considerably worse in liver transplant patients who developed ODS (Graff-Radford et al., 2011). Other recent series have also reported that good recovery from ODS is not uncommon and almost complete recovery from severe symptoms is possible (Menger and Jorg, 1999; Louis et al., 2012; Sai Kiran et al., 2014; Bernhardt et al., 2015).

Initially ODS was thought to have a consistently poor outcome because the diagnosis was only confirmed by necropsy. However, recent series have shown that good recovery from ODS is not infrequent and total, or near-total and at least remission of severe symptoms is possible (Singh et al., 2014; Sai Kiran et al., 2014).

NEUROLOGIC MANIFESTATIONS OF DISTURBANCES OF OTHER IONS

Calcium

Symptoms of hypocalcemia may be mild and nonspecific, including fatigue, hyperirritability, anxiety, and depression. Tetany, the hallmark of hypocalcemia, includes both sensory and muscular symptoms (Tohme and Bilezikian, 1993). It typically begins as perioral numbness, paresthesias of the hands and feet. Early motor symptoms include stiffness and clumsiness, myalgia, and cramps and may progress to include muscle spasms and focal or generalized seizures. In the hands the result is carpopedal spasm and when the glottis is involved the result is laryngismus stridulus, which can cause acute respiratory distress, cyanosis, and death (Chang et al., 2014). Patients in whom the onset of hypocalcemia is gradual tend to have fewer symptoms at the same serum calcium concentration (Tohme and Bilezikian, 1993). Occasionally hypocalcemia may present with generalized or focal seizures (Tohme and Bilezikian, 1993; Mrowka et al., 2004).

One of the classic findings in patients with tetany is the induction of carpopedal spasm by inflation of a sphygmomanometer above systolic blood pressure for 3 minutes (Trousseau's sign). Chvostek's sign consists of eliciting contraction of the ipsilateral facial muscles by tapping the facial nerve just anterior to the ear (Cooper and Gittoes, 2008).

Magnesium

Patients with chronic magnesium deficiency often present with complaints of neuromuscular hyperexcitability manifest by tremor, tetany, and weakness. They demonstrate positive Trousseau and Chvostek's signs, muscle spasms, and muscle cramps (Vallee et al., 1960; Shoback, 2000). In patients who are critically ill, weakness of the respiratory muscles may contribute to respiratory failure (Shiber and Mattu, 2002).

CNS symptoms involve alterations in sensorium, seizures, and involuntary movements. Mental status changes range from apathy and delirium to lethargy and finally coma. Generalized tonic-clonic seizures may occur but may also be multifocal motor in nature. Occasionally involuntary athetoid or choreiform movements are seen.

Magnesium concentration is maintained primarily by renal mechanisms. Diuretics, antibiotics, osmotic agents, sepsis, volume expansion, phosphate depletion, and metabolic acidosis promote renal magnesium loss. Cardiovascular manifestations include arrhythmias, hypotension, increased sensitivity to digitalis, and heart failure. Hypomagnesemia is linked to worse outcome in medical patients (Rubeiz et al., 1993) and correction improves outcome in models of head injury (McIntosh et al., 1989). Magnesium supplementation has been suggested in subarachnoid hemorrhage; however, several trials failed to demonstrate clinical benefit. (Dorhout Mees et al., 2015; Yamamoto et al., 2015).

Phosphate

Phosphate concentration is primarily controlled by intrarenal mechanisms. Factors that contribute to the development of hypophosphatemia include chronic alcoholism, diabetic ketoacidosis, intravenous alimentation with insufficient phosphate, urinary phosphate-wasting syndromes, chronic ingestion of antacids or other phosphate binders, hypomagnesemia, acidemia, hyperparathyroidism, infection, and renal tubular defects, parathyroid hormone, dextrose, digoxin, and chronic administration of corticosteroids.

The clinical expression of hypophosphatemia is primarily due to intracellular phosphate depletion. Manifestations include muscle weakness, tremor, altered mental status, respiratory insufficiency, rhabdomyolysis, and hemolysis (Weisinger and Bellorin-Font, 1998). Severe hypophosphatemia can lead to mild irritability, metabolic encephalopathy, paresthesias, delirium, generalized seizures, and coma (Subramanian and Khardori, 2000). Severe phosphate depletion may also contribute to the development of CPM (Turnbull et al., 2013). Neuromuscular manifestations include proximal myopathy and dysphagia.

REFERENCES

Adams RD, Victor M, Mancall EL (1959). Central pontine myelinolysis: a hitherto undescribed disease occurring in alcoholic and malnourished patients. AMA Arch Neurol Psychiatry 81: 154–172.

Arieff AI (1986). Hyponatremia, convulsions, respiratory arrest, and permanent brain damage after elective surgery in healthy women. N Engl J Med 314: 1529–1535.

Ayus JC, Krothapalli RK, Arieff AI (1985). Changing concepts in treatment of severe symptomatic hyponatremia. Rapid correction and possible relation to central pontine myelinolysis. Am J Med 78: 897–902.

Ayus JC, Krothapalli RK, Arieff AI (1987). Treatment of symptomatic hyponatremia and its relation to brain damage. A prospective study. N Engl J Med 317: 1190–1195.

Ayus JC, Wheeler JM, Arieff AI (1992). Postoperative hyponatremic encephalopathy in menstruant women. Ann Intern Med 117: 891–897.

Barth M, Capelle HH, Weidauer S et al. (2007). Effect of nicardipine prolonged-release implants on cerebral vasospasm and clinical outcome after severe aneurysmal subarachnoid hemorrhage: a prospective, randomized, double-blind phase IIa study. Stroke 38: 330–336.

Berl T, Anderson RJ, McDonald KM et al. (1976). Clinical disorders of water metabolism. Kidney Int 10: 117–132.

Bernhardt M, Pflugrad H, Goldbecker A et al. (2015). Central nervous system complications after liver transplantation: common but mostly transient phenomena. Liver Transpl 21: 224–232.

Boon AP, Carey MP, Salmon MV (1988). Central pontine myelinolysis not associated with rapid correction of hyponatraemia. Lancet 2: 458.

Brunner JE, Redmond JM, Haggar AM et al. (1990). Central pontine myelinolysis and pontine lesions after rapid correction of hyponatremia: a prospective magnetic resonance imaging study. Ann Neurol 27: 61–66.

Chang WT, Radin B, McCurdy MT (2014). Calcium, magnesium, and phosphate abnormalities in the emergency department. Emerg Med Clin North Am 32: 349–366.

Chung HM, Kluge R, Schrier RW et al. (1986). Postoperative hyponatremia. A prospective study. Arch Intern Med 146: 333–336.

Coenraad MJ, Meinders AE, Taal JC et al. (2001). Hyponatremia in intracranial disorders. Neth J Med 58: 123–127.

Cooper MS, Gittoes NJ (2008). Diagnosis and management of hypocalcaemia. BMJ 336: 1298–1302.

Cowley Jr AW (1985). Vasopressin-neural interactions in the control of cardiovascular function. In: RW Schrier (Ed.), Vasopressin. Raven Press, New York.

de Souza A (2013). Movement disorders and the osmotic demyelination syndrome. Parkinsonism Relat Disord 19: 709–716.

DiBona GF (1977). Neurogenic regulation of renal tubular sodium reabsorption. Am J Physiol 233: F73–F81.

DiBona GF (1985). Neural regulation of renal tubular sodium reabsorption and renin secretion. Fed Proc 44: 2816–2822.

Dorhout Mees SM, Algra A, Wong GK et al. (2015). Early magnesium treatment after aneurysmal subarachnoid hemorrhage: individual patient data meta-analysis. Stroke 46: 3190–3193.

Dunn FL, Brennan TJ, Nelson AE et al. (1973). The role of blood osmolality and volume in regulating vasopressin secretion in the rat. J Clin Invest 52: 3212–3219.

Ellis SJ (1995). Severe hyponatraemia: complications and treatment. QJM 88: 905–909.

Estol CJ, Faris AA, Martinez AJ et al. (1989). Central pontine myelinolysis after liver transplantation. Neurology 39: 493–498.

Gankam Kengne F, Nicaise C, Soupart A et al. (2011). Astrocytes are an early target in osmotic demyelination syndrome. J Am Soc Nephrol 22: 1834–1845.

Graff-Radford J, Fugate JE, Kaufmann TJ et al. (2011). Clinical and radiologic correlations of central pontine myelinolysis syndrome. Mayo Clin Proc 86: 1063–1067.

Hammond DN, Moll GW, Robertson GL et al. (1986). Hypodipsic hypernatremia with normal osmoregulation of vasopressin. N Engl J Med 315: 433–436.

Holliday MA, Kalayci MN, Harrah J (1968). Factors that limit brain volume changes in response to acute and sustained hyper- and hyponatremia. J Clin Invest 47: 1916–1928.

Lang RE, Tholken H, Ganten D et al. (1985). Atrial natriuretic factor – a circulating hormone stimulated by volume loading. Nature 314: 264–266.

Lappe RW, Dinish JL, Bex F et al. (1986). Effects of atrial natriuretic factor on drinking responses to central angiotensin II. Pharmacol Biochem Behav 24: 1573–1576.

Laragh JH (1985). Atrial natriuretic hormone, the renin–aldosterone axis, and blood pressure-electrolyte homeostasis. N Engl J Med 313: 1330–1340.

Leng G, Dyball R, Mason W (1985). Electrophysiology of osmoreceptors. In: W Schrier (Ed.), Vasopressin. Raven Press, New York.

Lester M, Nelson PB (1981). Neurological aspects of vasopressin release and the syndrome of inappropriate release of antidiuretic hormone. Neurosurg 8: 735–740.

Louis G, Megarbane B, Lavoue S et al. (2012). Long-term outcome of patients hospitalized in intensive care units with central or extrapontine myelinolysis. Crit Care Med 40: 970–972.

Massieu L, Montiel T, Robles G et al. (2004). Brain amino acids during hyponatremia in vivo: clinical observations and experimental studies. Neurochem Res 29: 73–81.

McDowell ME, Wolf AV, Steer A (1955). Osmotic volumes of distribution; idiogenic changes in osmotic pressure associated with administration of hypertonic solutions. Am J Physiol 180: 545–558.

McIntosh TK, Vink R, Yamakami I et al. (1989). Magnesium protects against neurological deficit after brain injury. Brain Res 482: 252–260.

McKee AC, Winkelman MD, Banker BQ (1988). Central pontine myelinolysis in severely burned patients: relationship to serum hyperosmolality. Neurology 38: 1211–1217.

McManus ML, Churchwell KB, Strange K (1995). Regulation of cell volume in health and disease. N Engl J Med 333: 1260–1266.

Melton JE, Patlak CS, Pettigrew KD et al. (1987). Volume regulatory loss of Na, Cl, and K from rat brain during acute hyponatremia. Am J Physiol 252: F661–F669.

Menger H, Jorg J (1999). Outcome of central pontine and extrapontine myelinolysis (n=44). J Neurol 246: 700–705.

Mrowka M, Knake S, Klinge H et al. (2004). Hypocalcemic generalised seizures as a manifestation of iatrogenic hypoparathyroidism months to years after thyroid surgery. Epileptic Disord 6: 85–87.

Narins RG (1986). Therapy of hyponatremia: does haste make waste? N Engl J Med 314: 1573–1575.

Norenberg MD, Leslie KO, Robertson AS (1982). Association between rise in serum sodium and central pontine myelinolysis. Ann Neurol 11: 128–135.

Ober KP (1991). Endocrine crises. Diabetes insipidus. Crit Care Clin 7: 109–125.

Oya S, Tsutsumi K, Ueki K et al. (2001). Reinduction of hyponatremia to treat central pontine myelinolysis. Neurology 57: 1931–1932.

Perianayagam A, Sterns RH, Silver SM et al. (2008). DDAVP is effective in preventing and reversing inadvertent overcorrection of hyponatremia. Clin J Am Soc Nephrol 3: 331–336.

Pullen RG, DePasquale M, Cserr HF (1987). Bulk flow of cerebrospinal fluid into brain in response to acute hyperosmolality. Am J Physiol 253: F538–F545.

Robertson GL (1976). The regulation of vasopressin function in health and disease. Recent Prog Horm Res 33: 333–385.

Robertson G, Athar S, Shelton R (1977). Osmotic control of vasopressin secretion. In: TE Andreoli, JJ Grantham, FC Rector (Eds.), Disturbances in body fluid osmolality. Williams & Wilkins, Baltimore.

Rossi NF, Cadnapaphornchai P (1987). Disordered water metabolism: hyponatremia. Crit Care Clin 3: 759–777.

Rubeiz GJ, Thill-Baharozian M, Hardie D et al. (1993). Association of hypomagnesemia and mortality in acutely ill medical patients. Crit Care Med 21: 203–209.

Rymer MM, Fishman RA (1973). Protective adaptation of brain to water intoxication. Arch Neurol 28: 49–54.

Sai Kiran NA, Mohan D, Sadashiva Rao A et al. (2014). Reversible extrapyramidal symptoms of extrapontine myelinolysis in a child following surgery for craniopharyngioma. Clin Neurol Neurosurg 116: 96–98.

Samson WK (1987). Atrial natriuretic factor and the central nervous system. Endocrinol Metab Clin North Am 16: 145–161.

Sanghvi SR, Kellerman PS, Nanovic L (2007). Beer potomania: an unusual cause of hyponatremia at high risk of complications from rapid correction. Am J Kidney Dis 50: 673–680.

Schrier RW (1988a). Pathogenesis of sodium and water retention in high-output and low-output cardiac failure, nephrotic syndrome, cirrhosis, and pregnancy (1). N Engl J Med 319: 1065–1072.

Schrier RW (1988b). Pathogenesis of sodium and water retention in high-output and low-output cardiac failure, nephrotic syndrome, cirrhosis, and pregnancy (2). N Engl J Med 319: 1127–1134.

Schrier RW, Berl T, Anderson RJ (1979). Osmotic and nonosmotic control of vasopressin release. Am J Physiol 236: F321–F332.

Share L, Levy MN (1962). Cardiovascular receptors and blood titer of antidiuretic hormone. Am J Physiol 203: 425–428.

Shiber JR, Mattu A (2002). Serum phosphate abnormalities in the emergency department. J Emerg Med 23: 395–400.

Shoback DM (2000). Hypocalcemia: Diagnosis and Treatment. In: LJ De Groot, P Beck-Peccoz, G Chrousos et al. (Eds.), Endotext. MDText.com, Inc., South Dartmouth, MA.

Singh TD, Fugate JE, Rabinstein AA (2014). Central pontine and extrapontine myelinolysis: a systematic review. Eur J Neurol 21: 1443–1450.

Sterns RH (2015). Disorders of plasma sodium – causes, consequences, and correction. N Engl J Med 372: 55–65.

Sterns RH, Riggs JE, Schochet Jr SS (1986). Osmotic demyelination syndrome following correction of hyponatremia. N Engl J Med 314: 1535–1542.

Sterns RH, Cappuccio JD, Silver SM et al. (1994). Neurologic sequelae after treatment of severe hyponatremia: a multicenter perspective. J Am Soc Nephrol 4: 1522–1530.

Sterns RH, Hix JK, Silver S (2010). Treatment of hyponatremia. Curr Opin Nephrol Hypertens 19: 493–498.

Strange K (1992). Regulation of solute and water balance and cell volume in the central nervous system. J Am Soc Nephrol 3: 12–27.

Strange K, Jackson PS (1995). Swelling-activated organic osmolyte efflux: a new role for anion channels. Kidney Int 48: 994–1003.

Subramanian R, Khardori R (2000). Severe hypophosphatemia. Pathophysiologic implications, clinical presentations, and treatment. Medicine (Baltimore) 79: 1–8.

Tohme JF, Bilezikian JP (1993). Hypocalcemic emergencies. Endocrinol Metab Clin North Am 22: 363–375.

Turnbull J, Lumsden D, Siddiqui A et al. (2013). Osmotic demyelination syndrome associated with hypophosphataemia: 2 cases and a review of literature. Acta Paediatr 102: e164–e168.

Urso C, Brucculeri S, Caimi G (2014). Physiopathological, epidemiological, clinical and therapeutic aspects of exercise-associated hyponatremia. J Clin Med 3: 1258–1275.

Vallee BL, Wacker WE, Ulmer DD (1960). The magnesium-deficiency tetany syndrome in man. N Engl J Med 262: 155–161.

Vokes TJ, Robertson GL (1988). Disorders of antidiuretic hormone. Endocrinol Metab Clin North Am 17: 281–299.

Weisinger JR, Bellorin-Font E (1998). Magnesium and phosphorus. Lancet 352: 391–396.

Yamamoto T, Mori K, Esaki T et al. (2015). Preventive effect of continuous cisternal irrigation with magnesium sulfate solution on angiographic cerebral vasospasms associated with aneurysmal subarachnoid hemorrhages: a randomized controlled trial. J Neurosurg 1–9.

Chapter 39

Management of neuro-oncologic emergencies

J.T. JO AND D. SCHIFF*
Neuro-Oncology Center, University of Virginia, Charlottesville, VA, USA

Abstract

Patients with brain tumors and systemic malignancies are subject to diverse neurologic complications that require urgent evaluation and treatment. These neurologic conditions are commonly due to the tumor's direct effects on the nervous system, such as cerebral edema, increased intracranial pressure, seizures, spinal cord compression, and leptomeningeal metastases. In addition, neurologic complications can develop as a result of thrombocytopenia, coagulopathy, hyperviscosity syndromes, infection, immune-related disorders, and adverse effects of treatment. Patients may present with typical disease syndromes. However, it is not uncommon for patients to have more subtle, nonlocalizing manifestations, such as alteration of mental status, that could be attributed to other systemic, nonneurologic complications. Furthermore, neurologic complications are at times the initial manifestations of an undiagnosed malignancy. Therefore a high index of suspicion is essential for rapid assessment and management. Timely intervention may prolong survival and improve quality of life. In this chapter, we will discuss the common neuro-oncologic emergencies, including epidemiology, pathophysiology, clinical presentation, diagnosis, and treatment.

INTRODUCTION

The incidence and survival of patients with primary and metastatic brain tumors are increasing, attributable to better recognition and improved treatment strategies. It is essential for neurocritical care professionals to recognize and appropriately treat common neuro-oncologic emergencies associated with these malignancies. These emergencies can be the result of direct effects of brain tumors on the central nervous system (CNS) (e.g., increased intracranial pressure (ICP), seizures, and spinal cord compression), indirect effects (e.g., stroke and CNS infection), or complications of various treatment modalities. Prompt diagnosis and timely management may preserve neurologic function and may even be life-saving.

DIRECT EFFECTS OF CANCER

Cerebral edema and increased intracranial pressure

EPIDEMIOLOGY

Increased ICP is a consequence of an expanding mass lesion and associated cerebral edema. Cerebral edema is a prominent feature of high-grade primary brain tumors such as glioblastoma (GBM), brain metastases, and sometimes meningiomas. Large brain metastases are the most common cause of increased ICP in oncologic patients (Pater et al., 2014). GBM and certain brain metastases such as melanoma, renal cell carcinoma, thyroid carcinoma, and choriocarcinoma have a particularly high propensity to develop intratumoral hemorrhage,

---

*Correspondence to: David Schiff, MD, Neuro-Oncology Center, University of Virginia Health System, Box 800432, Charlottesville VA 22908-0432, USA. Tel: +1-434-982-4415, Fax: +1-434-982-4467, E-mail: ds4jd@virginia.edu

Table 39.1

Common primary and metastatic brain tumors (Barnholtz-Sloan et al., 2004; Ostrom et al., 2014)

Primary	Secondary/metastases
Malignant tumors	Lung
Glioblastoma	Melanoma
Diffuse astrocytoma	Renal
Oligodendroglioma	Breast
Lymphoma	Colorectal
Nonmalignant tumors	
Meningioma	
Pituitary tumors	
Nerve sheath tumor	

leading to worsening mass effect. Brain herniation is a grave consequence of rapid-incremental mass effect, often necessitating emergency intervention. Table 39.1 lists the common primary and metastatic brain tumors.

Neuropathology and pathophysiology

Cerebral edema associated with brain tumors is most often vasogenic rather than cytotoxic (Kaal and Vecht, 2004). Vascular endothelial growth factor (VEGF) plays a key role in tumor angiogenesis and vasogenic edema formation. The newly formed tumoral blood vessels are structurally and functionally abnormal, with a compromised blood–brain barrier, resulting in fluid leakage into the surrounding brain parenchyma (Jo et al., 2012). VEGF may also impair the function of occludin, induce fenestration of the endothelium, and lead to synthesis and release of nitric oxide, which in turn leads to opening of tight junctions and enhancement of capillary permeability (Davies, 2002; Kaal and Vecht, 2004). VEGF expression is correlated with malignant progression and vascular density in gliomas. Upregulation of VEGF has also been observed in metastases and certain meningiomas, including the transitional and meningotheliomatous types (Stummer, 2007).

Peritumoral edema causes mass effect, resulting in increased ICP, compromise of local blood supply, brain tissue displacement, and ultimately brain herniation. Based on the Monro–Kellie hypothesis, an increase in brain parenchyma volume forces depletion of either the cerebrospinal fluid (CSF) or blood compartments (Mokri, 2011). The slower the tumor growth and the more gradual the incremental mass effect exerted by edema, the more effective the compensatory mechanism. Clearance of extravasated fluid is accomplished by bulk flow of edema fluid to the ventricular space, absorption through subarachnoid space, and absorption by local capillaries (Stummer, 2007). A rapid rise in ICP increases the risk of brain herniation. Marked transient elevation of ICP, classically in excess of 40 mmHg and lasting for a minimum of 5 minutes, is referred to as a plateau wave, and results in reduced cerebral perfusion pressure. Prolonged elevation of ICP for more than 30–40 minutes leads to insufficient cerebral blood flow, causing permanent damage and unfavorable outcomes (Castellani et al., 2009; Pater et al., 2014).

Clinical presentation

Peritumoral edema leads to signs and symptoms related to focal mass effect and increased ICP. These clinical manifestations usually begin insidiously over weeks to months, but some tumors presents acutely from intratumoral hemorrhage or obstructive hydrocephalus. History and rapid physical examination, focusing on signs of impending brain herniation, are important in the recognition and timely management of increased ICP.

Headache is present in almost half of patients with primary and metastatic brain tumors. The classic early-morning headache is relatively uncommon. Headache is usually characterized as a nonthrobbing, pressure-like sensation similar to tension-type headaches, and almost 40% of patients describe it as "the worst pain of their life." The headache due to increased ICP in brain tumor patients is generally severe, not relieved by common analgesics, and often associated with nausea and vomiting (Forsyth and Posner, 1993; Damek, 2009).

Headache can also be episodic due to plateau waves from abrupt fluctuations of ICP for 5–20 minutes, frequently accompanied by nausea, vomiting, visual loss, ataxia, and loss of consciousness, typically triggered by changes in body position. This phenomenon is sometimes confused with seizures (Damek, 2009). Impaired level of consciousness occurs in a third of patients with brain tumors, ranging from lethargy to coma (Damek, 2009; Pater et al., 2014). Papilledema may be observed in almost half of patients with increased ICP, although a decline to less than 10% has been reported owing to earlier diagnosis (Forsyth and Posner, 1993; Wen, 1997).

The presence of focal deficits depends on the location of the mass. Herniation syndromes commonly occur in patients with new large intracranial mass lesions or edema. Warning signs of impending herniation include impaired consciousness, abnormal breathing patterns (e.g., Cheyne–Stokes respiration or central neurogenic hyperventilation), repetitive respiratory reflexes (i.e., hiccups or yawning), and Cushing's triad (hypertension, bradycardia, and irregular respiration or apnea). Temporal-lobe masses cause uncal herniation, typically presenting with ipsilateral cranial nerve III palsy, hemiparesis, and impaired consciousness. Supratentorial masses can lead to central herniation, causing occlusion of the ipsilateral posterior cerebral artery, with resultant infarction (Stevens et al., 2012; Pater et al., 2014).

Neuroimaging

Brain magnetic resonance imaging (MRI) is superior to computed tomography (CT) in evaluation of the size, location, and number of brain masses, as well as secondary effects of brain tumors, such as edema and mass effect. However, noncontrast head CT is the preferred initial test due to speed and availability (Quinn and DeAngelis, 2000). Acute hemorrhage, hydrocephalus, edema, and mass effect can be easily visualized using CT. Once the patient is stable, cranial MRI can be obtained for more detailed characterization of the lesion. High water content, as with CSF and cerebral edema, appears hyperintense on T2-weighted imaging sequences. Fluid-attenuated inversion-recovery (FLAIR) MRI sequences suppress the CSF signal, and are beneficial in depicting the full extent of tumor and surrounding edema, especially for lesions adjacent to sulci and ventricles. Lesions with disrupted blood–brain barrier such as metastases, high-grade gliomas or secretory meningiomas avidly enhance on gadolinium-enhanced T1-weighted sequences (Fig. 39.1). T1 hyperintensity or "T1 shortening" and gradient recall echo (GRE) or susceptibility-weighted imaging (SWI) MRI sequences reveal hemorrhage within tumors. This is most common in metastases from melanoma, renal, thyroid, or choriocarcinoma (Kaal and Vecht, 2004; Walker and Kapoor, 2007).

Hospital course and management

If there is a high clinical suspicion of increased ICP or impending herniation, immediate measures to lower ICP take precedence over emergency neuroimaging. Intubation and hyperventilation to target pCO_2 between 25 and 30 mmHg is the most rapid method to decrease ICP. Decreased pCO_2 causes cerebral vasoconstriction and a subsequent decline in cerebral blood volume and ICP. This effect occurs within as little as 30 seconds of lowering pCO_2, but is transient and may cause cerebral ischemia (Ropper et al., 2003). Consequently, additional adjunctive therapies should be simultaneously initiated, and the pCO_2 should subsequently be normalized. Corticosteroids play a crucial role in alleviating symptomatic vasogenic edema associated with brain tumors. Dexamethasone is preferred over other corticosteroids, as it has the best CNS penetration, the least mineralocorticoid effect, a long half-life, and is the least protein-bound (Chamberlain, 2010a). The recommended starting dose is 4–8 mg daily for symptomatic patients, while an initial dose of 10 mg intravenous (IV) followed by

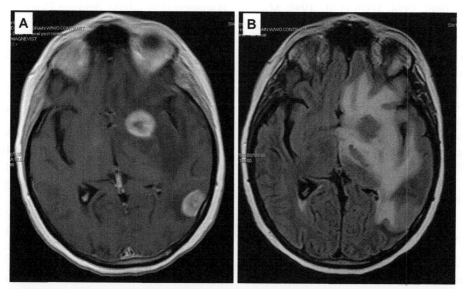

Fig. 39.1. Brain metastasis with vasogenic edema. Brain magnetic resonance imaging: postgadolinium study demonstrates peripherally enhancing mass in the left basal ganglia and left temporal lobe (**A**) with surrounding vasogenic edema on fluid-attenuated inversion recovery images (**B**) causing mass effect and left-to-right midline shift.

approximately 16 mg daily is advised for patients with acute signs of increased ICP (Kaal and Vecht, 2004). Therapeutic effect is usually reached within 24–72 hours.

Once clinical benefit is achieved, the dose should be titrated to the lowest possible in order to minimize adverse effects, such as immunosuppression, increased risk of opportunistic infection, and hyperglycemia. Prophylactic treatment for peptic ulcers is generally not recommended, except for patients with a history of previous ulcers, those concomitantly taking nonsteroidal anti-inflammatory drugs, or in the elderly (Wen et al., 2006). Hyperosmolar agents such as hypertonic saline solution (3–23.4% concentration) and 20–25% mannitol (0.5–1 g/kg) may reverse impending cerebral herniation and decrease ICP. These agents create an osmotic gradient that draws water across an intact blood–brain barrier to the higher osmolarity in the blood (Damek, 2009; Pater et al., 2014). Repeated doses may be required for clinical decline after initial improvement. However, frequent administration of hyperosmolar agents without time for clearance in between doses may lead to a rebound increase in ICP.

If the patient's condition continues to decline despite maximal medical management, neurosurgical intervention may be warranted for tumor debulking or hematoma evacuation, as well as ventriculostomy placement for obstructive hydrocephalus (Quinn and DeAngelis, 2000). Apart from rapidly reducing intracranial mass effect and permitting steroid tapering, resection of brain metastases in certain subgroups of patients provides local disease control and better quality of life when combined with whole-brain radiotherapy (RT) after surgery (Patchell et al., 1990, 1998; Vecht et al., 1993; Mintz et al., 1996). In patients with GBM, gross total resection along with RT and temozolomide improves overall survival (Stupp et al., 2005). Tumor location is an important factor, as lesions in the thalamus, basal ganglia, and brainstem are generally not amenable to resection. RT is an essential component of treatment for brain tumors. However, initiation of RT is not advisable in patients with significantly raised ICP, as this can further induce brain edema and herniation (Quinn and DeAngelis, 2000; Pater et al., 2014).

Tumor-associated epilepsy and status epilepticus

Epidemiology

Seizures are a frequent complication of primary and metastatic brain tumors. New-onset seizures are more common in slowly growing tumors than in more aggressive, rapidly growing ones. Seizures are either focal or secondarily generalized. Tumor-associated epilepsy usually occurs early in the course of the disease and the risk is inversely proportional to the World Health Organization tumor grade. In contrast, tumor-associated status epilepticus most often arises later in the course and may indicate tumor progression. This may be due to poor penetration of antiepileptic drugs (AEDs) into epileptogenic lesions, secondary to expression of multidrug-resistant proteins. In addition, there may be reduced AED efficacy due to drug interactions with concomitant chemotherapeutic agents used in the treatment of high-grade tumors (Goonawardena et al., 2015).

Among primary brain tumors, seizures most commonly occur with glioneuronal malignancies, such as dysembryoplastic neuroepithelial tumors (DNET) and gangliomas (80–100%), diffuse low-grade gliomas (60–88%), and GBM (40–60%). Among patients with low-grade gliomas, those with oligodendrogliomas and oligoastrocytomas are more prone to develop seizures than those with astrocytoma, owing to the more cortical location in the former and white-matter involvement in the latter (Ruda et al., 2012; Vecht et al., 2014). Seizures are less common with brain metastases, with an incidence of 20–40% (Maschio, 2012). This most commonly occurs in patients with melanoma (67%), presumably due to its tendency to involve the cerebral cortex and high propensity for hemorrhage. Metastases from primary lung cancer (29%) and gastrointestinal tumors (21%) have intermediate risk, while breast cancer has the lowest risk of seizures (16%) (Oberndorfer et al., 2002; van Breemen et al., 2007).

Status epilepticus is a neurologic emergency, with a reported 30-day mortality of more than 20% and a higher risk of death in patients with systemic malignancy (Cavaliere et al., 2006). Between 15 and 22% of patients with tumor-associated epilepsy develop status epilepticus. This is lower than the reported incidence of status epilepticus (30–40%) in patients with epilepsy in the general population (Goonawardena et al., 2015).

Neuropathology and pathophysiology

Tumors located in epileptogenic areas, such as the mesial temporal lobe, insula, and cortex, are more likely to cause seizures. Glioneuronal tumors are commonly located in the temporal lobe, explaining their strong tendency to trigger seizures (Baldwin et al., 2012).

Molecular factors are postulated to play a role in epileptogenesis of brain tumors. Isocitrate dehydrogenase 1 (IDH1) and 2 (IDH2), which are driver mutations in low-grade gliomas, are reported to be associated with an increased risk of seizures. IDH1 mutation is expressed in 70–80% of low-grade gliomas, as opposed to only 5–10% of high-grade gliomas (Cohen et al., 2013; Vecht et al., 2014). Mutated IDH1 catalyzes isocitrate to 2-hydroxyglutarate (2-HG), which is structurally similar to glutamate, instead of α-ketoglutarate in the Krebs cycle. The accumulation of 2-HG can activate N-methyl-D-aspartate (NMDA) receptors and promote tumor-associated epilepsy (Liubinas et al., 2014). Glutamate, a known "tumor growth factor" in gliomas, is also implicated in epileptogenesis. An increase in glutamate concentration and altered glutamate transporter expression are linked to tumor-associated seizures (Yuen et al., 2012; Pallud et al., 2013). In addition, GABAergic signaling is involved in tumor growth with perturbation of neuronal and tumor cell chloride homeostasis, potentially resulting in epileptiform activity (Pallud et al., 2013).

CLINICAL PRESENTATION

Specific seizure semiology reflects the tumor's location. Seizures in patients with brain tumors generally have characteristics of localization-related epilepsy, with typical subtypes including simple partial seizures (23–58%), complex partial seizures (7–31%), and focal seizures with secondary generalization (10–68%) (Vecht et al., 2014). Postictal or Todd's paralysis is common in patients with structural lesions, and may sometimes be prolonged (Damek, 2009). Status epilepticus can occur at tumor presentation (29%), during tumor progression (23%), and even when tumors are stable (23%) (Cavaliere et al., 2006). The clinical manifestations of nonconvulsive status epilepticus (NCSE) are nonspecific, and may include changes in personality, fluctuating mental status, focal myoclonic jerks, or abnormal ocular movements. NCSE is found to be the underlying etiology of altered mental status in up to 6% of cancer patients (Damek, 2009).

NEUROIMAGING

A thorough history and physical examination can establish a clinical diagnosis of seizures. However, other mimics, such as complex migraine, syncope, medication effects, extrapyramidal tremors, and posturing, should be ruled out. The first unprovoked seizure in adults warrants neuroimaging studies regardless of a known diagnosis of brain or systemic malignancies. CT is more readily available in the emergency setting and allows identification of some acute changes, such as overt hemorrhage or worsening intracranial mass effect. MRI is superior to CT in revealing number, size, and location of brain masses, secondary effects of tumors such as edema and mass effect, evidence of tumor progression, complications such as infarcts, as well as radiographic changes related to seizures, such as restricted diffusion in the cortical mantle of the epileptogenic focus. An electroencephalogram (EEG) is abnormal in approximately one-third of patients and may reveal epileptiform discharges localizing to the brain tumor region. An emergency EEG is essential in the diagnosis of NCSE. However, EEG is not necessarily routinely indicated for patients with witnessed seizures who have recovered (Damek, 2009; Baldwin et al., 2012).

HOSPITAL COURSE AND MANAGEMENT

Treatment of status epilepticus in patients with brain tumors is similar to standard management, which is discussed in Chapter 9. However, the choice of ongoing anti-epileptic drugs (AED) therapy is challenging due to the potential interaction between AEDs and chemotherapeutic agents. Older-generation AEDs, such as phenobarbital, phenytoin, and carbamazepine, induce cytochrome P450 enzymes, which can accelerate the metabolism of common chemotherapeutic agents and decrease their effectiveness (Baldwin et al., 2012). Newer AEDs, including levetiracetam, lamotrigine, and lacosamide, are preferred due to their low probability of pharmacokinetic interactions and a more favorable tolerability profile (Usery et al., 2010; Saria et al., 2013; Rossetti et al., 2014).

In a prospective study including 176 patients with newly diagnosed gliomas with epilepsy, 91% of patients treated with levetiracetam were seizure-free with no laboratory abnormalities observed with concomitant chemotherapy (Rosati et al., 2010). Levetiracetam is recommended for initial monotherapy due to its efficacy, tolerability, and pharmacokinetic properties. It has both oral and intravenous formulations and may be initiated at therapeutic dose.

If an additional AED is required, lacosamide is a reasonable second-line agent (Schiff et al., 2015). Valproic acid is another appropriate choice, with a possible additional advantage of some antitumor effect (Vecht and Wilms, 2010). In one study, patients with newly diagnosed GBM who received valproic acid either for seizure prophylaxis or treatment appeared to derive more survival benefit from the combination of temozolomide and RT than patients receiving enzyme-inducing AEDs or patients not receiving any AED. One potential mechanism for its antitumor effects is that valproic acid is a histone deacetylase inhibitor (Duenas-Gonzalez et al., 2008). However, valproic acid is also a CYP450 inhibitor and can potentially increase toxicity from chemotherapy (Schiff et al., 2015). Patients treated with valproic acid develop serious thrombocytopenia and leukopenia more often than other patients (Weller et al., 2011). Evidence has failed to show benefit of prophylactic AEDs in decreasing the incidence of new-onset seizures, and is therefore not recommended (Glantz et al., 2000). A randomized, placebo-controlled trial using lacosamide for seizure prophylaxis in patients with high-grade gliomas is currently under way (NCT01432171).

AED refractoriness appears to be more prevalent in tumor-associated epilepsy than in other patients with epilepsy, with as high as 60% failure with first-line AEDs (Goonawardena et al., 2015). AED resistance seems to be directly proportional to the tumor grade, with more than 60% of patients with high-grade glioma requiring polytherapy in one study, compared with only 36% and 29% in patients with low-grade gliomas or other brain tumors, respectively (van Breemen et al., 2009). These data support treating seizures in patients with high-grade gliomas relatively aggressively, with a low threshold for concomitantly using multiple AEDs (Goonawardena et al., 2015). Paradoxically, tumor-associated status epilepticus appears to be more responsive to simple AEDs than tumor-associated epilepsy or epilepsy in the general population (Cavaliere et al., 2006; Neligan and Shorvon, 2010; Goonawardena et al., 2015). First-line drug treatment (usually with benzodiazepines and phenytoin) terminates status epilepticus in the majority of patients (Lowenstein and Alldredge, 1993; Cavaliere and Schiff, 2007). The mechanisms accounting for this responsiveness remain elusive (Cavaliere and Schiff, 2007; Goonawardena et al., 2015). Treatment directed at the tumor, including surgery, radiation, or systemic therapy (i.e., chemotherapy and targeted therapy), may also significantly contribute to seizure control.

Epidural spinal cord compression

Epidemiology

Epidural spinal cord compression (ESCC) affects approximately 3% of patients with cancer per year. ESCC is by far the most common presentation of spinal cord metastases, and often occurs in patients with disseminated disease. Lung cancer is the most common primary source, followed by breast and prostate cancer, lymphoma, and multiple myeloma (Spinazze et al., 2005; Mak et al., 2011). The distribution of the spinal segments involved reflects the number and volume of vertebral bodies in each segment, with the thoracic spine most commonly involved (70%), followed by lumbar (20%) and cervical (10%) regions. In 10–40% of cases, multiple noncontiguous lesions are encountered (Klimo and Schmidt, 2004; Spinazze et al., 2005).

ESCC is a medical and/or surgical emergency, as neurologic deterioration may progress rapidly, potentially leading to irreversible deficits such as paralysis and sphincter dysfunction. Early recognition and urgent treatment are required to improve the chance of neurologic recovery and/or stabilization. The patient's pretreatment neurologic status is the most important prognostic factor for functional recovery, with about 70% of initially ambulant patients, 30% of paraparetic patients, and 5% of paraplegic patients retaining or regaining the ability to ambulate (Findlay, 1984; Bach et al., 1990). A prospective study of all newly diagnosed patients with ESCC demonstrated significant deterioration of motor or bladder function with a median delay from onset of symptoms of spinal cord compression to treatment of 14 (range 0–840) days. Of the total delays, 3 (range 0–300) days were accounted for by patients, 3 (0–330) days by general practitioners, 4 (0–794) days by district general hospitals, and 0 (0–114) days by the treatment unit (Husband, 1998). The median survival following diagnosis of metastatic ESCC is 3–6 months. The patient's neurologic status at the time of diagnosis and the tumor type are the strongest factors affecting survival. Nonambulatory patients, and those with bowel and bladder dysfunction tend to have the worst prognosis (Spinazze et al., 2005).

Neuropathology and pathophysiology

Tumor cells reach the epidural space by hematogenous spread to vertebral bodies in about 85% of cases, while 15% of the time there is spread through an intervertebral foramen. The latter mechanism is especially common in patients with lymphoma and neuroblastoma. Hematogenous dissemination occurs either via Batson's plexus, which serves as a pathway to transmit metastatic cells from thoracic, abdominal, and pelvic organs to the vertebral column, or through the arterial circulation to vertebral bodies. The metastatic tumor cells cause destruction to the vertebral bone, and expansion within the vertebral body, in turn leading to outgrowth into the epidural space. Direct hematogenous spread to the epidural space is relatively rare. Spinal cord injury can also occasionally result from direct mechanical injury to axons and myelin, and from compression of 2the epidural venous plexuses and spinal arteries, resulting in spinal cord edema and infarction (Schiff, 2003; Perrin and Laxton, 2004; Sun and Nemecek, 2010; Hammack, 2012).

Clinical presentation

Back pain is the most common presenting symptom (95%), followed by motor (60–85%) and sensory deficits (60%). Autonomic manifestations, such as bowel and bladder dysfunction, sexual disturbance, and orthostatic hypotension, usually occur later in the course of disease (Schiff, 2003; Hammack, 2012).

Pain is usually described as sharp, shooting, deep, or burning. It may be initially confined to the involved vertebral bodies, owing to stretching of the periosteum and adjacent pain-sensitive structures. However, the location of pain does not always correspond to the site of compression. C7 compression may cause referred pain to the midscapular region, and T12 compression to the sacroiliac or hip joints. Over half of patients with T1–T6 compression complain of lumbosacral pain (Levack et al., 2002, Abrahm, 2004). When nerve roots are compressed, patients present with radicular pain or a tight band sensation around the chest or abdomen. A Valsalva maneuver, or lying supine, may precipitate pain, which is suggestive of venous plexus distention. Back pain typically evolves over several weeks and gradually increases in severity. Acute pain may indicate a pathologic compression fracture. Spinal cord compression above L1 and L2 vertebral bodies produces an upper motor neuron pattern of weakness, with quadri- and paraparesis when there is cervical and thoracic cord involvement, respectively, as well as increased deep tendon reflexes. A sensory level is consistent with cord compression, while a dermatomal deficit implies nerve root compression. Compression of dorsal columns in the cervical and thoracic cord can produce l'Hermitte's phenomenon. If the cauda equina is compressed, a lower motor neuron pattern of weakness; diminished sensation over the buttocks, posterior superior thighs and perianal regions; urinary retention with overflow incontinence; and decreased anal sphincter tone are common presenting manifestations (Schiff, 2003; Abrahm, 2004; Sun and Nemecek, 2010; Hammack, 2012).

Neuroimaging

MRI is the gold standard for detecting an epidural disorder and consequent spinal cord compression, with overall accuracy of at least 95%. The entire spine should be imaged, since multilevel involvement is not uncommon in patients with ESCC. Vertebral metastases are typically hypointense on T1-weighted MRI sequences, hyperintense on T2-weighted sequences, and enhance with gadolinium (Fig. 39.2). Increased T2 signal within the cord suggests venous congestion or ischemia. CT myelogram is an acceptable alternative when MRI is contraindicated. Spinal CT, plain radiographs, and bone scans are not sufficient for diagnosis, as these do not accurately depict the tumor, the paraspinal region, and the extent of spinal cord involvement (Schiff et al., 1998; Schiff, 2003; Hammack, 2012).

Hospital course and management

The main treatment goals are pain control, preservation or recovery of neurologic function, and avoidance of complications. Treatment begins immediately with corticosteroid administration, followed by more definitive

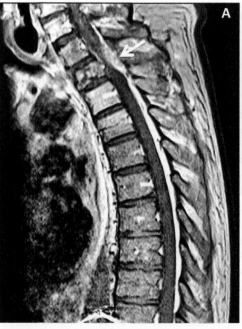
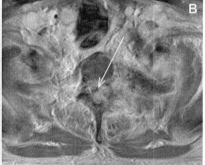

Fig. 39.2. Epidural spinal cord compression. Thoracic spine magnetic resonance imaging: (**A**) sagittal and (**B**) axial postgadolinium images demonstrate bone marrow infiltration and vertebral body collapse of T2, with circumferential epidural mass surrounding the thecal sac, notably within the posterior aspect of thecal sac, causing severe mass effect on the thoracic cord.

treatment with surgery, radiation therapy, chemotherapy, or a combination of these modalities.

Glucocorticoids, typically dexamethasone, usually improve pain through their antiprostaglandin effects, stabilize or restore neurologic function by reducing vasogenic edema, and may have a cytotoxic effect in lymphoma and multiple myeloma. A bolus dexamethasone dose of 8–10 mg followed by 16 mg/day in divided doses is advisable for patients with minimal or nonprogressive weakness. For patients with paraplegia, higher loading dose of up to 100 mg and maintenance doses of 96 mg/day can be given, although a recent Cochrane review did not find any advantage over lower-dose corticosteroids (George et al., 2015). Rapid steroid taper is important to prevent risk of complications, such as myopathies, psychosis, opportunistic infection, and peptic ulcer disease (Sun and Nemecek, 2010; Loblaw et al., 2012; Ribas and Schiff, 2012). Patients with back pain without signs of myelopathy and no massive invasion of the spine on imaging studies may forgo steroids (Maranzano et al., 1996).

Prompt surgical consultation is recommended following radiologic diagnosis to prevent neurologic decline. Surgery can provide pain relief, spinal cord decompression and restoration of neurologic status, re-establishment of spine stability, correction of deformity, as well as histologic diagnosis (Loblaw et al., 2012; Ribas and Schiff, 2012) (Patchell et al., 2005). Prior to availability of RT, simple laminectomy was the only form of definitive treatment for ESCC. With the advent of RT, several retrospective studies and small, randomized trials were conducted and demonstrated no benefit in combined laminectomy plus RT versus RT alone. However, since most metastases involve the vertebral body, laminectomy alone is not an optimal surgical approach. Tumor removal and immediate circumferential decompression are better achieved by anterior-approach vertebrectomy (Siegal et al., 1985). A randomized, nonblinded trial demonstrated superiority of combined direct decompressive surgery plus postoperative RT to RT alone for patients with spinal cord compression due to metastatic cancer. Of the 101 patients included, 84% were able to walk after combined treatment compared to 57% of patients who received RT alone; 62% in the surgery group regained the ability to walk compared to only 19% in the RT group; those treated with surgery retained the ability to walk significantly longer than those in the RT group (median days 122 vs. 13 days). In addition, corticosteroid and opioid analgesic requirements were significantly lower in the surgical group. Surgery resulted in significant benefit in maintenance of continence, muscle strength, and functional ability, while also increasing survival time (Patchell et al., 2005). Patients who presented more than 48 hours after total paraplegia were excluded in this study. Meaningful recovery is rarely attained after prolonged compression where irreversible vascular injury has already transpired. Patients with spinal cord compression at a single level presenting with incomplete motor deficits within 48 hours generally have a good prognosis and benefit from immediate surgery. Patients with life expectancy of less than 3–6 months are poor surgical candidates (L'Esperance et al., 2012).

All patients who are not candidates for surgery should receive RT. RT has been shown to reduce back pain in about 60%, and to maintain or restore ambulation and continence in 70% and 90% of patients, respectively (Maranzano et al., 2005; Loblaw et al., 2012). The patient's performance status at the start of RT, degree of tumor radiosensitivity, and speed of onset of the neurologic deficits are important predictors of RT response (Hammack, 2012). External-beam RT confers durable local control, particularly in radiosensitive malignancies such as lymphoma, multiple myeloma, and germ cell tumors. Various fractionation schedules have been used. Short-course therapy (8 Gy in one fraction or 16 Gy in two fractions) has the advantage of being faster and more convenient to the patient. However, the risk of tumor recurrence or progression within the field of radiation is lower in patients who receive a more protracted course (e.g., 30 Gy in 10 fractions). For patients who have limited life expectancy, due to either poor functional status or extensive systemic metastases, short-course therapy is recommended, since these patients are unlikely to survive long enough to develop radiation toxicity (Hammack, 2012; L'Esperance et al., 2012; Loblaw et al., 2012).

Stereotactic radiosurgery (SRS) allows precise delivery of high-dose radiation to metastases within or adjacent to the vertebral body and surrounding spinal cord. Since it has a steep dose gradient, it minimizes toxicity to the surrounding tissues. SRS appears to provide pain relief, with an overall improvement rate of 85–100%, and radiographic tumor control in 90% of patients (Gerszten et al., 2007; Sohn and Chung, 2012). Spinal column stabilization and epidural tumor resection followed by SRS result in high local tumor control at 1 year (Moussazadeh et al., 2014). This approach is effective regardless of histologic diagnosis and can be utilized in previously irradiated patients. However, high-quality supportive data are still lacking (Sohn and Chung, 2012).

Chemotherapy is considered for patients with chemosensitive tumors such as lymphoma and seminoma, who have minimal or no neurologic deficits. However, its role in emergency treatment of ESCC is limited (Hammack, 2012).

Leptomeningeal metastases

EPIDEMIOLOGY

Leptomeningeal metastases (LM) are a rare, but debilitating, complication of malignancy, with dissemination of cancer cells to the CSF, and pia and arachnoid mater. It occurs in 5–15% of hematologic malignancies, including lymphoma and leukemia; 1–5% of patients with solid tumors, most often breast or lung cancer, and melanoma; and 1–2% of patients with primary brain tumor such as high-grade gliomas, medulloblastoma, ependymoma, and pineoblastomas (Wassertrom et al., 2006; Chamberlain, 2010b; Groves, 2011). LM often occur in the setting of advanced disease. At autopsy, LM are detectable in almost 20% of cancer patients with neurologic signs and symptoms (Glass et al., 1979). The median survival after diagnosis of LM is only 4–6 weeks for untreated patients and 4–8 months with treatment. LM from hematologic tumors tends to have better prognosis than from solid tumors (median survival 4.7 vs. 2.3 months) (Chamberlain, 2010b; Groves, 2011; Clarke, 2012).

NEUROPATHOLOGY AND PATHOPHYSIOLOGY

Routes of leptomeningeal seeding include hematogenous spread via Batson's plexus or arterial dissemination, direct extension from adjacent structures, choroid plexus metastases into the CSF, and retrograde invasion along perineural or perivascular spaces (Clarke, 2012). Leptomeningeal involvement is associated with spread of tumor cells throughout the subarachnoid space, resulting in multifocal seeding, most prominent in the skull base, dorsal surface of spinal cord, and cauda equina. Hydrocephalus and increased ICP may result from tumor deposition and CSF flow obstruction (Chamberlain, 2010b; Groves, 2011; Clarke, 2012).

CLINICAL PRESENTATION

The majority of patients with LM present with multifocal symptoms reflecting multilevel nervous system involvement. Patients with cerebral involvement can manifest with headache, encephalopathy, nausea, vomiting, dizziness, and seizures. Cranial neuropathies may result in diplopia, facial motor and sensory impairment, dysarthria, dysphagia, hoarseness, and sensorineural hearing loss. Spinal symptoms include radiculopathy, back or neck pain, weakness or numbness in the anatomic distribution of involved levels, and bowel and bladder dysfunction (Groves, 2011; Clarke, 2012).

NEUROIMAGING

Clinical presentation, CSF cytology, and gadolinium-enhanced MRI are methods used to diagnose LM. MRI of the entire neuraxis is warranted to evaluate the extent of involvement and to plan treatment. It is positive in approximately 95% of cases (Clarke et al., 2010; Passarin et al., 2015). MRI findings suggestive of LM include linear or nodular enhancement in the cerebral sulci, cerebellar folia, basal cisterns, and cauda equina (Fig. 39.3), as well as enhancement of the subependyma, cranial and spinal nerves, and hydrocephalus (Groves, 2011).

Identification of malignant cells in the CSF remains the gold standard for diagnosis, with specificity of about 95%, but low sensitivity of less than 50% (Enting, 2005). To minimize false-negative cytology studies, withdrawing at least 10-mL samples, obtaining CSF from the site of known LM, immediate processing, and repeated CSF samplings are recommended. The yield increases from 71% with a single sample to 86% with a second, and up to 93% with a third (Glantz et al., 1988). CSF flow cytometry is more sensitive than cytology in patients with hematologic malignancies. With the exception of nonseminomatous germ cell tumors, CSF tumor markers generally have unproven clinical value (Chamberlain et al., 2014). According to the National Comprehensive Cancer Network guideline, the presence of any of the following in a patient with cancer is sufficient to diagnose LM: positive CSF cytology, positive radiologic findings with supportive clinical findings consistent with LM disease, or signs and symptoms consistent with LM disease and nonspecific (but abnormal) CSF studies (i.e., high white blood cell count, low glucose, and high protein) (Nabors et al., 2013).

HOSPITAL COURSE AND MANAGEMENT

The first dilemma in the management of LM disease is deciding which patients need aggressive treatment and which are better treated only with supportive care. Patients with low tumor burden, good functional performance status, lack of major neurologic deficits, no evidence of bulky disease on imaging studies, absence of CSF flow block using radioisotope imaging, expected survival > 3 months and limited extraneural metastatic disease are ideal patients to receive aggressive treatments (Chamberlain et al., 2014).

If LM-directed treatment is considered, the second challenge is determining the mode of treatment, including RT, systemic chemotherapy, or intrathecal (IT) chemotherapy. The goals of treatment are to improve or

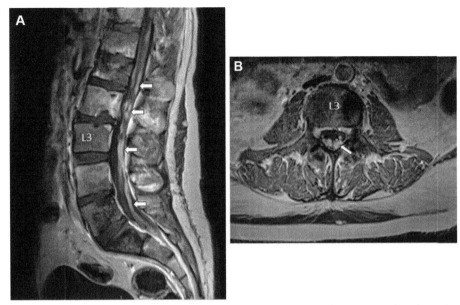

Fig. 39.3. Leptomeningeal carcinomatosis involving cauda equina. Lumbar magnetic resonance imaging: (**A**) sagittal and (**B**) axial postgadolinium images demonstrate diffuse nodularity of the cauda equina nerve roots (arrows) and bone marrow involvement of lumbar vertebral bodies with pathologic compression fracture at L1 and L2.

stabilize neurologic function, maintain quality of life, and possibly extend survival (Chamberlain, 2010b).

RT directed at the bulky or symptomatic disease sites provides palliation of symptoms and restoration of CSF flow. Whole-brain RT, in some cases followed by chemotherapy, is utilized for widespread CSF treatment. Cranial irradiation also targets focal collections of tumor cells causing noncommunicating hydrocephalus. Craniospinal irradiation is rarely recommended due to its significant adverse effects, including gastrointestinal toxicity, mucositis, and bone marrow suppression, and the lack of significant survival benefit (Chamberlain, 2010b; Clarke, 2012; Kak et al., 2015). Chemotherapy can be given either systemically or intrathecally. Systemic chemotherapy provides uniform drug distribution, can be administered to patients with bulky disease, is safe in patients with CSF flow block, and obviates the need for surgical placement of a reservoir (Glantz et al., 1998). Certain chemotherapeutic agents achieve cytotoxic CSF concentrations when given at high doses, including methotrexate ($3–8\ g/m^2$) or cytarabine ($3\ g/m^2$). Capecitabine, thiothepa, and temozolomide also cross the blood–brain barrier. The majority of chemotherapy, and targeted agents such as lapatinib, trastuzumab, and rituximab, do not penetrate the intact blood–brain barrier in sufficient concentrations to be effective (Chamberlain, 2010b; Clarke, 2012; Chamberlain et al., 2014).

IT delivery, either via lumbar puncture or intraventricular (Ommaya) reservoir, has the advantage of circumventing the blood–brain barrier, reducing the overall dosage, and decreasing systemic adverse effects (Grewal et al., 2012). Methotrexate, cytarabine, longer-acting liposomal cytarabine, and (less often) thiotepa can be given by the IT route. Several randomized controlled trials suggest modest benefit from IT treatment, but the median survival remains dismal (3 months), and often associated with treatment-related toxicity. No specific IT agent or regimen has shown superior efficacy in the treatment of LM, except for liposomal cytarabine in lymphomatous meningitis. One randomized controlled trial found no survival advantage in using single- versus multiagent treatment of LM (Jaeckle, 2006; Chamberlain et al., 2014). Furthermore, IT chemotherapy did not show a significant clinical response or improved median survival compared to systemic chemotherapy and CNS-directed RT (Boogerd et al., 2004). IT administration of trastuzumab, an antihuman epidermal growth factor receptor (HER-2) monoclonal antibody, has been shown to be safe and in some cases effective for treatment of meningeal carcinomatosis in HER-2-positive metastatic breast cancer (Zagouri et al., 2013). A phase I trial has identified a safe, well-tolerated IT trastuzumab dose (Raizer et al., 2014) and a phase II study is under way (https:/clinicaltrials.gov/ct2/show/NCT01325207).

IT rituximab, an anti-CD20 monoclonal antibody, may have a role in recurrent CNS non-Hodgkin's lymphoma (Schulz et al., 2004; Rubenstein et al., 2007, 2013). Intra-CSF interleukin-2 (IL-2) has been investigated in patients with LM from disseminated melanoma (Mitchell, 1989; Samlowski et al., 1993; Herrlinger et al., 1998).

INDIRECT COMPLICATIONS OF CANCER TO THE NERVOUS SYSTEM

Cerebrovascular disease

EPIDEMIOLOGY

Cancer patients are at increased risk for cerebrovascular disease (CVD). An autopsy study found pathologic evidence of CVD in 15% of patients with cancer, with about half having clinical manifestations of stroke (Graus et al., 1985). Intracerebral hemorrhage (ICH) is the most common form of stroke in patients with hematologic malignancies, while there is an approximately even distribution of ICH and ischemic stroke with solid tumors (Graus et al., 1985). Multiple embolic events are the predominant pattern of ischemic stroke in cancer patients (Hong et al., 2009). Vascular risk profiles are typically similar in stroke patients with or without cancer (Cetari et al., 2004; Zhang et al., 2006). However, approximately 40% of stroke patients with cancer lack a conventional stroke mechanism and 18% are found to have cryptogenic stroke (Kim et al., 2010). Patients with stroke and cancer have worse prognosis compared to the noncancer population (Zhang et al., 2006; Taccone et al., 2008).

NEUROPATHOLOGY AND PATHOPHYSIOLOGY

Classic CVD risk factors, such as hypertension, diabetes mellitus, atrial fibrillation, hyperlipidemia, carotid disease, and smoking, are still the main risk factors for stroke in cancer patients. In those without conventional stroke risk factors, cancer-related and treatment-induced mechanisms must be considered (Grisold et al., 2009; Bang et al., 2011).

Cancer-related mechanisms for ischemic stroke include: hypercoagulability via tumor production of mucin, leading to formation of platelet-rich thrombi; release of procoagulant molecules such as tissue factor and cancer procoagulant; and production of procoagulant cytokines such as tumor necrosis factor-alpha, IL-1, and IL-6, which in concert potentiate the clotting cascade (most commonly encountered in patients with adenocarcinoma); hyperviscosity leading to obstruction of small end vessels (usually seen in multiple myeloma, polycythemia vera, Waldenström's macroglobulinemia or Bing–Neel syndrome, acute myelogenous leukemia, or chronic lymphocytic leukemia); and direct tumor effect either from tumor compression of blood vessels through invasion or edema, such as those seen in brain metastases, primary CNS tumors, and with LM.

Direct vascular occlusion by tumor cells can also cause ischemic stroke through several mechanisms. Nonbacterial thrombotic endocarditis (NBTE), which is commonly seen in patients with adenocarcinoma, results from deposition of acellular aggregates of fibrin and platelet on normal heart valves, in turn leading to emboli. Direct tumor embolism can occur in patients with myxoma and lung tumors. Tumor infiltration of the blood vessel wall, as seen in patients with intravascular diffuse large B-cell lymphoma, can lead to vessel wall irregularities and subsequent arterial occlusion or embolism, resulting in multifocal cerebral infarction.

Hemorrhagic stroke can occur due to tumoral hemorrhagic conversion (common especially with glioblastoma and brain metastases from melanoma, renal cell carcinoma, and choriocarcinoma); as a complication of thrombocytopenia or coagulation disturbances; or by rupture of a neoplastic aneurysm (e.g., with choriocarcinoma).

Cerebral venous thrombosis can occur as a consequence of hypercoagulability and by direct compression or invasion of cerebral sinuses from dural or calvarial metastases and meningioma (Rogers, 2003; Nguyen and DeAngelis, 2006; Grisold et al., 2009; Bang et al., 2011; Dearborn et al., 2014). Pituitary apoplexy is a rare, potentially life-threatening complication, caused by hemorrhage and/or infarction of pituitary adenomas. The rapid increase in the contents of the sella results in direct compression of pituitary tissue and interruption of its blood supply. Acute alterations of the balance between pituitary tumor perfusion and metabolism (e.g., with hypotension or pregnancy) may precipitate pituitary apoplexy (Johnston et al., 2015; Oldfield and Merrill, 2015). Table 39.2 summarizes the patterns of stroke, causes, mechanisms, and associated malignancies.

CLINICAL PRESENTATION

Clinical presentations of acute ischemic stroke in patients with cancer range from silent disease to diffuse vascular encephalopathies and vascular events with focal deficits. Transient ischemic attacks can occur in patients with leptomeningeal disease owing to arterial vasospasm (Rogers, 2003; Grisold et al., 2009; Dearborn et al., 2014). Signs of increased ICP such as headache, vomiting, and papilledema may occur in patients with ICH and cerebral venous thrombosis. Patients with intravascular lymphoma usually present with diffuse encephalopathy or multifocal cerebral infarcts. Acute expansion of the pituitary gland causes sudden visual field and acuity defects, and various cranial neuropathies (Damek, 2009; Johnston et al., 2015).

Table 39.2

Cerebrovascular disease classification, cancer-related mechanism, and associated malignancies

Vascular events	Causes	Mechanisms	Associated malignancies
Ischemic (thomboembolic)	Hypercoagulability*	Microthrombi formation Release of tissue factor and cancer procoagulant Release of procoagulant cytokines (TNF-alpha, IL-1, and IL-6)	Adenocarcinoma (pancreas, colon, breast, lung, prostate and ovary) GBM Hematologic malignancies
	Hyperviscosity	Vessel obstruction by neoplastic cells	Multiple myeloma Bing–Neel syndrome AML CLL
	Direct tumor effect	Tumor invasion of vessels Compression of large vessels	Brain metastases Primary CNS tumors LM
	NBTE	Cerebral emboli from sterile vegetations on heart valves	Adenocarcinoma, particularly pancreas
	Tumor emboli	Tumor cells enter pulmonary circulation and distributed to the circulation	Cardiac myxoma Lung cancer Lung metastases Heart metastases
	Angioinvasive	Proliferation of tumor cells with vessel wall	Intravascular B-cell lymphoma
Hemorrhagic	Intratumoral hemorrhage	Hemorrhagic conversion of vascular tumors	GBM Brain metastases (melanoma, RCC, choriocarcinoma, papillary thyroid carcinoma)
	Coagulation disorders	Thrombocytopenia Hyperleukocytosis syndrome DIC Primary fibrinolysis Protein synthesis deficiency from liver injury	Hematologic malignancy, particularly AML, APML, CLL, lymphoma Multiple myeloma Mucin-producing adenocarcinomas
	Ruptured aneurysm	Aneurysm formation from neoplastic infiltration of arteries	Choriocarcinoma Lung cancer Cardiac myxoma
Cerebral venous thrombosis	Direct tumor effects on venous sinus	Compression or invasion of veins or sinuses	Dural or calvarial metastases Meningioma
	Hypercoagulability	As above*	As above*
Pituitary apoplexy	Direct compression and interruption of blood supply	Infarction and/or hemorrhage of pituitary tumor from precipitating factors	Pituitary macroadenoma (prolactinoma and growth hormone-secreting adenoma)

NBTE, nonbacterial thrombotic endocarditis; TNF-α, tumor necrosis factor-α; IL-1, interleukin-1; DIC, disseminated intravascular coagulopathy; GBM, glioblastoma; AML, acute myelogenous leukemia; CLL, chronic lymphocytic leukemia; CNS, central nervous system; LM, leptomeningeal metastasis; RCC, renal cell carcinoma; APML, acute promyelocytic leukemia.
*Hypercoagulability mechanisms and associated malignancies.

Neuroimaging

All cancer patients presenting with a change in mental status warrant at least a cranial CT scan and, in most cases, a gadolinium-enhanced MRI. Multiple lesions not respecting vascular territories, of varying ages, are frequently encountered in cancer-related ischemic stroke compared to involvement of one arterial territory in conventional stroke patients (Navi and Segal, 2009; Kim et al., 2010; Bang et al., 2011). Atypical location of ICH, multiple parenchymal hemorrhages, and the presence of enhancing masses are suggestive of bleeding from brain tumors. A high prevalence of embolic signal is detected by transcranial Doppler ultrasound in cancer patients with ischemic stroke (58%) compared to patients without cancer (33%). Detection of embolic signal in transcranial Doppler may provide clues regarding cancer-specific mechanism related to hypercoagulopathy, and may be useful for treatment monitoring (Grisold et al., 2009; Seok et al., 2010). Echocardiography, preferably transesophageal, demonstrates vegetations in most patients with NBTE (Grisold et al., 2009), but may sometimes be unrevealing, suggesting intravascular clot formation causing artery-to-artery embolism (Bang, 2011).

Complete blood count, coagulation profile, and disseminated intravascular coagulopathy panel should be obtained in stroke patients without conventional risk factors and in cancer patients without identifiable stroke etiology (i.e., direct tumor effects or treatment effects) (Dearborn et al., 2014). A platelet level of $<10\ 000\ mm^3$ creates high risk for spontaneous ICH, while a level over $1\ 000\ 000\ mm^3$ is associated with ischemic infarction. Elevated D-dimer is a direct measure of activated coagulation; a high level in the presence of infarcts in multiple vascular territories is consistent with cancer-related stroke (Kim et al., 2010). Tissue factor or thromboplastin, a potent procoagulant, is frequently upregulated in patients with cancer. The factors and products of thrombogenesis are potential biomarkers, including D-dimer, lipoprotein (a), homocysteine, tissue plasminogen activator, plasminogen activator inhibitor-1, and tissue factor-bearing microparticles (Navi and Segal, 2009; Bang et al., 2011; Jo et al., 2014). CSF examination may be useful in the diagnosis of leptomeningeal disease (Rogers, 2003).

Hospital course and management

In patients with cancer and ischemic stroke with identifiable CVD risk factors, secondary prevention with antiplatelet agents or anticoagulation and risk factor modification is appropriate. The role of antiplatelet agents in secondary prevention for cancer-related stroke remains uncertain (Dearborn et al., 2014). Although there is no proven therapy for cancer-related stroke, treatment mainly involves anticoagulation (Grisold et al., 2009; Dearborn et al., 2014). Low-molecular-weight heparin (LMWH) and warfarin are indicated in the treatment of venous thromboembolism (VTE) in cancer patients, including in those with brain tumors (Lyman et al., 2013). Whether these recommendations can be extrapolated to arterial stroke in the setting of hypercoagulability requires further investigation. Anticoagulation is the standard treatment for patients with CVD (Grisold et al., 2009). LMWH products are preferred over warfarin due to better bioavailability, longer half-life, lack of drug–drug interactions, and more predictable response (Jo et al., 2014). The CLOT trial demonstrated LMWH was more effective in reducing recurrent VTE without increasing the risk of bleeding in cancer patients (Lee et al., 2003). Use of novel oral anticoagulants that directly inhibit thrombin or factor Xa is currently not recommended for patients with malignancy due to limited data. Nevertheless, these agents can be used in patients with heparin-induced thrombocytopenia (Lyman et al., 2013).

The treatment approach for intratumoral hemorrhage depends on the site and size of hemorrhage, clinical status of the patient, and presence of coagulopathy. Correction of coagulation abnormalities is paramount with goal platelet level $>50\ 000/\mu L$ and normal coagulation tests. Surgical evacuation of the clot and ventricular drainage may be indicated to decrease ICP and relieve hydrocephalus (Quinn and DeAngelis, 2000; Grisold et al., 2009). Corticosteroids may reduce surrounding vasogenic edema. In pituitary apoplexy, prompt surgical decompression is indicated (Damek, 2009; Grisold et al., 2009).

Central nervous system infections

Epidemiology

CNS infection is an important cause of neurologic morbidity and mortality in patients with cancer. CNS infection is most common with hematologic malignancies, but may also occur with primary tumors (Pruitt, 2003). Increased susceptibility in cancer patients is primarily due to immunologic compromise from the underlying malignancy or treatment thereof.

Neuropathology and pathophysiology

Infectious agents reach the CNS by hematogenous spread, metastasis from focal infection elsewhere, or direct extension from surrounding structures (Patchell and Posner, 1985). The potential pathogens can be narrowed depending on the patient's type of immunodeficiency as well as the duration and degree of immunosuppression (Table 39.3) (Pruitt, 2004). Neutropenia ($<500/mm^3$) secondary to intensive chemotherapy

Table 39.3

Central nervous system (CNS) infection, common pathogens, and treatment (Patchell and Posner, 1985; Pruitt, 2004, 2012)

CNS infection	Associated malignancies/conditions	Empiric treatment
Bacterial		
Streptococcus pneumoniae	Lymphoma Leukemia Multiple myeloma Solid tumors	PCN intermediate resistance: ceftriaxone or cefotaxime PCN-resistant: vancomycin
Staphylococci	Head and spine tumors Patients with VP shunt	Methicillin-sensitive: nafcillin+cefotaxime Methicillin-resistant: vancomycin ± intraventricular vancomycin
Gram-negative rods	Lymphoma Leukemia Solid tumors Patients with VP shunt Cranial surgeries HSCT Chemotherapy- or radiation-induced bone marrow suppression	Ceftriaxone+gentamycin Ceftazidime for *Pseudomonas aeruginosa*
Listeria monocytogenes	Lymphoma Chronic steroid use HSCT	Ampicillin+gentamicin
Nocardia asteroides	Lymphoma Chronic steroid use HSCT	Sulfadiazine
Viruses		
Herpes simplex virus	Lymphoma Leukemia Solid tumors HSCT Organ transplant Chronic steroid use	Acyclovir
Varicella-zoster	Lymphoreticular neoplasms Organ transplant Chronic steroid use	Acyclovir
Cytomegalovirus	Lymphoma Leukemia Solid tumors HSCT Chemotherapy or radiation-induced bone marrow suppression	Ganciclovir
Epstein–Barr virus (PTLD)	Lymphoreticular neoplasms Organ transplant Chronic steroid use	Acyclovir
Fungi		
Cryptococcus neoformans	Lymphoma Organ transplant Chronic steroid use	Induction therapy: amphotericin+flucytosine or amphotericin B lipid complex for 2 weeks Maintenance therapy: fluconazole for at least 6 months
Candida albicans	Lymphoma Organ transplant Chronic steroid use	Amphotericin+flucytosine

PCN, penicillin; VP, ventriculoperitoneal; HSCT, hematopoietic stem cell transplant; PTLD, posttransplantation lymphoproliferative disorder.

or bone marrow invasion increases the risk of bacterial and fungal infections (Pruitt, 2004, 2012). Impairment of B lymphocytes and immunoglobulin production in patients with blood dyscrasias (i.e., chronic lymphocytic leukemia and multiple myeloma) increases risk of infection with encapsulated organisms, such as *Haemophilus influenzae* and *Streptococcus pneumoniae*. Impaired T-cell-mediated immunity, as in patients with lymphoma, hematopoietic stem cell transplantation (HSCT) recipients, and patients receiving immunosuppressive therapy such as tacrolimus, mycophenolate, bortezomib, cyclosporine, and chronic corticosteroids, predisposes to a wide variety of opportunistic CNS pathogens, including viruses (cytomegalovirus, herpes simples virus, varicella-zoster virus, JC virus, and Epstein–Barr virus), fungi (*Cryptococcus neoformans* and *Candida albicans*), parasites (*Toxoplasma gondii* and *Strongyloides stercoralis*), and bacteria (*Listeria monocytogenes, Nocardia asteroides,* and *Mycobacterium tuberculosis*). Barrier disruption after cranial surgery, ventricular shunt placement, and skin injury from radiation therapy predispose to infections with *Staphylococcus aureus, S. epidermidis,* and *Propionibacterium acnes,* as well as *Candida* species (Pruitt, 2004, 2012).

CLINICAL PRESENTATION

Patients with cancer may not display the typical signs and symptoms of CNS infection, such as fever, nuchal rigidity, or headache, due primarily to their deficient immune status (Patchell and Posner, 1985; Baldwin et al., 2012). Altered mental status may be the only presentation in severely immunocompromised patients (Patchell and Posner, 1985). Patients with brain abscess may present with focal deficits.

NEUROIMAGING

Diagnosis of CNS infection in cancer patients requires a careful clinical history, neuroimaging studies, and ancillary tests. The presence of focal deficits suggests bacterial or fungal infection; menigoencephalitic presentation is seen with viruses, bacteria, or fungi; diffuse encephalitis is likely viral in etiology; and cerebral infarction raises suspicion for bacterial or fungal endocarditis (Damek, 2009; Baldwin et al., 2012; Pruitt, 2012).

Brain CT or MRI may be helpful in identifying focal lesions concerning for abscess and meningeal enhancement seen in meningitis. However, neuroimaging studies do not distinguish CNS infection from its mimics (Baldwin et al., 2012; Pruitt, 2012). CSF analysis is indicated in appropriate patients suspected of having CNS infection. A neuroimaging study is recommended prior to lumbar puncture to rule out mass lesions (Patchell and Posner, 1985; Baldwin et al., 2012). As many as one-third of neutropenic patients with culture-proven bacterial or fungal meningitis demonstrate a normal CSF white blood cell count (Pruitt, 2012). CSF Gram stain and culture, and if indicated, cytology should be obtained in all patients. Polymerase chain reaction analysis is widely used to detect viral pathogens (Pruitt, 2004). Histopathologic analysis obtained by biopsy distinguishes brain abscess from metastasis (Baldwin et al., 2012).

HOSPITAL COURSE AND MANAGEMENT

Empiric treatment is based on the patient's clinical presentation, suspected organisms, and initial laboratory findings. Treatment of CNS infections is described further in Chapter 19.

Paraneoplastic neurologic disorders

EPIDEMIOLOGY

Paraneoplastic disorders are nonmetastatic, immune-related conditions that result from cancer (Martel et al., 2014). They are relatively uncommon, affecting less than 1% of patients with cancer overall. However, certain paraneoplastic syndromes are well described, such as Lambert–Eaton myasthenic syndrome (LEMS), which affects up to 3% of patients with small-cell lung carcinoma, and myasthenia gravis (MG), which occurs in up to 15% of patients with thymoma (Darnell and Posner, 2003; Pittock et al., 2004). Subacute cerebellar degeneration and limbic encephalitis are the most commonly diagnosed paraneoplastic syndromes affecting the CNS (24% and 10%, respectively), while sensory neuronopathy is the most common condition involving the peripheral nervous system (23%) (Giometto et al., 2010). Recognition of the paraneoplastic nature of the disorder may be challenging, but is important to enable initiation of treatment before neurologic injury becomes irreversible (Greenlee, 2010).

NEUROPATHOLOGY AND PATHOPHYSIOLOGY

Paraneoplastic syndromes can be divided into two groups based on epitopic targets and immunopathogenic mechanisms (Table 39.4).

One group of patients are those in whom the antibody response is directed against intracellular neuronal or neuroglial proteins. Examples include paraneoplastic cerebellar degeneration or encephalomyelitis. The culprit onconeuronal antibodies are associated with malignancies that target intracellular epitopes. It is known that these autoantibodies can trigger the activation of T cells. However, the antigen itself has been proposed to be capable of triggering a T-cell or humoral immune response through different pathways (Dalmau and

Table 39.4

Paraneoplastic neurologic syndromes

Location	Syndromes	Clinical presentation	Antibodies	Associated malignancies
Brain	Paraneoplastic cerebellar degeneration*	Gait disturbance, nausea, dizziness, diplopia	Anti-Yo (PCA-1) Anti-Tri Anti-mGluR1	Gynecologic and breast CA Hodgkin's lymphoma Hodgkin's lymphoma NSCLC
	Limbic encephalitis*	Seizures, memory loss, change in personality, hallucinations, insomnia	Anti-Ma1 Anti-Ma2 Anti-Tri Anti-mGluR1 AMPAR (Anti-GluR1/2)	Testicular, germ cell tumors, NSCLC, other solid tumors Hodgkin's lymphoma Hodgkin's lymphoma Lung, breast thymus CA
	Encephalomyelitis*	Multifocal neurologic deficits, involving brainstem (diplopia, dysphagia, dysarthria); cerebellar (ataxia, dizziness); autonomic (orthostatic hypotension, arrhythmias)	Anti-Hu (ANNA-1) Anti-CV2 (anti-CRMP5)	Small-cell CA (lung), neuroendocrine tumors Small-cell CA (lung)
	Opsoclonus-myoclonus*	Ataxia, myoclonus, opsoclonus, dysarthria, encephalopathy	Anti-Ri (ANNA-2)	Breast CA
Retina	Cancer-associated retinopathy	Painless, often bilateral photosensitivity, loss of color vision, central scotomas, visual flashes	Anti-recoverin	SCLC
	Melanoma-associated retinopathy	Sudden onset of night blindness, photopsia	Antibipolar cells of retina	Melanoma
Spinal cord	Stiff-person syndrome	Progressive rigidity and intermittent painful spams (distal lower limbs, sphincters, cranial nerves are spared)	Antiamphiphysin Anti-GAD	Breast, SCLC Thymoma, RCC

Peripheral nerves	Sensory neuronopathy*	Multifocal, asymmetric numbness, severe pain and paresthesia, loss of deep tendon reflexes	Anti-Hu (ANNA-1) Anti-CV2 (anti-CRMP5)	SCLC Small-cell CA
	Chronic sensorimotor neuronopathy	Mild to moderate distal symmetric sensorimotor deficits	Anti-CV2 (anti-CRMP5)	Small-cell CA
	Autonomic neuropathy	Panautonomic neuropathy (sympathetic, parasympathetic, and enteric; orthostatic hypotension; gastrointestinal dysfunction; arrhythmia)	Anti-Hu (ANNA-1) Anti-CV2 (anti-CRMP5) Antiganglionic AChR	SCLC SCLC and thymoma Thyroid, lung, pancreas, bladder, rectal CA
	Autonomic neuropathy with pseudo-obstruction*	Persistent constipation & abdominal distention, dysphagia, nausea, vomiting	Anti-Hu (ANNA-1)	SCLC
	Peripheral nerve excitability (neuromyotonia)	Muscle cramps, stiffness, weakness, excessive sweating	Anti-VGKC	Thymoma, breast CA, lung CA
Neuromuscular junction	Lambert–Eaton myasthenic syndrome*	Progressive proximal muscle weakness, reduced deep tendon reflexes, autonomic dysfunction (late in course)	Anti-VGCC	SCLC
	Myasthenia gravis	Fatigable weakness of voluntary muscles	Anti-AChR	Thymoma
Muscles	Dermatomyositis*	Skin: heliotrope rash (rash of upper eyelids), Gottron papules (scaly plaques on bony surfaces); shawl sign (photosensitive skin rash on face, neck, chest, back and shoulder) Muscles: symmetric progressive proximal muscle weakness	Anti-TIF1-γ, anti-NPX2	Ovaries, lung, colorectal, breast, gastric CA

CA, cancer; NSCLC, nonsmall-cell lung cancer; SCLC, small-cell lung cancer; RCC, renal cell carcinoma; VGKC, voltage-gated potassium channel; VGCC, voltage-gated calcium channel; AChR, acetylcholine receptor.
*Classis paraneoplastic neurologic disorders.

Rosenfeld, 2008; Didelot and Honnorat, 2014; Martel et al., 2014). Onconeuronal antibodies include anti-Yo (APCA, PCA1), anti-Hu (ANNA-1), anti-Ri (ANNA-2), anti-Tr, anti-CV2 (CRMP5), antiamphiphysin, anti-Ma1, and anti-Ma2. The mGluR1 is expressed both inside the cell and in the cell membrane. The expected response to immunomodulation treatment in this group is poor due to irreversible damage to neuronal cells (Greenlee, 2010; Martel et al., 2014).

A second group of patients are those in whom antibodies react with cell surface antigens. Neuronal cell surface antibodies impair neuronal function by targeting antigens in the neuronal or neuromuscular junction membranes (Greenlee, 2010; Martel et al., 2014). This group may or may not have an underlying malignancy. Relevant antibodies include NMDA receptor antibodies associated with encephalitis, voltage-gated calcium channel antibodies in LEMS, acetylcholine receptor antibodies in MG, and anti-GAD antibodies in stiff-person syndrome. These groups of patients have favorable responses to immunomodulation (Greenlee, 2010; Didelot and Honnorat, 2014; Martel et al., 2014).

Clinical presentation and neurodiagnostics

Neurologic symptoms are the first manifestation of a tumor in more than two-thirds of patients with paraneoplastic disorders, such that a high index of suspicion is required for early diagnosis. The onset of symptoms is typically subacute, followed by a rapidly progressive course (Damek, 2009; Pelosof and Gerber, 2010).

Patients with paraneoplastic encephalitis have evidence of inflammation in the CSF, typically with lymphocytic pleocytosis, elevated protein concentration, high IgG index, and, in some cases, oligoclonal bands (Dalmau and Rosenfeld, 2008). High titers of onconeuronal antigens are highly specific for diagnosis of paraneoplastic disease, while detection of neuronal surface antigen antibodies may be associated with malignancy or idiopathic. When possible, serum and CSF antibody titers should be obtained. Tissue-based assays have higher yield for intracellular antigens (Gultekin, 2015).

MRI may demonstrate high signal intensity in affected areas on T2 and FLAIR sequences involving the mesial temporal lobe in paraneoplastic limbic encephalitis and transient diffuse cerebellar hemispheric enlargement or widespread abnormalities in encephalomyelitis can be seen in MRI studies. However, normal neuroimaging studies are not uncommon (Damek, 2009; Xia et al., 2010, Kaira et al., 2014). Cerebral or cerebellar atrophy can be seen in chronic cases (Damek, 2009; Gultekin, 2015). Proposed diagnostic criteria for paraneoplastic neurologic disorders are proposed and summarized in Table 39.5 (Graus et al., 2004).

Thorough screening for occult neoplasm is indicated when a paraneoplastic disorder is suspected. If chest, abdomen, and pelvis CT are negative, fluorodeoxyglucose-positron emission tomography (FDG-PET) may still reveal the tumor. In cases where a potential primary malignancy is suspected, a more specific test such as mammography, lower/upper gastrointestinal endoscopy, pelvic and transvaginal ultrasound, and testicular ultrasound may be diagnostic (Dalmau and Rosenfeld, 2008; Pelosof and Gerber, 2010; Titulaer et al., 2011; Gultekin, 2015). Positive screening is seen in 70–80% of patients (Dalmau and Rosenfeld, 2008; Damek, 2009). If the screening studies fail to

Table 39.5

Diagnostic criteria for paraneoplastic neurologic disorders (PND) (Graus et al., 2004)

Definite PND	Possible PND
1. A classic syndrome and cancer that develops within 5 years of the diagnosis of the neurologic disorder	1. A classic syndrome, no onconeuronal antibodies, no cancer but a high risk of having an underlying tumor
2. A nonclassic syndrome that resolves or significantly improves after cancer treatment without concomitant immunotherapy provided that the syndrome is not susceptible to spontaneous remission	2. A neurologic syndrome (classic or not) with partially characterized onconeuronal antibodies and no cancer
3. A nonclassic syndrome with onconeuronal antibodies (well characterized or not) and cancer that develops within 5 years of the diagnosis of neurologic disorder	3. A nonclassic syndrome, no onconeuronal antibodies, and cancer present within 2 years of diagnosis
4. A neurologic syndrome (classic or not) with well-characterized onconeuronal antibodies (anti-Hu, Yo, CV21, Ri, Ma2, or amphiphysin), and no cancer	

identify malignancies, repeat screening after 3–6 months, followed by every 6 months for several years, is recommended, except in LEMS, where 2 years is sufficient (Titulaer et al., 2011).

HOSPITAL COURSE AND MANAGEMENT

In patients with rapidly progressive symptoms highly suggestive of a paraneoplastic disorder, prompt initiation of treatment is necessary to prevent neurologic deterioration. Rapid administration of treatment directed towards the underlying malignancy has been the most successful strategy for improvement or stabilization of neurologic deficits (Damek, 2009; Gultekin, 2015). Beyond this, no evidence-based studies above class III have been reported for most paraneoplastic disorders, with the exception of class II evidence for MG, LEMS, and stiff-person syndrome (Greenlee, 2010). Plasma exchange is beneficial mostly among patients with antibodies directed at cell surface antigens; its utility in patients with antibodies directed at intracellular antibodies remains uncertain (Vernino et al., 2004; Greenlee, 2010). Immunomodulatory treatments with corticosteroids (e.g., methylprednisolone 1 g/day for 3–5 days), intravenous immunoglobulin (0.4 g/kg IV for 5 days to a total dose of 2 g/kg), and cyclophosphamide (e.g., 2 mg/kg/day administered by mouth), singly or in combination, are often beneficial in paraneoplastic disorders affecting the peripheral nervous system, neuromuscular junction, and muscles. CNS paraneoplastic disorders are usually more refractory to treatment, although anecdotal reports have demonstrated responses to corticosteroids or IVIG (Damek, 2009; Greenlee, 2010). Rituximab, an anti-CD20 monoclonal agent, has shown beneficial effects in patients with anti-Hu associated sensory neuropathy and gastric pseudo-obstruction, anti-GAD and anti-amphiphysin in stiff-person syndrome, and patients with anti-Yo antibodies (Baker et al., 2005; Shams'ili et al., 2006; Coret et al., 2009; Dupond et al., 2010; Greenlee, 2010). 3,4-Diaminopyridine for LEMS, benzodiazepines for stiff-person syndrome, and cholinesterase inhibitors such as pyridostigmine for MG have been shown to be effective as supportive treatment (Jani-Acsadi and Lisak, 2007; Greenlee, 2010).

IATROGENIC NEUROLOGIC EMERGENCIES
Radiation therapy

The central and peripheral nervous system are vulnerable to effects of radiation. Neurologic complications are categorized based on the timing of tissue injury, which is helpful in predicting reversibility of symptoms (Sheline, 1977).

ACUTE INJURY (LESS THAN 1 MONTH)

Capillary injury and leakiness leading to edema may occur during or immediately after RT. Acute encephalopathy characterized by somnolence, headache, nausea, vomiting, and exacerbation of pre-existing neurologic deficits can develop days after starting RT and is generally responsive to corticosteroids. There is no specific imaging finding associated with early injury to the brain (Ricard et al., 2012). Acute worsening after radiation to the spinal cord warrants investigation for intratumoral hemorrhage or tumor progression (Posner, 1995).

EARLY-DELAYED INJURY (1–6 MONTHS)

Complications occurring 1–6 months after completion of RT are due to edema and demyelination. Patients may present with somnolence syndrome consisting of drowsiness, fatigue, anorexia, irritability, and transitory cognitive disturbance. MRI findings may be unchanged or may show increased edema and contrast enhancement mimicking tumor progression, termed pseudoprogression. This is most commonly encountered in patients with glioblastoma who received concurrent RT and temozolomide. Symptoms during this period are often reversible with corticosteroids (Brandsma et al., 2008; Ricard et al., 2012). L'Hermitte's phenomenon is seen in patients with early-delayed RT-induced myelopathy with no distinctive imaging findings and typically resolves spontaneously. RT to head and neck areas may produce cranial neuropathies involving hypoglossal and vagus nerves when treatment was directed at the neck region, and optic, oculomotor, trochlear, trigeminal, abducens, and facial nerves following treatment of skull base tumors (Lin et al., 2002; Kargiotis and Kyritsis, 2012). Treatment of breast, lung, and pelvic cancers may cause transient brachial and lumbosacral plexopathies (Chi et al., 2008).

LATE INJURY (MORE THAN 6 MONTHS)

Injuries inflicted at this period are irreversible and are hypothesized to be due to a combination of small- and medium-sized vessel injuries, demyelination with loss of oligodendrocytes, and an allergic response from antigens released by damaged glial cells. Radiation necrosis is the pathologic end state (Crossen et al., 1994; Giglio and Gilbert, 2010). Focal radiation necrosis may cause seizures, focal neurologic symptoms, and signs of increased ICP. MRI reveals focal areas of necrosis that is difficult to distinguish from tumor progression. Surgery is the only way to definitively confirm the diagnosis.

Permanent diffuse white-matter changes or leukoencephalopathy can present as mild cognitive impairment or severe dementia. MRI demonstrates white-matter abnormalities and atrophy (Giglio and Gilbert, 2010; Ricard et al., 2012). Late-delayed myelopathy can present with Brown-Séquard syndrome, weakness and sensory loss of lower extremities, and bowel and bladder dysfunction, occurring abruptly or insidiously, and is usually irreversible. MRI of the spinal cord demonstrates increased T2 signal and contrast enhancement at the affected level. Corticosteroids have variable benefits in radiation necrosis (Giglio and Gilbert, 2010).

Bevacizumab, a monoclonal VEGF inhibitor, has shown beneficial effects in patients with cerebral radiation necrosis (Torcuator et al., 2009; Levin et al., 2011). Permanent cranial neuropathies and plexopathies are seen due to delayed radiation effects to the peripheral nerves. Myokymia on electromyography and hypointensity in T1- and T2-weighted MR images can be seen (Qayyum et al., 2000; Chi et al., 2008; Jaeckle, 2010). Pain management and physical therapy are recommended for RT-induced neuropathy (Rogers, 2012). Table 39.6 summarizes the classification of neurologic complications of RT.

Chemotherapy and biologic agents

Certain cytotoxic agents and biologic agents produce neurotoxicity. The symptoms may resemble other neurologic complications of cancer or metastases. The severity depends on the treatment dose, duration, comorbidities, and concomitant administration of other neurotoxic agents.

High-dose methotrexate, ifosfamide, and procarbazine can cause acute confusion, hallucination, seizures, and drowsiness during or a few days after treatment (Kwong et al., 2009; Lee et al., 2012a). High doses of cytarabine may produce cerebellar manifestations in as many as 20% of patients (Herzig et al., 1987). Aseptic meningitis, characterized by neck stiffness, headache, nausea, vomiting, fever, and transverse myelopathy, presenting as paraplegia, leg pain, sensory level and neurogenic bladder dysfunction, can occur after IT administration of methotrexate and cytarabine (Patchell and Posner, 1985). Oxaliplatin may produce acute cold-induced dysesthesia involving distal extremities, throat, mouth, or face. Vestibulocochlear toxicity has been associated with cisplatin (Sioka and Kyritsis, 2009; Giglio and Gilbert, 2010; Lee et al., 2012a).

Bevacizumab, an anti-VEGF agent, increases the risk of thromboembolic stroke and intracranial hemorrhage. It has also been reported to cause posterior reversible encephalopathy syndrome (PRES) (Armstrong et al., 2012). Cetuximab, an epidermal growth factor receptor inhibitor, is associated with aseptic meningitis (Baselga et al., 2000). Acute inflammatory reactions such as myopathy, aseptic meningitis, severe meningoradiculoneuritis, temporal arteritis, and Guillain–Barré syndrome are reported to occur from ipilimumab (anticytotoxic T-lymphocyte antigen-4) (Tarhini, 2013; Liao et al., 2014). Crizotinib (ALK and c-MET inhibitor) may cause visual disturbance (Bang, 2011). Interferon-α is associated with neuropsychiatric symptoms and seizures, while IL-2 may cause transient encephalopathy and neurocognitive dysfunction (Apfel, 2012). Painful, length-dependent sensory axonal neuropathy occurs in about 35% of patients who received bortezomib (a proteosome inhibitor) (Richardson et al., 2006; Lee et al., 2012b). Table 39.7 summarizes the common neurologic complications from chemotherapy and biologic agents.

Hematopoietic stem cell transplantation

Symptomatic neurologic complications occur in 10–40% of patients undergoing HSCT (Rosenfeld and Pruitt, 2006). Myeloablative doses of chemotherapy may result in neurotoxicity not usually seen with conventional doses. These complications include encephalopathy from ifosfamide, melphalan, etoposide, and thiotepa; seizures from busulfan; and psychiatric symptoms from ifosfamide and busulfan (Rosenfeld and Pruitt, 2006; Giglio and Gilbert, 2010). Severe thrombocytopenia from bone marrow suppression predisposes patients to ICH (Quant and Wen, 2008; Giglio and Gilbert, 2010). Calcineurin inhibitors (tacrolimus and cyclosporine), used for prevention of graft-versus-host disease, cause headache, altered mental status, seizures, cortical blindness, hallucinations, spasticity, paresis, ataxia, and PRES (Hinchey et al., 1996; Sklar, 2006). Neurologic complications usually improve after reduction or discontinuation of treatment. CNS infection affects 3–8% of patients after HSCT. Neutropenic patients are at risk for systemic bacterial, *Candida*, and *Aspergillus* infections, as well as herpes simplex virus. Deficient cellular immunity predisposes patients to fungal, cytomegalovirus, Gram-positive bacterial infection, and JC virus reactivation (Quant and Wen, 2008; Giglio and Gilbert, 2010).

CONCLUSION

Neuro-oncologic emergencies are relatively common in patients with cancer and associated with significant morbidity and mortality. At times, these are the initial manifestations of an undiagnosed malignancy and a high index of suspicion is therefore essential. Rapid recognition and prompt treatment can prolong the patient's life and improve functional status and quality of life.

Table 39.6

Classification of neurologic complications of radiotherapy

Onset	Reversibility	Clinical manifestations				Treatment
		Brain	Spinal cord	Cranial nerves	Peripheral nerves	
Acute (<1 month)	Reversible	Somnolence Headache Nausea Vomiting Exacerbation of pre-existing neurologic deficits	–	–	Paresthesia	Corticosteroids
Early-delayed (1–6 months)	Reversible	Drowsiness Fatigue Anorexia Irritability Transient cognitive disturbance	Lhermitte's phenomenon	Painless visual loss Tongue weakness Hearing loss Anosmia	Transient plexopathy	Corticosteroids Bevacizumab
Late (>6 months)	Permanent	Focal radiation necrosis: seizure, focal neurologic deficits, and signs of increased ICP. White-matter disease: mild cognitive impairment, severe dementia	Brown-Séquard syndrome Weakness Sensory level Bowel and bladder dysfunction	Hearing loss Visual loss Lower CN palsies	Irreversible plexopathy Pain	Corticosteroids may or may not be beneficial Supportive care

ICP, intracranial pressure; CN, cranial nerve.

Table 39.7

Common neurologic complications of chemotherapy and biologic agents

Complications	Agents	Treatment
Encephalopathy	Ifosfamide, HD-MTX, HD-cytarabine, procarbazine, vincristine, IL-2	Methylene blue (for ifosfamide-induced) Discontinue drugs
Aseptic meningitis	IT-MTX and cytarabine	Discontinue/reduce drugs Corticosteroids
	Cetuximab	Discontinue drugs
	Ipilimumab	Discontinue drugs, IV corticosteroids, IVIG, plasmapharesis
Cerebellar syndromes	HD-cytarabine	Discontinue drugs Prevention by avoiding high doses
Seizures	Cisplatin, cytarabine, cyclophosphamide, nelarabine, ifosfamide, vincristine, Interferon-α	Discontinue drugs
Stroke	Bevacizumab, imatinib	Discontinue drugs
PRES	Cisplatin, cyclophosphamide, gemcitabine, bevacizumab, rituximab (rare)	Discontinue drugs
Neuropsychiatric symptoms	Interferon-α	Discontinue drugs
Myelopathy	IT-MTX and cytarabine	Discontinue/reduce drugs
Guillain–Barré syndrome	Ipilimumab	Discontinue drugs, IV corticosteroids, IVIG, plasmapharesis
Acute sensory dysesthesia	Oxaliplatin, ifosfamide, cytarabine (rare)	Discontinue/reduce drugs Antiepileptics (carbamazepine, gabapentin, lamotrigine) Tricyclic antidepressants Serotonin-norepinephrine reuptake inhibitors

HD, high dose; MTX, methotrexate; IL-2, interleukin-2; IT, intrathecal; IV, intravenous; IG, immunoglobulin; PRES, posterior reversible encephalopathy syndrome.

REFERENCES

Abrahm JL (2004). Assessment and treatment of patients with malignant spinal cord compression. J Support Oncol 2 (377–388): 391. discussion 391–393, 398, 401.

Apfel SC (2012). Neurologic complications of immunomodulatory agents. In: P Wen, D Schiff, EQ Lee (Eds.), Neurologic complications of cancer therapy, Demos Medical Publishing, New York, NY.

Armstrong T, Wen P, Gilbert M et al. (2012). Management of treatment-associated toxicities of anti-angiogenic therapy in patients with brain tumors. Neuro Oncol 14: 1203–1214.

Bach F, Larsen BH, Rohde K et al. (1990). Metastatic spinal cord compression. Occurrence, symptoms, clinical presentations and prognosis in 398 patients with spinal cord compression. Acta Neurochir (Wien) 107: 37–43.

Baker MR, Das M, Isaacs J et al. (2005). Treatment of stiff person syndrome with rituximab. J Neurol Neurosurg Psychiatry 76: 999–1001.

Baldwin KJ, Zivkovic SA, Lieberman F (2012). Neurologic emergencies in patients who have cancer: diagnosis and management. Neurol Clin 30: 101–128.

Bang YJ (2011a). The potential for crizotinib in non-small cell lung cancer: a perspective review. Ther Adv Med Oncol 3: 279–291.

Bang OY, Seok JM, Kim SG et al. (2011). Ischemic stroke and cancer: stroke severely impacts cancer patients, while cancer increases the number of strokes. J Clin Neurol 7: 53–59.

Barnholtz-Sloan JS, Sloan AE, Davis FG et al. (2004). Incidence proportions of brain metastases in patients diagnosed (1973 to 2001) in the Metropolitan Detroit Cancer Surveillance System. J Clin Oncol 22: 2865–2872.

Baselga J, Pfister D, Cooper MR et al. (2000). Phase I studies of anti-epidermal growth factor receptor chimeric antibody C225 alone and in combination with cisplatin. J Clin Oncol 18: 904–914.

Boogerd W, Van Den Bent M, Koehler PJ et al. (2004). The relevance of intraventricular chemotherapy for leptomeningeal metastasis in breast cancer: a randomized study. Eur J Cancer 40: 2726–2733.

Brandsma D, Stalpers L, Taal W et al. (2008). Clinical features, mechanisms, and management of pseudoprogression in malignant gliomas. Lancet Oncol 9: 453–461.

Castellani G, Zweifel C, Kim DJ et al. (2009). Plateau waves in head injured patients requiring neurocritical care. Neurocrit Care 11: 143–150.

Cavaliere R, Schiff D (2007). Chemotherapy and cerebral metastases: misperception or reality? Neurosurg Focus 22. E6.

Cavaliere R, Farace E, Schiff D (2006). Clinical implications of status epilepticus in patients with neoplasms. Arch Neurol 63: 1746–1749.

Cetari DM, Weine DM, Panageas KS et al. (2004). Stroke in patients with cancer: incidence and aetiology. Neurology 64: 2025–2030.

Chamberlain MC (2010a). Brain metastases: a medical neuro-oncology perspective. Expert Rev Neurother 10: 563–573.

Chamberlain MC (2010b). Leptomeningeal metastasis. Curr Opin Oncol 22: 627–635.

Chamberlain M, Soffietti R, Raizer J et al. (2014). Leptomeningeal metastasis: a Response Assessment in Neuro-Oncology critical review of endpoints and response criteria of published randomized clinical trials. Neuro Oncol 16: 1176–1185.

Chi D, Behin A, Delattre JY (2008). Neurologic complications of radiation therapy. In: D Schiff, S Kesari, P Wen (Eds.), Cancer neurology in clinical practice. Neurologic complications of cancer and its treatment, 2nd edn. Springer, New Jersey.

Clarke JI (2012). Leptomeningeal metastasis from systemic cancer. Continuum Lifelong Learning Neurol 18: 328–342.

Clarke JL, Perez HR, Jacks LM et al. (2010). Leptomeningeal metastases in the MRI era. Neurology 74: 1449–1454.

Cohen AL, Holmen SL, Colman H (2013). IDH1 and IDH2 mutations in gliomas. Curr Neurol Neurosci Rep 13: 345.

Coret F, Bosca I, Fratalia L et al. (2009). Long-lasting remission after rituximab treatment in a case of anti-Hu-associated sensory neuronopathy and gastric pseudoobstruction. J Neurooncol 93: 421–423.

Crossen JR, Garwood D, Glatstein E et al. (1994). Neurobehavioral sequelae of cranial irradiation in adults: a review of radiation-induced encephalopathy. J Clin Oncol 12: 627–642.

Dalmau J, Rosenfeld MR (2008). Paraneoplastic syndromes of the CNS. Lancet Neurol 7: 327–340.

Damek DM (2009). Cerebral edema, altered mental status, seizures, acute stroke, leptomeningeal metastases, and paraneoplastic syndrome. Emerg Med Clin North Am 27: 209–229.

Darnell RB, Posner JB (2003). Paraneoplastic syndromes involving the nervous system. N Engl J Med 349: 1543–1554.

Davies DC (2002). Blood–brain barrier breakdown in septic encephalopathy and brain tumours. J Anat 200: 639–646.

Dearborn JL, Urrutia VC, Zeiler SR (2014). Stroke and cancer – a complicated relationship. J Neurol Transl Neurosci 2: 1039.

Didelot A, Honnorat J (2014). Paraneoplastic disorders of the central and peripheral nervous systems. In: J Biller, JM Ferro (Eds.), Handbook of Clinical Neurology, Elsevier, San Diego.

Duenas-Gonzalez A, Candelaria M, Perez-Plascencia C et al. (2008). Valproic acid as epigenetic cancer drug: preclinical, clinical and transcriptional effects on solid tumors. Cancer Treat Rev 34: 206–222.

Dupond JL, Essalmi L, Gil H et al. (2010). Rituximab treatment of stiff-person syndrome in a patient with thymoma, diabetes mellitus and autoimmune thyroiditis. J Clin Neurosci 17: 389–391.

Enting RH (2005). Leptomeningeal neoplasia: epidemiology, clinical presentation, CSF analysis and diagnostic imaging. Cancer Treat Res 125: 17–30.

Findlay GF (1984). Adverse effects of the management of malignant spinal cord compression. J Neurol Neurosurg Psychiatry 47: 761–768.

Forsyth PA, Posner JB (1993). Headaches in patients with brain tumors: a study of 111 patients. Neurology 43: 1678–1683.

George R, Jeba J, Ramkumar G et al. (2015). Interventions for the treatment of metastatic extradural spinal cord compression. Cochrane Database Syst Rev 9. CD006716.

Gerszten PC, Burton SA, Ozhasoglu C et al. (2007). Radiosurgery for spinal metastases: clinical experience in 500 cases from a single institution. Spine (Phila Pa 1976) 32: 193–199.

Giglio P, Gilbert MR (2010). Neurologic complications of cancer and its treatment. Curr Oncol Rep 12: 50–59.

Giometto B, Grisold W, Vitaliani R et al. (2010). Paraneoplastic neurologic syndrome in the PNS Euronetwork database: a European study from 20 centers. Arch Neurol 67: 330–335.

Glantz MJ, Cole BF, Glantz LK (1988). Cerebrospinal fluid cytology in patients with cancer: minimizing false-negative results. Cancer 82: 733–739.

Glantz MJ, Cole BF, Recht L et al. (1998). High-dose intravenous methotrexate for patients with nonleukemic leptomeningeal cancer: is intrathecal chemotherapy necessary? J Clin Oncol 16: 1561.

Glantz MJ, Cole BF, Forsyth PA et al. (2000). Practice parameter: anticonvulsant prophylaxis in patients with newly diagnosed brain tumors. Report of the Quality Standards Subcommittee of the American Academy of Neurology. Neurology 54: 1886–1893.

Glass JP, Melamed M, Chernik NL et al. (1979). Malignant cells in cerebrospi fluid (CSF): the meaning of a positive CSF cytology. Neurology 29: 1369–1375.

Goonawardena J, Marshman LA, Drummond KJ (2015). Brain tumour-associated status epilepticus. J Clin Neurosci 22: 29–34.

Graus F, Rogers LR, Posner JB (1985). Cerebrovascular complications in patients with cancer. Medicine (Baltimore) 64: 16–35.

Graus F, Delattre JY, Antoine JC et al. (2004). Recommended diagnostic criteria for paraneoplastic neurological syndromes. J Neurol Neurosurg Psychiatry 75: 1135–1140.

Greenlee JE (2010). Treatment of paraneoplastic neurologic disorders. Curr Treat Options Neurol 12: 212–230.

Grewal J, Saria MG, Kesari S (2012). Novel approaches to treating leptomeningeal metastases. J Neurooncol 106: 225–234.

Grisold W, Oberndorfer S, Struhal W (2009). Stroke and cancer: a review. Acta Neurol Scand 119: 1–16.

Groves MD (2011). Leptomeningeal disease. Neurosurg Clin N Am 22 (67–78): vii.

Gultekin SH (2015). Recent development in paraneoplastic disorders of the nervous system. Surg Pathol 8: 85–99.

Hammack JE (2012). Spinal cord disease in patients with cancer. Continuum (Minneap Minn) 18: 312–327.

Herrlinger U, Weller M, Schabet M (1998). New aspects of immunotherapy of leptomeningeal metastasis. J Neurooncol 38: 233–239.

Herzig RH, Hines JD, Herzig GP et al. (1987). Cerebellar toxicity with high-dose cytosine arabinoside. J Clin Oncol 5: 927–932.

Hinchey J, Chaves C, Appignani B et al. (1996). A reversible posterior leukoencephalopathy syndrome. N Engl J Med 334: 494–500.

Hong CT, Tsai LK, Jeng JS (2009). Patterns of acute cerebral infarcts in patients with active malignancy using diffusion-weighted imaging. Cerebrovasc Dis 28: 411–416.

Husband DJ (1998). Malignant spinal cord compression: prospective study of delays in referral and treatment. BMJ 317: 18–21.

Jaeckle KA (2006). Neoplastic meningitis from systemic malignancies: diagnosis, prognosis and treatment. Semin Oncol 33: 312–323.

Jaeckle KA (2010). Neurologic manifestations of neoplastic and radiation-induced plexopathies. Semin Neurol 30: 254–262.

Jani-Acsadi A, Lisak RP (2007). Myasthenic crisis: guidelines for prevention and treatment. J Neurol Sci 261: 127–133.

Jo J, Schiff D, Purow B (2012). Angiogenic inhibition in high-grade gliomas: past, present and future. Expert Rev Neurother 12: 733–747.

Jo J, Schiff D, Perry JR (2014). Thrombosis in brain tumors. Semin Thromb Hemost 40: 325–331.

Johnston PC, Hamrahian AH, Weil RJ et al. (2015). Pituitary tumor apoplexy. J Clin Neurosci 22: 939–944.

Kaal EC, Vecht CJ (2004). The management of brain edema in brain tumors. Curr Opin Oncol 16: 593–600.

Kaira K, Okamura T, Takahashi H et al. (2014). Small-cell lung cancer with voltage-gated calcium channel antibody-positive paraneoplastic limbic encephalitis: a case report. J Med Case Rep 8: 119.

Kak M, Nanda R, Ramsdale EE et al. (2015). Treatment of leptomeningeal carcinomatosis: current challenges and future opportunities. J Clin Neurosci 22: 632–637.

Kargiotis O, Kyritsis AP (2012). Radiation-induced peripheral nerve disorder. In: P Wen, D Schiff, EQ Lee (Eds.), Neurologic complications of cancer therapy. New York, NY, Demos Medical.

Kim SG, Hong JM, Kim HY et al. (2010). Ischemic stroke in cancer patients with and without conventional mechanisms: a multicenter study in Korea. Stroke 41: 798–801.

Klimo JR P, Schmidt MH (2004). Surgical management of spinal metastases. Oncologist 9: 188–196.

Kwong YL, Yeung DY, Chan JC (2009). Intrathecal chemotherapy for hematologic malignancies: drugs and toxicities. Ann Hematol 88: 193–201.

Lee AY, Levine MN, Baker RI et al. (2003). Low-molecular-weight heparin versus a coumarin for the prevention of recurrent venous thromboembolism in patients with cancer. N Engl J Med 349: 146–153.

Lee EQ, Arrillaga-Romany IC, Wen PY (2012a). Neurologic complications of cancer drug therapies. Continuum (Minneap Minn) 18: 355–365.

Lee EQ, Norden A, Schiff D et al. (2012b). Neurologic complications of targeted therapy. In: P Wen, D Schiff, EQ Lee (Eds.), Neurologic complications of cancer therapy, New York, NY, Demos Medical.

L'esperance S, Vincent F, Gaudreault M et al. (2012). Treatment of metastatic spinal cord compression: CEPO review and clinical recommendations. Curr Oncol 19: e478–e490.

Levack P, Graham J, Collie D et al. (2002). Don't wait for a sensory level – listen to the symptoms: a prospectie audit of the delays in diagnosis of malignant cord compression. Clin Oncol (R Coll Radiol) 14: 472–480.

Levin VA, Bidaut L, Hou P et al. (2011). Randomized double-blind placebo-controlled trial of bevacizumab therapy for radiation necrosis of the central nervous system. Int J Radiat Oncol Biol Phys 79: 1487–1495.

Liao B, Shroff S, Kamiya-Matsuoka C et al. (2014). Atypical neurological complications of ipilimumab therapy in patients with metastatic melanoma. Neuro Oncol 16: 589–593.

Lin YS, Jen YM, Lin JC (2002). Radiation-related cranial nerve palsy in patients with nasopharyngeal carcinoma. Cancer 95: 404–409.

Liubinas SV, D'abacco GM, Moffat BM et al. (2014). IDH1 mutation is associate with seizures and protoplasmic subtype in patients with low-grade gliomas. Epilepsia 55: 1438–1443.

Loblaw A, Mitera G, Ford M et al. (2012). A 2011 updated systemic review and clinical practice guideline for the management of malignant extradural spinal cord compression. Int J Radiat Oncol Biol Phys: 1–6.

Lowenstein DH, Alldredge BH (1993). Status epilepticus at an urban public hospital in the 1980s. Neurology 43: 483–488.

Lyman GH, Khorana AA, Kuderer NM et al. (2013). Venous thromboembolism prophylaxis and treatment in patients with cancer: American Society of Clinical Oncology clinical practice guideline update. J Clin Oncol 31: 2189–2204.

Mak KS, Lee LK, Mak RH et al. (2011). Incidence and treatment patterns in hospitalizations for malignant spinal cord compression in the United States, 1998–2006. Int J Radiat Oncol Biol Phys 80: 824–831.

Maranzano E, Latiini P, Beneventi S et al. (1996). Radiotherapy without steroids in selected metastatic spinal cord compression patients. A phase II trial. Am J Clin Oncol 19: 179–183.

Maranzano E, Bellavita R, Rossi R et al. (2005). Short-course versus split-course radiotherapy in metastatic spinal cord compression: results of a phase III, randomized, multicenter trial. J Clin Oncol 23: 3358–3365.

Martel S, De Angelis F, Lapointe E et al. (2014). Paraneoplastic neurologic syndromes: clinical presentation and management. Curr Probl Cancer 38: 115–134.

Maschio M (2012). Brain tumor-related epilepsy. Current Neuropharmacol 10: 124–133.

Mintz AH, Kestle J, Rathbone MP et al. (1996). A randomized trial to assess the efficacy of surgery in addition to radiotherapy in patients with a single cerebral metastasis. Cancer 78: 1470–1476.

Mitchell MS (1989). Relapse in the central nervous system in melanoma patients successfully treated with biomodulators. J Clin Oncol 7: 1701–1709.

Mokri B (2011). The Monro-Kellie hypothesis: application in CSF volume depletion. Neurology 56: 1746–1748.

Moussazadeh N, Laufer I, Yamada Y et al. (2014). Separation surgery for spinal metastases: effect of spinal radiosurgery on surgical treatment goals. Cancer Control 21: 168–174.

Nabors L, Ammirati M, Bierman PJ et al. (2013). Central nervous system cancer. J Natl Compr Canc Netw 11: 1114–1151.

Navi B, Segal AZ (2009). Cancer and stroke. In: JD Geyer, CR Gomez (Eds.), Stroke: A practical approach, Lippincott Williams & Wilkins, Philadelphia, pp. 48–55.

Neligan A, Shorvon SD (2010). Frequency and prognosis of convulsive status epilepticus of different causes: a systematic review. Arch Neurol 67: 931–940.

Nguyen T, Deangelis LM (2006). Stroke in cancer patients. Curr Neurol Neurosci Rep 6: 187–192.

Oberndorfer S, Schmal T, Lahrmann H et al. (2002). The frequency of seizures in patients with primary brain tumor or cerebral metastases. An evaluation from the Ludwig Boltzmann Institute of Neuro-oncology and the Department of Neurology, Kaiser Franz Josef hospital, Vienna. Wien Klin Wochenschr 114: 911–916.

Oldfield EH, Merrill MJ (2015). Apoplexy of pituitary adenomas: the perfect storm. J Neurosurg: 1–6.

Ostrom QT, Gittleman H, Liao P et al. (2014). CBTRUS statistical report: primary brain and central nervous system tumors diagnosed in the United States in 2007–2011. Neuro Oncol 16 (Suppl 4): iv1–iv63.

Pallud J, Capelle L, Huberfeld G (2013). Tumoral epileptogenicity: how does it happen? Epilepsia 54: 30–34.

Passarin MG, Sava T, Furlanetto J et al. (2015). Leptomeningeal metastasis from solid tumors: a diagnostic and therapeutic challenge. Neurol Sci 36: 117–123.

Patchell RA, Posner JB (1985). Neurologic complications of systemic cancer. Neurol Clin 3: 729–750.

Patchell RA, Tibbs PA, Walsh JW et al. (1990). A randomized trial of surgery in the treatment of single metastases to the brain. N Engl J Med 322: 494–500.

Patchell RA, Tibbs PA, Regine WF et al. (1998). Postoperative radiotherapy in the treatment of single metastases to the brain: a randomized trial. JAMA 280: 1485–1489.

Patchell RA, Tibbs PA, Regine WF et al. (2005). Direct decompressive surgical resection in the treatment of spinal cord compression caused by metastatic cancer: a randomised trial. Lancet 366: 643–648.

Pater K, Puskulluoglu M, Zygulska AL (2014). Oncological emergencies: increased intracranial pressure in solid tumor's metastatic brain disease. Przegl Lek 71: 91–94.

Pelosof LC, Gerber DE (2010). Paraneoplastic syndromes: an approach to diagnosis and treatment. Mayo Clin Proc 85: 838–854.

Perrin RG, Laxton AW (2004). Metastatic spine disease: epidemiology, pathophysiology, and evaluation of patients. Neurosurg Clin N Am 15: 365–373.

Pittock SJ, Kryzer TJ, Lennon VA (2004). Paraneoplastic antibodies coexist and predict cancer, not neurological syndrome. Ann Neurol 56: 715–719.

Posner J (1995). Side effects of radiation therapy. In: J Posner (Ed.), Neurologic complications of cancer, F.A. Davis, Philadelphia.

Pruitt AA (2003). Nervous system infections in patients with cancer. Neurol Clin 21: 193–219.

Pruitt AA (2004). Central nervous system infections in cancer patients. Semin Neurol 24: 435–452.

Pruitt AA (2012). CNS infections in patients with cancer. Continuum (Minneap Minn) 18: 384–405.

Qayyum A, Macvicar AD, Padhani AR et al. (2000). Symptomatic brachial plexopathy following treatment for breast cancer: utility of MR imaging with surface-coil techniques. Radiology 214: 837–842.

Quant EC, Wen P (2008). Neurological complications of hematopoietic stem cell transplantation. In: D Schiff, S Kesari, P Wen (Eds.), Cancer Neurology in clinical practice: Neurologic complications of cancer and its treatment, Humana Press, Totowa, NJ.

Quinn JA, Deangelis LM (2000). Neurologic emergencies in the cancer patient. Semin Oncol 27: 311–321.

Raizer J, Pentsova E, Omuro A et al. (2014). Phase I trial of intrathecal trastuzumab in HER2 positive leptomeningeal metastases. Neuro Oncol 16.

Ribas ES, Schiff D (2012). Spinal cord compression. Curr Treat Options Neurol 14: 391–401.

Ricard D, Psimaras D, Soussain C et al. (2012). Central nervous system complications of radiation therapy. In: P Wen, D Schiff, EQ Lee (Eds.), Neurologic complications of cancer therapy, Demos Medical, New York.

Richardson PG, Briemberg H, Jagannath S et al. (2006). Frequency, characteristics, and reversibility of peripheral neuropathy during treatment of advance multiple myeloma with bortezomib. J Clin Oncol 24: 3113–3120.

Rogers LR (2003). Cerebrovascular complications in cancer patients. Neurol Clin 21: 167–192.

Rogers LR (2012). Neurologic complications of radiation. Continuum (Minneap Minn) 18: 343–354.

Ropper AH, Gress DR, Diringer MN (2003). Management of intracranial pressure and mass effect. In: AH Ropper (Ed.), 4th edn. Lippincott Williams & Wilkins, Philadelphia.

Rosati A, Buttolo L, Stefini R et al. (2010). Efficacy and safety of levetiracetam in patients with glioma: a clinical prospective study. Arch Neurol 67: 343–346.

Rosenfeld MR, Pruitt A (2006). Neurologic complications of bone marros, stem cell, and organ transplantation in patients with cancer. Semin Oncol 33: 352–361.

Rossetti AO, Jeckelmann S, Novy J et al. (2014). Levetiracetam and pregabalin for antiepileptic monotherapy in patients with primary brain tumors. A phase II randomized study. Neuro Oncol 16: 584–588.

Rubenstein JL, Fridlyand J, Abrey L et al. (2007). Phase I study of intraventricular administration of rituximab in patients with recurrent CNS and intraocular lymphoma. J Clin Oncol 25: 1350–1356.

Rubenstein JL, LI J, Chen L et al. (2013). Multicenter phase 1 trial of intraventricular immunochemotherapy in recurrent CNS lymphoma. Blood 121: 745–751.

Ruda R, Bello L, Duffau H et al. (2012). Seizures in low-grade gliomas: natural history, pathogenesis, and outcome after treatments. Neuro Oncol 14 (Suppl 4): iv55–iv64.

Samlowski WE, Park KJ, Galinsky RE et al. (1993). Intrathecal administration of interleukin-2 for meningeal carcinomatosis due to malignant melanoma: sequential evaluation of intracranial pressure, cerebrospinal fluid cytology, and cytokine induction. J Immunother Emphasis Tumor Immunol 13: 49.

Saria MG, Corle C, Hu J et al. (2013). Retrospective analysis of the tolerability and activity of lacosamide in patients with brain tumors: clinical article. J Neurosurg 118: 1183–1187.

Schiff D (2003). Spinal cord compression. Neurol Clin 21: 67–86. vii.

Schiff D, O'neill BP, Wang CH et al. (1998). Neuroimaging and treatment implications of patients with multiple epidural spinal metastases. Cancer 83: 1593–1601.

Schiff D, Lee EQ, Nayak L et al. (2015). Medical management of brain tumors and the sequelae of treatment. Neuro Oncol 17: 488–504.

Schulz H, Pels H, Schmidt-Wolf I et al. (2004). Intraventricular treatment of relapsed central nervous system lymphoma with the anti-CD20 antibody rituximab. Haematologica 89: 753–754.

Seok JM, Kim SG, Kim JW et al. (2010). Coagulopathy and embolic signal in cancer patients with ischemic stroke. Ann Neurol 68: 213–219.

Shams'ili S, De Beukelaar J, Gratama JW et al. (2006). An uncontrolled trial of rituximab for antibody associated paraneoplastic neurological syndromes. J Neurol 253: 16–20.

Sheline GE (1977). Radiation therapy of brain tumors. Cancer 39: 873–881.

Siegal T, Tiqva P, Siegal T (1985). Vertebral body resection for epidural compression by malignant tumors. Results of forty-seven consecutive operative procedures. J Bone Joint Surg Am 67: 375–382.

Sioka C, Kyritsis AP (2009). Central and peripheral nervous system toxicity of common chemotherapeutic agents. Cancer Chemother Pharmacol 63: 761–767.

Sklar EM (2006). Post-transplant neurotoxicity: what role do calcineurin inhibitors actually play? AJNR 27: 1602–1603.

Sohn S, Chung CK (2012). The role of stereotactic radiosurgery in metastasis to the spine. J Korean Neurosurg Soc 51: 1–7.

Spinazze S, Caraceni A, Schrijvers D (2005). Epidural spinal cord compression. Crit Rev Oncol Hematol 56: 397–406.

Stevens RD, Huff JS, Duckworth J et al. (2012). Emergency neurological life support: intracranial hypertension and herniation. Neurocrit Care 17 (Suppl 1): S60–S65.

Stummer W (2007). Mechanisms of tumor-related brain edema. Neurosurg Focus 22. E8.

Stupp R, Mason WP, Van Den Bent MJ et al. (2005). Radiotherapy plus concomitant and adjuvant temozolomide for glioblastoma. N Engl J Med 352: 987–996.

Sun H, Nemecek AN (2010). Optimal management of malignant epidural spinal cord compression. Hematol Oncol Clin North Am 24: 537–551.

Taccone FS, Jeangette SM, Blecic SA (2008). First-ever stroke as initial presentation of systemic cancer. J Stroke Cerebrovasc Disc 17: 169–174.

Tarhini A (2013). Immune-mediated adverse events associated with ipilimumab ctla-4 blockade therapy: the underlying mechanisms and clinical management. Scientifica (Cairo) 2013: 857519.

Titulaer MJ, Soffietti R, Dalmau J et al. (2011). Screening for tumours in paraneoplastic syndromes: report of an EFNS task force. Eur J Neurol 18: 19-e3.

Torcuator R, Zuniga R, Mohan YS et al. (2009). Initial experience with bevacizumab treatment for biopsy confirmed cerebral radiation necrosis. J Neurooncol 94: 63–68.

Usery JB, Michael LM, Sills AK et al. (2010). A prospective evaluation and literature review of levetiracetam use in patients with brain tumors and seizures. J Neurooncol 99: 251–260.

Van Breemen MS, Wilms EB, Vecht CJ (2007). Epilepsy in patients with brain tumours: epidemiology, mechanisms, and management. Lancet Neurol 6: 421–430.

Van Breemen MS, Rijsman RM, Taphoorn MJ et al. (2009). Efficacy of anti-epileptic drugs in patients with gliomas and seizures. J Neurol 256: 1519–1526.

Vecht CJ, Wilms EB (2010). Seizures in low- and high-grade gliomas: current management and future outlook. Expert Rev Anticancer Ther 10: 663–669.

Vecht CJ, Haaxma-Reiche H, Noordijk EM et al. (1993). Treatment of single brain metastasis: radiotherapy alone or combined with neurosurgery? Ann Neurol 33: 583–590.

Vecht CJ, Kerkhof M, Duran-Pena A (2014). Seizure prognosis in brain tumors: new insights and evidence-based management. Oncologist 19: 751–759.

Vernino S, O'neill BP, Marks RS et al. (2004). Immunomodulatory treatment trial for paraneoplastic neurological disorders. Neuro Oncol 6: 55–62.

Walker MT, Kapoor V (2007). Neuroimaging of parenchymal brain metastases. Cancer Treat Res 136: 31–51.

Wassertrom WR, Glass JP, Posner J (2006). Diagnosis and treatment of leptomeningeal metastases from solid tumors; experience with 90 patients. Cancer 49: 759–772.

Weller M, Gorlia T, Cairncross J et al. (2011). Prolonged survival with valproic acid use in the EORTC/NCIC temozolomide trial for glioblastoma. Neurology 77: 1156–1164.

Wen P (1997). Diagnosis and management of brain tumors, Blackwell Science, Cambridge, MA.

Wen PY, Schiff D, Kesari S et al. (2006). Medical mangement of patients with brain tumors. J Neurooncol 80: 313–332.

Xia Z, Mehta BP, Ropper AH et al. (2010). Paraneoplastic limbic encephalitis presenting as a neurological emergency: a case report. J Med Case Reports 4: 95.

Yuen TI, Morokoff AP, Bjorksten A et al. (2012). Glutamate is associated with a higher risk of seizures in patients with gliomas. Neurology 79: 883–889.

Zagouri F, Sergentanis TN, Bartsch R et al. (2013). Intrathecal administration of trastuzumab for the treatment of meningeal carcinomatosis in HER2-positive metastatic breast cancer: a systematic review and pooled analysis. Breast Cancer Res Treat 139: 13–22.

Zhang YY, Chan DKY, Cordato D et al. (2006). Stroke risk factor, pattern and outcome in patients with cancer. Acta Neurol Scand 114: 378–383.

Chapter 40

Management of neurologic complications of coagulopathies

J.D. VANDERWERF[1] AND M.A. KUMAR[2]*

[1]Department of Neurology, Perelman School of Medicine, Hospital of the University of Pennsylvania, Philadelphia, PA, USA

[2]Departments of Neurology, Neurosurgery, Anesthesiology and Critical Care, Perelman School of Medicine, Hospital of the University of Pennsylvania, Philadelphia, PA, USA

Abstract

Coagulopathy is common in intensive care units (ICUs). Many physiologic derangements lead to dysfunctional hemostasis; these may be either congenital or acquired. The most devastating outcome of coagulopathy in the critically ill is major bleeding, defined by transfusion requirement, hemodynamic instability, or intracranial hemorrhage. ICU coagulopathy often poses complex management dilemmas, as bleeding risk must be tempered with thrombotic potential. Coagulopathy associated with intracranial hemorrhage bears directly on prognosis and outcome. There is a paucity of high-quality evidence for the management of coagulopathies in neurocritical care; however, data derived from studies of patients with intraparenchymal hemorrhage may inform treatment decisions.

Coagulopathy is often broadly defined as any derangement of hemostasis resulting in either excessive bleeding or clotting, although most typically it is defined as impaired clot formation. Abnormalities in coagulation testing without overt clinical bleeding may also be considered evidence of coagulopathy. This chapter will focus on acquired conditions, such as organ failure, pharmacologic therapies, and platelet dysfunction that are associated with defective clot formation and result in, or exacerbate, intracranial hemorrhage, specifically spontaneous intraparenchymal hemorrhage and traumatic brain injury.

INTRODUCTION

Coagulopathy is common in intensive care units (ICUs). Many physiologic derangements lead to dysfunctional hemostasis; these may be either congenital or acquired. The most devastating outcome of coagulopathy in the critically ill is major bleeding, defined by transfusion requirement, hemodynamic instability, or intracranial hemorrhage (ICH). ICU coagulopathy often poses complex management dilemmas, as bleeding risk must be tempered with thrombotic potential. Coagulopathy associated with ICH bears directly on prognosis and outcome. There is a paucity of high-quality evidence for the management of coagulopathies in neurocritical care; however, data derived from studies of patients with intraparenchymal hemorrhage (IPH) may inform treatment decisions.

Coagulopathy is often broadly defined as any derangement of hemostasis resulting in either excessive bleeding or clotting, although most typically it is defined as impaired clot formation. Abnormalities in coagulation testing without overt clinical bleeding may also be considered evidence of coagulopathy. This chapter will focus on acquired conditions, such as organ failure, pharmacologic therapies, and platelet dysfunction that are associated with defective clot formation and result in, or exacerbate, ICH, specifically spontaneous IPH and traumatic brain injury (TBI).

EPIDEMIOLOGY

Coagulopathies are common in intensive care populations. The incidence of coagulopathy ranges widely (13–66%) and depends on the definition employed

---

*Correspondence to: Monisha A. Kumar, MD, Departments of Neurology, Neurosurgery, Anesthesiology and Critical Care, Perelman School of Medicine, Hospital of the University of Pennsylvania, 3 West Gates, Department of Neurology, 3400 Spruce Street, Philadelphia PA 19104, USA. E-mail: Monisha.Kumar@uphs.upenn.edu

(Chakraverty et al., 1996; Walsh et al., 2010). A prospective study of 235 patients in an adult medical-surgical ICU found that 14% had clinical bleeding secondary to coagulopathy, defined as bleeding unexplained by local or surgical factors (Chakraverty et al., 1996). Mortality was associated with clinical bleeding in 6% of these patients, with ICH representing one-third of these deaths.

Derangement of laboratory coagulation assays occurs in an even greater number of patients. An elevation of prothrombin time (PT) >1.5 times normal occurs in 30–66% of ICU patients, and 20–38% demonstrate thrombocytopenia (Chakraverty et al., 1996; Walsh et al., 2010). Presence of coagulopathy is associated with increased mortality, especially when it occurs after admission to the ICU (Walsh et al., 2010). There is limited information describing the incidence of coagulopathy in neurocritical care units.

Antithrombotic-associated intracranial hemorrhage

ICH is the most feared complication of antithrombotic therapy. The incidence of antithrombotic-associated ICH is 3.7–4.9 per 100 000 population (Nicolini et al., 2002). IPH represents the vast majority of these bleeds (Hart et al., 1995; Nicolini et al., 2002). IPH affects approximately 67 000 people in the USA each year, and antithrombotic use is associated with 12–41% of cases (Nilsson et al., 2000; Kissela et al., 2004; Lovelock et al., 2007; van Asch et al., 2010).

Vitamin K antagonist (VKA)-related ICH occurs in less than 1% of patients receiving VKA therapy. Seventy percent of these patients present with IPH, and the remainder have primarily subdural hematomas (Hart et al., 1995, 2005). Although the risk of ICH in patients treated with VKAs increases dramatically with an international normalized ratio (INR) >4.5, the majority of VKA-associated ICH occurs with an INR <3 (Hart et al., 1995; The Stroke Prevention in Reversible Ischemia Trial (SPIRIT) Study Group, 1997; Flaherty et al., 2007). This risk is magnified in older patients with hypertension, probably in part because of an increasing prevalence of atrial fibrillation in this population, with a consequent need for anticoagulation (Hart et al., 1995; Flaherty et al., 2007).

The incidence of ICH associated with unfractionated heparin (UFH) and low-molecular-weight heparin (LMWH) depends on the condition requiring treatment, route of administration, dosage of medication, and comorbid conditions (Schulman et al., 2008). The overall rate of ICH with UFH compared to LMWH is similar (Gould et al., 1999; Dolovich et al., 2000; Schulman et al., 2008). Therapeutic doses of either UFH or LMWH within 2 weeks of ischemic stroke carry approximately twice the risk of hemorrhagic conversion compared to aspirin or placebo (Schulman et al., 2008). In the International Stroke Trial, the rate of IPH over 14 days while on heparin 5000 units twice daily was 0.7%, compared with 1.8% while on heparin 12 500 units twice daily, and 0.3% in controls (International Stroke Trial Collaborative Group, 1997). In the setting of acute myocardial infarct (AMI), there is a nonsignificant increase in risk of ICH with either UFH or LMWH (Schulman et al., 2008). In one of the largest trials of LMWH in AMI, the risk of ICH was 0.3% compared to 0.1% with placebo (Schulman et al., 2008). There is an increased risk of bleeding with therapeutic doses of intermittent intravenous UFH compared to continuous infusion; however the risk is similar between continuous infusion and subcutaneous injection (Schulman et al., 2008). There is evidence that the risk of ICH rises with increasing dosage, age, and renal impairment (Schulman et al., 2008). A meta-analysis of 20 studies found a relative risk of 1.7 for major bleeding when LMWH was administered in patients with a glomerular filtration rate <60 mL/min, despite dose adjustments (Hoffmann and Keller, 2012).

Compared to VKAs, the target-specific oral anticoagulants (TSOACs) carry less risk of ICH (relative risk 0.49, 95% confidence interval 0.36–0.66) (Miller et al., 2012). In the Re-LY trial, dabigatran use was associated with a 0.3% risk of ICH compared to 0.8% with warfarin (Hart et al., 2012). In the ROCKET-AF trial, rivaroxaban use was associated with a 0.5% risk of ICH compared to 0.7% with warfarin (Patel et al., 2011). In the ARISTOTLE trial, apixaban was associated with a 0.3% risk of ICH compared to 0.8% with warfarin (Hylek et al., 2014). The average follow-up period for these trials was about 2 years. A meta-analysis of 11 randomized trials involving over 100 000 patients treated with TSOACs found a 47% odds reduction for fatal hemorrhage compared with VKA use (Caldeira et al., 2015).

The absolute risk of ICH associated with antiplatelet agents is low. Aspirin (acetylsalicylic acid: ASA) use results in an absolute increase of 0.1% per year in the risk of ICH compared to control (He et al., 1998). The "number needed to harm" (to cause one ICH) on ASA is 833 patients. In the CAPRIE trial, Plavix did not result in any significant difference in rate of ICH compared to ASA (0.3% vs. 0.5%) (CAPRIE Steering Committee, 1996).

Newer, more potent antiplatelet agents are increasingly used in the setting of acute coronary syndromes. Prasugrel, a thienopyridine, is associated with a higher risk of bleeding compared to clopidigrel; however, the risk of ICH is reported to be similar (Wiviott et al., 2007). Ticagrelor, a nonthienopyridine agent, which reversibly binds the P2Y12 platelet receptor, conveys a nonsignificantly increased risk of ICH compared to clopidigrel (Becker et al., 2011). Cangrelor is another nonthienopyridine, which is administered intravenously and likewise shows no increased risk of ICH compared to clopidigrel (Bhatt et al., 2009; Harrington et al., 2009). Although the risk of bleeding is low with single antiplatelet use, dual antiplatelet therapy conveys a similar rate of major bleeding compared to anticoagulants (Connolly et al., 2006; Hansen et al., 2010).

The use of recombinant tissue plasminogen activator (rtPA) for AMI and acute ischemic stroke also carries a risk of ICH. When rtPA is used in the setting of AMI, the incidence of ICH is 0.7% (Carlson et al., 1988). When rtPA is used for acute ischemic stroke, the incidence is somewhat higher, ranging from 3% to 9% (Broderick et al., 2007).

Thrombocytopenia

Thrombocytopenia is common in the ICU (20–38%) and its cause is often multifactorial. The risk of ICH varies depending on the etiology of thrombocytopenia and the presence of comorbidities. For example, idiopathic thrombocytopenic purpura is associated with only a 1% risk of ICH in adults (Neunert et al., 2015). Most spontaneous ICH occurs with platelet counts <20 000/mm^3 (Blanchette and Carcao, 2000). This risk is higher in those undergoing craniotomy. A small retrospective study of patients with modest perioperative thrombocytopenia undergoing cranial neurosurgery found that 40% developed postoperative ICH (Chan et al., 1989). All patients with a platelet count <100 000/μL who failed to respond to platelet transfusion developed postoperative ICH (Chan et al., 1989).

Disseminated intravascular coagulation

Disseminated intravascular coagulation (DIC) is a derangement of hemostasis consisting of widespread production of thrombin, which in turn leads to microvascular thrombosis, organ failure, and a consumptive coagulopathy. DIC is associated with several underlying conditions including sepsis, trauma, malignancy, and obstetric complications. Overall, sepsis is the most common cause of DIC; however trauma and malignancy represent other common causes in the neurocritical care unit (Hunt, 2014). DIC may lead to a thrombotic phenotype or a hemorrhagic phenotype, although it most often presents as hemorrhage (Hunt, 2014).

Coagulopathy after traumatic brain injury

TBI and polytrauma are often complicated by coagulopathy. Worldwide, there are 10 million reported cases of TBI per year resulting in hospitalization or death (Langlois et al., 2006). In the USA, there are at least 275 000 hospitalizations and over 50 000 deaths related to TBI yearly (Langlois et al., 2006; Coronado et al., 2011). Acute coagulopathy occurs in a high percentage of trauma patients and correlates with injury severity. The reported rate of coagulopathy observed in isolated TBI varies widely depending on severity of injury, type of clotting assays used, and timing of measurements from onset of injury. Diagnostic criteria for coagulopathy vary greatly in the literature. A meta-analysis found the mean incidence to be 33% (Harhangi et al., 2008). A more recent prospective study found evidence of coagulopathy (defined as at least one of thrombocytopenia, elevated INR, or elevated partial thromboplastin time (PTT)) in a similar percentage of patients (34%) with isolated TBI (Talving et al., 2009). There is evidence that injury severity is an important risk factor in predicting coagulopathy. Over 60% of patients with severe TBI have coagulopathy compared to <1% with mild head injury (Gómez et al., 1996; Hoyt, 2004).

Coagulopathy of malignancy

Bleeding complications are common in the setting of malignancy, especially leukemias. In a large retrospective series of autopsy reports from Memorial Sloan-Kettering Cancer Center from 1970 to 1981, 6% of cancer patients had IPH at autopsy and 2% were said to be from coagulopathy associated with malignancy (Graus et al., 1985). The incidence of IPH was much higher (16%) for patients with leukemia. When stratified by type of leukemia, it was found that 22% of acute myelocytic leukemia patients had IPH, with acute promyelocytic leukemia (APML) accounting for the majority of these cases. Over 60% of APML who died had IPH, which was often the presenting diagnosis.

Coagulopathy of liver failure

While chronic liver failure with cirrhosis occurs in about 15 per 100 000 person-years, there are only 1–6 cases of acute liver failure (ALF) per million persons per year in developed nations (Fleming et al., 2008; Bernal et al., 2010). Liver failure, whether chronic or acute, is associated with decreased synthesis of clotting factors. Although decreased synthesis likely affects procoagulant and anticoagulant proteins alike, the derangements in coagulation assays suggest a bleeding tendency. Assays of hypercoagulability are not readily available. The degree to which these derangements contribute to clinical hemorrhage, especially ICH, is not known. Recent studies using viscoelastic assays suggest that patients with ALF do not have an impaired ability to form clot *in vivo* despite abnormalities of standard coagulation assays and platelet count (Lisman and Leebeek, 2007).

Uremia

The prevalence of chronic kidney disease and use of renal replacement therapy is rising (Roderick et al., 2004). Uremia is associated with increased risk of hemorrhage secondary to platelet dysfunction, which often occurs in combination with thrombocytopenia. Uremic bleeding typically manifests as ecchymoses, epistaxis, gastrointestinal hemorrhage, or bleeding from puncture sites. Patients with chronic kidney disease have an increased risk of ICH, which may be 10 times greater in the setting of chronic hemodialysis than in the general population (Pavord and Myers, 2011). However, the incidence of major hemorrhages has decreased with increasing use and effectiveness of hemodialysis (Rabiner, 1972).

NEUROPATHOLOGY

Cell-based model of hemostasis

In order to appropriately discuss the pathology of coagulopathies, it is important to understand the process of hemostasis *in vivo* under normal conditions. In the 1960s, the "cascade" model of hemostasis was proposed after the discovery of several procoagulant proteins that were related to each other in a series of proteolytic reactions (Davie and Ratnoff, 1964; Macfarlane, 1964). This model describes a separate "intrinsic" and "extrinsic" pathway of proteolytic reactions, both of which appeared to be capable of forming thrombin, leading to fibrin clot. Originally the intrinsic pathway was believed to be the main contributor to hemostasis, but it later became evident that the extrinsic pathway played the major role *in vivo* (Hoffman and Monroe, 2001). This model enabled the development of the screening coagulation laboratory tests used most frequently in practice, namely the PT and activated PTT.

Since the 1960s, certain clinical observations have called into question the accuracy of the cascade model. These included the severity of bleeding in hemophiliacs from deficiency of "intrinsic factors," and the lack of clinical bleeding from deficiency of factor XII (FXII), high-molecular-weight kininogen, or prekallikrein. Furthermore, it has been shown that cellular elements (platelets, tissue factor-bearing cells, and erythrocytes) have a significant impact on coagulation (Roberts and Lozier, 1992; Hoffman and Monroe, 2001). Such observations led to the formation of a new perspective on hemostasis called the cell-based model, which provides a more accurate model of hemostasis *in vivo* (Roberts and Lozier, 1992).

The cell-based model holds that hemostasis is controlled by activity of cellular components more so than simple protein kinetics. Clot formation is divided into three stages: initiation, amplification, and propagation. The cell-based model of hemostasis has several implications, one of which is that PT and activated PTT (aPTT) values may not always give an accurate representation of the status of *in vivo* hemostasis, as these tests assess only early steps in the generation of thrombin. This helps explain why certain disease processes have a normal or enhanced ability to form clot (based on viscoelastic assays) despite abnormal PT or aPTT levels. Conversely, normalization of PT or aPTT levels in patients with coagulopathy may not imply normalized *in vivo* hemostasis. Awareness of the cell-based model when approaching a patient with coagulopathy is therefore important, as this can help guide clinical decisions.

Antithrombotic-associated IPH

IPH occurs when small penetrating arteries rupture, often in the area of the basal ganglia, thalamus, pons, or cerebellum. The pathology underlying antithrombotic-associated IPH is thought to be similar to spontaneous IPH. The use of antithrombotics exacerbates existing risk factors for IPH (Hart et al., 1995). The location of antithrombotic-associated IPH is not different from spontaneous IPH (Nilsson et al., 2000). Cerebral amyloid angiopathy increases in frequency with older age and affects superficial cortical vessels. A prospective case-control study suggests that lobar IPH in elderly patients with warfarin use is often related to underlying cerebral amyloid angiopathy (Rosand et al., 2000). Similarly, leukoaraiosis has also been shown to increase the risk of IPH in patients taking warfarin following ischemic stroke. Leukoaraiosis was present in 92% of warfarin users with IPH compared to 48% of warfarin users without IPH (Smith et al., 2002).

Hematoma expansion is a risk factor for poor outcome after IPH. Hematoma expansion is thought to result from ongoing bleeding from ruptured vessels, and possibly from additional vessels which are disrupted by the hematoma itself (Qureshi et al., 2001). Coagulopathy may increase the risk of expansion, which is likely to be one of the mechanisms which increases morbidity and mortality (Steiner et al., 2006).

Thrombocytopenia

The causes of thrombocytopenia are many and include sepsis, blood loss, hemodilution, DIC, mechanical fragmentation, hypersplenism, medications, immune-mediated disorders, and bone marrow suppression (Hunt, 2014). The most common cause in critically ill patients is sepsis (Greinacher and Selleng, 2010). It is beyond the scope of this chapter to discuss the unique pathology underlying each cause of thrombocytopenia. However the importance of the platelet in the amplification and propagation phases of hemostasis should be underscored.

Disseminated intravascular coagulation

With DIC, widespread microthrombosis is caused by the upregulation of tissue factor, downregulation of thrombomodulin, and increased phospholipid availability (Levi et al., 2009). This leads to massive fibrin deposition, followed by a consumptive coagulopathy, hyperfibrinolysis, and hemorrhage (Levi et al., 2009; Hunt, 2014).

TBI-associated coagulopathy

TBI-associated coagulopathy is common and is associated with an unfavorable prognosis (Harhangi et al., 2008; Talving et al., 2009). Coagulopathy in TBI involves a dysregulation of hemostasis leading to both hyper- and hypocoagulability. Studies suggest that an early tendency to bleed is followed by a later tendency to form clot, although these entities may coexist, complicating treatment (Laroche et al., 2012). The exact pathophysiology of coagulopathy is not fully understood, but several mechanisms have been postulated, including massive release of tissue factor, DIC, platelet dysfunction, and hypoperfusion, leading to activated protein C malfunction (Laroche et al., 2012). Despite a lack of clarity regarding the mechanism of coagulopathy, it is clear that the brain is highly enriched in tissue factor, mostly expressed in the membranes of astrocytes (Drake et al., 1989; Eddleston et al., 1993). It is postulated that widespread injury to brain parenchyma leads to a massive release of tissue factor, which promotes DIC (Keimowitz and Annis, 1973; Pathak et al., 2005). Release of procoagulant microparticles from activated platelets provides a surface for coagulation, and may promote both hyper- and hypocoagulability, with activation of coagulation followed by factor consumption (Morel et al., 2008; Laroche et al., 2012; Kumar, 2013). Another contributing mechanism is platelet activation and dysfunction, which can occur despite a normal platelet count (Nekludov et al., 2007). Additionally, hypoperfusion leads to activation of protein C. Activated protein C inhibits FVa, FVIIIa, and plasminogen activator inhibitor-1, leading to propensity to hemorrhage by causing hyperfibrinolysis (Cohen et al., 2007). This is then followed by depletion of activated protein C, which leads to a prolonged propensity for thromboembolism (Laroche et al., 2012). TBI-associated coagulopathy is a complex and fluid process associated with significant hemostatic derangements, the mechanisms of which remain to be further elucidated.

Leukemic coagulopathy

Malignancy is often associated with maladaptive hemostasis, sometimes leading to bleeding. Leukemia, especially APML, is associated with the highest risk of bleeding. Patients often have bleeding on presentation, with IPH being the most feared complication. The pathophysiology of coagulopathy caused by APML is unique. Leukemic promyelocytes have increased expression of tissue factor and cancer procoagulant, which leads to a hypercoagulable state, further exaggerated by apoptosis of leukemic cells, especially during chemotherapy (Zhu et al., 1999; Wang et al., 2001; Kwaan et al., 2002). Widespread activation of coagulation can lead to a consumptive coagulopathy and DIC, which in turn increases bleeding tendency (Kwaan and Cull, 2014). Concomitantly, patients with APML have increased levels of plasminogen activators tPA and uPA (Kwaan and Cull, 2014). There is also evidence of increased plasminogen cell surface receptors annexin A2 and S100A10 (O'Connell et al., 2011). In total, these derangements of hemostasis lead to hyperfibrinolysis, which causes a predilection for IPH. It has been hypothesized that IPH, as opposed to bleeding at other sites, is specifically increased because annexin A2 is highly expressed in brain endothelium (Kwaan et al., 2004).

Liver failure-associated coagulopathy

Liver failure is accompanied by derangements of hemostasis, long assumed to cause a bleeding tendency. Due to derangements in platelet count, PT/INR, and aPTT, bleeding events have been attributed to coagulopathy from a lack of hepatic synthetic function. However, major bleeding complications may also be related to portal hypertension rather than coagulopathy in patients with cirrhosis (Lisman and Leebeek, 2007).

Mounting evidence demonstrates a "rebalanced hemostasis" in patients with hepatic dysfunction due to a matched decrement in the synthesis of both procoagulant and anticoagulant proteins. Thrombocytopenia, impaired platelet function, increased nitrous oxide and prostacyclin, and decreased vitamin K-dependent coagulation factors are balanced by increased von Willebrand factor (vWF) and FVIII, and decreased protein C, S, and antithrombin III (Lisman and Leebeek, 2007). In fact, Tripodi et al. (2005) demonstrated that patients with cirrhosis have normal thrombin generation despite abnormal conventional coagulation assays. Because of large derangements in platelet count, PT/INR, and aPTT, it has likewise been assumed that patients with ALF have a propensity to hemorrhage. Nonetheless, there is convincing evidence that patients with ALF have a normal ability to form clot and may even be prothrombotic.

A prospective cohort study found that over 60% of patients with ALF had normal whole-blood hemostasis as measured by thromboelastography (TEG), and 8% were hypercoagulable (Stravitz et al., 2012). Similarly, another prospective study found normal thrombin generation and decreased fibrinolysis in ALF, and another found a 19-fold higher concentration of procoagulant microparticles compared to healthy controls (Lisman et al., 2012; Stravitz et al., 2013; Habib et al., 2014). The presence of hepatic encephalopathy was associated with hypercoagulable TEG parameters compared to acute liver injury without encephalopathy. Interestingly, thrombotic complications were almost twice as common as significant hemorrhage (Stravitz et al., 2012).

Uremia

Uremia interferes with clotting via impaired platelet function and abnormal platelet interaction with the endothelium (Pavord and Myers, 2011). Increased urea and other solutes reduce Gp1b receptors and affinity for vWF, and decrease platelet activation and aggregation in response to agonists (Pavord and Myers, 2011). Increased nitrous oxide and prostacyclin also inhibit platelet activation, aggregation, and adhesion to the vessel wall (Pavord and Myers, 2011). Anemia is common in chronic kidney disease and leads to decreased platelet interactions with the vessel wall due to altered rheology. Red blood cells normally flow in the center of the vessel lumen and displace platelets towards the endothelium. However, when the hematocrit is decreased, the radial dispersion of platelets is altered, which increases the distance between platelets and the vessel wall, thus limiting platelet adhesion (Pavord and Myers, 2011). Lastly, thrombocytopenia is also relatively common in patients on hemodialysis.

CLINICAL PRESENTATION

Major hemorrhage is the foremost complication of coagulopathy. In the neurocritical care unit, major bleeding may present as hemorrhagic shock, retroperitoneal bleeding after endovascular interventions, hemothorax, or bleeding from other extracranial sites. The most typical manifestation of major bleeding in the neurocritical care setting is IPH. IPH characteristically causes sudden onset of a focal neurologic deficit, followed by progressive deterioration in level of consciousness (Goldstein and Simel, 2005). As compared to ischemic stroke, headache, vomiting, and extreme elevations in blood pressure are more common with IPH (Panzer et al., 1985; Goldstein and Simel, 2005). Additionally, patients with IPH are at a high risk of early neurologic and cardiopulmonary deterioration in the first several hours from onset (Hemphill et al., 2015).

DIAGNOSTICS AND NEUROIMAGING

Imaging

Imaging, usually noncontrast computed tomography (CT), reveals the presence, size, and location of ICH, as the clinical examination cannot accurately distinguish the exact location of bleeding (epidural, subdural, subarachnoid, or parenchymal). Magnetic resonance imaging (MRI) reliably identifies acute ICH, but its use in emergent settings for critically ill patients is limited (Hemphill et al., 2015). CT imaging is also useful for evaluating associated skull fractures in the case of trauma. When reviewing neuroimaging in ICH, one should also make note of the presence and degree of intraventricular hemorrhage (IVH), hydrocephalus, perihematomal edema, mass effect, and herniation. Coagulopathic ICH is suggested by a distinct pattern of a fluid–fluid interface on CT imaging (Hart et al., 1995). Liquefied blood separates in a gravity-dependent nature with heavier, congealed cellular components on the bottom and less dense fluid towards the top. The presence of this sign on CT imaging may be the first clue that a patient has a coagulopathic ICH. Additionally, the presence of concomitant hemorrhage and thrombosis may be evident on MRI by areas of restricted diffusion and increased susceptibility, as might be seen in DIC or malignancy (Fig. 40.1).

Laboratory and point-of-care testing

Routine laboratory evaluation of any patient with ICH should include testing for coagulopathy. This should, as a minimum, include a complete blood count, PT or INR, and aPTT. These tests may suggest effects of warfarin, UFH, and/or screen for thrombocytopenia and anemia. However, patients may have a clinically significant coagulopathy in the setting of normal PT/INR or aPTT. Conversely, a mild elevation of these assays may not equate to a clinical bleeding tendency (Naidech et al., 2014a). In cases where coagulopathy is suspected, further testing such as point-of-care tests (POCT), viscoelastic assays, and others may be warranted.

PT/INR

The PT is a laboratory test developed to assess the "extrinsic pathway," according to the cascade model of hemostasis. Calcium and tissue factor are added to citrated blood and the time to coagulation is measured. The INR was developed to correct for differences in tissue factor preparations used across laboratories, and thereby standardize the measurement. The INR is only useful for monitoring the effect of warfarin therapy, as it has been standardized to accurately reflect warfarin

concentration is nonlinear and does not accurately reflect supratherapeutic levels.

PLATELET FUNCTION ASSAYS

Platelet function may be altered with the use of antiplatelet medications and in certain coagulopathies, despite a normal platelet count. In this instance, platelet function tests can detect occult platelet dysfunction and monitor recovery of function (Naidech et al., 2014a). Historically, the bleeding time was used for this purpose, but is no longer recommended because it is insensitive and operator-dependent (Collyer et al., 2009).

Light transmission aggregometry in platelet-rich plasma is considered the gold standard for measuring platelet function (Sibbing et al., 2008). It does this by measuring the response of platelets to agonists such as adenosine diphosphate (ADP). This method was used in the original dose-finding studies for antiplatelet medications such as clopidogrel (Sibbing et al., 2008). Unfortunately, platelet aggregometry is poorly standardized, logistically demanding, and time consuming, which makes it impractical for regular clinical use (Sibbing et al., 2008).

Commercially available point-of-care platelet function assays overcome the obstacles seen with platelet aggregometry. The VerifyNow-ASA (Accumetrics, CA), VerifyNow-P2Y12 assay (Accumetrics, San Diego, CA), and PFA-100 (Siemens AG, Germany) can detect dysfunction secondary to the use of antiplatelet medications (Goldenberg et al., 2005; Naidech et al., 2009a). Similarly, TEG with platelet mapping (TEG-PM) (Haemoscope Corporation, Niles, IL) is a specific viscoelastic hemostatic assay that has been shown to correlate with platelet aggregometry (Collyer et al., 2009). It also shows an ability to detect platelet dysfunction due to antiplatelet drugs and other coagulopathies. Each of these tests uses a slightly different method of measuring platelet activity in the presence of platelet inducers (Goldenberg et al., 2005; Collyer et al., 2009; Naidech et al., 2009a).

VISCOELASTIC HEMOSTATIC ASSAYS

Viscoelastic hemostatic assays, such as TEG and rotational thromboelastometry, were developed in Germany in the 1940s (Ganter and Hofer, 2008). They measure viscoelastic properties of blood at the bedside and more accurately reflect whole-blood clotting potential than routine coagulation assays (Naidech et al., 2014a). Viscoelastic hemostatic assays provide information on all stages of developing and resolving clot. For decades these studies have been used in cardiac surgery, liver transplantation, and multisystem trauma to guide transfusions and factor replacement. They show promise in guiding therapy in neurocritical care populations.

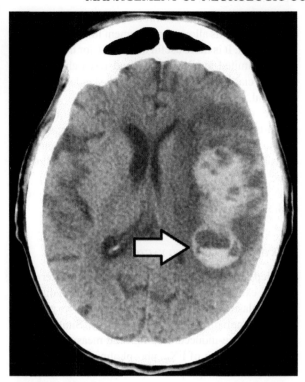

Fig. 40.1. Unenhanced head computed tomography of a patient on a heparin infusion who developed an acute intraparenchymal hemorrhage. Note the fluid level indicated by the arrow. Fluid levels within hematomas are a sign of coagulopathy, because anticoagulated blood separates, creating a horizontal interface between hyperdense settled blood (clumps of red blood cells) and the hypodense bloody serum above.

effect internationally (Hirsh and Poller, 1994). The PT is more sensitive than the aPTT for monitoring the effect of FXa inhibitors such as rivaroxaban or apixaban (Samama and Guinet, 2011). However, a normal PT does not reliably exclude ongoing antithrombotic effect (Samama and Guinet, 2011). It is often abnormal in other derangements of hemostasis caused by medications, liver failure, DIC, and trauma. However its clinical utility varies in each situation. For instance, an abnormal PT in liver failure may not lead to clinical bleeding tendency.

aPTT

The aPTT was developed to assess the "intrinsic pathway" of hemostasis. Calcium, phospholipid, and an activator such as kaolin are added to citrated blood, and time to coagulation is measured. The aPTT is most useful for monitoring UFH. Conversely, the effect of LMWH is not reliably reflected. In the setting of dabigatran use, aPTT is more sensitive than other routine coagulation assays. If normal, it is unlikely that the patient is still coagulopathic from medication effect (Lindahl et al., 2011). However, the relationship between aPTT and drug

TEG may be useful in predicting hematoma expansion in IPH. A prospective study using TEG within 6 hours of onset of spontaneous IPH found that patients with hematoma expansion had significantly slower clot formation compared to patients without hematoma expansion (Kawano-Castillo et al., 2014). TEG has also been used to monitor dabigatran effect in IPH, but has yet to be fully validated (Naidech et al., 2014a). Additionally, TEG may be useful for monitoring the effect of thrombolytic therapy and determining risk of hemorrhagic transformation after thrombolysis for acute ischemic stroke (Elliott et al., 2015).

TBI-associated coagulopathy may cause abnormalities in routine coagulation assays, but TEG has been shown to be much more sensitive at identifying early coagulopathic states (Windeløv et al., 2011; Kunio et al., 2012; Folkerson et al., 2015). Likewise, TEG with platelet mapping may reveal early platelet dysfunction in severe TBI (Nekludov et al., 2007). A prospective study assessing serial TEG profiles over the first 5 days found that patients with TBI developed a progressive and delayed hypercoagulable state compared with controls (Massaro et al., 2015). This information may help guide potential use of early hemostatic therapy and delayed antithrombotic therapy.

Although liver failure causes obvious derangements of routine coagulation assays and platelet count, these laboratory abnormalities do not accurately reflect clinical bleeding tendency. Use of TEG reveals that many patients have normal ability to form clot and may even be hypercoagulable (Lisman et al., 2012; Stravitz et al., 2012). Therefore TEG may be used to more accurately evaluate risk of clinical bleeding prior to invasive procedures, such as placement of intracranial pressure (ICP) monitors in patients with ALF.

Assays for measuring thrombin inhibition

The thrombin time (TT) directly measures thrombin activity in plasma and is very sensitive for detecting the effects of direct thrombin inhibitors such as dabigatran (van Ryn et al., 2010). It tends to be readily available, but is not standardized across separate laboratories. If normal, it reliably rules out an ongoing effect of dabigatran. It tends to be too sensitive to be very useful in emergency situations (van Ryn et al., 2010).

Dilution of a blood sample helps dampen the inflated sensitivity of TT assays. For example, the hemoclot diluted thrombin inhibitor assay (HYPHEN BioMed, France) is able to quickly and accurately assess the effect of dabigatran, but has limited availability (Stangier and Feuring, 2012).

The ecarin clotting time (ECT) uses snake venom to produce a thrombin intermediary, meizothrombin, which is also inhibited by direct thrombin inhibitors (van Ryn et al., 2010). This allows measurement for thrombin inhibition effect. It is more useful than aPTT for this purpose, but has limited availability beyond research purposes. Commercial kits are not validated or standardized (van Ryn et al., 2010).

The ecarin chromogenic assay (ECA) also uses ecarin-induced prothrombin activation, but instead of measuring the clotting of meizothrombin, changes in optical density induced by activity towards a chromogenic substrate are measured (Lange et al., 2003). The ECA is more sensitive than ECT for detecting the effects of direct thrombin inhibitors (Lange et al., 2003).

Assays for measuring factor Xa inhibition

The HepTest measures time to clot formation after addition of exogenous FXa, calcium, brain cephalin, and bovine plasma rich in FV and fibrinogen to patients' plasma samples (Samama et al., 2010). Similarly, the prothombinase-induced clotting time measures clotting time with addition of FXa, phospholipids, FV activator, and calcium to sample plasma (Samama et al., 2010). The chromogenic anti-FXa assay measures changes in optical density after addition of a chromogenic substrate to plasma incubated with exogenous FXa (Samama et al., 2010). There are also specific assays for each oral FXa inhibitor. All of these assays are more sensitive to ongoing FXa inhibition than PT and may be useful for measuring the effect of FXa inhibitors (Naidech et al., 2014a). Similarly, if prior use of LMWH is suspected, but not known from clinical history, anti-FXa activity can be measured (Hirsh and Raschke, 2004). Serial assessment of anti-FXa levels in patients treated with LMWH is warranted at extremes of age, during pregnancy, and with morbid obesity. However, emergent availability of these assays is currently limited.

Fibrinogen and D-dimer

In cases of hyperfibrinolysis, as with DIC, TBI-associated coagulopathy, and tPA use, measurement of fibrinogen and D-dimer levels can be useful. Fibrinolytic-associated ICH theoretically may cause hypofibrinogenemia and elevated D-dimer from clot lysis (Broderick et al., 2007). However, in a retrospective study of 20 patients with fibrinolytic-associated ICH, none had a fibrinogen level less than 100 mg/dL (Goldstein et al., 2010). There are no evidence-based guidelines, and very little observational data to guide management of these patients. Therefore fibrinogen levels and D-dimer should be checked and, if abnormal, serially followed for normalization with treatment.

HOSPITAL COURSE AND MANAGEMENT

The emergent treatment of ICH requires early diagnosis and treatment of coagulopathy, which may minimize hematoma expansion and improve outcome. Significant hematoma expansion (>33%) occurs in about one-third of patients with spontaneous IPH (Brott et al., 1997; Steiner et al., 2006). Many other patients have lesser degrees of growth. The vast majority of hematoma expansion occurs within the first 6 hours after initial hemorrhage, and most of the remainder within the first 24 hours (Kazui et al., 1996). In contrast, hematoma expansion occurs with a higher frequency and over a longer period of time in coagulopathic IPH (possibly up to 1–2 days after initial IPH) (Flibotte et al., 2004). It is therefore paramount that any coagulopathy be corrected as soon as possible in patients with IPH. The general management of IPH and TBI is beyond the scope of this chapter (Broderick et al., 2007). The remainder of this section will review the emergent management of various coagulation disturbances.

Treatment of antithrombotic-associated ICH

VITAMIN K ANTAGONISTS

Therapies to reverse warfarin are aimed at coagulation factor replacement. Vitamin K, fresh frozen plasma (FFP), prothrombin complex concentrates (PCC), and recombinant FVIIa (rFVIIa), used alone or in combination, are the current existing options (Broderick et al., 2007).

Vitamin K is usually given intravenously (IV) as a 10-mg dose (Holbrook et al., 2012). It takes several hours to take effect, and should never be used alone in emergent treatment of ICH (Broderick et al., 2007). It is crucial to administer vitamin K with other treatments because the half-life of vitamin K is longer than other therapies and it can durably maintain a normalized INR. The infusion carries a very small risk of anaphylaxis (<0.1%), which is increased with more rapid infusion (Choonara et al., 1985; Riegert-Johnson and Volcheck, 2002).

FFP is commonly used with vitamin K to reverse the effects of warfarin. It can effectively replete coagulation factors and normalize INR, but it has several limitations. Treatment with FFP requires considerable thaw time and infusion volumes (15–20 mL/kg) that can delay treatment and precipitate heart failure (Broderick et al., 2007). Additionally, it often takes hours to complete the infusion and it can have an unpredictable effect due to variable concentrations of coagulation factors in each bag of plasma (Broderick et al., 2007). Owing to these limitations, other methods of factor replacement are gaining favor. Hemorrhages can expand quickly and patients may arrive with larger hemorrhage with or without reversal of INR (Fig. 40.2)

PCC contains either three (II, IX, and X) or four factors (II, VII, IX, and X) (Broderick et al., 2007). It has the advantage of faster infusion, smaller volume, and more rapid INR correction when compared to FFP (Bechtel et al., 2011). A multicenter randomized trial compared the efficacy of four-factor PCC to FFP for treatment of major bleeding in patients receiving VKAs (Sarode

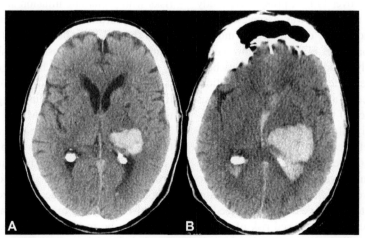

Fig. 40.2. Initial hemorrhage (**A**) and hematoma expansion (**B**) of a 76-year-old man on warfarin for atrial fibrillation admitted to the neurointensive care unit with a left thalamic intraparenchymal hemorrhage. He was in his normal state of health until 6 a.m., when he noticed acute right-sided weakness. In his local Emergency Department the initial head computed tomography (**A**), performed by 7 a.m., revealed an acute intraparenchymal hemorrhage. Afterwards he became more lethargic, with shallow respirations, and was intubated for airway protection prior to transfer to our intensive care unit. Repeat head computed tomography (**B**) at 5 p.m. showed significant hematoma expansion with intraventricular hemorrhage.

et al., 2013). It found that PCC is superior to FFP in correcting the INR within 30 minutes from the start of treatment (62% vs. 10%), with a similar rate of thromboembolic events. Volume overload was more common with FFP (13% vs. 5%). PCC was found to be noninferior to FFP as measured by an objective hemostatic efficacy scale. However, as few patients with ICH were included, it remains unclear whether this translates into improved patient outcomes. In a subsequent, larger clinical trial, 181 patients needing rapid VKA reversal prior to surgery or an invasive procedure were randomized to receive either four-factor PCC or FFP (Goldstein et al., 2015). Almost no patients with ICH were included in this study. The primary endpoint was effective hemostasis during surgery, and this was achieved in 90% of patients treated with PCC compared with 75% of those who received plasma ($p = 0.01$). Rapid INR correction was achieved in 55% of PCC-treated patients compared with 10% of plasma-treated patients ($p < 0.0001$). The risk of thromboembolic adverse events was not increased in the PCC group.

Observational studies assessing PCC and FFP in warfarin-associated IPH have demonstrated mixed results; however, some have found improved survival and less hematoma expansion with PCC (Huttner et al., 2006; Hanger et al., 2013; Frontera et al., 2014; Majeed et al., 2014). A retrospective study of 55 patients with warfarin-associated IPH found significantly less hematoma expansion with PCC, but this difference became nonsignificant if the INR was corrected within 2 hours by FFP (Huttner et al., 2006). Similarly, a retrospective study of 88 patients found that earlier administration of PCC led to improved survival (Hanger et al., 2013). This suggests that faster INR correction is the main advantage of PCC over FFP. A prospective study of 64 patients found that PCC was associated with improved mortality and severe disability at 3 months compared to FFP (Frontera et al., 2014). However, a larger retrospective study found no 30-day mortality benefit with PCC after adjusting for hematoma size, location, and patient age (Majeed et al., 2014). A meta-analysis observed a 1% incidence of thromboembolic complications with PCC (Dentali et al., 2011). Further prospective studies are needed to better evaluate the putative benefits of PCC in warfarin-associated IPH.

rFVIIa has also been shown to correct the INR more rapidly than FFP. However, whether the normalization of lab testing correlates with cessation of bleeding remains unclear. Furthermore, increased risk of thromboembolic complications limits its use (Broderick et al., 2007). Two large randomized trials evaluated the use of rFVIIa in spontaneous IPH (Mayer et al., 2006, 2008). Although treatment with an 80 μg/kg dose resulted in a significant reduction in hematoma expansion compared to placebo in the phase IIb trial, the larger phase III trial failed to show a benefit for functional outcome or mortality. Additionally, there were significantly more arterial thrombotic events compared to placebo (9% vs. 4%). This has led to concern about its use, especially in patients with risk factors for thromboembolic complications, such as pre-existing heart disease or atrial fibrillation. A comparative observational study of 45 patients with warfarin-associated IPH did not find an increase in clinically significant myocardial infarction or thromboembolism with rFVIIa compared to standard therapy (H-Y et al., 2012). However, caution in patients with risk factors is warranted until more evidence becomes available. Very limited data exist comparing PCC to rFVIIa. A mouse model of warfarin-associated IPH found that rFVIIa failed to prevent hematoma expansion compared to both PCC and FFP (Illanes et al., 2011). Current guidelines recommend against the routine use of rFVIIa alone for the reversal of VKAs (Hemphill et al., 2015).

HEPARIN AND HEPARINOIDS

Protamine sulfate is used to reverse the anticoagulant effect of UFH. Because the half-life of UFH is less than an hour, the amount of protamine required for emergent reversal depends upon the time elapsed since the infusion was discontinued (Hirsh and Raschke, 2004; Broderick et al., 2007). Immediate reversal requires 1 mg of protamine for every 100 U of UFH administered within the last 2–3 hours (Schulman et al., 2008). The maximum dose should not exceed 50 mg of protamine (Schulman et al., 2008). The main risks with protamine infusion are systemic hypotension and bradycardia, which can be mitigated via slow infusion. The aPTT can be followed to monitor reversal and guide redosing if necessary (Hirsh and Raschke, 2004).

LMWH has a longer half-life compared to UFH and is only partially reversed by protamine (about 60%), likely owing to poor binding by protamine to the small molecular moieties of LMWH (Hirsh and Raschke, 2004). The clinical significance of this is unclear, and there is no other currently available method of reversing LMWH. Therefore, it is recommended that 1 mg of protamine be given per 1 mg of enoxaparin, within 8 hours of the last dose. If bleeding is not stopped, a second dose of 0.5 mg protamine per 1 mg of enoxaparin may be required. A similar algorithm is recommended for dalteparin at a ratio of 1 mg protamine per 100 IU of dalteparin, which may be more effective due to variable sulfonation (Hirsh and Raschke, 2004). Unlike UFH, the aPTT does not reliably reflect the ongoing effect of LMWH, and anti-FXa assays must be used instead.

Direct thrombin inhibitors

The half-life of dabigatran is about 12 hours in patients with normal renal function (van Ryn et al., 2010). Dabigatran can be removed by intermittent hemodialysis, but fluid shifts may prove deleterious to patients with large mass lesions such as ICH. Animal models have shown a possible benefit with rFVIIa and PCC for reversing the dabigatran effect on aPTT, rat tail bleeding times, and hematoma expansion (van Ryn et al., 2010; Illanes et al., 2011). However, in a randomized study of 12 healthy volunteers who were given dabigatran, PCC failed to normalize aPTT, ECT, or TT (Eerenberg et al., 2011). Owing to this, some studies recommend the use of activated PCC over four-factor PCC for reversing dabigatran (Siegal and Cuker, 2013). No patients had thrombotic complications from activated PCC in a small case series of patients with TSOAC-associated IPH (Dibu et al., 2016). However, in clinical practice, concern over theoretical thrombotic complications of activated PCC may lead some to prefer rFVIIa for reversing dabigatran until more prospective data are available (Jauch et al., 2015).

A monoclonal antibody fragment against dabigatran, idarucizumab, has been shown to quickly normalize the anticoagulant effect of dabigatran. In a prospective cohort of 90 patients on dabigatran who had serious bleeding or required an urgent procedure, 5 grams of IV idarucizumab normalized the dilute TT and ECT in 88–98% of patients within minutes (Pollack et al., 2015). There was one reported thrombotic event within 72 hours of administration of the drug. Further prospective studies are required to determine the clinical efficacy of idarucizumab.

Factor Xa inhibitors rivaroxaban, apixaban, and edoxaban

Similar to dabigatran, limited preclinical *in vitro* data involving healthy volunteers and animal studies suggest that PCC, activated PCC (aPCC), and rFVIIa may be useful for reversing the effect of FXa inhibitors (Eerenberg et al., 2011; Perzborn et al., 2013). In one study, 12 healthy male volunteers were given rivaroxaban 20 mg twice daily then given either PCC or saline (Eerenberg et al., 2011). Their PT and endogenous thrombin potential were measured before and after treatment. The PT and endogenous thrombin potential were immediately normalized after administration of PCC. Another study looked at the effectiveness of PCC, aPCC, and rFVIIa for normalizing coagulation laboratory assays in rats and baboons (Perzborn et al., 2013). In rats, PCC, aPCC, and rFVIIa significantly decreased bleeding times. In baboons, aPCC returned bleeding time to baseline, while rFVIIa reduced bleeding time to 1.7-fold over baseline. PT was decreased by all treatments in both animals, but thrombin-antithrombin levels were only increased by PCC or aPCC and not rFVIIa. This suggests that PCC and aPCC may be more effective than rFVIIa.

Andexanet alfa is a modified FXa protein, which binds to and inactivates FXa inhibitors. It has been studied with rivaroxaban, apixaban, and LMWH (Ansell, 2016). In phase II studies and a randomized, placebo-controlled trial of healthy elderly volunteers given either rivaroxaban or apixaban, a continuous infusion of andexanet was able to quickly normalize several laboratory assays (Siegal et al., 2015). However, shortly after stopping the infusion, anti-FXa activity returned, suggesting it may require prolonged infusion. It has no intrinsic coagulant activity and was well tolerated; however it did reduce tissue factor pathway inhibitor levels, which may imply a prothrombotic effect (Ansell, 2016).

Nonspecific reversal agents

The synthetic compound ciraparantag may prove to be an effective universal reversal agent. It has been shown to normalize the effect of UFH, LMWH, and TSOACs. It was able to reduce bleeding in animal models and normalize coagulation assays in healthy volunteers given edoxaban and dabigatran with a single IV dose (Ansell et al., 2014; Ansell, 2016). There was no evidence of a prothrombotic effect. Results of further studies and Food and Drug Administration review are pending.

Fibrinolytic agents

Since only a small number of patients have IPH associated with thrombolysis, data guiding management in this situation are very limited. Clot lysis may cause hypofibrinogenemia and release D-dimer, which may bind to platelet fibrinogen receptor and cause an antiplatelet effect (Goldstein et al., 2010). There is therefore a theoretic basis for treatment with cryoprecipitate (which contains fibrinogen) and platelet infusions, while following fibrinogen and D-dimer levels for therapeutic effect (Broderick et al., 2007). The clinical benefit of this approach is not yet known. A small retrospective study of 20 patients found that no patients actually had a fibrinogen level <100 mg/dL (Goldstein et al., 2010). Furthermore, there was a wide variation in treatment strategy, and there was no improved outcome associated with the use of procoagulant therapy. The antifibrinolytic aminocaproic acid is not included in the American Heart Association guidelines given the concern for thrombotic complications. In the absence of evidence, various protocols have been adopted for management with the hope of expediting treatment (Rasler, 2007). This involves emergent infusion of 6–10 U cryoprecipitate with 6–8

U platelets while measuring fibrinogen and D-dimer levels for effect. Some may give thawed plasma or PCC in addition, based on severity of the clinical situation (Rasler, 2007). If fibrinolytics were administered during neuroendovascular procedure, it is important to consider protamine therapy to reverse any effect of UFH, which may have been used in conjunction.

ANTIPLATELET AGENTS

The absolute increase in risk of ICH while on antiplatelet agents is very small. Numerous studies have evaluated the effect of prior antiplatelet use on patient outcomes in IPH, but have had conflicting results. One study used the placebo arm of a large prospective randomized trial to evaluate the effect of antiplatelet use at time of IPH (Sansing et al., 2009). It included 282 patients, a quarter of whom were receiving antiplatelet agents. There was no correlation between antiplatelet use and hematoma size, edema volume, hematoma expansion, or 90-day functional outcome. Conversely, a separate observational study of 252 patients, about a quarter of whom were on antiplatelet therapy while suffering an IPH, found a correlation between antiplatelet use and hematoma expansion, need for emergent evacuation of hematoma, and death (Toyoda et al., 2005). It also found that those on antiplatelet therapy tended to be older and have a prior history of stroke and heart disease. A systematic review found no evidence of an effect of antiplatelet use on functional outcome (Thompson et al., 2010). Another systematic review found that the empiric administration of platelet transfusions or platelet activating agents failed to consistently show benefit (Nishijima et al., 2012).

The randomized controlled multicenter PATCH trial assessed the efficacy of platelet transfusion in patients with IPH receiving antiplatelet drugs. Patients were randomized within 6 hours of IPH and received one dose (5 units; if receiving a cyclooxygenase inhibitor), two doses (10 units; if receiving an ADP receptor inhibitor) or standard medical therapy alone. The vast majority of patients were receiving monotherapy with aspirin; fewer than 5% were treated with ADP receptor inhibitors. The chance of hematoma expansion was not reduced, and the proportion of patients with poor neurologic outcomes was actually significantly higher in the group receiving platelet transfusion.

Patients with platelet dysfunction are not always those treated with antiplatelet medications. Some observational studies suggest that treatment of platelet dysfunction, defined by POCTs, may improve outcomes (Naidech et al., 2009b). Functional platelet activity is restored to some degree after transfusion or desmopressin use (Naidech et al., 2012, 2014b). However, further studies specifically involving patients with platelet dysfunction are required to demonstrate clinical benefit (Table 40.1).

Thrombocytopenia

Patients with thrombocytopenia may require platelet transfusion depending on the clinical situation. In those without clinical bleeding, guidelines recommend a transfusion threshold of $<10\,000/\mu L$ (Slichter, 2007). Patients undergoing neurosurgery may be at risk of ICH with lower platelet counts (Chan et al., 1989). Because of the high morbidity and mortality in ICH, guidelines recommend a transfusion threshold of $<100\,000/\mu L$ in any patient with ICH or undergoing a neurosurgical procedure (Slichter, 2007). This recommendation is based largely on expert opinion, since data are limited.

Disseminated intravascular coagulation

The basic tenet in treatment of DIC is management of the underlying condition (Levi et al., 2009). There are no data from randomized trials concerning the efficacy of platelet transfusions or coagulation factor replacement. Therefore, it is not recommended to transfuse platelets or replace factors based upon laboratory changes alone (Levi et al., 2009). However, if a patient is actively bleeding, or is at high risk of bleeding, then it may be reasonable to transfuse platelets and replace factors. There is no evidence that factor replacement exacerbates DIC by adding "fuel to the fire" (Levi et al., 2009). If a patient is actively bleeding and has a platelet count $<50\,000/\mu L$, then platelet transfusion should be considered (Levi et al., 2009). Likewise, if a patient has active bleeding and prolonged PT, aPTT, or decreased fibrinogen, then giving FFP, PCC, and/or cryoprecipitate is reasonable (Levi et al., 2009).

Traumatic coagulopathy

A large multicenter randomized controlled trial evaluated the effect of tranexamic acid compared to placebo within 8 hours in trauma patients with bleeding or at risk of severe bleeding (Shakur et al., 2010). Tranexamic acid reduced all-cause mortality and death due to bleeding without a significant increase in thromboembolic complications. A substudy of this trial evaluated 280 patients with TBI and found a nonsignificant decrease in hemorrhagic contusion expansion, new focal cerebral ischemia, and mortality (CRASH-2 Collaborators Intracranial Bleeding Study, 2011). Following this, another trial evaluated use of tranexamic acid within 8 hours of moderate to severe TBI in 238 patients (Yutthakasemsunt et al., 2013). It also found a nonsignificant reduction in IPH expansion or mortality. Of note, this trial excluded patients with coagulopathy as measured by routine coagulation assays. Overall

Table 40.1
Management of antithrombotic-associated coagulopathy

Antithrombotic agent	Laboratory assessment	Reversal agent	Comments
Vitamin K antagonist	INR	Four-factor PCC and vitamin K 10 mg IV	PCC dose should be INR- and weight-based rFVIIa may be considered if PCC unavailable
Unfractionated heparin	PTT	1 mg protamine per 100 U UFH given in last 2–3 hours (maximum dose 50 mg)	Monitor for bradykinin reaction: hypotension
Low-molecular-weight heparin	Agent-specific chromogenic anti-FXa assay	1 mg protamine per 1 mg LMWH given in last 8 hours (maximum dose 50 mg)	Protamine reverses 66% of enoxaparin, but up to 95% of other LMWH (e.g., tinzaparin) Andexanet alfa (or ciraparantag) may be potential options
Direct thrombin inhibitors	PTT, TT, DTI, ECT, ECA	Oral agents (dabigatran) – idarucizamab 5 g IV IV: aPCC (FEIBA) 50 units/kg IV or four-factor PCC 50 units/kg IV	Consider activated charcoal if within 2–3 hours of ingestion and low risk of aspiration
Direct FXa inhibitors	HepTest, PiCT, Agent-specific chromogenic anti-FXa	aPCC (FEIBA) 50 units/kg IV or four-factor PCC 50 units/kg IV	Consider activated charcoal if within 2–3 hours of ingestion and low risk of aspiration Andexanet alpha or ciraparantag may be future treatment options
Thrombolytics	Fibrinogen, D-dimer	6–10 U cryoprecipitate and 6–8 U platelets (consider thawed plasma if delay in obtaining cryoprecipitate)	Reverse concomitant UFH given during endovascular procedures per above
Antiplatelets	Plt count, VerifyNow-ASA, VerifyNow-P2Y12, PFA-100, TEG-PM	Platelet transfusion or desmopressin 0.3–0.4 µg/kg IV	Platelet transfusion is not routinely recommended, but may be considered for patients who may require neurosurgical intervention

INR, international normalized ratio; PCC, prothrombin complex concentrate; PTT, partial thromboplastin time; UFH, unfractionated heparin; FXa, factor Xa; LMWH, low-molecular-weight heparin; TT, thrombin time; DTI, dilute thrombin time; ECT, ecarin clotting time; ECA, ecarin chromogenic assay; IV, intravenously; aPCC, activated prothrombin complex concentrate; PiCT, prothombinase-induced clotting time; Plt, platelet; TEG-PM, thromboelastography platelet mapping.

these data suggest that there may be a role for early antifibrinolytic therapy in moderate to severe TBI, but further studies in this population are required.

Limited data exist regarding the use of rFVIIa and PCC in TBI coagulopathy. A substudy of 30 patients from a randomized trial found there was no increased risk of adverse events with the use of rFVIIa in TBI (Kluger et al., 2007). Another study of 97 patients with TBI found a nonsignificant reduction in IPH expansion, without increased adverse events (Narayan et al., 2008). A single-center retrospective study of 85 patients found significantly lower mortality when PCC was used to treat TBI-associated coagulopathy (Joseph et al., 2013). Further prospective studies are needed to evaluate the therapeutic benefit of these treatments. TEG-PM may be considered to guide early hemostatic therapy in moderate to severe TBI (Walsh et al., 2011).

Coagulopathy associated with systemic disease

A leading cause of death from ALF is diffuse cerebral edema. Treatment of intracranial hypertension guided by ICP monitoring is used at some academic centers as a bridge to transplant. Retrospective studies of patients with ALF and ICP monitoring observed an ICH rate of 7–10%, only half of which were symptomatic (Vaquero et al., 2005; Karvellas et al., 2014). This rate varies depending on which factor replacements were given prior to the procedure, if any. The risk may be lower with use of epidural catheters; however these are not as accurate or reliable for assessing ICP compared to ventricular or parenchymal monitors (Blei et al., 1993). Since ALF causes large derangements of routine coagulation assays such as INR, many centers use procoagulant therapy to normalize INR prior to placing ICP monitors. rFVIIa effectively and quickly normalizes INR, and may be more effective compared to FFP (Krisl et al., 2011). Use of rFVIIa at 20–40 μg/kg, 30 minutes prior to the procedure, has been observed to be safe in case series (Krisl et al., 2011). Whether or not this strategy reduces clinical bleeding or alters outcome is not clear. It seems reasonable to consider use of viscoelastic hemostatic assays to determine whether patients are coagulopathic before empiric treatment with procoagulant therapies.

Many of the studies informing treatment of uremic bleeding were performed about 30 years ago and are of low quality (Hedges et al., 2007). Bleeding time was considered the most appropriate test to evaluate uremic coagulopathy. Erythropoietin, cryoprecipitate, conjugated estrogens, desmopressin, and tranexamic acid have all been shown to decrease bleeding time in uremic patients (Hedges et al., 2007).

Desmopressin is the most commonly used agent in uremic bleeding. It is given as a single dose of 0.3–0.4 μg/kg via the subcutaneous or intravenous route (Hedges et al., 2007). It has a rapid onset of action but short duration of therapeutic effect. A randomized trial comparing a 0.3 μg/kg infusion of desmopressin to placebo in patients with uremic coagulopathy found that desmopressin decreased bleeding time, increased FVIII activity, and created larger vWF-FVIII multimers in all patients (Mannucci et al., 1983). The effect lasted an average of 4 hours. A second trial compared a 0.4 μg/kg subcutaneous injection of desmopressin to placebo in patients with uremic bleeding (Köhler et al., 1989). Desmopressin levels peaked by 1 hour postinjection, and the mean half-life was about 3 hours. This study also showed decreased bleeding times and increased FVIII activity. Because desmopressin has a short half-life, use of another agent such as cryoprecipitate or tranexamic acid should be considered in patients with ongoing life-threatening bleeding. Additionally, hematocrit should be maintained around 30% with red blood cell transfusion when necessary, in order to improve rheology of blood and platelet interaction with the endothelium (Pavord and Myers, 2011) (Table 40.2).

CLINICAL TRIALS AND GUIDELINES

Guidelines published by the American Heart Association/American Stroke Association for management of spontaneous IPH and by the American College of Chest Physicians on antithrombotic and thrombolytic therapy provide input into management of antithrombotic-associated ICH (Schulman et al., 2008; Holbrook et al., 2012; Hemphill et al., 2015). However management of coagulopathies in the neurologic ICU is not the focus of these guidelines.

The Neurocritical Care Society (NCS) published guidelines in 2016 specifically on the reversal of antithrombotics in ICH (Frontera et al., 2016). These are the first such guidelines from the perspective of neurointensivists. Because of this unique perspective, these guidelines should be consulted when treating critically ill

Table 40.2

Management of coagulopathy associated with trauma or systemic disease

Disease process	Laboratory assessment	Hemostatic strategy	Comments
Traumatic coagulopathy	TEG	Tranexamic acid if multisystem trauma with extracranial bleeding. Consider TEG-guided treatment if abnormal	Routine lab assays may be less sensitive than TEG
Acute liver failure	TEG	Consider rFVIIa prior to ICP monitor placement but unclear benefit	Routine lab assays may be abnormal but may not reflect *in vivo* coagulation potential
DIC	INR, PTT, platelet count, fibrinogen, D-dimer	Treatment based on which laboratory assays are abnormal and presence of clinical bleeding	Focus treatment on the underlying condition
Uremia	VerifyNOW, PFA-100, TEG-PM	Desmopressin 0.3–0.4 μg/kg SC or IV, may require repeat doses due to short half-life (3–4 hours)	

TEG, thromboelastography; rFVIIa, recombinant factor VIIa; ICP, intracranial pressure; DIC, disseminated intravascular coagulation; INR, international normalized ratio; PTT, partial thromboplastin time; TEG-PM, thromboelastography platelet mapping; SC, subcutaneously; IV, intravenously.

patients who have neurologic complications of coagulopathy. Below is a brief summary of major recommendations. The full guideline manuscript should be referred to for complete recommendations.

For ICH attributable to VKAs, administering intravenous vitamin K in conjunction with four-factor PCC is recommended. The dose of PCC should be based on weight and admission INR. INR values should be followed serially, with FFP given if the INR remains elevated in the next 1–2 days. The NCS recommends against using rFVIIa for reversal of VKAs.

For ICH in patients receiving UFH, use of protamine at a dose of 1 mg per 100 U of heparin given in the last 2–3 hours, with repeat dosing at 0.5 mg per 100 U if necessary to normalize aPTT, is recommended. For enoxaparin, protamine is recommended at a dose of 1 mg per 1 mg of enoxaparin given in the last 8 hours and 0.5 mg per 1 mg for enoxaparin given 8–12 hours ago.

For ICH in patients receiving direct thrombin inhibitors, activated charcoal is recommended for those at low risk of aspiration who have ingested drug in the past 2 hours. Idarucizumab is recommended for ingestions of dabigatran in the last 3–5 half-lives, when renal impairment is absent. Activated PCC or four-factor PCC is recommended when idarucizumab is unavailable. The NCS recommends against using rFVIIa for reversal of direct thrombin inhibitors in ICH. For direct FXa inhibitors ingested in the last 2 hours, activated charcoal is recommended for those at low risk of aspiration. Activated PCC or four-factor PCC is recommended for ingestions in the last 3–5 half-lives.

The NCS recommends 10 U of cryoprecipitate to reverse thrombolytics given in the last 24 hours. If cryoprecipitate is unavailable or contraindicated, then tranexamic acid or aminocaproic acid are recommended alternatives. The NCS offers no recommendation for or against platelet transfusion use in this context.

The NCS recommends against the use of platelet transfusions in antiplatelet-associated ICH unless a patient is undergoing a neurosurgical procedure. If the patient is undergoing a neurosurgical procedure, it is recommended to assess platelet function, whenever available, to aid in making a decision about platelet transfusion. Additionally, transfusions should be given 1 unit at a time. Consideration should also be given to use of a single dose of desmopressin in antiplatelet-associated ICH regardless of planned neurosurgical intervention.

COMPLEX CLINICAL DECISIONS

Derangements of hemostasis with a tendency toward bleeding often coexist with a predilection to form clot. This is evident in traumatic coagulopathy, malignancy, liver disease, uremia with chronic kidney disease, and other conditions. Similarly, patients with coagulopathic bleeding from antithrombotic medications are prescribed those medications in the setting of an underlying prothrombotic condition. In light of this, it is difficult to find the perfect balance of acute reversal of coagulopathy in ICH with concomitant prevention of thromboembolic complications among patients who are at risk of these. The decision regarding whether and when to restart anticoagulation is also complex.

All patients with ICH are at increased risk for thromboembolic complications. About 7% of patients with IPH will suffer a thromboembolic complication during their hospital stay (Goldstein et al., 2009). Three percent suffer from ischemic stroke (Goldstein et al., 2009). Interestingly, a large prospective study found no difference in risk of these complications between patients with anticoagulant-related IPH and those with spontaneous IPH (Goldstein et al., 2009). It is still recommended to always reverse coagulopathy emergently in the acute phase of IPH, when the risk of harm from hematoma expansion outweighs risk of thrombosis (Broderick et al., 2007). After the first 24–72 hours, the risk of hematoma expansion decreases sharply, while the risk of thrombotic complications steadily increases. It is therefore not surprising that a clinical trial showed that it is safe to start prophylactic doses of subcutaneous UFH 48 hours from onset of IPH (Boeer et al., 1991). Guidelines recommend low-dose chemical thromboembolic prophylaxis starting 1–4 days after onset with documented cessation of bleeding on repeat CT scan (Hemphill et al., 2015).

The more complex decision involves whether and when to restart therapeutic anticoagulation for conditions such as atrial fibrillation, mechanical heart valves, or recent deep-vein thrombosis or pulmonary embolism. Unfortunately, there is no population-based study on the incidence of recurrent ICH specifically in patients on anticoagulation. A systematic review found an aggregate rate of recurrent IPH of 2% per patient-year in all primary IPH (Bailey et al., 2001). The risk depended partly on location of IPH, with lobar hemorrhages having a recurrence of 4% compared to 2% for deep hemorrhages. This likely reflects the increased risk of recurrent hemorrhage in patients with cerebral amyloid angiopathy. Another marker for this condition is the presence of cortical cerebral microbleeds on MRI.

The risk of recurrent ICH must be weighed against the risk of thromboembolism. The risk of embolic stroke in atrial fibrillation without prior ischemic stroke is about 5% per year; this increases to about 12% per year in patients with atrial fibrillation and prior ischemic stroke (Broderick et al., 2007). Patients with mechanical heart valves have about a 4% per year risk of embolic stroke (Broderick et al., 2007).

A study of 141 patients with high risk of embolism who had a warfarin-associated IPH were followed for incidence of stroke and recurrent IPH (Phan et al., 2000). It was estimated that the risk of stroke at 30 days for patients with mechanical heart valves was 2.9%, 2.6% for those with atrial fibrillation and prior stroke, and 4.8% for those with recurrent transient ischemic attacks or stroke. None of the patients who restarted warfarin after an average of 10 days had a recurrent IPH during their hospitalization. Another study of warfarin-associated IPH directly compared patients who restarted warfarin to those who did not and followed them for an average of 3.5 years (Claassen et al., 2008). It found that patients who restarted warfarin had increased risk of recurrent IPH, and patients who did not restart warfarin had increased risk of stroke. There was no therapeutic benefit to either strategy.

Given the lack of population-based data on the risk of IPH recurrence while on long-term anticoagulation, and the variable results of observational studies, the clinician is left to make decisions on an individual basis. The American Heart Association guidelines recommend restarting anticoagulation after 4 weeks, except in those at especially high risk of embolism (i.e., mechanical heart valves), in whom a time-frame of 7–10 days may be more appropriate (Broderick et al., 2007; Hemphill et al., 2015). A patient with a lobar hemorrhage who is elderly or has evidence of cortical cerebral microbleeds on MRI likely should not restart anticoagulation, even in the setting of a mechanical heart valve, given a high reported rate of recurrent IPH. Conversely, a patient with a deep hemorrhage and high risk of future embolism (such as atrial fibrillation and prior stroke or mechanical heart valve) would likely benefit more from restarting anticoagulation. Additionally, in the absence of contraindications such as kidney disease, novel oral anticoagulants should be chosen, given the documented lower risk of ICH compared to warfarin.

OUTCOME PREDICTION

It is always difficult for physicians to provide accurate outcome prediction to patients and families, but clinical tools are available to aid in this process. Numerous validated scores aim to predict mortality and functional outcome in both ICH and TBI. These include the ICH, Graeb, Leroux, IVH, and FUNC scores for ICH and the IMPACT, CRASH, Marshall, and Rotterdam scores for TBI (Marshall et al., 1992; Hemphill et al., 2001, 2009; Clarke et al., 2004; Hukkelhoven et al., 2005; Maas et al., 2005; Rost et al., 2008; Saatman et al., 2008; Steyerberg et al., 2008; Hwang et al., 2012). Each of these scores is described in detail elsewhere. Of note, none of these scores account for the presence of coagulopathy. However, it is known that mortality is higher in patients with coagulopathic IPH compared to spontaneous IPH (Steiner et al., 2006). A study found a correlation between admission hemoglobin, platelet count, and PT and 6-month outcome in TBI based on the IMPACT database (Van Beek et al., 2007). Scoring systems based on large populations of patients should be used with caution to predict outcome in any individual patient. Careful discussions with family members are required and should focus on the wishes of the patient, personal values, and what a "good outcome" means to that individual.

CONCLUSION

Coagulopathy is common in critical illness and poses complex management decisions. ICH is the major neurologic complication of coagulopathy. Although there is a paucity of data regarding the management of coagulopathy in the neurocritical care population, data derived from studies of general critical care populations, ICH, TBI, hematology, and cardiology can inform treatment decisions. Emergently identifying and correcting coagulopathy in ICH and TBI is crucial to decrease hematoma expansion and hopefully improve outcomes. Although widely available, routine coagulation assays do not reliably reflect true *in vivo* clotting potential. In certain situations, specific POCTs or viscoelastic assays are required to identify and/or monitor the resolution of coagulopathy. Currently available approaches to reversing coagulopathy include primarily factor replacement, which may not be efficacious for reversing TSOACs. As the use of TSOACs becomes more widespread, novel class-specific reversal agents may provide safe and effective means for reversing coagulopathy.

REFERENCES

Ansell JE (2016). Universal, class-specific and drug-specific reversal agents for the new oral anticoagulants. J Thromb Thrombolysis 41: 248–252.

Ansell JE, Bakhru SH, Laulicht BE et al. (2014). Use of PER977 to reverse the anticoagulant effect of edoxaban. N Engl J Med 371: 2141–2142.

Bailey RD, Hart RG, Benavente O et al. (2001). Recurrent brain hemorrhage is more frequent than ischemic stroke after intracranial hemorrhage. Neurology 56: 773–777.

Bechtel BF, Nunez TC, Lyon JA et al. (2011). Treatments for reversing warfarin anticoagulation in patients with acute intracranial hemorrhage: a structured literature review. Int J Emerg Med 4: 40.

Becker RC, Bassand JP, Budaj A et al. (2011). Bleeding complications with the P2Y12 receptor antagonists clopidogrel and ticagrelor in the PLATelet inhibition and patient outcomes (PLATO) trial. Eur Heart J 32: 2933–2944.

Bernal W, Auzinger G, Dhawan A et al. (2010). Acute liver failure. Lancet 376: 190–201.

Bhatt DL, Lincoff AM, Gibson CM et al. (2009). Intravenous platelet blockade with cangrelor during PCI. N Engl J Med 361: 2330–2341.

Blanchette V, Carcao M (2000). Approach to the investigation and management of immune thrombocytopenic purpura in children. Semin Hematol 37: 299–314.

Blei AT, Olafsson S, Webster S et al. (1993). Complications of intracranial pressure monitoring in fulminant hepatic failure. Lancet 341: 157–158.

Boeer A, Voth E, Henze T et al. (1991). Early heparin therapy in patients with spontaneous intracerebral haemorrhage. J Neurol Neurosurg Psychiatry 54: 466–467.

Broderick J, Connolly S, Feldmann E et al. (2007). Guidelines for the management of spontaneous intracerebral hemorrhage in adults: 2007 update: a guideline from the American Heart Association/American Stroke Association Stroke Council, High Blood Pressure Research Council, and the Quality of Care and Outcomes in Research Interdisciplinary Working Group. Stroke 38: 2001–2023.

Brott T, Broderick J, Kothari R et al. (1997). Early hemorrhage growth in patients with intracerebral hemorrhage. Stroke 28: 1–5.

Caldeira D, Rodrigues FB, Barra M et al. (2015). Non-vitamin K antagonist oral anticoagulants and major bleeding-related fatality in patients with atrial fibrillation and venous thromboembolism: a systematic review and meta-analysis. Heart 101: 1204–1211.

CAPRIE Steering Committee (1996). A randomised, blinded, trial of clopidogrel versus aspirin in patients at risk of ischaemic events (CAPRIE). Lancet 348: 1329–1339.

Carlson SE, Aldrich MS, Greenberg HS et al. (1988). Intracerebral hemorrhage complicating intravenous tissue plasminogen activator treatment. Arch Neurol 45: 1070–1073.

Chakraverty R, Davidson S, Peggs K et al. (1996). The incidence and cause of coagulopathies in an intensive care population. Br J Haematol 93: 460–463.

Chan KH, Mann KS, Chan TK (1989). The significance of thrombocytopenia in the development of postoperative intracranial hematoma. J Neurosurg 71: 38–41.

Choonara IA, Scott AK, Haynes BP et al. (1985). Vitamin K1 metabolism in relation to pharmacodynamic response in anticoagulated patients. Br J Clin Pharmacol 20: 643–648.

Claassen DO, Kazemi N, Zubkov AY et al. (2008). Restarting anticoagulation therapy after warfarin-associated intracerebral hemorrhage. Arch Neurol 65: 1313–1318.

Clarke JL, Johnston SC, Farrant M et al. (2004). External validation of the ICH score. Neurocrit Care 1: 53–60.

Cohen MJ, Brohi K, Ganter MT et al. (2007). Early coagulopathy after traumatic brain injury: the role of hypoperfusion and the protein C pathway. J Trauma 63: 1254–1261. discussion 1261-1252.

Collyer TC, Gray DJ, Sandhu R et al. (2009). Assessment of platelet inhibition secondary to clopidogrel and aspirin therapy in preoperative acute surgical patients measured by thrombelastography platelet mapping. Br J Anaesth 102: 492–498.

Connolly S, Pogue J, Hart R et al. (2006). Clopidogrel plus aspirin versus oral anticoagulation for atrial fibrillation in the Atrial fibrillation Clopidogrel Trial with Irbesartan for prevention of Vascular Events (ACTIVE W): a randomised controlled trial. Lancet 367: 1903–1912.

Coronado VG, Xu L, Basavaraju SV et al. (2011). Surveillance for traumatic brain injury-related deaths – United States, 1997–2007. MMWR Surveill Summ 60: 1–32.

CRASH-2 Collaborators Intracranial Bleeding Study (2011). Effect of tranexamic acid in traumatic brain injury: a nested randomised, placebo controlled trial (CRASH-2 Intracranial Bleeding Study). BMJ 343: d3795.

Davie EW, Ratnoff OD (1964). Waterfall sequence for intrinsic blood clotting. Science 145: 1310–1312.

Dentali F, Marchesi C, Pierfranceschi MG et al. (2011). Safety of prothrombin complex concentrates for rapid anticoagulation reversal of vitamin K antagonists. A meta-analysis. Thromb Haemost 106: 429–438.

Dibu JR, Weimer JM, Ahrens C et al. (2016). The role of FEIBA in reversing novel oral anticoagulants in intracerebral hemorrhage. Neurocrit Care 24: 413–419.

Dolovich LR, Ginsberg JS, Douketis JD et al. (2000). A meta-analysis comparing low-molecular-weight heparins with unfractionated heparin in the treatment of venous thromboembolism: examining some unanswered questions regarding location of treatment, product type, and dosing frequency. Arch Intern Med 160: 181–188.

Drake TA, Morrissey JH, Edgington TS (1989). Selective cellular expression of tissue factor in human tissues. Implications for disorders of hemostasis and thrombosis. Am J Pathol 134: 1087–1097.

Eddleston M, de la Torre JC, Oldstone MB et al. (1993). Astrocytes are the primary source of tissue factor in the murine central nervous system. A role for astrocytes in cerebral hemostasis. J Clin Invest 92: 349–358.

Eerenberg ES, Kamphuisen PW, Sijpkens MK et al. (2011). Reversal of rivaroxaban and dabigatran by prothrombin complex concentrate: a randomized, placebo-controlled, crossover study in healthy subjects. Circulation 124: 1573–1579.

Elliott A, Wetzel J, Roper T et al. (2015). Thromboelastography in patients with acute ischemic stroke. Int J Stroke 10: 194–201.

Flaherty ML, Kissela B, Woo D et al. (2007). The increasing incidence of anticoagulant-associated intracerebral hemorrhage. Neurology 68: 116–121.

Fleming KM, Aithal GP, Solaymani-Dodaran M et al. (2008). Incidence and prevalence of cirrhosis in the United Kingdom, 1992–2001: a general population-based study. J Hepatol 49: 732–738.

Flibotte JJ, Hagan N, O'Donnell J et al. (2004). Warfarin, hematoma expansion, and outcome of intracerebral hemorrhage. Neurology 63: 1059–1064.

Folkerson LE, Sloan D, Cotton BA et al. (2015). Predicting progressive hemorrhagic injury from isolated traumatic brain injury and coagulation. Surgery 158: 655–661.

Frontera JA, Gordon E, Zach V et al. (2014). Reversal of coagulopathy using prothrombin complex concentrates is associated with improved outcome compared to fresh frozen plasma in warfarin-associated intracranial hemorrhage. Neurocrit Care 21: 397–406.

Frontera JA, Lewin JJ, Rabinstein AA et al. (2016). Guideline for reversal of antithrombotics in intracranial hemorrhage. Neurocrit Care 24: 6–46.

Ganter MT, Hofer CK (2008). Coagulation monitoring: current techniques and clinical use of viscoelastic point-of-care coagulation devices. Anesth Analg 106: 1366–1375.

Goldenberg NA, Jacobson L, Manco-Johnson MJ (2005). Brief communication: duration of platelet dysfunction after a 7-day course of Ibuprofen. Ann Intern Med 142: 506–509.

Goldstein LB, Simel DL (2005). Is this patient having a stroke? JAMA 293: 2391–2402.

Goldstein JN, Fazen LE, Wendell L et al. (2009). Risk of thromboembolism following acute intracerebral hemorrhage. Neurocrit Care 10: 28–34.

Goldstein JN, Marrero M, Masrur S et al. (2010). Management of thrombolysis-associated symptomatic intracerebral hemorrhage. Arch Neurol 67: 965–969.

Goldstein JN, Refaai MA, Milling TJ et al. (2015). Four-factor prothrombin complex concentrate versus plasma for rapid vitamin K antagonist reversal in patients needing urgent surgical or invasive interventions: a phase 3b, open-label, non-inferiority, randomised trial. Lancet 385 (9982): 2077–2087.

Gómez PA, Lobato RD, Ortega JM et al. (1996). Mild head injury: differences in prognosis among patients with a Glasgow Coma Scale score of 13 to 15 and analysis of factors associated with abnormal CT findings. Br J Neurosurg 10: 453–460.

Gould MK, Dembitzer AD, Doyle RL et al. (1999). Low-molecular-weight heparins compared with unfractionated heparin for treatment of acute deep venous thrombosis. A meta-analysis of randomized, controlled trials. Ann Intern Med 130: 800–809.

Graus F, Rogers LR, Posner JB (1985). Cerebrovascular complications in patients with cancer. Medicine (Baltimore) 64: 16–35.

Greinacher A, Selleng K (2010). Thrombocytopenia in the intensive care unit patient. Hematology Am Soc Hematol Educ Program 2010: 135–143.

Habib M, Roberts LN, Patel RK et al. (2014). Evidence of rebalanced coagulation in acute liver injury and acute liver failure as measured by thrombin generation. Liver Int 34: 672–678.

Hanger HC, Geddes JA, Wilkinson TJ et al. (2013). Warfarin-related intracerebral haemorrhage: better outcomes when reversal includes prothrombin complex concentrates. Intern Med J 43: 308–316.

Hansen ML, Sørensen R, Clausen MT et al. (2010). Risk of bleeding with single, dual, or triple therapy with warfarin, aspirin, and clopidogrel in patients with atrial fibrillation. Arch Intern Med 170: 1433–1441.

Harhangi BS, Kompanje EJ, Leebeek FW et al. (2008). Coagulation disorders after traumatic brain injury. Acta Neurochir (Wien) 150: 165–175. discussion 175.

Harrington RA, Stone GW, McNulty S et al. (2009). Platelet inhibition with cangrelor in patients undergoing PCI. N Engl J Med 361: 2318–2329.

Hart RG, Boop BS, Anderson DC (1995). Oral anticoagulants and intracranial hemorrhage. Facts and hypotheses. Stroke 26: 1471–1477.

Hart RG, Tonarelli SB, Pearce LA (2005). Avoiding central nervous system bleeding during antithrombotic therapy: recent data and ideas. Stroke 36: 1588–1593.

Hart RG, Diener HC, Yang S et al. (2012). Intracranial hemorrhage in atrial fibrillation patients during anticoagulation with warfarin or dabigatran: the RE-LY trial. Stroke 43: 1511–1517.

He J, Whelton PK, Vu B et al. (1998). Aspirin and risk of hemorrhagic stroke: a meta-analysis of randomized controlled trials. JAMA 280: 1930–1935.

Hedges SJ, Dehoney SB, Hooper JS et al. (2007). Evidence-based treatment recommendations for uremic bleeding. Nat Clin Pract Nephrol 3: 138–153.

Hemphill JC, Bonovich DC, Besmertis L et al. (2001). The ICH score: a simple, reliable grading scale for intracerebral hemorrhage. Stroke 32: 891–897.

Hemphill JC, Farrant M, Neill TA (2009). Prospective validation of the ICH score for 12-month functional outcome. Neurology 73: 1088–1094.

Hemphill JC, Greenberg SM, Anderson CS et al. (2015). Guidelines for the management of spontaneous intracerebral hemorrhage: a guideline for healthcare professionals from the American Heart Association/American Stroke Association. Stroke 46: 2032–2060.

Hirsh J, Poller L (1994). The international normalized ratio. A guide to understanding and correcting its problems. Arch Intern Med 154: 282–288.

Hirsh J, Raschke R (2004). Heparin and low-molecular-weight heparin: the Seventh ACCP Conference on Antithrombotic and Thrombolytic Therapy. Chest 126: 188S–203S.

Hoffman M, Monroe DM (2001). A cell-based model of hemostasis. Thromb Haemost 85: 958–965.

Hoffmann P, Keller F (2012). Increased major bleeding risk in patients with kidney dysfunction receiving enoxaparin: a meta-analysis. Eur J Clin Pharmacol 68: 757–765.

Holbrook A, Schulman S, Witt DM et al. (2012). Evidence-based management of anticoagulant therapy: antithrombotic therapy and prevention of thrombosis, 9th ed: American College of Chest Physicians Evidence-Based Clinical Practice Guidelines. Chest 141: e152S–e184S.

Hoyt DB (2004). A clinical review of bleeding dilemmas in trauma. Semin Hematol 41: 40–43.

Hukkelhoven CW, Steyerberg EW, Habbema JD et al. (2005). Predicting outcome after traumatic brain injury: development and validation of a prognostic score based on admission characteristics. J Neurotrauma 22: 1025–1039.

Hunt BJ (2014). Bleeding and coagulopathies in critical care. N Engl J Med 370: 847–859.

Huttner HB, Schellinger PD, Hartmann M et al. (2006). Hematoma growth and outcome in treated neurocritical care patients with intracerebral hemorrhage related to oral anticoagulant therapy: comparison of acute treatment strategies using vitamin K, fresh frozen plasma, and prothrombin complex concentrates. Stroke 37: 1465–1470.

Hwang BY, Bruce SS, Appelboom G et al. (2012). Evaluation of intraventricular hemorrhage assessment methods for predicting outcome following intracerebral hemorrhage. J Neurosurg 116: 185–192.

Hylek EM, Held C, Alexander JH et al. (2014). Major bleeding in patients with atrial fibrillation receiving apixaban or warfarin: the ARISTOTLE trial (Apixaban for Reduction in Stroke and Other Thromboembolic Events in Atrial Fibrillation): predictors, characteristics, and clinical outcomes. J Am Coll Cardiol 63: 2141–2147.

H-Y CS, Xuemei C, G KR et al. (2012). Thromboembolic risks of recombinant factor VIIa Use in warfarin-associated intracranial hemorrhage: a case-control study. BMC Neurol 12: 158.

Illanes S, Zhou W, Schwarting S et al. (2011). Comparative effectiveness of hemostatic therapy in experimental warfarin-associated intracerebral hemorrhage. Stroke 42: 191–195.

International Stroke Trial Collaborative Group (1997). The International Stroke Trial (IST): a randomised trial of aspirin, subcutaneous heparin, both, or neither among 19435 patients with acute ischaemic stroke. Lancet 349: 1569–1581.

Jauch EC, Pineda JA, Claude Hemphill J (2015). Emergency neurological life support: intracerebral hemorrhage. Neurocrit Care 23 (Suppl 2): 83–93.

Joseph B, Hadjizacharia P, Aziz H et al. (2013). Prothrombin complex concentrate: an effective therapy in reversing the coagulopathy of traumatic brain injury. J Trauma Acute Care Surg 74: 248–253.

Karvellas CJ, Fix OK, Battenhouse H et al. (2014). Outcomes and complications of intracranial pressure monitoring in acute liver failure: a retrospective cohort study. Crit Care Med 42: 1157–1167.

Kawano-Castillo J, Ward E, Elliott A et al. (2014). Thrombelastography detects possible coagulation disturbance in patients with intracerebral hemorrhage with hematoma enlargement. Stroke 45: 683–688.

Kazui S, Naritomi H, Yamamoto H et al. (1996). Enlargement of spontaneous intracerebral hemorrhage. Incidence and time course. Stroke 27: 1783–1787.

Keimowitz RM, Annis BL (1973). Disseminated intravascular coagulation associated with massive brain injury. J Neurosurg 39: 178–180.

Kissela B, Schneider A, Kleindorfer D et al. (2004). Stroke in a biracial population: the excess burden of stroke among blacks. Stroke 35: 426–431.

Kluger Y, Riou B, Rossaint R et al. (2007). Safety of rFVIIa in hemodynamically unstable polytrauma patients with traumatic brain injury: post hoc analysis of 30 patients from a prospective, randomized, placebo-controlled, double-blind clinical trial. Crit Care 11: R85.

Köhler M, Hellstern P, Tarrach H et al. (1989). Subcutaneous injection of desmopressin (DDAVP): evaluation of a new, more concentrated preparation. Haemostasis 19: 38–44.

Krisl JC, Meadows HE, Greenberg CS et al. (2011). Clinical usefulness of recombinant activated factor VII in patients with liver failure undergoing invasive procedures. Ann Pharmacother 45: 1433–1438.

Kumar MA (2013). Coagulopathy associated with traumatic brain injury. Curr Neurol Neurosci Rep 13: 391.

Kunio NR, Differding JA, Watson KM et al. (2012). Thrombelastography-identified coagulopathy is associated with increased morbidity and mortality after traumatic brain injury. Am J Surg 203: 584–588.

Kwaan HC, Cull EH (2014). The coagulopathy in acute promyelocytic leukaemia – what have we learned in the past twenty years. Best Pract Res Clin Haematol 27: 11–18.

Kwaan HC, Wang J, Boggio LN (2002). Abnormalities in hemostasis in acute promyelocytic leukemia. Hematol Oncol 20: 33–41.

Kwaan HC, Wang J, Weiss I (2004). Expression of receptors for plasminogen activators on endothelial cell surface depends on their origin. J Thromb Haemost 2: 306–312.

Lange U, Nowak G, Bucha E (2003). Ecarin chromogenic assay – a new method for quantitative determination of direct thrombin inhibitors like hirudin. Pathophysiol Haemost Thromb 33: 184–191.

Langlois JA, Rutland-Brown W, Wald MM (2006). The epidemiology and impact of traumatic brain injury: a brief overview. J Head Trauma Rehabil 21: 375–378.

Laroche M, Kutcher ME, Huang MC et al. (2012). Coagulopathy after traumatic brain injury. Neurosurgery 70: 1334–1345.

Levi M, Toh CH, Thachil J et al. (2009). Guidelines for the diagnosis and management of disseminated intravascular coagulation. British Committee for Standards in Haematology. Br J Haematol 145: 24–33.

Lindahl TL, Baghaei F, Blixter IF et al. (2011). Effects of the oral, direct thrombin inhibitor dabigatran on five common coagulation assays. Thromb Haemost 105: 371–378.

Lisman T, Leebeek FW (2007). Hemostatic alterations in liver disease: a review on pathophysiology, clinical consequences, and treatment. Dig Surg 24: 250–258.

Lisman T, Bakhtiari K, Adelmeijer J et al. (2012). Intact thrombin generation and decreased fibrinolytic capacity in patients with acute liver injury or acute liver failure. J Thromb Haemost 10: 1312–1319.

Lovelock CE, Molyneux AJ, Rothwell PM et al. (2007). Change in incidence and aetiology of intracerebral haemorrhage in Oxfordshire, UK, between 1981 and 2006: a population-based study. Lancet Neurol 6: 487–493.

Maas AI, Hukkelhoven CW, Marshall LF et al. (2005). Prediction of outcome in traumatic brain injury with computed tomographic characteristics: a comparison between the computed tomographic classification and combinations of computed tomographic predictors. Neurosurgery 57: 1173–1182. discussion 1173–1182.

Macfarlane RG (1964). An enzyme cascade in the blood clotting mechanism, and its function as a biochemical amplifier. Nature 202: 498–499.

Majeed A, Meijer K, Larrazabal R et al. (2014). Mortality in vitamin K antagonist-related intracerebral bleeding treated with plasma or 4-factor prothrombin complex concentrate. Thromb Haemost 111: 233–239.

Mannucci PM, Remuzzi G, Pusineri F et al. (1983). Deamino-8-D-arginine vasopressin shortens the bleeding time in uremia. N Engl J Med 308: 8–12.

Marshall LF, Marshall SB, Klauber MR et al. (1992). The diagnosis of head injury requires a classification based on computed axial tomography. J Neurotrauma 9 (Suppl 1): S287–S292.

Massaro AM, Doerfler S, Nawalinski K et al. (2015). Thromboelastography defines late hypercoagulability after TBI: a pilot study. Neurocrit Care 22: 45–51.

Mayer SA, Brun NC, Broderick J et al. (2006). Recombinant activated factor VII for acute intracerebral hemorrhage: US phase IIA trial. Neurocrit Care 4: 206–214.

Mayer SA, Brun NC, Begtrup K et al. (2008). Efficacy and safety of recombinant activated factor VII for acute intracerebral hemorrhage. N Engl J Med 358: 2127–2137.

Miller CS, Grandi SM, Shimony A et al. (2012). Meta-analysis of efficacy and safety of new oral anticoagulants (dabigatran, rivaroxaban, apixaban) versus warfarin in patients with atrial fibrillation. Am J Cardiol 110: 453–460.

Morel N, Morel O, Petit L et al. (2008). Generation of procoagulant microparticles in cerebrospinal fluid and peripheral blood after traumatic brain injury. J Trauma 64: 698–704.

Naidech AM, Bernstein RA, Levasseur K et al. (2009a). Platelet activity and outcome after intracerebral hemorrhage. Ann Neurol 65: 352–356.

Naidech AM, Jovanovic B, Liebling S et al. (2009b). Reduced platelet activity is associated with early clot growth and worse 3-month outcome after intracerebral hemorrhage. Stroke 40: 2398–2401.

Naidech AM, Liebling SM, Rosenberg NF et al. (2012). Early platelet transfusion improves platelet activity and may improve outcomes after intracerebral hemorrhage. Neurocrit Care 16: 82–87.

Naidech AM, Kumar MAParticipants in the International Multidisciplinary Consensus Conference on Multimodality Monitoring (2014a). Monitoring of hematological and hemostatic parameters in neurocritical care patients. Neurocrit Care 21 (Suppl 2): 168–176.

Naidech AM, Maas MB, Levasseur-Franklin KE et al. (2014b). Desmopressin improves platelet activity in acute intracerebral hemorrhage. Stroke 45: 2451–2453.

Narayan RK, Maas AI, Marshall LF et al. (2008). Recombinant factor VIIA in traumatic intracerebral hemorrhage: results of a dose-escalation clinical trial. Neurosurgery 62: 776–786. discussion 786–778.

Nekludov M, Bellander BM, Blombäck M et al. (2007). Platelet dysfunction in patients with severe traumatic brain injury. J Neurotrauma 24: 1699–1706.

Neunert C, Noroozi N, Norman G et al. (2015). Severe bleeding events in adults and children with primary immune thrombocytopenia: a systematic review. J Thromb Haemost 13: 457–464.

Nicolini A, Ghirarduzzi A, Iorio A et al. (2002). Intracranial bleeding: epidemiology and relationships with antithrombotic treatment in 241 cerebral hemorrhages in Reggio Emilia. Haematologica 87: 948–956.

Nilsson OG, Lindgren A, Ståhl N et al. (2000). Incidence of intracerebral and subarachnoid haemorrhage in southern Sweden. J Neurol Neurosurg Psychiatry 69: 601–607.

Nishijima DK, Zehtabchi S, Berrong J et al. (2012). Utility of platelet transfusion in adult patients with traumatic intracranial hemorrhage and preinjury antiplatelet use: a systematic review. J Trauma Acute Care Surg 72: 1658–1663.

O'Connell PA, Madureira PA, Berman JN et al. (2011). Regulation of S100A10 by the PML-RAR-α oncoprotein. Blood 117: 4095–4105.

Panzer RJ, Feibel JH, Barker WH et al. (1985). Predicting the likelihood of hemorrhage in patients with stroke. Arch Intern Med 145: 1800–1803.

Patel MR, Mahaffey KW, Garg J et al. (2011). Rivaroxaban versus warfarin in nonvalvular atrial fibrillation. N Engl J Med 365: 883–891.

Pathak A, Dutta S, Marwaha N et al. (2005). Change in tissue thromboplastin content of brain following trauma. Neurol India 53: 178–182.

Pavord S, Myers B (2011). Bleeding and thrombotic complications of kidney disease. Blood Rev 25: 271–278.

Perzborn E, Gruber A, Tinel H et al. (2013). Reversal of rivaroxaban anticoagulation by haemostatic agents in rats and primates. Thromb Haemost 110: 162–172.

Phan TG, Koh M, Wijdicks EF (2000). Safety of discontinuation of anticoagulation in patients with intracranial hemorrhage at high thromboembolic risk. Arch Neurol 57: 1710–1713.

Pollack CV, Reilly PA, Eikelboom J et al. (2015). Idarucizumab for dabigatran reversal. N Engl J Med 373: 511–520.

Qureshi AI, Tuhrim S, Broderick JP et al. (2001). Spontaneous intracerebral hemorrhage. N Engl J Med 344: 1450–1460.

Rabiner SF (1972). Bleeding in uremia. Med Clin North Am 56: 221–233.

Rasler F (2007). Emergency treatment of hemorrhagic complications of thrombolysis. Ann Emerg Med 50: 485.

Riegert-Johnson DL, Volcheck GW (2002). The incidence of anaphylaxis following intravenous phytonadione (vitamin K1): a 5-year retrospective review. Ann Allergy Asthma Immunol 89: 400–406.

Roberts HR, Lozier JN (1992). New perspectives on the coagulation cascade. Hosp Pract (Off Ed) 27 (97–105): 109–112.

Roderick P, Davies R, Jones C et al. (2004). Simulation model of renal replacement therapy: predicting future demand in England. Nephrol Dial Transplant 19: 692–701.

Rosand J, Hylek EM, O'Donnell HC et al. (2000). Warfarin-associated hemorrhage and cerebral amyloid angiopathy: a genetic and pathologic study. Neurology 55: 947–951.

Rost NS, Smith EE, Chang Y et al. (2008). Prediction of functional outcome in patients with primary intracerebral hemorrhage: the FUNC score. Stroke 39: 2304–2309.

Saatman KE, Duhaime AC, Bullock R et al. (2008). Classification of traumatic brain injury for targeted therapies. J Neurotrauma 25: 719–738.

Samama MM, Guinet C (2011). Laboratory assessment of new anticoagulants. Clin Chem Lab Med 49: 761–772.

Samama MM, Martinoli JL, LeFlem L et al. (2010). Assessment of laboratory assays to measure rivaroxaban–an oral, direct factor Xa inhibitor. Thromb Haemost 103: 815–825.

Sansing LH, Messe SR, Cucchiara BL et al. (2009). Prior antiplatelet use does not affect hemorrhage growth or outcome after ICH. Neurology 72: 1397–1402.

Sarode R, Milling TJ, Refaai MA et al. (2013). Efficacy and safety of a 4-factor prothrombin complex concentrate in patients on vitamin K antagonists presenting with major bleeding: a randomized, plasma-controlled, phase IIIb study. Circulation 128: 1234–1243.

Schulman S, Beyth RJ, Kearon C et al. (2008). Hemorrhagic complications of anticoagulant and thrombolytic treatment: American College of Chest Physicians Evidence-Based Clinical Practice Guidelines (8th edition). Chest 133: 257S–298S.

Shakur H, Roberts I, Bautista R et al. (2010). Effects of tranexamic acid on death, vascular occlusive events, and blood transfusion in trauma patients with significant haemorrhage (CRASH-2): a randomised, placebo-controlled trial. Lancet 376: 23–32.

Sibbing D, Braun S, Jawansky S et al. (2008). Assessment of ADP-induced platelet aggregation with light transmission aggregometry and multiple electrode platelet aggregometry before and after clopidogrel treatment. Thromb Haemost 99: 121–126.

Siegal DM, Cuker A (2013). Reversal of novel oral anticoagulants in patients with major bleeding. J Thromb Thrombolysis 35: 391–398.

Siegal DM, Curnutte JT, Connolly SJ et al. (2015). Andexanet alfa for the reversal of factor xa inhibitor activity. N Engl J Med 373: 2413–2424.

Slichter SJ (2007). Evidence-based platelet transfusion guidelines. Hematology Am Soc Hematol Educ Program 172–178.

Smith EE, Rosand J, Knudsen KA et al. (2002). Leukoaraiosis is associated with warfarin-related hemorrhage following ischemic stroke. Neurology 59: 193–197.

Stangier J, Feuring M (2012). Using the HEMOCLOT direct thrombin inhibitor assay to determine plasma concentrations of dabigatran. Blood Coagul Fibrinolysis 23: 138–143.

Steiner T, Rosand J, Diringer M (2006). Intracerebral hemorrhage associated with oral anticoagulant therapy: current practices and unresolved questions. Stroke 37: 256–262.

Steyerberg EW, Mushkudiani N, Perel P et al. (2008). Predicting outcome after traumatic brain injury: development and international validation of prognostic scores based on admission characteristics. PLoS Med 5: e165. discussion e165.

Stravitz RT, Lisman T, Luketic VA et al. (2012). Minimal effects of acute liver injury/acute liver failure on hemostasis as assessed by thromboelastography. J Hepatol 56: 129–136.

Stravitz RT, Bowling R, Bradford RL et al. (2013). Role of procoagulant microparticles in mediating complications and outcome of acute liver injury/acute liver failure. Hepatology 58: 304–313.

Talving P, Benfield R, Hadjizacharia P et al. (2009). Coagulopathy in severe traumatic brain injury: a prospective study. J Trauma 66: 55–61. discussion 61–52.

The Stroke Prevention in Reversible Ischemia Trial (SPIRIT) Study Group (1997). A randomized trial of anticoagulants versus aspirin after cerebral ischemia of presumed arterial origin. Ann Neurol 42: 857–865.

Thompson BB, Béjot Y, Caso V et al. (2010). Prior antiplatelet therapy and outcome following intracerebral hemorrhage: a systematic review. Neurology 75: 1333–1342.

Toyoda K, Okada Y, Minematsu K et al. (2005). Antiplatelet therapy contributes to acute deterioration of intracerebral hemorrhage. Neurology 65: 1000–1004.

Tripodi A, Salerno F, Chantarangkul V et al. (2005). Evidence of normal thrombin generation in cirrhosis despite abnormal conventional coagulation tests. Hepatology 41: 553–558.

van Asch CJ, Luitse MJ, Rinkel GJ et al. (2010). Incidence, case fatality, and functional outcome of intracerebral haemorrhage over time, according to age, sex, and ethnic origin: a systematic review and meta-analysis. Lancet Neurol 9: 167–176.

Van Beek JG, Mushkudiani NA, Steyerberg EW et al. (2007). Prognostic value of admission laboratory parameters in traumatic brain injury: results from the IMPACT study. J Neurotrauma 24: 315–328.

van Ryn J, Stangier J, Haertter S et al. (2010). Dabigatran etexilate – a novel, reversible, oral direct thrombin inhibitor: interpretation of coagulation assays and reversal of anticoagulant activity. Thromb Haemost 103: 1116–1127.

Vaquero J, Fontana RJ, Larson AM et al. (2005). Complications and use of intracranial pressure monitoring in patients with acute liver failure and severe encephalopathy. Liver Transpl 11: 1581–1589.

Walsh TS, Stanworth SJ, Prescott RJ et al. (2010). Prevalence, management, and outcomes of critically ill patients with prothrombin time prolongation in United Kingdom intensive care units. Crit Care Med 38: 1939–1946.

Walsh M, Thomas SG, Howard JC et al. (2011). Blood component therapy in trauma guided with the utilization of the perfusionist and thromboelastography. J Extra Corpor Technol 43: 162–167.

Wang J, Weiss I, Svoboda K et al. (2001). Thrombogenic role of cells undergoing apoptosis. Br J Haematol 115: 382–391.

Windeløv NA, Welling KL, Ostrowski SR et al. (2011). The prognostic value of thrombelastography in identifying neurosurgical patients with worse prognosis. Blood Coagul Fibrinolysis 22: 416–419.

Wiviott SD, Braunwald E, McCabe CH et al. (2007). Prasugrel versus clopidogrel in patients with acute coronary syndromes. N Engl J Med 357: 2001–2015.

Yutthakasemsunt S, Kittiwatanagul W, Piyavechvirat P et al. (2013). Tranexamic acid for patients with traumatic brain injury: a randomized, double-blinded, placebo-controlled trial. BMC Emerg Med 13: 20.

Zhu J, Guo WM, Yao YY et al. (1999). Tissue factors on acute promyelocytic leukemia and endothelial cells are differently regulated by retinoic acid, arsenic trioxide and chemotherapeutic agents. Leukemia 13: 1062–1070.

Chapter 41

Prognosis of neurologic complications in critical illness

M. VAN DER JAGT AND E.J.O. KOMPANJE*
Department of Intensive Care, Erasmus MC University Medical Center, Rotterdam, The Netherlands

Abstract

Neurologic complications of critical illness require extensive clinical and neurophysiologic evaluation to establish a reliable prognosis. Many sequelae of intensive care unit (ICU) treatment, such as delirium and ICU-acquired weakness, although highly associated with adverse outcomes, are less suitable for prognostication, but should rather prompt clinicians to seek previously unnoticed persisting underlying illnesses. Prognostication can be confounded by drug administration particularly because its clearance is abnormal in critical illness. Some neurological complications are severe, and can last for months or years after discharge from ICU. The most important ethical aspects regarding neurologic complications in critically ill patients are prevention, recognition, and identification, and prevention of self-fulfilling prophecies. This chapter summarizes the tool of prognostication of major neurological complications of critical illness.

INTRODUCTION

Three fundamental questions can be addressed in the management of critically ill patients: (1) What is the diagnosis?; (2) What is the most appropriate treatment?; and (3) What is the prognosis?

In most patients in an intensive care unit (ICU), the first two questions are seldom troublesome, but predicting outcome is often very difficult. Often we cannot predict the course of the disease and whether or not complications will occur, but an estimate of prognosis provides guidance for continuation, withholding, or withdrawal of life-sustaining measures. Relatives of the patient expect from us a prediction of survival and outcome, but at the same time, predicting the future may engender strong emotions among relatives and even physicians and nurses themselves. Prognostication is an essential part of daily care in the ICU, but is also one of the most elaborate tasks (Christakis, 1999). Prognostication may become even more difficult when the original critical illness is complicated by medical events that have a strong impact on prognosis.

A medical complication is an unintended, harmful occurrence or condition resulting from a diagnostic, prophylactic, or therapeutic intervention, or an accidental injury occurring in the hospital setting (Rubins and Moskowitz, 1990). Critically ill patients are especially vulnerable to complications that arise from the underlying disease or comorbidities, as well as from advanced intensive care treatments. Well-known are complications resulting from mechanical ventilation and catheter-related blood stream infections, among many others. Acute and chronic neurologic complications can occur both in neurologic and nonneurologic critically ill patients, of which many are perceived as highly detrimental to survival and outcome. Survivors of critical illness are often left with neurologic and cognitive disabilities (Ortega-Guitierrez et al., 2009; Desal et al., 2011), even without evident neurologic complications during the course of their ICU stay. In a sense, it may seem that this is the price we pay for advances in critical care medicine, engaging us to keep patients alive simply because we can, in an increasingly elderly, multimorbid population prone to long-term (neurologic) sequelae.

Some neurologic complications are severe, and can last for months or years after discharge from ICU. These neurologic disabilities form a significant reason for decreased quality of life after ICU discharge (Guerra et al., 2012). Long-term cognitive and other neurologic

---

*Correspondence to: Dr. Erwin J.O. Kompanje, PhD, Department of Intensive Care, Erasmus MC University Medical Center Rotterdam, P.O. Box 2040, 3000CA Rotterdam, The Netherlands. Tel: +31-6-53837655, E-mail: erwinkompanje@me.com

disabilities resulting from critical illness constitute a personal, social, and economic burden for patients, their relatives, and society, and in most cases will result in high costs on resource consumption, such as sick leave, hospitalizations, outpatient care, drugs, social services, early retirement, and premature death. In many cases, neurologic complications after ICU admission form the basis of disability-adjusted life years. Although not many robust clinical studies exist that have included premorbid cognitive and functional status at ICU admission, the scarce data published to date have clearly indicated that the ICU phase may contribute to neurologic impairments.

Medically or surgically critically ill patients are at risk for acute neurologic and long-term cognitive complications. This is a particular concern among patients admitted with severe sepsis with multiple organ failure (MOF); immunocompromised patients, including those undergoing solid-organ or bone marrow transplantation or those who are treated with neurotoxic immunosuppressive agents; patients with severe metabolic, electrolyte, and acid–base disturbances, leading to encephalopathy; and those requiring long-term sedation. Common neurologic syndromes, such as delirium and ICU-acquired weakness (ICUAW), occur regularly in these patients, and have strong prognostic significance.

Prognostication of neurologic complications seems highly dependent on the etiology and severity of the primary condition, age of the patient, and comorbidities. This chapter deals with the prognosis of neurologic complications in critically ill patients developing neurologic manifestations. This information is aimed primarily at physicians and nurses who regularly deal with critically ill patients. The summaries of prognostic knowledge of the main neurologic complications of critical illness are intended to provide guidance in: (1) deciding when to continue treatment versus transition towards end-of-life care; (2) informing families of critically ill patients about what they may or may not expect in the short and long term; and (3) defining knowledge gaps in prognostic information (where currently only clinical experience and intense multidisciplinary consultations are available to estimate prognosis).

We will also explore the ethics behind prognostication: what is the personal and societal burden of neurologic impairment resulting from critical illness? What are the ethical issues raised by the use of the prognostic information?

PROGNOSIS OF ICU-ACQUIRED ENCEPHALOPATHY

Patients admitted to the ICU can suffer from several types of encephalopathy, including drug-induced, hepatic, hypoxic-ischemic, uremic, Wernicke's, and hypertensive encephalopathy. Some of these can be seen as complications of critical care treatment, since they develop during admission to the ICU, whether iatrogenic (e.g., drug-induced encephalopathy), as a manifestation of the condition the patient is admitted for, or as a consequence of underlying comorbidity (hepatic encephalopathy (HE), uremic encephalopathy). Treating the underlying cause may reverse the symptoms, but in some cases, structural changes are irreversible. Prognosis of encephalopathy depends on many factors, including age, comorbidity, recognition of the underlying cause, and degree of organ dysfunction.

Drug-induced encephalopathy

Administration of certain drugs in patients with impaired cerebral metabolism can lead to serious encephalopathy. Usually, this is not attributed to structural cerebral lesions or diseases, but the drug-induced encephalopathy can form the basis of structural lesions, with disturbances of consciousness, personality, and cerebral function, and can give rise to clinical signs and symptoms (e.g., seizures), with subsequent impact on mortality and morbidity. Since many severely ill patients admitted to an ICU have impaired cerebral metabolism, and because of the widespread use of certain drugs, drug-induced encephalopathy is not uncommon, although often underrecognized.

A variety of commonly used drugs are neurotoxic and can cause drug-induced encephalopathies, among which are analgesics, sedatives, antibiotics, antivirals, anticonvulsants, chemotherapeutics, immunosuppressants, and neuroleptics. We focus on three groups of drugs in which encephalopathy is described as a complication specifically in critically ill patients: antibiotics, antiviral agents, and analgesics.

Most drug-induced encephalopathies have a good prognosis when the cause is recognized and reversed. Although fatal drug-induced encephalopathy is described (Cossaart et al., 2003), we were unable to find such a complication in the treatment of critically ill patients in the ICU.

Several antibiotics cross the blood–brain barrier, making them good treatment options for serious central nervous system (CNS) infections. On the other hand, these agents can, for the same reason, cause neurotoxicity, complicating treatment of infections in the ICU (Grill and Maganti, 2011). The toxic effects on the CNS are often initially unrecognized, and the signs and symptoms are mistakenly seen as manifestations of other neurologic conditions and complications. Since many ICU patients suffer from primary or nosocomial infections, which necessitate use of broad-spectrum antibiotics, the ICU population is at high risk of neurotoxicity associated with various groups of antibiotics (Grill and Maganti, 2011). Severe infections, comorbidities, polypharmacy, and advanced age give rise to neurologic deterioration and hamper the diagnosis of drug-induced encephalopathy. Besides this, drug-induced encephalopathy is often ascribed to metabolic disturbances and MOF, such that the incidence is underestimated. As most antibiotic-induced encephalopathies are reversible after the drug is stopped, early recognition is essential.

Metronidazole

Administration of the antibiotic metronidazole can lead to a variety of neurologic adverse effects, including peripheral neuropathy, seizures, cerebellar dysfunction, and encephalopathy (Kusumi et al., 1980; Ahmed et al., 1995; Horlen et al., 2000; Woodruff et al., 2002; Heaney et al., 2003; Seok et al., 2003; Kim et al., 2007, 2011; Mulcahy and Chaddha, 2008; Graves et al., 2009; Sarna et al., 2009; Huang et al., 2012). Metronidazole is a nitroimidazole antibiotic commonly used as treatment for anaerobic-related infections, *Clostridium difficile* colitis, and in patients with HE (to remove the nitrogenous load in the gastrointestinal tract). After discontinuation of the medication, patients in case reports usually made a rapid recovery. The hyperintense lesions on T2-weighted and fluid-attenuated inversion recovery (FLAIR) images were found to be completely or partially reversible (Horlen et al., 2000; Cecil et al., 2002; Heaney et al., 2003; Kim et al., 2007; Mulcahy and Chaddha, 2008). Metronidazole-induced encephalopathy must be differentiated from osmotic demyelination and other transient T2-hyperintense lesions, as these will lead to different therapeutic strategies. When adequately recognized using brain magnetic resonance imaging (MRI), prognosis of metronidazole-induced encephalopathy is favorable, disappearing a few days to weeks after discontinuation of the medication.

Cephalosporins

Cephalosporin-induced neurotoxicity may manifest as encephalopathy, among other neurologic signs and symptoms. Elderly patients, those with renal failure, and patients with prior neurologic conditions are especially prone to cephalosporin-induced encephalopathy (Abanades et al., 2004; Lam and Gomolin, 2006; Grill and Maganti, 2008). Cefepime is a parenterally administered fourth-generation cephalosporin antibiotic used for the treatment of Gram-positive and Gram-negative infections in critically ill patients. Approximately 3% of patients treated with cefepime experienced adverse neurologic reactions (Neu, 1996). Cefepime-induced status epilepticus has been reported in more than 25 cases (Primavera et al., 2004; Maganti et al., 2006; Thabet et al., 2009). Sonck and colleagues (2008) described 8 patients with renal failure showing slowly progressive neurologic symptoms after initiation of cefepime. The mortality in this series was 100%, with death occurring shortly after neurologic deterioration in 3 cases. Three patients who survived longer showed neurologic improvement after drug discontinuation. Partial or complete recovery after early recognition and withdrawal of cefepime appear to be the most common outcome (Jallon et al., 2000; Martinez-Rodriguez et al., 2000; Chattelier et al., 2002; Chow et al., 2003; Ferrara et al., 2003).

Ceftriaxone has also been described as inducing a reversible encephalopathy (Klion et al., 1994; Herishanu et al., 1998; Martinez-Rodriguez et al., 2000; Dakdouki and Al-Awar, 2004; Roncon-Albuquerque et al., 2009; Grill and Maganti, 2011; Kim et al., 2012; Sharma et al., 2012; Sadafi et al., 2014). As with cefepime, toxicity is more common with previous CNS disease and renal failure (Denysenko and Nicolson, 2011; Kim et al., 2012; Sadafi et al., 2014). Latency of ceftriaxone-induced encephalopathy is 1–10 days after drug initiation, and regression of all neurologic symptoms usually follows within 2–7 days following drug suspension (Roncon-Albuquerque et al., 2009; Kim et al., 2012; Sharma et al., 2012).

Other cephalosporins (cefuroxime, ceftazidime, cefazolin) have been described to cause encephalopathy as well (Schwankhaus et al., 1985; Josse et al., 1987; Geyer et al., 1988; Pascual et al., 1990; Ortiz et al., 1991; Jackson and Berkovic, 1992; Herishanu et al., 1998). After withdrawal or reduction in dosage, most patients show recovery of signs and symptoms of encephalopathy. In some cases, symptoms persisted for more than a week after withdrawal of the antibiotics (Jackson and Berkovic, 1992).

Clarithromycin

Clarithromycin is an antibiotic of the macrolide family, used widely in respiratory infections. Clarithromycin may give rise to psychiatric symptoms (Négrin-González et al., 2014), but may also induce encephalopathy, appearing 1–10 days after drug intake (Bandettin di Poggio et al., 2011). Early detection of clarithromycin-induced encephalopathy and discontinuation of the drug result in full recovery.

Cycloserine

Cycloserine is a broad-spectrum antibiotic used for the treatment of drug-resistant tuberculosis. Excessive doses can give rise to serious neurologic complications (Kwon et al., 2008; Kim et al., 2014). A characteristic finding is reversible cytotoxic edema in the dentate nuclei, visible on MRI. Discontinuation of cycloserine usually gives complete resolution.

Linezolid

Linezolid is an antibiotic used for the treatment of serious infections caused by Gram-positive bacteria that are resistant to other antibiotics, including vancomycin-resistant *Enterococcus* species and methicillin-resistant *Staphylococcus aureus*. In the ICU, it is often used in the treatment of life-threatening pneumonia and serious skin infections. Linezolid has been described in a small number of patients as inducing peripheral and central neurotoxicity (Ferry et al., 2005; Fletcher et al., 2010), among which is posterior reversible encephalopathy syndrome (PRES) (Hinchey et al., 1996; Nagel et al., 2007). In one series of 4 patients, linezolid-induced peripheral neuropathy resulted in persistent neurologic damage, while central neurotoxicity was transient after discontinuation of linezolid (Ferry et al., 2005).

Penicillin

Penicillin encephalopathy has been, since 1952 (Bateman et al., 1952), described in many patients (Conway et al., 1968). As with other antibiotic-induced encephalopathies, it is often associated with renal failure, which leads to drug accumulation. In every case where penicillin has been stopped or reduced, considerable improvement, usually resulting in complete recovery of encephalopathic symptoms, occurred (Conway et al., 1968).

Antiviral drugs

In critically ill patients with renal failure, neurotoxicity of acyclovir leading to encephalopathy is described in several case reports (Revankar et al., 1995; Ernst and Franey, 1998; Delerme et al., 2002; Dulluc et al., 2004; Carlon et al., 2005; Onuigbo et al., 2009; Van Kan et al., 2009). Prevention is especially important in patients with renal failure. With the prompt initiation of hemodialysis, neurotoxicity is reversible, making prognosis favorable in many cases (Laskin et al., 1982; Almond et al., 1995). However, Van Kan et al. (2009) could not exclude, however unlikely, the contribution of acyclovir neurotoxicity in the death of a critically ill patient.

Analgesics

Opioids can induce neurotoxicity, especially in severely ill patients secondary to dehydration, infection, or drug interactions (Gallagher, 2007). Morphine-induced encephalopathy is described as a complication in patients with severe liver dysfunction, which interferes with the metabolism of morphine (Hasselström et al., 1990; Dumont et al., 1998). Therefore, it is a challenge to ensure adequate analgesia and pain treatment in patients with liver failure. Shinagawa et al. (2008) described a case of morphine-induced encephalopathy where the excess of ammonia due to constitutional constipation in this patient was exacerbated by the use of morphine. The patient regained consciousness by receiving aminoleban and a suppository for constipation.

PROGNOSIS OF HEPATIC ENCEPHALOPATHY IN ACUTE LIVER FAILURE

Acute liver failure (ALF) is a feared condition in critically ill patients, resulting in rapid and severe decline in liver function (impaired synthetic parameters, such as international normalized ratio $(INR) > 1.5$) and HE, the latter resulting from reduced detoxification. ALF may be observed as a complication of critical illness. Although ALF is most commonly drug-induced and starts outside the ICU, it can also result from severe shock, MOF, viruses, vascular origin, or autoimmunity. Rare causes of ALF with HE include pregnancy and Reye's syndrome. In more than 80% of cases, the cause of ALF can be determined. ALF is always life-threatening and may affect relatively young patients. Liver transplantation is often the only possible therapeutic option impacting survival, without which patients may die within days, despite all supportive ICU treatments. Before the era of liver transplantation, mortality of ALF was as high as 85%, but in the last several decades, survival rates have risen to 60–80%. Recovery with only standard intensive care treatment is seen in more than half of the patients admitted with ALF due to an overdose with acetaminophen and pregnancy-related (fatty) liver failure. On the other end of the spectrum, poor prognosis is often seen in hepatitis B infection, other drug intoxications (e.g., ecstasy), and autoimmune liver failure.

HE is an important factor for predicting survival in ALF. HE is characterized by decrease in level of consciousness, asterixis, myoclonus and may progress to coma with extensor motor responses, and a variety of precipitating factors that can affect the prognosis. Renal failure with electrolyte and acid–base imbalance, respiratory failure, and sepsis are common. Neurologic features of HE in ALF include the presence of raised intracranial pressure, which can complicate the outcome, even with complete recovery of liver function or after liver transplantation. The grade of HE is an important prognostic factor. Low-grade HE (I–II) is associated with spontaneous recovery in up to 70%, but recovery is below 20% in grade IV HE (O'Grade et al., 1989).

In the acute phase, the outcome of ALF is often unpredictable, making prognostication very difficult. Age, the etiology of ALF, severity of illness, long periods of raised intracranial pressure, and comorbidities are independent factors in prognostication.

Several prognostic scoring systems have been developed and are in use for screening before liver transplantation (García-Martínez et al., 2011). The most commonly used prediction model is the King's College criteria, which were developed using data from 588 patients with ALF. The model separates acetaminophen-induced ALF from other causes. In acetaminophen-induced ALF, criteria for transplantation include an arterial pH < 7.3 (irrespective of grade of HE) or all three of serum creatinine > 3.4 mg/dL, prothrombin time (PT) > 100 seconds (INR > 6.5), and grade III–IV HE. In patients with other causes of ALF, transplantation is indicated in patients with PT > 100 seconds (INR > 6.5, irrespective of the grade of HE) or three of the following five variables: idiosyncratic drug reactions, > 7 days' jaundice prior to onset of HE, PT > 50 seconds (INR > 3.5), age < 10 or > 40 years, and serum bilirubin > 18 mg/dL.

PROGNOSIS OF ICU-ACQUIRED UREMIC ENCEPHALOPATHY

Uremic encephalopathy may complicate severe acute and chronic renal failure. Especially in patients with acute renal failure, the symptoms are generally more pronounced and progress more rapidly (Van Dijck et al., 2012). Renal failure is fatal if left untreated. Uremic encephalopathy reflects the severity of renal failure. If the renal failure is not treated, or cannot be treated any more, uremic encephalopathy progresses to coma and death. On the other hand, uncomplicated uremic encephalopathy is reversible, making prompt recognition and treatment (hemodialysis) important. An important note is the fact that there is scarce literature on the effects of treatment of uremic encephalopathy with renal replacement therapies (RRT) on clinically relevant outcomes. Although patients may certainly improve in level of consciousness after RRT has been initiated, robust prospective studies are lacking.

PROGNOSIS OF ICU-ACQUIRED WEAKNESS

ICUAW is the most common modern terminology for the syndrome of weakness that is also referred to by other names, such as "critical illness polyneuromyopathy." The many terms used to identify this syndrome have hampered robust outcome research. Only relatively recently proposed criteria have been published for ICUAW, which may be regarded as a starting point for more robust short- and long-term outcome studies, now that definitions have become more uniform (Stevens et al., 2009; Fan et al., 2014; Sharshar et al., 2014). Nevertheless, ICUAW in its previous diverse connotations is a well-recognized complication since the first description in the 1980s (Bolton et al., 1984), which has been associated with long-term morbidity and mortality in survivors (van der Jagt, 2010; Kress and Hall, 2014). Current studies on prognosis, many with intrinsic limitations with regard to definitions and terminology, will be summarized in this section.

In general it is difficult to truly separate the influence of ICUAW from outcomes from other prognostic factors, such as age, cognitive status, and various comorbidities. Moreover, the fact that ICUAW is heterogeneous in itself (polyneuropathic ICUAW seems to have a worse prognosis than predominantly myopathic ICUAW (Koch et al., 2014; Hermans and Van den Berghe, 2015)) complicates matters further. Finally, a plethora of interventions in the ICU seem to impact on the occurrence and course of ICUAW (e.g., medications such as corticosteroids, early mobilization practices, and nutrition), although true prospective randomized studies that reveal cause-and-effect relationships are scarce (Schweickert et al., 2009; Farhan et al., 2016). Therefore, at present, it is unclear whether ICUAW is merely a marker of severity of illness, or that it independently contributes to eventual adverse clinical outcomes, although the truth most probably lies somewhere in between (Hermans and Van den Berghe, 2015). Nonetheless, ICUAW and cognitive dysfunction together represent the most important long-term ICU sequelae (Kress and Hall, 2014), and therefore deserve our attention. In other words, there is more to ICU-related outcomes than just "discharge from ICU alive." In ICU survivors, functional outcome is of major interest. In addition, ICU length of stay (ICU-LOS) is of importance both from the perspective of the patient and cost effectiveness.

Earlier investigations have reported highly variable mortality risk associated with ICUAW, ranging from 7 to 61%, depending on the target population and the time frame of evaluation after ICU admission (van der Jagt, 2010). Adjustment for confounders has shown a statistically independent relationship between ICUAW and 30-day mortality, which may be fivefold higher compared to patients without ICUAW. In addition, ICU-LOS may double after a diagnosis of ICUAW (Ali et al., 2008). Importantly, easy-to-perform bedside tests may assist in the evaluation of muscle strength in suspected ICUAW and unveil an increased risk of death (Ali et al., 2008; Lee et al., 2012). More recent investigations by separate research groups have confirmed the strong and independent association between ICUAW and extended ICU and hospital LOS, as well as long-term risk of death, suggesting a causal relation (Hermans et al., 2014). The finding enhances this notion that there seems to be a dose–response relationship between the severity of muscle weakness at ICU discharge and mortality (Hermans et al., 2014). These studies suggest that scrutiny is indicated in the follow-up of these patients after the ICU period because of their higher risk for adverse outcomes. However, studies evaluating the benefit of follow-up programs have not been performed.

Critical illness portends long-term functional disabilities in those who are sensitive to the adversities of both the critical illness itself and the aggressive treatment approaches. Long-term neurophysiologic abnormalities (e.g., nerve conduction abnormalities) found in survivors of critical illness up to 5 years after ICU discharge illustrate the potential long-term impact of treatment in the ICU (Fletcher et al., 2003). These often subtle abnormalities may go unnoticed for ICU physicians, who are frequently focused primarily on "saving lives," but may have an important impact on functional recovery, activities of daily living, and quality of life after ICU discharge.

Previous health status seems to have an important impact on the relationship between ICUAW and long-term functional outcome, with younger, relatively healthy patients suffering least from the consequences of ICUAW (Semmler et al., 2013). After discharge from ICU, mortality risk in ICU survivors with ICUAW may differ less significantly compared to those without ICUAW, but in this subset of ICUAW patients, lower physical functioning has been found as an important outcome difference (Wieske et al., 2015). In spite of the significant functional impairments months after ICU discharge, there is also a prolonged potential for ongoing recovery when patients' physical reserve is sufficient (De Jonghe et al., 2002). Although observational studies confirm an association between early mobilization and improved functional outcomes, it remains to be seen in future studies whether such mobilization programs improve physical functioning for sure (TEAM Study Investigators et al., 2015). Further, current randomized and nonrandomized studies on early rehabilitation in critically ill patients that include a control group are difficult to interpret, due to varying interventions and outcome definitions, although a signal seems to be present in favor of early rehabilitation, at least with regard to walking distance at discharge from the hospital (Castro-Avila et al., 2015). Ongoing controlled studies on early rehabilitation will likely provide more insight into efficacy of specific early interventions on functional outcomes after ICU.

There is a need for structured outcome prediction of ICUAW for several reasons: (1) early rehabilitation may benefit those destined for adverse functional outcomes the most, and better prognostic tools may aid in the design of future intervention studies; and 2) prognostic information helps patients and next of kin to have realistic expectations on the time course of the recovery process. To develop prediction models for outcome, internal and external validations are needed in large, preferably prospective datasets examining various outcomes of interest. For this purpose, pooling data from prospective intervention studies has been shown to be appropriate in other settings (Walgaard et al., 2010; Roozenbeek et al., 2012), and should be considered for future studies of ICUAW.

Risk factors have been identified for the development of ICUAW in several studies and include sepsis, disease severity at admission, poor nutritional status, treatment with corticosteroids or neuromuscular blocking agents, prolonged sedation, mechanical ventilation, hyperglycemia, and female sex (De Jonghe et al., 2002; Hermans et al., 2014). Recently, a prediction model has been reported for ICUAW, with fair discriminative performance (area under the curve for model of >0.70, compared with 0.64 and 0.66, respectively, for Acute Physiology and Chronic Health Evaluation (APACHE-IV) compared with the Sequential Organ Failure Assessment (SOFA) score) (Wieske et al., 2014). However, external validity has not yet been demonstrated.

Although ICUAW is undoubtedly a relevant clinical entity, there are no known interventions that are supported by rigorous evidence that improve recovery from it. Nevertheless, recognition of ICUAW is of utmost importance, as has been shown in a landmark paper by Latronico et al. (1996). This study showed that, in patients who were comatose after sustaining different kinds of brain injuries, the presence of ICUAW might severely confound prognostic estimations, because of underestimating the motor response as part of the Glasgow Coma Scale. This may lead to unreasonably pessimistic prognostication that may even result in inadvertent withdrawal of treatment orders when there is in fact a salvageable brain. Therefore, it seems especially pertinent to establish protocols for the standardized evaluation of ICUAW in ICUs where neurologic patients are cared for (Latronico and Bolton, 2011; Farhan et al., 2016). In a more general sense, it seems sensible to avoid oversedating patients without a strict indication, in order to avoid muscle wasting and catabolism. Ideally, intensivists should strive for awake, but comfortable, patients who are evaluated on a daily basis for suitability for early physiotherapy and mobilization (Hermans and Van den Berghe, 2015). However, such practices are often hampered by implementation issues and scarce resources (Trogrlic et al., 2015). Further, ICUAW has been associated with extubation failure, which highlights that diagnosing ICUAW is clinically relevant (Jung et al., 2016), although obviously ICUAW is not the only relevant factor (Thille et al., 2015).

PROGNOSIS OF POSTERIOR REVERSIBLE ENCEPHALOPATHY SYNDROME

PRES, also known as reversible posterior leukoencephalopathy syndrome or hypertensive encephalopathy, was first described in 1996 (Hinchey et al., 1996). Signs and symptoms include confusion, depressed consciousness, visual loss, and headache. Despite its name, PRES is seldom isolated only to the posterior parts of the brain. On MRI, areas of edema are visible mostly in the occipital and parietal regions, but also frontal, temporal, cerebellar, and brainstem. PRES can, especially in cases related to a severe hypertensive crisis, lead to cardiac stunning (Banuelos et al., 2008).

The pathophysiology of PRES is largely unknown, but hypotheses include impaired autoregulation of the brain, endothelial dysfunction, and vasospasm resulting in ischemia. Underlying conditions are eclampsia, malignant hypertension, cytotoxic medications, sepsis, MOF, renal failure, low serum magnesium, immunosuppressive therapy (e.g., tacrolimus, cyclosporine), and chemotherapy. Acute kidney injury has been reported in 10–15% of patients with PRES (Ni et al., 2011). Autoimmune disease is present in more than 40% of cases (Fugate et al., 2010).

The diagnosis is clinical, and in critical care patients it is often suspected based on MRI findings. As many patients admitted to the ICU with serious illness are unconscious or pharmacologically sedated, the diagnosis may go unrecognized. In these unrecognized cases, seizures may deteriorate into status epilepticus (Riggael et al., 2013).

Since many patients admitted to an ICU have conditions that are risk factors for PRES, the diagnosis is not uncommon. PRES can be viewed as a complication of underlying conditions that are often treated on ICUs.

Recently, Muhammed et al. (2016) described an illustrative case of PRES in a patient admitted with subarachnoid hemorrhage (SAH), who was treated with induced hypertension for cerebral vasospasm. Besides their own case, the authors found seven articles describing 10 patients with PRES after hypertensive therapy of vasospasm and delayed cerebral ischemia. The time to development of PRES after starting hypertensive treatment was 7.8 ± 3.8 days (range 1–13 days). In all patients, the clinical symptoms reversed within days of normalization of the blood pressure.

As more than 60% of patients with SAH develop vasospasm, often treated with induced hypertension, in some cases maintaining mean arterial pressure above 110 mmHg, PRES is probably sometimes overlooked, given the rarity of published case reports. Because the prognosis of PRES complicating induced hypertension is generally favorable after normalization of the blood pressure, recognition is paramount. Therefore, in patients with SAH and symptomatic vasospasm in whom clinical symptoms worsen after induced hypertension, PRES should be considered.

The prognosis of PRES depends on the cause, but in most cases clinical signs and symptoms resolve within several weeks after controlling the underlying condition. Findings on MRI imaging usually also resolve within days to weeks. However, in other cases, signs and symptoms may be long-lasting and give rise to prolonged neurologic disturbances, mostly visual (Roth and Ferber, 2011). In critically ill patients, especially when there is an inadequately treated underlying condition, PRES can potentially be life-threatening. This is especially described in initially unrecognized cases in which PRES causes ischemia and infarction. Recurrent episodes of PRES have been reported in the literature, especially in patients requiring dialysis (Ergün et al., 2008; Hobson et al., 2012). MRI provides not only a means of diagnosing PRES, but also prognostic information. Covarrubias et al. (2002) have found that some patients develop foci of high signal intensity in the cortex on diffusion-weighted images, consistent with infarction. Apparent diffusion coefficient values in these areas are normal or only slightly elevated, perhaps because of pseudonormalization, resulting from intravoxel averaging of values in cytotoxic and vasogenic edema. The extent of T2 and diffusion-weighted imaging signal intensity correlates well with outcome and can provide guidance in decision making.

PROGNOSIS OF DELIRIUM IN CRITICALLY ILL PATIENTS

Delirium in critically ill patients is regarded as a highly lethal form of vital organ failure and is highly prevalent, especially in mechanically ventilated patients (Ely et al., 2004). Emergence or presence of delirium has implications for prognosis, which will be exacerbated with prolonged duration (Pisani et al., 2009). Therefore delirium may be viewed as a complication of critical illness, although it is certainly not exclusive to critically ill patients alone.

Delirium has been firmly linked to several adverse clinical outcomes in critically ill patients (Salluh et al., 2015).

Delirium has been reported to increase the risk of death approximately threefold compared to nondelirious critically ill patients in a seminal study by Ely and coworkers (2004), when statistical adjustments were incorporated into the analyses to diminish the potential for confounding. It was further shown that delirium arising after a state of coma (often due to sedatives in a patient receiving mechanical ventilation) was still associated with death, whereas such a comatose state that was not followed by a delirious state did not have a worse prognosis. This signifies that comatose state and delirium may not be viewed as a continuum of impaired brain function, but rather that delirium and coma are different both in a pathophysiologic and prognostic sense (Skrobik et al., 2013). The strong association of delirium with mortality in mechanically ventilated patients has been corroborated by others, who have reported even higher risks associated with delirium (Lin et al., 2004). However, since there are essentially three states of consciousness in critically ill patients (coma, delirium, no delirium), it is important to scrutinize whether, in studies on associations of delirium and adverse outcomes, the presence of coma has been accounted for. For instance, delirium may occur more frequently when sedation practices are mitigated towards less sedation use because this may result in fewer patients being comatose due to sedatives and more delirium simply because more patients can be assessed for it. Therefore, some studies have focused on "acute brain dysfunction" as a composite risk factor (i.e., delirium or coma), which still is associated with mortality (Almeida et al., 2014).

Recently, two studies from the same research group have challenged the independent association of delirium with mortality in ICU survivors using advanced statistical methods to adjust for time-varying variables (Klein Klouwenberg et al., 2014; Wolters et al., 2014). In the latter study, the association with mortality could not be established for a single day of delirium, but was still present for delirium persisting for 2 or more days.

In spite of the vast number of studies in recent years, a pathophysiologic explanation for the association of delirium with death is still elusive (few ICU physicians will have ever seen a patient acutely die from delirium), and measures that reduce delirium in controlled trials have not been consistently shown to reduce mortality (Al-Qadheeb et al., 2014). Therefore it seems more likely that focusing on preventive measures that act in concert to decrease both mortality and delirium will prove beneficial, rather than solely focusing on decreasing delirium as a syndrome of solitary organ failure (Al-Qadheeb et al., 2014; van der Jagt et al., 2014).

Emerging evidence links ICU delirium to significant cognitive impairment in ICU survivors (Pandharipande et al., 2013). Especially delirium duration is an important variable predicting future cognitive impairment. The degree of cognitive impairment up to 1 year after ICU stay may, in severe cases, be comparable to that seen in mild forms of Alzheimer's disease, or mild traumatic brain injury. This may be true even in younger patients. Although one may argue that the effects of delirium on cognition may simply represent additional pathologic "hits" on those with already vulnerable brains before their ICU stay, several studies suggest that a bidirectional interaction exists, linking cognitive decline to susceptibility towards infection and critical illness on the one hand, but also infection or sepsis to further decline of cognitive function on the other hand (Robinson et al., 2012; Shah et al., 2013).

Delirium has also been linked to functional impairment and reduced quality of life in ICU survivors, although it is hard to truly separate cognitive and functional impairment (Abelha et al., 2013; Svenningsen et al., 2014). Functional impairment is not limited to frail elderly patients, but may also be very significant in younger patients who were healthy at baseline (Iwashyna et al., 2010). Such functional impairments, especially in elderly ICU survivors, may result in increased rates of discharge to long-term care facilities (Balas et al., 2009).

Severity of delirium seems to predict the risk for adverse outcomes, in terms of mortality risk, functional outcome, or place of discharge (home vs. other) (Ouimet et al., 2007). Therefore, delirium should be regarded as a syndrome with a disease spectrum, rather than a type of organ failure that is either present or absent. Practically, the Intensive Care Delirium Screening Checklist (ICDSC), composed of a scoring form listing specific delirium symptoms, is more suitable to "grade" delirium severity than the Confusion Assessment Method for the ICU (CAM-ICU), which regards delirium as a dichotomous syndrome being either present or absent (Devlin et al., 2007). Further, some data indicate that the hypoactive subtype of delirium, rather than the hyperactive or mixed subtype, predicts worse prognosis (Robinson et al., 2011; Stransky et al., 2011). A distinction between 1-day versus prolonged delirium seems to be prognostically and pathophysiologically sound, since prolonged delirium as opposed to 1-day delirium seems more firmly linked to mortality (Klein Klouwenberg et al., 2014). Finally, sedation-associated delirium manifestations that resolve after halting the sedatives (so-called rapidly reversible, sedation-related delirium) have been found to be more "benign" compared with delirium symptoms that persist after sedatives have been stopped (Patel et al., 2014). Obviously, delirium comes in different "sizes" and also here the adage "one size fits all" does not entirely apply.

PROGNOSIS OF COMA AFTER CARDIOPULMONARY RESUSCITATION

Patients who remain comatose after cardiopulmonary resuscitation (CPR) and return of spontaneous circulation (ROSC) pose the dilemma to intensivists whether or not return of consciousness will occur. This issue usually becomes relevant after the first 24 hours of ICU management, which generally includes targeted temperature management (TTM) with sedation and analgesia. Several prognostic factors are known, such as initial heart rhythm, underlying disease and severity, previous medical history, and whether the resuscitation took place in an out-of-hospital or in-hospital setting. However, such prognostic factors can, on their own, not be used to determine with certainty whether individual patients will have a favorable or poor outcome. Therefore, recent guidelines have adopted a multimodal approach to prognostication (Nolan et al., 2015). Estimation of prognosis is useful to inform next of kin and discuss appropriateness of medical treatments or limitations thereof.

The European Resuscitation Council (ERC) guidelines have recently been updated (Nolan et al., 2015) and the International Liaison Committee on Resuscitation (ILCOR) Advanced Life Support Task Force published its consensus in parallel (Callaway et al., 2015).

Since the 2010 guidelines (Deakin et al., 2010), which stressed increased uncertainties rather than fixed truths with regard to prognostication due to the uncertain effects of widespread use of therapeutic hypothermia as compared with the 2005 guidelines (Nolan et al., 2005), outcome assessment has now become a multimodal approach, using several clinical and electrophysiologic variables to guide clinical judgment with greater confidence. A practical guide to prognostication is elaborated on in the next paragraph, but some general remarks apply. In general, decisions on halting or limiting treatment because of estimated poor prognosis are made in a multidisciplinary setting involving intensivists and neurologists, and such decisions may also involve other consulting specialists depending on the initial or underlying disease (e.g., cardiologists and nurses). It is of utmost importance that a locally implemented protocol is in use, which is strictly adhered to with the goal of minimizing variability in clinical practice. This implies that intensivists and neurologists, as the most pertinent specialists with regard to decision making based on estimated prognosis, are well versed with regard to mutual practices. For instance, when benzodiazepines and/or muscle relaxants have been administered during the initial TTM period, this may imply residual sedation, which should render consulting neurologists cautious in their prognostic assessments.

In this multidisciplinary setting, the protocol should be strictly followed, because even small deviations may complicate clinical decision making, especially when family members perceive inappropriate suggestions indicating dismal prognosis. For instance, early electroencephalography (EEG) to assess the presence of seizures (<72 hours after ROSC) in a patient with daily improving Glasgow Coma Scale motor scores, but persistent unresponsiveness, may yield a low-voltage EEG, suggesting poor prognosis, when in fact the clinical course signifies that progressive brain functioning is occurring, justifying daily clinical follow-up, rather than use of information from electrophysiologic tests to establish a definitive prognostic verdict. In general, it is wise not to use information from tests that were executed for an indication other than prognostic assessment, since this may complicate prognostication, rather than being helpful. This is also in line with the explicit admonition in the newest guidelines (Callaway et al., 2015), stating that any doubts about prognosis should be met with a low threshold for treatment continuation to allow more time for follow-up on neurologic status, rather than institution of tests to confirm poor prognosis.

Figure 41.1 from the ERC guideline publication shows a prognostication algorithm (Nolan et al., 2015). Importantly, patients eligible for prognostication according to the algorithm are those with ROSC and persistent coma, but the guidelines do not further specify with regard to specific subgroups, such as in-hospital versus out-of-hospital cardiac arrest. However, the robust evidence focuses mainly on out-of-hospital witnessed cardiac arrest victims, such that the management advice described in the guidelines is supported more for these patients than those who had cardiac arrest in different circumstances. In this section, for the purpose of uniformity, the situation of a comatose survivor is presumed to have been one that was managed with TTM, in line with Figure 41.1. However, we acknowledge that some patients will still be managed without TTM on an individual basis and depending on the organization of local care.

A comatose survivor of cardiac arrest will most commonly be managed with 24 hours of TTM between 32 and 36°C, after which rewarming will occur slowly. However, the whole process of cooling and rewarming will take less than 48 hours after ROSC, after which the first question should be whether there could be a confounder present for the comatose state. The most common confounders include residual sedation or metabolic disturbances, which should be firmly ruled out before concluding too soon that severe brain damage is present. No uniform guidelines are available to assess such potential confounders. A pharmacist may be helpful in cases of possible residual sedation, if there is doubt regarding its presence. When confounders are excluded,

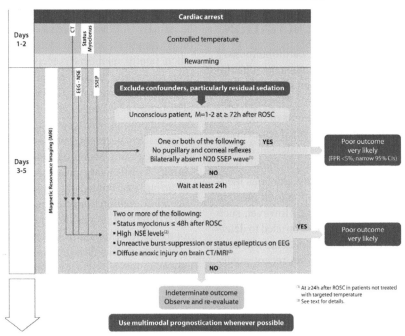

Fig. 41.1. Algorithm for prognostication in coma after cardiac arrest from the 2015 European Resuscitation Counsil (ERC) Guidelines. CT, computed tomography; EEG-NSE, electroencephalogram-neuron-specific enolase; SSEP, somatosensory evoked potential; ROSC, return of spontaneous circulation; FPR, false positive rate; Cis, confidence intervals.

and the patient is still comatose after 72 hours of ROSC (Fig. 41.1), there should be assessment of the pupillary and corneal reflexes. When both are absent, poor outcome is extremely likely (this is also true when only the pupillary reflexes are absent, only very slightly less so). Otherwise, bilaterally absent somatosensory evoked potentials (SSEPs) indicate a poor prognosis. When bilateral SSEPs and brainstem reflexes are present, more time is needed to determine prognosis, as indicated by the algorithm. Although the algorithm subsequently suggests including two or more of several variables (status myoclonus <48 hours after ROSC, high neuron-specific enolase levels, nonreactive burst suppression, status epilepticus on EEG, or diffuse anoxic injury on brain computed tomography (CT) or MRI), it should be noted that this advice is based on expert opinion, without very robust evidence. Apart from the low level of evidence for this latter part of the algorithm, applicability of this advice may be hampered by several issues depending on local settings: (1) implementation problems (e.g., neuron-specific enolase assessments may not be widely available); and (2) external validity (inherent to the lower level of evidence, these variables have been based on low numbers, from only a few hospitals, and applicability to other settings is disputable until further studies confirm the robustness of the predictive value of the variables).

Absence of EEG reactivity, a burst suppression pattern, and very low voltage seem to be relatively good indicators of poor prognosis after 72 hours of normothermia being reached and in the absence of confounders, although the guideline mentions a study reporting 3 patients who awoke in spite of absent EEG reactivity (Bouwes et al., 2012b). However, no further details were provided on these patients (e.g., could there have been residual sedation?). Status epilepticus also is not invariably a sign of poor prognosis and may be a reason to start a trial of aggressive treatment with antiepileptic drugs (AEDs). Whether this is indeed useful is currently being investigated (Ruijter et al., 2014), since many more patients with status epilepticus in this setting die than survive. More data on the prognostic value of EEG patterns are becoming available, and it is expected that their prognostic value will gain importance; not only to predict poor outcomes but also to help identify patients with good outcomes at an earlier stage (Hofmeijer et al., 2015; Westhall et al., 2016).

A vexing problem can be the patient with severe myoclonic jerking, with positive SSEPs and brainstem reflexes, which can only be managed with continuous intravenous sedatives to obtain an acceptable level of suppression of the myoclonus. In such a patient it may be best to treat with several AEDs, potentially in high dosages (e.g., valproate, levetiracetam, and when necessary, one or two other AEDs), to try and diminish the myoclonus, yet also abolish the sedatives to be able to assess the neurologic status. When this regimen is unsuccessful, most physicians will feel the inclination to stop treatment aimed at recovery, but theoretically early severe myoclonus may constitute Lance–Adams syndrome, which is compatible with awakening.

The guideline mentions that any doubt on ultimate prognosis should lead to consideration of prolonged treatment. Clearly, absence of any subsequent improvement of neurologic status suggests an ultimately worse prognosis. Reference is made to several studies reporting "late awakening" (up to 25 days after ROSC), but these cases did not provide any detail on the presence or absence of important confounders (Rittenberger et al., 2012) and, in one unexpected case of late awakening, an early EEG showed reactivity, which is a strong reason to suspect brain recovery, and in that sense the positive outcome may not have been so unexpected (Greer, 2013).

The ERC/European Society of Intensive Care Medicine guidelines (Nolan et al., 2015) mention the very small possibility of awakening in spite of absent SSEPs (i.e., false-positive results) with referral to three publications (Young et al., 2005; Bouwes et al., 2012a; Dragancea et al., 2015). In the first study, reference details on the single patient who awoke in spite of absent SSEPs were not provided (Young et al., 2005), and the study concerned patients who were not managed with therapeutic hypothermia. In the second study, the discussion refers to the risk of technically indeterminate SSEP results, but the results of the study itself did not include false positives in patients with SSEPs after rewarming (Bouwes et al., 2012a). In the third study (Dragancea et al., 2015), there was mention of one SSEP that proved to be false positive, but this was done in a patient who was actually following commands, and the SSEP itself was reported to be technically insufficient (in addition to the fact that it was obviously not indicated). Based on these insufficient data, one may therefore still argue that a technically well-performed SSEP, in a patient in whom there is a strong clinical suspicion of neurologic damage compatible with persistent comatose state (i.e., motor score 4 or less at least 72 hours after ROSC) and no residual sedation, muscle relaxants, or other potential confounders such as uremic encephalopathy, is a reliable tool to confirm poor prognosis, indicating that further treatment is inappropriate. Thus, it is important that a multidisciplinary team, including at least a clinical neurophysiologist, neurologist, and intensivist, establishes this verdict, and that intensivists should decide on withdrawal of treatment only after having certified that the other team members have corroborated and confirmed their assessment in the medical file.

There has been a point of discussion as to whether absent SSEPs truly indicate irreversible brain damage, or that SSEPs, being an intricate part of decision making, actually constitute a real risk with regard to a self-fulfilling prophecy. Proponents of the dangers of the self-fulfilling prophecy indicate that treatment limitations based on bilaterally absent SSEPs may result in potentially premature treatment cessation in patients who would still have a small chance of recovery. However, inappropriate withdrawal of life support with negative SSEPs only seems possible when technical artifacts of the SSEP and/or residual sedation are present; otherwise extubation may give a spontaneously breathing patient the opportunity to show neurologic improvement in the course of the following days, at least when sedatives are withheld. If unanticipated awakenings in such situations had been described in the medical literature, the fear of the self-fulfilling prophecy would be substantiated, but such reports are not known, and should not be mixed up with observational series reporting late awakenings (e.g., more than 3 days after rewarming), in which details of sedation practices and protocols for withdrawal of treatment practices have not been extensively described (Gold et al., 2014).

One may argue that a patient who cannot sustain patency of the airway in spite of intact brainstem function and absence of residual sedation must have sustained damage to the cortical structures so severe that it is essentially incompatible with a meaningful recovery, but studies on natural disease course of patients who have been certified to fulfill these criteria are also absent. However, it is very unlikely that a trial will ever be conducted in patients with absent SSEPs comparing standard practice with a strategy of sustained intensive care treatment, which will be arguably the only way to definitely establish whether fear of a self-fulfilling prophecy is justified.

Although chances of survival of in-hospital (nonperioperative) cardiac arrest and coma are lower than those associated with out-of-hospital cardiac arrest (Nolan et al., 2007), this knowledge does not generally impact on prognostic decision making. To a certain extent, the cerebral prognosis may be considered as separate from other organ failures, because the first consideration of prognostication focuses on whether or not the patient will regain consciousness. This means that, when the patient will not regain consciousness, treatment may be stopped independently of other organ failures, whereas when cerebral prognosis is regarded as indeterminate or good, the extent of organ failure may be much more important in the assessment of overall prognosis. In line with this, the new guidelines only stratify in the prognostication process based on whether patients have been managed with TTM or not, and do not stress the distinction between in- or out-of-hospital cardiac arrest patients, or initial rhythms as crucial factors for prognostication.

In conclusion, clinical variables other than those indicated by the algorithm in Figure 41.1 and paragraph 4.2.2

of the ERC guidelines should generally not be used for prognostication with regard to cerebral short-term outcomes. An exception may exist for in-hospital cardiac arrest survivors, since a very large population-based validation study proposed practical criteria to predict favorable outcome, which may be usable for clinical and "family-shared" decision making on treatment continuation or limitation (Chan et al., 2012). However, further external validation seems necessary.

We conclude with a case which exemplifies that the absence of clinical indicators of dismal prognosis, such as bilaterally absent SSEPs, warrants extreme caution, especially when potential confounders for neurologic assessment are present.

Illustrative case

A 24-year old woman was admitted to an academic ICU after prolonged cardiac arrest and CPR (90 minutes) in another institution caused by massive pulmonary embolism after recent start of oral contraceptives. The patient had been cannulated for venoarterial extracorporeal life support by a mobile extracorporeal membrane oxygenation (ECMO) team due to lack of persistent ROSC in spite of ongoing resuscitation efforts, but had some instances of nonsustained spontaneous output during the whole resuscitation period. She had been treated with thrombolytics, with no apparent effect. After transfer for ECMO support to an academic ICU, a laparotomy was performed because of abdominal compartment syndrome due to intra-abdominal bleeding caused by liver rupture as a result of prolonged chest compressions and intense anticoagulant therapies. After surgery, temperature control with avoidance of fever was initially chosen as a treatment strategy rather than hypothermia.

The subsequent course was as follows:

- Initially, she had absent pupillary and corneal reflexes, and status myoclonus.
- SSEPs were bilaterally present.
- EEG at 72 hours after ROSC was not low-voltage (>20 μV), but was nonreactive, although the patient was still receiving sedatives (opiate and propofol) in order to attenuate the severe myoclonus.
- The course was complicated by septic shock with acute kidney injury, which further hampered detailed assessment of her neurologic status, although pupillary and corneal reflexes returned.
- Some members of the treatment team doubted whether further treatment was indicated because of perceptions of a very poor prognosis, but others favored continuing treatment since important confounders were present.
- Over the subsequent week, after RRT had normalized the serum urea, the motor response improved to withdrawal, and the patient seemed to open her eyes to voice, but still did not track. A repeat EEG showed return of reactivity. She still required low dosages of sedatives because of frequent myoclonus, which had not disappeared despite treatment with two AEDs (valproate and levetiracetam).
- Although there was still some doubt about the prognosis, a tracheostomy was performed.
- Over the next week, her neurologic status improved, and she followed commands after sedation was tapered. It was concluded that the severe myoclonus was a combination of Lance–Adams syndrome and possibly the uremia earlier in the course.

According to the ERC algorithm (Fig. 41.1), one might argue that only status myoclonus was present, but a second criterion for poor prognosis was absent, indicating a strategy to "observe and re-evaluate." This strategy proved justified in this case.

PROGNOSIS OF NEUROLOGIC COMPLICATIONS OF LIVER TRANSPLANTATION

Orthotopic liver transplantation is a life-saving procedure for liver failure. All patients with ALF, and all patients after orthotopic liver transplantation are treated in ICUs.

Neurologic complications after liver transplantation surgery are reported in 13–47% of patients, especially in those receiving cadaveric grafts. Key topics in this setting include: immunosuppressive neurotoxicity; seizures; osmotic demyelination; neuromuscular complications; cerebrovascular complications; and CNS infections (Guarino et al., 2011).

The most commonly used immunosuppressive drugs after liver transplantation are cyclosporine and tacrolimus, which are well known for their complicating neurotoxicity (Guarino et al., 2006; Saner et al., 2007). Other, more recently introduced agents, such as sirolimus and everolimus, lack this neurotoxicity. Neurotoxicity can become evident soon after initiation of these medications, even while patients are still in the ICU. Guarino et al. (2011) distinguish "minor" events, (e.g., headache, tremor, insomnia, and paresthesias) and "major" events (e.g., encephalopathy, seizures, akinetic mutism, and polyneuropathy). Severe complications are often preceded by hypertension, hypomagnesemia, and HE (Saner et al., 2007). The prognosis of immunosuppressive neurotoxicity is favorable in most cases, but is highly dependent on timely discontinuation and a switch to nonneurotoxic agents. Irreversible neurologic damage is possible if the culprit drug is not stopped.

The postoperative course of liver transplantation is complicated by tonic-clonic seizures in up to 40% of patients (Guarino et al., 2006, 2011; Saner et al.,

2007). Seizures are observed especially in the immediate postoperative period, and result from metabolic derangements, adverse effects of medications, hypoxic-ischemic injury, but especially because of neurotoxicity of immunosuppressive medications (Guarino et al., 2006). Prognosis of seizures after liver transplantation is dependent on their cause. Seizures due to metabolic derangements or immunosuppressive neurotoxicity have a favorable outlook, but seizures attributable to sepsis, acute transplant rejection, or cerebrovascular events may have a more guarded prognosis.

Central pontine and extrapontine myelinolysis (CPEPM), first described in 1959, is a feared complication, which may be associated with rapid correction of hyponatremia (>15 mmol/L in 24 hours or 18 mmol/L in 48 hours when the sodium concentration was < 120 mmol/L) (Abbasoglu et al., 1998; Yu et al., 2004). It was first described after liver transplantation in 1978 (Starzl et al., 1978). The incidence of CPEPM after liver transplantation was reported to be 5–10% (Yu et al., 2004), but in a recent series was found in only 1.4% of 1378 consecutive patients (Morard et al., 2014).

CPEPM is the most serious neurologic complication after liver transplantation. In the ICU, CPEPM occurs in malnourished, usually alcoholic patients and in those with cirrhosis. Other risk factors are chronic renal failure necessitating hemodialysis and adrenal insufficiency (Morais et al., 2009). After liver transplantation, CPEPM is seen in patients with cirrhosis with pretransplantation serum hyponatremia, where large volumes of intravenous fluids were given in the operating theatre, resulting in an increase in plasma osmolality. Effective treatment modalities in the acute phase of CPEPM are lacking. Despite the absence of any compelling evidence, treatment has been attempted with corticosteroids, plasmapheresis, and intravenous immunoglobulin.

The prognosis of CPEPM has traditionally been considered to be poor. In one series, CPEPM occurred 3–18 days after liver transplantation, and had a 100% case fatality rate, with a median survival of 25 days (Yu et al., 2004). In a series described by Abbasoglu et al. (1998), all patients expired within 3 months. Others report a mortality rate > 50% in the first 2 weeks and 90% after 6 months (Morais et al., 2009). However, some recent reports are less pessimistic. Musana (2005) reported that all 6 of their patients had some clinical improvement over time. In another series of 25 patients, 11 had a favorable outcome, of which 7 had a full recovery, and the remaining 4 were independent in activities of daily living with some mild cognitive or extrapyramidal deficits (Kallakatta et al., 2011). Because most published series have shown a grim prognosis, there may also be a degree of self-fulfilling prophecy, since patients who undergo withdrawal of life support measures usually die. In a recent series of 36 patients with CPEPM (not as a complication after liver transplantation), 11 (31%) were dead 1 year after withdrawal of life-sustaining measures, whereas 14 (56%) of the survivors were still alive with a Rankin score less than 1 (Louis et al., 2012). The reason for withdrawal of life-sustaining measures in the 11 patients was, in all cases, severe motor deficits. Improvements in neurologic function may be seen over the course of months after onset of CPEPM. When individuals are offered time to recover, many show gradual improvement, and can even recover completely (Niehaus et al., 2001), while other surviving patients have residual deficits, ranging from minor functional and cognitive difficulties to locked-in syndrome, quadriplegia, and pseudobulbar paralysis (Newell and Kleinschmidt-DeMasters, 1996; Brown, 2000). Patients with CPEPM often have other complications, such as MOF, infection, and gastrointestinal hemorrhage, making the prognosis more unfavorable (Brito et al., 2006). Kallakatta et al. (2011) showed that three factors were significantly correlated with poor prognosis: sodium concentration < 115 mmol/L, associated hypokalemia, and Glasgow Coma Scale < 10. Prevention of CPEPM, recognizing a patient at risk with slow correction of hyponatremia, is paramount (Abbasoglu et al., 1998; Yu et al., 2004).

The diagnosis, treatment, and prognosis of muscular weakness, stroke, and CNS infections after liver transplantation are similar to other settings, and will not be discussed here.

ETHICAL CONSIDERATIONS

Critically ill patients are subject to many complications connected with life-sustaining treatment and measures required for their serious conditions. As complications can worsen outcome and even cause death, they are of ethical concern. High-quality medical care can be defined as evidence-based care based on the results of well-conducted research and delivered by well-trained clinicians. A complication is usually defined as an unintended, harmful occurrence or condition resulting from a diagnostic, prophylactic, or therapeutic intervention, or an accidental injury occurring in a hospital setting. However, when a complication is a result of medical care that was not indicated, or not well applied, it can be labeled as unethical. The three most important ethical aspects regarding complications in critically ill patients are prevention, identification, and avoidance of self-fulfilling prophecies.

Efforts to prevent complications of intensive care should be vigorously pursued. This includes adequate monitoring of signs and symptoms, educational programs for physicians and nurses, and the provision of adequate staffing. Due to the severity of the medical conditions of patients, their multiple comorbidities, and the complexity of the intensive care environment, many complications can never be completely prevented. For example, although some ICU patients develop delirium due to a single preventable cause, delirium more often occurs when a vulnerable patient with multiple predisposing risk factors encounters a serious course of illness (Brummel and Girard, 2013). Early identification and treatment of conditions leading to MOF, avoidance of deep sedation, treatment of hyperglycemia, promotion of early mobilization, and carefully weighing the administration of corticosteroids might help reduce the incidence and severity of ICUAW (De Jonghe et al., 2009).

Correct identification of neurologic complications in critically ill patients can be challenging, but can have a significant impact on outcomes. For example, initially unrecognized cases of PRES can progress to cerebral infarction. Early recognition of risk factors and symptoms is paramount and can, for some complications, make the difference between a good or bad outcome. Physicians and nurses should have knowledge of the possible risk factors associated with neurotoxicity of drugs. Monitoring of neurologic signs and symptoms potentially associated with the administration of drugs should be routine, but also correct identification that the symptoms are, in fact, drug-related, and not a sign of underlying neurologic disease (George et al., 2010).

Prognostication in critically ill patients can raise additional ethical concerns. As with all prognostication in medicine, how sure are we when we predict outcome? This is especially important in cases where a perceived poor prognosis leads to withholding or withdrawing of life-sustaining measures. Predictions of poor outcome may become self-fulfilling if life-sustaining measures, such as mechanical ventilation, are subsequently withheld or withdrawn on the basis of that prediction (Wilkinson, 2009). Predictions of poor outcome may also affect the perception of patients, relatives, and healthcare providers. Physicians make decisions on the basis of available evidence. If the evidence is not based on the most relevant knowledge, or is based on the presumption that the signs and symptoms are those of the underlying disease, and not of a (reversible) complication, this can lead to unethical decisions. It is accepted that it is ethically permissible to withhold or withdraw life-sustaining measures in the face of uncertainty in critical care medicine based on clinical signs and symptoms pointing to a poor prognosis, but it is unethical to do this based on wrong assumptions.

REFERENCES

Abanades S, Nolla J, Rodriguez-Campello A et al. (2004). Reversible coma secondary to cefepime neurotoxicity. Ann Pharmacother 38 (4): 606–608.

Abbasoglu O, Goldstein RM, Vodapally MS et al. (1998). Liver transplantation in hyponatremic patients with emphasis on central pontine myelinolysis. Clin Transplant 12: 263–269.

Abelha FJ, Luis C, Veiga D et al. (2013). Outcome and quality of life in patients with postoperative delirium during an ICU stay following major surgery. Crit Care 17: R257.

Ahmed A, Loes DJ, Bressler EL (1995). Reversible resonance imaging findings in metronidazole-induced encephalopathy. Neurology 45: 588–589.

Ali NA, O'Brien Jr JM, Hoffmann SP et al. (2008). Acquired weakness, handgrip strength, and mortality in critically ill patients. Am J Respir Crit Care Med 178: 261–268.

Almeida IC, Soares M, Bozza FA et al. (2014). The impact of acute brain dysfunction in the outcomes of mechanically ventilated cancer patients. PLoS One 9:e85332.

Almond MK, Fan S, Dhillon S et al. (1995). Avoiding acyclovir neurotoxicity in patients with chronic renal failure undergoing haemodialysis. Nephron 69: 428–432.

Al-Qadheeb NS, Balk EM, Fraser GL et al. (2014). Randomized ICU trials do not demonstrate an association between interventions that reduce delirium duration and short-term mortality: a systematic review and meta-analysis. Crit Care Med 42: 1442–1454.

Balas MC, Happ MB, Yang W et al. (2009). Outcomes associated with delirium in older patients in surgical ICUs. Chest 135: 18–25.

Bandettin di Poggio M, Anfosso S, Andenino D et al. (2011). Clarithromycin-induced neurotoxicity in adults. J Clin Neurosci 18: 313–318.

Banuelos PA, Temes R, Lee VH (2008). Neurogenic stunned myocardium associated with reversible posterior leukoencephalopathy syndrome. Neurocrit Care 9: 108–111.

Bateman JC, Barberio JR, Grice P et al. (1952). Fatal complications of intensive antibiotic therapy in patients with neoplastic disease. Arch Intern Med 90: 763.

Bolton CF, Gilbert JJ, Hahn AF et al. (1984). Polyneuropathy in critically ill patients. J Neurol Neurosurg Psychiatry 47: 1223–1231.

Bouwes A, Binnekade JM, Kuiper MA et al. (2012a). Prognosis of coma after therapeutic hypothermia: a prospective cohort study. Ann Neurol 71: 206–212.

Bouwes A, van Poppelen D, Koelman JH et al. (2012b). Acute posthypoxic myoclonus after cardiopulmonary resuscitation. BMC Neurol 12: 63.

Brito AR, Vasconcelos MM, Cruz LC et al. (2006). Central pontine and extrapontine myelinolysis: report of a case with a tragic outcome. J Pediatr 82: 157–160.

Brown WD (2000). Osmotic demyelination disorders: central pontine and extrapontine myelinolysis. Curr Opin Neurol 13: 691–697.

Brummel NE, Girard TD (2013). Preventing delirium in the intensive care unit. Crit Care Clin 29: 51–65.

Callaway CW, Soar J, Aibiki M et al. (2015). Part 4: Advanced Life Support: 2015 International Consensus on Cardiopulmonary Resuscitation and Emergency Cardiovascular Care Science With Treatment Recommendations. Circulation 132: S84–S145.

Carlon R, Possamai C, Corbanese U (2005). Acute renal failure and neurotoxcity following valacyclovir. Intensive Care Med 31: 1593.

Castro-Avila AC, Seron P, Fan E et al. (2015). Effect of early rehabilitation during intensive care unit stay on functional status: systematic review and meta-analysis. PLoS One 10: e0130722.

Cecil KM, Halsted MJ, Schapiro M et al. (2002). Reversible MR imaging and MR spectroscopy abnormalities in association with metronidazole therapy. J Comput Assist Tomo 26: 948–951.

Chan PS, Spertus JA, Krumholz HM et al. (2012). A validated prediction tool for initial survivors of in-hospital cardiac arrest. Arch Intern Med 172: 947–953.

Chattelier D, Jourdain M, Mangalaboyi J et al. (2002). Cefepime-induced neurotoxicity: an underestimated complication of antibiotherapy in patients with acute renal failure. Intensive Care Med 28: 214–217.

Chow KM, Szeto CC, Hui AC et al. (2003). Retrospective review of neurotoxicity induced by cefepime and ceftazidime. Pharmacotherapy 23: 369–373.

Christakis NA (1999). Death foretold. Prophecy and prognosis in medical care. University of Chicago Press, Chicago.

Conway N, Beck E, Somerville J (1968). Penicillin encephalopathy. Postgrad Med J 44: 891–897.

Cossaart N, SantaCruz KS, Preston D et al. (2003). Fatal chemotherapy-induced encephalopathy following high-dose therapy for metastatic breast cancer: a case report and review of the literature. Bone Marrow Transplant 31: 57–60.

Covarrubias DJ, Luetmer PH, Campeau NG (2002). Posterior reversible encephalopathy syndrome: prognostic itility of quantitative diffusion-weigthed MR Images. Am J Neuroradiol 23: 1038–1048.

Dakdouki CK, Al-Awar GN (2004). Cefepime-induced encephalopathy. Int J Infect Dis 8: 59–61.

De Jonghe B, Sharshar T, Lefaucheur JP et al. (2002). Paresis acquired in the intensive care unit: a prospective multicenter study. JAMA 288: 2859–2867.

De Jonghe B, Lacherade JC, Sharshar T et al. (2009). Intensive care unit-acquired weakness: risk factors and prevention. Crit Care Med 37: S309–S315.

Deakin CD, Nolan JP, Soar J et al. (2010). European Resuscitation Council Guidelines for Resuscitation 2010 Section 4. Adult advanced life support. Resuscitation 81: 1305–1352.

Delerme S, De Jonghe B, Proost O et al. (2002). Acyclovir-induced coma in a young patient without preexisting renal impairment. Intensive Care Med 28: 661–662.

Denysenko L, Nicolson SE (2011). Cefoxitin and ciprofloxacin neurotoxicity and catatonia in a patient on hemodialysis. Psychosomatics 52: 379–383.

Desal SV, Law TJ, Needham DM (2011). Long-term complications of critical care. Crit Care Med 39: 371–379.

Devlin JW, Fong JJ, Fraser GL et al. (2007). Delirium assessment in the critically ill. Intensive Care Med 33: 929–940.

Dragancea I, Horn J, Kuiper M et al. (2015). Neurological prognostication after cardiac arrest and targeted temperature management 33 degrees C versus 36 degrees C: Results from a randomised controlled clinical trial. Resuscitation 93: 164–170.

Dulluc AMY, Latour P, Goas JY (2004). Encephalopathy and acute renal failure during acyclovir treatment. Rev Neurol 160: 704–706.

Dumont L, Picard V, Marti RA et al. (1998). Use of remifentanil in a patient with chronic hepatic failure. Br J Anaesth 81: 265–267.

Ely EW, Shintani A, Truman B et al. (2004). Delirium as a predictor of mortality in mechanically ventilated patients in the intensive care unit. JAMA 291: 1753–1762.

Ergün T, Lakadamyali H, Yilmaz A (2008). Recurrent posterior reversible encephalopathy syndrome in a hypertensive patient with end-stage renal disease. Diagn Intrev Radiolo 14: 182–185.

Ernst ME, Franey RJ (1998). Acyclovir- and ganciclovir-induced neurotoxicity. Ann Pharmacother 32: 111–113.

Fan E, Cheek F, Chlan L et al. (2014). An official American Thoracic Society clinical practice guideline: the diagnosis of intensive care unit-acquired weakness in adults. Am J Respir Crit Care Med 190: 1437–1446.

Farhan H, Moreno-Duarte I, Latronico N et al. (2016). Acquired muscle weakness in the surgical intensive care unit: nosology, epidemiology, diagnosis, and prevention. Anesthesiology 124: 207–234.

Ferrara N, Abete P, Giordano M et al. (2003). Neurotoxicity induced by cefepime in a very old hemodialysis patient. Clin Nephrol 59: 388–390.

Ferry T, Ponceau B, Simon M et al. (2005). Possibly linezolid-induced peripheral and central neurotoxicity: report of four cases. Infection 33: 151–154.

Fletcher SN, Kennedy DD, Ghosh IR et al. (2003). Persistent neuromuscular and neurophysiologic abnormalities in long-term survivors of prolonged critical illness. Crit Care Med 31: 1012–1016.

Fletcher J, Aykroyd LE, Feucht EC et al. (2010). Early onset probable linezolid induced encephalopathy. J Neurol 257: 433–435.

Fugate JE, Claassen DO, Cloft HJ et al. (2010). Posterior reversible encephalopathy syndrome: associated clinical and radiologic findings. Mayo Clin Proc 85: 427–432.

Gallagher R (2007). Opioid-induced neurotoxicity. Can Fam Physician 53: 426–427.

García-Martínez R, Símon-Talero M, Córdoda J (2011). Prognostic assessment in patients with hepatic encephalopathy. Dis Markers 31: 171–179.

George EL, Henneman EA, Tasato FJ (2010). Nursing implications for prevention of adverse drug events in the intensive care unit. Crit Care Med 38: S136–S144.

Geyer J, Höffler D, Demers HG et al. (1988). Cephalosporin-induced encephalopathy in uremic patients. Nephron 48: 237.

Gold B, Puertas L, Davis SP et al. (2014). Awakening after cardiac arrest and post resuscitation hypothermia: are we pulling the plug too early? Resuscitation 85: 211–214.

Graves TD, Condon M, Loucaidou M et al. (2009). Reversible metronidazole-induced cerebellar toxicity in a multiple transplant recipient. J Neurol Sci 285: 238–240.

Greer DM (2013). Unexpected good recovery in a comatose post-cardiac arrest patient with poor prognostic features. Resuscitation 84: e81–e82.

Grill MF, Maganti RK (2008). Cephalosporin-induced neurotoxicity: clinical manifestations, potential pathogenic mechanisms, and the role of electroencephalographic monitoring. Ann Pharmacother 42: 1843–1850.

Grill MF, Maganti RK (2011). Neurotoxic effects associated with antibiotic use: management consideartions. Br J Clin Pharmacol 72: 381–393.

Guarino M, Benito-León J, Decruyenaere J et al. (2006). EFNS guidelines on management of neurological problems in liver transplantation. Eur J Neurol 13: 2–9.

Guarino M, Benito-León J, Decruyenaere J et al. (2011). Neurological problems in liver transplantation. In: NE Gilhus, MP Barnes, M Brainin (Eds.), European Handbook of Neurological Management, 2nd Edn. Vol. 1. Blackwell Publishing, Oxford, pp. 491–499.

Guerra C, Linde-Zwirble WT, Wunsch H (2012). Risk factors for dementia after critical illness in elderly Medicare beneficiaries. Crit Care 16: R233.

Hasselström J, Eriksson S, Persson A et al. (1990). The metabolism and bioavailability of morphine in patients with severe liver cirrhosis. Br J Clin Pharmac 29: 289–297.

Heaney CJ, Campeau NG, Lindell EP (2003). MR imaging and diffuse-weigthed imaging changes in metronidazole (flagyl)-induced cerebellar toxicity. Am J Neuroradiol 24: 1615–1617.

Herishanu YO, Ziotnik M, Mostoslavsky M et al. (1998). Ceforoxime-induced encephalopathy. Neurology 50: 1873–1875.

Hermans G, Van den Berghe G (2015). Clinical review: intensive care unit acquired weakness. Crit Care 19: 274.

Hermans G, Van Mechelen H, Clerckx B et al. (2014). Acute outcomes and 1-year mortality of intensive care unit-acquired weakness. A cohort study and propensity-matched analysis. Am J Respir Crit Care Med 190: 410–420.

Hinchey J, Chaves C, Appignanni B et al. (1996). A reversible posterior leukoencepahlopathy syndrome. NEJM 334: 494–500.

Hobson EV, Craven I, Blank SC (2012). Posterior reversible encephalopathy syndrome: a truly treatable neurologic illness. Perit Dial Int 32: 590–594.

TEAM Study Investigators Hodgson C, Bellomo R et al. (2015). Early mobilization and recovery in mechanically ventilated patients in the ICU: a bi-national, multi-centre, prospective cohort study. Crit Care 19: 81.

Hofmeijer J, Beernink TM, Bosch FH et al. (2015). Early EEG contributes to multimodal outcome prediction of postanoxic coma. Neurology 85: 137–143.

Horlen CK, Seifert CF, Malouf CS (2000). Toxic metronidazole-induced MRI changes. Ann Pharmacother 34: 1273–1275.

Huang YT, Chen LA, Cheng SJ (2012). Metronidazole-induced encephalopathy: case report and review literature. Acta Neurol Taiwan 21: 74–78.

Iwashyna TJ, Ely EW, Smith DM et al. (2010). Long-term cognitive impairment and functional disability among survivors of severe sepsis. JAMA 304: 1787–1794.

Jackson GD, Berkovic SF (1992). Ceftazidime encephalopathy: absence status and toxic hallucinations. J Neurol Neurosurg Psychiatry 55: 333–334.

Jallon P, Frankhauser I, Du Pasqueir R et al. (2000). Severe but reversible encephalopathy associated with cefepime. Neurophysiol Clin 30: 383–386.

Josse S, Godin M, Fillastre JP (1987). Cefazolin induced encephalopathy in a uremic patient. Nephron 45: 72.

Jung B, Moury PH, Mahul M et al. (2016). Diaphragmatic dysfunction in patients with ICU-acquired weakness and its impact on extubation failure. Intensive Care Med 42: 853–861.

Kallakatta RN, Radhakrishnan A, Fayaz RK et al. (2011). Clinical and functional outcome and factors predicting prognosis is osmotic demyelination syndrome (central pontine and/or extrapontine myelinolysis) in 25 patients. J Neurol Neurosurg Psychiatry 82: 326–331.

Kim E, NA DG, Kim EY et al. (2007). MR Imaging of metronidazole-induced encephalopathy: lesion distribution and diffusion-weighted imaging findings. AJNR 28: 1652–1658.

Kim H, Lim Y, Kim SR et al. (2011). Metronidazole-induced encephalopathy in a patient with infectious colitis: a case report. J Med Case Reports 5: 63.

Kim KB, Lim SM, Park W et al. (2012). Ceftriaxone-induced neurotoxicity: case report, pharmakinetic considerations, and literature review. J Korean Med Sci 27: 1120–1123.

Kim S, Kang M, Cho JH et al. (2014). reversible magnetic resonance imaging findings in cycloserine-induced encephalopathy: a case report. Neurol Asia 19: 417–419.

Klein Klouwenberg PM, Zaal IJ, Spitoni C et al. (2014). The attributable mortality of delirium in critically ill patients: prospective cohort study. BMJ 349: g6652.

Klion AD, Kallsen J, Cowl CT et al. (1994). Ceftazidime-related nonconvulsive status epilepticus. Arch Intern Med 154: 586–589.

Koch S, Wollersheim T, Bierbrauer J et al. (2014). Long-term recovery In critical illness myopathy is complete, contrary to polyneuropathy. Muscle Nerve 50: 431–436.

Kress JP, Hall JB (2014). ICU-acquired weakness and recovery from critical illness. N Engl J Med 370: 1626–1635.

Kusumi RK, Plouffe JF, Wyatt RH et al. (1980). Central nervous system toxicity associated with metronidazole therapy. Ann Intern Med 93: 59–60.

Kwon HM, Lim HK, Cho J et al. (2008). Cycloserine-induced encephalopathy: evidence on brain MRI. Eur J Neurol 15: e60–e61.

Lam S, Gomolin IH (2006). Cefepime neurotoxicity: case report, pharmacokinetic considerations, and literature review. Pharmacotherapy 26: 1169–1174.

Laskin OL, Longstreth JA, Whelton A et al. (1982). Acyclovir kinetics in end-stage renal disease. Clin Pharmacol Ther 31: 594–601.

Latronico N, Bolton CF (2011). Critical illness polyneuropathy and myopathy: a major cause of muscle weakness and paralysis. Lancet Neurol 10: 931–941.

Latronico N, Fenzi F, Recupero D et al. (1996). Critical illness myopathy and neuropathy. Lancet 347: 1579–1582.

Lee JJ, Waak K, Grosse-Sundrup M et al. (2012). Global muscle strength but not grip strength predicts mortality and length of stay in a general population in a surgical intensive care unit. Phys Ther 92: 1546–1555.

Lin SM, Liu CY, Wang CH et al. (2004). The impact of delirium on the survival of mechanically ventilated patients. Crit Care Med 32: 2254–2259.

Louis G, Megarbane B, Lavoué S et al. (2012). Long-term outcome of patients hospitalized in intensive care units with central or extrapontine myelinolysis. Crit Care Med 40: 970–972.

Maganti R, Jolin D, Rishi D et al. (2006). Nonconvulsive status epilepticus due to cefepime in a patient with normal renal function. Epilepsy Behav 8: 312–314.

Martinez-Rodriguez JE, Barriga FJ, Santamaria J et al. (2000). Nonconvulsive status epilepticus associated with cephalosporins in patients with renal failure. Am J Med 111: 115–119.

Morais BS, Cameiro FS, Araújo RM et al. (2009). Central pontine myelinolysis after liver transplantation: is sodium the only villain? Case Report. Rev Bras Anestesiol 59: 344–349.

Morard I, Gasche Y, Kneteman M et al. (2014). Identifying risk factors for central pontine and extrapontine myelionlysis after liver transplantation: a case-control study. Neurocrit Care 20: 287–295.

Muhammed S, Güresir Á, Greschus S et al. (2016). Posterior reversible encephalopathy syndrome as an overlooked complication of induced hypertension for cerebral vasospasm. Stroke 47: 519–522.

Mulcahy H, Chaddha SKB (2008). MRI of metronidazole-induced encephalopathy. Radiology Case Reports 3: 1–4.

Musana AK (2005). Central pontine myelinolysis: case series and review. Wisconsin Med J 104: 56–60.

Nagel S, Kohrmann M, Huttner HB et al. (2007). Linezolid-induced posterior reversible leukoencephalopathy syndrome. Arch Neurol 64: 748.

Négrin-González J, Peralto Filpo G, Carrasco JL (2014). Psychiatric adverse reaction induced by Clarithromycin. Eur Ann Allergy Clin Immunol 46: 114–115.

Neu HC (1996). Safety of cefepime: a new extended spectrum perenteral cephalosporin. Am J Med 100: 68S–75S.

Newell KL, Kleinschmidt-DeMasters BK (1996). Central pontine myelinolysis at autopsy: a twelve year retrospective analysis. J Neurol Sci 142: 1394–1399.

Ni J, Zhou LX, Hao HL et al. (2011). The clinical and radiological spectrum of posterior reversible encephalopathy syndrome: a retrospective series of 24 patients. J Neuroimaging 21: 219–224.

Niehaus L, Kulozik A, Lehmann R (2001). Reversible central pontine and extrapontine myelinolysis in a 16-year old girl. Childs Nerv Syst 17: 294–296.

Nolan JP, Deakin CD, Soar J et al. (2005). European Resuscitation Council guidelines for resuscitation 2005. Section 4. Adult advanced life support. Resuscitation 67 (Suppl 1): S39–S86.

Nolan JP, Laver SR, Welch CA et al. (2007). Outcome following admission to UK intensive care units after cardiac arrest: a secondary analysis of the ICNARC Case Mix Programme Database. Anaesthesia 62: 1207–1216.

Nolan JP, Soar J, Cariou A et al. (2015). European Resuscitation Council and European Society of Intensive Care Medicine Guidelines for Post-resuscitation Care 2015: Section 5 of the European Resuscitation Council Guidelines for Resuscitation 2015. Resuscitation 95: 202–222.

O'Grade JG, Alexander GJ, Hayllar KM et al. (1989). Early indicators of prognosis in fulminant hepatic failure. Gastroenterology 97: 439–445.

Onuigbo MAC, Nye D, Iloanya PC (2009). Drug-induced encephalopathy secondary to non renal dosing of common medications in two dialysis patients. Adv Peritoneal Dialysis 25: 89–91.

Ortega-Guitierrez S, Wolge T, Panya DJ et al. (2009). Neurologic complications in non-neurological intensive care units. Neurologist 15: 254–267.

Ortiz A, Martin-Llonch N, Garron MP et al. (1991). Cefazolin-induced encephalopathy in uremic patients. Rev Infect Dis 13: 772–773.

Ouimet S, Riker R, Bergeron N et al. (2007). Subsyndromal delirium in the ICU: evidence for a disease spectrum. Intensive Care Med 33: 1007–1013.

Pandharipande PP, Girard TD, Jackson JC et al. (2013). Long-term cognitive impairment after critical illness. N Engl J Med 369: 1306–1316.

Pascual J, Liano F, Ortuno J (1990). Cefotaxime-induced encephalopathy in an uremic patient. Nephron 54: 92.

Patel SB, Poston JT, Pohlman A et al. (2014). Rapidly reversible, sedation-related delirium versus persistent delirium in the intensive care unit. Am J Respir Crit Care Med 189: 658–665.

Pisani MA, Kong SY, Kasl SV et al. (2009). Days of delirium are associated with 1-year mortality in an older intensive care unit population. Am J Respir Crit Care Med 180: 1092–1097.

Primavera A, Cocito L, Audenino D (2004). Nonconvulsive status epilepticus during cephalosporin therapy. Neuropsychobiology 49: 218–222.

Revankar SG, Applegate AL, Markovitz DM (1995). Delirium associated with acyclovir treatment in a patient with renal failure. Clin Infect Dis 21: 435–436.

Riggael BD, Waked CS, Okun MS (2013). Diagnosis and treatment of altered mental status. In: AJ Layon, A Gabrielli, WA Friedman (Eds.), Textbook of Neurointensive Care. Springer Verlag, London, pp. 521–540.

Rittenberger JC, Popescu A, Brenner RP et al. (2012). Frequency and timing of nonconvulsive status epilepticus in comatose post-cardiac arrest subjects treated with hypothermia. Neurocrit Care 16: 114–122.

Robinson TN, Raeburn CD, Tran ZV et al. (2011). Motor subtypes of postoperative delirium in older adults. Arch Surg 146: 295–300.

Robinson TN, Wu DS, Pointer LF et al. (2012). Preoperative cognitive dysfunction is related to adverse postoperative outcomes in the elderly. J Am Coll Surg 215: 12–17. discussion 17–18.

Roncon-Albuquerque R, Pires I, Martins K et al. (2009). Ceftriaxone-induced reversible encephalopathy in a patient treated for a urinary tract infection. North J Med 67: 72–75.

Roozenbeek B, Lingsma HF, Lecky FE et al. (2012). Prediction of outcome after moderate and severe traumatic brain injury: external validation of the International Mission on Prognosis and Analysis of Clinical Trials (IMPACT) and Corticoid Randomisation After Significant Head injury (CRASH) prognostic models. Crit Care Med 40: 1609–1617.

Roth C, Ferbert A (2011). The posterior reversible encephalopathy syndrome: what's certain, what's new? Pract Neurol 11: 136–144.

Rubins HB, Moskowitz MA (1990). Complications of care in a medical intensive care unit. J Gen Intern Med 5 (2): 104–109.

Ruijter BJ, van Putten MJ, Horn J et al. (2014). Treatment of electroencephalographic status epilepticus after cardiopulmonary resuscitation (TELSTAR): study protocol for a randomized controlled trial. Trials 15: 433.

Sadafi S, Mao M, Dillon JJ (2014). Ceftriaxone-induced acute encephalopathy in a peritoneal dialysis patient. Case Reports in Nephrology 108185.

Salluh JI, Wang H, Schneider EB et al. (2015). Outcome of delirium in critically ill patients: systematic review and meta-analysis. BMJ 350: h2538.

Saner FH, Sotiropoulos GC, Gu Y et al. (2007). Severe neurological events following liver transplantation. Arch Med Res 38: 75–79.

Sarna JR, Brownell AK, Furtado S (2009). Cases: reversible cerebellar syndrome caused by metronidazole. CMAJ 181: 611–613.

Schwankhaus JD, Masucci EF, Kurtze JF (1985). Cefazolin-induced encephalopathy in a uremic patient. Ann Neurol 17: 211.

Schweickert WD, Pohlman MC, Pohlman AS et al. (2009). Early physical and occupational therapy in mechanically ventilated, critically ill patients: a randomised controlled trial. Lancet 373: 1874–1882.

Semmler A, Okulla T, Kaiser M et al. (2013). Long-term neuromuscular sequelae of critical illness. J Neurol 260: 151–157.

Seok JI, Yi H, Song YM, Lee WY (2003). Metronidazole-induced encephalopathy and inferior olivary hypertrophy: lesion analysis with diffuse-weighted imaging and apparent diffusion coefficient maps. Arch Neurol 60: 1796–1800.

Shah FA, Pike F, Alvarez K et al. (2013). Bidirectional relationship between cognitive function and pneumonia. Am J Respir Crit Care Med 188: 586–592.

Sharma N, Batish S, Gupta A (2012). Ceftriaxone-induced acute reversible encephalopathy in apatient with enteric fever. Drug Watch 44: 124–125.

Sharshar T, Citerio G, Andrews PJ et al. (2014). Neurological examination of critically ill patients: a pragmatic approach. Report of an ESICM expert panel. Intensive Care Med 40: 484–495.

Shinagawa J, Hashimoto Y, Ohmae Y (2008). A case of hepatic encephalopathy induced by adverse effect of morphine sulfate. Gan To Kaguka Ryoho 35: 1025–1027.

Skrobik Y, Leger C, Cossette M et al. (2013). Factors predisposing to coma and delirium: fentanyl and midazolam exposure; CYP3A5, ABCB1, and ABCG2 genetic polymorphisms; and inflammatory factors. Crit Care Med 41: 999–1008.

Sonck J, Laureys G, Verbeelen D (2008). The neurotoxicity and safety of treatment with cefepime in patients with renal failure. Nephrol Dial Transplant 23: 966–970.

Stevens RD, Marshall SA, Cornblath DR et al. (2009). A framework for diagnosing and classifying intensive care unit-acquired weakness. Crit Care Med 37: S299–S308.

Stransky M, Schmidt C, Ganslmeier P et al. (2011). Hypoactive delirium after cardiac surgery as an independent risk factor for prolonged mechanical ventilation. J Cardiothorac Vasc Anesth 25: 968–974.

Starzl TE, Schneck SA, Mazzoni G et al. (1978). Acute neurological complications after liver transplantation with particular reference to intraoperative cerebral air embolus. Ann Surg 187: 236–240.

Svenningsen H, Tonnesen EK, Videbech P et al. (2014). Intensive care delirium – effect on memories and health-related quality of life – a follow-up study. J Clin Nurs 23: 634–644.

Thabet F, Al Maghrabi M, Al Barraq A et al. (2009). Cefepime-induced nonconvulsive status epilepticus: case report and review. Neurocrit Care 10: 347–351.

Thille AW, Boissier F, Ben Ghezala H et al. (2015). Risk factors for and prediction by caregivers of extubation failure in ICU patients: a prospective study. Crit Care Med 43: 613–620.

Trogrlic Z, van der Jagt M, Bakker J et al. (2015). A systematic review of implementation strategies for assessment, prevention, and management of ICU delirium and their effect on clinical outcomes. Crit Care 19: 157.

van der Jagt M (2010). Intensive care unit-acquired weakness. Crit Care Med 38: 1617–1619. author reply 1619.

van der Jagt M, Trogrlic Z, Ista E (2014). Untangling ICU delirium: is establishing its prevention in high-risk patients the final frontier? Intensive Care Med 40: 1181–1182.

Van Dijck A, Van Daele W, De Deyn PP (2012). Uremic encephalopathy. In: R Tanasescu (Ed.), Misellanea on Encephalopathies – A second look. InTech, Available from: http://www.intechopen.com/books/miscellanea-on-encepahlopathies-a-second-look/uremic-encephalopathy.

van Kan HJM, De Jonge E, Bijleveld YA et al. (2009). Viral encephalitis masking acyclovir neurotoxicity? A case report. Neth J Crit Care 13: 96–98.

Walgaard C, Lingsma HF, Ruts L et al. (2010). Prediction of respiratory insufficiency in Guillain-Barré syndrome. Ann Neurol 67: 781–787.

Westhall E, Rossetti AO, van Rootselaar AF et al. (2016). Standardized EEG interpretation accurately predicts prognosis after cardiac arrest. Neurology 86: 1482–1490.

Wieske L, Witteveen E, Verhamme C et al. (2014). Early prediction of intensive care unit-acquired weakness using easily available parameters: a prospective observational study. PLoS One 9:e111259.

Wieske L, Dettling-Ihnenfeldt DS, Verhamme C et al. (2015). Impact of ICU-acquired weakness on post-ICU physical functioning: a follow-up study. Crit Care 19: 196.

Wilkinson D (2009). The self-fulfilling prophecy in intensive care. Theor Med Bioeth 30: 401–410.

Wolters AE, van Dijk D, Pasma W et al. (2014). Long-term outcome of delirium during intensive care unit stay in survivors of critical illness: a prospective cohort study. Crit Care 18: R125.

Woodruff BK, Wijdicks EF, Marshall WF (2002). Reversible metronidazole-induced lesions of the cerebellar dentate nuclei. N Engl J Med 346: 68–69.

Young GB, Doig G, Ragazzoni A (2005). Anoxic-ischemic encephalopathy: clinical and electrophysiological associations with outcome. Neurocrit Care 2: 159–164.

Yu J, Zheng SS, Liang TB et al. (2004). Possible causes of central pontine myelinolysis after liver transplantation. World J Gastroenterol 10: 2540–2543.

Index

NB: Page numbers in *italics* refer to figures, tables and boxes.

A

Abscess, spinal epidural
  hospital course and management of 331
  neuropathology of 320
Absence status epilepticus 136
Abulia 444–445
Access, care in comatose patient *127*
Accidental hypothermia 695–699
  clinical presentation of 696–697
  clinical trials and guidelines of 699
  complex clinical decisions of 699
  epidemiology of 695, *695*
  hospital course and management of 698–699
  neurodiagnostics and imaging of 698
  neuropathology of 695–696, *696*
  neurorehabilitation of 699
  outcome prediction of 699
Acephalgic migraine 660
Acetaminophen, in shivering 625–626
Acetylcholine neurotransmitter system, in delirium 451–452
Acetylcholinesterase inhibitors, for cognitive impairment 584
Acid-base disorders *698*
Acid-base status, TH and 626
Acidemic hypocarbia 24
Acidosis
  in convulsive status epilepticus 135
  metabolic, in comatose patient 124–125
Acinetobacter 366–367
Acute basilar artery embolus, acute ischemic stroke and 169–170
Acute basilar artery thrombosis, in acute ischemic stroke 169–170
Acute brain failure 449–450
Acute brain injury
  airway management and mechanical ventilation in 15–32
  problems after 23
  decompressive craniectomy in 299–318
    complications of 311–313
    pathophysiologic rationale and impact of 300–302
    perceived indications for *300*
    quality of life after 313
Acute central neurotoxicologic syndromes 485–498
  anticholinergic toxicity in *492*, 493–496
  dissociative agents in 489–490
  opioids in 487–489, *488*
  sedative-hypnotic toxicity in 485–487, *486*

Acute central neurotoxicologic syndromes *(Continued)*
  sedative-hypnotic withdrawal in 490–493
  sympathomimetic toxicity in 496–498
Acute central neurotoxicology, of drugs of abuse 485–506
Acute confusional state 444–445
Acute encephalitis 337–348
  cerebrospinal diagnostic testing for 340
  clinical presentation of 339–340
  common agents of *339*
  diagnostic criteria for *338*
  EEG for 340
  epidemiology of 338–339
  flavivirus encephalitides in 342–343
  herpes encephalitis in 340–342
  hospital course of 340–345
  hospital management of 345
  immune-mediated encephalitis in 344–345
  Japanese encephalitis virus in 343
  neurodiagnostics and neuroimaging of 340
  neurorehabilitation for 345–346
  rabies and 343–344, *344*
  tick-borne encephalitis in 343
  West Nile virus in 342–343, *342*
Acute encephalopathy 449
Acute environmental injuries, neurologic complications of 685–704
  injuries related to electricity 689–691
  injuries related to extremes of temperature 695–699
  injuries related to fire 685–689
  injuries related to water 691–695
Acute hemorrhage 717
Acute hydrocephalus 213–214
  management of *210*
Acute hyponatremia 709
Acute ischemic stroke
  critical care in 153–176
    clinical presentation of 156–158
    clinical trials and guidelines in 168
    complex clinical decisions in 168–170
    hospital course and management of 161–164
    indications for 154–155
    neurodiagnostics and neuroimaging of 158–161
    neuropathology of 155–156
    neurorehabilitation in 170–171

Acute ischemic stroke *(Continued)*
    patient, intensive care of 164–168
  in critically ill pregnant patients 667
Acute liver failure
  coagulopathies and 755–756
  hepatic encephalopathy in, prognosis of 768–769
Acute myocardial infarct 744
Acute neuromuscular disorders
  clinical features of 231
  diagnostic tests of 233–234
  epidemiology of 230
  hospital course and further management 234–235
  management of 229–238
  neuromuscular respiratory failure 231–233, *233*
  neurorehabilitation 235
  pathophysiology of 230–231
Acute neuromuscular respiratory failure, causes of *230*
Acute neurovascular disease, prognosis of 55
Acute Physiology and Chronic Health Evaluation 6
Acute Physiology and Chronic Health Evaluation II score, in ICU delirium 450, *451*
Acute respiratory distress syndrome 25, 39–40, *39*
  in opioid-poisoned patients 488
  therapeutic strategies leading 40
Acute structural brain injury, delirium in 450
Acute weakness, rare causes of respiratory failure *230*
Acyclovir, for herpes encephalitis 341
Adrenocorticotrophic hormone, for superrefractory status epilepticus 522
Age 381–382
  intensive care unit-acquired weakness and 534
  as risk factor of cerebral aneurysms and SAH *201–203*
Agitation, traumatic brain injury and 253–254, 261
Air emboli 639–640, *639*
Air embolism, after transplantation 549
Airway
  breathing, and circulation, assessment of, for seizure patients 512–513
  difficult

Airway *(Continued)*
    approach to 21
    preparation of 18–19
    monitoring of, in seizures 517
    neuropathology of 16–18
Airway compromise, in acute ischemic stroke 154
Airway management
    in acute ischemic stroke 154
    and mechanical ventilation in acute brain injury 15–32
    clinical trials in 20–21
    complex clinical decisions 21–25
    guidelines in 20–21
Airway pressure, intracranial pressure and 25
Airway-protective reflexes, impaired 17
Akinetic mutism, after transplantation 552
Alberta Stroke Program Early CT Score 159, *159*
Alcohol intake, as risk factor of cerebral aneurysms and SAH *201–203*
Alcohol withdrawal syndrome
    delirium secondary to 642
    diagnosis of 643
    treatment for 645
Alcoholic hallucinosis 491
Alcoholic tremulousness 491
Alcoholic withdrawal seizures 491
Aldosterone 707
Alkalemic hypocarbia 24
Alkalosis, respiratory 16
Allogeneic bone marrow transplant, PRES following 468
Allograft dysfunction, primary 414
Allopregnanolone, for seizures 524
Alpha-II spectrin 385
Alteplase Thrombolysis for Acute Noninterventional Therapy in Ischemic Stroke trial 161
Altered mental status 444, 641–645
    clinical presentation of 643
    clinical trials and guidelines in 643–644
    complex clinical decisions in 644–645, *644*
    differential diagnosis for *642*
    epidemiology of 641–642
    hospital course and 644–645, *644*
    neurodiagnostics and imaging in 643
    neuropathology of 642–643
    neurorehabilitation/outcomes in 645
Amebic meningoencephalitis, after transplantation 564
American Clinical Neurophysiology Society 115
    indications for EEG *647*
American College of Surgeons Advanced Trauma Life Support guidelines 277
American Heart Association/American Stroke Association *162*, 168
    guidelines
        for coagulopathy, management of 183, *183*
        for intracerebral hemorrhage, management of 182–183

American Heart Association-International Liaison Committee on Resuscitation, 2015 *604*
American Spinal Injury Association scoring system, of spinal cord injuries 278, *279*
Amniotic fluid embolism 662
Amphetamines 496–497
Ampicillin 368
Amygdala 50, *50*
Anaerobic glycolysis 372
Analeptic agents, for sedative-hypnotic toxicity 487
Analgesics, drug-induced encephalopathy and 768
Ancillary testing, in brain death 422, *425–427*
Andexanet alfa 753
Anemia *216–221*
    in acute ischemic stroke 167–168
    management of 167–168
Anesthetic drips, for seizures 518
Anesthetics, NCSE and
    in comatose patients 143
    with preservation of consciousness 143
Aneurysm
    acute hemorrhage from *198*
    formation and rupture 197–203
    growth, as risk factor of cerebral aneurysms and SAH *201–203*
    location, as risk factor of cerebral aneurysms and SAH *201–203*
    repair 209–210
    shape, as risk factor of cerebral aneurysms and SAH *201–203*
    size, as risk factor of cerebral aneurysms and SAH *201–203*
Aneurysmal subarachnoid hemorrhage 195
    acute hydrocephalus of 213–214
    aneurysm formation and rupture 197–203
    aneurysm repair 209–210
    clinical presentation of 205–206
    clinical trials and guidelines of 214–215
    complex clinical decisions of 215
    complications and secondary treatment 211
    computed tomography of 206, *207*
    in critically ill pregnant patients 666
    delayed cerebral ischemia 204–205, 211–213
    diagnosis *216–221*
        of cause of 207–209
    early brain injury 204
    epidemiology of 195–196
    etiology of 196
    grading, initial assessment *216–221*
    hospital course and management of 209–214
    increased intracranial pressure and 214
    lumbar puncture 207
    magnetic resonance imaging 206–207
    management of 195–228
    natural history of 195–196

Aneurysmal subarachnoid hemorrhage *(Continued)*
    neurodiagnostics and imaging of 206–209
    neuropathology of 196–205
    neurorehabilitation 222
    outcome prediction of 215–222
    seizures 214
    subtypes of 196
    summary of guidelines from different countries and committees *216–221*
    weather and climate in 196
Angioedema, in acute ischemic stroke 163
Angiographic vasospasm 204–205, 212–213
Angiography
    complications of, of subarachnoid hemorrhage *212*
    in PRES 474
Anoxic-ischemic encephalopathy 386–387
    continuous EEG of 110–111, *112*
Anterior spinal artery syndrome 580
Antiangiogenic drugs, in PRES 470
Antibiotics
    for bacterial CNS infections *357*
    drug-induced encephalopathy and 766–767
    intraventricular administration of 373
Anticholinergic drugs, in delirium 451–452, 457
Anticholinergic toxicity 493–496
    clinical course for 495
    clinical presentation in 494–495, *494*
    complex clinical decisions for 495–496
    epidemiology of 493
    laboratory imaging and testing in 495
    management of 495
    pharmacology of *492*, 493–494
Anticholinergic xenobiotics 493
Anticholinesterase medications 661
Anticoagulation, after ischemic stroke 168–169
Antidiuretic hormone, circulating 707
Antiepileptic drugs 520–523, 660, 718
    dosing and pharmacokinetics
        for refractory status epilepticus *519*
        for seizures *516*
    fosphenytoin 520–521
    ketamine 522
    lacosamide 521–522
    levetiracetam 521
    for neurologic complications of transplantation 553
    phenobarbital 522
    phenytoin 520–521
    SUDEP and 56
    for traumatic brain injury 254–255
    valproic acid 521
Antifibrinolytic drugs 209
Antihypertensive Treatment of Acute Cerebral Hemorrhage II 182
Antimicrobial therapy, for community-acquired bacterial meningitis *358*
Antimuscarinic agents 494

Antiplatelet agents, intracranial hemorrhage and 754
Antiplatelet management, in acute ischemic stroke 166–167
Antipsychotics, in delirium 457–458
Antithrombotic-associated intracranial hemorrhage 744–745, 751–754
Antiviral drugs, drug-induced encephalopathy and 768
Apical ballooning syndrome, status epilepticus and 135
Apixaban 753
  in intracerebral hemorrhage 179
Apnea, central sleep 37
Apneusis 37–38
Apoptosis 677
Arboviruses 339
Arbovirus-mediated encephalitis 340
Arginine vasopressin 705–706
Arrhythmia *698*
  categories of 55
Arterial blood pressure 71
  ICP and, pulse transmission of 79–80
Arterial blood volume, ICP and 67–68
Arterial dissection 635–637
  clinical presentation of 635
  clinical trials and guidelines in 636, *636*
  complex clinical decisions in 636–637
  epidemiology of 635
  hospital course and 636–637
  neurodiagnostics and imaging in 635–636
  neuropathology of 635
Arterial gas embolism 692
Arterial injury, spinal cord injuries and 290
Artery of Adamkiewicz 320, 577–578
Artifacts, in electroencephalography 115
Ascending reticular arousal system 381
ASIA Impairment Scale 278, *280*
Aspergillosis, after transplantation *561*
Aspirin
  in acute ischemic stroke 166–167
  in intracerebral hemorrhage 179
  for ischemic stroke 582–583
Assays for measuring thrombin inhibition 750
Astrocytic, function and microglia activation 677
Ataxic breathing 38
Atelectasis, inspiratory load determined by several factors and mechanisms of *232*
Atrial natriuretic factor 707
Atrioventricular node 49–50, *50*
Attention disturbances, in delirium 452
Autoimmune disease, PRES and *468*, 470
Automaticity, respiratory
  impaired 16–17
  neuropathology of 16–18
Autonomic dysreflexia, management of 332–333
Autonomic nervous system *52*
  cardiovascular system and 49–50
  parasympathetic branch of 50
  sympathetic branch of 50

Autoregulation 469
  of cerebral blood flow 68–69, *68*
  cerebrovascular pressure reactivity and 74
Autoregulatory failure, in PRES 469–470
Autosomal dominant polycystic kidney disease, as risk factor of cerebral aneurysms and SAH *201–203*
Awaken, failure to, after transplantation 549–551, *549*
  evaluation for 550–551, *550*

B
Baboon animal model 53
Back pain 720
Bacteremia 369–370
Bacterial infections, of central nervous system 349–364
  clinical presentation of 351–352
  clinical trials and guidelines for 359
  complex clinical decisions in 360
  epidemiology of 349–351
  hospital course and management of 355–359
  neurodiagnostics and neuroimaging of 352–355
  neuropathology of 351
  neurorehabilitation for 361
  outcome prediction of 360–361, *361*
Bacterial meningitis 337–338, *339*
  coma and 120
  community-acquired 349–350
    antimicrobial therapy for *358*
    clinical presentation of 351–352
    clinical trials and guidelines for 359
    complex clinical decisions in 360
    epidemiology of 349–350
    hospital course and management of 355–359
    neurodiagnostics of 352–354
    neuropathology of 351
    outcome prediction in 360–361, *361*
    recommendations for initial antibiotic therapy for *357*
  hospital-acquired 350
    clinical presentation of 352
    clinical trials and guidelines for 359
    epidemiology of 350
    hospital course and management of 358
    neurodiagnostics and neuroimaging of 354
    neuropathology of 351
    recommendations for initial antibiotic therapy for *357*
Bacterial ventriculitis 371–372
Barbiturate coma, ICP and 84–85
Baroreceptors 706–707
Barthel index 381
Batson's plexus 720
Bedrest, and intensive care unit-acquired weakness 533
Bedside multimodality monitoring, for traumatic brain injury 249–251, *249*, *252*
Bedside pulmonary function tests 233

Benzodiazepine-refractory ethanol withdrawal 493
Benzodiazepines 720
  in anticholinergic toxicity 495
  in delirium 450, 456, 460
  in ICU 644
  for intracranial hypertension 82–83
  for neurologic complications of transplantation 553
  responsiveness, loss of 510–512, 517–518
  in sedative-hypnotic toxicity 487
  for seizures 498, 515
  for status epilepticus 140
  for traumatic brain injury 252–253
  trial, for nonconvulsive status epilepticus 514–515, *515*
BEST-TRIP trial, in ICP monitoring 95
Beta-blockers, for neurogenic pulmonary edema 38–39
Bifrontal craniectomy 301
Bioinformatics, multimodal neurologic monitoring 100–101
Biologic agents 734
Biomarkers, and intensive care unit-acquired weakness 536
Biopsy
  extensive, and muscle weakness 533
  in PRES 471
Biot's breathing 38
Bladder, care in comatose patient *127*
Bladder dysfunction 720
Blast theory 39–40
Blood cultures
  for diagnosing bacterial meningitis 354
  in intracerebral hemorrhage 180
Blood diseases *196*
Blood gases *697*
Blood glucose targets, traumatic brain injury and 254
Blood pressure management, in acute ischemic stroke 164–165
Blood vessel rupture, in intracerebral hemorrhage 179–180
Blood-based biomarkers
  of brain injury 597–598
  in outcome prediction, of CPR 605
Blood-brain barrier breakdown 677
Blood-brain barrier damage, in acute ischemic stroke 156
Blunt carotid-vertebral injury 635
Blunt cervical trauma-associated vascular injuries, management of 289–290, *289*
Bolus dexamethasone 722
Bone flap resorption, following cranioplasty 312–313
Bone marrow transplant, PRES following *468*, 470
BOOST 2 trial, in brain tissue oxygen monitoring 97
Borrelia burgdorferi, intramedullary infectious myelopathies and *323*
Bötzinger complex 36–37

Bowel care regimen, in comatose patient 127
Bowel dysfunction 720
Bowell *698*
Boyle's law 692
Brachial plexopathy 581
Brachial plexus injury *649*
  clinical deficits from 648
  incidence of 648
  operative repair of 650
Bradford Hill criteria, for causality 461, *461*
Bradyarrhythmias 55
Brain abscess 374–375
  clinical presentation of 352
  complex clinical decisions in 360
  epidemiology of 350–351
  hospital course and management of 358–359
  neurodiagnostics and neuroimaging of 354–355, *355*
  neuropathology of 351
  recommendations for initial antibiotic therapy for *357*
Brain cerebral blood flow, normal 155
Brain death 404
  acceptance of, by families 430
  ancillary testing in 422, *425–427*
  clinical testing in 421–422
  concept and controversies 430–431
  definition of 420–421
  determination of 421–422, *423–424*
  epidemiology of 428
  ethical considerations in 430–431
  maternal 669
  organ support in 428–429
  pathophysiology of 422–428
  systemic complications in *428*
Brain edema, ICP and 67–68
Brain herniation 715–716
Brain imaging, in PRES 471–473
Brain infarction, hypotension and 22–23
Brain injury
  acute, decompressive craniectomy in 299–318
  airway management and mechanical ventilation in 15–32
  early 204
  post cardiac arrest 596
  respiratory complications and 15
  traumatic
    abnormalities in vasomotor tone and reactivity of 244–245
    balancing second-tier therapies for 259
    basic intensive care for 252–255
    bedside neurodiagnostics and monitoring for 248–251, *249, 252*
    biomarkers in 247–248, *248*
    brain swelling in 242–244, *242, 244*
    clinical decisions for 257–260
    clinical neuropathology of 240–245
    clinical presentation of 245
    clinical trials and guidelines for 257
    coma and 119–120

Brain injury *(Continued)*
    critical care management of 239–274
    dysautoregulation in 242–244
    in elderly patient 258–259
    energy failure in 242–244, *243*
    epidemiology of 240
    focal pathology of 240–241, *240*
    genetic modulation of disease course and outcome in 241–242
    hemostatic abnormalities after 257–258
    hospital course and management of 251–257
    imaging for 246–247, *246–247*
    molecular mechanisms of 242
    multiple trauma in 260
    neurorehabilitation for 261
    in nonassessable patient, monitoring of 259
    nonfocal (diffuse) pathology of *240*, 241
    osmotic therapy for 255
    outcome of 262, *262*
    pathophysiology of 240–245, *240*
    prediction for 262–263
    rescue therapies for 255
    routine ICU interventions for 255
    sedation for 252–254, *253*
    seizure prophylaxis for 254–255
    temperature management for 254
    thromboprophylaxis for 258
    ventilatory strategies for 252
Brain ischemia, hypotension and 22–23
Brain metabolism, cerebral microdialysis and 98–100
Brain metastases 715–716
Brain natriuretic peptide 707
Brain oxygenation, traumatic brain injury and 249
Brain swelling, in traumatic brain injury 242–244, *242, 244*
Brain tissue microtransducers 70
Brain tissue oxygen pressure-reactivity index 97
Brain tissue oxygen tension 97
Brain tissue oxygenation 96–98
  intraparenchymal cerebral oxygen monitoring 97
  jugular venous bulb oximetry 96–97
  near-infrared spectroscopy 97–98
Brain Trauma Foundation, for intracranial pressure monitoring 95
Brain tumors 664
  in critically ill pregnant patients 666–667
Brain-specific serum markers 385–386
Brainstem, lesions in 16–17
Brainstem edema, in PRES 476, *477*
Brainstem injury, after transplantation 549
Brainstem reflexes 121, *123*
  in Full Outline of Unresponsiveness score *122*
Breathing
  after stroke 36
  ataxic 38

Breathing *(Continued)*
  Biot's 38
  disordered 36–38
  patterns, abnormal 16
Bronchoalveolar lavage 367
B-type natriuretic peptide, use of 57
Burn encephalopathy 685–687
Burn-induced delirium 685–686
Burn-related peripheral neuropathies 686
Burns 685
Burst suppression *111*
Buspirone, in shivering 625–626
Busulfan, drug toxicity and 554–555

C

C. glabrata 371
Calcineurin inhibitor-induced pain syndrome 556
Calcineurin inhibitors, drug toxicity of 555, *555–556*
  diagnostic criteria for *558*
  symptoms of 555
Calcium, neurologic manifestations of disturbances 710–711
Canadian C-spine Rule 281
Candida albicans 371, 727–729
Candidiasis, after transplantation *561*
Cangrelor 744–745
Capecitabine 724
Carbamazepine 719
Carbapenem 368
Carbon dioxide pressure, effects on brain physiology 17–18
Carboxyhemoglobin 686
Cardiac arrest 593
  blood-based biomarkers in 597–598
  clinical examination in 597, *598*
  clinical presentation in 595–596
  clinical trials and guidelines in 602–603, 622–623
  complex clinical decisions in 603–604
  epidemiology in 593–594
  hospital course and management 599–602
  neurodiagnostics and imaging in 597–599
  neuroimaging in 598–599
  neuropathology in 594–595
  neurophysiologic testing in 598
  outcome prediction of 604–607
    blood-based biomarkers in 605
    clinical examination in 604–605
    multimodal prognostication in 606–607, *608–609*
    neuroimaging in 606, *606–607*
    neurophysiologic testing in 605
    neurorehabilitation in 607–608
  TH and 620
Cardiac care, acute ischemic stroke and 155
Cardiac dysfunction 57–58
Cardiac function, TH and 626
Cardiac histopathology 52–53
Cardiac plexus 50, *50*

Cardiac surgery, neurologic complications
of 573–592
clinical presentation of 579–581
clinical trials and guidelines for 585
complex clinical decisions 585
epidemiology of *574*, 575–577
hospital course and management of
582–585
neurodiagnostics and imaging 581–582,
582
neuropathology of 577–579
neurorehabilitation for 587
outcome prediction 585–587
Cardiopulmonary complications
216–221
Cardiopulmonary resuscitation
cardiac arrest and, neurology of 593
blood-based biomarkers in 597–598
clinical examination in 597, *598*
clinical presentation in 595–596
clinical trials and guidelines in
602–603
complex clinical decisions in
603–604
epidemiology in 593–594
hospital course and management
599–602
neurodiagnostics and imaging in
597–599
neuroimaging in 598–599
neuropathology in 594–595
neurophysiologic testing in 598
outcome prediction of 604–607
coma after, prognosis of 773–776
neurology of 593–618
Cardiovascular autonomic dysfunction 52
Cardiovascular system *698*
ANS and 49–50
CNS and 50
Carotid artery surgery, consultation for 445
Carotid endarterectomy 574–575
stroke etiology after 577
Carotid Revascularization Endarterectomy
versus Stenting Trial 575
Carotid stenting, in acute ischemic stroke
155
Caspofungin 372–373
Catecholamine, status epilepticus and 135
Catecholamine reuptake inhibition 490
Catha edulis 496
Catheter angiography, for cerebrovascular
injury 290
Catheter-associated UTIs 370–371
Catheter-related bacteremia, risk for 369
Cathinones *494*, 496–497
synthetic 497
Cats, electric stimulation and 51
Caudate 178, *178*
Cefepime 368, 770
Ceftriaxone, reversible encephalopathy
and 767
Cell-based model of hemostasis 746, *749*
Cellular injury, during cardiac arrest 594
Centers for Disease Control and
Prevention 365

Centers for Medicare and Medicaid
Services, comatose patient and
127–128
Central alveolar hypoventilation
syndromes 37
Central nervous system
abstract of 49
bacterial infections of
clinical presentation of 351–352
clinical trials and guidelines for 359
complex clinical decisions in 360
epidemiology of 349–351
hospital course and management of
355–359
management of 349–364
neurodiagnostics and neuroimaging
of 352–355
neuropathology of 351
neurorehabilitation for 361
outcome prediction in 360–361, *361*
infections of, after transplantation
563–566
donor-derived 564
immune reconstitution inflammatory
syndrome 566
reactivation of 564–566, *564*
Central nervous system infections 727
Central neurogenic hyperventilation 37
Central pontine and extrapontine
myelinolysis 777
Central pontine myelinolysis 709–710
as neurologic complication of
transplantation 560–562, *562*
Central sleep apnea 37
Central venous catheters 369
Cephalosporin 368
drug-induced encephalopathy and 767
Cerebellar edema, in PRES 476, *477*
Cerebellar hemisphere infarction, acute
ischemic stroke and 154
Cerebellum 178, *178*
Cerebral adaptation
to hypernatremia 708, *709*
to hyponatremia 708
Cerebral amyloid angiopathy, in
intracerebral hemorrhage
178–179, *179*
Cerebral aneurysm, risk factors for
201–203
Cerebral aspergillosis, after
transplantation 565–566
Cerebral biochemistry 80
Cerebral blood flow 67, 80, 100
autoregulation of 68–69, *68*
control of 69, *69*
traumatic brain injury and 244
Cerebral blood volume 68
barbiturate coma for 84–85
fever and 82
Cerebral edema 602, 715–718
acute ischemic stroke and 156
clinical presentation 716–717
epidemiology in 715–716, *716*
fever and 82
hospital course and 717–718

Cerebral edema *(Continued)*
management and 717–718
neuroimaging 717, *717*
neuropathology and 716
pathophysiology and 716
Cerebral electric activity 80
Cerebral hyperperfusion syndrome
clinical presentation of 580
epidemiology of 576
hospital course and management of
584–585
neuropathology of 578
outcome prediction 587
Cerebral hypoxia 372
brain injury and 96
Cerebral ischemia
delayed 204–205, 211–213
angiographic vasospasm 204–205
cortical spreading depolarization/
ischemia 205
microthromboembolism 205
pathophysiology of *204*
prophylaxis and treatment of
212–213
hyperglycemia and 82
seizures and 83
Cerebral microdialysis 80, 98–100
for traumatic brain injury 249–250, *249*
Cerebral oxygenation *74*, 80
Cerebral Performance Category scale 381
Cerebral perfusion, intubation in impaired
22–23
Cerebral perfusion pressure 67
attempts to measure 70–71
cerebrovascular resistance and 68–69
maintaining of 81–82
management of 187
monitoring 96
optimal *68*, 74–78
pressure-reactivity index and 74, *77*
therapeutic targets for *78*
in traumatic brain injury
management protocols of 255–257,
256
monitoring of 254
Cerebral saccular aneurysm, pathogenesis
of *199–200*
Cerebral swelling, in acute ischemic stroke
154
Cerebral venous sinus thrombosis 658
in critically ill pregnant patients
667–668
Cerebral venous thrombosis 725
Cerebritis 374
Cerebrospinal diagnostic testing, for acute
encephalitis 340
Cerebrospinal fluid
analysis, in GBS 234
culture, for diagnosing bacterial
meningitis 353
spinal cord emergencies and *323*, *328*,
329
Cerebrospinal fluid pleocytosis, herpes
encephalitis and 343
Cerebrovascular disease 725–727, *726*

Cerebrovascular injury, catheter angiography for 290
Cerebrovascular pressure reactivity, autoregulation and 74
Cervical collars, for spinal cord injuries 277
Cervical spine
  intubation in unstable 23
  reduction, closed, for spinal cord injuries 283–286, *284*
Cessation of statin use, in intracerebral hemorrhage 179
CHANCE trial 167
Chemotherapeutic drugs, in PRES *468*, 470
Chemotherapy 734
Chest physiotherapy 42
Cheyne-Stokes respiration 716–717
China Antihypertensive Trial in Acute Ischemic Stroke 165
Chinese Acute Stroke Trial 166–167
Cholesterol, as risk factor of cerebral aneurysms and SAH *201–203*
Cholinergic drugs, in delirium 451–452
Cholinesterase inhibitors 495
Choriocarcinoma 662, 715–716
Chronic hydrocephalus, management of 210
Chronic hyponatremia 709
Chvostek's sign 711
"Circulatory determination of death," 409–410
  donation after 411–420
Cisatracurium 600
Clarithromycin, drug-induced encephalopathy and 767
Clinical institute withdrawal of alcohol scale *492*, 493
Clinical syndrome, and neurologic complications of transplantation 548
Clonidine, for cerebral hyperperfusion syndrome 584–585
Clopidogrel
  in acute ischemic stroke 167
  in intracerebral hemorrhage 179
Closed cervical spine reduction, for spinal cord injuries 283–286, *284*
Coagulation 698
  hyponatremia and *697*
Coagulopathies
  after traumatic brain injury 745, 747
  associated with systemic disease 755–756, *756*
  clinical presentation of 748
  clinical trials and guidelines 756–757
  complex clinical decisions in 757–758
  epidemiology of 743–746
  hospital course and management of 751–756, *751*, *755*
  imaging 748, *751*
  in intracerebral hemorrhage 179, 183
  laboratory and point-of-care testing in 748–750

Coagulopathies *(Continued)*
  leukemic 747
  liver failure of 745, 747–748
  malignancy of 745
  management of 183–184, *183*, *185*
  neurodiagnostics 748–750
  neurologic complications of 743–764
  neuropathology of 746–748
  outcome prediction in 758
  in traumatic brain injury 255
Coagulopathy-related neurovascular phenomena 637–638
Cocaine 496–497
Cognition 380–381
Cognitive impairment
  clinical presentation of 580
  diagnosis of 581
  epidemiology of 576
  hospital course and management of 584
  neuropathology of 578
  outcome prediction 586
Cognitive interventions, for delirium 583–584
"Cold diuresis," 626
Cold ischemia time 414
Cold preservation 415
Cold storage, machine perfusion *versus* 415–416
Colistin 373
Color density spectral array 114–115, *114*
Coma 380, 596
  see also Comatose patient
  after cardiopulmonary resuscitation, prognosis of 773–776
  causes of 118, *119*
  classification of *119*
  neuropathophysiology in 118–120
  nonconvulsive status epilepticus and 135–136
  outcome prediction in 128–129
  prevalence of 118
  of uncertain etiology, questions to ask in 118, *118*
Comatose cardiac arrest survivors
  clinical case in, practical application of prognostic knowledge in *776*
  in-hospital, management of 775
  practical guidance for prognostication in (2015 guidelines) *774*
Comatose patient
  daily concerns in care in *127*
  laboratory tests for 124–125
  management of 117–130
  with NCSE 143
  neuroimaging in 124–125
  neurologic examination of 120–124
  neurologic findings of 123–124, *124*
  neurorehabilitation for 127–128
Combined pain, agitation, and delirium protocols 458–459
Communication, in NCCU
  general principles of 398–400
  strategies of 400–401
Community-acquired bacterial meningitis
  antimicrobial therapy for *358*

Community-acquired bacterial meningitis *(Continued)*
  clinical presentation of 351–352
  clinical trials and guidelines for 359
  complex clinical decisions in 360
  epidemiology of 349–350
  hospital course and management of 355–359
  neurodiagnostics of 352–354
  neuropathology of 351
  outcome prediction in 360–361, *361*
  recommendations for initial antibiotic therapy for *357*
Complement cascade activation 230–231
Complete background suppression, of seizures 520
Compound muscle action potential, as marker of nerve or muscle dysfunction 535
Computed tomography
  of aneurysmal subarachnoid hemorrhage 206, *207*
  brain 679
  for brain abscess 354–355
  of spinal cord injuries 281–282
  of traumatic brain injury *244*, 246, *246–247*
Computed tomography angiography
  in brain death *425–427*
  for spinal cord injuries 290
  for traumatic brain injury 246
Computed tomography myelogram, for extramedullary lesions 324–325
Conflict, and conflict resolutions 403
Confusion Assessment Method for the Intensive Care Unit 453, *454*, 551, 772
Conjunctivitis, comatose patient and 126
Consciousness
  abnormal level of 661
  focal nonconvulsive status epilepticus and
    with 136
    without 136
  NCSE with, anesthetics in 143
  recovery of 380
Consults, in ICU 444
  neurologic
    complexity of *446*
    essentials of *445*
  surgical and trauma 445
Continue or Stop PostStroke Antihypertensives Collaborative Study trial 165
Continuous electroencephalography monitoring
  clarifying nature of movements 109
  controversies and future endeavors 115
  detection and management of seizures in 107–109, *108*, *109*
  grading severity of encephalopathy in 109–110, *110–111*, *111*
  indications for 107
  in intensive care unit 107–116

Continuous electroencephalography (Continued)
    monitoring depth of sedation in 109
    prognostication in 110–113, 112
    standardization in 115
    technical and logistic considerations in 113–115
    for seizures 514
    for traumatic brain injury 250
Continuous positive airway pressure 26, 422
Continuous-infusion AEDs, for seizures 518
Contraction band necrosis 52–53
    baboon animal model of 53
    presence of 53
Contrast extravasation, in intracerebral hemorrhage 181, *181*
"Controlled" DCD 411
    identification of candidates of 413–414
    protocols and practice guidelines in 416–417, *418*
Conventional angiography
    in intracerebral hemorrhage 181–182
    for spinal cord injuries 290
Conventional cerebral angiogram, in brain death 425–427
Convulsive status epilepticus 134–135
Cooling 620, 626–627
Corneal erosion, comatose patient and 126
Coronary artery bypass grafting 573–574
Coronary artery disease, factors of 54
Coronary artery stenting 574–575
Cortical spreading depolarization/ischemia 205
Cortical spreading depression, traumatic brain injury and 243
Corticosteroid Randomisation After Significant Head Injury 386
Corticosteroids 717–718
    herpes encephalitis and 341
    for immune-mediated encephalitis 345
    intensive care unit-acquired weakness and 534
    for metastatic epidural spinal cord compression 330–331
Cough 17
    ICP and 82–83
Cranial computed tomographic scanning 382
Cranial irradiation 724
Cranial nerve deficits
    clinical presentation of 581
    epidemiology of 576
    hospital course and management of 585
    neuropathology of 578–579
    outcome prediction 587
Cranial nerves
    assessment of 597
    neurologic evaluation of, in comatose patient 121–122
Craniectomy
    decompressive 301–302
        in acute brain injury 299–318
        combined with hypothermia 310–311

Craniectomy (Continued)
    complications of 311–313, *311*
    in elderly 309–310
    for hemispheric stroke 304–310
    for middle cerebral artery infarction, trials of 306–309, *307*
    pathophysiologic rationale and impact of 300–302
    perceived indications for *300*
    quality of life after 313
    for trauma 302–304
    for intracranial hypertension 85
Cranioplasty
    bone flap resorption following 312–313
    complication of, following decompressive craniectomy 312–313
Critical care medicine 3
    see also Critical care neurology; Intensive care; Neurocritical care
Critical care neurology. *see also* Critical care medicine; Intensive care; Neurocritical care
Critical care nurses, in early beginnings of neurocritical care 7
Critical ICP 68, 74–78
Critical illness
    delirium and 451
    ethical considerations in 777–778
    neurologic complications in
        identification 778
        prevention of 777–778
        prognosis of 765–784
        self-fulfilling prophecies in 778
    neurology of
        consult in 445–446
        field of 444, *444*
        scope of 441–448
Critical illness myopathy 532–533
Critical illness polyneuropathy 531–532
    axonal damage in 533
Critically ill patients, delirium in, prognosis of 771–772
Critically ill pregnant patients, neurologic complications in 657–674
    abnormal level of consciousness 661
    clinical presentation of 659–661
    clinical trials and guidelines 667–668
    complex clinical decisions 668–669
    epidemiology of 657–658
    generalized weakness 661
    hospital course and management 665–667
    increased intracranial pressure 661
    neurodiagnostics and imaging 664–665
    neuropathology of 658–659
    neurorehabilitation 670
    new focal neurologic signs of 660
    outcome prediction of 669
    pregnancy-related clinical syndromes 661–664
    seizures 659–660
    serious headaches 660, *660*
    treatment of specific conditions 666–667

Cryptococcus, after transplantation 561, 565–566
Cryptococcus neoformans 727–729
CSF cytology 723
CSF lactate 372
CT angiography
    acute ischemic stroke and 160
    in arterial dissection 635
    in intracerebral hemorrhage 181–182
CT perfusion, acute ischemic stroke and 160
CT scan, in comatose patient 125
CT without contrast, in intracerebral hemorrhage 180
Cushing, Harvey 7–8
Cushing's triad 716–717
Cyanide exposure 688
Cycloserine, drug-induced encephalopathy and 767–768
Cyclosporine, drug toxicity and 554–555
Cytokine-mediated inflammation 40
Cytokines, in traumatic brain injury *248*
Cytomegalovirus infection, after transplantation *561*
Cytotoxic drugs, in PRES 470
Cytotoxic edema
    in acute ischemic stroke 156
    traumatic brain injury and 242–243

D

Dabigatran 753
Daclizumab, drug toxicity and 554–555
Davson's equation 68
DCD-N score 413
D-dimer 727, 750
"Dead donor rule," 409–410
Death
    brain
        acceptance of, by families 430
        ancillary testing in 422, *425–427*
        clinical testing in 421–422
        concept and controversies 430–431
        definition of 420–421
        determination of 421–422, *423–424*
        epidemiology of 428
        ethical considerations in 430–431
        future trends in 431–432
        organ support in 428–429
        pathophysiology of 422–428
        systemic complications in *428*
    definition of 410
    determination 411–413
    early, in stroke 398
    organ recovery prior to 417–419
    in PRES 477
Decerebrate posturing, comatose patient and 123
Decompression illness 691–695
    clinical presentation of 692–693, *692*
    clinical trials and guidelines of 694
    complex clinical decisions of 694
    epidemiology of 691
    hospital course and management of 693–694
    neurodiagnostics and imaging of 693

Decompression illness *(Continued)*
  neuropathology of 692
  neurorehabilitation of 695
  outcome prediction of 694–695
Decompressive craniectomy 301–302
  in acute brain injury 299–318
  combined with hypothermia 310–311
  complications of 311–313, *311*
    on CSF flow and dynamics 312
    hemorrhagic 311–312
    infectious and inflammatory 312
  in elderly 309–310
  for hemispheric stroke 304–310
  for middle cerebral artery infarction, trials of 306–309, *307*
  pathophysiologic rationale and impact of 300–302
  perceived indications for *300*
  quality of life after 313
  for trauma 302–304
Decompressive Craniectomy in Diffuse Traumatic Brain Injury study 302–304
  critique of 303–304
Decompressive Craniectomy in Malignant Cerebral Artery Infarction 307–308
  decompressive hemicraniectomy and 169
  pooled analysis of 308–309
Decompressive hemicraniectomy, acute ischemic stroke and 169
Decompressive Surgery for the Treatment of Malignant Infarction of the Middle Cerebral Artery 308
  decompressive hemicraniectomy and 169
  II trial 309–310
  pooled analysis of 308–309
DEcompressive surgery Plus hypoTHermia for Space-Occupying Stroke trial 310–311
Decorticate posturing, comatose patient and 123
Deep-vein thrombosis 167
  prophylaxis
    in acute ischemic stroke 167
    for spinal cord injuries 290–291
Delayed cerebral ischemia 204–205, 211–213
  angiographic vasospasm 204–205
  cortical spreading depolarization/ischemia 205
  diagnosis of 211–212
  endovascular management of *216–221*
  hemodynamic management of *216–221*
  microcirculatory dysfunction 205
  microthromboembolism 205
  monitoring for *216–221*
  other prophylaxis for *216–221*
  pathophysiology of *204*
  prophylaxis and treatment of 212–213
Delayed graft function 414

Delirium
  after transplantation 551–552
  clinical presentation of 452, 580
  clinical trials and guidelines for 459–460, *459*
  complex clinical decisions in 460
  criteria for 450
  in critically ill patients 449–466
    prognosis of 771–772
  diagnosis of 581
  diagnostic workup *644*
  encephalopathy and 449
  epidemiology of 450, 576
  frequency of 450
  hospital course and management of 454–459, *455*, 583–584
    combined pain, agitation, and delirium protocols in 458–459
    nonpharmacologic prevention and treatment in 455–456
    pharmacologic prevention in 456–457
    prediction in 454
  hypoactive 444–445
  imaging in 452–453, 643
  long-term cognitive and functional impairment after 772
  mortality after 772
  neurodiagnostics in 452–453, 643
  neuropathology of 450–452, 578
  outcome prediction in 460–461, 586
  prognostic literature of 771
  risk factors for 450, *450–451*
  screening for 453, *454*
  (sub)types of, prognostic differences of 772
  in trauma patients 641, *642*
Delirium tremens 491
Dementia, in ICU delirium 450
Denver classification system 635, *635*
Depth of sedation, monitoring 109
Desmopressin 756
Deterioration, causes of subarachnoid hemorrhage *212*
Dexamethasone 717–718
  for cerebral lesions 83
Dexmedetomidine 493
  for delirium 457–458, 583–584
  in ICU 644
  for traumatic brain injury 252–253
Dextromethorphan abuse 489
Diagnostic and Statistical Manual of Mental Disorders, 5th edition criteria, for delirium 450
Diarrhea, in comatose patient 127
Diazepam, for seizures 515–517
Diffuse axonal injury, traumatic brain injury and 241
Diffuse low-grade gliomas 718
Diffusion tensor imaging 383, 595
Diffusion-weighted imaging, in PRES 473, *474*
Diffusion-weighted magnetic resonance imaging, for traumatic brain injury 247

Dipyridamole, in acute ischemic stroke 167
Direct brain tissue damage, in intracerebral hemorrhage 179–180
Direct muscle stimulation, and intensive care unit-acquired weakness 535–536
Direct thrombin inhibitors 753
Disability Rating Scale 128, 381
Disordered breathing 36–38
Disorders of consciousness 380
Disseminated intravascular coagulation
  epidemiology of 745
  management of 754
  neuropathology of 747
Dissociative agents 489–490
  clinical presentation of 490
  complex clinical decisions for 490
  epidemiology of 489–490
  laboratory testing in 490
  management of 490
  pharmacology of 490
Donation after brain death 411
  future trends in 431–432
  identification of candidates in 421
  pathophysiology of 422–428
  protocols and guidelines in 429–430, *431*
Donation after circulatory death 411–420
  category I, donor 411, *412*
  category II, donor 411, *412*
  category III, donor 411, *412*
  category IV, donor 411, *412*
  category V, donor 411, *412*
  "controlled," 411
    identification of candidates of 413–414
    protocols and practice guidelines in 416–417, *418*
  epidemiology in 415
  ethical considerations in 417–420
  future trends in 420
  organ support in 415–416
  pathophysiology in 414–415
  protocols and guidelines in 416–417
  uncontrolled 411
    continuation of resuscitation efforts *versus* 420
    pathophysiology of 414
    protocols and guidelines in 417, *419*
Donors, categories of 411
Do-not-resuscitate orders 398
Dopamine
  in brain-dead donors 429
  in delirium 451–452
Dopaminergic agents, for comatose patient 128
"Drip and ship" transfers 158
Drug metabolism *697*
Drug toxicity
  after transplantation 554–559
  differential diagnosis *554*
  management of 557–558
  neuromuscular syndromes 558–559
  risk factors for 557
  stroke syndrome 558, *559*

Drug-induced encephalopathy 766–768
Drugs, antifibrinolytic 209
Dysautonomia
　with BD 428
　rabies and 343–344
Dysautoregulation, in traumatic brain injury 242–244
Dysembryoplastic neuroepithelial tumors 718
Dysfunction, microcirculatory 205
Dysphagia 711
　treatment of 170–171

E

E4 allele, in traumatic brain injury 242
Early brain injury 204
Early mobilization, for delirium 583–584
EARLY trial 167
Eastern Association for the Surgery of Trauma 281
Eaton-Lambert syndrome 231
Ecarin chromogenic assay 750
Ecarin clotting time 750
Echocardiography 639
　acute ischemic stroke and 160
Eclampsia
　in critically ill pregnant patients 667
　definition of 662
　PRES and *468*, 470
　seizure and 659–660
　status epilepticus and 145
Edema 717
　after ischemia 620
　cerebral 602
　comatose patient and 119
　cytotoxic 594–595
Edoxaban 753
Elderly
　decompressive craniectomy in 309–310
　traumatic brain injury in 258–259
Electrocardiogram *698*
　changes *697*
Electrocorticographic detection, for traumatic brain injury 243, *244*, 250
Electrodes, for electroencephalography 114
Electroencephalography 92–95, 384, 645–646, 664, 719
　for acute encephalitis 340
　artifacts in 115
　in brain death *425–427*
　burst suppression, of seizures 520
　in cerebral electric activity 80
　continuous, for seizures 514
　electrodes for 114
　montages 113–114
　in neurophysiologic testing 605
　in PRES 471
　quantitative 109
　raw EEG *vs*. quantitative displays *113–114*, 114–115
　in status epilepticus 136–139, 142–143
　subhairline 113
　TH and 621

Electrolyte abnormalities, major
　adaptive response to sodium disturbances 707–708
　neurologic manifestations of 705–714
　of disturbances of other ions 710–711
　osmolality and 705–707
　pathophysiology response to sodium disturbances 707–708
　physiologic regulation of sodium and 705–707
Electrolyte disturbances *706*
Electrolyte levels, reduced, TH and 626
Electrolytes, hypothermia and *697*
Electromyography 650, 664
Electrophysiologic evaluation, for nonconvulsive seizures 559
Electrophysiologic testing, TH and 621
Electroporation 689
Embolic strokes 638–640
　clinical presentation in 639
　clinical trials and guidelines in 640
　complex clinical decisions in 640
　drug toxicity and 558
　epidemiology of 638
　hospital course and 640
　neurodiagnostics and imaging in 639–640
　neuropathology of 638–639
Emergency department, acute ischemic stroke and 156, *157*
Emergency medical services personnel, for spinal cord injuries 276, *276*
Emergency Neurological Life Support 10
Emergency room, seizure patients presenting to 513–514
Emergent control therapy, for seizures 515–517
Empiric antibiotic therapy 374–375
Encephalitis 337
　acute 337–348
　autoimmune, immunosuppression in 144–145
　endemic 338
　epidemic 338–339
　immune-mediated 344–345
　Japanese encephalitis virus in 343
　tick-borne 343
Encephalomyelitis 337–338
Encephalopathy 444
　acute 548
　delirium and 449
　diagnostic workup *644*
　drug-induced 766–768
　grading severity of 109–110, *110–111*, *111*
　hepatic, in acute liver failure, prognosis of 768–769
　hypertensive 659, 771
　ICU-acquired, prognosis of 766–768
　and neurologic complications of transplantation 548
　penicillin 768
　septic 675–683

Encephalopathy *(Continued)*
　clinical presentation and neurodiagnostics 678–679
　definition of 676
　epidemiology of 675–676
　hospital course and management 679–680
　neuropathology of 676–678
　signs and symptoms of *678*
Endemic encephalitis 338
Endocrine function *216–221*
Endocrine support, in organ donation 428–429
Endocrine system *697*
Endothelial cells, aneurysm and *199–200*
Endothelial dysfunction, in PRES 470–471
Endotoxins, for septic encephalopathy 676
Endotracheal intubation
　for airway protection 487
　for traumatic brain injury 252
Endovascular therapy, patient management after 164
Energy failure, in traumatic brain injury 242–244, *243*
Enterobacter 366–367
Enterovirus, intramedullary infectious myelopathies and *323*
Epidemic encephalitis 338–339
Epidemiology-Based Mortality Score in Status Epilepticus 146
Epidural hematoma, traumatic brain injury and 240–241
Epidural sensors 70
Epidural space 320
Epidural spinal cord compression 720–722, *721*
　metastatic
　　etiologies and clinical pearls of *322–323*
　　hospital course and management of 330–331
　　neuropathology of 320
Epidural spinal hematoma, neuropathology of 320
Epilepsia partialis continua 136, 508
Epilepsy, sudden cardiac death and 56
Epileptogenic areas 718
Epstein-Barr virus infection, after transplantation 561
ERC/European Society of Intensive Care Medicine guidelines 775
ESCAPE 164
Escherichia coli 366–367
Esmolol, for cerebral hyperperfusion syndrome 584–585
Ethanol, use of 491
Ethnicity, as risk factor of cerebral aneurysms and SAH *201–203*
Etomidate, in intracerebral hemorrhage 182
European Cooperative Acute Stroke Study II trial 161
EuroTherm trial 623

INDEX

Eurotransplant Organization 411
Evoked potentials, in brain death 425–427
Excitotoxicity, brain injury and 204
Exclusion criteria, in NINDS trial 161, *162*
Exercise rehabilitation programs, for intensive care unit-acquired weakness 539
Extended Glasgow Outcome scale 381
EXTEND-IA *164*
External-beam RT 722
Extracorporeal membrane oxygenation 573, 575, 776
  mortality rate 586
  periprocedural stroke rate 575
  stroke etiology 577
  for traumatic brain injury 260
Extramedullary lesions, neuroimaging for 324–325, *325*
Eye movements, in comatose patient 121, *123*
Eye response, in Full Outline of Unresponsiveness score *122*

F

Facial nerve, injury to 578–579, 581
Factor Xa 727
Factor X_A inhibition 750
Familial intracranial aneurysm study 200
Family history, as risk factor of cerebral aneurysms and SAH *201–203*
Fast Assessment of Stroke and Transient Ischemic Attack to Prevent Early Recurrence trial 167
Fat emboli 640
Fat embolism syndrome 638
  clinical manifestations of 640
  clinical triad of 639
  treatment of 640
Fatal cerebrogenic arrhythmias, SUDEP and 56
Fever *216–221*
  in acute ischemic stroke 156
  cerebral edema and 82
  in comatose patients 126
  management of 184–185
  sepsis and 677–678
Fiberoptic bronchoscopy 41
Fibrinogen 750
Fibrinolytic agents 753–754
Field triage, for spinal cord injuries 276
Filamentary keratopathy, in comatose patient 126
"Flapping tremor," 678
Flavivirus encephalitides 342–343
Fluid-attenuated inversion recovery 383, 717
Flumazenil 487
Fluorodeoxyglucose-positron emission tomography 383–384
Fluoroquinolone 368
Focal motor status 508, *509–510*
Focal signs 660
Foscarnet, for HHV6 341

Fosphenytoin 520–521, 667
  for seizures *516*
  for status epilepticus 140–141
FOUR score 382
Fracture-dislocation injury, closed cervical reduction of 283, *284*
Fresh frozen plasma 751
  in intracerebral hemorrhage 183
Frontotemporoparietal hemicraniectomy 301–302, *302*
Full Outline of UnResponsiveness score, (FOUR score) 121, *122*, 597, *598*
  for traumatic brain injury 245
Functional independence, recovery of 381
Functional Independence Measure 293
Functional neuroimaging 383–384

G

GABAergic signaling 718–719
Gabapentin, for superrefractory status epilepticus 523
Gadolinium, in PRES 473, *475*
Gadolinium-enhanced MRI 723
Gamma-aminobutyric acid, in delirium 451–452
Gamma-hydroxybutyrate toxicity 487
Ganciclovir, drug toxicity and 554–555
Gangliomas 718
Gardner-Wells tongs, for spinal cord injuries 283, *285*
Gastrointestinal system *697–698*
  care in comatose patient *127*
Gastrotomy, percutaneous 125–126, *125*
Generalized convulsive status epilepticus 507
  in adults 507–508
Genetics, as risk factor of cerebral aneurysms and SAH *201–203*
Gentamicin 373
Glasgow Coma Scale 121, 180, *206*, 380, 597, *598*
  for traumatic brain injury 245
Glial fibrillary acidic protein 385
  in traumatic brain injury *248*
Glioblastoma multiforme 715–716
Glioneuronal tumors 718
Glossopharyngeal nerve, injury to 581
Glucocorticoids 722
  cerebral lesions and 83
Glucose *216–221*
  in cerebral microdialysis 98
Glutamate 718–719
  in cerebral microdialysis 98
Glycemic management, in acute ischemic stroke 165–166
Glycerol, in cerebral microdialysis 98
Glycine 594–595
Gradient recall echo 717
Graft pancreatitis 415
Graft-*versus*-host disease 470
  after transplantation 562–563
Gram staining, for diagnosing bacterial meningitis 353

Guillain-Barré syndrome 230
  cerebrospinal fluid analysis in 234
  diagnosis of 231
  drug toxicity and 558–559
  immunopathogenesis of 230–231
  mortality in 235
  respiratory failure in 232–233
  specific treatment for 234
Gyriform signal enhancement, in PRES 473, *475*

H

Haemophilus influenzae 366–367, 727–729
Halo ring, and vest, for spinal cord injuries *285*
Haloperidol, for delirium 456, 458
HAMLET trial 308
  decompressive hemicraniectomy and 169
  pooled analysis of 308–309
Headache 686–687, 716
  migraine 660
  in PRES 471
  serious 660, *660*
HEADDFIRST 306–307
Hearing loss, bacterial meningitis and 361
Heart rate *697*
Heart rate variability 52, *52*
Heat stroke 695–699
  clinical presentation of 696–697
  clinical trials and guidelines of 699
  complex clinical decisions of 699
  epidemiology of 695, *695*
  hospital course and management of 698–699
  neurodiagnostics and imaging of 698
  neuropathology of 695–696, *696*
  neurorehabilitation of 699
  outcome prediction of 699
Hematologic effects, TH and 627
Hematologic system *698*
Hematoma, spinal epidural
  hospital course and management of 331
  neuropathology of 320
Hematoma expansion 746, *751*
  in intracerebral hemorrhage 179–180
Hematopoietic stem cell transplantation 545–547, 734
  morbidity after 548
  procedures for *546*
Hemicraniectomy, frontotemporoparietal 301–302, *302*
Hemicraniectomy After Middle Cerebral Artery infarction with Life-threatening Edema Trial 308
Hemicraniectomy and Durotomy Upon Deterioration From Infarction-Related Swelling Trial 306–307
Hemispheric index 100
Hemispheric stroke, decompressive craniectomy for 304–310

Hemoclot diluted thrombin inhibitor assay 750
Hemodynamic changes 658
Hemodynamic management, of spinal cord injuries, acute 287–288
Hemodynamic support, in organ donation 429
Hemodynamics *698*
Hemorrhage 664
   in comatose patient 119–120
   subarachnoid 35–36
Hemorrhagic CSF 371–372
Hemorrhagic stroke 725
   drug toxicity and 558
   outcome prediction 585–586
Hemorrhagic transformation 156
   infarction with 154–155
Hemostasis, cell-based model of 746, *749*
Hemostatic agents, in intracerebral hemorrhage 184
Heparin 636, 752
   associated with, epidural hematoma 331
Heparinoids 752
Hepatic encephalopathy 551–552
   in acute liver failure, prognosis of 768–769
HER-2 724
Herniation 374
Herniation syndromes 716–717
Herpes encephalitis 340–342
Herpes simplex, encephalitis and 340–341
Herpes zoster, encephalitis and 341
Herpesviruses, and CNS infections 564
Heterogeneous pattern, in intracerebral hemorrhage 181–182
Higher neuropsychologic processing, recovery of 380–381
High-frequency centroid 79
High-mobility group box protein 1 34–35
Histone deacetylase inhibitor 719–720
Horner's syndrome 579
Hospital and system of care *216–221*
Hospital-acquired bacterial meningitis
   clinical presentation of 352
   clinical trials and guidelines for 359
   epidemiology of 350
   hospital course and management of 358
   neurodiagnostics and neuroimaging of 354
   neuropathology of 351
   recommendations for initial antibiotic therapy for *357*
Hospital-acquired infections 365
Human herpes virus 6 341
   after transplantation *561*, 564
Hydration, for delirium 583–584
Hydrocephalus *216–221*, 717
   acute and chronic, management of *210*
   complications of decompressive craniectomy *311*
Hyperactive subtype, of delirium 452
Hyperammonemia, idiopathic, after transplantation 563
Hyperbaric oxygen therapy 688

Hyperdense right middle cerebral artery 158–159, *159*
Hyperemesis gravidarum 661
Hyperglycemia
   in acute ischemic stroke 155–156, 165
   cerebral ischemia and 82
   cerebral microdialysis and 99
   and intensive care unit-acquired weakness 533–534
   management of 185
Hypernatremia 708
   traumatic brain injury and 255
Hyperosmolar agents 718
Hyperosmolar therapy, for ICP 83–84
Hyperosmolar urea 705–706
Hyperoxia
   effects on brain physiology 17
   exacerbate primary injury 24–25
Hyperperfusion theory 469
Hyperpyrexia 677–678
Hypertension
   in intracerebral hemorrhage 178, *178*, 182
   management of 182–183
   in PRES 469–470, 475
   as risk factor of cerebral aneurysms and SAH *201–203*
Hypertensive encephalopathy 659, 771
   treatment for 475
Hyperthermia
   in acute ischemic stroke 155–156, 166
   complications of *698*
   definition of 695
Hypertonic saline 718
   for ICP 83
Hyperventilation 84
   central neurogenic 37
   effects on brain physiology 23–24
   hypocarbia and 18
   ICP and 67–68
   for intracranial pressure 24
   pathophysiologic central 16
   physiologic central 17
   respiratory alkalosis and 16
   subarachnoid hemorrhage and 35–36
Hypoactive delirium 444–445, 452
Hypocalcemia, symptoms of 710–711
Hypocapnia, ICP and 82
Hypocarbia
   acidemic 24
   alkalemic 24
   hyperventilation and 18, 24
Hypoglossal nerve, injury to 578–579, 581
Hypoglycemia, in acute ischemic stroke 165
Hypomagnesemia 711
Hyponatremia 707–708
   traumatic brain injury and 255
   volume status and *216–221*
Hypoperfusion-related cerebral ischemia 640–641
Hypophosphatemia 711
Hypotension, traumatic brain injury and 245
Hypothalamus 50, *50*

Hypothermia 662
   complications of *697*
   decompressive craniectomy combined with 310–311
   ICP and 84
   induced 621
   induction phase of 622
   intraischemic 620
   maintenance phase of 622
   rewarming phase of 622
   for superrefractory status epilepticus 522–523
   therapeutic 619–632
   for traumatic brain injury 254
Hypoventilation
   central alveolar 37
   effects on brain physiology 23–24
   from hypercarbia 17–18
Hypovolemia *697*
Hypoxemia, ictal 38
Hypoxia
   cellular, in brain 594–595
   effects on brain physiology 17
   exacerbate primary injury 24–25
   injured brain and 82
   with neurogenic pulmonary edema 51
   traumatic brain injury and 245
Hypoxic-ischemic encephalopathy 549, *606–607*
Hypoxic-ischemic injury, coma and 119

I

Ictal hypoxemia 38
ICU-acquired uremic encephalopathy, prognosis of 769
ICU-acquired weakness
   functional outcomes of 770
   prediction of 770
   prognosis of 769–771
   prognostication and, practical implications of 770–771
Idarucizumab 753
Idiopathic hemorrhage *196*
Idiopathic hyperammonemia, after transplantation 563
Idiopathic thrombocytopenic purpura 745
Illness trajectories 403
Immune function, impaired, TH and 626–627
Immune reconstitution inflammatory syndrome 566
Immune-mediated encephalitis 344–345
Immunochromatography antigen testing, for community-acquired bacterial meningitis 354
Immunomodulators
   in PRES *468*
   for superrefractory status epilepticus 522
Immunosuppressive drugs, in PRES *468*, 470
Incidence, of VAP 366
Inclusion criteria, in NINDS trial 161
Increased intracranial pressure 214, 661
   after ischemia 620

Induced hypertension 663–664
Induced hypothermia 621
Induction agents, drug toxicity and 554–555
Induction phase, of hypothermia 622
Infarction
 brain, hypotension and 22–23
 with hemorrhagic transformation, acute ischemic stroke and 154–155
Infections
 in comatose patient 126
 hyponatremia and *697*
 PRES and *468*, 470
 SAH and *196*
 in traumatic brain injury 33–34
 urinary tract, in comatose patient 127
Infectious Disease Society of America 367
Infectious myelopathies, management of 332
Inflammation
 in intracerebral hemorrhage 180
Inflammatory demyelinating myelitis, hospital course and management of 332
Inflammatory mediators, for septic encephalopathy 676–677
Inflammatory response, of ischemic brain tissue 156
Infusion, of cold fluids, in cooling process 621
In-hospital cardiac arrest 593–594
Insula *50*, 51
 nonlacunar stroke syndromes and 52
Insulin resistance, TH and 626
Intensive Blood Pressure Reduction in Acute Cerebral Hemorrhage 2 trial 182
Intensive care 3
 see also Critical care medicine
 early beginnings of 4–7
 monitoring of the patient 10
 statistical models used in 10
Intensive Care Delirium Screening Checklist 453, *454*, 551, 772
Intensive care unit 443
 acute ischemic stroke and 156–158, *158*
 closed 11
 consults in 444
  complexity of, neurologic *446*
  essentials of, neurology *445*
  surgical and trauma 445
 delirium in 450
  hospital course and management of *455*
  precipitating factors in pathogenesis of *451*
  predisposing factors in pathogenesis of *451*
 neurologic conditions in *446*
 neurologic models 11
 open 11
 posterior reversible encephalopathy syndrome 467–484

Intensive care unit *(Continued)*
 precursor of 6
 weakness in 445
Intensive care unit-acquired weakness 531–544
 clinical presentation of 534–535
 clinical trials and guidelines 537–539, *538*
 complex clinical decisions 539
 differential diagnosis *539*
 epidemiology of 531–532
 hospital course and management of 536–537
 models for 533
 neurodiagnostics 535–536
 neuropathology 532–533, *532*
 neurorehabilitation for 539
 outcome prediction 539
 risk factors for 533–534
 terms for 532
Intensive physical therapy, for cognitive impairment 584
Interleukin-2 724–725
Internal mammary artery dissection 579
International Encephalitis Consortium, and diagnosis of encephalitis 340
International Mission for Prognosis and Analysis of Clinical Trials 386
International normalized ratio 748–749
International Stroke Trial 164–167
Internuclear ophthalmoplegia, and calcineurin inhibitors 556
Intervention, triggers for *216–221*
Intoxication, coma due to 120
Intra-arrest therapeutic hypothermia 599
Intra-arterial thrombectomy, acute ischemic stroke and 163–164, *164*
Intra-arterial thrombolysis
 acute ischemic stroke and 163
 for ischemic stroke 583
Intracerebral hemorrhage 49, 197, 725
 CBN and 53
 clinical trials and guidelines in 624
 common neurocardiac sequelae of 54
 epidemiology of 575–576
 hospital course and management of 583
 management of 177–194, *210*
  clinical presentation of 180
  clinical trials or large studies of *185*
  complex clinical decisions in 186–188
  definition of 177
  epidemiology of 177–178
  guidelines to 182–186
  hospital course and management of 182–186
  introduction of 177
  neurodiagnostics and imaging of 180–182, *181*
  neuropathology of 178–180
  neurorehabilitation in 189
  outcome prediction of 188–189, *188*
 neuropathology of 578
 patients with 51, 57

Intracranial hemorrhage
 antithrombotic-associated 744–745, 751–754
 fibrinolytic-associated 750
 in PRES *476*
 thromboembolic complications of 757
 traumatic brain injury and 240–241
Intracranial hypertension
 cerebral autoregulation and 68–69, *68*
 essential principles of 67–68
 neuropathology of 67–69
 optimal CCP and *68*
 pathophysiology of 67–69
 prevention of 81–83, *81–82*
 semiquantitative relationship and 67–68
 traumatic brain injury and 245
Intracranial pressure 67, 204
 airway pressure and 25
 amplitude of pulsatile component of 69, *69*
 clinical presentation and neurodiagnostics of 70–80
  attempts to measure 70–71
  autoregulation and 74
  cerebrovascular pressure reactivity and 74
  critical *68*, 74–78
  ICP analysis, methods of 79–80
  measurement, methods of 70
  pressure-volume compensatory reserve and *69*, 78–79, *79*
  raised ICP, consequences of 80
  typical waves and trends in 71–74, *72*
 dynamic components of 68
 hospital course and management of 81–86
  acute exacerbations of, treatment of 85
  continuous measurement of 85–86
  hyperosmolar treatment for 83–84
  raised ICP, treatment of 81–86, *81*, *86*
  therapies for 84
  treatment threshold of 81
 hyperventilation for 24
 increased 661, 715–718
  clinical presentation 716–717
  epidemiology in 715–716, *716*
  hospital course and 717–718
  management and 717–718
  neuroimaging 717, *717*
  neuropathology and 716
  pathophysiology and 716
 intubation for elevated 22
 management of 187
 monitoring of 67–90
 peaks of 71–72, *73*
 plateau waves of 72–73, *74*
 slow vasogenic waves of 72, *73*
 power of 80
 therapeutic targets for *78*
 in traumatic brain injury 34
  management protocols of 255–257, *256*
  monitoring of 254
 treatment of 67–90

Intracranial pressure *(Continued)*
   value of 85–86
   variations of 70, *70*
   waveforms
      arterial pressure and 71
      pulse, morphologic analysis of 80
Intracranial pressure monitoring 95–96
Intramedullary lesions, neuroimaging for 325–328, *326–329*
Intraoperative Hypothermia for Aneurysm Surgery Trial 624
Intraparenchymal cerebral oxygen monitoring 97
Intraparenchymal hemorrhage
   antithrombotic-associated 746
   in PRES 473
   recurrence of 758
Intraparenchymal microtransducers 70
Intrathecal antibiotics 373
Intravenous immunoglobulin 661
   for West Nile virus 343
Intravenous thrombolysis trials, acute ischemic stroke and 161–162, *162*
Intraventricular hemorrhage 197
   management of 187–188, *188*
Intubation
   alternatives to 19
   contraindications to 19
   in elevated intracranial pressure 22
   in impaired cerebral perfusion 22–23
   peri-, reducing risk 20
   preparation of 18–20
   in unstable cervical spine 23
Iodinated contrast dye, in acute ischemic stroke 160
Iron lung *4*
Ischemia 372, 595
   brain, hypotension and 22–23
   brain injury and 96
   cerebral delayed 204–205, 211–213
      angiographic vasospasm 204–205
      cortical spreading depolarization/ischemia 205
      microthromboembolism 205
      pathophysiology of *204*
      prophylaxis and treatment of 212–213
   in delirium 452
   traumatic brain injury and 241
Ischemic penumbra, acute ischemic stroke and 155–156
Ischemic stroke 49
   CBN and 53
   common neurocardiac sequelae of 54
   consequences of 55–56
   in critically ill pregnant patients 666
   epidemiology of 575
   hospital course and management of 582–585
   neuropathology of 577–578
   outcome prediction 585–586
   patients with 51
      cardiac ischemia in 53–54
   TH and 624–625
Isocitrate dehydrogenase 1 718–719

Isocitrate dehydrogenase 2 718–719
Isoflurane, for superrefractory status epilepticus 522
Isotonic solutions, in intracerebral hemorrhage 182
IT rituximab 724–725
IV tPA treatment (tissue plasminogen activator), in acute ischemic stroke 156
   after, patient management 162–163
   prior to, patient management 162
IV vitamin K, in intracerebral hemorrhage 183

J
Japanese encephalitis virus 338–339, 343
JC virus, and CNS infections 564
Jugular venous bulb oximetry 96–97
Jugular venous bulb oxygen saturation 80
Jugular venous saturation 96

K
Kainic acid-induced NCSE 134
"Keraunoparalysis," 689
Ketamine 22, 489–490, 522
   for refractory status epilepticus *519*
Kidney dysfunction, TH and 626
Kidneys
   pathophysiology of 414
   PCAS and 596
Klebsiella pneumoniae 366–367
Kölliker fuse nuclei 36–37

L
Labetalol, for cerebral hyperperfusion syndrome 584–585
Lacosamide 521–522, 719
   for neurologic complications of transplantation 553–554
   for seizures *516*
Lactate/pyruvate ratio 98
Lamotrigine 719
Large cerebral hemisphere, acute ischemic stroke and 154
Large hematoma volume, in intracerebral hemorrhage 181–182
Laryngeal nerve, injury to 579
Laryngismus stridulus 710–711
Laser Doppler flowmetry 100
Lassen's autoregulation curve 68–69, *68*
Latex agglutination, for community-acquired bacterial meningitis 354
LEMON mnemonic 18
Length of stay, in ICU, mortality and 769–770
Leptomeningeal metastases 723–725, *724*, *728*
Leukemic coagulopathy 747
Leukemic promyelocytes 747
Leukoaraiosis 746
Levetiracetam 521, 719
   for neurologic complications of transplantation 553–554
   for seizures *516*, 517

Lidocaine
   for cognitive impairment 584
   for superrefractory status epilepticus 522
Life-sustaining interventions, de-escalation of
   in neurocritical care 397–408
   neurologic disorders requiring 397–398
   prognostic models and factors associated with 401–403
Light microscopy, CBN and 53
Light transmission aggregometry 749
Lightning 689–691
   clinical presentation of 690, *690*
   clinical trials and guidelines of 691
   complex clinical decisions of 691
   epidemiology of 689
   hospital course and management of 691
   neurodiagnostics and imaging of 690
   neuropathology of 689
   neurorehabilitation of 691
   outcome prediction of 691
Limbic encephalitis 729
Linezolid, drug-induced encephalopathy and 768
Lipohyalinosis, in intracerebral hemorrhage 178
Lipopolysaccharides 676
Lipoprotein 727
Listeria monocytogenes 350, 372, 727–729
Listeria monocytogenes meningitis 565
Listeriosis, after transplantation *561*, 565
Liver *697–698*
   pathophysiology of 415
   PCAS and 596
Liver failure
   acute, coagulopathies and 755–756
   of coagulopathies 745, 747–748
Liver transplantation
   neurologic complications of, prognosis of 776–777
   PRES in 468, 470
"Locked-in" state 169–170
Long term cognitive decline 578
Long-latency evoked potentials 384–385
Lorazepam
   for intracranial hypertension 82–83
   for seizures 515–517, *516*
   IV, administration of 517
   for status epilepticus 140
Low tidal volumes, lung-protective ventilation utilizing 41
Low-molecular-weight heparin 727, 752
   intracranial hemorrhage and 744
   for spinal cord injuries 291
Lucid interval, in traumatic brain injury 240–241
Ludwig Guttmann 8
Lumbar puncture
   of aneurysmal subarachnoid hemorrhage 207
   for bacterial meningitis 352–353
   for seizure patients 513–514
Lumbar spine, MRI of 328

Lumbosacral plexus injury 648, *649*
  clinical deficits from 648
  from hip dislocation 648
  management of 650
Lund concept, for traumatic brain injury 41
Lung cancer 720
Lung dysfunction
  cytokine-mediated inflammation in 40
  goal of management of 41
  pathogenesis of 34–35
  subarachnoid hemorrhage in 35
Lung-protective ventilation 25
  utilizing low tidal volumes 41
Lungs
  care in comatose patient *127*
  pathophysiology of 414

M

Maastricht classification, of nonheart-beating donors 411, *412*
Machine perfusion, cold storage *versus* 415–416
Magnesium
  infusion, in shivering 625–626
  neurologic manifestations of disturbances 711
  for superrefractory status epilepticus 523
Magnesium sulfate 666
  in PRES 475
Magnetic resonance angiography, in brain death *425–427*
Magnetic resonance imaging
  for acute encephalitis 340
  of aneurysmal subarachnoid hemorrhage 206–207
  in arterial dissection 635–636
  brain 679
    acute ischemic stroke and 160
  for brain abscess *355*
  in comatose patient 125
  of head 328
  of lumbar spine 328
  in PRES 473
  for seizures 514
  for spinal cord emergencies 324–325, *325–326*
  of spinal cord injuries 282
  in status epilepticus 140
  for subdural empyema edema *356*
  for traumatic brain injury 246–247, *246–247*
  ultrasound and noncontrasted 665
  without contrast, in intracerebral hemorrhage 180
Magnocellular neurons 706–707
Maintenance phase, of hypothermia 622
Malignant MCA stroke 169
Mannitol 708, 718
  for ICP 83
Mask ventilation, preparation of 18–19
Maternal brain death 669
Maximal inspiratory pressure 233

Mechanical ventilation 41, 601
  in acute brain injury, airway management and 15–32
    problems after 23
  in acute ischemic stroke 154
  discussion of, with families 403
  for traumatic brain injury 252
Medical Research Council scale 535
Medication, care in comatose patient *127*
Melanoma 715–716
Memantine, for cognitive impairment 584
Meningiomas 715–716
Meningitis 337–338
  bacterial, coma and 120
  complications of decompressive craniectomy *311*
Meropenem 373
Mesial temporal-lobe epilepsy 52
Metabolic acidosis, in comatose patient 124–125
Metabolic disorders 579
Metabolic neuroimaging 383–384
Metabolism
  of CSF leukocytes 372
  disorders of, PRES and *468*
Metastatic epidural spinal cord compression
  etiologies and clinical pearls of *322–323*
  hospital course and management of 330–331
  neuropathology of 320
Methicillin-resistant *S. aureus* 366–367
Methylamine 708
Methylprednisone, for superrefractory status epilepticus 522
Metronidazole, drug-induced encephalopathy and 767
Microcirculation, for septic encephalopathy 677
Microcirculatory dysfunction 205
Microglial cells 677
Microtubule-associated protein 385
Midazolam
  infusion, seizures during *108*
  for intracranial hypertension 82–83
  intramuscular, for seizures 517
  for refractory status epilepticus 518, *519*
  status epilepticus and 140, 142
Middle cerebral artery 70–71
Middle cerebral artery infarction, randomized controlled trials of decompressive craniectomy for 306–309, *307*
Migraine 660
Minimally conscious state 380
Minimally Invasive Surgery and rt-PA for Intracerebral Hemorrhage Evacuation II trial 187
Mitochondrial DNA haplotype K, in traumatic brain injury 241–242
Mitochondrial dysfunction 677
  in traumatic brain injury 242–243
MOANS mnemonic 18
Mobilization, for intensive care unit-acquired weakness 536

Modified Rankin Scale *158*
  in intracerebral hemorrhage 182
Modulators, during seizures 510
Monocyte chemoattractant protein 1 *199–200*
Monro-Kellie doctrine 67, 299
Monro-Kellie hypothesis 716
Montages, electroencephalography 113–114
Mortality
  after delirium 772
  ICU-LOS and 769–770
Motility *697*
Motor response, in Full Outline of Unresponsiveness score *122*
Motor system, testing of, in comatose patient 122
Motor unit action potentials, and myopathy 536
Motor vehicle collisions 635
Movements, nature of 109
MR angiography, acute ischemic stroke and 160
MR CLEAN 163–164, *164*
MR perfusion, acute ischemic stroke and 160
Multidetector computed tomography, for traumatic brain injury 246
Multifactorial metabolic encephalopathy, in critical illness 444
Multifocal myoclonus 552–553, 678
Multimodal brain monitoring, ICP with 80
Multimodal neurologic monitoring 91–106, *93–94*
  bioinformatics 100–101
  brain metabolism and cerebral microdialysis 98–100
  brain tissue oxygenation 96–98
  cerebral blood flow 100
  electroencephalography for 92–95
  intracranial pressure monitoring 95–96
  monitoring modalities recommendations for 92
  transcranial Doppler ultrasonography 100
Multimodal prognostication, after cardiac arrest 606–607, *608–609*
Multiple sclerosis 322–323, 327
Multiple trauma, in traumatic brain injury 260
Multiple-organ failure, and intensive care unit-acquired weakness 533–534
Muscle immobilization, and intensive care unit-acquired weakness 533
Muscle specific tyrosine kinase antibodies 233
Muscle ultrasound, and intensive care unit-acquired weakness 536
Muscle weakness 532, *532*
Muscular system *698*
Mutism, akinetic, after transplantation 552
Myasthenia gravis 230
  respiratory failure in 233
  specific treatment for 234
Myasthenic crisis 231

Mycobacterium tuberculosis 727–729
  intramedullary infectious myelopathies and 323
Myelopathies, autoimmune, demyelinating, and other inflammatory 321
Myocardial CBN 52–53
Myocardial dysfunction, post cardiac arrest 596
Myoclonic status epilepticus 135, 523–524, *523*, 552–553
Myoclonus, multifocal 552–553
Myocytolysis 52–53
  intracranial pressure associated with 53
Myofibrillary degeneration 52–53

N

N20 signals 384
Naloxone 489
National Acute Brain Injury Study: Hypothermia II 623
National Comprehensive Cancer Network guideline 723
National Emergency X-Radiography Utilization criteria 281
National Healthcare Safety Network 366
National Institutes of Health Stroke Scale *157*
Near-infrared spectroscopy 97–98
  for traumatic brain injury 250
Neck trauma, penetrating, spinal cord injuries and 277
Necrosis, intracranial pressure associated with 53
Needle biopsy, and intensive care unit-acquired weakness 536
Needle cricothyroidotomy, for difficult airway 21–22
Neisseria meningitidis, in bacterial meningitis 350
Neoplasms, SAH and *196*
Neostigmine 549–550
Nerve conduction studies 664
  for intensive care unit-acquired weakness 535
  for neurologic complications of transplantation 560
Nerve/plexus injuries 648–650
  clinical presentation in 648, *649*
  clinical trials and guidelines in 650
  complex clinical decisions in 650
  epidemiology of 648
  hospital course of 650
  neurodiagnostics and imaging in 648–650
  neuropathology of 648
  neurorehabilitation of 650
Neural autoantibodies, for spinal cord emergencies 330
Neurocardiology 49–66
  abstract of 49
  anatomy and physiology of 49–53, *50*
  introduction of 49
  neurocritically ill patients, cardiac disease in 53–61

Neurocardiology (Continued)
  cardiac dysfunction and 57–58
  subarachnoid hemorrhage and 57–58
  tako-tsubo cardiomyopathy and 59–61, *59–60*
Neurocatastrophes
  definition of 379
  determinants of prognosis in 379–396
  etiologic classification of 380
  multivariable models 386–387
  neurobiology 381
  phenotypes of recovery 380–381
  prognostic variables 381–386
Neurocritical care. *see also* Critical care medicine; Critical care neurology; Intensive care
  history of 1–14
    in 21st century 12
    early beginnings of 4–7
    interactions with other specialties *11*
    major landmarks in development of *9*
    monitoring of the patient 10
    statistical models used in 10
Neurocritical Care Society 9
  recommendations for multimodal monitoring 91–92, *92*
Neurocritical care unit
  brain death in 404
  communication in 398–400
    general principles of 398–400
    strategies of 400–401
  conflict, and conflict resolutions in 403
  illness trajectories and 403
  introduction to 397
  life-sustaining interventions in, de-escalation of 397–408
  neurologic disorders requiring 397–398
  prognostic models and factors associated with 401–403
  mechanical ventilation in, discussion of, with families 403
  organ donation in 404
  religious, and spiritual support, of family choices 404
  surrogate decision makers in 398
Neurocritically ill patients, cardiac disease in 53–61
Neurodiagnostics, of intensive care unit-acquired weakness 535–536
Neurofilament proteins, in traumatic brain injury *248*
Neurogenic pulmonary edema 38–39, *39*
  hypoxia with 51
  management of 41
  subarachnoid hemorrhage and 35
Neurogenic shock 324
Neuroimaging, in cardiac arrest 598–599, 606, *606–607*
Neurointensive care programs 230
Neurointensive care unit
  in acute ischemic stroke 154
  organ donation and 411

Neurointensive care unit-related infections
  bacteremia 369–370
  conclusion 375
  introduction in 365
  management of 365–378
  pneumonia 365–369, *366*
  postneurosurgical wound 373–375
  urinary tract 370–371
  ventriculitis 371–373
Neurointensivists, practice of 12
Neuroleptic malignant syndrome 552
Neurologic complications, in critically ill pregnant patients 657–674
  abnormal level of consciousness 661
  clinical presentation of 659–661
  clinical trials and guidelines 667–668
  complex clinical decisions 668–669
  epidemiology of 657–658
  generalized weakness 661
  hospital course and management 665–667
  increased intracranial pressure 661
  neurodiagnostics and imaging 664–665
  neuropathology of 658–659
  neurorehabilitation 670
  new focal neurologic signs of 660
  outcome prediction of 669
  pregnancy-related clinical syndromes 661–664
  seizures 659–660
  serious headaches 660, *660*
  treatment of specific conditions 666–667
Neurologic consults, in ICU
  complexity of 445–446, *446*
  essentials of *445*
Neurologic disorders, requiring de-escalation of life-sustaining interventions 397–398
Neurologic presentation 382
Neurologic symptoms, causes of sudden onset of *659*
Neurological system *697–698*
Neuromuscular blockade 600–601
Neuromuscular disorders, acute
  clinical features of 231
  diagnostic tests of 233–234
  epidemiology of 230
  hospital course and further management 234–235
  management of 229–238
  neuromuscular respiratory failure 231–233, *233*
  neurorehabilitation 235
  pathophysiology of 230–231
Neuromuscular respiratory failure 231–233, *233*
  acute, causes of *230*
Neuromuscular syndromes, and drug toxicity 558–559
Neuromyelitis optica spectrum disorder 321, *322–323, 327*
Neuronal cell death 428
Neuronal nitric oxide synthase 594–595

Neuron-specific enolase 385, 597–598
  in traumatic brain injury 248
Neuro-oncologic emergencies
  direct effects of cancer 715–725
  iatrogenic neurologic emergencies 733–734
  indirect complications of 725–733
  introduction of 715
  management of 715–742, 735–736
Neuropathology
  of aneurysmal subarachnoid hemorrhage 196–205
  of intensive care unit-acquired weakness 532–533, 532
Neurophysiologic testing 384–385
  in cardiac arrest 598, 605
Neuropsychologic evaluation, long-term care, rehabilitation 216–221
Neuropulmonology 33–48
  introduction to 33
  neurocritical disorders associated with pulmonary disease 33–40
Neurorehabilitation 235
  of aneurysmal subarachnoid hemorrhage 222
  in cardiac arrest 607–608
  for comatose patient 127–128
  for intensive care unit-acquired weakness 539
  neurologic complications, in critically ill pregnant patients 670
  for neurologic complications of cardiac and vascular surgery 587
  for PRES 479
  for traumatic brain injury 261
Neurotransmitters
  after ischemia 620
  during seizures 510
New-onset RSE 132
  autoimmune encephalitis and 144–145
Nicardipine, for cerebral hyperperfusion syndrome 584–585
Nimodipine 216–221
NINDS trial (National Institute of Neurologic Disorders and Stroke) 161
Nitric oxide 716
Nitric oxide synthase 620
Nitroprusside, for cerebral hyperperfusion syndrome 584–585
N-methyl-D-aspartate receptors 594–595
Nocardia asteroides 727–729
Nocardiosis, after transplantation 561, 565
Nomogram, for time and death following WLST 413
Nonaneurysmal perimesencephalic SAH 196
Nonbacterial thrombotic endocarditis 725
Noncontrast head CT 717
  acute ischemic stroke and 158–159, 159
Nonconvulsive seizures, incidence of 107–108
Nonconvulsive status epilepticus 132, 507, 552, 719
  in coma 134–136

Nonconvulsive status epilepticus (Continued)
  comatose patients with 143
  criteria for 514–515, 515
  diagnosis of 139
    benzodiazepine trial for 514–515, 515
  focal
    with impaired consciousness 136
    without impaired consciousness 136
  incidence of 107–108, 109
  kainic acid-induced 134
  mechanisms of 510
  patterns in critical illness 143
  phenotypes of 508
  with preservation of consciousness 143
  semiologic spectrum of 512
  without coma 136
Nondepolarizing neuromuscular paralytics, in intracerebral hemorrhage 182
Nonfatal drowning 691–695
  clinical presentation of 692–693, 692
  clinical trials and guidelines of 694
  complex clinical decisions of 694
  epidemiology of 691
  hospital course and management 693–694
  neurodiagnostics and imaging of 693
  neuropathology of 692
  neurorehabilitation of 695
  outcome prediction of 694–695
Noninvasive positive-pressure ventilation 19, 665
Noninvasive ventilation, contraindications to 19
Nonlacunar stroke syndromes, insula and 52
Nonspecific reversal agents 753
Normal respiration 16
Normal ventilation 36
Nurses, critical care 7
Nursing care, in early beginnings of neurocritical care 7
Nutrition, for comatose patient 125, 127

O
Obstructive sleep apnea, stroke and 36
Obtundation 708
Oligoastrocytomas 718
Oligodendrogliomas 718
Onconeural antibodies 729–732
Opioids 487–489
  clinical course for 488–489
  clinical presentation in 488
  complex clinical decisions for 489
  in delirium 457
  drug-induced encephalopathy and 768
  epidemiology of 487
  laboratory testing and imaging in 488
  management of 488–489
  pharmacology of 487–488
  receptors of 487–488, 488
  for traumatic brain injury 253–254

Opportunistic infections, and neurologic complications of transplantation 548, 561
Optimal CPP 68, 74–78
Organ
  preservation 419–420
  recovery, prior to death 417–419
  support 428–429
Organ donation 404
  after brain determination of death/heart-beating donors 420–432
  background of 409–411
  deceased 409–410
  demand and supply mismatch in 410
  donation after brain death 411
  donation after circulatory death 411–420
  neurointensive care unit and 411
  organ and
    preservation of 419–420
    recovery of 417–419
    support of 415–416, 428–429
  protocols 409–440
Organ Procurement and Transplantation Network 421
Organ procurement organizations 430
Organ transplant 545
  procedures for 546
Orthostatic hypotension 720
Osmolytes 708
Osmoreceptors 705–706
Osmotherapy, for ICP 67–68
Osmotic demyelination syndrome 557, 709–710
Osmotic therapy, for traumatic brain injury 255
Ottawa SAH rule 205–206
Out-of-hospital cardiac arrest 593
  score 386
Oxygenation, in subarachnoid hemorrhage 36

P
P300 wave 385
Pain 720–721
Pancreas 697
  pathophysiology of 415
Pancreatitis, graft 415
Papilledema 716
Paracetamol (Acetaminophen) in Stroke trial 166
Paraneoplastic neurologic disorders 729–733, 730–732
Paraplegia, in surgical ICU 445
Parasites, intramedullary infectious myelopathies and 323
Paroxysmal sympathetic hyperactivity syndrome, in comatose patient 126
Partial thromboplastin time, active 749
Patent foramen ovale 693
Patients in persistent vegetative state, diagnosis of 129
$P_{bt}O_2$, traumatic brain injury and 249

Penetrating neck trauma, spinal cord injuries and 277
Penicillin, drug-induced encephalopathy and 768
Pentobarbital
 ICP and 84–85
 for refractory status epilepticus 518, *519*
 complications of 518
Penumbral brain tissue, in acute ischemic stroke 155
Peptic ulcers, prophylactic treatment for 718
Perceptual disturbances, in delirium 452
Percussion peak, ICP and 71–72, *73*
Percutaneous tracheostomy 27
Perihematomal edema, in intracerebral hemorrhage 180
Perihematomal ischemic penumbra, in intracerebral hemorrhage 182
Peripheral nervous system, and graft-*versus*-host disease 563
Peripheral neuropathies 686
 clinical presentation of 581
 epidemiology of 577
 hospital course and management of 585
 intraoperative monitoring for 581–582
 neuropathology of 579
 outcome prediction 587
Peritumoral edema 716
Petechial hemorrhage, in acute ischemic stroke 154–155
pH, effects on brain physiology 17–18
Pharmacologic therapy, for spinal cord injuries 288–289
Pharmacoresistance, of seizures 510–512
Pharmacotherapy, status epilepticus and *137–139*, 140–142
Phencyclidine 489
Phenobarbital 522, 719
 for seizures *516*
Phenylethylamines 496
Phenytoin 520–521, 719
 for intracranial hypertension 83
 for neurologic complications of transplantation 553
 for seizures *516*
Phosphate 711
Phrenic nerve, injury to 579, 581
Physical examination, with MRC scale 535
Physiotherapy, for traumatic brain injury 261
Physostigmine 495
Piperacillin-tazobactam 368
Piracetam, for cognitive impairment 584
Pituitary apoplexy 664
Plasma exchange 661
Plasmapheresis, for superrefractory status epilepticus 522
Plateau waves 67–68, 716
Platelet function assays 749
Platelets *698*
Pneumonia 365–369
 aspiration and 367
 bronchial cultures and 367

Pneumonia *(Continued)*
 causative organisms 366–367
 diagnosis 367
 stroke and 36
 treatment in 368–369, *368*
Poisoning, coma due to 120
Poliomyelitis, epidemics 4
 in Denmark 4–6, *5*
 in Minnesota 6–7
Poliovirus, intramedullary infectious myelopathies and *323*
Polyalcohols 708
Polycystic kidney disease, autosomal dominant, as risk factor of cerebral aneurysms and SAH *201–203*
Polymerase chain reaction
 for acute encephalitis 340
 analysis 727–729
 for diagnosing bacterial meningitis 353–354
Polymorphisms, in traumatic brain injury 241–242
Polytrauma
 altered mental status in 641–645
 definition of 633
 electroencephalography and 645–646
 neurologic complications of 633–656, *634*
 neuromuscular complications in 648–650
 neurovascular complications of 635–641
  arterial dissection 635–637
  coagulopathy-related neurovascular phenomena 637–638
  embolic strokes 638–640
  hypoperfusion-related cerebral ischemia 640–641
 other vascular complications in 641
 seizures and 645–646, *646*
 spinal complications in 646–648
  clinical presentation of 647
  clinical trials and guidelines of 647
  complex clinical decisions of 647–648
  epidemiology of 646
  hospital course and 647–648
  neurodiagnostics and imaging of 647
  neuropathology of 646–647
  neurorehabilitation/outcomes of 648
Pons 178, *178*
Pontine lesions, in comatose patient 122
Positive end-expiratory pressure 41
Positron emission tomography 213
 for spinal cord emergencies 328–329, *328*
 for traumatic brain injury 244
Post cardiac arrest syndrome 595–596
Post intensive care syndrome 531
Postarrest therapeutic hypothermia 599
Postendovascular care, acute ischemic stroke and 155
Postepidural headache 660

Posterior reversible encephalopathy syndrome 555, *555–556, 557*, 771
 autoregulatory failure in 469–470
 clinical presentation of 471, *471*
 clinical trials for 478
 complex clinical decisions for 478–479
 conditions and medications associated with *468*
 diagnosis of 556
  proposed algorithm for *479*
 differential diagnosis in *472*
 endothelial dysfunction in, other causes of 470–471
 epidemiology of 468
 gadolinium enhancement in *475*
 guidelines for 478
 histopathology in 471
 hospital course and management for 474–476, *477*
 hypertension in 469–470
 imaging in 472–474, *473, 476*
 in intensive care unit 467–484
 neurodiagnostics in 471–474
 neuropathology of 468–471
 neurorehabilitation for 479
 outcome prediction in 476–478
 pathophysiology of 468–471
 recurrent 477–478
 restricted diffusion in *474*
Posterior reversible syndrome, prognosis of 771
Postictal 718–719
Postneurosurgical wound infections 373–375
Postoperative complications, of subarachnoid hemorrhage 212
Postoperative encephalopathy, after transplantation 551–552, *551*
Postpartum woman, differential diagnosis for seizure in 659
Postreperfusion syndrome 549
Postthrombolysis transfers, acute ischemic stroke and 158
Posttransplantation lymphoproliferative disorder 566, *566*
Posttraumatic cardioembolic strokes 638
Posttraumatic cerebral infarction 637
Posturing reflexes, comatose patient and 123
Power lines, accidents with 689–691
 clinical presentation of 690, *690*
 clinical trials and guidelines 691
 complex clinical decisions of 691
 epidemiology of 689
 hospital course and management of 691
 neurodiagnostics and imaging of 690
 neuropathology of 689
 neurorehabilitation of 691
 outcome prediction of 691
Practical guidance for prognostication in comatose cardiac arrest survivors (2015 guidelines) 773, *774*
Prasugrel 744–745
Pre-Bötzinger complex 36–37

INDEX

Pre-eclampsia 658–659, 661–664
  incidence of 657
  PRES and *468*, 470
Pregnancy 658
  emergency and pathologic causes of headache during *668*
  in status epilepticus 145
Pregnancy-related clinical syndromes 661–664
Prehospital management, in intracerebral hemorrhage 182
Preintubation neurologic evaluation 19–20
  standard elements of *20*
Preoxygenation, in intubation 22
Pressure ulcers, spinal cord injuries and 277
Pressure-reactivity index 74, 96
  of traumatic brain injury 250, *251*
  variability in 77
Pressure-volume compensatory reserve 69, 78–79, *79*
Prevalence, of VAP 366
Primary graft dysfunction 414
Prognosis on Admission of Aneurysmal Subarachnoid Hemorrhage scales 206, *206*
Prognostic models and factors, de-escalation of life-sustaining interventions associated with 401–403
Prognostication
  cerebral microdialysis for 98–99
  in continuous EEG 110–113, *112*
  electroencephalography for 94
Progressive multifocal encephalopathy 557
Progressive multifocal leukoencephalopathy, after transplantation *561*, 564–565, *565*
Proinflammatory cytokines 677
Proinflammatory signaling pathways, activate *199–200*
Prophylaxis, care in comatose patient *127*
Propionibacterium acnes 727–729
Propofol 602
  in ICU 644
  in intracerebral hemorrhage 182
  for intracranial hypertension 82–83
  for refractory status epilepticus 518–520, *519*
  for traumatic brain injury 252–253
Propofol infusion syndrome 82–83, 518–520
  status epilepticus and 142
Protamine, for heparin-associated epidural hematoma 331
Protamine sulfate 752
  in intracerebral hemorrhage 184
Proteus 366–367
Prothrombin complex concentrates 751–752
  in intracerebral hemorrhage 183
Prothrombin time 748–749
Prothrombotic state, of pregnancy 658
Proximal myopathy 711

Pseudomonas aeruginosa 366–367
Puerperium, emergency and pathologic causes of headache during *668*
Pulmonary edema, neurogenic 38–39, *39*
Pulmonary infiltrates *39*
  in subarachnoid hemorrhage 35
Pulmonary injuries, traumatic brain injury and 260
Pulsatility index 100
Pulse oximetry 234
Pupillary light reflex 382
Putamen 178, *178*
Pyridoxine, for superrefractory status epilepticus 523
Pyruvate, in cerebral microdialysis 98
Pyuria 371

Q

Quantitative EEG 109

R

Rabies 339, 343–344, *344*
  intramedullary infectious myelopathies and *323*
Radcliffe respirator *7*
Radiation therapy 733–734
Radiculopathy 723
Radiographic analysis, of spinal cord injuries 278–283, *281–283*
Radionuclide cerebral perfusion scan, in brain death 425–427
Ramping up *20*
Rankin scale 381
Rapid-sequence intubation, in intracerebral hemorrhage 182
Rebleeding
  initial management and prevention of 209
  preventing *216–221*
Receptor for advanced glycation end products 34–35
Recombinant activated factor VIIa, in intracerebral hemorrhage 183
Recombinant FVIIa 752
Recurrent PRES 477–478
Refractory status epilepticus 132, 142, 508–510
  antiepileptic drug dosing and pharmacokinetics for *519*
  hospital course and management of 518–520
  mortality rate 525
Regional cerebral oximetry, for cognitive impairment 584
Regional wall motion abnormalities 58
Regular physical exercise, as risk factor of cerebral aneurysms and SAH *201–203*
Rejection encephalopathy, after transplantation 549–550
Relative hyponatremia 709
Religious, and spiritual support, of family choices 404
Renal, hypothermia and *697*
Renal cell carcinoma 715–716

Renal failure, PRES and 470–471
Renal system *698*
Renin-angiotensin-aldosterone system 707
Reperfusion injury 595
  acute ischemic stroke and 156
RESCUEicp trial 304, *305*
Respiration 16
  in Full Outline of Unresponsiveness score *122*
Respiratory alkalosis 16
  hyperventilation and 16
Respiratory complications, brain injury and 15
Respiratory depression, in dextromethorphan 490
Respiratory distress syndrome, acute 25, 39–40, *39*
  traumatic brain injury and 260
Respiratory failure
  rare causes of acute weakness and *230*
  spinal cord injuries and 287
  spinal shock and 330
Respiratory intoxication 685–689
  clinical presentation of 686–687, *687*
  complex clinical decisions of 688
  epidemiology of 685
  hospital course and management of 688
  neurodiagnostics and imaging of 687–688
  neuropathology of 685–686
  neurorehabilitation 689
  outcome prediction of 688–689
Respiratory management, for spinal cord injuries 286–288
Respiratory muscle weakness 232
Respiratory rhythm generator mechanism 16–17
Respiratory system *697–698*
Respiratory weakness 534
Restricted diffusion, in PRES 473, *474*
Return of spontaneous circulation 599
Reversal agents, nonspecific 753
Reversible cerebral vasoconstriction syndrome, PRES and 474
Reversible posterior leukoencephalopathy 555
Reversible posterior leukoencephalopathy syndrome. *see* Posterior reversible encephalopathy syndrome
Rewarming phase, of hypothermia 622
Rhabdomyolysis 497–498
Risk factors, for bacteremia 369
Risperidone, for delirium 583–584
Rivaroxaban 753
ROCKET-AF trial 744
Rocuronium 600
Rupture, cerebral saccular aneurysm and *199–200*
Ruptured aneurysm repair *216–221*

S

S100-B, in traumatic brain injury *248*
Saphenous nerve, injury to 579, 581

Sarcoidosis, of spinal cord 321, *322–323*, 327
Scalp electrodes, for electroencephalography 95
Scandinavian Candesartan Acute Stroke Trial 165
Schistosomiasis, intramedullary infectious myelopathies and *323*
Secondary brain injury, cerebral microdialysis for 99
Sedation
  depth of, monitoring 109
  for traumatic brain injury 252–254, *253*
Sedative-hypnotic toxicity 485–487
  agents in, abused in 485–486, *486*
  clinical course for 487
  clinical presentation in 486
  complex clinical decisions for 487
  epidemiology of 485–486
  laboratory imaging and testing in 486–487
  management of 487
  pharmacology of 486
Sedative-hypnotic withdrawal 490–493
  alcoholic hallucinosis in 491
  alcoholic tremulousness in 491
  alcoholic withdrawal seizures in 491
  clinical course for 491–493
  clinical presentation of 491
  complex clinical decisions for 493
  delirium tremens in 491
  epidemiology of 490–491
  laboratory testing and imaging in 491
  management of 491–493
  pharmacology of 491
Seizures 38, *216–221*, 645–646, *646*, 716
  activity 384
  acute ischemic stroke and 155
  after transplantation 552–553
    evaluation management of, algorithm for *553*
  of aneurysmal subarachnoid hemorrhage 214
  cerebral ischemia and 83
  clinical presentation of 580, *580*
  in critically ill 507–530
    clinical presentation of 512
    clinical trials and guidelines for 524
    complex clinical decisions in 524–525
    epidemiology of 508–510
    hospital course and management of 515–524
    imaging 512–515
    neurodiagnostics 512–515
    neuropathology of 510–512
    neurorehabilitation for 525
    nonconvulsive, semiologic spectrum of *512*
    outcome prediction in 525
    pharmacoresistance in 510–512
    prophylaxis in 524
    substances causing 513–514
  in critically ill pregnant patients 659–660, 666

Seizures *(Continued)*
  detection and management of 107–109, *108, 109*
  epidemiology of 576
  hospital course and management of 583
  ICP and 83
  management of 184
  neuropathology of 578
  nonconvulsive 552
    electrophysiologic evaluation for 559
  outcome prediction 586
  in pregnant woman, differential diagnosis for 659
  in PRES 471, 474–475
  prophylaxis, for traumatic brain injury 254–255
  SUDEP and 56
  traumatic brain injury and 92–93, 245
Sepsis
  clinical presentation and neurodiagnostics 678–679
  epidemiology of 675–676
  hospital course and management 679–680
  intensive care unit-acquired weakness and 533–534
  neurologic complications of 675–683
  PRES and *468*, 470
  signs of 369
Sepsis-associated delirium 676
Septic encephalopathy 675–683
  clinical presentation and neurodiagnostics 678–679
  definition of 676
  epidemiology of 675–676
  hospital course and management 679–680
  neuropathology of 676–678
  signs and symptoms of *678*
Serratia marcescens 366–367
Serum inflammatory markers, for diagnosing bacterial meningitis 354
Serum S-100B protein 597–598
SETscore 27
Sexual disturbances 720
SGE-102, for seizures 524
Shivering
  management of 600–601
  TH and 625–626
Shock ward 6
Short-term cognitive decline 578
Signs/symptoms, of CAUTI 370–371
"Sinking skin flap syndrome," 312
Sinoatrial node 49–50, *50*
Sirolimus, for drug toxicity 557–558
Skin
  care in comatose patient *127*
  hyponatremia and *697*
Sleep-disordered breathing, stroke and 36
Sleep-wake cycle, in delirium 457
SLUDGE acronym 231
Smoking, as risk factor of cerebral aneurysms and SAH *201–203*

Society of Critical Care Medicine 6
Sodium, perturbations of 705
Sodium channel blockade, in cocaine 498
Sodium regulation 707
Solitaire Flow Restoration device 163
Somatic support, in pregnant women 669
Somatosensory evoked potentials 112–113, 384
  in neurophysiologic testing 605
  reliability of 775
  TH and 621
Sonography, transcranial Doppler 70–71
αII-Spectrin breakdown products, in traumatic brain injury 248
Sphincter dysfunction 720
Spinal clearance, for spinal cord injuries 276
Spinal cord compression 720–721
Spinal cord emergencies
  clinical presentation of 321–324, *322–323*
  diagnosis of 319–336
  epidemiology of 319
  examination patterns in 324
  hospital course and management of 330–333
  management of 319–336
  neurodiagnostics and imaging in 324–330
  neuropathology of 320–321
  neurorehabilitation for 333
  outcome prediction of 333
Spinal cord infarction 575
  aortic dissection-associated 331
  etiologies and clinical pearls of *322–323*
  hospital course and management of 331
  neuropathology of 320
Spinal cord injuries 646, 720
  clinical trials and guidelines in 625
  history of care of 8
  traumatic 275–298
    blunt cervical trauma-associated vascular injuries, management of 289–290, *289*
    clinical presentation of 276–278
    clinical trials and guidelines for 291, *292*
    closed cervical spine reduction for 283–286, *284*
    complex decision making in 289–291
    deep venous thrombosis prophylaxis for 290–291
    definitive clinical assessment of 291–293
    epidemiology of 275–276, *276*
    field triage and spine clearance for 276
    hemodynamic management of 287–288
    hospital course and management of 278–289
    initial hospital evaluation of 278, *278, 279*

Spinal cord injuries *(Continued)*
  initial radiographic analysis of 278–283, *281–283*
  pharmacologic therapy for 288–289
  prehospital management of 276–278
  respiratory management for 286–288
  spinal immobilization in 277
  transportation of patient with 277–278
Spinal epidural abscess
  hospital course and management of 331
  neuropathology of 320
Spinal epidural hematoma, hospital course and management of 331
Spinal immobilization
  complications of 277
  methods of 277
Spinal injuries, traumatic brain injury and 260
Spinal shock
  hospital course and management of 330
  neuropathology of 324
Spondylotic compressive myelopathies, management of 332
Spontaneous breathing trials 26, *26*
"Spot sign," 382–383
Staphylococcus aureus 366–367, 727–729
Static cold storage 415–416
Statins *216–221*
  in delirium 457
Status epilepticus 131, 384, 507, 660, 718–720
  absence 136
  causes of *133*
  clinical presentation of 134–136
  complex clinical decisions in 143
  convulsive 134–135
  in critically ill pregnant patients 667
  defined 507–508
  diagnostic tests and imaging in *513*
  epidemiology of 132–133
  etiology of *514*
  hospital course for 140–143
  imaging in 136–140
  management of 131–151, *646*
  myoclonic 135
  neurodiagnostics and 136–140, *139*
  neuropathology of 133–134
  nonconvulsive 132
    in coma 134–136
    comatose patients with 143
    patterns in critical illness 143
    with preservation of consciousness 143
    without coma 136
  outcome prediction in 145–146
  pharmacotherapy and *137–139*, 140–142
  in pregnancy, management of 145
  in PRES 474–475
  refractory 132, 142, 508–510
    new-onset 132, 144–145
  subtle 134–136
  superrefractory 132, 143–144, *144*, 508–510

Status Epilepticus Severity Score 146, 525
Stereotactic radiosurgery 722
Steroids, for superrefractory status epilepticus 522
Streptococcus pneumoniae 366–367, 727–729
  in bacterial meningitis 349–350
Stress-induced cardiomyopathy, status epilepticus and 135
Stress-related cardiomyopathy 49
  SAH and 57
Stroke 36, 580
  acute, NCCU population and 398
  causes of 659
  hemispheric, decompressive craniectomy for 304–310
  ischemic, TH and 624–625
  in pregnancy 660, 662
"Stroke code" teams 156
Stroke syndrome, drug toxicity and 558, *559*
Strongyloides stercoralis 727–729
Structural brain injury, acute, delirium in 450
Subacute cerebellar degeneration 729
Subacute encephalopathy 449
Subarachnoid blood 204
Subarachnoid hemorrhage 35–36, 49, 57–58
  algorithm for investigation of patients with *208*
  aneurysmal, cardiac enzyme with 56
  causes of *196*
    deterioration after *212*
  CBN associated with 52–53
  characteristics of 58
  clinical trials and guidelines in 624
  common neurocardiac sequelae of 54
  computed tomography of different types of *197, 207*
  diagnosis of cause of 207–209
  differentiating from traumatic lumbar puncture *208*
  early pathophysiology of *198*
  history of *201–203*
  ICH and 52–53
  IS and 53
  management of 58–59
  pathogenesis of 58
  pathology studies in 53
  patients with 51
  prevention of *216–221*
  risk factors for *201–203, 216–221*
  risk of 663
  RWMA in, diverse pattern of 58
  secondary to *212*
  stress-related CMO and 57
  TH and 621
Subarachnoid space 320
Subdural empyema 373–374
  clinical presentation of 352
  complex clinical decisions in 360
  epidemiology of 351
  hospital course and management of 359

Subdural empyema *(Continued)*
  neurodiagnostics and neuroimaging of 355, *356*
  recommendations for initial antibiotic therapy for *357*
Subdural hematomas, traumatic brain injury and 240–241
Subdural hygroma, complications of decompressive craniectomy *311*
Subdural space 320
Subhairline EEG 113
Substance P, during seizures 510
Succinylcholine 22
Sudden cardiac arrest 402–403
Sudden cardiac death 49
  definition of 55
  epilepsy and 56
  tachyarrhythmias associated with 55
Sudden unexplained death in epilepsy 38, 49
  definition of 56
Superrefractory status epilepticus 132, 143–144, *144*, 508–510
  alternative treatments for 522–523
  defined 520
  hospital course and management of 520–523
Supratentorial masses 716–717
Surface cooling systems 621–622
Surface echocardiography, acute ischemic stroke and 160–161
Surface-cooling systems 599–600
Surgery, for ICP 67–68
Surgical ICUs, consults in 445
Surgical tracheostomy 27
Surgical Trial in Intracerebral Hemorrhage I and II 186
Surrogate decision makers 398
Surviving Sepsis Campaign, further adjunctive therapeutic measures according to guidelines of *679*
Susceptibility-weighted imaging 717
SWIFTPRIME *164*
Sympathetic nerve injury 579
Sympathetic nervous system 707
Sympathetic stimulation, in autoregulation 469
Sympathomimetic toxicity 496–498
  amphetamines in 496–497
  cathinones in *494*, 496–497
  clinical presentation in 497
  cocaine in 496–497
  complex clinical decisions for 498
  epidemiology of 496
  laboratory and imaging in 497–498
  management of 498
  pharmacology of 496–497
  phenylethylamines in 496
  synthetic cathinones in 497
Symptomatic aneurysm, as risk factor of cerebral aneurysms and SAH *201–203*
Symptomatic hemorrhagic transformation, in acute ischemic stroke 162
Syndrome of the trephined *311*, 312

Synthetic cathinones 497
Systemic disease, coagulopathies associated with 755–756, 756
Systemic inflammatory response syndrome
  and intensive care unit-acquired weakness 533–534
  in subarachnoid hemorrhage 35–36

T

Tachyarrhythmias 55
Tako-tsubo cardiomyopathy 59–61, 59–60
  status epilepticus and 135
Targeted temperature management 382, 599
  for comatose survivors 602–603
  neuroprognostication after cardiac arrest using 608–609
  provision of 603
Telemedicine, in intracerebral hemorrhage 182
Temozolomide 718
Temperature management, in acute ischemic stroke 166
Temperature modulation techniques, TH and 621–622
Temperatures, core body 626
Temporal lobe masses 716–717
Tetany 710–711
Thalamus 50, 50, 178, 178
"The Milwaukee protocol," 343–344
Therapeutic hypothermia 601–602
  clinical trials and guidelines in 622–625
  complex clinical decisions in 625–627
  description of 619
  hospital course and management of 621–622
  intra-arrest 599
  neurodiagnostics and imaging in 621
  neuropathology in 619
  outcome prediction in 627
  postarrest 599
  protocols 619–632
Thermal burns 685–689
  clinical presentation of 686–687, 687
  complex clinical decisions and 688
  epidemiology of 685–689
  hospital course and management of 688
  neurodiagnostics and imaging of 687–688
  neuropathology 685–686
  neurorehabilitation of 689
  outcome prediction of 688–689
Thermoregulation 697
Thiamine deficiency 487
Thoracoabdominal asynchrony 232
Thrombectomy, for ischemic stroke 583
Thrombin inhibition, assays for measuring 750
Thrombin time 750
Thrombocytopenia
  epidemiology of 745
  management of 754
  neuropathology of 745–747
Thromboelastography 750

Thrombolysis, for spinal cord infarction 331
Thromboprophylaxis, for traumatic brain injury 258
Thyroid carcinoma 715–716
Ticagrelor 744–745
Tick-borne encephalitis 338–339, 343
Time to treatment, and seizures 517
Tissue plasminogen activator administration 161–162, 162
Todd's paralysis 719
Tonic-clonic seizures 711
Topiramate, for superrefractory status epilepticus 522
Toxemia 658–659
Toxic-metabolic encephalopathy 340
Toxicology 485
  panel, for seizure patients 513–514
Toxidrome-oriented physical examination, in unknown ingestion 499–500
Toxins, SAH and 196
Toxoplasma, after transplantation 561
Toxoplasma gondii 727–729
Toxoplasmosis, after transplantation 564
T-piece trials 26
Trach collar trials 26
Tracheobronchial aspiration 367
Tracheostomy 27, 126, 126
  alternative to intubation 19
  in early beginnings of neurocritical care 4–6
  for spinal cord injuries 287
Tracheotomy, in acute ischemic stroke 171
Transcranial Doppler
  Aaslid's description of 70–71
  in brain death 425–427
Transcranial Doppler sonography 70–71, 212
Transcranial Doppler ultrasonography 100
Transcranial NIRS 80
Transesophageal echocardiograms, acute ischemic stroke and 160–161
Transfusion, blood, for traumatic brain injury 259
Transfusion targets 216–221
Transient global ischemia 198
Transplant recipient, consultation in 445
Transplantation
  neurologic complications of 545–572
    akinetic mutism as 552
    central pontine myelinolysis as 560–562
    clinical presentation 548–553
    CNS infections as 563–566
    drug toxicity 554–559, 554
    epidemiology of 547–548
    failure to awaken as 549–551, 549–550
    graft-versus-host disease as 562–563
    guidelines for 567
    idiopathic hyperammonemia as 563
    incidence of, contemporary studies evaluating 547
    neurodiagnostics and imaging of 559–560, 560

Transplantation (Continued)
    neuropathology 548
    outcomes of 560
    in pediatric populations 548
    postoperative encephalopathy and delirium 551–552
    posttransplantation lymphoproliferative disorder as 566, 566
    risk factors of 548
    seizures as 552–553
    timing of 548
    treatment of 553–554
    white matter diseases 557
  procedures for 546
Transthoracic echocardiogram 639
  acute ischemic stroke and 160–161
Transthoracic echocardiography 599
Trastuzumab 724
Trauma
  decompressive craniectomy for 302–304
    DECRA trial 302–304
    RESCUEicp trial 304, 305
  ICUs, consults in 445
  SAH and 196
Traumatic axonal injury 382
  traumatic brain injury and 241
Traumatic brain injury 33–35, 385, 398, 402
  abnormalities in vasomotor tone and reactivity of 244–245
  balancing second-tier therapies for 259
  basic intensive care for 252–255
  bedside neurodiagnostics and monitoring for 248–251, 249, 252
  biomarkers in 247–248, 248
  brain swelling in 242–244, 242, 244
  clinical decisions for 257–260
  clinical neuropathology of 240–245
  clinical presentation of 245
  clinical trials and guidelines for 257, 623–624
  coagulopathy after 745, 747
  coma and 119–120
  computed tomography of 93
  critical care management of 239–274
  dysautoregulation in 242–244
  in elderly patient 258–259
  energy failure in 242–244, 243
  epidemiology of 240
  focal pathology of 240–241, 240
  genetic modulation of disease course and outcome in 241–242
  hemostatic abnormalities after 257–258
  hospital course and management of 251–257
  imaging for 246–247, 246–247
  intracranial pressure monitoring for 95
  Lund concept for 41
  molecular mechanisms of 242
  multimodal monitoring for, time-synchronized data of 94
  multiple trauma in 260
  neurorehabilitation for 261

Traumatic brain injury (Continued)
  in nonassessable patient, monitoring of 259
  nonfocal (diffuse) pathology of 240, 241
  organ dysfunction following 40
  osmotic therapy for 255
  outcome of 262, 262
  pathophysiology of 240–245, 240
  prediction for 262–263
  rescue therapies for 255
  routine ICU interventions for 255
  sedation for 252–254, 253
  seizure prophylaxis for 254–255
  seizures and 92–93
  skull X-ray of 93
  temperature management for 254
  TH and 620–621
  thromboprophylaxis for 258
  ventilatory strategies for 252
Traumatic intracerebral hemorrhage, traumatic brain injury and 240–241
Traumatic lumbar puncture, differentiating from subarachnoid hemorrhage 208
Traumatic spinal cord injuries, acute 275–298
  blunt cervical trauma-associated vascular injuries, management of 289–290, 289
  clinical presentation of 276–278
  clinical trials and guidelines for 291, 292
  closed cervical spine reduction for 283–286, 284
  complex decision making in 289–291
  deep venous thrombosis prophylaxis for 290–291
  definitive clinical assessment of 291–293
  epidemiology of 275–276, 276
  field triage and spine clearance for 276
  hemodynamic management of 287–288
  hospital course and management of 278–289
  initial hospital evaluation of 278, 278, 279
  initial radiographic analysis of 278–283, 281–283
  pharmacologic therapy for 288–289
  prehospital management of 276–278
  respiratory management for 286–288
  spinal immobilization in 277
  transportation of patient with 277–278
Trephined, syndrome of 311, 312
Treponema pallidum, intramedullary infectious myelopathies and 323
Troponin I 57
Truncal ataxia 693
Tuberculosis, after transplantation 561
"Tumor growth factor," 718–719
Tumor location 718
Tumor necrosis factor-alpha 725
Tumor-associated epilepsy 718–720

U
Ubiquitin C-terminal hydrolase-L1 385
UCH-L1, in traumatic brain injury 248
UK Glucose Insulin in Stroke Trial 166
Ulcers, pressure, spinal cord injuries and 277
Unconsciousness, prolonged, in comatose patient 128–129
Uncontrolled DCD 411
  continuation of resuscitation efforts *versus* 420
  pathophysiology of 414
  protocols and guidelines in 417, 419
Unfractionated heparin
  intracranial hemorrhage and 744
  for spinal cord injuries 290–291
Uniform Determination of Death Act 409–410
United Council of Neurologic Specialties 9–10
United Network for Organ Sharing 410
  criteria 413
University of Wisconsin DCD Evaluation Tool 413
Unknown ingestion, patient with, approach to 498–500
  central nervous system imaging in 500
  history in 498–499
  laboratory imaging in 500
  physical examination in
    general 499
    toxidrome-oriented 499–500
Unresponsive wakefulness syndrome 380
Unruptured aneurysms 216–221
Uremia
  after transplantation 549–550
  coagulopathy and 756
  epidemiology of 746
  neuropathology of 748
Uremic encephalopathy, ICU-acquired, prognosis of 769
Urinary tract infections 370–371
  in comatose patient 127
Urine drug-of-abuse testing, reflexive use of 500

V
Vagus nerve, injury to 578–579, 581
Valproic acid 521, 719–720
  for seizures 516
  for status epilepticus 140–141
Valsalva maneuver 720–721
Vancomycin 371
Varicella zoster virus infection, after transplantation 561
Vascular compromise 372
Vascular endothelial growth factor 716
Vascular hemorrhage 196
Vascular malformations
  etiologies and clinical pearls of 322–323
  hospital course and management of 331–332
  neuropathology of 320–321
Vascular occlusion 725

Vascular surgery, neurologic complications of 573–592
  clinical presentation of 579–581
  clinical trials and guidelines for 585
  complex clinical decisions 585
  epidemiology of 574, 575–577
  hospital course and management of 582–585
  neurodiagnostics and imaging 581–582, 582
  neuropathology of 577–579
  neurorehabilitation for 587
  outcome prediction 585–587
Vasodilatatory cascade 72–73
Vasogenic edema
  in acute ischemic stroke 156
  traumatic brain injury and 243–244
Vasomotor tone, traumatic brain injury and 244–245
Vasopressors, in intracerebral hemorrhage 182
Vasospasm
  angiographic 204–205, 212–213
  in PRES 474
  TCD ultrasound and 100
Vecuronium 600
Vegetative state 380
Venography, in intracerebral hemorrhage 181–182
Venous outflow obstruction, ICP and 67–68
Venous thromboembolism 216–221, 727
  management of 186
Ventilation
  mechanical
    airway management and 15–32
    monitoring of, in seizures 517
    noninvasive, contraindications to 19
    problems after acute brain injury 23
Ventilator weaning 26–27
Ventilator-associated pneumonia 41–42, 365–369, 367
Ventilatory support, in organ donation 429
Ventral respiratory group 36–37
Ventricular arrhythmias, pathophysiology of 52
Ventriculitis 371–373
  decompressive craniectomy, complications of 311
VerifyNow-ASA 749
Vertebral metastases 721
Veterans Affairs Cooperative Study 517
Video intubation 21
Viral encephalitis 339
Viruses, intramedullary infectious myelopathies and 323
Viscoelastic hemostatic assays 749–750
Visual disturbances, in PRES 471
Vitamin K, intravenous, for warfarin-associated epidural hematoma 331
Vitamin K antagonists 751–752
  intracranial hemorrhage and 744
Volume replacement, in comatose patient 126–127

W

Warfarin
  associated with, epidural hematoma 331
  in intracerebral hemorrhage 179
Warm ischemia time 414
Water regulation 705–707
Waterhouse-Friderichsen syndrome 680
Weakness
  ICU-acquired
    functional outcomes of 770
    prediction of 770
    prognosis of 769–771
    prognostication and, practical implications of 770–771
  intensive care unit-acquired 531–544
    clinical presentation of 534–535

Weakness *(Continued)*
    clinical trials and guidelines 537–539, *538*
    complex clinical decisions 539
    differential diagnosis *539*
    epidemiology of 531–532
    hospital course and management of 536–537
    models for 533
    neurodiagnostics 535–536
    neuropathology 532–533, *532*
    neurorehabilitation for 539
    risk factors for 533–534
    terms for 532
    respiratory 534
Weaning, ventilator 26–27

Weather and climate, in aneurysmal subarachnoid hemorrhage 196
Wernicke's encephalopathy 642
  triad of 643
West Nile virus 342–343, *342*
  after transplantation *561*
  intramedullary infectious myelopathies and *323*
Withdrawal of life-sustaining therapies 413
World Federation of Neurologic Surgeons, clinical grading scales include Hunt and Hess *206*

X

Xanthochromia 207
Xenobiotics *492*